Excursions in Modern Mathematics

Excursions in Modern Mathematics

Peter Tannenbaum
Robert Arnold
CALIFORNIA STATE UNIVERSITY—FRESNO

PRENTICE HALL
UPPER SADDLE RIVER NJ 07458

Library of Congress Cataloging-in-Publication Data

Tannenbaum, Peter (date)
Excursions in modern mathematics / Peter Tannenbaum, Robert Arnold.--3rd ed.
p. cm.
Includes bibliographical references and index.
ISBN 0-13-598335-5
1. Mathematics. I. Arnold, Robert, mathematician. II. Title.
QA36.T35 1997
519--dc21 97-17313
CIP

Editorial Director: *Tim Bozik*
Editor-in-Chief: *Jerome Grant*
Acquisitions Editor: *Sally Denlow*
Production Editors: *Richard DeLorenzo and Barbara Mack*
Managing Editor: *Linda Behrens*
Executive Managing Editor: *Kathleen Schiaparelli*
Assistant Vice President of Production and Manufacturing: *David W. Riccardi*
Marketing Manager: *John Tweeddale*
Creative Director: *Paula Maylahn*
Art Director: *Sheree Goodman Design, Inc.*
Interior Design: *Brand X Studios*
Cover Art and Design: *Hothouse Design, Inc.*
Art Manager: *Gus Vibal*
Photo Editor: *Lori Morris-Nantz*
Photo Research: *Kim Moss and Teri Stratford*
Manufacturing Buyer: *Alan Fischer*
Manufacturing Manager: *Trudy Pisciotti*
Supplements Editor/Editorial Assistant: *April Thrower*

Simon & Schuster/A Viacom Company
Upper Saddle River, New Jersey 07458

Printed in the United States of America
10 9 8 7 6 5 4 3 2 1

ISBN 0-13-598335-5

Prentice-Hall International (UK) Limited, *London*
Prentice-Hall of Australia Pty. Limited, *Sydney*
Prentice-Hall Canada Inc., *Toronto*
Prentice-Hall Hispanoamericana S.A., *Mexico City*
Prentice-Hall of India Private Limited, *New Delhi*
Prentice-Hall of Japan, Inc., *Tokyo*
Simon & Schuster Asia Pte. Ltd., *Singapore*
Editora Prentice-Hall do Brasil, Ltda., *Rio de Janeiro*

To my parents, Nicholas and Anna
and my wife Sally

PT

To my wife Rachael,
and my son Craig

RA

Contents

PART 3 Growth and Symmetry

Preface

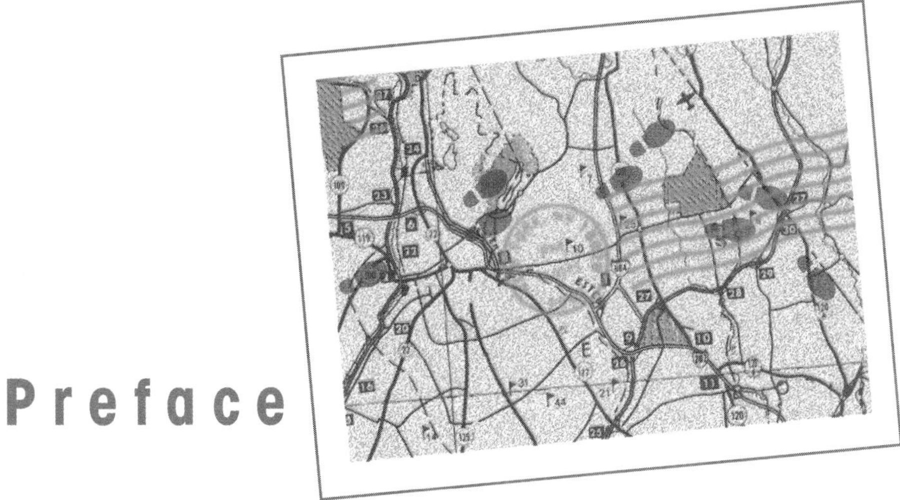

To most outsiders, modern mathematics is unknown territory. Its borders are protected by dense thickets of technical terms; its landscapes are a mass of indecipherable equations and incomprehensible concepts. Few realize that the world of modern mathematics is rich with vivid images and provocative ideas.

Ivars Peterson , The Mathematical Tourist

Excursions in Modern Mathematics is, as we hope the name might suggest, a collection of "trips" into that vast and alien frontier that many people perceive mathematics to be. While the purpose of this book is quite conventional--it is intended to serve as a textbook for a college-level liberal arts mathematics course--its contents are not. We have made a concerted effort to introduce the reader to a view of mathematics entirely different from the traditional algebra-geometry-trigonometry-finite math curriculum which so many people have learned to dread, fear, and occasionally abhor. The notion that general education mathematics must be dull, unrelated to the real world, highly technical, and deal mostly with concepts that are historically ancient is totally unfounded.

The excursions in this book represent a collection of topics chosen to meet a few simple criteria.

Applicability.

The connection between the mathematics presented here and down-to-earth, concrete real-life problems is direct and immediate. The often heard question, "What is this stuff good for?" is a legitimate one and deserves to be met head on. The often heard answer, "Well, you need to learn the material in Math 101 so that you can understand Math 102 which you will need to know if you plan to take Math 201 which will teach you the real applications," is less than persuasive and in many cases reinforces students' convictions that mathematics is remote, labyrinthine, and ultimately useless to them.

Accessibility.

Interesting mathematics need not always be highly technical and built on layers upon layers of concepts. As a general rule, the choice of topics in this book is such that a heavy mathematical infrastructure is not needed: We have found Intermediate Algebra to be an appropriate and sufficient prerequisite. (In the few instances in which more advanced concepts are unavoidable we have endeavored to provide enough background to make the material self-contained.) A word of caution--this does not mean that the material is easy! In mathematics, as in many other walks of life, simple and straightforward is not synonymous with easy and superficial.

Age.

Much of the mathematics in this book has been discovered in this century, some as recently as 20 years ago. Modern mathematical discoveries do not have to be only within the grasp of experts.

Aesthetics.

The notion that there is such a thing as beauty in mathematics is surprising to most casual observers. There is an important aesthetic component in mathematics and, just as in art and music (which mathematics very much resembles), it often surfaces in the simplest ideas. A fundamental objective of this book is to develop an appreciation for the aesthetic elements of mathematics. It is not necessary that the reader love everything in the book—it is sufficient that he or she find one topic about which they can say, "I really enjoyed learning this stuff!" We believe that anyone coming in with an open mind almost certainly will.

Outline

The material in the book is divided into four independent parts. Each of these parts in turn contains four chapters dealing with interrelated topics.

Part 1

(Chapters 1 through 4). ***The Mathematics of Social Choice.*** This part deals with mathematical applications in social science. How do groups make decisions? How are elections decided? What is power? How can power be measured? What is fairness? How are competing claims on property resolved in a fair and equitable way?

Part 2

(Chapters 5 through 8). ***Management Science***. This part deals with methods for solving problems involving the organization and management of complex activities—that is, activities involving either a large number of steps and/or a large number of variables (routing the delivery of packages, landing a spaceship on Mars, organizing a banquet, scheduling classrooms at a big university, etc.). Efficiency is the name of the game in all these problems. Some limited or precious

resource (time, money, raw materials) must be managed in such a way that waste is minimized. We deal with problems of this type (consciously or unconsciously) every day of our lives.

Part 3

(Chapters 9 through 12). Growth and Symmetry. This part deals with nontraditional geometric ideas. How do sunflowers and seashells grow? How do animal populations grow? What are the symmetries of a snowflake? What is the symmetry type of a wallpaper pattern? What is the geometry of a mountain range? What kind of symmetry lies hidden in our circulatory system?

Part 4

(Chapters 13 through 16). Statistics. In one way or another, statistics affects all of our lives. Government policy, insurance rates, our health, our diet, and public opinion are all governed by statistical laws. This part deals with some of the most basic aspects of statistics. How should statistical data be collected? How is data summarized so that it is intelligible? How should statistical data be interpreted? How can we measure the inherent uncertainty built into statistical data? How can we draw meaningful conclusions from statistical information? How can we use statistical knowledge to predict patterns in future events?

Exercises

We have endeavored to write a book that is flexible enough to appeal to a wide range of readers in a variety of settings. The exercises, in particular, have been designed to convey the depth of the subject matter by addressing a broad spectrum of levels of difficulty—from the routine drill to the ultimate challenge. For convenience (but with some trepidation) we have classified them into three levels of difficulty:

Walking.

These exercises are meant to test a basic understanding of the main concepts, and they are intended to be within the capabilities of students at all levels.

Jogging.

These are exercises that can no longer be considered as routine—either because they use basic concepts at a higher level of complexity, or they require slightly higher order critical thinking skills, or both.

Running.

This is an umbrella category for problems that range from slightly unusual or slightly above average in difficulty to problems that can be a real challenge to even the most talented of students. This category also includes an occassional open-ended problem suitable for a project.

Teaching Extras Available with the Third Edition

The New York Times Mathematics and Statistics Supplement

selected by Peter Tannenbaum and others

Prentice Hall and the **New York Times** jointly sponsor "A Contemporary View", a collection of articles taken from the pages of the New York Times. These articles deal with mathematically significant breakthroughs or with recent news stories which illustrate the importance, relevance, and currency of mathematics in real life. The articles may be used to prompt classroom discussion, to suggest topics for term papers and research projects, or merely to show the connections between class material and the world that surrounds the student.

"*A Contemporary View*" is published once a year and is available exclusively through Prentice Hall. Through this program, at the instructor's request, each student will receive a free copy of this supplement. Instructors are encouraged to call their Prentice Hall representative for more information.

ABC News Videos and Accompanying Projects

edited and written by Kim Query

Segments are taken from Nightline, World News Tonight and This Week with David Brinkley covering various current events through which mathematics is used. Also includes a manual with extended group projects and discussion questions to accompany the video clips.

Prentice Hall Companion Web Page

Features Internet projects with links that complement the activities in the text, "Net Tutors" for on-line tutoring, and links to related math resources available on the Web.
http://www.prenhall.com/tannenbaum

Instructor's Manual

The Instructor's Manual contains notes and comments of a general nature, and a test bank consisting of approximately 500 multiple choice questions and worked out solutions to all exercises in the text.
ISBN: 0-13-746959-4

Student's Solutions Manual

The Student's Solutions Manual contains the solutions to the odd numbered problems in the text and other helpful hints and suggestions.
ISBN: 0-13-746967-5

Prentice Hall Custom Test (IBM and MAC)

A fully editable test generator which features an instructor's grade book and on-line testing.
IBM: 0-13-747123-8
MAC: 0-13-747081-9

Life on the Internet - Mathematics

Guides students and instructors through the complexity of the Internet, offering navigation strategies, practice exercises, and lists of resources. Contact your local Prentice Hall representative for the latest copy.

THE THIRD EDITION

This third edition of Excursions in Modern Mathematics retains the topics and organization of previous editions, in a more attractive and hopefully more user friendly package. Most chapters have been rewritten, and new examples and applications have been added throughout. In addition, some chapters have undergone substantive changes in organization(Chapters 5, 6, and 7), or coverage (Chapters 11, 13, 15, and 16). New topics not in previous editions are: the classification of finite shapes by their symmetry types (Chapter 11); the capture-recapture method for estimating the size of a population (Chapter 13); permutations and combinations (Chapter 15); and an introduction to statistical inference (Chapter 16).

A FINAL WORD

This book grew out of the conviction that a liberal arts mathematics course should teach students more than just a collection of facts and procedures. The ultimate purpose of this book is to instill in the reader an overall appreciation of mathematics as a discipline and an exposure to the subtlety and variety of its many facets: problems, ideas, methods, and solutions. Last, but not least, we have tried to show that mathematics can be fun.

ACKNOWLEDGMENTS

The following mathematicians reviewed previous editions of the book and made many invaluable suggestions:

CARMEN ARTINO, *College of Saint Rose;*
DONALD BEATON, *Norwich University;*
MARGRET BOS, *St. Lawrence University;*
THOMAS A. CARNEVALE, *Shawnee State University*
TERRY L. CLEVELAND, *New Mexico Military Institute;*
LESLIE COBAR, *University of New Orleans;*
RONALD CZOCHOR, *Rowan College of New Jersey;*
KATHRYN E. FINK, *Moorpark College;*
JOSEPHINE GUGLIELMI, *Meredith College;*
WILLIAM S. HAMILTON, *Community College of Rhode Island;*
GLENDA HAYNIE, *North Carolina State University;*

HAROLD JACOBS, *East Stroudsburg University;*
TOM KILEY, *George Mason University;*
JEAN KRICHBAUM, *Broome Community College;*
CHRISTOPHER MCCORD, *University of Cincinnati;*
THOMAS O'BRYAN, *University of Wisconsin-Milwaukee;*
DANIEL E. OTERO, *Pennsylvania State University;*
DENNIS D. PENCE, *Western University;*
MATTHEW PICKARD, *University of Puget Sound;*
LANA RHOADS, *William Baptist College;*
DAVID E. RUSH, *University of California at Riverside;*
KATHLEEN C. SALTER, *Eastern New Mexico University;*
THERESA M. SANDIFER, *Southern Connecticut State University;*
PAUL SCHEMBARI, *East Stroudsburg University of Pennsylvania;*
CONNIE S. SCHROCK, *Emporia State University*
WILLIAM W. SMITH, *University of North Carolina at Chapel Hill;*
DAVID STACY, *Bellevue Community College;*
JOHN WATSON, *Arkansas Technical University;*
TAMELA WILLETT, *McHenry County College.*

We would like to extend special thanks to *Professor Benoit Mandelbrot* of Yale University who read the manuscript for Chapter 12 and made several valuable suggestions. We gratefully acknowledge *Vahack Haroutunian, Ronald Wagoner, Carlos Valencia, and L. Taylor Ullmann,* all of whom made significant contributions to the exercise sets.

Excursions in Modern Mathematics

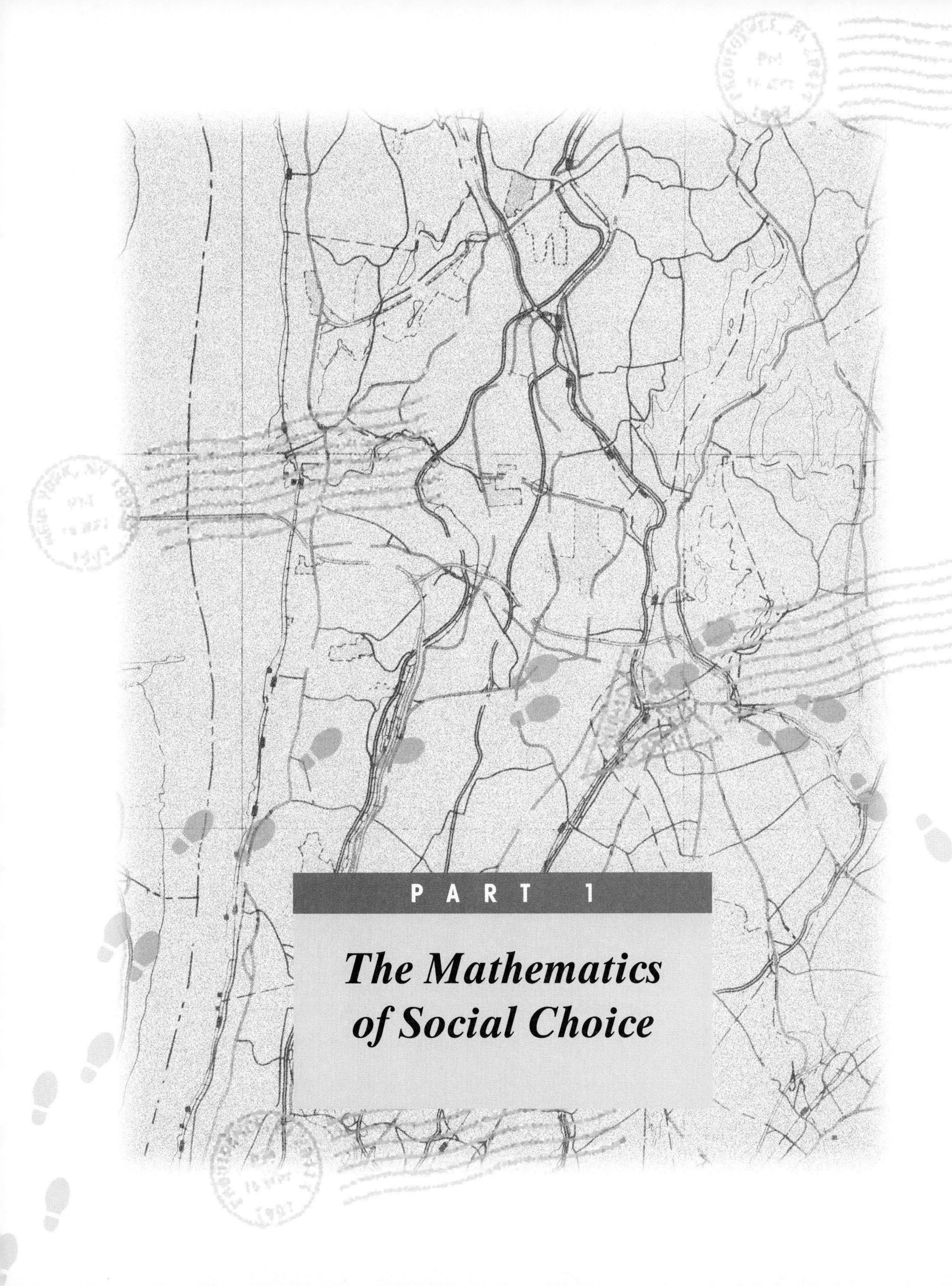

PART 1

The Mathematics of Social Choice

It's not the voting that's democracy; it's the counting.

Tom Stoppard

The Mathematics of Voting

The Paradoxes of Democracy

CHAPTER 1

We are constantly reminded, especially at election time, that voting is the backbone of democracy—the right and the duty of every adult citizen. That every citizen is entitled to vote is an issue no longer in question, but why should voting be a duty? In these times of increasing cynicism about politics and politicians, many citizens fail to vote in local, state, or even national elections—usually because of the conviction that single votes do not make a difference. In a democracy, does the voice of one person really count? If so, how?

The answer to these questions can best be understood when we look at the notion of *voting* in the broadest possible sense. Voting is much more than dropping a ballot in a box and choosing a president, or a senator, or even a member of the school board. There is voting behind many kinds of public events, ranging in importance from the trite to the critical, depending on one's priorities: Which automobile gets selected *Automobile of the Year*? Which college football player wins the Heisman trophy? Which movie wins the Oscar for "best movie"? Who gets that executive job you are applying for? In subtle and often misunderstood ways elections are happening all around us, invisibly shaping the world we live in.

The reason we have elections is that we don't all think alike. In a democracy, an election is a *process*—in fact, the only acceptable process—by which the many and often conflicting choices of individual voters are sifted so that a single choice representing the collective voice of the group can emerge. Understanding the intricacies behind this process is what **voting theory** is about.

But wait just a second! Voting theory? Why do we need a fancy theory? It all sounds pretty simple: We have an election; we count the ballots. Based on the ballots we decide the outcome of the election using a procedure that is consistent and fair. Surely, there must be a reasonable way to accomplish all of this! Surprisingly, there isn't.

In 1952, the mathematical economist Kenneth Arrow discovered a remarkable fact: For elections involving more than two choices, *there is no consistently fair democratic method for choosing a winner*. In fact, Arrow demonstrated that *a voting method that is democratic and always fair is a mathematical impossibility*. This, the most famous fact in voting theory, is known as **Arrow's impossibility theorem.**[1]

The primary purpose of this chapter is to explore some of the basic ideas in the mathematical theory of voting and to clarify the meaning and significance of Arrow's impossibility theorem. In so doing, we will take a brief tour through some of the most commonly used **voting methods**—how they work, what their implications are, and most importantly, how they stack up when we put them to some basic tests of *fairness*.

Preference Ballots and Preference Schedules

We kick off our discussion of voting theory with a simple but important example, which we will revisit several times throughout the chapter. You may want to think of this example as a mathematical parable—its importance being not in the story itself, but in what lies hidden behind it.

Example 1 The Math Club Election.

The Math Appreciation Society (MAS) is a student organization dedicated to an unsung but worthy cause—that of fostering the enjoyment and appreciation of mathematics among college students. The Tasmania State University chapter of MAS is holding its annual election for president. There are four candidates running for president: Alisha, Boris, Carmen, and Dave (*A, B, C,* and *D* for short). Each of the 37 members of the club votes by means of a ballot indicating his or her first, second, third, and fourth choice (ties are not allowed in a ballot). The 37 ballots submitted are shown in Fig. 1-1. Once the ballots are in, it's decision time. Who should be the *winner*[2] of the election? Why? ■

---

[1] In 1972 Arrow was awarded the Nobel Prize in Economics (there is no Nobel Prize in Mathematics) for his pioneering work in what is now known as *social-choice theory,* a discipline that combines aspects of mathematics, economics, and political science.

[2] In this particular example, there can only be one club president, so this election can have only one winner. Every once in a while, however, there are situations where an election can have two or more winners. In 1968, for example, Katherine Hepburn and Barbra Streisand both won the Oscar for Best Actress. While it is the exception rather than the rule, an election can have more than one winner, and in our usage of the word "winner" we will allow for this possibility.

Ballot	Ballot	Ballot	Ballot	Ballot	Ballot	Ballot	Ballot	Ballot	Ballot	Ballot
1st *A*	1st *B*	1st *A*	1st *C*	1st *B*	1st *C*	1st *A*	1st *B*	1st *C*	1st *A*	1st *C*
2nd *B*	2nd *D*	2nd *B*	2nd *B*	2nd *D*	2nd *B*	2nd *B*	2nd *D*	2nd *B*	2nd *B*	2nd *B*
3rd *C*	3rd *C*	3rd *C*	3rd *D*	3rd *C*	3rd *D*	3rd *C*	3rd *C*	3rd *D*	3rd *C*	3rd *D*
4th *D*	4th *A*	4th *D*	4th *A*	4th *A*	4th *A*	4th *D*	4th *A*	4th *A*	4th *D*	4th *A*

Ballot	Ballot	Ballot	Ballot	Ballot	Ballot	Ballot	Ballot	Ballot	Ballot	Ballot	Ballot	Ballot
1st *D*	1st *A*	1st *A*	1st *C*	1st *A*	1st *C*	1st *D*	1st *C*	1st *A*	1st *D*	1st *D*	1st *C*	1st *C*
2nd *C*	2nd *B*	2nd *B*	2nd *B*	2nd *B*	2nd *B*	2nd *C*	2nd *B*	2nd *B*	2nd *C*	2nd *C*	2nd *B*	2nd *B*
3rd *B*	3rd *C*	3rd *C*	3rd *D*	3rd *C*	3rd *D*	3rd *B*	3rd *D*	3rd *C*	3rd *B*	3rd *B*	3rd *D*	3rd *D*
4th *A*	4th *D*	4th *D*	4th *A*	4th *D*	4th *A*	4th *A*	4th *A*	4th *D*	4th *A*	4th *A*	4th *A*	4th *A*

Ballot	Ballot	Ballot	Ballot	Ballot	Ballot	Ballot	Ballot	Ballot	Ballot	Ballot	Ballot	Ballot
1st *D*	1st *A*	1st *D*	1st *C*	1st *A*	1st *D*	1st *B*	1st *A*	1st *C*	1st *A*	1st *A*	1st *D*	1st *A*
2nd *C*	2nd *B*	2nd *C*	2nd *B*	2nd *B*	2nd *C*	2nd *D*	2nd *B*	2nd *D*	2nd *B*	2nd *B*	2nd *C*	2nd *B*
3rd *B*	3rd *C*	3rd *B*	3rd *D*	3rd *C*	3rd *B*	3rd *C*	3rd *C*	3rd *B*	3rd *C*	3rd *C*	3rd *B*	3rd *C*
4th *A*	4th *D*	4th *A*	4th *A*	4th *D*	4th *A*	4th *A*	4th *D*	4th *A*	4th *D*	4th *D*	4th *A*	4th *D*

FIGURE 1-1
The 37 anonymous ballots for the MAS election

Before we try to answer these two deceptively simple questions, let's discuss the nature of the ballots shown in Fig. 1-1. Ballots in which a voter is asked to rank all the candidates in order of preference are called **preference ballots.** While a preference ballot is not the only (or even the most common) form a ballot can take,[3] it is the best way of getting information from a voter, since it gives a detailed accounting of how the voter feels about the relative merits of all the candidates.

In the MAS election, as is often the case, there are many repetitions among the ballots submitted. The repetitions represent nothing more than the fact that different voters can end up with the same exact ranking of the candidates. Thus, a logical way to organize the ballots is to group together identical ballots (Fig. 1-2), and this leads in a rather obvious way to Table 1-1, which is called the **preference schedule** for the election. The preference schedule is the simplest and most compact way to completely summarize the balloting in an election. (If we want to think of an election as a formal process, then the preference schedule is the **input** to the process.)

Ballot	Ballot	Ballot	Ballot	Ballot
1st *A*	1st *C*	1st *D*	1st *B*	1st *C*
2nd *B*	2nd *B*	2nd *C*	2nd *D*	2nd *D*
3rd *C*	3rd *D*	3rd *B*	3rd *C*	3rd *B*
4th *D*	4th *A*	4th *A*	4th *A*	4th *A*
14	10	8	4	1

FIGURE 1-2
The 37 MAS election ballots organized into neat little piles

Table 1-1 Preference Schedule for the MAS Election

Number of voters	**14**	**10**	**8**	**4**	**1**
1st choice	*A*	*C*	*D*	*B*	*C*
2nd choice	*B*	*B*	*C*	*D*	*D*
3rd choice	*C*	*D*	*B*	*C*	*B*
4th choice	*D*	*A*	*A*	*A*	*A*

[3] In most elections for public office, for example, a ballot asks for just one choice—our top choice.

Transitivity and Elimination of Candidates

There are two important facts that we need to keep in mind when we work with preference ballots. The first goes by the name of *the transitivity of individual preferences*. It basically says that if we know that a voter prefers A to B and B to C, then it follows automatically that this voter must prefer A to C. A useful consequence of this observation is this: *If we need to know which candidate a voter would vote for if it came down to a choice between just two candidates, all we have to do is look at which candidate was placed higher on that voter's ballot.* We will use this fact throughout the chapter.

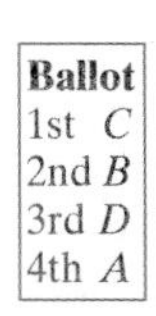

FIGURE 1-3

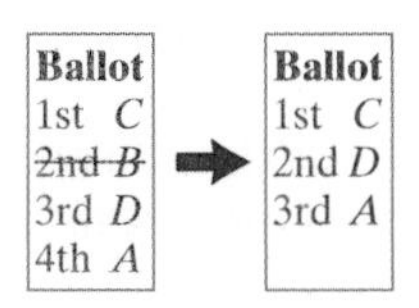

FIGURE 1-4

The other important fact is that the relative preferences of a voter are not affected by the elimination of one or more of the candidates. Take, for example, the ballot shown in Fig. 1-3 and pretend that, due to an emergency, candidate B drops out of the race right before the ballots are submitted. How would this voter now rank the remaining three candidates? As Fig. 1-4 shows, the relative positions of the remaining candidates are unaffected: C remains the first choice, D moves up to the second choice, and A moves up to the third choice.

Let's now return to the business of deciding the outcome of elections in general and the MAS election (Example 1) in particular.

The Plurality Method

Perhaps the best known and most commonly used method for finding a winner in an election is the **plurality method.** Essentially this method says that the candidate (or candidates, if there is more than one) with the *most* first-place votes wins. Notice that in the plurality method, the only thing that matters is first-place votes, and that we don't actually need the entire preference ballot, just the first choice.

When we apply the plurality method to the Math Appreciation Society election, this is what we get:

A gets 14 first-place votes.

B gets 4 first-place votes.

C gets 11 first-place votes.

D gets 8 first-place votes.

The Math Appreciation Society News

ALISHA ELECTED PRESIDENT OF MAS!

In this case, the results of the election are clear—the winner is Alisha!

The popularity of the plurality method stems not only from its simplicity but also from the fact it is a natural extension of the **majority rule:** In any democratic election between *two* candidates, the one with the majority (more than half) of the votes wins.

When there are three or more candidates, the majority rule cannot always be applied: In the MAS election, 19 first-place votes (out of 37) are needed for a majority, but none of the candidates received 19 first-place votes, so no one has the required majority. Alisha, with 14 first-place votes, has more than anyone else, so she has a *plurality*.

The Majority Criterion

While a plurality does not imply a majority, a majority does imply a plurality: A candidate that has more than half of the first-place votes must automatically have more first-place votes than any other candidate. The implications are simple but important: A candidate that has a majority of the first-place votes is automatically the winner under the plurality method.

The notion that having a majority of the first-place votes should automatically guarantee the win in an election makes good sense and is an important requirement for a fair and democratic election. In fact, it is important enough to have a name: **the majority criterion.**

The Majority Criterion. If there is a choice that has a majority of the first-place votes in an election, then that choice should be the winner of the election.

We already know that if a candidate happens to have a majority of first-place votes, then that candidate is guaranteed to win under the plurality method. A fancy way to say this is that *the plurality method satisfies the majority criterion.*

The majority criterion is something that in a democracy we tend to think of as a given. We will soon see that this need not be the case: There are important and widely used voting methods where a candidate could have a majority of the first-place votes and yet another candidate could end up winning the election.

What's Wrong with the Plurality Method?

In spite of its widespread use, the plurality method has many flaws and is usually a rather bad method for choosing the winner of an election when there are more than two candidates. Its principal weakness is that it fails to take into consideration the voters' preferences other than first choice, and in so doing can lead to some very bad election results.

To underscore the point, consider the following example.

Example 2.

Tasmania State University has a superb marching band. They are so good that this coming New Year they have been invited to march at five different bowl games: The Rose Bowl (R), the Hula Bowl (H), the Cotton Bowl (C), the Orange Bowl (O), and the Sugar Bowl (S). An election is held among the 100 members of the band to decide in which of the five bowl games they will march. A preference schedule giving the results of the election is shown in Table 1-2.

If the plurality method is used, the winner of the election is the Rose Bowl, with 49 first-place votes. Note, however, that the Hula Bowl (H) which has 48

Table 1-2 Preference Schedule for the Band Election

Number of voters	**49**	**48**	**3**
1st choice	R	H	C
2nd choice	H	S	H
3rd choice	C	O	S
4th choice	O	C	O
5th choice	S	R	R

first-place votes, also has 52 second-place votes. Simple common sense tells us that the Hula Bowl is a much better choice. In fact, we can make the following persuasive argument in favor of the Hula Bowl: If we compare the Hula Bowl to any other bowl on a *head-to-head* basis, the Hula Bowl is always the preferred choice. Take, for example, a comparison between the Hula Bowl and the Rose Bowl: The Hula Bowl would get 51 votes (48 from the second column plus the 3 votes in the last column) versus 49 votes for the Rose Bowl. Likewise, a comparison between the Hula Bowl and the Cotton Bowl would result in 97 votes for the Hula Bowl (first and second columns) and 3 votes for the Cotton Bowl. And when the Hula Bowl is compared to either the Orange Bowl or the Sugar Bowl, it wins by a landslide. ■

We can now summarize the problem as follows: Although *H* wins in a head-to-head comparison between it and any other choice, the plurality method fails to choose *H* as the winner. In the language of voting theory, we say that the plurality method *violates* a basic requirement of fairness called the **Condorcet[4] criterion.**

The Condorcet Criterion. If there is a choice that in a head-to-head comparison is preferred by the voters over every other choice, then that choice should be the winner of the election.

Before we go on, a word about how to interpret some of the terminology we have just introduced. When we say that the plurality method *violates the Condorcet criterion,* we mean that it is possible to come across examples of elections in which there is a candidate that wins in a head-to-head comparison against every other candidate and yet, using the plurality method, a different candidate wins the election. The band election is one such example. We should not conclude, however, that the problem must occur in every election—it doesn't! (Think about it this way: To violate the speeding laws we do not have to drive above the speed limit all the time—it is enough that we do it once in a while.)

When there is a candidate that is preferred by the voters in each head-to-head comparison with the other candidates, we will call such a candidate a **Condorcet candidate.** Using this wording, the Condorcet criterion has a nice ring to it: It simply says that if in an election there happens to be a Condorcet candidate, that candidate should be the winner of the election. Of course, in many cases there is no Condorcet candidate in the election. In these cases the Condorcet criterion does not apply.

We will return to the idea of head-to-head comparisons between the candidates shortly. In the meantime, we conclude this section by discussing another important weakness of the plurality method: The ease with which **insincere voting** can affect the results of the election. (A voter who changes the true order of his or her preferences in the ballot in an effort to influence the outcome of the election against a certain candidate is said to vote *insincerely*.) With the plurality method, this can be easily accomplished. As an example, consider once again Table 1-2 and look at the

---

[4] Named after Marie Jean Antoine Nicolas Caritat, Marquis de Condorcet (1743–1794). Condorcet was a French aristocrat, mathematician, philosopher, economist, and social scientist. As a member of a group of liberal thinkers (the *encyclopédistes*), his ideas were instrumental in leading the way to the French Revolution. Unfortunately, his ideas eventually fell into disfavor and he died in prison.

last column of the preference schedule, which represents the ballots of three specific band members, which we might add are dead set against the Rose Bowl (allergies!). Assuming that they have some idea of how the election is likely to turn out and that their first choice (the Cotton Bowl) has no chance of winning the election, their best strategy is to vote *insincerely,* moving their second choice (the Hula Bowl) to first and in so doing effectively changing the outcome of the election.

In real-world elections insincere voting can have serious and unexpected consequences. Take, for example, the overwhelming tendency for a two-party system in American politics. Why is the two-party system so entrenched in the United States? Partly, it is because the plurality method encourages insincere voting. It is well known that third-party candidates have a hard time getting their just share of the votes: Many voters who actually prefer the third-party candidate end up reluctantly voting for one of the two major-party candidates for fear of "wasting" their vote. Allegedly, this occurred in the 1992 presidential election when many voters who were inclined to vote for Ross Perot ended up voting (insincerely) for either Bill Clinton or George Bush.

The Borda Count Method

An entirely different approach to finding the winner in an election is the **Borda count method.**[5] In this method each place on a ballot is assigned points. In an election with N candidates we give 1 point for last place, 2 points for second from last place, . . . , and N points for first place. The points are tallied for each candidate separately, and the candidate with the highest total is the winner.

Let's use the Borda count method to choose the winner of the Math Appreciation Society election. Table 1-3 shows the point values under each column based on first place worth 4 points; second place worth 3 points; third place worth 2 points; and fourth place worth 1 point.

Table 1-3 Borda Points for the MAS Election

Number of voters	14	10	8	4	1
1st choice: 4 points	*A:* 56 pts	*C:* 40 pts	*D:* 32 pts	*B:* 16 pts	*C:* 4 pts
2nd choice: 3 points	*B:* 42 pts	*B:* 30 pts	*C:* 24 pts	*D:* 12 pts	*D:* 3 pts
3rd choice: 2 points	*C:* 28 pts	*D:* 20 pts	*B:* 16 pts	*C:* 8 pts	*B:* 2 pts
4th choice: 1 point	*D:* 14 pts	*A:* 10 pts	*A:* 8 pts	*A:* 4 pts	*A:* 1 pt

When we tally the points,

A gets $56 + 10 + 8 + 4 + 1 = 79$ points

B gets $42 + 30 + 16 + 16 + 2 = 106$ points

C gets $28 + 40 + 24 + 8 + 4 = 104$ points

D gets $14 + 20 + 32 + 12 + 3 = 81$ points,

we find that the winner is Boris!

The Math Appreciation Society News

BORIS ELECTED PRESIDENT OF MAS!

[5] This method is named after the Frenchman Jean-Charles de Borda (1733–1799). Borda was a military man—a cavalry officer and naval captain—who wrote on such diverse subjects as mathematics, physics, the design of scientific instruments, and voting theory.

What's Wrong with the Borda Count Method?

In contrast to the plurality method, the Borda count method takes into account *all* the information provided by the voters' preferences and produces as a winner the *best compromise candidate*. This is good! The real problem with the Borda count method is that a candidate with a majority of first-place votes can lose the election—in other words, *it violates the majority criterion*. The next example illustrates how this can happen.

Example 3.

The last principal at George Washington Elementary School has just retired and a new principal must be hired by the School Board. The four finalists for the job are Mrs. Amaro, Mr. Burr, Mr. Castro, and Mrs. Dunbar (*A, B, C,* and *D,* respectively). After interviewing the four finalists, the eleven members of the school board vote, each member ranking each of the four finalists, and it is agreed that the winner will be decided using the Borda count method. The results of the voting are shown in Table 1-4.

Table 1-4 Preference Schedule for Example 3

Number of voters	6	2	3
1st choice	*A*	*B*	*C*
2nd choice	*B*	*C*	*D*
3rd choice	*C*	*D*	*B*
4th choice	*D*	*A*	*A*

It is a simple matter of arithmetic (which we leave to the reader to verify) that under the Borda count method, Mr. Burr gets the principal's job, with a total of 32 points. This happens in spite of the fact that Mrs. Amaro has 6 out of the 11 first-place votes and therefore a majority. What we have here is a violation of the majority criterion. ■

Here is another problem with the Borda count method: Since any violation of the majority criterion is an automatic violation of the Condorcet criterion as well (see Exercise 49), we can use Example 3 to show that the Borda count method also *violates the Condorcet criterion*.

In spite of these drawbacks, the Borda count method is widely used in a variety of important real-world elections, especially when there is a large number of candidates. The winner of the Heisman award; the winners of various music awards (Country Music Vocalist of the Year, etc.); the hiring of school principals and university presidents; and a host of other jobs, awards, and distinctions are decided using the Borda count method.

The Plurality-with-Elimination Method

The plurality-with-elimination method is the electoral version of the principle of *survival of the fittest*. The basic idea is to keep eliminating the most "unfit" candidates, one by one, until there is a winner left. The criterion for "fitness" is the number of first-place votes (thus the name *plurality with elimination*).

A more formal description of the process (we do have to be careful with some of the details) goes like this:

- **Round 1.** Count the first-place votes for each candidate, just as you would in the plurality method. If a candidate has a majority of first-place votes, that candidate is automatically declared the winner. Otherwise, eliminate the candidate (or candidates if there is a tie) with the fewest first-place votes.
- **Round 2.** Cross out the name(s) of the candidates eliminated from the preference schedule and recount the first-place votes. If a candidate has a majority of first-place votes, declare that candidate the winner. Otherwise, eliminate the candidate with the fewest first-place votes.
- **Rounds 3, 4, etc.** Repeat the process, each time eliminating one or more candidates, until there finally is a candidate with a majority of first-place votes, which is then declared the winner.

Let's apply the plurality-with-elimination method to the Math Appreciation Society election. For the reader's convenience Table 1-5 shows the preference schedule again—it is exactly the same as Table 1-1.

Table 1-5 Preference Schedule for the MAS Election

Number of voters	14	10	8	4	1
1st choice	*A*	*C*	*D*	*B*	*C*
2nd choice	*B*	*B*	*C*	*D*	*D*
3rd choice	*C*	*D*	*B*	*C*	*B*
4th choice	*D*	*A*	*A*	*A*	*A*

- **Round 1.**

Candidate	*A*	*B*	*C*	*D*
Number of first-place votes	14	4	11	8

Since *B* has the fewest first-place votes, he is eliminated first.

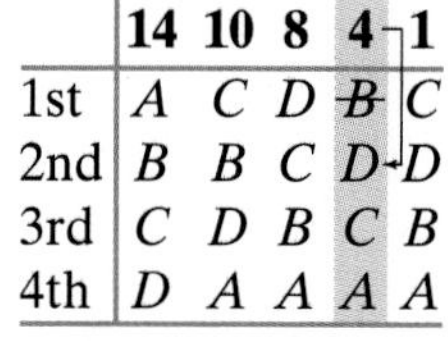

	14	10	8	4	1
1st	*A*	*C*	*D*	~~*B*~~	*C*
2nd	*B*	*B*	*C*	*D*	*D*
3rd	*C*	*D*	*B*	*C*	*B*
4th	*D*	*A*	*A*	*A*	*A*

FIGURE 1-5

- **Round 2.** Once *B* is eliminated, the four votes that originally went to *B* in round 1 will now go to *D*, the next-best candidate in the opinion of these four voters (see Fig. 1-5). The new tally is

Candidate	*A*	*B*	*C*	*D*
Number of first-place votes	14		11	12

In this round *C* has the fewest first-place votes and is eliminated.

- **Round 3.** The 11 votes that went to *C* in round 2 now go to *D* (just check the relative positions of *D* and *A* in those columns). This gives

Candidate	*A*	*B*	*C*	*D*
Number of first-place votes	14			23

The Math Appreciation Society News

DAVE ELECTED PRESIDENT OF MAS!

We now have a winner, and lo and behold, its neither Alisha nor Boris. The winner of the election, with 23 first-place votes, is Dave!

For the reader who likes straight answers to simple questions, what's been happening with the Math Appreciation Society election may be somewhat disconcerting. The question, "Who is the winner of the MAS election?" is beginning to look quite ambiguous: The answer depends not just on the ballots themselves but on the way we choose to interpret those ballots!

Winner	Voting method
Alisha	Plurality
Boris	Borda count
Dave	Plurality-with-elimination

Applying the Plurality-with-Elimination Method

The next two examples are intended primarily to illustrate some subtleties that can come up when applying the plurality-with-elimination method. To speed things up, we describe each election by simply showing the preference schedule.

Example 4.

Table 1-6 shows the preference schedule for an election between five candidates *A*, *B*, *C*, *D*, and *E*.

Table 1-6 Preference Schedule for Example 4

Number of voters	10	5	2	1	4	4
1st choice	*A*	*B*	*C*	*C*	*D*	*E*
2nd choice	*B*	*D*	*A*	*E*	*C*	*D*
3rd choice	*C*	*E*	*E*	*B*	*A*	*C*
4th choice	*D*	*C*	*B*	*A*	*E*	*A*
5th choice	*E*	*A*	*D*	*D*	*B*	*B*

From the preference schedule, we can determine that the number of voters is $10 + 5 + 2 + 1 + 4 + 4 = 26$, and therefore 14 or more votes are needed for a majority. Let's use the plurality-with-elimination method to find a winner.

- **Round 1.**

Candidate	*A*	*B*	*C*	*D*	*E*
Number of first-place votes	10	5	3	4	4

Here *C* has the fewest number of first-place votes and is eliminated first.

- **Round 2.** Of the three votes originally going to *C*, now two go to *A* (look at the third column of the preference schedule) and one goes to *E* (from the fourth column of the preference schedule).

Candidate	*A*	*B*	*C*	*D*	*E*
Number of first-place votes	12	5		4	5

In this round *D* has the fewest first-place votes and is eliminated.

- **Round 3.** The four votes originally going to *D* would next go to *C* (look at the fifth column of the preference schedule), but *C* is out of the picture at this point, so we dip further down into the column and find A, the top candidate in that column still among the living. The four votes go to A.

Candidate	A	B	C	D	E
Number of first-place votes	16	5			5

At this point we can stop, as there is no need to go on! Candidate A has a majority of the first-place votes and is the winner of the election. ■

Example 5.

Table 1-7 shows the result of an election among the four candidates W, X, Y, and Z.

Table 1-7 Preference Schedule for Example 5

Number of voters	8	6	2	19
1st choice	W	X	Y	Z
2nd choice	X	Z	Z	X
3rd choice	Z	Y	W	W
4th choice	Y	W	X	Y

The number of voters in this election is $8 + 6 + 2 + 19 = 35$, so it takes 18 or more votes for a majority. But notice that candidate Z has 19 first-place votes right out of the gate. This means that we are done—Z is automatically the winner! ■

There is a simple but important lesson to be learned from Example 5: *The plurality-with-elimination method satisfies the majority criterion.*

What's Wrong with the Plurality-with-Elimination Method?

The main problem with the plurality-with-elimination method is quite subtle and is illustrated by the next example.

Example 6.

Three cities, Athens (A), Babylon (B), and Carthage (C), are competing to host the next Summer Olympic Games. The final decision is made by a secret vote of the 29 members of the Executive Council of the International Olympic Committee, and the winner is chosen using the plurality-with-elimination method. Two days before the actual election is to be held, a straw vote (just to see how things stand) is conducted by the Executive Council. The results of the straw poll are shown in Table 1-8.

Table 1-8 Preference Schedule in Straw Vote Two Days Before the Actual Election

Number of voters	7	8	10	4
1st choice	A	B	C	A
2nd choice	B	C	A	C
3rd choice	C	A	B	B

Based on the straw vote we can predict the results of the election: In the first round Athens would have 11 votes, Babylon would have 8, and Carthage would have 10, which means that Babylon would be eliminated first. In the second round, Babylon's 8 votes would go to Carthage (see the second column of Table 1-8), so Carthage ends up with 18 votes, more than enough to lock up the election.

Although the results of the straw poll are supposed to be secret, the word gets out that unless some of the voters turn against Carthage, Carthage is going to win the election. Not surprisingly (everybody loves a winner), what ends up happening in the actual election is that even more first-place votes are cast for Carthage than in the straw poll. Specifically, the four voters in the last column of Table 1-8 decide as a block to switch their first-place votes from Athens to Carthage. Surely, this is just frosting on the cake for Carthage, but just to be sure, let's go through the motions of checking the results of the election.

Table 1-9 shows the preference schedule for the actual election. [Table 1-9 is the result of just switching A and C in the last column of Table 1-8 and combining columns 3 and 4 (they are now the same) into a single column.]

Table 1-9 Preference Schedule for the Actual Election

Number of voters	7	8	14
1st choice	A	B	C
2nd choice	B	C	A
3rd choice	C	A	B

When we apply the plurality-with-elimination method to Table 1-9, Athens (with 7 first-place votes) is eliminated first, and the 7 votes originally going to Athens now go to Babylon, giving it 15 votes *and the win!* How could this happen? How could Carthage lose an election it had locked up simply because some voters moved Carthage from second to first choice? To the people of Carthage this was surely the result of an evil Babylonian plot, but checking and double-checking the figures makes it clear that everything is on the up and up—Carthage is just the victim of a twisted quirk in the plurality-with-elimination method: The possibility that you can actually do worse by doing better! In the language of voting theory this is known as the *violation of the monotonicity criterion*. ■

The Monotonicity Criterion. If choice X is a winner of an election and, in a reelection, the only changes in the ballots are changes that only favor X, then X should remain a winner of the election.

We now know that the plurality-with-elimination method *violates the monotonicity criterion*. We leave it as an exercise for the reader to verify that plurality with elimination also *violates the Condorcet criterion* (see Exercise 51).

In spite of its flaws, the plurality-with-elimination method is used in many real-world situations, usually in elections in which there are few candidates (typically three or four, rarely more than six). While Example 6 was just a simple dramatization, it is a fact that when choosing which city gets to host the Olympic Games, the International Olympic Committee uses the plurality-with-elimination method. (For details as to how the 2000 Summer Olympics were awarded to Sydney, the reader is referred to Appendix 2.)

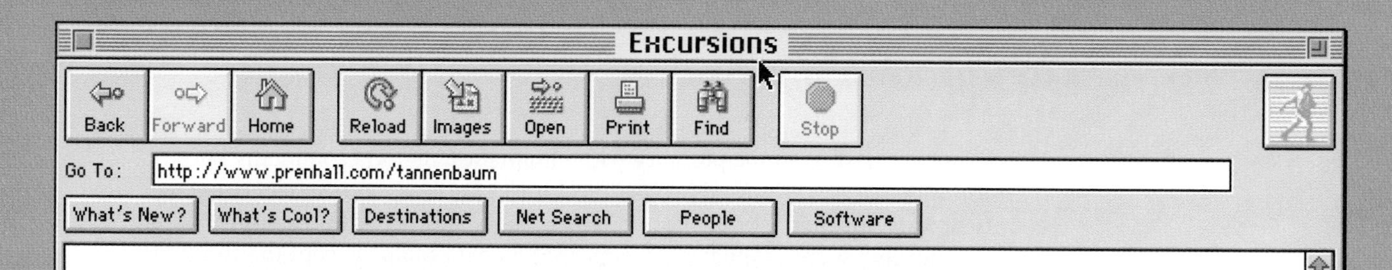

Get Out The Vote!

Many U.S. city councils are elected using the extended (or "at-large") plurality method, in which each voter casts as many votes as there are open seats, and the winning candidates are the top vote-getters. This method is easy for voters to understand. However, once a candidate has enough votes to win, any further votes for that candidate make no difference. And none of the votes for a losing candidate make any difference either. This can lead many voters to feel that they needn't have bothered voting at all.

For this reason, some people favor a system in which more of the ballots really matter. One city using such a system is Cambridge, Massachusetts (home of Harvard University). The type used there is called the single transferable vote method, or STV for short. An explanation of this method, and an interactive demonstration, can be found in the Tannenbaum Website under Chapter 1.

Questions:

1. Why is the "quota," $\frac{\text{voters}}{\text{seats} + 1} + 1$, defined the way it is?

Consider a race among candidates *A, B, C,* and *D* for two open seats on a city council, and imagine that there are only five voters: 1, 2, 3, 4, and 5. Suppose the voters' preferences among the candidates are as follows:

	Voter				
Candidate	1	2	3	4	5
A	1st	1st	1st	3rd	3rd
B	2nd	2nd	2nd	4th	4th
C	3rd	3rd	3rd	1st	1st
D	4th	4th	4th	2nd	2nd

Clearly there are two voting blocks: voters 1–3, and voters 4 and 5.

2. In an extended plurality vote, each voter would cast a vote for each of his or her two favorite candidates. Which candidates would win?
3. Use the demo to find the winners under the STV method.
4. How many voters' ballots matter under each method?
5. STV is often called a type of proportional representation, because it tends to give groups of voters power in proportion to their relative numbers. How does the above scenario illustrate this?

A simple variation of the plurality-with-elimination method commonly known as *plurality with a runoff* is used in elections for local political office (city councils, county boards of supervisors, school boards, etc.). Plurality with a runoff works just like plurality with elimination except that *all* candidates except the top two get eliminated in the first round (see Exercise 54).

The Method of Pairwise Comparisons

So far, all three voting methods we have discussed violate the Condorcet criterion, but this is not an insurmountable problem. It is reasonably easy to come up with a voting method that satisfies the Condorcet criterion. Our next method, commonly known as **the method of pairwise comparisons,**[6] illustrates how this can be done.

The method of pairwise comparisons is like a *round-robin tournament* in which every candidate is matched *one-on-one* with every other candidate. Each of these one-on-one matchups is called a **pairwise comparison.** In a pairwise comparison between candidates X and Y each vote is assigned to either X or Y, *the vote going to whichever of the two candidates is higher on the ballot*. The winner of the pairwise comparison is the one with the most votes, and as in an ordinary tournament, a win is worth 1 point (a loss is worth nothing!). In case of a tie each candidate gets $\frac{1}{2}$ point. Needless to say, the winner of the election is the candidate with the most points after all the pairwise comparisons are tabulated. In case of a tie (ties are common under this method) we can either have more than one winner (if multiple winners are permitted) or use a predetermined tie-breaking procedure.

Once again, we will illustrate the method of pairwise comparisons using the Math Appreciation Society election.

Let's start with a pairwise comparison between A and B. Looking at Table 1-10, we can see that the 14 votes in the first column of the preference schedule go to A,

Table 1-10 Comparing Candidates A and B

Number of voters	14	10	8	4	1
1st choice	(A)	C	D	(B)	C
2nd choice	B	(B)	C	D	D
3rd choice	C	D	(B)	C	(B)
4th choice	D	A	A	A	A

but the remaining 23 votes go to B. Consequently, the winner of the pairwise comparison between A and B is B. We summarize this result by

A versus B (14 to 23): B wins (B gets 1 point; A gets 0 points).

Let's next look at the pairwise comparison between C and D (Table 1-11). In this one, the 14 votes in the first column, the 10 votes in the second column, and the 1 vote in the last column go to C, for a total of 25 votes. Clearly, this one goes to C.

[6] This method is sometimes known as *Copeland's method* and attributed to A. H. Copeland.

Table 1-11 Comparing Candidates C and D

Number of voters	14	10	8	4	1
1st choice	A	C	D	B	C
2nd choice	B	B	C	D	D
3rd choice	C	D	B	C	B
4th choice	D	A	A	A	A

C versus *D* (25 to 12): *C* wins (*C* gets 1 point).

If we continue in this manner, comparing in all possible ways two candidates at a time, we end up with the following scoreboard:

A versus *B* (14 to 23): *B* wins (*B* gets 1 point);
A versus *C* (14 to 23): *C* wins (*C* gets 1 point);
A versus *D* (14 to 23): *D* wins (*D* gets 1 point);
B versus *C* (18 to 19): *C* wins (*C* gets 1 point);
B versus *D* (28 to 9): *B* wins (*B* gets 1 point);
C versus *D* (25 to 12): *C* wins (*C* gets 1 point).

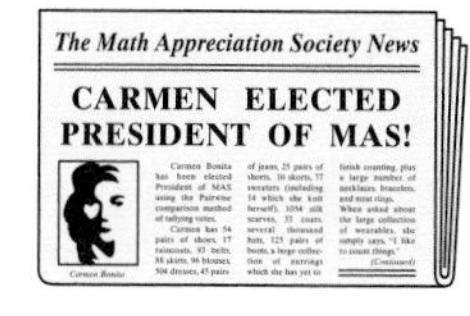
The Math Appreciation Society News

CARMEN ELECTED PRESIDENT OF MAS!

The final tally gives *A*, 0 points; *B*, 2 points; *C*, 3 points, and *D*, 1 point. Can it really be true? Yes! The winner of the election is Carmen!

Winner	Voting method
Alisha	Plurality
Boris	Borda count
Dave	Plurality with elimination
Carmen	Pairwise comparisons

It is easy to see that the method of pairwise comparisons *satisfies the Condorcet criterion*. After all, in an election with N candidates, the maximum number of points a candidate can get under the method of pairwise comparisons is $N - 1$, and that happens when that candidate wins every one of its pairwise comparisons—in other words, when that candidate is a Condorcet candidate. Thus, a Condorcet candidate always wins the election under the method of pairwise comparisons.

It is also true that the method of pairwise comparisons *satisfies both the majority criterion and the monotonicity criterion* (see Exercises 48 and 57). Hmmm . . . this is beginning to look promising.

So, What's Wrong with the Method of Pairwise Comparisons?

Unfortunately, the method of pairwise comparisons is not without flaws. The next example illustrates the most serious one.

Example 7.

As the newest expansion team in the NFL, the Los Angeles Web Surfers will be getting the number-one choice in the upcoming draft of college football players. After narrowing the list of candidates to five players (Allen, Byers, Castillo, Dixon, and

Evans), the coaches and team executives meet to discuss the candidates and eventually have a vote, a decision of major importance to both the team and the chosen player. According to team rules, the final decision must be made using the method of pairwise comparisons. Table 1-12 shows the preference schedule after all the ballots are turned in.

Table 1-12 Preference Schedule for LA's Draft Choices

Number of voters	2	6	4	1	1	4	4
1st choice	*A*	*B*	*B*	*C*	*C*	*D*	*E*
2nd choice	*D*	*A*	*A*	*B*	*D*	*A*	*C*
3rd choice	*C*	*C*	*D*	*A*	*A*	*E*	*D*
4th choice	*B*	*D*	*E*	*D*	*B*	*C*	*B*
5th choice	*E*	*E*	*C*	*E*	*E*	*B*	*A*

We leave it to the reader to verify that the results of the ten possible pairwise comparisons are:

A versus *B* (7 to 15): *B* wins;
A versus *C* (16 to 6): *A* wins;
A versus *D* (13 to 9): *A* wins;
A versus *E* (18 to 4): *A* wins;
B versus *C* (10 to 12): *C* wins;
B versus *D* (11 to 11): tie (*B* gets $\frac{1}{2}$ point; *D* gets $\frac{1}{2}$ point);
B versus *E* (14 to 8): *B* wins;
C versus *D* (12 to 10): *C* wins;
C versus *E* (10 to 12): *E* wins;
D versus *E* (18 to 4): *D* wins.

The final tally is *A*, 3 points; *B*, $2\frac{1}{2}$ points; *C*, 2 points; *D*, $1\frac{1}{2}$ points, and *E*, 1 point. It looks like Allen(*A*) is the lucky young man who will make millions of dollars playing for the Los Angeles Web Surfers.

The interesting twist to the story surfaces when it is discovered that one of the other players (Castillo) has accepted a Rhodes scholarship to go to graduate school in England, and has never intended to play professional football. Of course this fact should have no effect on the choice of Allen as the top draft choice. Or should it?

Suppose we were to eliminate Castillo from the original election, which we can easily do by crossing *C* from the preference schedule shown in Table 1-12 and thus getting the preference schedule shown on Table 1-13.

Table 1-13 Preference Schedule for LA's Draft Choices After *C* Is Eliminated

Number of voters	2	6	4	1	1	4	4
1st choice	*A*	*B*	*B*	*B*	*D*	*D*	*E*
2nd choice	*D*	*A*	*A*	*A*	*A*	*A*	*D*
3rd choice	*B*	*D*	*D*	*D*	*B*	*E*	*B*
4th choice	*E*	*E*	*E*	*E*	*E*	*B*	*A*

The results of the six possible pairwise comparisons would now be:

A versus B (7 to 15): B wins;
A versus D (13 to 9): A wins;
A versus E (18 to 4): A wins;
B versus D (11 to 11): tie (B gets $\frac{1}{2}$ point; D gets $\frac{1}{2}$ point);
B versus E (14 to 8): B wins;
D versus E (18 to 4): D wins.

In this new scenario A would have 2 points, B would have $2\frac{1}{2}$ points, D would have $1\frac{1}{2}$ points, and E would have 0 points, and the winner would be Byers. In other words, if the election had been conducted with the knowledge that Castillo was not really a candidate, then Byers, and not Allen would have been the winner, and the millions of dollars that are going to go to Allen would have gone instead to Byers! On its surface, this outcome seems grossly unfair to Byers. ■

The strange happenings in Example 7 help illustrate an important fact: The method of pairwise comparisons may satisfy all of our previous fairness criteria, but unfortunately, it violates another basic requirement of fairness known as *the independence-of-irrelevant-alternatives criterion.*

The Independence-of-Irrelevant-Alternatives Criterion. If candidate or alternative X is a winner of an election and one (or more) of the other candidates or alternatives is removed and the ballots recounted, then X should still be a winner of the election.

A second problem with the method of pairwise comparisons is that sometimes it can produce an outcome in which everyone is a winner.

Example 8.

The Icelandia State University varsity hockey team is on a road trip. An important decision needs to be made: Where to go for dinner? In the past, this has led to some heated arguments, so this time they decide to hold an election. The choices boil down to three restaurants: Hunan (H); Pizza Palace (P), and Danny's (D). The decision is to be made using the method of pairwise comparisons. Table 1-14 shows the results of the voting by the 11 players on the squad.

Table 1-14 The Preference Schedule for Example 8

Number of voters	4	2	5
1st choice	H	P	D
2nd choice	P	D	H
3rd choice	D	H	P

Here H beats P (9 to 2), P beats D (6 to 5), and D beats H (7 to 4). This results in a three-way tie for first place. What now? In this particular example it is unrealistic to declare the result of the election a three-way tie and have everybody go to the restaurant of their choice. Here, as in most situations, it becomes necessary to break the tie.

In general, there is no set way to break a tie, and in practice, it is important to establish the rules as to how ties are to be broken ahead of time. Otherwise, consider what might happen. Those who want to eat at Danny's could argue, not unreasonably, that the tie should be broken by counting first-place votes. In this case, Danny's would win. On the other hand, those who want to eat at the Hunan could make an equally persuasive argument that the tie should be broken by counting total points (Borda count). In this case the Hunan gets 24 points and Danny's gets 23, so the Hunan would win. As the reader can see, it would have been smart to think about these things before the election. For a more detailed discussion of ties and how to break them, the reader is encouraged to look at Appendix 1 at the end of this chapter. ■

How Many Pairwise Comparisons?

One practical difficulty with the method of pairwise comparisons has to do with the amount of work required to come up with a winner. The reader may have noticed that there seem to be a lot of pairwise comparisons to check out. Exactly how many? Since comparisons are based on taking two candidates at a time, the answer obviously depends on the number of candidates. We already saw that with *four* candidates there are *six* pairwise comparisons possible, and in Example 7 we saw that with *five* candidates there are *ten* possible pairwise comparisons. To make it a little more challenging, let's say that we have an election with 12 candidates. How many pairwise comparisons are possible? Let's try to systematically count the comparisons making sure that we don't count any comparison twice.

- The first candidate must be compared with each of the other 11 candidates—*11 pairwise comparisons.*
- Compare the second candidate with each of the other candidates except the first one, since that comparison has already been made—*10 pairwise comparisons.*
- Compare the third candidate with each of the other candidates except the first and second candidates, since those comparisons have already been made—*9 pairwise comparisons.*

  ⋮

- Compare the eleventh candidate with each of the other candidates except the first 10 candidates, since those comparisons have already been made. In other words, compare the eleventh candidate with the twelfth candidate—*1 pairwise comparison.*

We see that the total number of pairwise comparisons is

$$1 + 2 + 3 + 4 + 5 + 6 + 7 + 8 + 9 + 10 + 11 = 66.$$

How many pairwise comparisons are there in an election with 100 candidates? Well, using an argument similar to the preceding one, it's not hard to convince oneself that the total number of pairwise comparisons is

$$1 + 2 + 3 + 4 + \cdots + 99,$$

and in a more generic vein we can say that if there are N candidates, the total number of pairwise comparisons is

$$1 + 2 + 3 + 4 + \cdots + (N - 1).$$

Note that the last number added is one less than the number of candidates.

Next, we are going to learn about a very useful mathematical formula. Let's go back to the case of 100 candidates. How much is $1 + 2 + 3 + \cdots + 99$? One could always add up these numbers, but that is not very imaginative, to say nothing of the fact that it is also a lot of work.

Let's try a different way to count the comparisons. Suppose that before a pairwise comparison between two candidates takes place, each candidate gives the other one his or her business card. Then, clearly, each candidate would end up with the business card of every other candidate, and there would be a total of $99 \times 100 = 9900$ cards handed out (each of the 100 candidates would hand out 99 cards, one to each of the other candidates). But since each comparison resulted in two cards being handed out, the total number of comparisons must be half as many as the number of cards. Consequently,

$$1 + 2 + 3 + 4 + \cdots + 99 = \frac{99 \times 100}{2} = 4950.$$

Similar arguments show that if there are N candidates, the number of pairwise comparisons needed is

$$1 + 2 + 3 + 4 + \cdots + (N - 1) = \frac{(N - 1)N}{2}.$$

Because the number of comparisons grows quite fast in relation to the number of candidates, the method of pairwise comparisons is cumbersome and time consuming when there are more than a handful of candidates and is seldom used in practice as a method for deciding elections.

Rankings

Quite often it is important not only to know who wins the election but also to know who comes in second, third, etc. Let's consider once again the Math Appreciation Society election. Suppose now that instead of electing just the president we need to elect a board of directors consisting of a president, a vice president, and a treasurer. The club's bylaws state that rather than having separate elections for each office, the winner of the election gets to be the president, the second-place candidate gets to be the vice president, and the third-place candidate gets to be the treasurer. In a situation like this, we need a voting method that gives us not just a winner but also a second place, a third place, etc.—in other words, a **ranking** of the candidates.

Extended Ranking Methods

Each of the four voting methods we discussed earlier in this chapter has a natural extension that can be used to produce a ranking of the candidates.

Let's start with the plurality method and see how we might extend it to produce a ranking of the four candidates in the Math Appreciation Society election. For the reader's convenience, the preference schedule is shown again in Table 1-15.

Table 1-15 Preference Schedule for the MAS Election

Number of voters	14	10	8	4	1
1st choice	*A*	*C*	*D*	*B*	*C*
2nd choice	*B*	*B*	*C*	*D*	*D*
3rd choice	*C*	*D*	*B*	*C*	*B*
4th choice	*D*	*A*	*A*	*A*	*A*

The count of first-place votes is

A: 14 first-place votes

B: 4 first-place votes

C: 11 first-place votes

D: 8 first-place votes.

We know that, using the plurality method, *A* is the winner. Who should be second? The answer seems obvious: *C* has the second most first-place votes (11), so we declare *C* to come in second. By the same token, we declare *D* to come in third (8 votes) and *B* last. In short, the *extended plurality method* gives us a complete ranking of the candidates, shown in Table 1-16.

Table 1-16 Ranking the Candidates in the MAS Election Using the Extended Plurality Method

Office	Place	Candidate	First-place votes
President	1st	*A*	14
Vice president	2nd	*C*	11
Treasurer	3rd	*D*	8
	4th	*B*	4

Ranking the candidates using the extended Borda count method is equally simple. In the MAS election, for example, the point totals under the Borda count method were

A: 79 Borda points

B: 106 Borda points

C: 104 Borda points

D: 81 Borda points.

The resulting ranking, based on the *extended Borda count method,* is shown in Table 1-17.

Table 1-17 Ranking the Candidates in the MAS Election Using the Extended Borda Count Method

Office	Place	Candidate	Borda points
President	1st	*B*	106
Vice president	2nd	*C*	104
Treasurer	3rd	*D*	81
	4th	*A*	79

Ranking the candidates using the *extended plurality-with-elimination method* is a bit more subtle: We rank them in reverse order of elimination (the first candidate eliminated is ranked last, the second candidate eliminated is ranked next to last, etc.).[7] Table 1-18 shows the results of ranking the candidates in the MAS election using the *extended plurality-with-elimination method.*

Table 1-18 Ranking the Candidates in the MAS Election Using the Extended Plurality-with-Elimination Method

Office	Place	Candidate	Eliminated in
President	1st	*D*	
Vice president	2nd	*A*	3rd round
Treasurer	3rd	*C*	2nd round
	4th	*B*	1st round

Last, we can rank the candidates using the *extended method of pairwise comparisons* according to the number of pairwise comparisons won (recall that we count a tie as $\frac{1}{2}$ point). In the case of the MAS election, *C* won 3 pairwise comparisons, *B* won 2 pairwise comparisons, *D* won 1 pairwise comparison, and *A* won none. The results of ranking the candidates under the *extended method of pairwise comparisons* are shown in Table 1-19.

Table 1-19 Ranking the Candidates in the MAS Election Using the Extended Method of Pairwise Comparisons

Office	Place	Candidate	Points
President	1st	*C*	3
Vice president	2nd	*B*	2
Treasurer	3rd	*D*	1
	4th	*A*	0

A summary of the results of the MAS election using the different extended ranking methods is shown in Table 1-20.

Table 1-20 Ranking the Candidates in the MAS Election: A Tale of Four Methods

	Ranking			
Method	**1st**	**2nd**	**3rd**	**4th**
Extended plurality	*A*	*C*	*D*	*B*
Extended Borda count	*B*	*C*	*D*	*A*
Extended plurality with elimination	*D*	*A*	*C*	*B*
Extended pairwise comparisons	*C*	*B*	*D*	*A*

The most striking thing about Table 1-20 is the wide discrepancy of results. While it is somewhat frustrating to see this much equivocation, it is important to keep things in context: This is the exception rather than the rule. One purpose of the MAS example is to illustrate how crazy things can get in some elections, but in most real-life elections there tends to be much more consistency among the various methods.

[7] In cases where a candidate gets a majority of first-place votes before the ranking of all the candidates is complete, we continue the process of elimination to rank the remaining candidates.

Recursive Ranking Methods

We will now discuss a different, somewhat more involved strategy for ranking the candidates, which we will call the **recursive** approach. The basic strategy here is the same regardless of which voting method we choose—only the details are different.

Let's say we are going to use some voting method X and the recursive approach to rank the candidates in an election. We first use method X to find the winner of the election. So far, so good. We then remove the name of the winner on the preference schedule and obtain a new, modified preference schedule with one less candidate on it. We apply method X once again to find the "winner" based on this new preference schedule, and this candidate is ranked second. (This makes a certain amount of sense: What we're saying is that, after the winner is removed, we run a brand-new race, and the best candidate in that race is the second-best candidate overall.) We repeat the process again (cross out the name of the last winner, calculate the new preference schedule, and apply method X to find the next winner, who is then placed next in line in the ranking) until we have ranked as many of the candidates as we want.

We will illustrate the basic idea of recursive ranking with a couple of examples, both based on the Math Appreciation Society election.

Example 9.

Suppose we want to rank the four candidates in the MAS election using the *recursive plurality method*. The preference schedule, once again, is given in Table 1-21.

Table 1-21 Preference Schedule for the MAS Election

Number of voters	14	10	8	4	1
1st choice	*A*	*C*	*D*	*B*	*C*
2nd choice	*B*	*B*	*C*	*D*	*D*
3rd choice	*C*	*D*	*B*	*C*	*B*
4th choice	*D*	*A*	*A*	*A*	*A*

- **Step 1.** (Choose the winner using plurality.) We already know the winner is A with 14 first-place votes.
- **Step 2.** (Choose second place.) First we remove A from the original schedule—this gives us a "new" preference schedule to work with (Table 1-22).

Table 1-22 Preference Schedule for the MAS Election after *A* Has Been Removed

Number of voters	14	10	8	4	1
1st choice	*A*	*C*	*D*	*B*	*C*
2nd choice	*B*	*B*	*C*	*D*	*D*
3rd choice	*C*	*D*	*B*	*C*	*B*
4th choice	*D*	*A*	*A*	*A*	*A*

➡

Number of voters	14	10	8	4	1
1st choice	*B*	*C*	*D*	*B*	*C*
2nd choice	*C*	*B*	*C*	*D*	*D*
3rd choice	*D*	*D*	*B*	*C*	*B*

In this schedule the winner using plurality is B, with 18 first-place votes. Thus, *second place goes to B*.

- **Step 3.** (Choosing third place.) We now remove B from the preceding schedule. The resulting schedule is shown on the right in Table 1-23.

Table 1-23 Preference Schedule for the MAS Election after *A* and *B* Have Been Removed

Number of voters	14	10	8	4	1
1st choice	*B*	*C*	*D*	*B*	*C*
2nd choice	*C*	*B*	*C*	*D*	*D*
3rd choice	*D*	*D*	*B*	*C*	*B*

➡

Number of voters	25	12
1st choice	*C*	*D*
2nd choice	*D*	*C*

Using plurality, the winner for this schedule is *C* with 25 first-place votes. This means that third place goes to *C* and, needless to say, last place goes to *D*. The final ranking of the candidates under the *recursive plurality method* is shown in Table 1-24.

Table 1-24 Ranking the Candidates in the MAS Election Using the Recursive Plurality Method

Office	Place	Candidate
President	1st	*A*
Vice president	2nd	*B*
Treasurer	3rd	*C*
	4th	*D*

It is worth noting how different this ranking turned out to be when compared with the earlier ranking obtained using the *extended plurality method*. In fact, except for first place (which will always be the same), all the other positions turned out to be different. ■

Example 10.

For our second example, we will apply the *recursive plurality-with-elimination* method to rank the candidates in the MAS election.

To help the reader understand how this method works, we will make a semantic distinction. In running the plurality-with-elimination method, candidates are "eliminated" until there is a winner left. Having locked up a place in the ranking, this winner is then "removed" so that the election can be rerun for the next place in the ranking. Here is how it works:

- **Step 1.** We apply the plurality-with-elimination method to the original preference schedule and get a winner: *D*. (We did all the busy work earlier.)
- **Step 2.** We now remove the winner *D* from the preference schedule (Table 1-25).

Table 1-25 Preference Schedule for the MAS Election after the Winner *D* Has Been Removed

Number of voters	14	10	8	4	1
1st choice	*A*	*C*	*D*	*B*	*C*
2nd choice	*B*	*B*	*C*	*D*	*D*
3rd choice	*C*	*D*	*B*	*C*	*B*
4th choice	*D*	*A*	*A*	*A*	*A*

➡

Number of voters	14	10	8	4	1
1st choice	*A*	*C*	*C*	*B*	*C*
2nd choice	*B*	*B*	*B*	*C*	*B*
3rd choice	*C*	*A*	*A*	*A*	*A*

Once again, we apply the plurality-with-elimination method to the revised schedule. *B* is eliminated first, and *A* second (the reader should verify all the details), leaving *C* as the winner. This means that second place in the original election goes to *C*.

- **Step 3.** We now remove *C* from the last preference schedule (Table 1-26).

Table 1-26 Preference Schedule for the MAS Election after *D* and *C* Have Been Removed

Number of voters	14	10	8	4	1
1st choice	*A*	*C*	*C*	*B*	*C*
2nd choice	*B*	*B*	*B*	*C*	*B*
3rd choice	*C*	*A*	*A*	*A*	*A*

➡

Number of voters	14	10	8	4	1
1st choice	*A*	*B*	*B*	*B*	*B*
2nd choice	*B*	*A*	*A*	*A*	*A*

The winner of this election under plurality with elimination is *B*. This means that third place goes to *B*.

The final ranking of the candidates under the *recursive-plurality-with elimination method* is shown in Table 1-27.

Table 1-27 Ranking the Candidates in the MAS Election Using the Recursive Plurality-with-Elimination Method

Office	Place	Candidate
President	1st	*D*
Vice president	2nd	*C*
Treasurer	3rd	*B*
	4th	*A*

While somewhat more complicated than the extended ranking methods, the recursive ranking methods are of both practical and theoretical importance. We encourage the reader to give them a try (Exercises 31 through 40). ■

Conclusion: Fairness and Arrow's Impossibility Theorem

When is a voting method fair? Throughout this chapter we have introduced several standards of fairness known as *fairness criteria*.[8] Let's review what they are.

- **Majority Criterion.** If there is a candidate or alternative that has a majority of the first-place votes, then that candidate or alternative should be the winner of the election.

- **Condorcet Criterion.** If there is a candidate or alternative that is preferred by the voters over any other candidate or alternative, then that candidate or alternative should be the winner of the election.

[8] Singular: criterion; plural: criteria.

- **Monotonicity Criterion.** If candidate or alternative X is a winner of an election and, in a reelection, all the voters who change their preferences do so in a way that is favorable only to X, then X should still be a winner of the election.
- **Independence-of-Irrelevant-Alternatives Criterion.** If candidate or alternative X is a winner of an election, and one (or more) of the other candidates or alternatives is removed and the ballots recounted, then X should still be a winner of the election.

Each of the above four criteria represents a basic standard of fairness, and it is reasonable to expect that a fair voting method ought to satisfy all of them. Surprisingly, none of the four voting methods we discussed in this chapter does. The question remains: Is there a democratic voting method that satisfies all four of the fairness criteria—if you will, a perfect voting method? For elections involving more than two alternatives the answer is No! *No perfect voting method exists.*

At first glance, this fact seems a little surprising. Given the obvious importance of elections in a democracy and given the collective intelligence and imagination of social scientists and mathematicians, how is it possible that no one could come up with a voting method satisfying all of the fairness criteria? Up until the early 1950s this was one of the most challenging questions in social-choice theory. Finally, in 1952, Kenneth Arrow demonstrated the now famous **Arrow's impossibility theorem:** *It is mathematically impossible for a democratic voting method to satisfy all four of the fairness criteria.* No matter how hard we look for it, there can be no perfect voting method. Ironically, total and consistent fairness is inherently impossible in a democracy.

Key Concepts

Arrow's impossibility theorem
Borda count method
Condorcet candidate
Condorcet criterion
extended rankings methods
independence-of-irrelevant-alternatives criterion
insincere voting
majority criterion
method of pairwise comparisons
monotonicity criterion
plurality method
plurality-with-elimination method
preference ballot
preference schedule
rankings
recursive ranking methods

Exercises

Walking

1. The management of the XYZ Corporation has decided to treat their office staff to dinner. The choice of restaurants is The Atrium (A), Blair's Kitchen (B), The Country Cookery (C), and Dino's Steak House (D). Each of the 12 staff members is asked to submit a preference ballot listing his or her first, second, third, and fourth choices among these restaurants. The resulting preference ballots are shown in the figure.

Ballot	Ballot	Ballot	Ballot	Ballot	Ballot	Ballot	Ballot	Ballot	Ballot	Ballot	Ballot
1st A	1st C	1st B	1st C	1st C	1st C	1st A	1st C	1st A	1st A	1st C	1st A
2nd B	2nd B	2nd D	2nd B	2nd B	2nd B	2nd B	2nd B	2nd B	2nd B	2nd B	2nd B
3rd C	3rd D	3rd C	3rd A	3rd A	3rd D	3rd C	3rd A	3rd C	3rd C	3rd D	3rd C
4th D	4th A	4th A	4th D	4th D	4th A	4th D	4th D	4th D	4th D	4th A	4th D

(a) Write out the preference schedule for this election.

(b) Is there a majority winner?

(c) Find the winner of the election using the plurality method.

(d) Find the winner of the election using the Borda count method.

2. The Latin Club is holding an election to choose its president. There are three candidates, Arsenio, Beatrice, and Carlos (*A, B,* and *C* for short). The other 11 members of the club (the candidates are not allowed to vote) vote as shown below.

Voter	Sue	Bill	Tom	Pat	Tina	Mary	Alan	Chris	Paul	Kate	Ron
1st choice	*C*	*A*	*C*	*A*	*B*	*C*	*A*	*A*	*C*	*B*	*A*
2nd choice	*A*	*C*	*B*	*B*	*C*	*B*	*C*	*C*	*B*	*C*	*B*
3rd choice	*B*	*B*	*A*	*C*	*A*	*A*	*B*	*B*	*A*	*A*	*C*

(a) Write out the preference schedule for this election.

(b) Is there a majority winner?

(c) Find the winner of the election using the plurality method.

(d) Find the winner of the election using the Borda count method.

Exercises 3 through 6 refer to the following: A math class is asked by the instructor to vote among four possible times for the final exam—A (December 15, 8:00 A.M.), B (December 20, 9:00 P.M.), C (December 21, 7:00 A.M.), and D (December 23, 11:00 A.M.). The class preference schedule is given below:

Number of voters	3	4	9	9	3	5	8	2	12
1st choice	*A*	*A*	*A*	*B*	*B*	*B*	*C*	*C*	*D*
2nd choice	*B*	*B*	*C*	*C*	*A*	*C*	*D*	*A*	*C*
3rd choice	*C*	*D*	*B*	*D*	*C*	*A*	*B*	*D*	*A*
4th choice	*D*	*C*	*D*	*A*	*D*	*D*	*A*	*B*	*B*

3. (a) How many students in the class voted?

(b) Find the winner of the election using the plurality method.

4. Find the winner of the election using the Borda count method.

5. Find the winner of the election using the plurality-with-elimination method.

6. Find the winner of the election using the method of pairwise comparisons.

Exercises 7 through 13 refer to an election with 5 candidates (A, B, C, D, and E), 21 voters, and the preference schedule given below.

Number of voters	5	3	5	3	2	3
1st choice	*A*	*A*	*C*	*D*	*D*	*B*
2nd choice	*B*	*D*	*E*	*C*	*C*	*E*
3rd choice	*C*	*B*	*D*	*B*	*B*	*D*
4th choice	*D*	*C*	*A*	*E*	*A*	*C*
5th choice	*E*	*E*	*B*	*A*	*E*	*A*

7. Find the winner of the election using the Borda count method.

8. Find the winner of the election using:

(a) the plurality method;

(b) the plurality-with-elimination method;

(c) the method of pairwise comparisons.

9. Find the ranking of the candidates using the extended plurality-with-elimination method.

10. Find the ranking of the candidates using the extended plurality method.

11. Find the ranking of the candidates using the extended pairwise comparisons method.

12. Find the ranking of the candidates using the extended Borda count method.

13. Suppose that, for some unexplained reason, the votes in this election have to be recounted, but before this is done, candidate E drops out.

(a) Write out the preference schedule after E drops out.

(b) Find the winner of the new election using the Borda count method.

Exercises 14 through 20 refer to an election with 5 alternatives (R, H, C, O, and S), 100 voters, and the preference schedule given below.

Number of voters	**45**	**40**	**15**
1st choice	H	R	C
2nd choice	O	C	O
3rd choice	S	O	S
4th choice	C	S	R
5th choice	R	H	H

14. Find the winner of the election using the Borda count method.

15. Find the winner of the election using the plurality-with-elimination method.

16. Find the winner of the election using the method of pairwise comparisons.

17. Find the ranking of the alternatives using the extended plurality method.

18. Find the ranking of the alternatives using the extended plurality-with-elimination method.

19. Find the ranking of the alternatives using the extended Borda count method.

20. Find the ranking of the alternatives using the extended pairwise comparisons method.

Exercises 21 through 24 refer to an election with 5 candidates (A, B, C, D, and E), 24 voters, and the preference schedule given by the following table.

Number of voters	**8**	**7**	**6**	**2**	**1**
1st choice	A	D	D	C	E
2nd choice	B	B	B	A	A
3rd choice	C	A	E	B	D
4th choice	D	C	C	D	B
5th choice	E	E	A	E	C

21. (a) Find the winner of the election using the Borda count method.

(b) Does this example illustrate a violation of the majority criterion? Why or why not?

22. (a) Find the winner of the election using the plurality-with-elimination method.

(b) Does the example illustrate a violation of the Condorcet criterion? Why or why not?

23. Find the ranking of the candidates using the extended plurality method.

24. Find the ranking of the candidates using the extended Borda count method.

Exercises 25 through 27 refer to an election with 5 candidates (A, B, C, D, and E), 23 voters, and the preference schedule given below.

Number of voters	8	3	1	5	6
1st choice	*C*	*D*	*E*	*A*	*E*
2nd choice	*B*	*C*	*A*	*B*	*A*
3rd choice	*D*	*B*	*B*	*D*	*C*
4th choice	*A*	*E*	*C*	*E*	*D*
5th choice	*E*	*A*	*D*	*C*	*B*

25. Find the ranking of the candidates using the extended pairwise comparisons method.

26. (a) Find the winner of the election using the Borda count method.

(b) Does this example illustrate a violation of the majority criterion? Why or why not?

(c) Does this example illustrate a violation of the Condorcet criterion? Why or why not?

27. (a) Find the ranking of the candidates using the extended plurality-with-elimination method.

(b) Does this example illustrate a violation of the Condorcet criterion? Why or why not?

Exercises 28 through 31 refer to an election with 4 candidates (A, B, C, and D), 27 voters, and the preference schedule given below.

Number of voters	4	1	9	8	5
1st choice	*A*	*B*	*C*	*A*	*D*
2nd choice	*C*	*A*	*D*	*B*	*C*
3rd choice	*B*	*D*	*A*	*D*	*B*
4th choice	*D*	*C*	*B*	*C*	*A*

28. (a) Find the winner of the election using the plurality method.

(b) Does this example illustrate a violation of the Condorcet criterion? Why or why not?

29. (a) Find the winner of the election using the Borda count method.

(b) Does this example illustrate a violation of the majority criterion? Why or why not?

30. Find the ranking of the candidates using the extended plurality-with-elimination method.

31. Find the ranking of the candidates using the recursive plurality-with-elimination method.

Exercises 32 through 34 refer to the MAS election. For the reader's convenience the preference schedule is given below.

Number of voters	14	10	8	4	1
1st choice	*A*	*C*	*D*	*B*	*C*
2nd choice	*B*	*B*	*C*	*D*	*D*
3rd choice	*C*	*D*	*B*	*C*	*B*
4th choice	*D*	*A*	*A*	*A*	*A*

32. Find the ranking of the candidates using the recursive Borda count method.

33. Find the ranking of the candidates using the recursive pairwise comparisons method.

34. Find the ranking of the candidates using the recursive plurality method.

Exercises 35 through 37 refer to an election with 5 candidates (A, B, C, D and E), 24 voters, and the preference schedule given below.

Number of voters	8	6	2	3	5
1st choice	*A*	*B*	*C*	*D*	*E*
2nd choice	*B*	*D*	*A*	*E*	*A*
3rd choice	*C*	*E*	*E*	*A*	*D*
4th choice	*D*	*C*	*B*	*C*	*B*
5th choice	*E*	*A*	*D*	*B*	*C*

35. Find the ranking of the candidates using the recursive plurality-with-elimination method.

36. Find the ranking of the candidates using the recursive pairwise comparisons method.

37. Find the ranking of the candidates using the recursive Borda count method.

Exercises 38 through 40 refer to the same election as Exercises 25 through 27.

38. Find the ranking of the candidates using the recursive Borda count method.

39. Find the ranking of the candidates using the recursive plurality-with-elimination method.

40. Find the ranking of the candidates using the recursive pairwise comparisons method.

Jogging

41. **(a)** Suppose that 50 players sign up for a round-robin tennis tournament (everyone plays everyone else). How many tennis matches must be scheduled?

(b) If there are 50 people in a room and everyone kisses everyone else (on the cheek of course), how many kisses take place?

42. Show that the four voting methods we discussed in this chapter give the same winner when there are only two candidates and that the winner is just determined by straight majority.

43. **The mystery election problem.** You are given the following information about an election. There are 5 candidates and 21 voters. The preference schedule for the election has been lost—the only thing you know is that there were only *two* columns in the schedule. (As you can imagine, this is the key piece of information in the problem.)

(a) Explain why there must be a majority winner in this election.

(b) Explain why the argument given in (a) still works with any number of candidates and any odd number of voters.

44. An election is held among 4 candidates (*A*, *B*, *C*, and *D*) using the Borda count method. There are 11 voters. Suppose that after the ballots are in and the points are tallied, *B* gets 32 points, *C* gets 29 points, and *D* gets 18 points. How many points does *A* get, and why?

45. An election involving 5 candidates and 21 voters is held, and the results of the election are to be determined using the Borda count method. Unfortunately, the elections committee completely botches up the addition of the points. After a complaint is lodged by one of the candidates, the results of the election are recomputed using 4 points for a first-place vote, 3 points for a second-place vote, 2 points for a third-place vote, 1 point for a fourth-place vote, and 0 points for a fifth-place vote. Explain why computing the results this way gives the same election outcome as using the traditional Borda count method.

46. Explain why the plurality method satisfies the monotonicity criterion.

47. Explain why the plurality-with-elimination method satisfies the majority criterion.

48. Explain why the method of pairwise comparisons satisfies the majority criterion.

49. Explain why any method that violates the majority criterion must also violate the Condorcet criterion.

50. The preference schedule below shows the results of an election with 4 candidates (A, B, C, and D) and 13 voters.

Number of voters	**7**	**4**	**2**
1st choice	A	B	D
2nd choice	B	D	A
3rd choice	C	C	C
4th choice	D	A	B

(a) Does this election have a candidate with a majority of the first-place votes?

(b) Does this election have a Condorcet candidate?

(c) Find the winner of this election using the Borda count method.

(d) Suppose that candidate C drops out of the race. Who among the remaining candidates wins the election using the Borda count method?

(e) Based on (a) through (d), which of the four fairness criteria are violated in this election? Explain.

51. The preference schedule below shows the results of an election with 4 candidates (A, B, C, and D) and 27 voters.

Number of voters	**10**	**6**	**5**	**4**	**2**
1st choice	A	B	C	D	D
2nd choice	C	D	B	C	A
3rd choice	D	C	D	B	B
4th choice	B	A	A	A	C

(a) Does this election have a candidate with a majority of the first-place votes?

(b) Does this election have a Condorcet candidate?

(c) Find the winner of this election using the plurality-with-elimination method.

(d) Suppose that candidate D drops out of the race. Who among the remaining candidates wins the election using the plurality-with-elimination method?

(e) Based on (a) through (d), which of the four fairness criteria are violated in this election? Explain.

52. Give an example of an election decided under the Borda count method in which the Condorcet criterion is violated but the majority criterion is not.

53. The table below shows the three top-ranked college football teams at the end of the 1993 football season according to the CNN/*USA Today* coaches poll.

Team	Points	Number of first-place votes
1. Florida State	1523	36
2. Notre Dame	1494	25
3. Nebraska	1447	1
⋮	⋮	⋮

This poll is based on the votes of 62 coaches, each one of whom ranks the top 25 teams. (The remaining 22 teams are not shown because they are irrelevant to this exercise.) A team gets 25 points for each first-place vote, 24 points for each second-place vote, 23 points for each third-place vote, etc.

(a) Based on the information give in the table it is possible to conclude that all 62 coaches had Florida State, Notre Dame, and Nebraska in some order as their top three choices. Explain why this is true.

(b) Find the number of second- and third-place votes for each of the 3 teams.

54. Plurality with a runoff. This is a simple variation of the plurality-with-elimination method. Here, if a candidate has a majority of the first-place votes, then that candidate wins the election; otherwise we eliminate all candidates except the two with the most first-place votes. The winner is then chosen between these two by recounting the votes in the usual way.

(a) Use the MAS election to show that plurality with a runoff can produce a different outcome than plurality-with-elimination.

(b) Give an example that shows that plurality with a runoff violates the monotonicity criterion.

(c) Give an example that shows that plurality with a runoff violates the Condorcet criterion.

Running

55. The Coombs method. This method is just like the plurality-with-elimination method except that in each round we eliminate the candidate with the *largest number of last-place votes* (instead of the one with the fewest first-place votes).

(a) Find the winner of the MAS election using the Coombs method.

(b) Give an example showing that the Coombs method violates the Condorcet criterion.

(c) Give an example showing that the Coombs method violates the monotonicity criterion.

56. Give an example (do not use one given in the book) of an election decided under the plurality-with-elimination method in which the Condorcet criterion is violated.

57. Show that the method of pairwise comparisons satisfies the monotonicity criterion.

58. Show that if, in an election with an odd number of voters, there is no Condorcet candidate, then any ranking of the candidates based on the extended pairwise comparisons method must result in at least two candidates ending up tied in the rankings.

59. The Pareto criterion. The following fairness criterion was proposed by the Italian economist Vilfredo Pareto (1848–1923): *If every voter prefers alternative X over alternative Y, then a voting method should not choose Y as the winner.* Show that all four voting methods discussed in the chapter satisfy the Pareto criterion. (A separate analysis is needed for each of the four methods.)

60. Suppose the following was proposed as a fairness criterion: *If a majority of the voters prefer alternative X to alternative Y, then the voting method should rank X above Y.* Give an example to show that all four of the extended voting methods discussed in the chapter can violate this criterion. (Hint: Consider an example with no Condorcet candidate.)

61. Consider the following fairness criterion: *If a majority of the voters prefer every alternative over alternative X, then a voting method should not choose alternative X as the winner.*

(a) Give an example to show that the plurality method an violate this criterion.

(b) Give an example to show that the plurality-with-elimination method can violate this criterion.

(c) Explain why the method of pairwise comparisons satisfies this criterion.

(d) Explain why the Borda count method satisfies this criterion.

62. The Condorcet loser criterion. *If there is an alternative that loses in a one-to-one comparison to each of the other alternatives, then that alternative should not be the winner of the election.* (This fairness criterion is a sort of mirror image of the regular Condorcet criterion.)

(a) Give an example that shows that the plurality method can violate the Condorcet loser criterion.

(b) Explain why the plurality-with-elimination method violates the Condorcet loser criterion.

(c) Explain why the Borda count method satisfies the Condorcet loser criterion.

63. Consider a variation of the Borda count method in which a first-place vote in an election with N candidates is worth F points, (where $F > N$), and all other places in the ballot are the same as in the ordinary Borda count: $N - 1$ points for second place, $N - 2$ points for third place, . . . , 1 point for last place. By making F large enough, this variation of the Borda count method can be made to satisfy the majority criterion. Find the smallest value of F for which this happens.

APPENDIX 1: Breaking Ties

By and large, most of the examples given in the chapter were carefully chosen to avoid tied winners, but of course in the real world ties are bound to occur.

In this appendix we will discuss very briefly the problem of how to break ties when necessary. Tie-breaking methods can raise some fairly complex issues, and our purpose here is not to study such methods in great detail but rather to make the reader aware of the problem and give some inkling as to possible ways to deal with it.

For starters, consider the election with preference schedule shown in Table A-1. If we look at this preference schedule carefully, we can see that there is complete *symmetry* in the positions of the three candidates. Essentially, this means that we could interchange the names of the candidates and the preference schedule would not change. Given the complete symmetry of the preference schedule, it is clear that no rational voting method could choose one candidate as the winner over the other two. In this situation, a tie is inevitable regardless of the voting method used. We call this kind of tie an **essential tie.** Essential ties cannot be broken using a rational tie-breaking procedure, and we must rely on some sort of outside intervention such as chance (flip a coin, draw straws, etc.), a third party (the judge, mom, etc.), or even some outside factor (experience, age, etc.).

Table A-1 A Three-Way Essential Tie

Number of voters	7	7	7
1st choice	*A*	*B*	*C*
2nd choice	*B*	*C*	*A*
3rd choice	*C*	*A*	*B*

Most ties are not essential ties, and they can often be broken in a more rational way: either by implementing some tie-breaking rule or by using a different voting method to break the tie. To illustrate some of these ideas let's consider as an example the election with preference schedule shown in Table A-2.

Table A-2 A Tie That Could Be Broken

Number of voters	5	3	5	3	2	4
1st choice	*A*	*A*	*C*	*D*	*D*	*B*
2nd choice	*B*	*B*	*E*	*C*	*C*	*E*
3rd choice	*C*	*D*	*D*	*B*	*B*	*A*
4th choice	*D*	*C*	*A*	*E*	*A*	*C*
5th choice	*E*	*E*	*B*	*A*	*E*	*D*

If we decide this election using the method of pairwise comparisons, we have:

A versus *B* (13 to 9): *A* wins

A versus *C* (12 to 10): *A* wins

A versus *D* (12 to 10): *A* wins

A versus *E* (10 to 12): *E* wins

B versus *C* (12 to 10): *B* wins

B versus *D* (12 to 10): *B* wins

B versus *E* (17 to 5): *B* wins

C versus *D* (14 to 8): *C* wins

C versus *E* (18 to 4): *C* wins

D versus *E* (13 to 9): *D* wins

In this election *A* and *B*, with three wins each, tie for first place. How could we break this tie? Here is just a sampler of the many possible ways:

1. Use the results of a pairwise comparison between the winners. In the above example, since *A* beats *B* 13 votes to 9, the tie would be broken in favor of *A*.
2. Use the total point differentials. For example, since *A* beats *B* 13 to 9, the point differential for *A* is $+4$, and since *A* lost to *E* 10 to 12, the point differential for *A* is -2. Computing the total point differentials for *A* gives $4 + 2 + 2 - 2 = 6$. Likewise, the total point differential for *B* is $2 + 2 + 12 - 4 = 12$. In this case the point differentials favor *B*, so *B* would therefore be declared the winner.
3. Use first-place votes. In the example, *A* has 8 and *B* has 4. With this method, the winner would be *A*.
4. Use Borda count points to choose between the two winners. Here

   A has $(5 \times 8) + (3 \times 4) + (2 \times 7) + (1 \times 3) = 69$ points;

   B has $(5 \times 4) + (4 \times 8) + (3 \times 5) + (1 \times 5) = 72$ points;

   and the tie would be broken in favor of *B*.

By now we should not be at all surprised that different tie-breaking methods produce different winners and that there is no single *right* method for breaking ties. In retrospect, flipping a coin might not be such a bad idea!

APPENDIX 2: A Sampler of Elections in the Real World

Olympic Venues. The selection of the city that gets to host the Olympic Games has tremendous economic and political impact for the cities involved, and it goes without saying that it always generates a fair amount of controversy. The process of selection is carried out by means of an election very much like some of the ones we studied in this chapter (see Example 6). The voters are the members of the International Olympic Committee, and the actual voting method used to select the winner is the plurality-with-elimination method with a minor twist: Instead of indicating their preferences all at once, the voters let their preferences be known one round at a time. Here are the actual details of how Sydney, Australia, was chosen to host the 2000 Summer Olympic Games.

Sydney, Australia, site of the 2000 Olympics. Olympics host cities are chosen using the plurity-with-elimination method.

On September 23, 1993, the 89 members of the International Olympic Committee met in Monte Carlo, Monaco, to vote on the selection of the site for the 2000 Summer Olympics. Five cities made bids: Beijing (China), Berlin (Germany), Istanbul (Turkey), Manchester (England), and Sydney (Australia). In each round, the delegates voted for just one city, and the city with the fewest votes was eliminated. The voting went as follows:

- **Round 1.**

City	Beijing	Berlin	Istanbul	Manchester	Sydney
Votes	32	9	7	11	30

Istanbul is eliminated in round 1.

- **Round 2.**

City	Beijing	Berlin	Manchester	Sydney
Votes	37	9	13	30

Berlin is eliminated in round 2.

- **Round 3.**

City	Beijing	Manchester	Sydney	Abstentions
Votes	40	11	37	1

Manchester is eliminated in round 3

- **Round 4.**

City	Beijing	Sydney	Abstentions
Votes	43	45	1

Beijing is eliminated in round 4; Sydney gets the gold!

The Academy Awards. The Academy of Motion Picture Arts and Sciences gives its annual Academy Awards ("Oscars") for various achievements in connection with motion pictures (best picture, best director, best actress, etc.). The winner in each category is chosen by means of an election held among the eligible members of the Academy. The election process varies slightly from award to award and is quite complicated. For the sake of brevity we will describe the election process for best picture. (The process is almost identical for each of the major awards.) The election takes place in two stages: (1) the nomination stage, in which the five top pictures are nominated, and (2) the final balloting for the winner.

We describe the second stage first because it is so simple: Once the five top pictures are nominated, each eligible member of the Academy is asked to vote for one picture, and the winner is chosen by simple plurality. Because the number of voters is large (somewhere between 4000 and 5000), ties are not likely to occur, but if they do, they are not broken. Thus, it is possible for two candidates to share an award.

The process for selecting the five nominations is considerably more complicated and is based on a voting method called **single transferable voting.** Each eligible member of the Academy is asked to submit a preference ballot with the names of their top five choices ranked from first to fifth. Based on the total number of valid ballots submitted, the minimum number of votes needed to get a nomination (called the **quota**) is established, and any picture with enough first-place votes to make the quota is automatically nominated.

The quota is always chosen to be a number that is over one-sixth (16.66%) but not more than one-fifth (20%) of the total number of valid ballots cast. (Setting the quota this way ensures that it is impossible for six or more pictures to get automatic nominations.) While in theory it is possible for five pictures to make the quota right off the bat and get an automatic nomination (in which case the nomination process is over), this has never happened in practice. In fact, what usually happens is that there are no pictures that make the quota automatically. Then, the picture with the fewest first-place votes (say X) is eliminated, and on all the ballots that originally had X as the first choice, X's name is crossed off the top and all the other pictures are moved up one spot. The ballots are then counted again. If there are still no pictures that make the quota, the process of elimination is repeated. Eventually, there will be one or more pictures that make the quota and are nominated.

The moment that one or more pictures are nominated, there is a new twist: Nominated pictures "give back" to the other pictures still in the running (not nominated but not eliminated either) their "surplus" votes. This process of giving back votes (called a **transfer**) is best illustrated with an imaginary example. Suppose that the quota is 400 (a nice, round number) and at some point a picture (say Z) gets 500 first-place votes, enough to get itself nominated. The surplus for Z is $500 - 400 = 100$ votes, and these are votes that Z doesn't really need. For this reason the 100 surplus votes are taken away from Z and divided fairly among the second-place choices on the 500 ballots cast for Z. The way this is done may seem a little bizarre, but it makes perfectly good sense. Since there are 100 surplus votes to be divided into 500 equal shares, each second-place vote on the 500 ballots cast for Z is worth $\frac{100}{500} = \frac{1}{5}$ vote. While one-fifth of a vote may not seem like much, enough of these fractional votes can make a difference and help some other picture or pictures make the quota. If that's

the case, then once again the surplus or surpluses are transferred back to the remaining pictures following the procedure described above; otherwise, the process of elimination is started up again. Eventually, after several possible cycles of eliminations, transfers, eliminations, transfers, . . . , five pictures get enough votes to make the quota and be nominated, and the process is over.

The method of single transferable voting is not unique to the Academy Awards. It is used to elect officers in various professional societies as well as the members of the Irish Senate.

Corporate Boards of Directors. In most corporations and professional societies, the members of the Board of Directors are elected by a method called **approval voting.** In approval voting, a voter does not cast a preferential ballot but rather votes for as many candidates as he or she wants. Each of these votes is simply a yes vote for the candidate, and it means that the voter approves of that candidate. The candidate with the most approval votes wins the election.

Table A-3 shows an example of a hypothetical election based on approval voting.

Table A-3 An Election Based on Approval Voting

	Voters							
Candidates	**Sue**	**Bill**	**Tito**	**Prince**	**Tina**	**Van**	**Devon**	**Ike**
A	Yes		Yes	Yes	Yes		Yes	Yes
B			Yes		Yes	Yes		
C	Yes				Yes	Yes	Yes	

The results of this election are as follows: Winner, *A* (6 approval votes); second place, *C* (4 approval votes); last place, *B* (3 approval votes). Note that a voter can cast anywhere from no approval votes at all (such as Bill did above) to approval votes for all the candidates (such as Tina did above). It is somewhat ironic that the effect of Tina's vote is exactly the same as that of Bill's.

In the last few years a strong case has been made suggesting that for political elections, approval voting is a big improvement over the more traditional voting methods. In particular, approval voting encourages voter turnout. The reason for this is psychological: Voters are more likely to vote when they feel they can make intelligent decisions, and unquestionably it is easier for a voter to give an intelligent answer to the question, Do you approve of this candidate—yes or no? than it is to the question, Which candidate is your first choice, second choice, etc.? The latter requires a much deeper knowledge of the candidates, and in today's complex political world it is a knowledge that very few voters have.

References and Further Readings

1. Arrow, Kenneth J., *Social Choice and Individual Values.* New York: John Wiley & Sons, Inc., 1963.
2. Brams, Steven J., and Peter C. Fishburn, *Approval Voting.* Boston: Birkhäuser, 1982.
3. Dummett, M., *Voting Procedures.* New York: Oxford University Press, 1984.
4. Farquharson, Robin, *Theory of Voting.* New Haven, CT: Yale University Press, 1969.
5. Fishburn, Peter C., and Steven J. Brams, "Paradoxes of Preferential Voting," *Mathematics Magazine,* 56 (1983), 207–214.
6. Gardner, Martin, "Mathematical Games (From Counting Votes to Making Votes Count: The Mathematics of Elections)," *Scientific American,* 243 (October 1980), 16–26.
7. Guinier, Lani, *The Tyranny of the Majority: Fundamental Fairness in Representative Democracy.* New York: Free Press, 1994.
8. Kelly, J., *Arrow Impossibility Theorems.* New York: Academic Press, 1978.
9. Merrill, S., *Making Multicandidate Elections More Democratic.* Princeton, NJ: Princeton University Press, 1988.
10. Niemi, Richard G., and William H. Riker, "The Choice of Voting Systems," *Scientific American,* 234 (June 1976), 21–27.
11. Nurmi, H., *Comparing Voting Systems.* Dordretch, Holland: D. Reidel, 1987.
12. Saari, D. G., *The Geometry of Voting.* New York: Springer-Verlag, 1994.
13. Straffin, Philip D., Jr., *Topics in the Theory of Voting,* UMAP Expository Monograph. Boston: Birkhäuser, 1980.
14. Taylor, Alan, *Mathematics and Politics: Strategy, Voting, Power and Proof.* New York: Springer-Verlag, 1995.

Each state shall appoint . . . a Number of Electors, equal to the whole Number of Senators and Representatives to which the State may be entitled in the Congress. . . .

Article II, Section 1, Constitution of the United States

Weighted Voting Systems

The Power Game

CHAPTER 2

In a democracy we take many things for granted, not the least of which is the idea that we are all equal. When it comes to individual rights, the principle of *one person–one vote* is the core principle of a democratic society. But is the principle of *one person–one vote* always justified? Should it also apply when the *voters* are something other than individuals, such as organizations, states, and even countries? Shouldn't differences between *voters* sometimes be recognized?

In a diverse society, it is in the very nature of things that voters—be they individuals or institutions—are not equal, and sometimes it is actually desirable to recognize their differences by giving them different amounts of say over the outcome of the election. What we are talking about here is the exact philosophical opposite of the principle of *one voter–one vote,* a principle naturally described as *one voter–x votes,* but more formally known as **weighted voting.**

A classic example of weighted voting is built into the process for electing the president of the United States—the controversial and much-maligned Electoral College. In the Electoral College, the *voters* are the 50 states plus the District of Columbia, and, as we all know, they are far from being equal: Whereas California gets 54 votes to choose the president, Montana only gets three! (A state gets as many votes as it has delegates in Congress, and the District of Columbia gets three.)

Other examples of weighted voting can be found in regional and local governments such as county boards of supervisors and school boards; international legislative bodies such as the United Nations Security Council; corporate shareholders' elections (each shareholder controls as many votes as the number of shares he or she owns); and even at home, where it often seems to be the case that mom (OK, sometimes it's dad) has more votes than anyone else.

In this chapter we will analyze and discuss **weighted voting systems,** that is to say voting situations in which voters are not necessarily equal in terms of the number of votes they control. (The name makes sense once we learn that the word **weight** is used to describe the number of votes controlled by a player.)

An issue of primary interest when studying weighted voting systems is that of **power**: How much *power* over the outcome of the election does each voter have? Of course, understanding who has the power and how much of it they have is important and useful in almost any walk of life. In this chapter we will develop a mathematical understanding of how to measure power.

Weighted Voting Systems

To keep things simple, in this chapter we will consider voting between only *two* alternatives or candidates. Any vote involving only two choices can be thought of as a *yes-no* vote and is generally referred to as a **motion.** The great advantage of dealing with motions is that we don't have to worry about the choice of voting method, because all reasonable voting methods boil down to the same rule: *majority wins.*

We will start by introducing some terminology and illustrating the basic elements of every weighted voting system.

Terminology

Every weighted voting system is characterized by three elements: the *players,* the *weights* of the players, and the *quota.* The **players** are just the voters themselves. (From now on we will stick to the usual convention of using "voters" when we are dealing with a *one person–one vote* situation as in Chapter 1, and "players" in the case of a weighted voting system.) We will use the letter N to represent the number of players and the symbols $P_1, P_2, \ldots, P_N$ to represent the names of the players—it is a little less personal but a lot more convenient than using Archie, Betty, Jughead, etc. In a weighted voting system, each player controls a certain number of votes, and this number is called the player's **weight.** We will use the symbols $w_1, w_2, \ldots, w_N$ to represent the weights of $P_1, P_2, \ldots, P_N$, respectively. Finally, there is the **quota,** the minimum number of votes needed to pass a motion. We will use the letter q to denote the quota.

It is important to note that the quota q can be something other than a strict majority of the votes. There are many voting situations in which a majority of the votes is not enough to pass a motion—the rules may stipulate a different definition of what is needed for passing. Take, for example, the rules in the U.S. Senate. To pass an ordinary law, a strict majority of the votes is sufficient, but when the Senate is attempting to override a presidential veto, the rules state that two-thirds of the votes are needed. In other organizations the rules may stipulate that 75% of the votes are needed or 83% (why not?) or even 100% (unanimous consent). In fact, any number can be a reasonable choice for the quota q, as long as *it is more than half of the total number of votes but not more than the total number of votes.* To put it somewhat more formally,

$$\frac{w_1 + w_2 + \cdots + w_N}{2} < q \leq w_1 + w_2 + \cdots + w_N.$$

Notation and Examples

A convenient way to describe a weighted voting system is

$$[q: w_1, w_2, \ldots, w_N].$$

The quota is always given first, followed by a colon and then the respective wights of the individual players. (It is customary to write the weights in numerical order, starting with the highest, and we will adhere to this convention throughout the chapter.)

Example 1.

Consider [25: 8, 6, 5, 3, 3, 3, 2, 2, 1, 1, 1, 1].

This is a weighted voting system with 12 players ($P_1, P_2, \ldots, P_{12}$). P_1 has 8 votes, P_2 has 6 votes, P_3 has 5 votes, etc. The total number of votes is 36. The quota is 25 (informally, this is described as *over two-thirds required to pass a motion*). ■

Example 2.

Consider [7: 5, 4, 4, 2].

This weighted voting system doesn't make any sense because *the quota (7) is less than half of the total number of votes (15)*! If P_1 and P_4 voted yes and P_2 and P_3 voted no, both groups would win. This is a mathematical version of anarchy, and we will not consider this to be a legal weighted voting system. ■

Example 3.

Consider [17: 5, 4, 4, 2].

Here the quota is too high. In this weighted voting system no motion could ever pass. We can't allow this to pass either! ■

Example 4.

Consider the weighted voting system [11: 4, 4, 4, 4, 4].

In this weighted voting system all 5 players are equal. To pass a motion at least 3 out of the 5 players are needed. Note that if the quota ($q = 11$) were changed to 12, the situation would still remain the same—at least 3 out of the 5 players would be needed. What we really have here, somewhat in disguise, is a *one person–one vote situation with strict majority needed for passing a motion.*

In terms of how it works, this weighted voting system is equivalent to the weighted voting system [3: 1, 1, 1, 1, 1]. ■

Example 5.

Consider the weighted voting system [15: 5, 4, 3, 2, 1].

Here we have 5 players with a total of 15 votes. Since the quota is 15, the *only way a motion can pass is by unanimous consent of the players.* How does this voting system differ from the voting system [5: 1, 1, 1, 1, 1]? Well, the latter also has 5 players, and the only way a motion can pass is by unanimous consent of the players. So, in terms of how they work, [15: 5, 4, 3, 2, 1] and [5: 1, 1, 1, 1, 1] are equivalent weighted voting systems. ■

The surprising conclusion of Example 5 is that the weighted voting system [15: 5, 4, 3, 2, 1] describes a one person–one vote situation in disguise. This seems like a contradiction only if we think of a *one person–one vote* situation as implying that all players have an *equal number of votes rather than an equal say in the outcome of the election.* Apparently, these two things are not the same! As the example makes abundantly clear, just looking at the number of votes a player controls can be very deceptive.

Power; More Terminology; More Examples

Let's look at a few more examples of weighted voting systems and start to informally focus on the notion of power.

Example 6.

Consider the weighted voting system [11: 12, 5, 4].

Here is a situation in which a single player (P_1) controls enough votes to pass any measure single-handedly. Such a player has all the power, and, not surprisingly, we call such a player a *dictator.* ■

In general, we will say that a player is a **dictator** if the player's weight is bigger than or equal to the quota. Notice that whenever there is a dictator, all the other players, regardless of their weights, have absolutely no power. A player without power is called a **dummy.**

Example 7.

Consider the weighted voting system [12: 9, 5, 4, 2].

Here we have a situation in which player P_1, while not a dictator, has the power to obstruct by preventing any motion from passing. This happens because even if all the remaining players were to vote together, they wouldn't have the votes to pass a motion against the will of P_1. ■

A player that is not a dictator but can single-handedly prevent the rest of the players from passing a motion is said to have **veto power.**

Example 8.

Consider the weighted voting system [101: 99, 98, 3].

How is power distributed in this weighted voting system? At first glance it appears that P_1 and P_2 have lots of power while P_3 has very little power (if any). On closer inspection, however, we notice that it takes two of the players to pass a motion, and in fact, any two can do so. It seems appropriate, therefore, to claim that P_3, with a measly three votes, has as much power as either of the other two players. While hard to believe, this is in fact the case: The three players have equal power in this weighted voting system. ■

The Banzhaf Power Index

We are almost ready to formally introduce our first mathematical interpretation of power for weighted voting systems. This particular definition of power was suggested by John Banzhaf[1] in 1965.

Let's analyze the weighted voting system [101: 99, 98, 3] (Example 8) in a little more detail. Although this example itself is fairly simple, we will use it to introduce some important concepts.

Which sets of players could join forces and, voting together, carry a motion? Looking at the numbers, we can see that there are four such sets:

Set 1: P_1 and P_2 (this group controls 197 votes)
Set 2: P_1 and P_3 (this group controls 102 votes)
Set 3: P_2 and P_3 (this group controls 101 votes, just enough to win)
Set 4: P_1, P_2, and P_3 (this group controls all the votes).

From now on we will adhere to the standard language of voting theory and call any set of players that might join forces to vote together a **coalition.** (We use the word "coalition" in a rather generous way and will allow for even single-player coalitions.) The total number of votes controlled by a coalition is called the **weight of the coalition.** Of course, some coalitions have enough votes to win and some don't. Quite naturally, we call the former **winning coalitions,** and the latter **losing coalitions.**

Since coalitions are just sets of players, the most convenient way to describe coalitions mathematically is to use set notation. For example, the coalition consisting of players P_1 and P_2 can be written as the set $\{P_1, P_2\}$, the coalition consisting of just player P_2 by itself can be written as the set $\{P_2\}$, and so on. Table 2-1 summarizes the situation in Example 8.

Table 2-1 The Seven Possible Coalitions for Example 8

	Coalition	Coalition Weight	Win or Lose
1	$\{P_1\}$	99	Lose
2	$\{P_2\}$	98	Lose
3	$\{P_3\}$	3	Lose
4	$\{P_1, P_2\}$	197	Win
5	$\{P_1, P_3\}$	102	Win
6	$\{P_2, P_3\}$	101	Win
7	$\{P_1, P_2, P_3\}$	200	Win

[1] Banzhaf, who was a lawyer and not a mathematician, was mostly concerned with issues of equity and fairness in state and local systems of government.

If we now analyze the winning coalitions in Table 2-1, we notice the following: In coalitions 4, 5, and 6 both players are *critical* for the win (if either player were to leave the coalition, the coalition would no longer have the votes to carry a motion); in coalition 7 no *single* player is critical to the win (even if a player were to desert the coalition, the coalition would have enough votes to carry a motion).

We will look for players whose desertion turns a winning coalition into a losing coalition, and we will call such a player a **critical player** for the coalition. Notice that a winning coalition can have more than one critical player, and occasionally a winning coalition has no critical players. Losing coalitions never have critical players.

The critical-player concept is the basis for the definition of the **Banzhaf power index.** Banzhaf's key idea is that a player's power is proportional to the number of coalitions for which that player is critical, so that the more often the player is critical, the more power he or she holds.

We now know that in Example 8 each player is critical twice, so they all have equal power. Since there are three players, we can say that each player holds one-third of the power.

We can now formalize our approach for finding the Banzhaf power index of any player in a generic weighted voting system with N players.

Finding the Banzhaf Power Index of Player *P*

- **Step 1.** Make a list of all possible coalitions.
- **Step 2.** Determine which of them are winning coalitions.
- **Step 3.** In each winning coalition, determine which of the players are *critical* players.
- **Step 4.** Count the total number of times player P is critical. (Let's call this number B.)
- **Step 5.** Count the total number of times all playes are critical. (Let's call this number T.)

The Banzhaf power index of player P is then given by the fraction B/T.

Example 9.

Foreman & Sons is a family-owned corporation. Three generations of Foremans (George I, George II, and George III) are involved in its management, but, their names notwithstanding, the Foremans are not all the same. When it comes to making final decisions, George I has 3 votes, George II has 2 votes, and George III has 1 vote. A majority of 4 (out of the 6 possible votes) is needed to carry a motion. How is the power divided among the three Georges?

What we have here is the weighted voting system [4: 3, 2, 1]. To find the Banzhaf power index of each player we follow the five steps described above. (For consistency, we will use P_1 for George I, P_2 for George II, and P_3 for George III.)

- **Step 1.** There are 7 possible coalitions. They are
  $\{P_1\}, \{P_2\}, \{P_3\}, \{P_1, P_2\}, \{P_1, P_3\}, \{P_2, P_3\}, \{P_1, P_2, P_3\}$
- **Step 2.** The winning coalitions are
  $\{P_1, P_2\}, \{P_1, P_3\}$, and $\{P_1, P_2, P_3\}$.

- **Step 3.**

Winning coalitions	Critical players
$\{P_1, P_2\}$	P_1 and P_2
$\{P_1, P_3\}$	P_1 and P_3
$\{P_1, P_2, P_3\}$	P_1 only

- **Step 4.**

  P_1 is critical three times.

  P_2 is critical one time.

  P_3 is critical one time.

- **Step 5.**
  $T = 5$.

The Banzhaf power index of each of the players is

P_1: $\frac{3}{5}$

P_2: $\frac{1}{5}$

P_3: $\frac{1}{5}$

Note the surprising fact that P_2 and P_3 have the same Banzhaf power indexes. One has to wonder, Why does George II put up with this arrangement? (Could it be that he doesn't quite fully understand the mathematics of power?) ■

We will refer to the complete listing of the Banzhaf power indexes as the **Banzhaf power distribution** of a weighted voting system. It is a common practice to write power indexes as percentages, rather than fractions. Percentagewise, the Banzhaf power distribution of the weighted voting system in Example 9 is

P_1: 60%

P_2: 20%

P_3: 20%

Example 10.

Among the most important decisions a professional basketball team must make is the drafting of college players. In many cases the decisions as to whether to draft a specific player is made through weighted voting. Take, for example, the case of the Akron Flyers. In their system, the head coach (HC) has 4 votes, the general manager (GM) has 3 votes, the director of scouting operations (DS) has 2 votes, and the team psychiatrist (TP) has 1 vote. A simple majority of 6 votes is required for a yes vote on a player. In essence, the Akron Flyers operate as the weighted voting system [6: 4, 3, 2, 1].

We will now find the Banzhaf power distribution of this weighted voting system. Table 2-2 shows the 15 possible coalitions, which ones are winning and which are losing coalitions, and, for each winning coalition, the critical players (underlined).

Table 2-2 The 15 Coalitions for the Akron Flyers Management Team with Critical Players Underlined

Coalition	Weight	Win or Lose
$\{HC\}$	4	Lose
$\{GM\}$	3	Lose
$\{DS\}$	2	Lose
$\{TP\}$	1	Lose
$\{\underline{HC}, \underline{GM}\}$	7	Win
$\{\underline{HC}, \underline{DS}\}$	6	Win
$\{HC, TP\}$	5	Lose
$\{GM, DS\}$	5	Lose
$\{GM, TP\}$	4	Lose
$\{DS, TP\}$	3	Lose
$\{\underline{HC}, GM, DS\}$	9	Win
$\{\underline{HC}, \underline{GM}, TP\}$	8	Win
$\{\underline{HC}, \underline{DS}, TP\}$	7	Win
$\{\underline{GM}, \underline{DS}, \underline{TP}\}$	6	Win
$\{HC, GM, DS, TP\}$	10	Win

All we have to do now is count the number of times each player is underlined and divide by the total number of underlines. The Banzhaf power distribution is

HC: $\frac{5}{12} = 41\frac{2}{3}\%$

GM: $\frac{3}{12} = 25\%$

DS: $\frac{3}{12} = 25\%$

TP: $\frac{1}{12} = 8\frac{1}{3}\%$

(Note that the power indexes always add up to 1. This fact provides a useful check on your calculations.) ■

How Many Coalitions?

Before we go on to the next example, let's take a brief detour and consider the following mathematical question: For a given number of players, how many different coalitions are possible? Here, our identification of coalitions with sets will come in particularly handy. Except for the empty subset { }, we know that every other subset of the set of players can be identified with a different coalition. This means that we can count the total number of coalitions by counting the number of subsets and subtracting one. So, how many subsets does a set have?

A careful look at Table 2-3 shows us that each time we add a new element we are doubling the number of subsets—the same subsets we had before we added the element plus an equal number consisting of each of the above but with the new element thrown in.

Since each time we add a new player we are doubling the number of subsets, we will find it convenient to think in terms of powers of 2. Table 2-4 summarizes what we have learned.

Table 2-3 The Subsets of a Set

Set	$\{P_1, P_2\}$	$\{P_1, P_2, P_3\}$	$\{P_1, P_2, P_3, P_4\}$	$\{P_1, P_2, P_3, P_4, P_5\}$
Number of Subsets	4	8	16	32
	$\{\ \}$ $\{P_1\}$ $\{P_2\}$ $\{P_1, P_2\}$	$\{\ \}$ $\{P_3\}$ $\{P_1\}$ $\{P_1, P_3\}$ $\{P_2\}$ $\{P_2, P_3\}$ $\{P_1, P_2\}$ $\{P_1, P_2, P_3\}$	$\{\ \}$ $\{P_4\}$ $\{P_1\}$ $\{P_1, P_4\}$ $\{P_2\}$ $\{P_2, P_4\}$ $\{P_1, P_2\}$ $\{P_1, P_2, P_4\}$ $\{P_3\}$ $\{P_3, P_4\}$ $\{P_1, P_3\}$ $\{P_1, P_3, P_4\}$ $\{P_2, P_3\}$ $\{P_2, P_3, P_4\}$ $\{P_1, P_2, P_3\}$ $\{P_1, P_2, P_3, P_4\}$	The 16 subsets from the previous column along with each of these with P_5 thrown in.

Table 2-4 The Number of Possible Coalitions

Players	Number of Subsets	Number of Coalitions
$\{P_1, P_2\}$	$4 = 2^2$	$2^2 - 1 = 3$
$\{P_1, P_2, P_3\}$	$8 = 2^3$	$2^3 - 1 = 7$
$\{P_1, P_2, P_3, P_4\}$	$16 = 2^4$	$2^4 - 1 = 15$
$\{P_1, P_2, P_3, P_4, P_5\}$	$32 = 2^5$	$2^5 - 1 = 31$
...	...	...
$\{P_1, P_2, \ldots, P_N\}$	2^N	$2^N - 1$

Example 11.

The disciplinary committee at George Washington High School has five members: the principal (P_1), the vice principal (P_2), and three teachers (P_3, P_4, and P_5). When voting on a specific disciplinary action the principal has three votes, the vice principal has two votes, and each of the teachers has one vote. A total of five votes are needed for a motion to carry. We can describe this voting system as [5: 3, 2, 1, 1, 1].

We now know that with five players there are 31 possible coalitions. Rather than plow straight ahead and list them all, we can sometimes cut corners by figuring out directly which are winning coalitions. Table 2-5 shows the winning coalitions only, with the critical players in each coalition underlined. We leave it to the reader to verify the details, which, while important, are not the main point of this example (the main point being that sometimes we can save a lot of work by directly zeroing in on the winning coalitions).

The Banzhaf power distribution of the disciplinary committee is

Principal (P_1): $\frac{11}{25} = 44\%$

Vice principal (P_2): $\frac{5}{25} = 20\%$

Teacher (P_3): $\frac{3}{25} = 12\%$

Teacher (P_4): $\frac{3}{25} = 12\%$

Teacher (P_5): $\frac{3}{25} = 12\%$ ■

Table 2-5 Winning Coalitions for Example 11 with Critical Players Underlined

Winning Coalitions	Comments
$\{\underline{P_1}, \underline{P_2}\}$	Only possible winning two-player coalition.
$\{\underline{P_1}, \underline{P_2}, P_3\}$	Winning three-player coalitions must contain P_1.
$\{\underline{P_1}, \underline{P_2}, P_4\}$	
$\{\underline{P_1}, \underline{P_2}, P_5\}$	
$\{\underline{P_1}, \underline{P_3}, \underline{P_4}\}$	
$\{\underline{P_1}, \underline{P_3}, \underline{P_5}\}$	
$\{\underline{P_1}, \underline{P_4}, \underline{P_5}\}$	
$\{\underline{P_1}, P_2, P_3, P_4\}$	Any four-player coalition wins
$\{\underline{P_1}, P_2, P_3, P_5]$	
$\{\underline{P_1}, P_2, P_4, P_5\}$	
$\{\underline{P_1}, P_3, P_4, P_5\}$	
$\{\underline{P_2}, \underline{P_3}, \underline{P_4}, \underline{P_5}\}$	
$\{P_1, P_2, P_3, P_4, P_5\}$	The **grand coalition** (all players).

Example 12.

The Tasmania State University Promotion and Tenure committee consists of five members: the dean (D) and four other faculty members of equal standing (F_1, F_2, F_3, and F_4). In this committee motions are carried by strict majority, but the dean never votes except to break a tie. How is power distributed in this voting system?

While in this example we are not given the weights of the various players, we can still proceed in the usual manner. Table 2-6 shows the winning coalitions with the critical players underlined.

Table 2-6 Winning Coalitions for Example 12 with Critical Players Underlined

Winning coalitions without the dean	Winning coalitions with the dean
$\{\underline{F_1}, \underline{F_2}, \underline{F_3}\}$	$\{\underline{D}, \underline{F_1}, \underline{F_2}\}$
$\{\underline{F_1}, \underline{F_2}, \underline{F_4}\}$	$\{\underline{D}, \underline{F_1}, \underline{F_3}\}$
$\{\underline{F_1}, \underline{F_3}, \underline{F_4}\}$	$\{\underline{D}, \underline{F_1}, \underline{F_4}\}$
$\{\underline{F_2}, \underline{F_3}, \underline{F_4}\}$	$\{\underline{D}, \underline{F_2}, \underline{F_3}\}$
$\{F_1, F_2, F_3, F_4\}$	$\{\underline{D}, \underline{F_2}, \underline{F_4}\}$
	$\{\underline{D}, \underline{F_3}, \underline{F_4}\}$

The Banzhaf power distribution in this committee is

D: $\frac{6}{30} = 20\%$

F_1: $\frac{6}{30} = 20\%$

F_2: $\frac{6}{30} = 20\%$

F_3: $\frac{6}{30} = 20\%$

F_4: $\frac{6}{30} = 20\%$

Surprise! All the members (including the dean) have the same amount of power. ■

An interesting variation of Example 12 occurs in the U.S. Senate, where the vice president of the United States votes only to break a tie. An analysis similar to the one in Example 12 would show that, assuming all 100 senators are voting, he has exactly the same amount of power as any other member of the senate.

Applications of the Banzhaf Power Index

The Nassau County Board of Supervisors, New York. John Banzhaf first introduced the Banzhaf power index in 1965 in an analysis of how power was distributed in the Board of Supervisors of Nassau County, New York. Although Banzhaf was a lawyer, it was his mathematical analysis of power in the Nassau County Board that provided the legal basis for a series of lawsuits[2] involving the mathematics of weighted voting systems and its implications regarding the "equal protection" guarantee of the Fourteenth Amendment.

Nassau County is divided into six different districts, and, based on 1964 population figures, a total of 115 votes were allocated to the districts. Table 2-7 shows the names of the districts and their allocation of votes:

Table 2-7

District	Votes in 1964
Hempstead #1	31
Hempstead #2	31
Oyster Bay	28
North Hempstead	21
Long Beach	2
Glen Cove	2

The number of votes needed to pass a motion was 58, so that the Nassau County Board in effect was the weighted voting system [58: 31, 31, 28, 21, 2, 2]. So far, so good, but what about the power of each district? In his lawsuit, Banzhaf argued that in this instance, all the power in the County Board was concentrated in the hands of the top three players—Hempstead #1, Hempstead #2, and Oyster Bay. After a moment's reflection we can see why this is so: No winning coalition is possible without two of the top three players in it, and since any two of the top three already form a winning coalition, none of the last three play-

[2] For students of the law, here are the case references: *Graham v. Board of Supervisors* (1966): *Franklin v. Krause* (1974); *Bechtle v. Board of Supervisors* (1981); *League of Women Voters v. Board of Supervisors* (1983); and *Jackson v. Board of Supervisors* (1991). All of the above lawsuits involved the Nassau County Board of Supervisors. Other important legal cases involving the Banzhaf power index are *Ianucci v. Board of Supervisors* (1967) and *Morris v. Board of Estimate* (U.S Supreme Court, 1989).

ers can ever be critical players. (We leave it to the reader to verify all the details—see Exercise 26.) The long and the short of it was that, as Banzhaf successfully argued, this County Board was in practice a three-member board, with Hempstead #1, Hempstead #2, and Oyster Bay each having one-third of the power, and North Hempstead, Glen Cove, and Long Beach having absolutely no power at all!

Based on Banzhaf's analysis, the number of votes allocated to each district was changed, and has indeed been changed several times since 1965. Since 1994, the Nassau County Board operates as the weighted voting system [65: 30, 28, 22, 15, 7, 6] (see Exercise 27).

The United Nations Security Council The main body responsible for maintaining the international peace and security of nations is the Security Council of the United Nations. At present the composition of the Security Council is as follows: There are 5 **permanent** members of the council (the United Kingdom, France, the People's Republic of China, Russia, and the United States), plus 10 additional **nonpermanent** slots filled by other countries on a rotating basis.[3] According to the voting rules of the Security Council, each of the permanent members has veto power, so that a resolution cannot pass unless each permanent member votes for it. In addition, at least 4 of the 10 nonpermanent members must also vote yes. It turns out that these rules are equivalent to giving 7 votes to each permanent member and 1 vote to each nonpermanent member, with a requirement of at least 39 votes to carry a motion (see Exercise 54). Thus, in our language, the U.N. Security Council is nothing more than the weighted voting system [39: 7, 7, 7, 7, 7, 1, 1, 1, 1, 1, 1, 1, 1, 1, 1]. Once the Security Council is described this way, it is possible to compute the Banzhaf power index of each member. (Since the calculations are a bit involved, we will omit them.) The long and the short of it is that the Banzhaf power index of each permanent member is 848/5080 $\approx$ 16.7%, while the Banzhaf power index of each nonpermanent member is 84/5080 $\approx$ 1.65%. These numbers raise some interesting political questions: Is it reasonable that a perma-

The U.N. Security Council: a weighted voting system in which five players have most of the power and the rest have hardly any power.

[3] When the original Security Council was set up by the United Nations in 1945, the number of nonpermanent members was only 6. The number was increased to 10 in 1963.

nent member have more than 10 times as much power as a nonpermanent member? Was this the original intent in the United Nations charter? If not, should the voting rules of the Security Council be reconsidered?

The Electoral College. As we should all know, the president of the United States is chosen using an institution called the *electoral college.* In choosing the president, each state is allowed to cast a certain number of votes, equal to the total number of members of Congress (senators plus representatives) from that state. The votes are cast by individuals called *electors,* who are chosen to represent the citizens of their respective states. The general rule is that all the electors from a particular state vote the same way (for the presidential candidate who wins a plurality of the votes in that state). While there have been challenges to the constitutionality of this rule (known as the *unit rule* or *winner-take-all rule*), and in a few instances the rule has been violated by individual electors, it is currently standard procedure in the electoral college.

A second important point is the fact that under America's strong two-party system, most presidential elections boil down to a choice between just two viable candidates.

Under the unit rule and in an election between only two viable candidates, the electoral college represents one of the most important examples of a weighted voting system (and one of the most unusual—the United States is the only country in the world with such a system). The players in this voting system are the states (actually the 50 states plus the District of Columbia), and the weight of a state is the number of senators plus representatives from that state (the weight of the District of Columbia is set at 3). The quota is defined by a strict majority of the electoral vote. Since 1964, the total number of electoral votes has been set at 538 and the quota at 270. The appendix at the end of this chapter shows, among other things, the electoral votes for each state based on the 1990 census and the Banzhaf power index of each state. (The calculations for the power indexes require both sophisticated mathematical methods and a powerful computer.)

The Shapley-Shubik Power Index

In this section we will discuss a different approach to measuring power, first proposed jointly by Lloyd Shapley and Martin Shubik[4] in 1964. The key difference between the Shapley-Shubik interpretation of power and Banzhaf's centers around the concept of a *sequential coalition.* According to Shapley and Shubik, coalitions are formed sequentially: Every coalition starts with a first player, who may then be joined by a second player, then a third, and so on. Thus, to an already complicated situation we are adding one more wrinkle—the question of the order in which the players joined the coalition.

Let's illustrate the difference with a simple example. According to the Banzhaf interpretation of power, the coalition $\{P_1, P_2, P_3\}$ represents the fact that P_1, P_2, and P_3 have joined forces and will vote together. We don't care who joined

[4] Lloyd S. Shapley was a mathematician at the Rand Corporation and Martin Shubik an economist at Yale University.

the coalition when. According to the Shapley-Shubik interpretation of power, the same three players can form six different sequential coalitions: $\langle P_1, P_2, P_3\rangle$ (this means that P_1 started the coalition, then P_2 joined in, and last came P_3); $\langle P_1, P_3, P_2\rangle$; $\langle P_2, P_1, P_3\rangle$; $\langle P_2, P_3, P_1\rangle$; $\langle P_3, P_1, P_2\rangle$; $\langle P_3, P_2, P_1\rangle$.

Note the change in notation: From now on the notation $\langle\ \rangle$ will indicate that we are dealing with a sequential coalition; that is, we care about the order in which the players are listed.

Factorials

It is now time to consider another one of those *How many?* questions: For a given number of players *N, how many sequential coalitions containing the N players are there?* We have just seen that with three players there are six sequential coalitions: $\langle P_1, P_2, P_3\rangle$, $\langle P_1, P_3, P_2\rangle$, $\langle P_2, P_1, P_3\rangle$, $\langle P_2, P_3, P_1\rangle$, $\langle P_3, P_1, P_2\rangle$, and $\langle P_3, P_2, P_1\rangle$. What happens if we have four players? We could try to write down all of the sequential coalitions, a tedious and unimaginative task. Instead, let's argue as follows: To fill the first slot in a coalition we have 4 choices (any one of the 4 players); to fill the second slot we have 3 choices (any one of the players except the one in the first slot); to fill the third slot we have only 2 choices, and to fill the last slot we have only 1 choice. We can now combine these choices by multiplying them. Thus, the total number of possible sequential coalitions with four players turns out to be $4 \times 3 \times 2 \times 1 = 24$.

The one question that may still remain unclear is, Why did we multiply? The answer lies in a basic rule of mathematics called the **multiplication rule**: *If there are m different ways to do X and n different ways to do Y, then X and Y together can be done in m × n different ways.* For example, if an ice cream shop offers 2 different types of cones and 3 different flavors of ice cream, then according to the multiplication rule there are $2 \times 3 = 6$ different cone/flavor combinations. Figure 2-1 shows why this is so. We will discuss the multiplication rule and its uses in greater detail in Chapter 15. Meanwhile, back to the issue at hand.

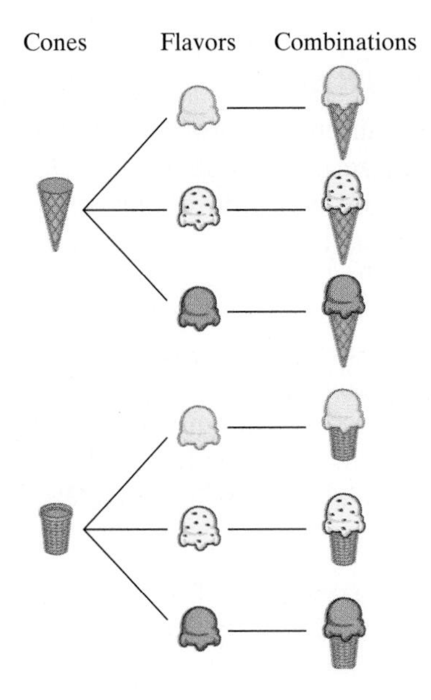

FIGURE 2-1

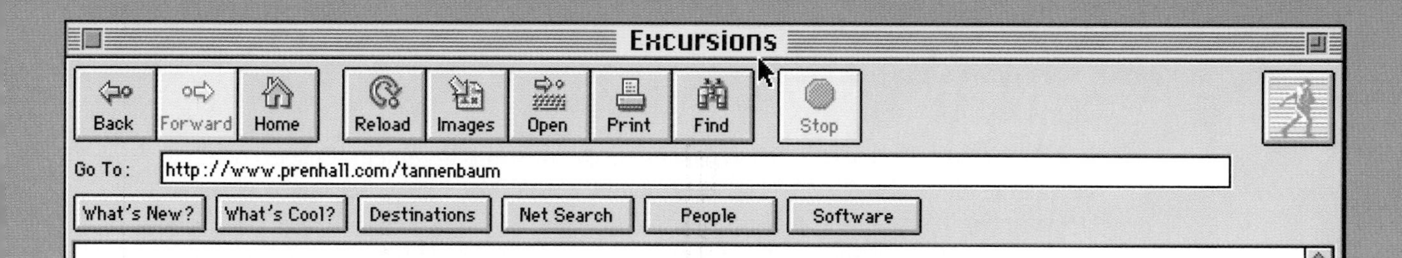

Why Send A Stranger To College?

Every four years we elect a president. And every four years we're told, long after the election is over, that the president has been officially voted into office by a group of people called the Electoral College. Strictly speaking, voters at the polls aren't voting for the president at all; they're choosing electors, who then cast the votes that actually decide the presidential race.

This process seems undemocratic, and for many years people have talked of getting rid of it. However, physicist Alan Natapoff has recently argued that the system, instead of taking power out of the hands of individual voters, actually gives them more power than a direct vote would! You can see the argument for yourself, in a very readable article, in the Tannenbaum Website under Chapter 2.

Questions:

1. Why do all of the examples in the article use an odd number of voters?
2. According to Natapoff, voters' power should be not only equal, but also as large as possible. What would Natapoff say about the Banzhaf and Shapley-Shubik power indices?
3. The Banzhaf power index is $\frac{\text{number of times player } P \text{ is critical}}{\text{number of times all players are critical}}$. Why might $\frac{\text{number of times player } P \text{ is critical}}{\text{number of coalitions player } P \text{ belongs to}}$ be called the Natapoff power index? What kinds of elections would Natapoff say it works for?
4. Use the Natapoff index from Question 3 to verify Natapoff's calculation of the power one voter has in a direct-vote, dead-even contest with a nine-voter population. To avoid writing out all the coalitions, use the fact that there are $n!/[r! \cdot (n-r)!]$ ways to select r objects from a group of n objects without regard to order. Many calculators can calculate this number using the nCr function.
5. If you have a calculator with the nCr function, verify Natapoff's calculation for the power of one voter in a direct-vote, dead-even contest with 135 voters.

If we have 5 players, following up on our previous argument, we can count on a total of $5 \times 4 \times 3 \times 2 \times 1 = 120$ sequential coalitions. In a more general vein, the number of sequential coalitions with N players is $1 \times 2 \times 3 \times \cdots \times N$.

Numbers of the form $1 \times 2 \times 3 \times \cdots \times N$ are some of the most important numbers in mathematics and will show up several times in this book. The number $1 \times 2 \times 3 \times \cdots \times N$ is called the **factorial** of N and is written in the shorthand form $N!$. The factorial of 5, for example, is written 5! and equals 120 ($1 \times 2 \times 3 \times 4 \times 5 = 120$), while $10! = 3{,}628{,}800$ (check it out!). To summarize,

> The number of sequential coalitions with N players is
>
> $N! = 1 \times 2 \times 3 \times \cdots \times N$.

Back to the Shapley-Shubik Power Index

Suppose that we have a weighted voting system with N players. We know from the preceding discussion that there is a total of $N!$ different sequential coalitions containing *all* the players. In each of these coalitions there is one player that tips the scales—the moment that player joins the coalition, the coalition changes from a losing to a winning coalition (see Fig. 2-2). We call such a player a **pivotal player** for the sequential coalition. The underlying principle of the Shapley-Shubik approach is that the pivotal player deserves special recognition. After all, the players who came before the pivotal player did not have enough votes to carry a motion, and the players who came after the pivotal player are a bunch of Johnny-come-latelies. (Note that we can talk about "before" and "after" only because we are considering sequential coalitions.) According to Shapley and Shubik, a player's power depends on the total number of times that player is pivotal in relation to all other players.

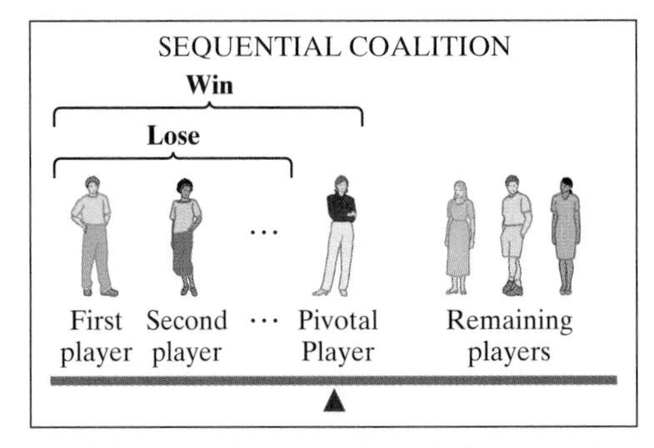

FIGURE 2-2 The pivotal player tips the scales.

The formal description of the procedure for finding the Shapley-Shubik power index of any player in a generic weighted voting system with N players is as follows:

Finding the Shapley-Shubik Power Index of Player *P*

- **Step 1.** Make a list of all sequential coalitions containing all N players. There are $N!$ of them.
- **Step 2.** In each sequential coalition determine *the* pivotal player. There is one in each sequential coalition.
- **Step 3.** Count the total number of times P is pivotal and call this number S.

The Shapley-Shubik power index of P is then given by the fraction $S/N!$.

A listing of the Shapley-Shubik power indexes of all the players gives the **Shapley-Shubik power distribution** of the weighted voting system.

Example 13.

Let's consider, once again, an analysis of power at Foreman & Sons, the company we first discussed in Example 9. (Remember George I, George II and George III?) This time we will use the Shapley-Shubik interpretation of power. Recall that we are dealing here with the weighted voting system [4: 3, 2, 1].

- **Step 1.**

There are $3! = 6$ sequential coalitions of the three players. They are

$\langle P_1, P_2, P_3 \rangle$

$\langle P_1, P_3, P_2 \rangle$

$\langle P_2, P_1, P_3 \rangle$

$\langle P_2, P_3, P_1 \rangle$

$\langle P_3, P_1, P_2 \rangle$

$\langle P_3, P_2, P_1 \rangle$

- **Step 2.**

Sequential coalition	Pivotal player
$\langle P_1, P_2, P_3 \rangle$	P_2
$\langle P_1, P_3, P_2 \rangle$	P_3
$\langle P_2, P_1, P_3 \rangle$	P_1
$\langle P_2, P_3, P_1 \rangle$	P_1
$\langle P_3, P_1, P_2 \rangle$	P_1
$\langle P_3, P_2, P_1 \rangle$	P_1

- **Step 3.**

P_1 is pivotal four times.

P_2 is pivotal one time.

P_3 is pivotal one time.

The Shapley-Shubik power distribution is

P_1: $\frac{4}{6} = 66\frac{2}{3}\%$

P_2: $\frac{1}{6} = 16\frac{2}{3}\%$

P_3: $\frac{1}{6} = 16\frac{2}{3}\%$

Note that the power distribution is different from the Banzhaf power distribution (P_1: 60%; P_2: 20%; P_3: 20%) obtained in Example 9. Under the Shapley-Shubik interpretation of power, George I has even more power—his son and grandson have each a little less. One fact that hasn't changed is that poor George II still has the same amount of power as his son! ■

Example 14.

We will now reconsider Example 10, the one about the Akron Flyers system for picking players in the draft. The weighted voting system in this example is [6: 4, 3, 2, 1], and we will now find its Shapley-Shubik power distribution.

There are 24 different sequential coalitions involving the 4 players. They are listed in Table 2-8 with the pivotal player underlined.

Table 2-8 The 24 Sequential Coalitions for Example 14 with Pivotal Players Underlined

$\langle HC, \underline{GM}, DS, TP\rangle$	$\langle GM, \underline{HC}, DS, TP\rangle$	$\langle DS, \underline{HC}, GM, TP\rangle$	$\langle TP, HC, \underline{GM}, DS\rangle$
$\langle HC, \underline{GM}, TP, DS\rangle$	$\langle GM, \underline{HC}, TP, DS\rangle$	$\langle DS, \underline{HC}, TP, GM\rangle$	$\langle TP, HC, \underline{DS}, GM\rangle$
$\langle HC, \underline{DS}, GM, TP\rangle$	$\langle GM, DS, \underline{HC}, TP\rangle$	$\langle DS, GM, \underline{HC}, TP\rangle$	$\langle TP, GM, \underline{HC}, DS\rangle$
$\langle HC, \underline{DS}, TP, GM\rangle$	$\langle GM, DS, \underline{TP}, HC\rangle$	$\langle DS, GM, \underline{TP}, HC\rangle$	$\langle TP, GM, \underline{DS}, HC\rangle$
$\langle HC, TP, \underline{GM}, DS\rangle$	$\langle GM, TP, \underline{HC}, DS\rangle$	$\langle DS, TP, \underline{HC}, GM\rangle$	$\langle TP, DS, \underline{HC}, GM\rangle$
$\langle HC, TP, \underline{DS}, GM\rangle$	$\langle GM, TP, \underline{DS}, HC\rangle$	$\langle DS, TP, \underline{GM}, HC\rangle$	$\langle TP, DS, \underline{GM}, HC\rangle$

The Shapley-Shubik power distribution is

HC: $\frac{10}{24} = 41\frac{2}{3}\%$

GM: $\frac{6}{24} = 25\%$

DS: $\frac{6}{24} = 25\%$

TP: $\frac{2}{24} = 8\frac{1}{3}\%$

It is worth mentioning that in this example the Shapley-Shubik power distribution turns out to be exactly the same as the Banzhaf power distribution (see Example 10). If nothing else, this shows that it is not impossible for the Banzhaf and Shapley-Shubik power distributions to agree. While it doesn't happen often, it can happen. ■

Example 15.

The city of Cleansburg operates under what is called a "strong-major" system. The strong-major system in Cleansburg works like this: There are five council members, namely the major and four "ordinary" council members. A motion can pass only if the major and at least two other council members vote for it, or alternatively, if all four of the ordinary council members vote for it.[5]

Common sense tells us that under these rules, the four ordinary council members have the same amount of power but the major has more. Exactly how much more is what we will now try to find out, and we will use the Shapley-Shubik interpretation of power to do so.

Since there are 5 players in this voting system, there are $5! = 120$ sequential coalitions to consider. Obviously, we will want to try some kind of a shortcut. We will first try to find the Shapley-Shubik power index of the major. In what position does the major have to be in order to be the pivotal player in a sequential coalition?

---

[5] This situation is usually described by saying that the *major has veto power but a unanimous vote of the other council members can override the major's veto.*

Does the major have to be in first place? No way! No player can be pivotal in the first position unless he or she is a dictator. Second? No—an ordinary council member plus the major are not enough to carry a motion. Third place? Yes! If the major is in third place, he is the pivotal player in that sequential coalition (see Fig. 2-3a). Likewise, if the major is in fourth place, he is the pivotal player in that sequential coalition, because the three preceding ordinary members are not enough to carry a motion (see Fig. 2-3b). Finally, when the major is in the last (fifth) place in a sequential coalition, he is not the pivotal player—the four ordinary members preceding him do have enough votes to carry the motion (see Fig. 2-3c).

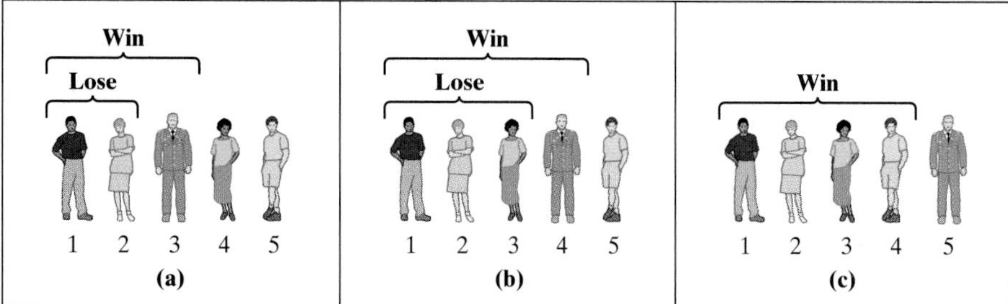

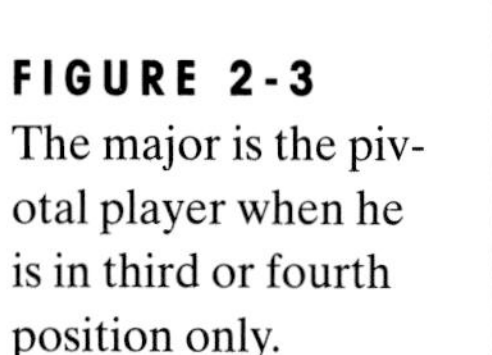

FIGURE 2-3
The major is the pivotal player when he is in third or fourth position only.

Now comes a critical question: In how many (of these 120) sequential coalitions is the major in first place? second place? . . . fifth place? The symmetry of the positions tells us that there should be just as many sequential coalitions in which the major is in first place as in any other place. It follows that the 120 sequential coalitions can be divided into 5 groups of 24—24 with the major in first place, 24 with the major in second place, etc.

We finally have a handle on the major (that's more than many people in Cleansburg can say!). The major is the pivotal player in all sequential coalitions in which he is either in the third or fourth position, and there are 24 of each. Thus, the Shapley-Shubik power index of the major is 48/120 = 2/5 (= 40%). Since the four ordinary council members must share the remaining 60% of the power equally, it follows that each of them must have a Shapley-Shubik power index of 15%. We are done! ■

(For the purposes of comparison, the reader is encouraged to calculate the Banzhaf power distribution of the Cleansburg city council [see Exercise 48].)

Applications of the Shapley-Shubik Power Index

The Electoral College Revisited. Calculating the Shapley-Shubik power index of the states in the electoral college is no small task. Checking every possible sequential coalition by hand would take literally thousands of years, so a direct approach is out of the question. There are, however, some sophisticated mathematical shortcuts, which, when coupled with a computer and the right kind of software, allow the calculations to be done quite efficiently. The appendix at the end of this chapter shows (in its last column) the Shapley-Shubik power index of each state. If we compare the Banzhaf and the Shapley-Shubik power indexes, we notice that there is a very small difference between the two. This example shows that in some situations the Banzhaf and Shapley-Shubik power indexes give essentially the same answer. The next example illustrates a very different situation.

The United Nations Security Council Revisited. As mentioned earlier in the chapter, under its present rules the U.N. Security Council can be formally described as the weighted voting system [39: 7, 7, 7, 7, 7, 1, 1, 1, 1, 1, 1, 1, 1, 1, 1], and the Banzhaf power index of the five permanent members (the ones with seven votes) is about 10 times larger than that of the nonpermanent members (16.7% against 1.65%). What if we use the Shapley-Shubik power index instead? Once again we skip the calculations, which are somewhat involved, but it turns out that the Shapley-Shubik power index of each of the five permanent members is 19.6%, while that of each of the nonpermanent members is about 0.2%. The disproportion between the power of the permanent and nonpermanent is much more pronounced here: According to the Shapley-Shubik interpretation of power, a permanent member has about 100 times more power than a nonpermanent member! Here is a situation where the Banzhaf power index and the Shapley-Shubik power index produce significantly different answers.

Conclusion

In any society, no matter how democratic, some individuals and groups have more power than others. This is simply a consequence of the fact that individuals and groups are not all equal. Diversity is the inherent reason why the concept of power exists.

Power itself comes in many different forms. We often hear expressions such as "In strength lies power" and "Money is power," and in our new information age, "Knowledge is power." In this chapter we discussed the notion of power as it applies to formal voting situations called *weighted voting systems* and saw how mathematical methods allow us to measure the power of an individual or group by means of a *power index.* In particular, we looked at two different kinds of power indexes: the *Banzhaf power index* and the *Shapley-Shubik power index.*

These indexes provide two different ways to measure power, and while they occasionally agree, they often differ significantly. Of the two, which one is closer to reality?

Unfortunately, there is no simple answer. Both of them are useful, and in some sense the choice is subjective. Perhaps the best way to evaluate them is to think of them as being based on a slightly different set of assumptions. The idea behind the Banzhaf interpretation of power is that players are free to come and go, negotiating their allegiance for power (somewhat like professional athletes since the advent of free agency). Underlying the Shapley-Shubik interpretation of power is the assumption that when a player joins a coalition, he or she is making a commitment to stay. In the latter case a player's power is generated by his ability to be in the right place at the right time.

In practice, the choice of which method to use for measuring power is based on which of the assumptions better fits the specifics of the situation. Mathematics, contrary to what we've often come to expect, does not give us the answer, just the tools that might help us make an informed decision.

Key Concepts

Banzhaf power distribution
Banzhaf power index
coalition
coalition weight
critical player
dictator
dummy
factorial
losing coalition
motion
multiplication rule
pivotal player
player
quota
sequential coalition
Shapley-Shubik power distribution
Shapley-Shubik power index
veto power
weighted voting system
weight
winning coalition

Exercises

Walking

1. Consider the weighted voting system [10: 6, 5, 4, 2]

(a) How many players are there?

(b) What is the quota?

(c) What is the weight of the coalition formed by P_1 and P_3?

(d) Write down all winning coalitions.

(e) Which players are critical in the coalition $\{P_1, P_2, P_3\}$?

(f) Find the Banzhaf power distribution of this weighted voting system.

2. Consider the weighted voting system [5: 3, 2, 1, 1].

(a) How many players are there?

(b) What is the quota?

(c) What is the weight of the coalition formed by P_1 and P_3?

(d) Which players are critical in the coalition $\{P_2, P_3, P_4\}$?

(e) Which players are critical in the coalition $\{P_1, P_3, P_4\}$?

(f) Write down all winning coalitions.

(g) Find the Banzhaf power distribution of this weighted voting system.

3. (a) Find the Banzhaf power distribution of the weighted voting system [6: 5, 2, 1].

(b) Find the Banzhaf power distribution of the weighted voting system [3: 2, 1, 1]. Compare your answers in (a) and (b).

4. (a) Find the Banzhaf power distribution of the weighted voting system [7: 5, 2, 1].

(b) Find the Banzhaf power distribution of the weighted voting system [5: 3, 2, 1]. Compare your answers in (a) and (b).

5. (a) Find the Banzhaf power distribution of the weighted voting system [10: 5, 4, 3, 2, 1]. (If possible, do it without writing down all coalitions—just the winning ones.)

(b) Find the Banzhaf power distribution of the weighted voting system [11: 5, 4, 3, 2, 1]. [*Hint*: Note that the only change from (a) is in the quota, and use this fact to your advantage.]

6. **(a)** Find the Banzhaf power distribution of the weighted voting system [9: 5, 5, 4, 2, 1].
   (b) Find the Banzhaf power distribution of the weighted voting system [9: 5, 5, 3, 2, 1].

7. Consider the weighted voting system [16: 9, 8, 7].
   (a) Write down all the sequential coalitions involving all 3 players.
   (b) In each of the sequential coalitions in (a), underline the pivotal player.
   (c) Find the Shapley-Shubik power distribution of this weighted voting system.

8. Consider the weighted voting system [8: 7, 6, 2].
   (a) Write down all the sequential coalitions involving all 3 players.
   (b) In each of the sequential coalitions in (a), underline the pivotal player.
   (c) Find the Shapley-Shubik power distribution of this weighted voting system.

9. Find the Shapley-Shubik power distribution of each of the following weighted voting systems.
   (a) [8: 8, 5, 1]
   (b) [8: 7, 5, 2]
   (c) [8: 7, 6, 1]
   (d) [8: 6, 5, 1]
   (e) [8: 6, 5, 3]

10. Find the Shapley-Shubik power distribution of each of the following weighted voting systems.
    (a) [6: 4, 3, 2, 1]
    (b) [7: 4, 3, 2, 1]
    (c) [8: 4, 3, 2, 1]
    (d) [9: 4, 3, 2, 1]
    (e) [10: 4, 3, 2, 1]

11. Consider the weighted voting system [5: 3, 2, 1, 1].
    (a) Find the Banzhaf power distribution of this weighted voting system.
    (b) Find the Shapley-Shubik power distribution of this weighted voting system.

12. Consider the weighted voting system [60: 32, 31, 28, 21].
    (a) Find the Banzhaf power distribution of this weighted voting system.
    (b) Find the Shapley-Shubik power distribution of this weighted voting system.

13. In each of the following weighted voting systems, determine which players (i) are dictators; (ii) have veto power; (iii) are dummies:
    (a) [6: 4, 2, 1]
    (b) [6: 7, 3, 1]
    (c) [10: 9, 9. 1]

14. In each of the following weighted voting systems, determine which players (i) are dictators; (ii) have veto power; (iii) are dummies:
    (a) [95: 95, 80, 10, 2]
    (b) [95: 65, 35, 30, 25]
    (c) [48: 32, 16, 8, 4, 2, 1]

15. In each of the following weighted voting systems, determine which players (i) are dictators; (ii) have veto power; (iii) are dummies:

(a) [19: 9, 7, 5, 3, 1]

(b) [15: 16, 8, 4, 1]

(c) [17: 13, 5, 2, 1]

(d) [25: 12, 8, 4, 2]

16. In each of the following weighted voting systems, determine which players (i) are dictators; (ii) have veto power; (iii) are dummies:

(a) [27: 12, 10, 4, 2]

(b) [22: 10, 8, 7, 2, 1]

(c) [21: 23, 10, 5, 2]

(d) [15: 11, 5, 2, 1]

17. Consider the weighted voting system [q: 10, 6, 5, 4, 2].

(a) What is the smallest value that the quota q can take?

(b) What is the largest value that the quota q can take?

(c) How many coalitions are there for this weighted voting system?

(d) How many sequential coalitions are there involving all the players?

18. Consider the weighted voting system [q: 5, 3, 2, 2, 1, 1].

(a) What is the smallest value that the quota q can take?

(b) What is the largest value that the quota q can take?

(c) How many coalitions are there for this weighted voting system?

(d) How many sequential coalitions are there involving all the players?

19. Consider the weighted voting system [q: 5, 3, 1]. Find the Shapley-Shubik power distribution of this weighted voting system when

(a) $q = 5$

(b) $q = 6$

(c) $q = 7$

(d) $q = 8$

(e) $q = 9$

20. Consider the weighted voting system [q: 5, 3, 1]. Find the Banzhaf power distribution of this weighted voting system when

(a) $q = 5$

(b) $q = 6$

(c) $q = 7$

(d) $q = 8$

(e) $q = 9$

21. This exercise is intended to help you develop a better understanding of factorials. If you have a fancy calculator with a factorial key, don't use it—use only the multiplication key!

(a) Calculate 6!

(b) Calculate 10!

(c) Calculate 11! [*Hint*: Think of a shortcut using the answer for (b).]

(d) Calculate 9! [*Hint*: Think of a shortcut using the answer for (b).]

(e) $12x = 12!$ Solve for x. (*Hint*: You already calculated the answer.)

22. This exercise should be done *without* a calculator. The amount of arithmetic is minimal. Remember that it is always better to do cancellations first and multiplications later.

(a) Calculate 10!/9!.

(b) Calculate 30!/28!.

(c) Calculate 9!/6!3!.

(d) Calculate 15!/11!4!.

23. A business firm is owned by 4 partners, *A, B, C,* and *D*. When making group decisions, each partner has one vote and the majority rules, except in the case of a 2-2 tie. Then, the coalition that contains *D* (the partner with the least seniority) loses. What is the Banzhaf power distribution in this partnership?

24. A business firm is owned by 4 partners, *A, B, C,* and *D*. When making group decisions, each partner has one vote and the majority rules, except in the case of a 2-2 tie. Then, the coalition that contains *A* (the senior partner) wins. What is the Shapley-Shubik power distribution in this partnership?

25. A business firm is owned by 5 partners: *A, B, C, D,* and *E*. When making group decisions, each partner has one vote and majority rules, except that *A* and *B* both have veto power and therefore must be in all winning coalitions. What is the Banzhaf power distribution in this partnership?

26. The 1964 Nassau County Board of Supervisors could be described by the weighted voting system [58: 31, 31, 28, 21, 2, 2].

(a) Describe all the winning coalitions.

(b) Find the Banzhaf power distribution for this weighted voting system.

Jogging

27. The 1994 Nassau County Board of Supervisors can be described by the weighted voting system [65: 30, 28, 22, 15, 7, 6]. Find the Banzhaf power distribution for this weighted voting system.

28. Consider the weighted voting system [21: 6, 5, 4, 3, 2, 1]. (Note that here the quota is the sum of all the weights of the players.)

(a) How many coalitions are there?

(b) Write down the winning coalitions only and underline the critical players.

(c) Find the Banzhaf power index of each player.

(d) Explain why in any weighted voting system $[q: w_1, w_2, \ldots, w_N]$, if $q = w_1 + \cdots + w_N$, then the Banzhaf power index of each player is $1/N$.

29. Consider the weighted voting system [21: 6, 5, 4, 3, 2, 1].

(a) How many different sequential coalitions are there?

(b) There is only one way in which a player can be pivotal in one of these sequential coalitions. Describe it.

(c) In how many sequential coalitions is P_6 pivotal?

(d) What is the Shapley-Shubik power index of P_6?

(e) What are the Shapley-Shubik power indexes of the other players?

(f) Explain why in any weighted voting system $[q: w_1, w_2, \ldots, w_N]$, if $q = w_1 + \cdots + w_N$, then the Shapley-Shubik power index of each player is $1/N$.

30. Give an example of a weighted voting system in which P_1 has twice as many votes as P_2 and

(a) the Banzhaf power index of P_1 is greater than twice the Banzhaf power index of P_2.

(b) the Banzhaf power index of P_1 is less than twice the Banzhaf power index of P_2.

(c) the Banzhaf power index of P_1 is equal to twice the Banzhaf power index of P_2.

(d) the Banzhaf power index of P_1 is equal to the Banzhaf power index of P_2.

31. Give an example of a weighted voting system in which P_1 has twice as many votes as P_2 and

(a) the Shapley-Shubik power index of P_1 is greater than twice the Shapley-Shubik power index of P_2.

(b) the Shapley-Shubik power index of P_1 is less than twice the Shapley-Shubik power index of P_2.

(c) the Shapley-Shubik power index of P_1 is equal to twice the Shapley-Shubik power index of P_2.

(d) the Shapley-Shubik power index of P_1 is equal to the Shapley-Shubik power index of P_2.

32. **(a)** Consider the weighted voting system [22: 10, 10, 10, 10, 1]. Are there any dummies? Explain your answer.

(b) Without doing any work [but using your answer for (a)], find the Banzhaf and Shapley-Shubik power distributions of this weighted voting system.

(c) Consider the weighted voting system $[q: 10, 10, 10, 10, 1]$. Find all the possible values of q for which P_5 is not a dummy.

33. Consider the weighted voting system $[q: 8, 4, 1]$.

(a) What are the possible values of q?

(b) Which values of q results in a dictator? (Who? Why?)

(c) Which values of q results in exactly one player with veto power? (Who? Why?)

(d) Which values of q results in more than one player with veto power? (Who? Why?)

(e) Which values of q results in one or more dummies? (Who? Why?)

34. Consider the weighted voting systems $[9: w, 5, 2, 1]$.

(a) What are the possible values of w?

(b) Which values of w results in a dictator? (Who? Why?)

(c) Which values of w results in a player with veto power? (Who? Why?)

(d) Which values of w results in one or more dummies? (Who? Why?)

35. **(a)** Verify that the weighted voting systems [12: 7, 4, 3, 2] and [24: 14, 8, 6, 4] result in exactly the same Banzhaf power distribution. (If you need to make calculations, do them for both systems side by side and look for patterns.)

(b) Based on your work in (a), explain why the two proportional weighted voting systems $[q: w_1, w_2, \ldots, w_N]$ and $[cq: cw_1, cw_2, \ldots, cw_N]$ always have the same Banzhaf power distribution.

36. **(a)** Verify that the weighted voting systems [12: 7, 4, 3, 2] and [24: 14, 8, 6, 4] result in exactly the same Shapley-Shubik power distribution. (If you need to make calculations, do them for both systems side by side and look for patterns.)

(b) Based on your work in (a), explain why the two proportional weighted voting systems $[q: w_1, w_2, \ldots, w_N]$ and $[cq: cw_1, cw_2, \ldots, cw_N]$ always have the same Shapley-Shubik power distribution.

37. **A dummy is a dummy is a dummy....** This exercises shows that a player that is a dummy is a dummy regardless of which interpretation of power is used.

(a) Show that if a player has a Banzhaf power index of 0 in a weighted voting system, then that player must also have a Shapley-Shubik power index of 0.

(b) Show that if a player has a Shapley-Shubik power index of 0 in a weighted voting system, then that player must also have a Banzhaf power index of 0.

38. Consider the weighted voting system $[q: 5, 4, 3, 2, 1]$.

(a) For what values of q is there a dummy?

(b) For what values of q do all players have the same power?

39. The weighted voting system [6: 4, 2, 2, 2, 1] represents a partnership among 5 people (P_1, P_2, P_3, P_4, and you!). You are the last player (the one with 1 vote), which in this case makes you a dummy! Not wanting to remain a dummy, you offer to buy 1 vote. Each of the other four partners is willing to sell you one of their votes, and they are all asking the same price. Which partner should you buy from in order to get as much power for your buck as possible? Use the Banzhaf power index for your calculations. Explain your answer.

40. The weighted voting system [27: 10, 8, 6, 4, 2] represents a partnership among 5 people (P_1, P_2, P_3, P_4, and P_5). You are P_5, the one with 2 votes. You want to increase your power in the partnership and are prepared to buy 1 share (1 share = 1 vote) from any of the other partners. P_1, P_2, and P_3 are each willing to sell cheap ($1000 for one share), but P_4 is not being quite as cooperative—she wants $5000 for 1 share. Given that you still want to buy 1 share, who should you buy it from? Use the Banzhaf power index for your calculations. Explain your answer.

41. The weighted voting system [18: 10, 8, 6, 4, 2] represents a partnership among 5 people (P_1, P_2, P_3, P_4, and P_5). You are P_5, the one with 2 votes. You want to increase your power in the partnership and are prepared to buy shares (1 share = 1 vote) from any of the other partners.

(a) Suppose that each partner is willing to sell 1 share and they are all asking the same price. Assuming that you decide to buy only 1 share, which partner should you buy from? Use the Banzhaf power index for your calculations.

(b) Suppose that each partner is willing to sell 2 shares and they are all asking the same price. Assuming that you decide to buy 2 shares from a single partner, which partner should you buy from? Use the Banzhaf power index for your calculations.

(c) If you have the money and the cost per share is fixed, should you buy 1 share or 2 shares (from a single person)? Explain.

42. Sometimes in a weighted voting system, 2 or more players decide to merge—that is to say, to combine their votes and always vote the same way. (Notice that a merger is different from a coalition—coalitions are temporary, mergers are permanent.) For example, if in the weighted voting system [7: 5, 3, 1] P_2 and P_3 were to merge, the weighted voting system would then become [7: 5, 4]. In this exercise, we explore the effects of mergers on a player's power.

(a) Consider the weighted voting system [4: 3, 2, 1]. In Example 9 we saw that P_2 and P_3 each have a Banzhaf power index of $\frac{1}{5}$. Suppose that P_2 and P_3 merge and become a single player P^*. What is the Banzhaf power index of P^*?

(b) Consider the weighted voting system [5: 3, 2, 1]. Find first the Banzhaf power indexes of players P_2 and P_3 and then the Banzhaf power index of P^* (the merger of P_2 and P_3). Compare.

(c) Same as (b) for [6: 3, 2, 1].

(d) What are your conclusions from (a), (b), and (c)?

Running

Exercises 43 through 48 refer to the concept of equivalent weighted voting systems. Two weighted voting systems are defined to be ***equivalent*** *if they have the same number of players and exactly the same winning coalitions.*

43. **(a)** Show that [8: 5, 3, 2] is equivalent to [2: 1, 1, 0].

(b) Without doing any further calculations, give the Banzhaf and Shapley-Shubik power distributions of this weighted voting system.

44. **(a)** Show that [6: 5, 3, 2] is equivalent to [3: 2, 1, 1].

(b) Without doing any further calculations, what can you say about the Banzhaf and Shapley-Shubik power indexes of the three players?

45. **(a)** Explain why a player who is critical in a particular winning coalition of a weighted voting system must be critical in that same coalition in any equivalent weighted voting system.

(b) Explain why a player must have the same Banzhaf power index in equivalent weighted voting systems.

46. **(a)** Explain why a player who is pivotal in a particular sequential coalition of a weighted voting system must be pivotal in that same sequential coalition in any equivalent weighted voting system.

(b) Explain why a player must have the same Shapley-Shubik power index in equivalent weighted voting systems.

47. A weighted voting system is called **minimal** if there is no equivalent weighted voting system with a smaller quota or with a smaller total weight. (Quotas and weights are always whole numbers.)

(a) Show that [2: 1, 1, 1] is minimal.

(b) Show that [3: 2, 1, 1] is minimal.

(c) Show that [4: 2, 1, 1] is not minimal. Find an equivalent minimal weighted voting system.

(d) Show that [4: 2, 2, 1] is not minimal. Find an equivalent minimal weighted voting system.

(e) Show that [8: 5, 3, 1] is not minimal. Find an equivalent minimal weighted voting system.

(f) Explain why a weighted voting system with a dictator is equivalent to the minimal weighted voting system $[1: 1, 0, \cdots, 0]$.

48. **(a)** Show that the 1994 Nassau County Board of Supervisors voting system: [65: 30, 28, 22, 15, 7, 6] is equivalent to [15: 7, 6, 5, 4, 2, 1].

(b) Show that the above system [15: 7, 6, 5, 4, 2, 1] is minimal (see Exercise 47). (*Hint*: First show that all six weights must be positive and all must be different. Then examine possible quotas that would give the correct results for the three coalitions $\{P_1, P_2, P_5\}$, $\{P_1, P_2, P_6\}$, and $\{P_2, P_3, P_4\}$. Conclude that the players' weights *cannot* be 6, 5, 4, 3, 2, 1. Use the same coalitions to conclude that $w_3 + w_4 > w_1 + w_6$, and finally that if $w_1 = 7$ then $w_4 = 4$.)

49. **(a)** Give an example of a weighted voting system with 4 players and such that the Shapley-Shubik power index of P_1 is $\frac{3}{4}$.

(b) Show that in any weighted voting system with 4 players, a player cannot have a Shapley-Shubik power index of more than $\frac{3}{4}$ unless he or she is a dictator.

(c) Show that in any weighted voting system with N players, a player cannot have a Shapley-Shubik power index of more than $(N - 1)/N$ unless he or she is a dictator.

(d) Give an example of a weighted voting system with N players and such that P_1 has a Shapley-Shubik power index of $(N - 1)/N$.

50. **(a)** Give an example of a weighted voting system with 3 players and such that the Shapley-Shubik power index of P_3 is $\frac{1}{6}$.

(b) Show that in any weighted voting system with 3 players, a player cannot have a Shapley-Shubik power index of less than $\frac{1}{6}$ unless he or she is a dummy.

51. **(a)** Give an example of a weighted voting system with 4 players and such that the Shapley-Shubik power index of P_4 is $\frac{1}{12}$.

(b) Show that in any weighted voting system with 4 players, a player cannot have a Shapley-Shubik power index of less than $\frac{1}{12}$ unless he or she is a dummy.

52. **(a)** Give an example of a weighted voting system with N players having a player with veto power who has a Banzhaf power index of $1/N$.

(b) Show that in any weighted voting system with N players, a player with veto power must have a Banzhaf power index of at least $1/N$.

53. **(a)** Give an example of a weighted voting system with N players having a player with veto power who has a Shapley-Shubik power index of $1/N$.

(b) Show that in any weighted voting system with N players, a player with veto power must have a Shapley-Shubik power index of at least $1/N$.

54. **The United Nations Security Council.** The U.N. Security Council is made up of 15 countries. There are 5 permanent members (the People's Republic of China, France, Russia, the United Kingdom, and the United States) and 10 nonpermanent members. All 5 of the permanent members have veto power. A winning coalition must consist of the 5 permanent members plus at least 4 nonpermanent members. Explain why the Security Council can formally be described by the weighted voting system [39: 7, 7, 7, 7, 7, 1, 1, 1, 1, 1, 1, 1, 1, 1, 1].

55. **The original United Nations Security Council.** The original Security Council set up in 1945 consisted of only 11 countries (5 permanent members plus 6 nonpermanent members). A winning coalition had to include all 5 permanent members plus at least 2 of the nonpermanent members. Give a formal description of the original Security Council as a weighted voting system.

56. The Cleansburg City Council. Find the Banzhaf power index of the Cleansburg city council. (See Example 15 for details.)

57. The Fresno City Council. In Fresno, California, the city council consists of 7 members (the mayor and 6 other council members). A motion can be passed by the mayor and at least 3 other council members, or by at least 5 of the 6 ordinary council members.

(a) Describe the Fresno City Council as a weighted voting system

(b) Find the Shapley-Shubik power distribution for the Fresno City Council.

(*Hint*: See Example 15 for some useful ideas.)

58. Suppose that in a weighted voting system there is a player A who hates another player P so much that he will always vote the opposite way of P, regardless of the issue. We will call A the **antagonist** of P.

(a) Suppose that in the weighted voting system [8; 5, 4, 3, 2], P is the player with 2 votes and his antagonist A is the player with 5 votes. What are the possible coalitions under these circumstances? What is the Banzhaf power distribution under these circumstances?

(b) Suppose that in a generic weighted voting system with N players there is a player P that has an antagonist A. How many coalitions are there under these circumstances?

(c) Give examples of weighted voting systems where a player A can:

(i) increase his Banzhaf power index by becoming an antagonist of another player,

(ii) decrease his Banzhaf power index by becoming an antagonist of another player,

(d) Suppose that the antagonist A has more votes than his enemy P. What is a strategy that P can use to gain power at the expense of A?

State	Number of Electoral Votes*	Percent of Electoral Votes	Percent of Power per Banzhaf Power Index	Percent of Power per Shapley-Shubik Power Index
California	54	10.04	11.14	10.81
New York	33	6.13	6.20	6.29
Texas	32	5.95	6.00	6.09
Florida	25	4.65	4.63	4.69
Pennsylvania	23	4.28	4.25	4.30
Illinois	22	4.09	4.06	4.11
Ohio	21	3.90	3.87	3.91
Michigan	18	3.35	3.30	3.33
New Jersey	15	2.79	2.75	2.76
North Carolina	14	2.60	2.56	2.57
Georgia	13	2.42	2.38	2.38
Virginia	13	2.42	2.38	2.38
Indiana	12	2.23	2.19	2.20
Massachusetts	12	2.23	2.19	2.20
Missouri	11	2.04	2.01	2.01
Tennessee	11	2.04	2.01	2.01
Washington	11	2.04	2.01	2.01
Wisconsin	11	2.04	2.01	2.01
Maryland	10	1.86	1.82	1.82
Minnesota	10	1.86	1.82	1.82
Alabama	9	1.67	1.64	1.64
Louisiana	9	1.67	1.64	1.64
Arizona	8	1.49	1.46	1.46
Colorado	8	1.49	1.46	1.46
Connecticut	8	1.49	1.46	1.46
Kentucky	8	1.49	1.46	1.46
Oklahoma	8	1.49	1.46	1.46
South Carolina	8	1.49	1.46	1.46
Iowa	7	1.30	1.28	1.27
Mississippi	7	1.30	1.28	1.27
Oregon	7	1.30	1.28	1.27
Arkansas	6	1.12	1.09	1.09
Kansas	6	1.12	1.09	1.09
Nebraska	5	0.93	0.91	0.90
New Mexico	5	0.93	0.91	0.90
Utah	5	0.93	0.91	0.90
West Virginia	5	0.93	0.91	0.90
Hawaii	4	0.74	0.73	0.72
Idaho	4	0.74	0.73	0.72
Maine	4	0.74	0.73	0.72
Nevada	4	0.74	0.73	0.72
New Hampshire	4	0.74	0.73	0.72
Rhode Island	4	0.74	0.73	0.72
Alaska	3	0.56	0.55	0.54
Delaware	3	0.56	0.55	0.54
Montana	3	0.56	0.55	0.54
North Dakota	3	0.56	0.55	0.54
South Dakota	3	0.56	0.55	0.54
Vermont	3	0.56	0.55	0.54
Wyoming	3	0.56	0.55	0.54
District of Columbia	3	0.56	0.55	0.54
Total	538	100	100	100

*Number of seats in Congress (2 senators plus number of members in the House of Representatives).

References and Further Readings

1. Banzhaf, John F., III, "Weighted Voting Doesn't Work," *Rutgers Law Review,* 19 (1965), 317–343.
2. Brams, Steven J., *Game Theory and Politics.* New York: Free Press, 1975, chap. 5.
3. Brams, Steven J., William F. Lucas, and Philip D. Straffin, *Political and Related Models.* New York: Springer-Verlag, 1983, chaps. 9 and 11.
4. Grofman, B., "Fair Apportionment and the Banzhaf Power Index," *American Mathematical Monthly,* 88 (1981), 1–5.
5. Hively, Will, "Math Against Tyranny," *Discover,* November 1986, 74–85.
6. Imrie, Robert W., "The Impact of the Weighted Vote on Representation in Municipal Governing Bodies of New York State," *Annals of the New York Academy of Sciences,* 219 (November 1973), 192–199.
7. Lambert, John P., "Voting Games, Power Indices and Presidential Elections," *UMAP Journal,* 3 (1988), 213–267.
8. Merrill, Samuel, "Approximations to the Banzhaf Index of Voting Power," *American Mathematical Monthly,* 89 (1982), 108–110.
9. Riker, William H., and Peter G. Ordeshook, *An Introduction to Positive Political Theory.* Englewood Cliffs, NJ: Prentice-Hall, Inc., 1973, chap. 6.
10. Shapley, Lloyd, and Martin Shubik, "A Method for Evaluating the Distribution of Power in a Committee System," *American Political Science Review,* 48 (1954), 787–792.
11. Straffin, Philip D., Jr., "The Power of Voting Blocs: An Example," *Mathematics Magazine,* 50 (1977), 22–24.
12. Straffin, Philip D., Jr., *Topics in the Theory of Voting, UMAP Expository Monograph.* Boston: Birkhäuser, 1980, chap. 1.
13. Tannenbaum, Peter, "Power in Weighted Voting Systems," *The Mathematica Journal,* 7 (1997), 58–63.
14. Taylor, Alan, *Mathematics and Politics: Strategy, Voting, Power and Proof.* New York: Springer-Verlag, 1995, chaps. 4 and 9.

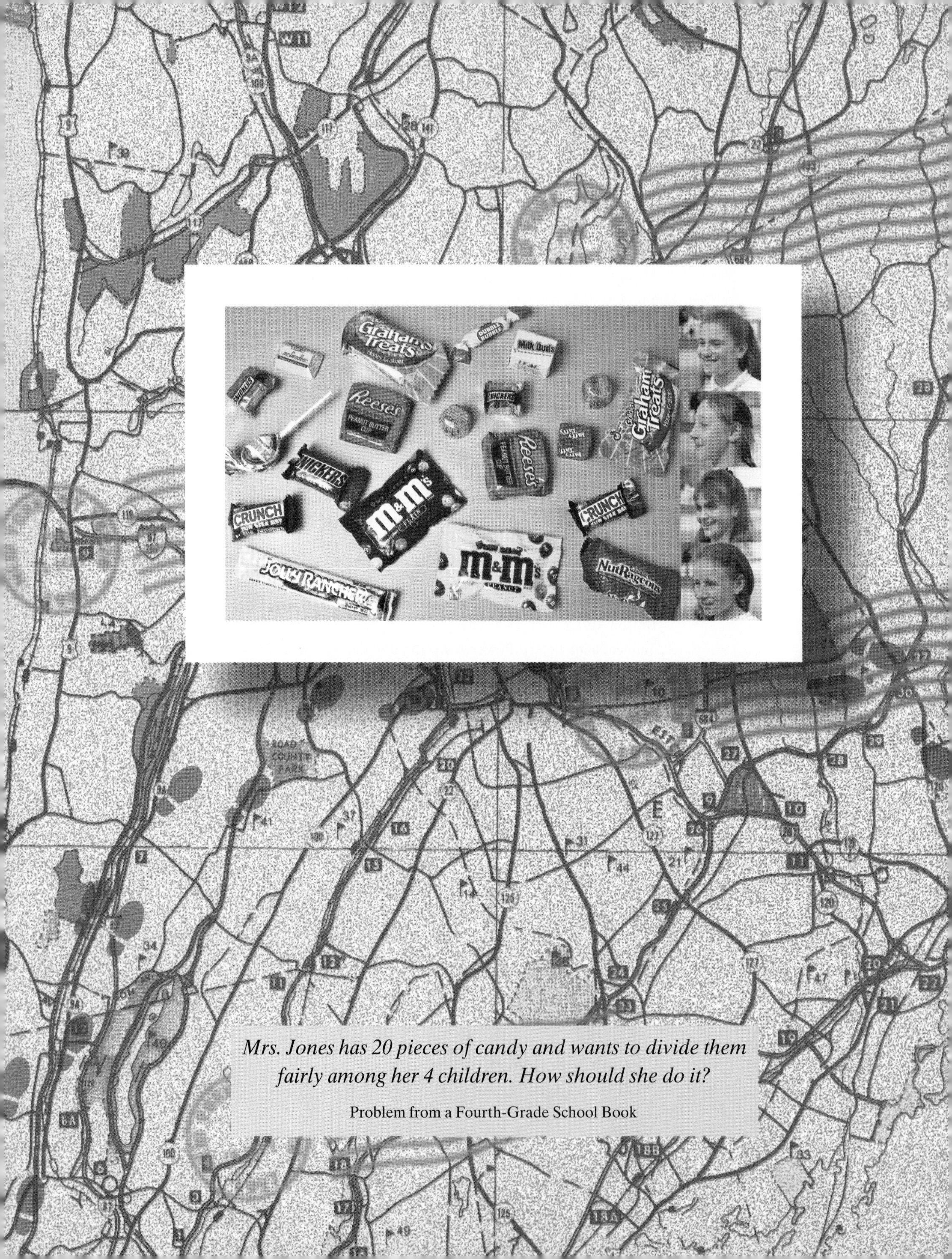

Mrs. Jones has 20 pieces of candy and wants to divide them fairly among her 4 children. How should she do it?

Problem from a Fourth-Grade School Book

Fair Division

The Slice Is Right

CHAPTER 3

Remember those old classic division problems in fourth grade? Well, they are back! Believe it or not, we are resurrecting the theme in this chapter. How come? When looked at in the right way, the problem of dividing candy among children is a paradigm for an important and surprisingly subtle category of problems known as *fair-division problems.*

The classic grammar school version of the problem is based on two unspoken assumptions: the pieces of candy are all *identical* and the children are *entitled to equal shares.* In this case we solve the problem by giving each child the same amount of candy—in Mrs. Jones' case, 5 pieces each. But what happens if the pieces of candy are not all identical? Let's imagine that Mrs. Jones is dividing a cornucopia of sweets: M&M's Reese's Pieces, Snickers, Milk Duds, caramels, bubblegum, etc. Should we stubbornly insist on giving each child 5 pieces? What about pieces that are more valuable than others? What about individual likes and dislikes? How do we take all of these things into account and yet divide the candy fairly? And by the way, what does *fairly* mean in this situation?

Before we try to answer some of these questions, an even more basic question must be addressed: Who cares? Who cares whether some little munchkin is happy or not with her share of candy? Why the big deal? Now close your eyes and imagine heirs to an estate instead of children; million-dollar paintings and fabulous jewelry instead of candy, and you might have your answer. A little imagination is all it takes to see that dividing candy among children is only a metaphor for a common and extremely important problem in the real world: How can something that must be shared by a set of competing parties be divided among them in a way that ensures that each party is satisfied with the outcome? We call this kind of a problem a *fair-division problem.*

Problems of fair division are as old as mankind. One of the best-known and best-loved biblical stories is built around a fair-division problem: two women, both claiming to be mothers of the same baby, make their case to King Solomon. The King proposes to cut the baby in half and give each woman a share, a division that is totally unacceptable to the true mother—she would rather see the baby go to the other woman than be slaughtered! The final settlement, of course, is that the baby is returned to its rightful mother.

The Judgment of Solomon, by Nicolas Poussin (1594–1665).

For more modern examples all we have to do is look at the news. Breaking up nations (Bosnia, 1995); dividing a country's airspace (Iraq, 1996); dividing the cost of cleaning environmental pollution (The Rio Grande and NAFTA, 1994) are all, at their core, problems of fair division. For some of these problems more than Solomonic wisdom is required, and yet, for many others, a little mathematics is all it takes. The settlement of an estate among heirs, the division of common property in a divorce proceeding, the subdivision of land among competing claimants, and, of course, the division of candy or cake among children are just a few examples of such problems. In this chapter we will learn how to identify and solve some of these types of problems, so that at the end, while not quite as wise as King Solomon, we might be wiser than before.

Fair-Division Problems and Fair-Division Schemes

Regardless of whether the problem is dividing fairly a bunch of candy among children, or an expensive art collection among the heirs to an estate, from a formal point of view **fair-division problems** all share the same essential elements.

1. A set of *goods* to be divided. These "goods" can be anything that has a potential value. Typically, the goods are tangible physical objects, such as candy, cake, pizza, jewelry, art, property (cars, boats, houses, land), etc. In more esoteric situations the goods may be intangible things such as rights (water rights, drilling rights, broadcast licenses, etc.).[1] We will use the symbol S to denote the objects (or object—it could be just one) to be divided (the "loot" if you will).
2. A set of *players*, who are the parties entitled to share the set S. For lack of better names, we will call the players $P_1, P_2, \ldots, P_N$. The players are usually persons, but they could also be countries, states, ethnic or political groups, and institutions. The one key characteristic that each of the players must have is a *value system*—the ability to assign value not only to the set S but to any part of it as well. Needless to say, each player has its own value system.

Given the set S, and the players $P_1, P_2, \ldots, P_N$, each with its own opinions about how S should be divided, here are some key questions:

- What does it mean for a player to get a *fair share* of S?
- Is it possible to divide S into shares (one for each player) in such a way that every player gets a *fair share?* If so, how?

Let's answer the first question first; it is the easiest. To us, a **fair share** will mean any share that *in the opinion of the player receiving it* has a value that is *at least one Nth* of the total value of S, N being the number of players. Notice that our definition of a fair share is recipient dependent. The only thing that matters in determining the fairness of a share is *what the player receiving the share thinks of it;* what the other players think is irrelevant.

Say, for example, that we have 5 players, and player X gets a share that in her opinion is worth at least $\frac{1}{5}$ (20%) of the total value of S, but the other four players think that player X's share is pretty much worthless. This is just fine! Player X is satisfied with her share, and the other players should be happy that X didn't get any of the "good" part.

The answer to the second question is the main theme of this chapter. In the 1940s the Polish mathematician Hugo Steinhaus, together with some of his colleagues and students, developed various **fair-division schemes**—systematic methods for dividing things among a set of players, with the wonderful property that, when properly followed, guarantee that each of the players ends up with a fair share. Another attraction of these methods is that they are *internal* to the players—in other words, they work without the need for outsiders such as a judge, lawyers, a referee, etc. The only requirements to accomplish this are that the play-

[1] To keep things simple we will stick to positive goods throughout our discussion, but it is also possible for the "goods" to have negative value (pollution, trash, chores, etc.) in which case one could call them "bads." With minor variations, all the methods discussed in this chapter can be used to divide "bads" just as well as "goods."

ers *act in a rational manner* (i.e., that their value systems conform to the basic laws of arithmetic); that the players have *no knowledge* about each others' value systems (just as, in a card game, you don't want the other players to see your hand, here you don't want any of the other players to have knowledge about your likes and dislikes). Finally, of course, the success of these methods requires the willingness of the players to abide by the rules and results of the game.

Types of Fair-Division Problems

Depending on the nature of the set of goods *S*, fair-division problems can be classified into three types: continuous, discrete, and mixed.

In a **continuous** fair-division problem the set *S* is divisible in infinitely many ways, and shares can be increased or decreased by arbitrarily small amounts. Typical examples of continuous fair-division problems are dividing a parcel of land, dividing commercial air time, dividing a cake, a pizza, ice cream, etc.

A fair-division problem is **discrete** when the set *S* is made up of objects that are indivisible (paintings, houses, cars, boats, jewelry, etc.). As far as candy is concerned, yes, it could be chopped up into smaller and smaller pieces, but nobody really does that (it makes for a big mess!), so let's agree that throughout this chapter we will think of candy as indivisible and therefore discrete (a semantic convenience).

A **mixed** fair-division problem is one in which some of the components are continuous and some are discrete. Dividing an estate consisting of a car, a house, and a parcel of land is a mixed division problem.

Fair-division schemes are classified according to the nature of the problem involved. Thus, there are *discrete fair-division schemes* (which are used to solve fair-division problems in which the set *S* is made up of indivisible, discrete objects); and *continuous fair-division schemes* (which are used to solve fair-division problems in which the objects are infinitely divisible, continuous objects). (Mixed fair-division problems can usually be solved by dividing the continuous and discrete parts separately, so we will not discuss them in this chapter.)

We will start our discussion with continuous fair-division problems.

Two Players: The Divider-Chooser Method

This is undoubtedly the best known of all continuous fair-division schemes. This scheme can be used anytime there is a continuous fair-division problem involving just two players. Most of us have unwittingly used it at some time or another, and informally it is best known as the *you cut—I choose method.* As this name suggests, one player (the *divider*) divides the *cake* (we will use the word "cake" as a convenient metaphor for a continuous set *S*) into two pieces, and the second player (the *chooser*) picks the piece he or she wants, leaving the other piece to the divider. When played honestly, this method guarantees that each player will get a share that he or she believes to be worth *at least one-half* of the total: The divider can guarantee this for herself in the mere act of dividing, and the chooser because of the simple fact that when anything is divided into two parts, one of the parts must be worth at least one-half or more of the total.[2]

---

[2] We must point out a hidden assumption here: *The value of the object S being divided does not diminish when the object is cut.* Thus, when cutting our theoretical cakes, there will be no crumbs.

Example 1.

On their first date, Damian and Cleo go to the county fair. With a $2 raffle ticket they win the chocolate-strawberry cake shown in Fig. 3-1(a). (Actually they had their hearts set on the red Corvette, but third prize is better than nothing.)

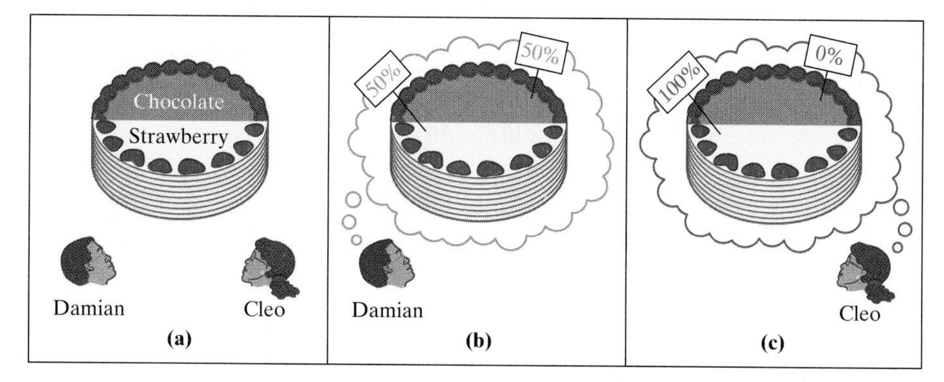

FIGURE 3-1 A chocolate-strawberry cake. The values are in the eyes of the beholder.

To Damian, chocolate and strawberry are equal in value—he has no preference for one over the other. Thus, in Damian's eyes, the value of the cake is distributed evenly between the chocolate and strawberry parts [Fig. 3-1(b)]. On the other hand, Cleo is allergic to chocolate so she won't eat any of the chocolate part. Thus, in Cleo's eyes the value of the cake is concentrated entirely in the strawberry half; the chocolate half has *zero value* [Fig. 3-1(c)]. Since this is their first date, neither one of them knows anything about the other's likes and dislikes.

Let's now see how Damian and Cleo might divide this cake using the divider-chooser method. Damian volunteers to go first and be the divider. His cut is shown in Fig. 3-2(a). There is no need to psychoanalyze the reasons for Damian's cut (granted, it is a little weird, but then, so is Damian). The important thing here is that, mathematically speaking, it is a perfectly logical cut based on Damian's value system. It is now Cleo's time to choose, and her choice is obvious—she will pick the piece having the most strawberry [Fig. 3-2(b)].

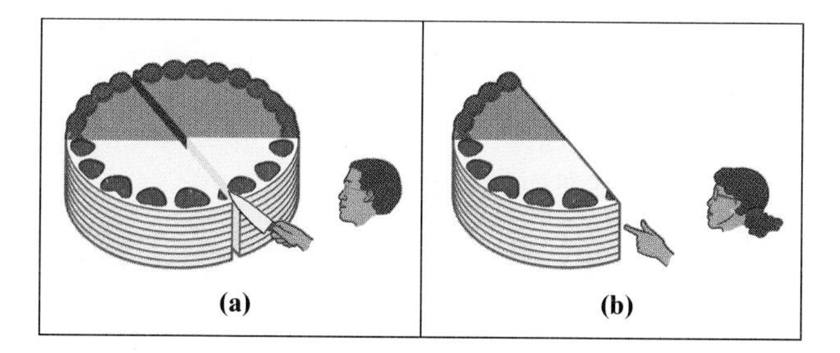

FIGURE 3-2 (a) Damian cuts (b) Cleo picks.

Notice that while Damian gets a share that (to him) is worth exactly one-half, Cleo ends up with a share that (to her) is worth much more than one-half. ■

Example 1 illustrates the fact that a fair-division method need not necessarily by *symmetric* (i.e., treat all the players equally). While it is true that each player gets a share that in his or her opinion is worth *at least one-half of the total*, there is a definite advantage to being the chooser. The simplest way to handle this problem is to randomly choose who gets to be the divider and who gets to be the chooser (toss a coin, draw straws, etc.).

The divider-chooser method can be extended for continuous fair-division problems involving more than two players in several ways. We will present three different schemes in this chapter: the *lone-divider method,* the *lone-chooser method,* and the *last-diminisher method.* In the lone-divider method, one of the players is the divider and all the rest are choosers; in the lone-chooser method, one of the players is the chooser and all the rest are dividers; in the last-diminisher method, each player has a chance to be both a divider and a chooser.

The Lone-Divider Method

For the sake of simplicity we describe the method for the case of three players, of which one will be the divider and the other two choosers. Since being the divider is somewhat of a disadvantage, the fairest way to decide which player is the divider is by random selection (roll a die, draw straws, draw cards from a deck, etc.). Let's call the divider D, and the choosers C_1 and C_2.

- **Step 1 (The Division).** The divider D divides the cake into three pieces (s_1, s_2, and s_3). D will get one of these pieces, but at this point she does not know which one. This forces her[3] to divide the cake in such a way that all three pieces have equal value, namely one-third of the value of the entire cake.
- **Step 2 (The Bids).** Each chooser declares (usually by writing it down on a slip of paper) which of the pieces cut up by the divider are, in his or her opinion, fair shares. We will call these the *choosers' bids.* It is important that the bids be done independently, without the choosers seeing each other's bids. A chooser *must* bid for any piece that he values to be worth one-third or more of the cake, not just the piece he likes the best. Thus, a chooser could bid for all three pieces (possible but unlikely—it would only happen if that chooser had exactly the same value system as the divider), or two pieces, or just one piece. Notice that it is logically *impossible for the chooser to consider all three pieces unfair,* so there always has to be at least one piece in every bid.
- **Step 3 (The Distribution).** Who gets which piece? The answer, of course, depends on the bids. For convenience, we will separate the pieces into two groups: the "bid-for" pieces (pieces that were bid for by one or both of the choosers), and the "unbid" pieces (pieces that neither chooser had any use for). We now consider two cases, depending on whether the "bid-for" group has several pieces or just one.

  Case 1. There are two or more pieces in the "bid-for" group. Here, it is possible to give each chooser a piece that he bid for and to give the divider the last remaining piece. Once this is done, every player has received a fair share, and our goal of fair division has been met.[4]

  Case 2. There is only one piece in the "bid-for" group. Now we are in trouble, because both choosers covet the same piece (let's call it the *C-piece*).

---

[3] Unless she is willing to take some chances.

[4] This does not preclude the possibility that each chooser may like the other chooser's piece better, in which case it is perfectly reasonable to let them swap their pieces. This would make each of them happier than they already were, and who could be against that?

Fortunately, there is a way out: We first combine the C-piece with *one* of the other two unbid-for pieces (we can pick either one, so the most direct way to decide which one is to flip a coin) to make a single "big" piece, which we will call the *B-piece* [see Figs. 3-3(a) and (b)]. The last remaining piece (call it the *L-piece*) goes to the divider, to whom all pieces are fair shares. Now what? First, a simple matter of arithmetic: the B-piece plus the L-piece equal the entire cake, and the value of the L-piece to both choosers is less than one-third of the total (it was unbid for). This means that in the eyes of both choosers, the B-piece is worth more than two-thirds of the value of the original cake. The trick now is to divide the B-piece fairly between the two choosers. If we can do this, we are done. But in fact, we do know a way to divide a piece fairly among *two* players: the *divider-chooser method.* Applying the divider-chooser method gives us a way to give each of the two choosers a fair share of the B-piece, and thus, a fair share of the original cake (a one-half or better share out of a two-thirds or better share is worth at least one-third).

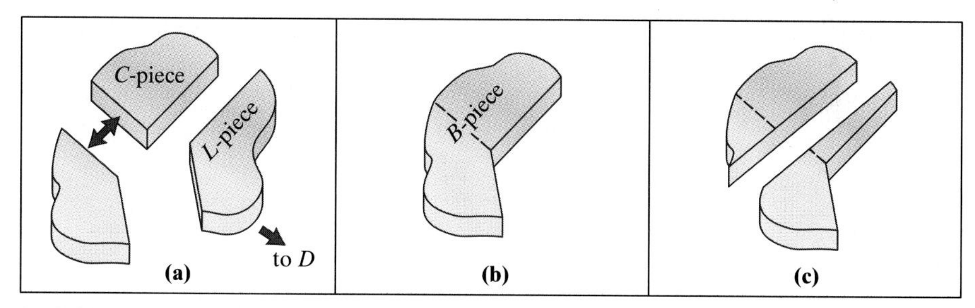

FIGURE 3-3
Case 2 in the lone-divider method (3 players). (a) One of the unwanted (by the choosers) pieces goes to D. (b) The other two pieces are recombined into the B-piece. (c) The B-piece is divided in 2 by the divider-chooser method.

We will illustrate the lone-divider method for three players with several examples. In all of these examples, we will assume that the divider D has already divided the cake into three pieces s_1, s_2, and s_3. In each example, the values that each of the three players assigns to the pieces, expressed as percentages of the total value of the cake, are shown in the form of a table. The reader should realize, however, that an individual player would not know what the other player's percentages are.

Example 2.

Table 3-1 shows the values of the three pieces in the eyes of each of the players. With three players, the threshold for an acceptable piece is $33\frac{1}{3}\%$. Looking at the table, we can see that C_1's bid should be $\{s_1, s_3\}$, and C_2's bid should also be $\{s_1, s_3\}$. The set of bid-for pieces being $\{s_1, s_3\}$, a couple of divisions are possible that meet our definition of fairness—for example, giving s_1, to C_1, s_2 to D, and s_3 to C_2. An

Table 3-1

	s_1	s_2	s_3
D	$33\frac{1}{3}\%$	$33\frac{1}{3}\%$	$33\frac{1}{3}\%$
C_1	35 %	10 %	55 %
C_2	40 %	25 %	35 %

even better division (from both choosers' point of view) would be to give s_3 to C_1, s_2 to D, and s_1 to C_3, but the players themselves have no way of knowing this, since the information available in the game is the bids (and not the percentages). After the division has been made, however, if two of the players want to swap pieces, it is perfectly permissible to let them. In this case, it is clear that both C_1 and C_2 would benefit by exchanging their pieces, and (remember, the players are *rational*) they will undoubtedly do so. ■

Example 3.

Table 3-2 shows the values of the three pieces in the eyes of each of the players. In this example, C_1's bid consists of just $\{s_2\}$; C_2's bid consists of just $\{s_1\}$. Here, the only possible fair division under the lone-divider method is to give s_2 to C_1, s_1 to C_2, and s_3 to D.

Table 3-2

	s_1	s_2	s_3
D	$33\frac{1}{3}\%$	$33\frac{1}{3}\%$	$33\frac{1}{3}\%$
C_1	30 %	40 %	30 %
C_2	60 %	15 %	25 %

■

Example 4.

Table 3-3 shows the values of the three pieces in the eyes of each of the players. Here we are in a case 2 situation: both choosers will bid only for s_3, which, in our terminology, becomes the C-piece. We must now decide which of the other two pieces to keep and which one to give to D. According to the rules, we should flip a coin. Let's suppose that the L-piece turns out to be s_1, and $s_2 + s_3$ becomes the B-piece. Note that the B-piece is worth 80% of the original cake in the eyes of C_1 and 90% of the original cake in the eyes of C_2. The B-piece will now be divided between C_1 and C_2, using the divider-chooser method, and regardless of how the details of that division work out, we know that C_1 will end up with a piece that is worth at least 40% of the original cake and C_2 will end up with a piece that is worth at least 45% of the original cake.

Table 3-3

	s_1	s_2	s_3
D	$33\frac{1}{3}\%$	$33\frac{1}{3}\%$	$33\frac{1}{3}\%$
C_1	20 %	30 %	50 %
C_2	10 %	20 %	70 %

■

Example 5.

In this example we will show what could happen to a player that tries to cheat. Table 3-4 shows the values of the three pieces in the eyes of each of the players. According to the rules, C_1's bid should be $\{s_1, s_2\}$, but C_1 is greedy and really wants s_1, so he decides to bid only for s_1. When the bids are opened, s_1 is the only bid-for piece,

Table 3-4

	s_1	s_2	s_3
D	$33\frac{1}{3}\%$	$33\frac{1}{3}\%$	$33\frac{1}{3}\%$
C_1	60 %	40 %	0 %
C_2	42 %	28 %	30 %

and after a flip of the coin s_2 goes to the divider D. The B-piece now becomes $s_1 + s_3$, to be divided between the choosers using the divider-chooser method. After a second flip of a coin, C_1 turns out to be the divider. Thus the best that C_1 can do is to get exactly half of the value of the B-piece, which is only 30%. Had C_1 not cheated, he could have assured himself a piece that was at least 40% of the cake. ■

The lone-divider method can be extended to any number of players N by picking one player to be the divider D, and the remaining $N - 1$ players to be choosers. The divider then proceeds to divide the cake into N pieces, and the choosers make their bids for the acceptable pieces. The final distribution depends on the bids. In general, the method is not difficult to carry out, but discussing the various possibilities can be a little involved, so we leave the details to the exercises (but to be fair, we give some helpful hints!) (see Exercises 12 through 19 and 50).

The Lone-Chooser Method

Once again, we start with a description of the method for the case of three players. Here we have one chooser and two dividers. As usual, we decide who is what by random selection. Let's say that C is the chooser and D_1 and D_2 the dividers.

- **Step 1 (The First Division).** D_1 and D_2 cut the cake [Fig. 3-4(a)] between themselves into *two* fair shares. To do this, they use the divider-chooser method. Let's say that D_1 gets s_1, and D_2 gets s_2 [Fig. 3-4(b)]. Each considers his slice worth at least one-half of the total.
- **Step 2 (The Second Division).** Each divider divides "his" piece into three equal shares. Thus, D_1 divides s_1 into three pieces, which we will call s_{1a}, s_{1b}, and s_{1c}. Likewise, D_2 divides s_2 into three pieces, which we will call s_{2a}, s_{2b}, and s_{2c}. [Fig. 3-4[c)].
- **Step 3 (The Selection).** The lone chooser C now comes riding upon the horizon. C gets to choose one of D_1's three pieces and one of D_2's three pieces (whichever she likes best). These two pieces make up C's final share. D_1 then gets to keep the remaining two pieces from s_1, and D_2 gets to keep the remaining two pieces from s_2 [Fig. 3-4[d)].

Why is this a fair division of the cake? D_1 ends up with exactly two-thirds of s_1, which was worth at least one-half of the original cake. The same argument applies to D_2. What about the chooser's share? We don't know what C thought of the first division between D_1 and D_2, but it really doesn't matter. Suppose that in C's eyes, s_1 was worth $x\%$ of the original cake. This automatically implies that s_2 was worth $(100 - x)\%$. Now C got a slice from s_1 worth at least of $x\%$, and another slice from s_2

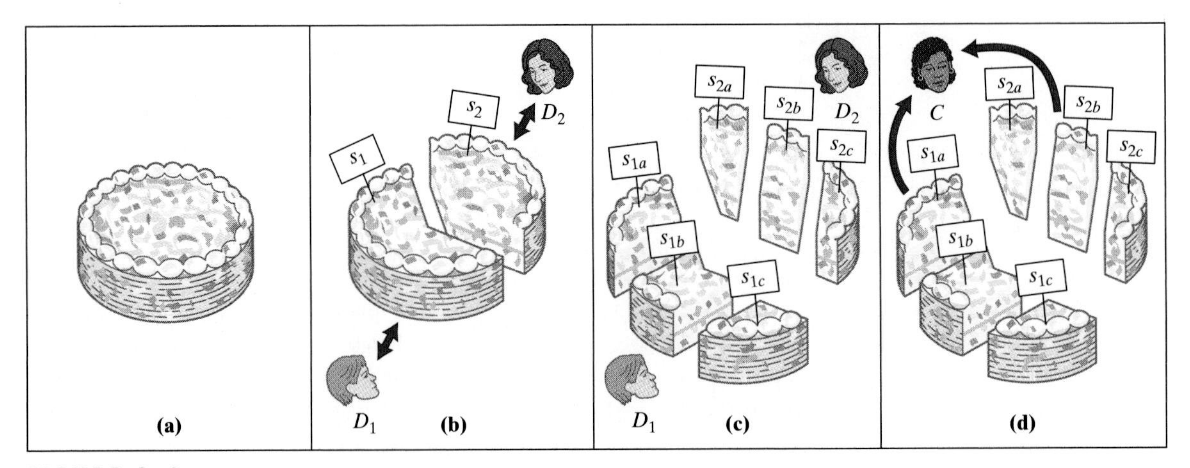

FIGURE 3-4
(a) The original cake (b) First Division (c) Second Division (d) Selection

worth at least one-third of $(100 - x)\%$. Between the two pieces, C got a share worth at least one-third of the cake.

The following example illustrates in detail how the lone-chooser method works.

Example 6.

David, Dinah, and Cher are planning to divide an orange-pineapple cake valued by each of them at $27 [Fig. 3-5(a)] using the *lone-chooser method.* They draw straws and Cher gets to be the chooser, so David and Dinah first divide the cake using the *divider-chooser method.* Since David drew a shorter straw than Dinah, he will be the one to cut the cake. Now, for their value systems:

- David: likes pineapple and orange the same ($P = O$). To him, value is synonymous with size, so in his eyes the cake looks like Fig. 3-5(b).
- Dinah: likes orange but hates pineapple ($P = 0$). In Dinah's eyes the entire value of the cake is concentrated in the orange half, so in her eyes the cake looks like Fig. 3-5(c).

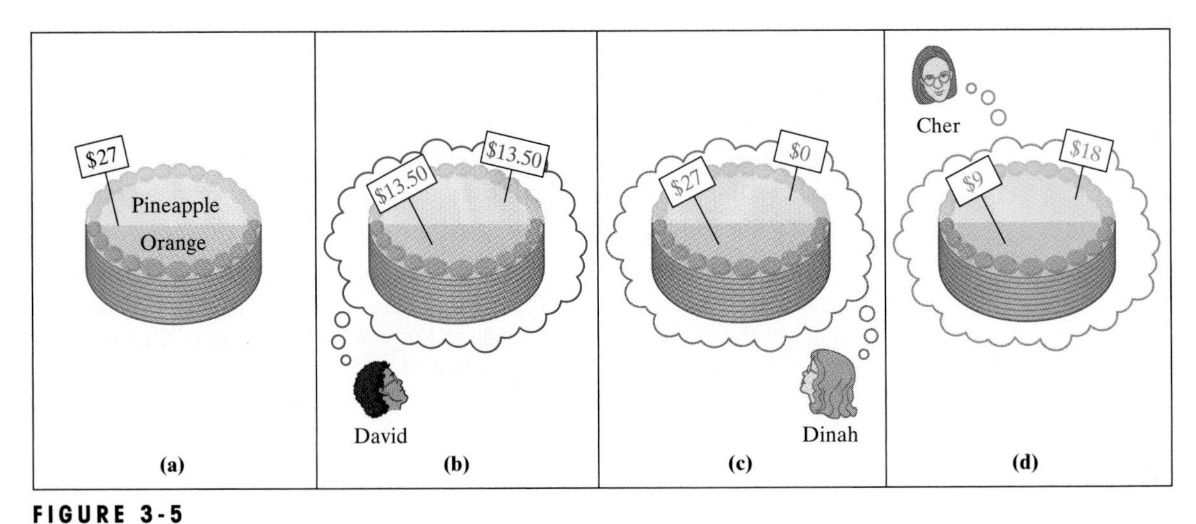

FIGURE 3-5
(a) The original cake (b) in David's eyes (c) in Dinah's eye (d) in Cher's eyes

- Cher likes pineapple twice as much as she likes orange ($P = 2O$). In Cher's eyes the cake looks like Fig. 3-5(d).
- **Step 1.** David starts by cutting the cake in 2 equal shares. His cut is shown in Fig. 3-6(a). Of course, since Dinah doesn't like pineapple, she will take the piece with the most orange. The value of the 2 pieces in each player's eyes is shown in Fig. 3-6.
- **Step 2.** David divides his piece into 3 pieces that in his opinion are of equal value. Of course to David this only means the volumes are equal, so he cuts the pieces as shown in Fig. 3-7(a). Dinah also divides her piece into 3 pieces that in her opinion are of equal value. Remember that Dinah hates pineapple, so she should cut in such a way as to have one-third of the orange in each of her pieces [as, for example, the cuts shown in Fig. 3-7(b)].

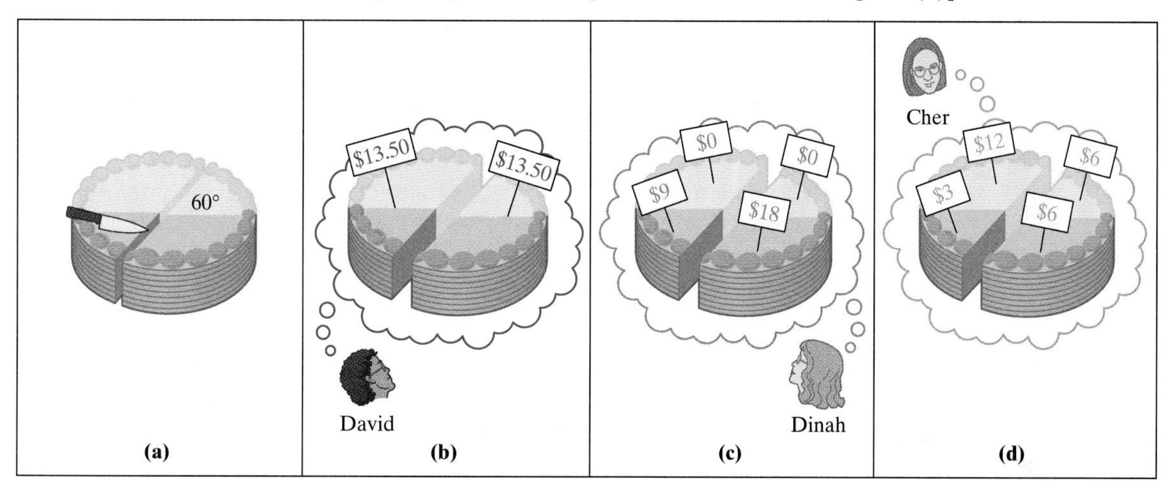

FIGURE 3-6
The first cut and the values of the pieces in the eyes of each player.

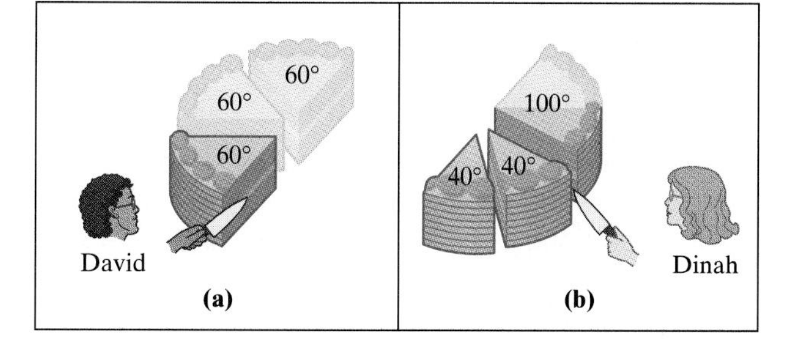

FIGURE 3-7
(a) David cuts "his" piece (b) Dinah cuts "her piece."

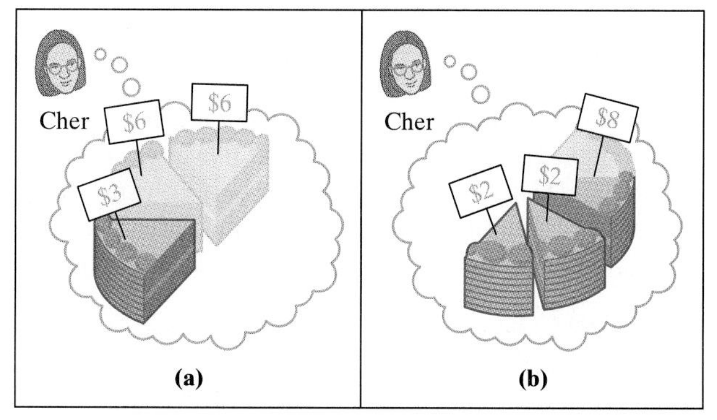

FIGURE 3-8
The values of the pieces in Cher's eyes.

- **Step 3.** It's now Cher's turn to choose 1 piece from David's 3 pieces and 1 piece from Dinah's 3 pieces. Figure 3-8 shows the value of the pieces in Cher's eyes.

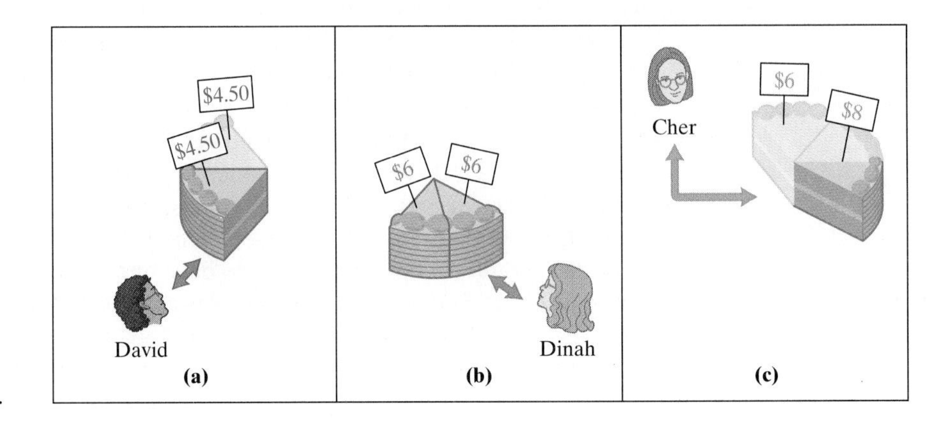

FIGURE 3-9 The share each player ends up with.

The final division of the cake is shown in Fig. 3-9.

Notice that each person has received a share that is worth at least one-third of the value of the cake. Notice also that David feels that Cher got a more valuable share than he did. Remember—a fair division only guarantees that each player will receive a fair share, not the best share. ■

The Last-Diminisher Method

We will describe the method for the general case of *N* players. The basic idea behind this method is that at any time throughout the game, the cake is divided into two pieces (which we will call the *C*-piece and the *R*-piece), and the players are divided into two groups: one player who is the "claimant" of the *C*-piece and all the other players, whom we will call the "nonclaimants." As the game goes on, each player gets the opportunity to become a claimant or a nonclaimant, and thus the *C*-piece, the *R*-piece, the claimant, and the nonclaimants all can change (this is what is going to keep the players honest!). Here are the details of exactly how it all works:

- **Preliminaries.** Before the game starts, the players are randomly assigned an order (P_1 first, P_2 second, . . . , P_N last), and the players will play in this order throughout the game. The game is played in rounds, and at the end of each round there is one less player and a smaller piece of cake to be divided.
- **Round 1.** The first player (P_1) starts by becoming the first claimant. P_1's job is to cut for herself a slice from the cake that she believes to be an *exact* fair share (one-*N*th) of the cake. This will be the *C*-piece, claimed by P_1. Since P_1 does not know if she will end up with this piece or not, she must be careful that her claim is neither too small (she may end up with that piece) nor too large (somebody else might end up with it). The next player (P_2) now has the right to become a claimant (*play*) or to remain a member of the nonclaimant group (*pass*) on the *C*-piece. P_2 should play only if he thinks that the *C*-piece is better than a fair share (worth more than one-*N*th) of the cake; otherwise he should pass and

remain a nonclaimant. If P_2 plays, he must do so by cutting out of the *C*-piece an appropriate sized sliver and *diminishing* the *C*-piece to the point where *it is a fair share* (in P_2's eyes). When this happens, P_2 becomes the claimant of the diminished *C*-piece, the sliver cut off from the old *C*-piece becomes a part of the *R*-piece, and P_1 happily (the *R*-piece got bigger) goes back to the nonclaimant group (see Fig. 3-10). It is now P_3's turn to pass or play on the *C*-piece, regardless of whether it is P_1's or P_2's *C*-piece. If P_3 passes, then nothing changes, and we move on to the next player. If P_3 plays (only if she thinks that the *C*-piece is better than a fair share), she cuts a sliver out of the *C*-piece so that it becomes a fair share, the sliver is added to the *R*-piece, and the previous claimant happily joins the nonclaimant group. We continue in this way until all the players in order have a chance to pass or play. The player who is the claimant at this point (*the last diminisher*) gets his *C*-piece and is out of the game. It is clear that if this player played honestly, he will end up with a fair share of the cake. What happens to the remaining players (the nonclaimants)? They move on to the next round.

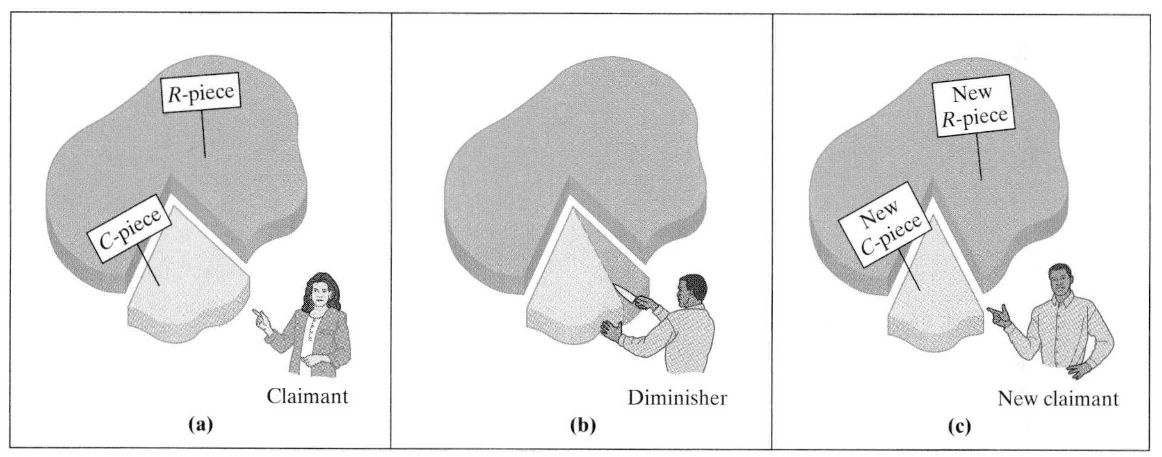

FIGURE 3-10
A diminisher becomes the new claimant.

- **Round 2.** The *R*-piece becomes the "new cake," to be divided fairly among the $N - 1$ remaining players, all of whom were nonclaimants in the previous round and therefore value the *R*-piece at more than $(N - 1) \cdot (1/N)$ of the original cake. In other words, since none of them thought that the old *C*-piece was all that great, they should all be happy to be in the position of having to divide this "new cake" fairly among $N - 1$ players. This is done by repeating the whole process (claimants, nonclaimants, *C*-pieces, and *R*-pieces), but remembering that now, with one less player, the threshold for a fair share is different [one-$(N - 1)$th of the total]. At the end of this round, the claimant to the *C*-piece gets it and is out of the game.
- **Round 3, 4, etc.** Repeat the process, each time with one less player and a smaller "cake," until there are just two players left. At this point divide the last piece of cake between the last two players using the *divider-chooser method.*

Example 7.

Five sailors are marooned on a lush, deserted tropical island. Knowing how to make the best out of a good situation, they decide to claim ownership of the island, divide it among themselves, and lead the good life there forever. Knowing something about fair division, they agree to do it using the last-diminisher method. Here pictures speak louder than words, so the whole story unfolds in Figs. 3-11 through 3-15.

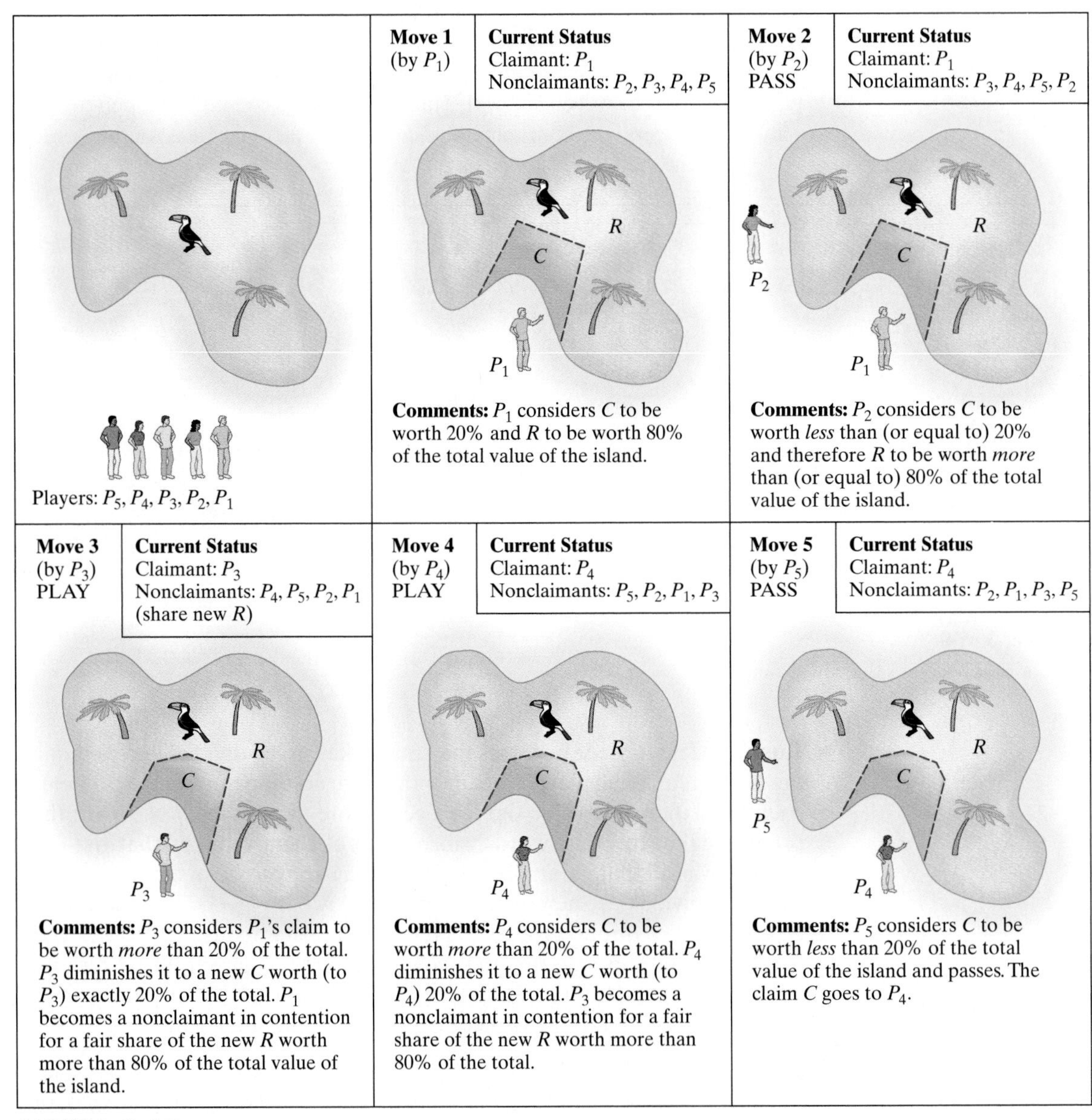

FIGURE 3-11
Example 7, Round 1.

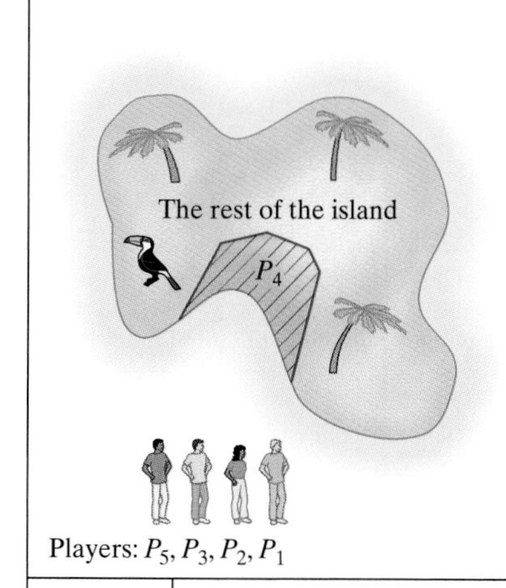

Players: P_5, P_3, P_2, P_1

Move 1 (by P_1)

Current Status
Claimant: P_1
Nonclaimants: P_2, P_3, P_5

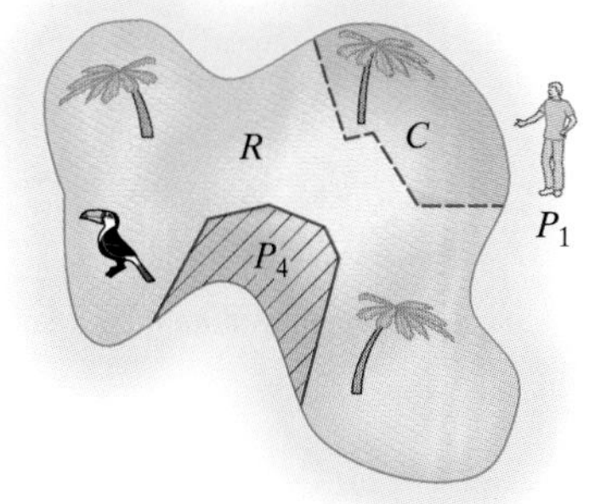

Comments: P_1 considers C to be worth 25% and R to be worth 75% of the value of the rest of the island.

Move 2 (by P_2) PASS
Move 3 (by P_3) PASS

Current Status
Claimant: P_1
Nonclaimants: P_5, P_2, P_3

R C P_1 P_2 P_4 P_3

Comments: P_2 and P_3 consider C to be worth *less* than or equal to 25% and therefore R to be worth *more* than or equal to 75% of the total value of the rest of the island.

Move 4 (by P_5) PLAY

Current Status
Claimant: P_3
Nonclaimants: P_2, P_3, P_1

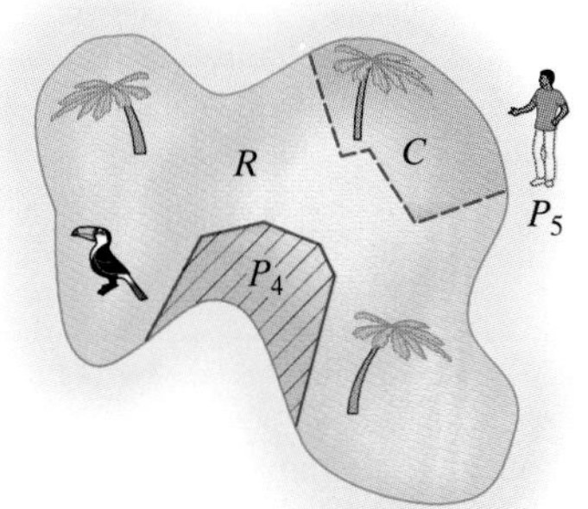

Comments: P_5 considers C claim to be worth *more* than 25%. P_5 diminishes it to a new C worth (to P_5) exactly 25% of the total. P_1 becomes a nonclaimant and the claim goes to P_5.

FIGURE 3-12 Example 7, Round 2. (4 players left)

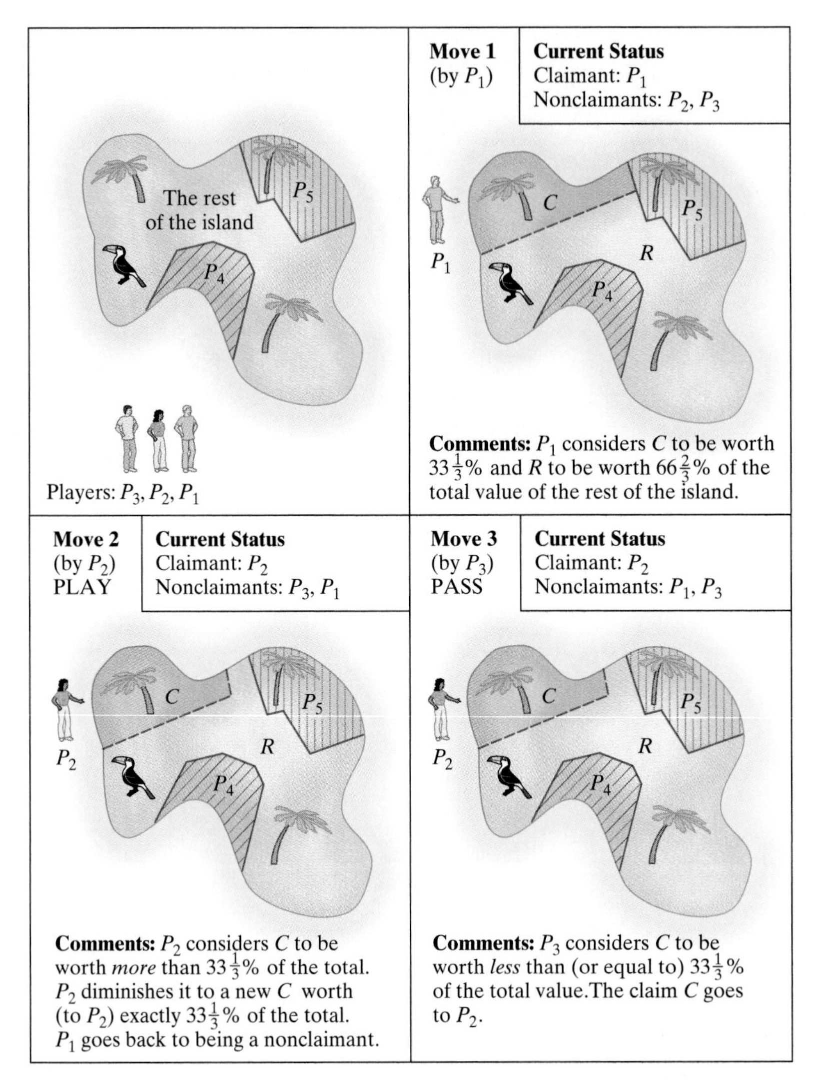

FIGURE 3-13
Example 7, Round 3. (3 players left)

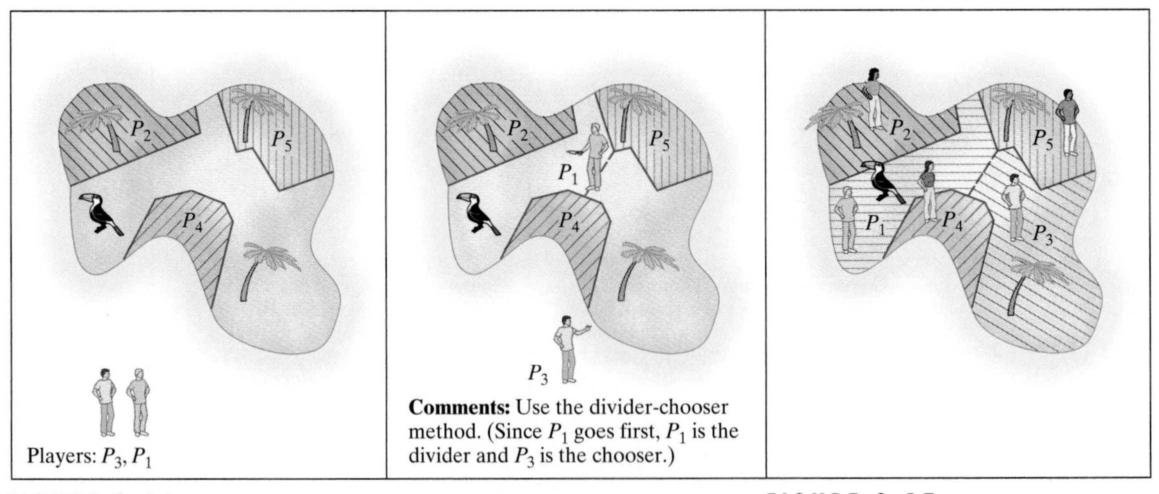

FIGURE 3-14
Example 7, last round (divider-chooser method).

FIGURE 3-15
The final division of the island. ■

We will now move on to *discrete* fair-division schemes, where the set S consists of objects that are indivisible (houses, cars, paintings, candy, etc.).

The Method of Sealed Bids

One of the most important discrete fair-division schemes is the **method of sealed bids**. The easiest way to illustrate how this method works is by means of an example.

Example 8.

In her last will and testament, Grandma plays a little joke on her four grandchildren (Art, Betty, Carla, and Dave) by leaving just three valuable items—a house, a Rolls Royce, and Picasso painting—with the stipulation that the items must remain with the grandchildren (not sold to outsiders) and must be divided fairly in equal shares among them. How could we possibly resolve this conundrum? The method of sealed bids will allow us to do this in a very elegant way.

Step 1 (The Bids). Each of the players is asked to make a bid for each of the items in the estate, giving his or her *honest*[5] assessment of the dollar value of each item. It is important that the bids are done independently, and no player should see another player's bids before making his or her own. Of course, the easiest way is for each player to submit his bid in a sealed envelope. When all the bids are in, they are opened. Table 3-5 shows each player's bid on each item in the estate.

Table 3-5 The Bids

	Art	Betty	Carla	Dave
House	220,000	250,000	211,000	198,000
Rolls Royce	40,000	30,000	47,000	52,000
Picasso	280,000	240,000	234,000	190,000

Step 2 (The Allocation). Each item goes to the highest bidder for that item. In this example, the house goes to Betty, the Rolls Royce goes to Dave and the Picasso painting goes to Art. Carla gets nothing. So far, this doesn't sound very fair!

Step 3 (The Payments). Now come the payments. Depending on what a player got, the player may owe money or be owed money by the estate. To calculate the numbers, we first calculate how much each player believes his or her share is worth. This is done by adding all the players' bids and dividing by the number of players. Table 3-6 shows the value of a fair share to each player (last row). If the

Table 3-6

	Art	Betty	Carla	Dave
Home	220,000	250,000	211,000	198,000
Rolls Royce	40,000	30,000	47,000	52,000
Picasso	280,000	240,000	234,000	190,000
Total	540,000	520,000	492,000	440,000
Fair share	135,000	130,000	123,000	110,000

[5] Players who make dishonest bids run the risk of coming out short at the end. As with every other fair-division method, the honesty of the players is built into the method itself.

total value of the items that the player gets is more than the value of that player's fair share, the player pays the estate the difference; if the total value of the items that the player gets is less than the value of the player's fair share, the player collects the difference in cash. Let's try it with each of our players.

- **Art.** Art's fair share is, by his own estimation, worth $135,000. By the same token, he is getting a Picasso painting he values at $280,000. This means that Art must pay to the estate $145,000 ($280,000 − $135,000). Notice that if Art was honest in his assessment of the value of each item, he is now getting a fair share of the estate.
- **Betty.** Betty's fair share is worth $130,000. She is getting the house, which she values at $250,000, so she must pay the estate the difference of $120,000. Once again, notice that Betty is now getting a fair share of the estate.
- **Carla.** Carla's fair share is, by her own estimation, $123,000. Since she is getting no items from the estate, she gets her full $123,000 in cash. Her fair share of the estate is now settled.
- **Dave.** Dave's assessment of the value of his fair share is $110,000. Now Dave is getting the Rolls, which he values at $52,000, so he has the balance of $58,000 coming to him in cash.

At this point each of the heirs has received a fair share, and we might consider our job done, except that now comes the fun part. If we add Art's and Betty's payments to the estate and subtract the payments made by the estate to Carla and Dave, we discover that something truly remarkable has happened: There is $84,000 left over ($145,000 + $120,000 coming in; $123,000 and $58,000 going out). This leads to the next move, where everybody wins!

Step 4 (Dividing the Surplus). The surplus money is divided equally among the four heirs. In our example each player's share of the $84,000 surplus is $21,000. This means that in the final settlement each player gets a fair share (in items plus or minus cash) plus an extra $21,000 in cash—everyone has to be tickled pink! ■

The method of sealed bids works so well because of a clever idea:[6] In most ordinary transactions there is a buyer and a seller, and the buyer knows the other party is the seller and vice versa. In a sense, this works to both parties' disadvantage. In the method of sealed bids, each player is simultaneously a buyer and a seller, without actually knowing which one until all the bids are opened. This keeps the players honest and, in the long run, works out to everyone's advantage. For the method to work, however, certain conditions must be satisfied.

1. Each player must have enough money to play the game. If a player is going to make honest bids on the items, he must be prepared to take some or all of them, which means that he may have to pay the estate certain sums of money. If the player does not have this money available, he is at a definite disadvantage in playing the game.
2. Each player must accept money (if it is a sufficiently large amount) as a substitute for any item. This means that no player can consider any of the items priceless (*I want Grandma's diamond ring, and no amount of money in the world is going to make me change my mind!* is not an attitude conducive to a good resolution of the problem).

[6] Like most ideas in this chapter, it comes from the work of the Polish mathematician Hugo Steinhaus in the 1940s.

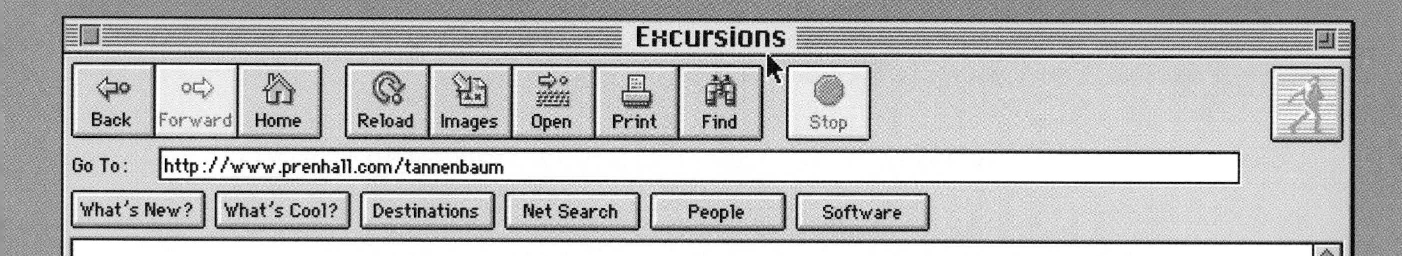

Choosing Sides

Stop by your typical playground and sooner or later you will witness an earnest and often comic playground ritual: a bunch of players trying to divide themselves into two supposedly "evenly matched" teams for the purpose of having a game. This ritual, commonly known as "choosing sides," is at heart just another *fair division problem* (the objects being divided are the players and the "players" in the fair division game are the captains doing the choosing). A significant example of "choosing sides" known as the *expansion draft* occurs in professional sports when new teams join an already existing league: The new teams are formed by the owners (or their representatives) drafting players from a designated free agent pool of players—choosing sides, if you will.

With just two teams, the most common approach for "choosing sides," and the one used in most expansion drafts, is the *single alternating scheme:* X chooses, then Y, then X again, and so on. It is clear that under this method, you want to be X. A better approach is the *double alternating scheme:* X chooses, then Y chooses twice, then X chooses twice, and so on (Y choosing once at the end). It's not quite as clear here whether you would want to be X or Y—it depends on the situation. Over the years, many other variations of these sequential approaches have been tried, in an effort to find the one that produces the fairest division of the talent pool.

Recently, Brian Dawson, a mathematician, proposed a new approach based on a simple variation of the divider-chooser method. This new approach generally gives a fairer result than any previously used drafting method. A description of Dawson's *draft protocol* and additional background information can be found under Chapter 3 in the Tannenbaum Website (http://www.prenhall.com/tannenbaum).

Questions:

1. How would you modify the single and double alternating schemes for the case in which the players must be divided into k teams ($k > 2$)?
2. Explain the purpose of Step 3 in Dawson's draft protocol.
3. Describe a scenario under which the "chooser" (Owner 2) would want to exchange one of his players with a player from the waiver pool.
4. Describe two different scenarios under which the "divider" (Owner 1) would want to exchange one of his players with a player from the waiver pool.

The method of sealed bids takes a particularly simple form in the case of two players and one item. Consider the following example:

Example 9.

Table 3-7

Al	Betty
$130,000	$142,000

Al and Betty are getting a divorce. The only common property of value is their house. Since the divorce is amicable and they are not particularly keen on going to court or hiring an attorney, they decide to divide the house using the method of sealed bids. The bids are shown in Table 3-7. Betty, being the highest bidder, gets the house but must pay the estate $71,000 (she is entitled to only half of the value of the house). Al's fair share is half of his bid, namely, $65,000. The surplus of $6000 is divided equally between Al and Betty, and the bottom line is that Betty gets the house but pays Al $68,000. Notice that this result is equivalent to assessing the value of the house as the value halfway between the two bids ($136,000) and splitting this value equally between the two parties, with the house going to the highest bidder and the cash to the other party (Exercise 40). ■

The Method of Markers

This is a discrete fair-division scheme that does not require the players to put up any of their own money. In this sense it has a definite advantage over the method of sealed bids. On the other hand, unlike the method of sealed bids, this method cannot be used effectively unless there are many more items to be divided than there are players.

In this method, we start with the items lined up in an *array* (a fixed sequence which cannot be changed). For convenience, think of the array as a string of objects. Each player independently bids for segments of consecutive items in the array (as many segments as there are players) by "cutting" the string. Notice that if we want to cut a string into N sections, all we need is $N - 1$ cuts. In practice, one way to make the "cuts" is to place markers in the places where the cuts are made (thus the name *method of markers*). Thus, each player can make her bids by placing $N - 1$ markers which divide the array into N segments, each of which represents an acceptable share of the entire set of items. This is important, because we will guarantee to each player that she will get one of her chosen segments, but we cannot guarantee which one. Needless to say, no player should see the markers of another player before laying down her own.

After all the players are through laying down their respective markers, the method allocates to each player one of their bid segments (a section between two consecutive markers). The easiest way to explain how this is done is with an example.

Example 10.

Four children—Alice, Bianca, Carla and Dana (A, B, C, and D)—are to divide the 20 pieces of candy shown in Fig. 3-16. Their teacher, Mrs. Jones, offers to divide the candy, but the children reply that they can do it themselves, thank you, using the method of markers. The 20 pieces are randomly arranged into the array shown in Fig. 3-17. For convenience, we will label the pieces of candy 1 through 20.

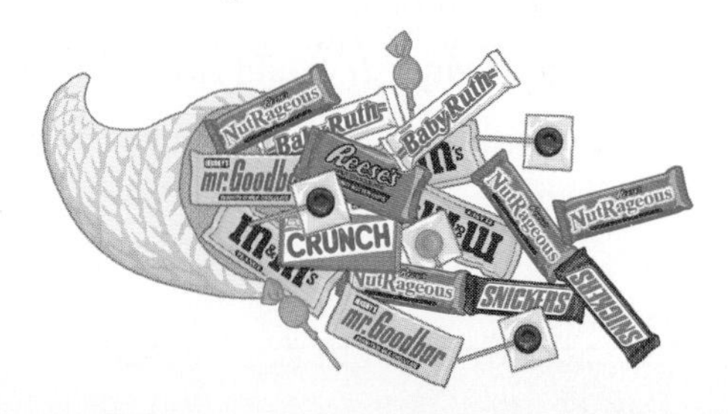

FIGURE 3-16
The "loot."

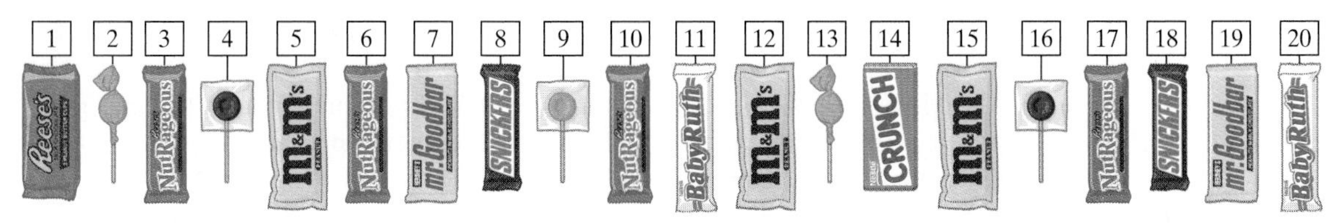

FIGURE 3-17

Step 1 (The Bids). Each child writes down independently on a piece of paper exactly where they want their 3 markers (4 players means 3 markers).[7] The bids are now opened, and the results are shown in Fig. 3-18.

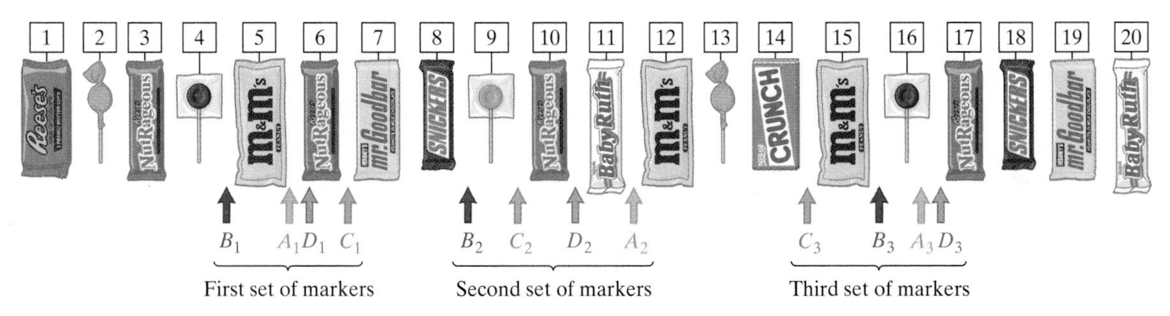

FIGURE 3-18
The results of the bidding.

Step 2 (The Allocations). We are now ready to allocate one segment of the array to each child. To do so we start scanning the array from left to right, until we find someone's first marker. Here the first *first marker* going from left to right is Bianca's (B_1), so we give Bianca her first segment (pieces 1 through 4, Fig. 3-19).

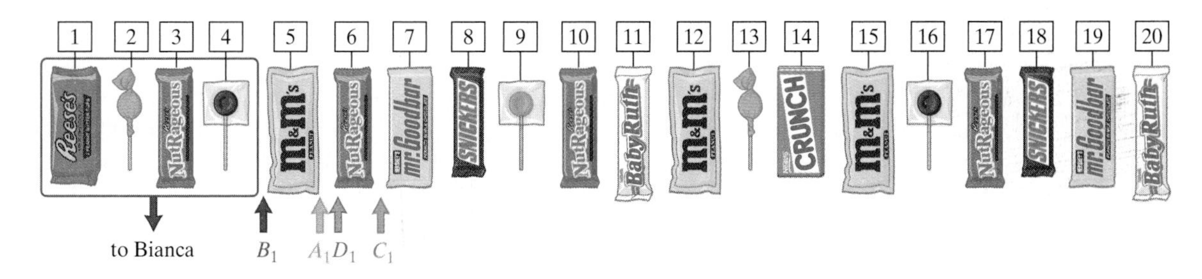

FIGURE 3-19
Bianca, the owner of the first first marker, gets her first segment.

Bianca has now received a fair share of the candy and is happily gone (and her markers removed, since they are no longer needed). We now continue scanning from left to right looking for the first *second marker.* This marker belongs to Carla (C_2). We now give to Carla her second segment, going from first marker to second

---

[7] This can be done by writing something like *first marker between items 3 and 4, second marker between . . . ,* etc.

marker (pieces 7 through 9, Fig. 3-20). Once Carla and her markers are out of the picture, we continue scanning from left to right until we find the first *third marker.* This is a tie between Alice and Dana (A_3, D_3), and we can break the tie randomly. After a coin toss, Alice ends up with her third segment (pieces 12 through 16, Fig. 3-21), and finally we give to the last player (Dana) her last segment (pieces 17 through 20, Fig. 3-22). Now everybody has gotten one of their chosen segments. The amazing part is that there is *leftover candy!*

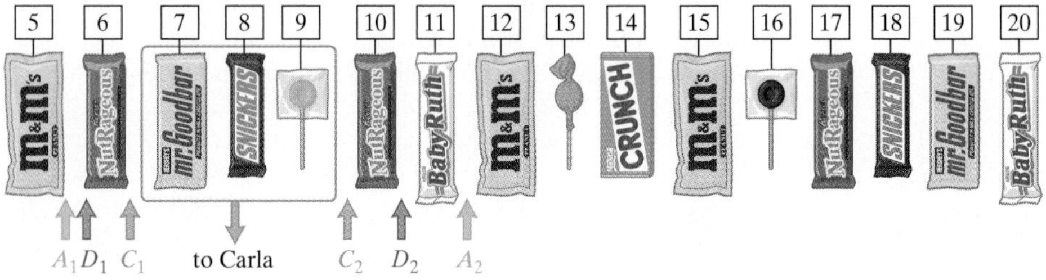

FIGURE 3-20 Carla, the owner of the first second marker (among the remaining players) gets his second segment.

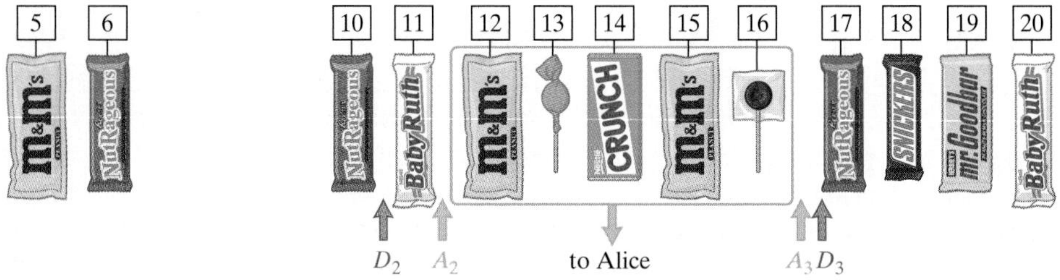

FIGURE 3-21 Alice and Dana both own the first third marker. After a coin toss, the third segment goes to Alice.

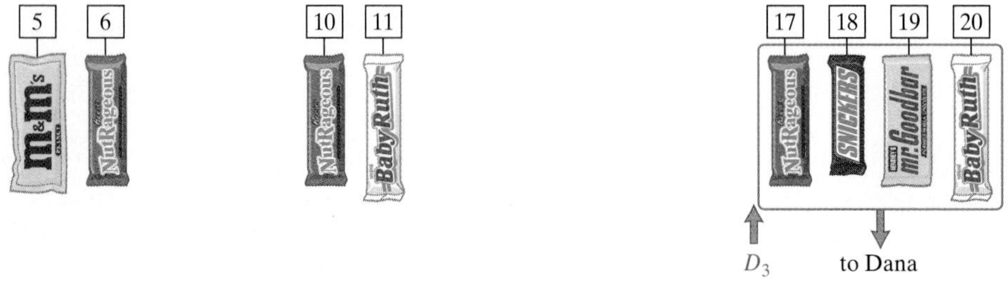

FIGURE 3-22 Dana is the last player left. She gets her last segment.

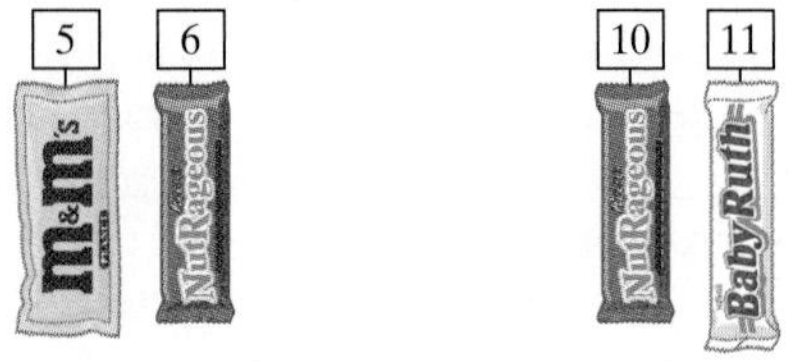

FIGURE 3-23 The leftovers (to be given randomly to the players one at a time) are a bonus.

Step 3 (Dividing the Leftovers). Usually there are just a few pieces of candy left over, not enough to play the game all over again, so the simplest thing to do is randomly draw lots and let the children go in order picking one piece at a time until there is no more candy left. Here the leftover pieces are 5, 6, 10, and 11. The players now draw lots; Carla gets to choose first and takes piece 11. Dana chooses next and takes piece 5. Bianca and Alice end up with pieces 6 and 10, respectively. ■

We now give the general description of the **method of markers** with N players and M items which are arranged into an array.

- **Step 1 (The Bids).** Each player independently divides the array into N acceptable segments (the player can do this by placing $N - 1$ markers).
- **Step 2 (The Allocations).** Scan the array from left to right until the first *first marker* is located. The player owning that marker gets to keep his first segment, and his markers are removed. In case of a tie, break the tie randomly. We continue moving from left to right, looking for the first *second marker.* The player owning it gets to keep her second segment. Continue this process until each player has received one of the segments.
- **Step 3 (Leftovers).** The leftover items can be divided among the players by some form of lottery, and, in the rare case that there are many more items than players, the method of markers could be used again.

In spite of its simple elegance, the method of markers can be used only under some fairly restrictive conditions. In particular, the method assumes that every player is able to divide the array of items into segments in such a way that each of the segments has approximately equal value. This is usually possible when the items are of small and homogeneous value, but almost impossible to accomplish when there are expensive items involved. (Imagine, for example, trying to divide fairly a bunch of pieces of candy plus a gold coin using the method of markers.)

Conclusion

The problem of dividing an object or set of objects among the members of a group is a practical problem that comes up regularly in our daily lives. When the object is a pizza, a cake, or a bunch of candy, we don't always pay a great deal of attention to the issue of fairness, but when the object is an estate, land, jewelry or some other valuable asset, dividing things fairly becomes a critical issue.

On the surface, problems of fairness seem far removed from the realm of mathematics. We are more likely to think of economics, political science, or law as being the proper fields for a discussion of this topic. It is surprising, therefore, that when certain basic conditions are satisfied, mathematics can provide fair-division methods that not only guarantee fairness but often turn out to actually do much better than that.

In this chapter we discussed several such methods, which we called *fair-division schemes.* The choice of which is the best fair-division scheme to use in a particular situation is not always clear, and in fact there are many situations in which a fair division is mathematically unattainable (we will discuss an important example of this in Chapter 4). At the same time, in a large number of everyday situations the fair-division schemes we described in this chapter (or simple variations thereof) will work. Remember these methods the next time you must divide an inheritance, a piece of real estate, or even some of the chores around the house. They may serve you well.

Key Concepts

continuous fair-division problem
discrete fair-division problem
divider-chooser method
fair-division problem
fair-division scheme
fair share
last-diminisher method
lone-chooser method
lone-divider method
method of markers
method of sealed bids

Exercises

Walking

1. Alex buys a chocolate-strawberry mousse cake [shown in (i) below] for $12. Alex values chocolate 3 times as much as he values strawberry.

(a) What is the value of the chocolate half of the cake to Alex?

(b) What is the value of the strawberry half of the cake to Alex?

(c) What is the value of a 60° strawberry wedge of the cake to Alex?

(d) What is the value of a 40° chocolate wedge of the cake to Alex?

(e) A piece of the cake is cut as shown in (ii). What is the value of the piece of cake to Alex?

(i)

(ii)

2. Jody buys a chocolate-strawberry mousse cake [shown in (i) below] for $13.50. Jody values strawberry 4 times as much as she values chocolate.

(a) What is the value of the chocolate half of the cake to Jody?

(b) What is the value of the strawberry half of the cake to Jody?

(c) What is the value of a 60° strawberry wedge of the cake to Jody?

(d) What is the value of a 40° chocolate wedge of the cake to Jody?

(e) A piece of the cake is cut as shown in (ii). What is the value of the piece of cake to Jody?

(i)

(ii)

3. Kala buys a chocolate-strawberry-vanilla cake [shown in (i) opposite page] for $12. Kala likes all 3 flavors but likes strawberry twice as much as vanilla and likes chocolate 3 times as much as vanilla.

(a) What is the value of the chocolate part of the cake to Kala?

(b) What is the value of the strawberry part of the cake to Kala?

(c) What is the value of the vanilla part of the cake to Kala?

(d) The cake is cut into 6 pieces equal in size (pieces 1 through 6) as shown in (ii). What is the value of each of the 6 pieces of cake to Kala?

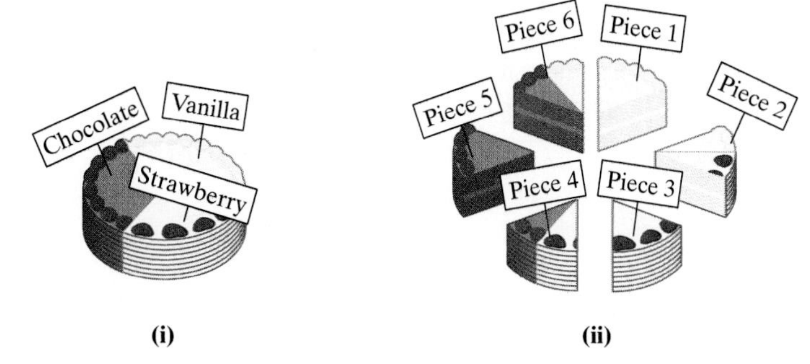

(i) (ii)

4. Malia buys a chocolate-strawberry-vanilla cake [shown in (i) below] for \$11.20. Malia likes all 3 flavors but likes strawberry twice as much as chocolate and likes chocolate twice as much as vanilla.

(a) What is the value of the chocolate part of the cake to Malia?

(b) What is the value of the strawberry part of the cake to Malia?

(c) What is the value of the vanilla part of the cake to Malia?

(d) The cake is cut into 6 pieces equal in size (pieces 1 though 6) as shown in (ii). What is the value of each of the 6 pieces of cake to Malia?

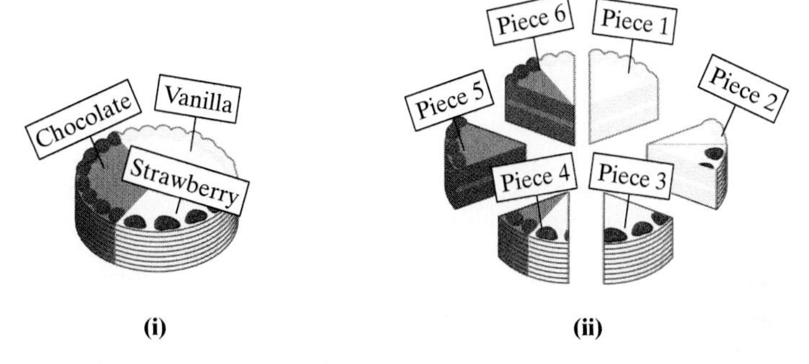

(i) (ii)

5. Three players (Ana, Ben, and Cara) must divide a cake among themselves. Suppose the cake is divided into 3 slices (s_1, s_2, and s_3). The value of the entire cake and of each of the 3 slices in the eyes of each of the players is shown in the table below. For each player, indicate which of the slices are fair shares.

	Whole cake	s_1	s_2	s_3
Ana	\$12.00	\$3.00	\$5.00	\$4.00
Ben	\$15.00	\$4.00	\$4.50	\$6.50
Cara	\$13.50	\$4.50	\$4.50	\$4.50

6. Three players (Ana, Ben, and Cara) must divide a cake among themselves. Suppose the cake is divided into 3 slices (s_1, s_2, and s_3). The table below shows the percentage of the value of the entire cake that each slice represents to each player. For each player, indicate which of the slices are fair shares.

	s_1	s_2	s_3
Ana	30 %	40%	30 %
Ben	35 %	25%	40 %
Cara	$33\frac{1}{3}\%$	50%	$16\frac{2}{3}\%$

7. Four players (Abe, Betty, Cory, and Dana) must divide a cake among themselves. Suppose the cake is divided into 4 slices (s_1, s_2, s_3 and s_4). The table below shows the percentage of the value of the entire cake that each slice represents to each player. For each player, indicate which of the slices are fair shares.

	s_1	s_2	s_3	s_4
Abe	30%	24%	20%	26%
Betty	35%	25%	20%	20%
Cory	25%	40%	15%	20%
Dana	20%	20%	20%	40%

8. Four players (Abe, Betty, Cory, and Dana) must divide a cake among themselves. Suppose the cake is divided into 4 slices (s_1, s_2, s_3 and s_4). The value of the entire cake and of each of the 4 slices in the eyes of each of the players is shown in the table below. For each player, indicate which of the slices are fair shares.

	Whole cake	s_1	s_2	s_3	s_4
Abe	$15.00	$3.00	$5.00	$5.00	$2.00
Betty	$18.00	$4.50	$4.50	$4.50	$4.50
Cory	$12.00	$4.00	$3.50	$1.50	$3.00
Dana	$10.00	$2.75	$2.25	$2.50	$2.50

9. Two friends (David and Paul) decide to divide the pizza shown in the accompanying figure using the divider-chooser method. David likes pepperoni, sausage, and mushrooms equally well but hates anchovies. Paul likes anchovies, mushrooms, and pepperoni equally well but hates sausage. Neither one knows anything about the other one's likes and dislikes (they are new friends).

(a) Suppose that David is the divider. Which of the cuts (i) through (iv) show a division of the pizza into fair shares according to David?

(b) For each of the cuts in (a), which piece should Paul choose?

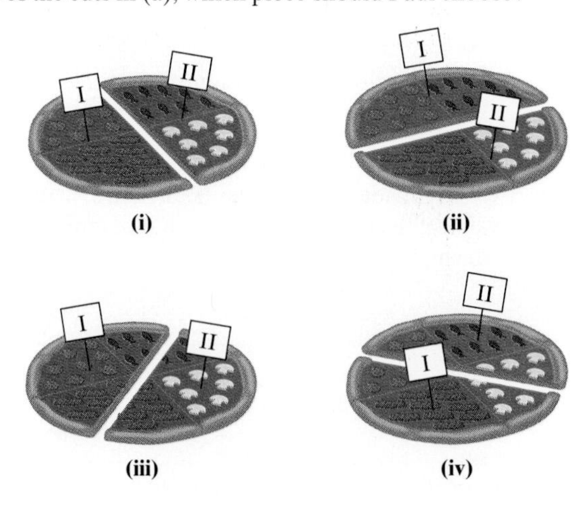

10. Raul and Karli want to divide a chocolate-strawberry mousse cake. Raul values chocolate 3 times as much as he values strawberry; Karli values chocolate twice as much as she values strawberry.

(a) If Raul is the divider, which of the following cuts are consistent with Raul's value system?

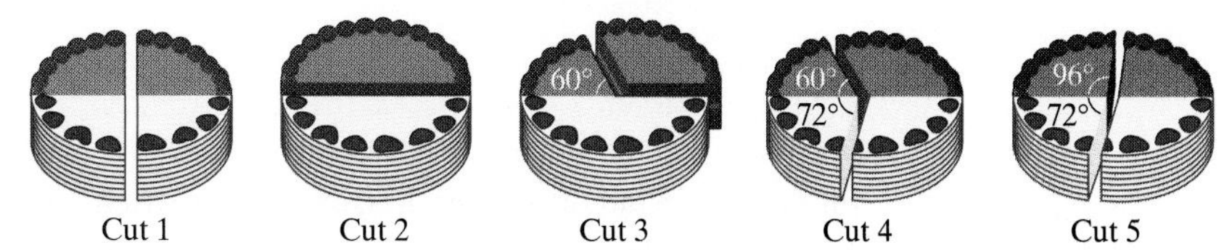

(b) For each of the cuts in (a), indicate which of the pieces is Karli's best choice.

(c) Suppose Karli is the divider. Draw three different cuts that are consistent with her value system.

(d) Explain why the following cut is consistent with any value system.

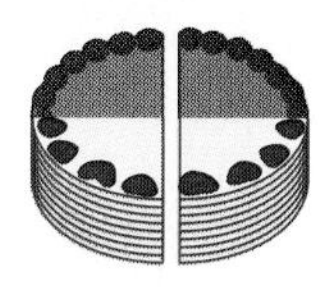

11. Three players want to divide a cake fairly using the lone-divider method. The divider cuts the cake into 3 slices (s_1, s_2, s_3).

(a) If the chooser declarations are

Chooser 1: $\{s_2, s_3\}$

Chooser 2: $\{s_1, s_3\}$

describe a possible fair division of the cake.

(b) If the chooser declarations are

Chooser 1: $\{s_1, s_2, s_3\}$

Chooser 2: $\{s_1\}$

describe a possible fair division of the cake.

(c) If the chooser declarations are

Chooser 1: $\{s_1\}$

Chooser 2: $\{s_2\}$

describe a possible fair division of the cake.

(d) If the chooser declarations are

Chooser 1: $\{s_1\}$

Chooser 2: $\{s_1\}$

describe how to proceed to obtain a possible fair division of the cake.

12. Four players want to divide a cake fairly using the lone-divider method. The divider cuts the cake into 4 slices (s_1, s_2, s_3, s_4), and the choosers make the following declarations:

Chooser 1: $\{s_2, s_3\}$

Chooser 2: $\{s_3, s_4\}$

Chooser 3: $\{s_4\}$.

(a) Describe a fair division of the cake.

(b) Explain why the answer in (a) is the only possible fair division of the cake.

13. Four players want to divide a cake fairly using the lone-divider method. The divider cuts the cake into 4 slices (s_1, s_2, s_3, s_4), and the choosers make the following declarations:

Chooser 1: $\{s_2, s_3\}$

Chooser 2: $\{s_1, s_3\}$

Chooser 3: $\{s_1, s_2\}$.

(a) Describe a fair division of the cake.

(b) Describe a fair division of the cake different from the one given in (a).

(c) Is it possible to find a fair division of the cake such that the divider doesn't get s_4? Explain your answer.

14. Four players want to divide a cake fairly using the lone-divider method. The divider cuts the cake into 4 slices (s_1, s_2, s_3, s_4) and the choosers make the following declarations:

Chooser 1: $\{s_1, s_2\}$

Chooser 2: $\{s_1, s_2\}$

Chooser 3: $\{s_2\}$.

Describe how to proceed to obtain a possible fair division of the cake.

15. Five players want to divide a cake fairly using the lone divider method. The divider cuts the cake into 5 slices (s_1, s_2, s_3, s_4, s_5), and the choosers make the following declarations:

Chooser 1: $\{s_2, s_4\}$

Chooser 2: $\{s_2, s_4\}$

Chooser 3: $\{s_2, s_3, s_4\}$

Chooser 4: $\{s_2, s_3, s_5\}$.

(a) Describe a fair division of the cake.

(b) Describe a fair division of the cake different from the one given in (a).

(c) Is it possible to find a fair division of the cake such that the divider doesn't get s_1? Explain your answer.

16. Five players want to divide a cake fairly using the lone-divider method. The divider cuts the cake into 5 slices (s_1, s_2, s_3, s_4, s_5), and the choosers make the following declarations:

Chooser 1: $\{s_2, s_5\}$

Chooser 2: $\{s_1, s_2, s_5\}$

Chooser 3: $\{s_1, s_4, s_5\}$

Chooser 4: $\{s_2, s_4\}$

(a) Describe a fair division of the cake.

(b) Describe a fair division of the cake different from the one given in (a).

(c) Is it possible to find a fair division of the cake such that the divider doesn't get s_3? Explain your answer.

17. Six players want to divide a cake fairly using the lone-divider method. The divider cuts the cake into 6 slices ($s_1, s_2, s_3, s_4, s_5, s_6$), and the choosers make the following declarations:

Chooser 1: $\{s_2, s_3, s_5\}$

Chooser 2: $\{s_1, s_5, s_6\}$

Chooser 3: $\{s_3, s_5, s_6\}$

Chooser 4: $\{s_2, s_3\}$

Chooser 5: $\{s_3\}$.

(a) Describe a fair division of the cake.

(b) Explain why the answer in (a) is the only possible fair division of the cake.

18. Six players want to divide a cake fairly using the lone-divider method. The divider cuts the cake into 6 slices ($s_1, s_2, s_3, s_4, s_5, s_6$), and the choosers make the following declarations:

Chooser 1: $\{s_1]$

Chooser 2: $\{s_2, s_3\}$

Chooser 3: $\{s_4, s_5\}$

Chooser 4: $\{s_4, s_5\}$

Chooser 5: $\{s_1\}$.

Describe how to proceed to obtain a fair division of the cake.

19. Four players want to divide a parcel of land valued at $120,000 fairly using the lone-divider method. The divider cuts the parcel into 4 slices (s_1, s_2, s_3, s_4) as shown in the figure.

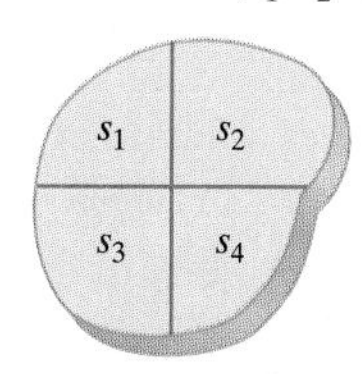

The choosers value (in thousands of dollars) these pieces as follows:

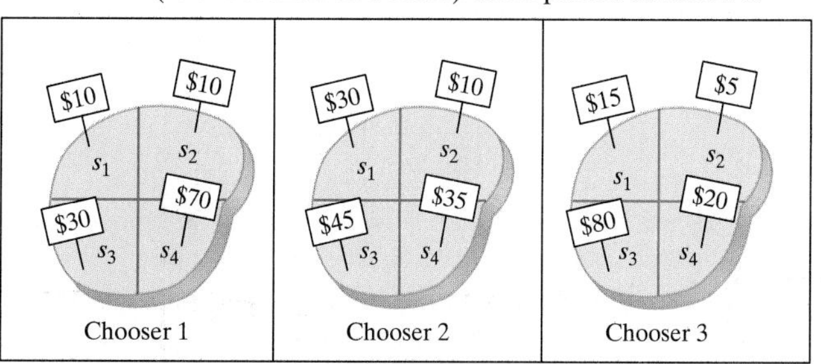

(a) What should the chooser declarations be?

(b) Describe a possible fair division of the land.

20. Three players (X, Y, and Z) decide to divide a $12 vanilla-strawberry cake using the lone-chooser method. The dollar amounts of the cake in each player's eyes are given in the following figure.

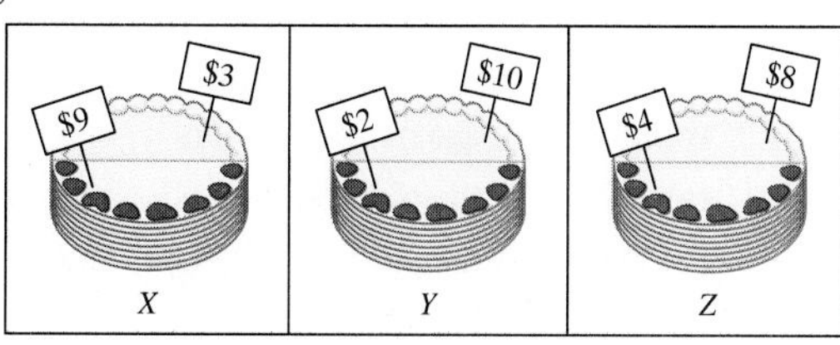

(a) Suppose that in the first division X cuts and Y chooses and that the player making the final selection is Z. Describe a possible fair division of the cake and for each player's final share give the dollar amounts of the share in the players' own eyes.

(b) Suppose that in the first division Z cuts and X chooses and that the player making the final selection is Y. Describe a possible fair division of the cake and for each player's final share give the dollar amounts of the share in the players' own eyes.

(c) Suppose that in the first division Y cuts and Z chooses and that the player making the final selection is X. Describe a possible fair division of the cake and for each player's final share give the dollar amounts of the share in the players' own eyes.

21. Three players (X, Y, and Z) decide to divide the cake shown in the accompanying figure using the lone-chooser method.

X likes chocolate and orange equally well, but hates strawberry and vanilla.

Y likes chocolate, and strawberry equally well, but hates orange and vanilla.

Z likes chocolate and vanilla equally well, but hates orange and strawberry.

(a) Suppose that in the first division X cuts and Y chooses and that the player making the final selection is Z. Describe a possible fair division of the cake, and for each player's final share give the value (as a percentage of the total) of the share in the players' own eyes.

(b) Suppose that in the first division Z cuts and X chooses and that the player making the final selection is Y. Describe a possible fair division of the cake, and for each player's final share give the value (as a percentage of the total) of the share in the player's own eyes.

(c) Suppose that in the first division Y cuts and Z chooses and that the player making the final selection is X. Describe a possible fair division of the cake, and for each player's final share give the value (as a percentage of the total) of the share in the player's own eyes.

22. A cake is to be divided among 4 people using the last-diminisher method. The 4 people are randomly ordered and are called P_1, P_2, P_3, and P_4. In round 1, P_1 cuts a piece s, P_2 and P_3 think it is fair and don't diminish it, and P_4 diminishes it.

(a) Is it possible for P_2 to end up with any part of s in his final share?

(b) Who gets a piece at the end of round 1?

(c) Who cuts the piece at the beginning of round 2?

(d) Who is the last person who has an opportunity to diminish the piece in round 2?

(e) How many rounds are required to divide the cake among the 4 people?

23. A cake is to be divided among 12 people using the last-diminisher method. The 12 people are randomly ordered and are called $P_1, P_2, P_3, \ldots, P_{12}$. In round 1, P_1 cuts a piece, and P_3, P_7, and P_9 are the only diminishers. In round 2, the only diminisher is P_5, and in round 3 there are no diminishers.

(a) Who gets the piece at the end of round 1?

(b) Who cuts the piece at the beginning of round 2?

(c) Who is the last person who has an opportunity to diminish the piece in round 2?

(d) Who gets the piece at the end of round 2?

(e) Who gets the piece at the end of round 3?

(f) Who cuts the piece at the beginning of round 4?

(g) Who is the last person who has an opportunity to diminish the piece in round 4?

(h) How many rounds are required to divide the cake among the 12 people?

24. A cake is to be divided among 6 people using the last-diminisher method. The 6 people are randomly ordered and are called $P_1, P_2, P_3, P_4, P_5, P_6$. In round 1, P_1 cuts a piece, and P_2, P_5, and P_6 are the only diminishers. In round 2 there are no diminishers. In round 3, everyone who can, diminishes.

(a) Who gets the piece at the end of round 1? Why?

(b) Who cuts the piece at the beginning of round 2?

(c) Who is the last person who has an opportunity to diminish the piece in round 2? Why?

(d) Who gets the piece at the end of round 2? Why?

(e) Who cuts the piece at the beginning of round 3? Why?

(f) Who diminishes in round 3?

(g) Who gets the piece at the end of round 3?

(h) Who is the first person who has an opportunity to diminish in round 4?

(i) Who is the last person who has an opportunity to diminish the piece in round 4?

(j) How many rounds are required to divide the cake among the 6 people?

25. Three sisters (*A*, *B*, and *C*) wish to divide up 4 pieces of furniture they shared as children using the method of sealed bids. Their bids on each of the items are given in the following table.

	A	*B*	*C*
Dresser	$150	$300	$275
Desk	180	150	165
Vanity	170	200	260
Tapestry	400	250	500

Describe the outcome of this fair-division problem.

26. Robert and Peter equally inherit their parents' old house and classic car. They decide to divide the 2 items using the method of sealed bids. Robert bids $29,200 on the car and $60,900 on the house. Peter bids $33,200 on the car and $65,300 on the house. Describe the outcome of this fair-division problem.

27. Bob, Ann, and Jane wish to dissolve their partnership using the method of sealed bids. Bob bids $240,000 for the partnership, Ann bids $210,000, and Jane bids $225,000.

(a) Who gets the business and for how much?

(b) What do the other two people get?

28. Three heirs (A, B, and C) wish to divide up an estate consisting of a house, a small farm, and a painting, using the method of sealed bids. The heirs' bids on each of the items are given in the following table:

	A	*B*	*C*
House	$150,000	$146,000	$175,000
Farm	430,000	425,000	428,000
Painting	50,000	59,000	57,000

Describe the outcome of this fair-division problem.

29. Three people (A, B, and C) wish to divide up 4 items using the method of sealed bids. Their bids on each of the items are given in the following table:

	A	*B*	*C*
Item 1	$20,000	$18,000	$15,000
Item 2	46,000	42,000	35,000
Item 3	3,000	2,000	4,000
Item 4	201,000	190,000	180,000

Describe the outcome of this fair-division problem.

30. Three people (A, B, and C) wish to divide up 5 items using the method of sealed bids. Their bids on each of the items are given in the following table:

	A	*B*	*C*
Item 1	$14,000	$12,000	$22,000
Item 2	24,000	15,000	33,000
Item 3	16,000	18,000	14,000
Item 4	16,000	16,000	18,000
Item 5	18,000	24,000	20,000

(a) What does A end up with? (Does A pay anything?)

(b) What does B end up with? (Does B pay anything?)

(c) What does C end up with? (Does C pay anything?)

31. Five heirs (A, B, C, D, and E) wish to divide up an estate consisting of 6 items using the method of sealed bids. The heirs' bids on each of the items are given in the following table:

	A	*B*	*C*	*D*	*E*
Item 1	$352	$295	$395	$368	$324
Item 2	98	102	98	95	105
Item 3	460	449	510	501	476
Item 4	852	825	832	817	843
Item 5	513	501	505	505	491
Item 6	725	738	750	744	761

Describe the outcome of this fair-division problem.

32. Three players (A, B, and C) agree to divide the 13 items shown by lining them up in order and using the method of markers. The players' bids are as indicated.

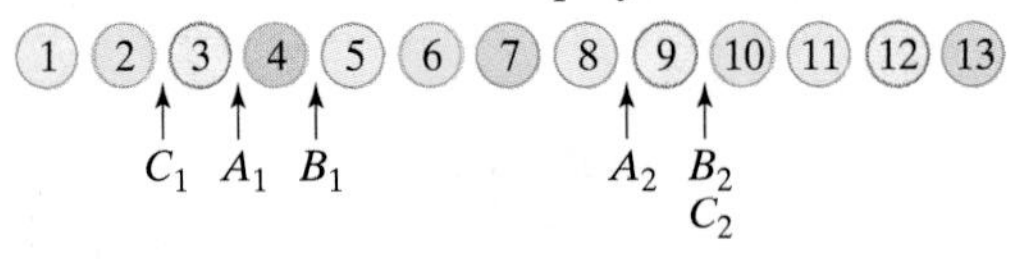

(a) Describe the allocation of items to each player.

(b) Which items are left over?

33. Three players (A, B, and C) agree to divide the 13 items shown by lining them up in order and using the method of markers. The players' bids are as indicated.

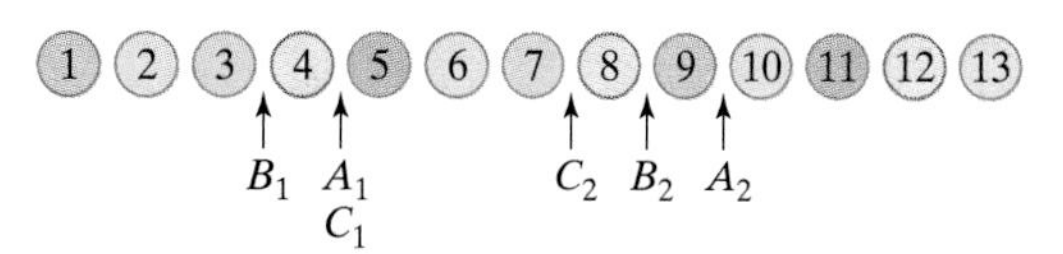

(a) Describe the allocation of items to each player.

(b) Which items are left over?

34. Two players (A and B) agree to divide the 12 items shown by lining them up in order and using the method of markers. The players' bids are as indicated.

(a) Describe the allocation of items to each player.

(b) Which items are left over?

35. Three players (A, B, and C) agree to divide the 12 items shown by lining them up in order and using the method of markers. The players' bids are as indicated.

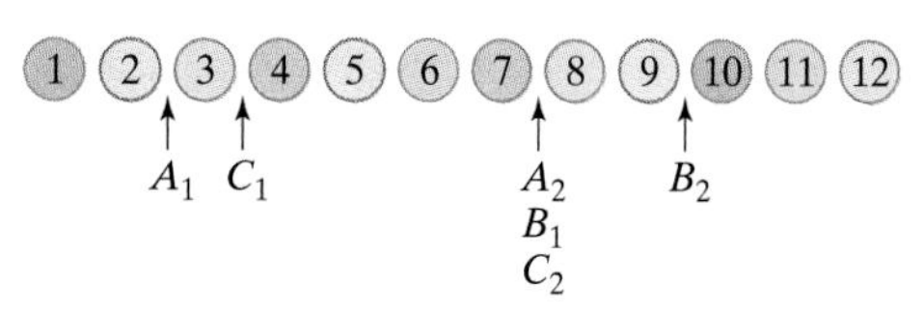

(a) Describe the allocation of items to each player.

(b) Which items are left over?

36. Three players (A, B, and C) agree to divide the 12 items shown by lining them up in order and using the method of markers. The players' bids are as indicated.

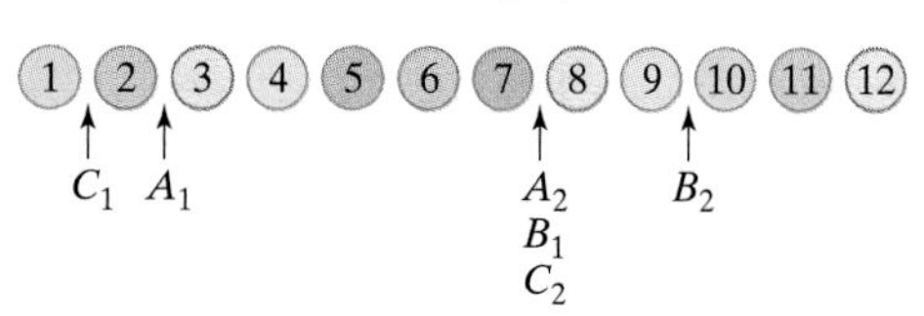

(a) Describe the allocation of items to each player.

(b) Which items are left over?

37. Five players (A, B, C, D, and E) agree to divide the 20 items shown by lining them up in order and using the method of markers. The players' bids are as indicated.

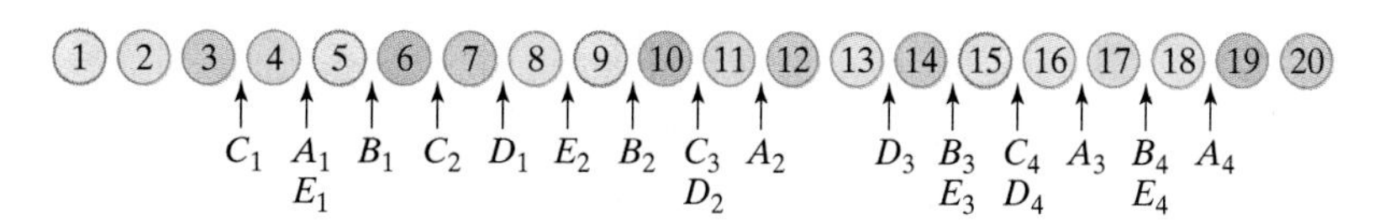

(a) Describe the allocation of items to each player.

(b) Which items are left over?

38. Four players (A, B, C, and D) agree to divide the 15 items shown below by lining them up in order and using the method of markers. The players' bids are as indicated.

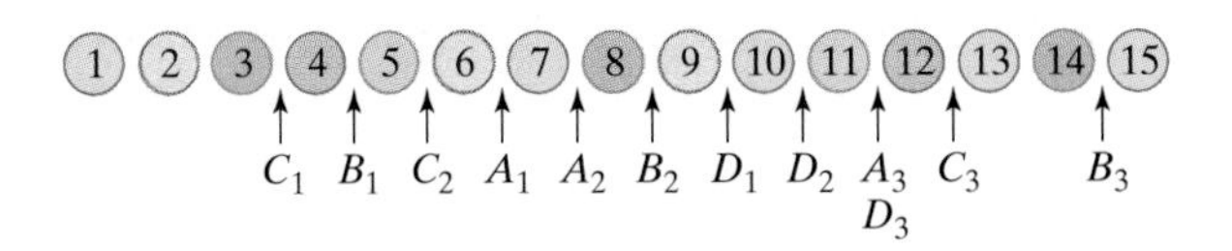

(a) Describe the allocation of items to each player.

(b) Which items are left over?

39. Four players (A, B, C, and D) agree to divide the 15 items shown by lining them up in order and using the method of markers. The players' bids are as indicated.

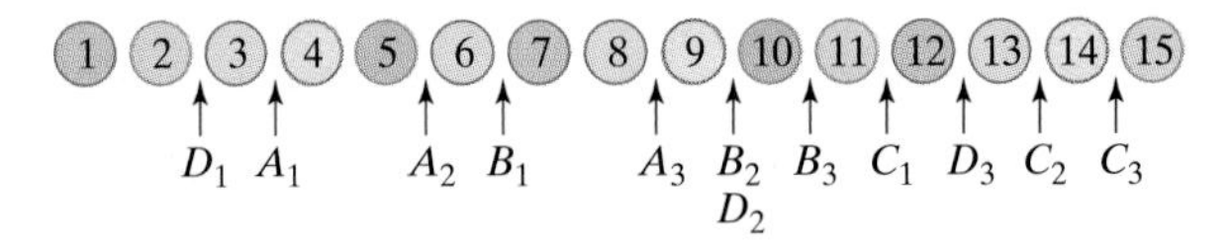

(a) Describe the allocation of items to each player.

(b) Which items are left over?

Jogging

40. Every Friday night, Marty's Ice Cream Parlor sells "Kitchen Sink Sundaes" (Fridaes?) for $6.00 each. A KiSS consists of 12 mixed scoops of whatever flavors Marty wants to get rid of. The customer has no choice. Three friends (Abe, Babe, and Cassandra) decide to share one. Abe wants to eat half of it and pays $3.00 while Babe and Cassie pay $1.50 each. They decide to divide it by the lone-divider method. Abe spoons the sundae onto four plates (P, Q, R, and S) and says that he will be satisfied with any two of them.

(a) If both Babe and Cassie find only Q and R acceptable, discuss how to proceed.

(b) If Babe finds only Q and R acceptable, and Cassie finds only P and S acceptable, discuss how to proceed.

(c) If Babe and Cassie both find only R acceptable, discuss how to proceed.

41. Three friends, Peter, Paul, and Mary, each pay $1.20 and plan to divide a $3.60 half-gallon brick of "Neopolitan" ice cream made of equal size bricks of strawberry, vanilla, and chocolate, using the lone-chooser method with Mary as the chooser. The value of the three flavors to each player is shown in the figure. (None of the players knows anything about the others' tastes and value systems.)

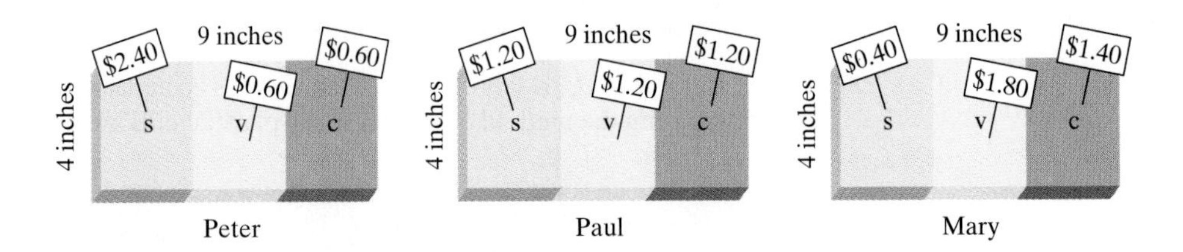

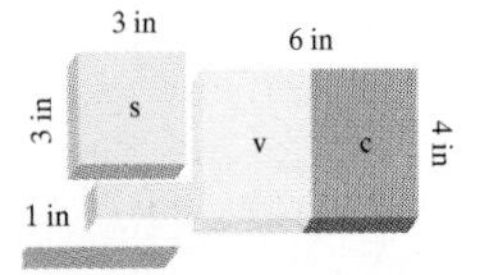

Peter starts by cutting the whole brick into 2 pieces as shown in the figure in the margin. Assuming that all players play honestly and that all of the remaining cuts are horizontal, describe how the rest of the division would proceed. Who gets what and how much is each player's share worth to the player receiving it.

42. Three players (P_1, P_2, and P_3) agree to divide the property shown below using the last-diminisher method. The order of the players is P_1, P_2, P_3. The first player to play, P_1, makes a claim C as shown in the figure. We know that both P_2's and P_3's value systems are the same and that they value the land uniformly.

Park Place

100 m

100 m

60 m

C

200 m

Baltic Avenue

300 m

(a) Give a geometric argument why P_2 and P_3 would both pass in round 1 and P_1 would end up with C.

(b) Describe a possible cut that the divider in round 2 might make.

(c) Suppose that, after round 1 is over, P_2 and P_3 discover that the city requires that the next cut be made parallel to Park Place. Describe a possible cut that the divider in round 2 might make in this case.

(d) Repeat (c) for a cut that must be made parallel to Baltic Avenue.

43. Three players (P_1, P_2, and P_3) agree to divide the property shown using the last-diminisher method. The order of the players is P_1, P_2, P_3. The first player to play, P_1, makes a claim C as shown in the figure.

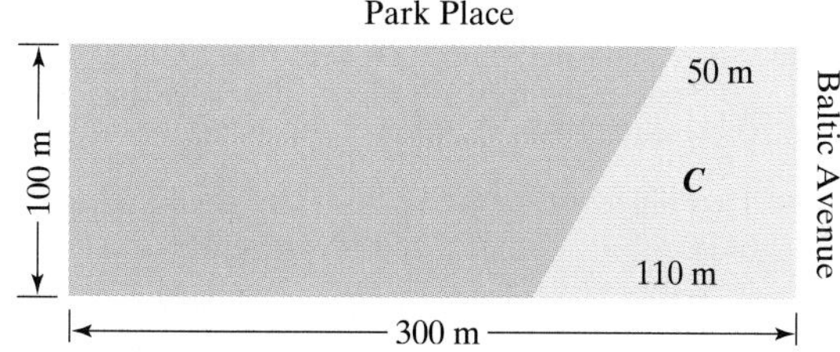

We know that both P_2's and P_3's value systems are the same and that they value the land uniformly.

(a) Give a geometric argument why P_2 and P_3 would both pass in round 1 and P_1 would end up with C.

(b) Describe a possible cut that the divider in round 2 might make.

(c) Suppose that, after round 1 is over, P_2 and P_3 discover that the city requires that the next cut be made parallel to Baltic Avenue. Describe a possible cut that the divider in round 2 might make in this case.

44. Three players (P_1, P_2, and P_3) agree to divide the property shown using the last-diminisher method. The order of the players is P_1, P_2, P_3. The first player to play, P_1, makes a claim C as shown in the figure.

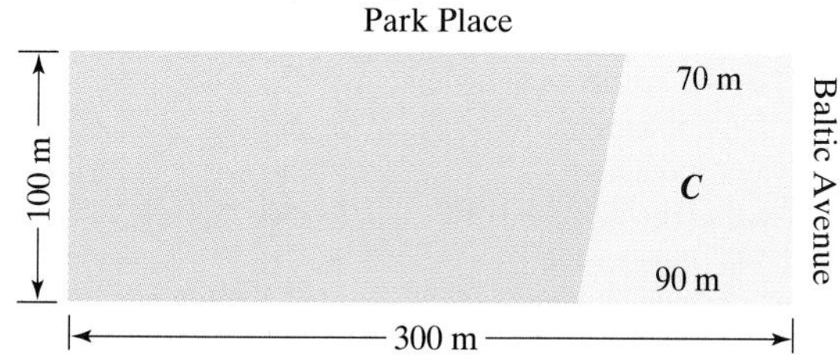

We know that both P_2's and P_3's value systems are the same and that they value the land uniformly except for the square 20 meter by 20 meter plot in the upper left corner of the property which is contaminated by an old, underground gas station tank that will cost twice as much to remove and clean up as that square plot would otherwise be worth.

(a) Give an argument why P_2 and P_3 would both pass in round 1 and P_1 would end up with *C*.

(b) Suppose that after round 1 is over P_2 and P_3 discover that the city requires that the next cut be made parallel to Baltic Avenue. Describe a possible cut that the divider in round 2 might make in this case.

45. Four friends (*E, F, G,* and *H* always playing in that order) divide a roasted chicken by the last-diminisher method. In round 1 *E* cuts off the left breast, *F* diminishes by removing the skin, *G* passes, and *H* diminishes. In round 2 both drumsticks are cut off as one piece. There are no diminishers. In round 3 the right breast is cut off and the chooser takes it.

(a) Who gets the left breast?

(b) Who gets the right breast?

(c) Who gets the back?

(d) Who cuts off the right breast?

(e) Who gets the skin of the left breast?

(f) Who had the last opportunity to diminish the drumsticks?

46. This exercise is based on Example 8. Suppose that in her will Grandma stipulates that the estate be divided among the 4 heirs as follows: Art, 25%; Betty, 35%; Carla, 30%; and Dave, 10%. Describe a variation of the method of sealed bids that will accomplish this.

47. Say that *N* players ($P_1, P_2, \ldots, P_N$) are heirs to an estate. According to the will, P_1 is entitled to r_1% of the estate, P_2 is entitled to r_2%, etc. ($r_1 + r_2 + \cdots + r_N = 100$). Describe a general variation of the method of sealed bids that gives a fair division for this estate. (You should try this exercise only after you have finished Exercise 46.)

48. Two players (*A* and *B*) wish to dissolve their partnership using the method of sealed bids. *A* bids $\$x$ and *B* bids $\$y$, where $x < y$.

(a) How much are *A* and *B*'s original fair shares worth?

(b) How much is the surplus after the original allocations are made?

(c) When all is said and done, how much must *B* pay *A* for *A*'s half of the partnership?

Exercises 49 and 50 show how the method of sealed bids can be used when some values are negative. If you offer to pay a bid (as for a purchase), the bid is listed as a positive amount. It follows that, if you offer to receive a bid (as for labor), the bid is listed as a negative amount. Regardless, the winning bid is the highest number.

49. Three women (Ruth, Sarah, and Tamara) share a house and wish to divide the chores: bathrooms, cooking, dishes, laundry, and vacuuming. For each chore, they privately write the least they are willing to receive monthly (their negative valuation) in return for doing that chore. The results are shown in the following table.

	Ruth	Sarah	Tamara
Clean bathrooms	\$−20	\$−30	\$−40
Do cooking	\$−50	\$−10	\$−25
Wash dishes	\$−30	\$−20	\$−15
Mow the lawn	\$−30	\$−20	\$−10
Vacuum and dust	\$−20	\$−40	\$−15

[Note: For doing the dishes, the high bid is \$−15.]

Divide the chores using the method of sealed bids. (Who does which chores? Who gets paid, and how much? Who pays, and how much?)

50. Four roommates are going their separate ways after graduation and wish to divide up their jointly owned furniture (equal shares) and the moving chores by the method of sealed bids. Their bids (in dollars) on the items are shown in the following table.

	Quintin	**Ramon**	**Stephone**	**Tim**
Stereo	300	250	200	280
Couch	200	350	300	100
Table	250	200	240	80
Desk	150	150	200	220
Cleaning the rugs	−80	−70	−100	−60
Patching nail holes	−60	−30	−60	−40
Repairing the window	−60	−50	−80	−80

(a) What is each roommate's estimate of his part of the total?

(b) How much surplus cash is there?

(c) What is the final outcome?

(d) What percentage of the total value (of everything) does each roommate get using the roommate's own valuation?

(e) If Tim is dishonest and sneaks a peek at the bid lists of the other 3 roommates before filling out his own, how could he adjust his bids (in whole dollars) so as to get the same furniture as before, but no chores, and also maximize his cash receipts? Explain your reasoning.

51. Four roommates (Quintin, Ramon, Stephone, and Tim) want to divide 18 small items by the method of markers. The items are lined up as shown.

❶ ❷ ❷ ❸ ❷ ❶ ❶ ❹ ❸ ❸ ❸ ❷ ❸ ❷ ❸ ❷ ❹ ❹

The (secret) values of the items by the roommates is shown in the following table.

	Quintin	**Ramon**	**Stephone**	**Tim**
Each ❶ is worth	$12	$ 9	$ 8	$5
Each ❷ is worth	$ 7	$ 5	$ 7	$4
Each ❸ is worth	$ 4	$ 5	$ 6	$4
Each ❹ is worth	$ 6	$11	$14	$7

(a) Show where each roommate would place his markers (use Q_1, Q_2, Q_3 for Quintin's markers, R_1, R_2, R_3 for Ramon's markers, etc.).

(b) Describe who gets what piece and which pieces are left over.

(c) How would you divide the leftovers?

52. Three players (A, B, and C) agree to divide some candy using the method of markers. The candy consists of 3 Nestle Crunch Bars, 6 Snickers Bars, and 6 Reese's Peanut Butter Cups. The players' value systems are as follows:

A loves Nestle Crunch Bars but does not like Snickers Bars or Reese's Peanut Butter Cups at all.

B loves Snickers Bars and Nestle Crunch Bars equally well (i.e., 1 Snickers Bar = 1 Nestle Crunch Bar) but does not like Reese's Peanut Butter Cups at all.

C loves Snickers Bars and Reese's Peanut Butter Cups equally well (i.e., 1 Snickers Bar = 1 Reese's Peanut Butter Cup) but is allergic to Nestle Crunch Bars.

The candy is lined up as shown.

(a) What bid would *A* make to ensure that she gets her fair share (according to her value system)?

(b) What bid would *B* make to ensure that he gets his fair share (according to his value system)?

(c) What bid would *C* make to ensure that she gets her fair share (according to her value system)?

(d) Describe the allocations to each player.

(e) What items are left over?

53. Repeat Exercise 52 with the candy lined up as follows.

54. Suppose that two players (*A* and *B*) buy a chocolate-strawberry mousse cake with a caramel swirl and assorted frostings for $10. Since *A* contributes $7 and *B* only $3, they both agree that a fair division of the cake is one in which *A* gets a piece that is worth (in *A*'s opinion) at least 70% of the cake and *B* gets a piece that is worth (in *B*'s opinion) at least 30% of the cake. Describe a variation of the lone-divider method that can be used in this situation. (*Hint*: Think of this problem as an ordinary fair-division problem with many players.)

55. This problem is a variation of Exercise 54. Three players are involved in buying a $10 cake (*A*, *B*, and *C*). *A* contributes $2.50, *B* contributes $3.50, and *C* contributes $4 toward the purchase of the cake. Describe a fair-division scheme for this problem.

56. Consider the following variation of the divider-chooser method for 2 players: After the divider cuts the cake in 2 pieces, the chooser (who is unable to see either piece) picks his piece randomly by flipping a coin. The divider, of course, gets the other piece.

(a) In this a fair-division scheme according to our definition? Explain your answer.

(b) Who would you rather be—divider or chooser? Explain.

Running

57. Alternative notion of a fair share. To explain this exercise we will need some additional terminology. We will say that a player is *fairly happy* if she receives a share that is (in her opinion) worth at least $1/N$ of the total (what we have been calling a fair share). We will say that a player is *ecstatic* if she receives a share that is worth (in her opinion) at least as much as anyone else's—in other words, the player would not want to trade her share with anyone else. All the fair-division schemes we presented in the chapter guarantee that all the players will be fairly happy but not that all the players will be ecstatic. Human nature being what it is, being fairly happy but not ecstatic might make some of the players unhappy. The purpose of this exercise is to discuss this alternative interpretation of fairness.

(a) Explain why the divider-chooser method for 2 players guarantees that each player will be ecstatic.

(b) Give an example of a fair division using the lone-divider method with 3 players in which at least one of the players ends up with a piece that makes him happy but not ecstatic.

(c) Give an example of a fair division using the lone-divider method with 3 players in which all 3 players are ecstatic.

58. Using the terminology of Exercise 57, describe a continuous fair-division method for 3 players that guarantees that all three players will be ecstatic.

59. **(a)** Explain why after the original allocation is made in the method of sealed bids, the surplus produced must be either positive or zero.

(b) Under what condition is the surplus zero?

60. The purpose of this exercise is to extend the ideas of the lone-divider method for 3 players to any number of players.

(a) Describe how the lone-divider method would work in the case of 4 players. (*Hint:* Consider several cases following the format for the case of 3 players. Also look at Exercises 12 through 19.)

(b) Describe the lone-divider method for the general case of N players.

References and Further Readings

1. Brams, Steven, and Alan Taylor, *Fair Division.* Cambridge, England: Cambridge University Press, 1996.
2. Demko, Stephen, and Theodore Hill, "Equitable Distribution of Indivisible Objects," *Mathematical Social Sciences,* 16 (1988), 145–158.
3. Dubins, L. E., "Group Decision Devices," *American Mathematical Monthly,* 84 (1977), 350–356.
4. Fink, A. M., "A Note on the Fair Division Problem," *Mathematics Magazine,* 37 (1964), 341–342.
5. Gardner, Martin, *aha! Insight.* New York: W. H. Freeman, 1978, 124.
6. Hill, Theodore, "Determining a Fair Border," *American Mathematical Monthly,* 90 (1983), 438–442.
7. Hively, Will, "Dividing the Spoils," *Discover,* 16 (1995), 49–57.
8. Kuhn, Harold W., "On Games of Fair Division," *Essays in Mathematical Economics,* Martin Shubik, ed. Princeton, NJ: Princeton University Press, 1967, 29–37.
9. Olivastro, Dominic, "Preferred Shares," *The Sciences,* March–April 1992, 52–54.
10. Steinhaus, Hugo, *Mathematical Snapshots.* New York: Oxford University Press, 1960, 65–71.
11. Steinhaus, Hugo, "The Problem of Fair Division," *Econometrica,* 16 (1948), 101–104.
12. Stewart, Ian, "Fair Shares for All," *New Scientist,* 146 (1982), 42–46.
13. Stromquist, Walter, "How to Cut a Cake Fairly," *American Mathematical Monthly,* 87 (1980), 640–644.
14. Weingartner, H. M., and B. Gavish, "How to Settle an Estate," *Management Science,* 39 (1993), 588–601.

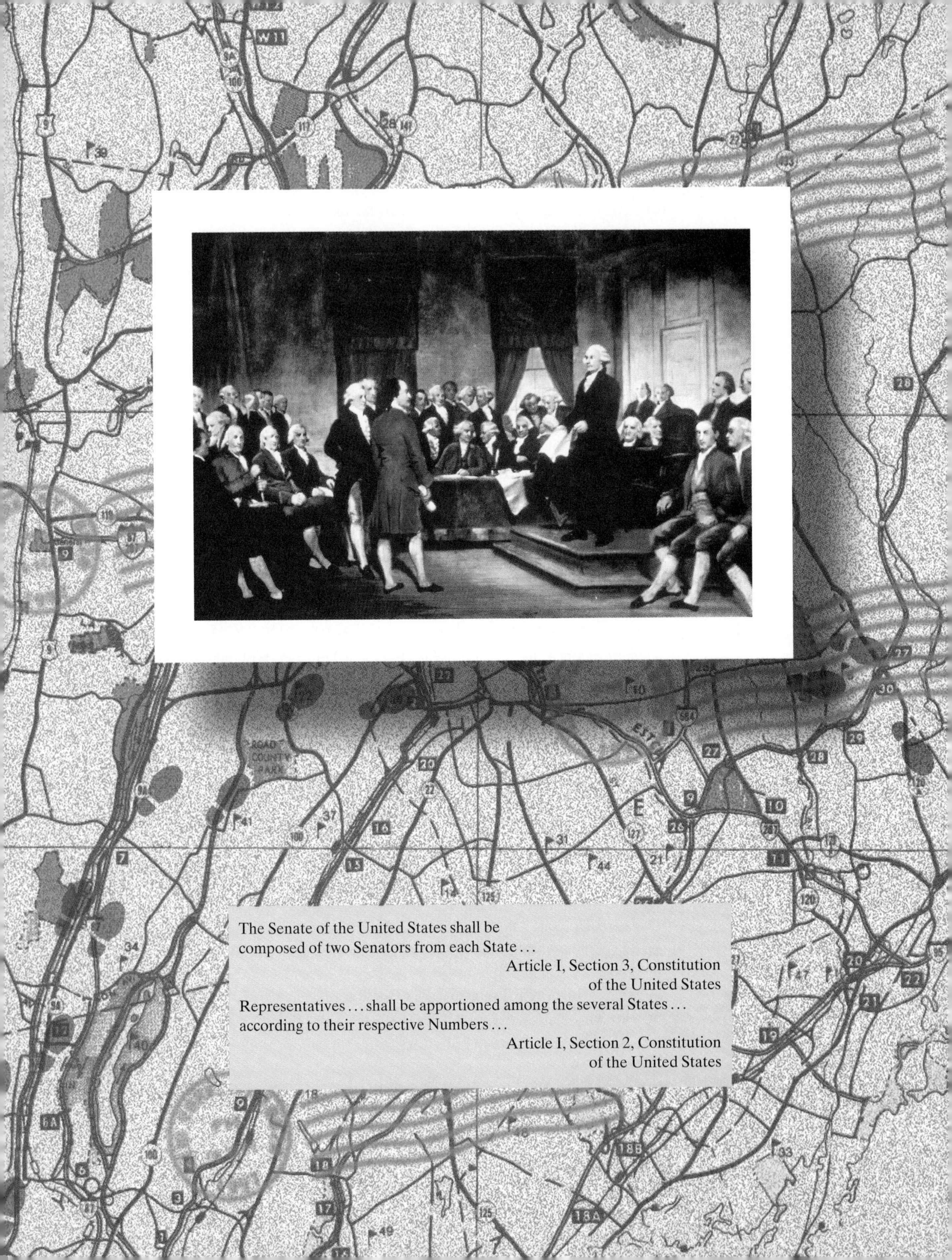
The Senate of the United States shall be composed of two Senators from each State . . .
Article I, Section 3, Constitution of the United States
Representatives . . . shall be apportioned among the several States . . . according to their respective Numbers . . .
Article I, Section 2, Constitution of the United States

The Mathematics of Apportionment

Making the Rounds

CHAPTER 4

In the stifling heat of the Philadelphia summer of 1787, delegates from the thirteen states met to draft a Constitution for a new nation. Except for Thomas Jefferson (then minister to France) and Patrick Henry (who refused to participate), all the main names of the American Revolution were there—George Washington, Ben Franklin, Alexander Hamilton, James Madison. Without a doubt, the most important and heated debate at the Constitutional Convention concerned the makeup of the legislature. The small states, led by New Jersey, wanted all states to have the same number of representatives. The larger states, led by Virginia, wanted some form of proportional representation. The final resolution of this dispute is all too familiar to us: a Senate, in which every state has two senators, and a House of Representatives, in which each state has a number of representatives that is a function of its population. This so-called *Great Compromise* was embodied in Article 1, Sections 2 and 3, of the Constitution of the United States.

One set of details was not fully spelled out in Article 1, Section 2: Other than the fact that state apportionments in the House of Representatives are based on populations (*Representatives . . . shall be apportioned among the several States . . . according to their respective numbers. . .*), how should the exact apportionments to each state be calculated? Undoubtedly, the delegates felt that this was a relatively minor detail—a matter of simple arithmetic that could be easily figured out and agreed upon by reasonable people. Certainly it was not the kind of thing to clutter a Constitution with, or spend time arguing over in the heat of the summer. What the Founding Fathers did not realize is that Article 1, Section 2, set the Constitution of the United States into a collision course with a mathematical iceberg known today (but certainly not then) as *the apportionment problem.*

What is an apportionment? What is the apportionment problem? Why is there a problem? Why is the problem so complicated? Why should anyone care? These, in essence, are the questions we will answer in this chapter. In so doing, we will learn some interesting mathematics and at the same time get a glimpse of a little-known but fascinating chapter of United States history. (When was the last time you saw the words mathematics and United States history uttered in the same breath?)

Apportionment Problems

What is generally now known as the **apportionment**[1] **problem** is really a special kind of *discrete fair-division problem*—a sort of mirror image of some of the problems we discussed in Chapter 3. As in Chapter 3, we have *indivisible objects* that we would like to divide fairly among a set of *players.* The difference is this: in Chapter 3, each player was entitled to an equal share but the objects were different; now *the objects are all going to be the same, but the players are going to be entitled to different-sized shares.*

The most important example of an apportionment problem is that of *proportional representation* in a legislative body, exactly the kind of problem faced by our Founding Fathers in 1787. Here, the identical, *indivisible objects* to be apportioned are slots (called *seats*) in the legislature, and the players are the *states* (or provinces, regions, etc.). The idea of proportional representation is that each state is entitled to a number of seats that is proportional to its population. Most of our discussion for this chapter will take place in the context of this particular type of apportionment problem, but it is important to realize that apportionment problems occur in many other guises as well. The point is best illustrated with a couple of examples.

Example 1. Kitchen Capitalism.

Mom has a total of exactly 50 identical, indivisible pieces of candy (let's say they are caramels) which she is going to divide among her five children. Like any good mom, she is intent on doing this fairly. Of course, the easiest thing to do would be to give each child ten caramels, but mom has loftier goals—she wants to teach her children about the value of work and about the relationship between work and

[1] **ap·pôr·tion:** to divide and assign in due and proper proportion or according to some plan (*Webster's New Twentieth Century Dictionary*).

reward. This leads her to the following idea: The candy is going to be *apportioned* among the children based strictly on the amount of time each child spends helping with the weekly kitchen chores.

Here we are, trying to divide candy once again! But now things are quite different from Example 10 in Chapter 3. We have 50 identical objects (the caramels) to be divided among 5 players (the kids), each of which is entitled to a different share of the total. How should this be done?

Table 4-1 shows the amount of work (in minutes) done by each child during the week.

Table 4-1 Amount of Work (In Minutes) Per Child

Child	Alan	Betty	Connie	Doug	Ellie	Total
Minutes worked	150	78	173	204	295	900

Once the figures are in, it is time to divide the candy. According to the ground rules, Alan, who worked 150 out of a total of 900 minutes, is entitled to $16\frac{2}{3}\%$ of the 50 pieces of candy $[(150/900) = 16\frac{2}{3}\%]$, namely to $8\frac{1}{3}$ pieces. Here comes the problem: Since the pieces of candy are indivisible, it is impossible for Alan to get the exact share he is entitled to—he can get 8 pieces (and get shorted) or he can get 9 pieces (and someone else will get shorted). A similar problem occurs with each of the other children: Betty's exact fair share should be $4\frac{1}{3}$ pieces, Connie's fair share should be $9\frac{11}{18}$ pieces, Doug's fair share should be $11\frac{1}{3}$ pieces, and Ellie's fair share should be $16\frac{7}{18}$ pieces (we leave it to the reader to verify these figures). Needless to say, none of these shares can be realized. Clearly, an absolutely fair apportionment of the candy is going to be impossible. What should mom do? (Do you have any suggestions?) ■

Our next example shows a more classical version of an apportionment problem.

Example 2. The Intergalactic Congress.

It is the year 2525 and all the planets in the galaxy have finally signed a peace treaty. Five of the planets (Alanos, Betta, Conii, Dugos, and Ellisium) decide to join forces and form an Intergalactic Federation. The Federation will be ruled by an Intergalactic Congress consisting of 50 delegates, and each of the 5 planets will be entitled to a number of delegates that is proportional to its population. The population data for each of the planets is shown in Table 4-2. How many delegates should each planet get?

Table 4-2 Intergalactic Federation: Population Figures (in billions) for 2525

Planet	Alanos	Betta	Conii	Dugos	Ellisium	Total
Population	150	78	173	204	295	900

Example 2 is not just another example of an apportionment problem, it is Example 1 revisited. When we compare Example 2 with Example 1, we see that the numbers are identical—it is only the setting that is changed. While the merits of the problem may be different, mathematically speaking, Examples 1 and 2 are one and the same! ■

Between the extremes of apportioning the seats in the Intergalactic Congress (important, but too far away!) and apportioning the caramels among the children (closer to home, but the world will not come to an end if some of the kids feel shorted!) fall many other applications of the apportionment problem that are both important and relevant: apportioning nurses to shifts in a hospital, apportioning telephone calls to switchboards in a network, apportioning subway cars to routes in a subway system, etc.

Our primary purpose in this chapter is to learn various methods for solving apportionment problems (**apportionment methods**), something that sounds reasonably simple but has many subtleties and surprises. In fact, our travels through this chapter will be somewhat reminiscent of our experiences with *voting methods* in Chapter 1.

Over the years, statesmen, politicians, and mathematicians have designed many ingenious apportionment methods, and we will study some of the best known in this chapter. Interestingly, the names associated with many of these apportionment methods one would expect to find in a history book, rather than a mathematics book: Alexander Hamilton, Thomas Jefferson, John Quincy Adams, and Daniel Webster. The reason, of course, is an accident of history—the *Great Compromise* of the Constitution. Here we have one of those rare subjects where history, politics, and mathematics become intertwined. Thus, before we start a detailed mathematical discussion of the various apportionment methods, we will find it illuminating to briefly look at the history of the apportionment problem in the United States.

A Little Bit of U.S. History

It didn't take long after the Constitutional Convention in 1787 for the controversy over apportionment to start. The very first time the House of Representatives was to be apportioned was after the census of 1790, and the method by which this was to be done was to be decided by Congress. Two very different methods were under consideration, one proposed by Alexander Hamilton, the other by Thomas Jefferson (we will learn the details of both soon). After considerable and sometimes heated debate, a bill was passed to use *Hamilton's method.* The bill was then submitted to President George Washington, who, after considering the pros and cons of each of the two methods, vetoed the bill (the first bill vetoed by a president in U.S. history!). Unable to override the veto and facing a damaging political stalemate, Congress ended going along with the president and adopting *Jefferson's method.*

Jefferson's method was used for five decades (until 1842) and then replaced by a method proposed by Daniel Webster. *Webster's method* was soon replaced by one equivalent to that proposed by Hamilton in 1790, which was then replaced again by Webster's method, which was eventually replaced (in 1941) by the current method of apportionment (called *Huntington-Hill's method*). Each change of method was preceded by considerable discussion and debate, often of a very nasty nature. There were decades in which there was no official apportionment method, making the very existence of the House of Representatives unconstitutional.[2]

---

[2] For a little more detail the reader is referred to Appendix 2 at the end of this chapter. A really detailed account of the history of the apportionment of the House of Representatives can be found in references 1 and 2.

George Washington and his Cabinet. From left to right: Washington, Henry Knox, Alexander Hamilton, Thomas Jefferson, and Edmund Randolph. (*1876 lithograph, by Currier and Ives. Courtesy of the Granger Collection.*)

The latest controversy over apportionment methods is of very recent vintage. In 1992, the state of Montana mounted a legal challenge to the apportionment method now being used, with the case ending up in the Supreme Court. (Primarily, Montana was concerned that, based on the census of 1990, it would have to give up one of its two seats in the House of Representatives to the state of Washington.) After hearing the case, the Supreme Court ruled against Montana and upheld the validity of the Huntington-Hill method. Good-bye, Montana; Hello, Washington!

The Mathematics of Apportionment: Basic Concepts

We are now ready to take on the systematic study of apportionment methods. Much of our discussion for the rest of the chapter will be centered around the following simple but important example.

Example 3. The Congress of Parador.

Parador is a new republic located in Central America. It is made up of six states: Azucar, Bahia, Cafe, Diamante, Esmeralda, and Felicidad (*A, B, C, D, E,* and *F* for short). According to the new constitution of Parador, the Congress will have 250 seats, divided among the six states according to their respective populations. The population figures for each state are given in Table 4-3.

Table 4-3 Republic of Parador (Population Data by State)

State	*A*	*B*	*C*	*D*	*E*	*F*	Total
Population	1,646,000	6,936,000	154,000	2,091,000	685,000	988,000	12,500,000

A natural starting point for our mathematical adventure is to calculate the ratio of national population to number of seats in Congress, a sort of national average of people per seat. Since the total population of Parador is 12,500,000 and the number of seats is 250, this average is given by 12,500,000/250 = 50,000. In general, *for any apportionment problem in which the total population of the country is P and the number of seats to be apportioned is M, the ratio P/M gives the number of people per seat in the legislature on a national basis.* We will call the number P/M the **standard divisor**.[3]

Using the standard divisor, we can calculate the *fraction of the total number of seats that each state would be entitled to if the seats were not indivisible.* This number, which we will call the **state's standard quota**,[4] is obtained by dividing the state's population by the standard divisor.

$$\text{Standard divisor} = \frac{\text{population}}{\text{total number of seats}}$$

$$\text{State's standard quota} = \frac{\text{state's population}}{\text{standard divisor}}$$

Table 4-4 shows the standard quota of each state in Parador, given to two decimal places (two is usually enough!). These were obtained by dividing each state's population by 50,000. (Notice that the sum of all the standard quotas is 250, the total number of seats to be divided.)

Table 4-4 Republic of Parador (Standard Quotas for Each State)

State	*A*	*B*	*C*	*D*	*E*	*F*	Total
Population	1,646,000	6,936,000	154,000	2,091,000	685,000	988,000	12,500,000
Standard quota	32.92	138.72	3.08	41.82	13.70	19.76	250

Associated with each state's standard quota are two whole numbers: the state's **lower quota** (the standard quota rounded down), and the state's **upper quota** (the standard quota rounded up). For example, state *A* (with a standard quota of 32.92) has a lower quota of 32 and an upper quota of 33. (In the unusual case that the state's quota is a whole number, then the lower and upper quotas are the same.)

The standard quotas represent each state's exact fair share of the 250 seats, and if the seats in the legislature could be chopped up into fractional parts, we would be done. Unfortunately, seats have to be given out whole, so it now becomes a question of how to *round the standard quotas into whole numbers.* At first glance, the obvious strategy would appear to be the traditional approach to rounding we learned in school, which we will call *conventional rounding:* Round down if the fractional part is less than 0.5, round up otherwise. Unfortunately, this approach is not guaranteed to work. Look at what happens when we try it with the standard quotas of the states in Parador (Table 4-5).

---

[3] Note that, although our definition is given in the context of Example 3, the concept of standard divisor applies to any apportionment problem, regardless of the context. In our original candy example, the standard divisor is $P/M = 900/50 = 18$, and it represents the average number of minutes of work needed to earn a single caramel.

[4] It is also known as the *exact quota.*

Table 4-5 Conventional Rounding Doesn't Always Work!

State	Population	Standard quota	Rounded to
A	1,646,000	32.92	33
B	6,936,000	138.72	139
C	154,000	3.08	3
D	2,091,000	41.82	42
E	685,000	13.70	14
F	988,000	19.76	20
Total	12,500,000	250.00	251

As the total in the last column of Table 4-5 shows, we have a slight problem: We are giving out 251 seats in Congress! Where is that extra seat going to come from? ■

The example of Parador's Congress illustrates the major problem with conventional rounding of the standard quotas—a seductive idea that doesn't always work. In this example, we ended up giving out more seats than we were supposed to; other times we could end up giving out less. Occasionally, by sheer luck, it works out just right. Clearly, it is not an apportionment method we can count on!

The fact that conventional rounding of the standard quotas does not work as an apportionment method is disappointing, but hardly a surprise at this point. After all, if this obvious and simple-minded approach worked all the time, the whole issue of apportionment would be mathematically trivial (and there wouldn't be a reason for this chapter to exist!). Given this, we will be forced to consider more sophisticated approaches to apportionment. Our strategy for the rest of this chapter will be to look at several important apportionment methods (important both historically and mathematically) and find out what is good and bad about each one. (Shades of Chapter 1?)

Hamilton's Method

While historically Hamilton's method did not come first, we will discuss it first because it is mathematically the simplest.[5]

Alexander Hamilton

Hamilton's Method

- **Step 1.** Calculate each state's standard quota.
- **Step 2.** Give to each state (for the time being) its *lower quota* (in other words, round each state's quota down).
- **Step 3.** Give the surplus seats (one at a time) to the states with the largest fractional parts until we run out of surplus seats.

[5] Hamilton's method is also known as the *method of largest remainders* and sometimes as *Vinton's method.* It is unclear as to whether Hamilton thought of the method himself or learned about it from someone else.

Example 4. Hamilton's Method Meets Parador's Congress.

Let's apply Hamilton's method to apportion Parador's Congress. Table 4-6 shows all the details and speaks for itself. (The reader is reminded that in Example 3 we found that the standard divisor is 50,000.)

Table 4-6 Republic of Parador: Apportionment Based on Hamilton's Method

State	Population	Standard quota (Step 1)	Lower quota (Step 2)	Fractional part	Surplus seats (Step 3)	Final apportionment
A	1,646,000	32.92	32	0.92	1 (1st)	33
B	6,936,000	138.72	138	0.72	1 (4th)	139
C	154,000	3.08	3	0.08		3
D	2,091,000	41.82	41	0.82	1 (2nd)	42
E	685,000	13.70	13	0.70		13
F	988,000	19.76	19	0.76	1 (3rd)	20
Total	12,500,000	250.00	246	4.00	4	250

■

Essentially, Hamilton's method can be described as follows: Every state gets at least its lower quota. As many states as possible get their upper quota, the one with highest fractional part having first priority, the one with second highest fractional part second, and so on.

Is Hamilton's a fair and reasonable apportionment method? Hamilton thought it was, and he lobbied quite forcefully to President Washington on its behalf. However, we can already see in Example 4 some hints of unfairness. Consider the sad fate of state *E*, next in line for a surplus seat with a hefty fractional part of 0.70 and yet getting none. Is state *B* (with a fractional part of 0.72) truly that much more deserving than state *E* in getting that last surplus seat? The answer is not quite clear. If we look at fractional parts in absolute terms, the answer is yes (0.72 is more than 0.70). However, when we look at the fractional part as a percentage of the entire state's population, state *E*'s 0.70 represents a much larger proportion of its quota (13.70) than state *B*'s 0.72 does of its quota (138.72). It would not be totally unreasonable to argue that state *E*, rather than *B*, should be getting that last surplus seat.[6]

While on the surface the rules of the game under Hamilton's method sound very fair, a little probing shows that Hamilton's method consistently works to the advantage of the larger states over the smaller ones, and there is some reason to suspect that Hamilton himself knew this.

The Quota Rule

The net effect of Hamilton's method is to separate the states into two groups: The *lucky* ones that get a surplus seat end up with a number of seats equal to their *upper quota*, and the *unlucky* ones getting no surplus seats—they end up with a

[6] The idea that relative rather than absolute fractional parts should determine the order in which the surplus seats are handed out is the basis for an apportionment method known as *Lowndes' method*. For details, the reader is referred to Exercise 40.

number of seats equal to their *lower quota.* By the same token, no state ends up with an apportionment that is off by more than one seat from the standard quota, which is our benchmark for fairness. This is good! An apportionment method that always does this is said to *satisfy* the **quota rule**.

The Quota Rule

A state's apportionment should always be either its *upper quota* or its *lower quota.*

Satisfying the quota rule seems like the least one could ask from a fair apportionment method. Since we cannot give out fractional apportionments, and we cannot always round the quotas to the nearest integer, let's at least *round the quotas to one of the two nearest integers.* Surprisingly, some of the most important apportionment methods (including the one currently used to apportion the House of Representatives) can violate the quota rule. We will study some of these methods later in the chapter, but for now let's return to Hamilton's method.

So far, Hamilton's method seems to have a few things going for it. It is easy to understand, it satisfies the quota rule, and—except for what appears to be a natural favoritism toward large states over small ones—it seems reasonably fair. Why isn't it the answer to our prayers? As you may have guessed, we are about to discover the dark side of Hamilton's method.

The Alabama Paradox

The most serious (in fact, the fatal) flaw of Hamilton's method is commonly known as the **Alabama paradox**. In essence, the Alabama paradox occurs when an *increase in the total number of seats, in and of itself, forces a state to lose one of its seats.* The best way to understand what this means is to look carefully at the following example.

Example 5.

A small country consists of 3 states: *A, B,* and *C*. Table 4-7 shows the apportionment under Hamilton's method when there are $M = 200$ seats to be apportioned. Table 4-8 shows the apportionment under Hamilton's method when there are $M = 201$ seats to be apportioned. The reader is encouraged to verify the necessary calculations. (Here is some help: The standard divisor when $M = 200$ is 100; the standard divisor when $M = 201$ is 99.5.)

Table 4-7 Apportionment Under Hamilton's Method for $M = 200$

State	Population	Standard quota when $M = 200$	Apportionment under Hamilton's method
A	940	9.4	10
B	9030	90.3	90
C	10,030	100.3	100
Total	20,000	200.0	200

Table 4-8 Apportionment Under Hamilton's Method for $M = 201$

State	Population	Standard quota when $M = 201$	Apportionment under Hamilton's method
A	940	9.45	9
B	9030	90.75	91
C	10,030	100.80	101
Total	20,000	201.00	201

Using Hamilton's method to apportion the seats, we can see that when there are $M = 200$ seats to be apportioned, A gets the only surplus seat and the final apportionment is: A gets 10 seats, B gets 90 seats, and C gets 100 seats (Table 4-7).

What happens when the number of seats to be apportioned increases to $M = 201$? Now there are two surplus seats and they go to B and C, so that the final apportionment is: A gets 9 seats, B gets 91 seats, and C gets 101 seats (Table 4-8). ■

The shocking conclusion of Example 5 is that it is possible for a state to get a smaller apportionment with a larger legislature than with a smaller one. Undoubtedly, this is a very unfair situation. The first serious instance of this problem occurred in 1880, when it was noted that, based on Hamilton's method, if the House of Representatives were to have 299 seats Alabama would get 8 seats, but if the House of Representatives were to have 300 seats Alabama would end up with 7 (see Table 4-9). This is how the name *Alabama paradox* came about.

Table 4-9 Hamilton's Method and the Alabama Paradox, 1880

State	Standard quota with $M = 299$	Apportionment with $M = 299$	Standard quota with $M = 300$	Apportionment with $M = 300$
Alabama	7.646	8	7.671	7
Texas	9.64	9	9.672	10
Illinois	18.64	18	18.702	19

Mathematically, the Alabama paradox is the result of some quirks of basic arithmetic. When we increase the number of seats to be apportioned, each state's standard quota goes up, but not in the same amount. Thus, the *priority order* for surplus seats used by Hamilton's method can get scrambled around, moving some states from the front of the priority order to the back, and vice versa. This can result in some state or states losing seats they already had. This is exactly what happened in Example 5 and in the Alabama fiasco of 1880.

More Problems with Hamilton's Method

The discovery of the Alabama paradox in 1880 turned out to be the kiss of death for Hamilton's method. Ironically, two other serious flaws of Hamilton's method were discovered later, when the method was no longer being used. We will briefly discuss these in this section, primarily because they are mathematically interesting, but also because they show that Hamilton's method has serious problems even if the Alabama paradox were not an issue.

The Population Paradox

Sometime around 1900 it was discovered that under Hamilton's method, *state X could lose seats to state Y even though the population of X had grown at a higher rate than that of Y.* Needless to say, this is quite unfair. The following example illustrates how this can actually happen.

Example 6.

We are going to revisit Example 2, the one about the Intergalactic Federation of 2525. Here are the population figures once again:

Table 4-10 Intergalactic Federation: Population Figures (in billions) for 2525

Planet	Alanos	Betta	Conii	Dugos	Ellisium	Total
Population	150	78	173	204	295	900

The total population for the 5 planets comes to 900 billion. If we divide this number by 50, we get a standard divisor of 18 billion. Using this standard divisor, we can get the standard quotas (column 3 of Table 4-11) and then carry out steps 2 and 3 of Hamilton's method, as shown in columns 4 and 5 of Table 4-11, respectively. The final apportionment is shown in column 6. We call the reader's attention to two planets that play a key role in this story: Betta (4 delegates), and Ellisium (17 delegates).

Table 4-11 Intergalactic Federation: Apportionment of 2525 (Hamilton's Method)

Planet	Population (in billions)	Standard quota (Population ÷ 18)	Lower quota (Step 2)	Surplus seats (Step 3)	Final apportionment
Alanos	150	$8.\overline{3}$	8		8
Betta	78	$4.\overline{3}$	4		4
Conii	173	$9.6\overline{1}$	9	1	10
Dugos	204	$11.\overline{3}$	11		11
Ellisium	295	$16.3\overline{8}$	16	1	17
Total	900	50.00	48	2	50

Intergalactic Federation. Part II. Ten years have gone by, and it is time to reapportion the Intergalactic Congress. Actually not much has changed (populationwise) within the Federation. Conii's population has increased by 8 billion (Coniians are very prolific), and Ellisium's population has increased by 1 billion. All other planets have stayed exactly the same (Table 4-12).

Table 4-12 Intergalactic Federation: Population Figures (in billions) for 2525

Planet	Alanos	Betta	Conii	Dugos	Ellisium	Total
Population	150	78	181	204	296	909

Since the total population is now 909 billion and the number of delegates is still 50, the standard divisor we now get is 909/50 = 18.18. Table 4-13 shows the steps for Hamilton's method based on this new standard divisor. Once again, the final apportionment is shown in column 5. Do you notice something terribly wrong with this apportionment? Ellisium, whose population went up by 1 billion, is losing a delegate to Betta, whose population did not go up at all! ■

Table 4-13 Intergalactic Federation: Apportionment of 2535 (Hamilton's Method)

Planet	Population (in billions)	Standard quota (Population ÷ 18.18)	Lower quota (Step 2)	Surplus seats (Step 3)	Final apportionment
Alanos	150	8.25	8		8
Betta	78	4.29	4	1	5
Conii	181	9.96	9	1	10
Dugos	204	11.22	11		11
Ellisium	296	16.28	16		16
Total	909	50.00	48	2	50

This is, in essence, the **population paradox**: *State X has a population growth rate higher than that of state Y and yet, when the apportionment is recalculated based on the new population figures, state X loses a seat to state Y.*

The New-States Paradox

Another paradox produced by Hamilton's method was discovered when Oklahoma became a state in 1907. Previously, the House of Representatives had 386 seats. Based on its population, Oklahoma was entitled to 5 seats, so the size of the House of Representatives was changed from 386 to 391. The obvious intent in adding the extra 5 seats was to leave all the other states' apportionments unchanged. However, when the apportionments were recalculated under Hamilton's method, using the same population figures but adding Oklahoma's population, and using 391 seats instead of 386, something truly bizarre took place: Maine's apportionment went up (from 3 to 4 seats) and New York's went down (from 38 to 37 seats). The mere addition of Oklahoma (with its fair share of seats) to the Union would force New York to give a seat to Maine! The perplexing fact that *the addition of a new state with its fair share of seats can, in and of itself, affect the apportionments of other states,* is called the **new-states paradox**.

The following example gives a simple illustration of the new-states paradox. For a change of pace, we will try something other than legislatures.

Example 7.

Central School District has two high schools: North High has an enrollment of 1045 students; South High has an enrollment of 8955. The school district is allocated a counseling staff of 100 counselors, who are to be apportioned between the two schools using Hamilton's method. This results in an apportionment of 10 counselors to North High and 90 counselors to South High. The computation is summarized in Table 4-14.

November 16, 1907: Oklahoma becomes the 45th state.

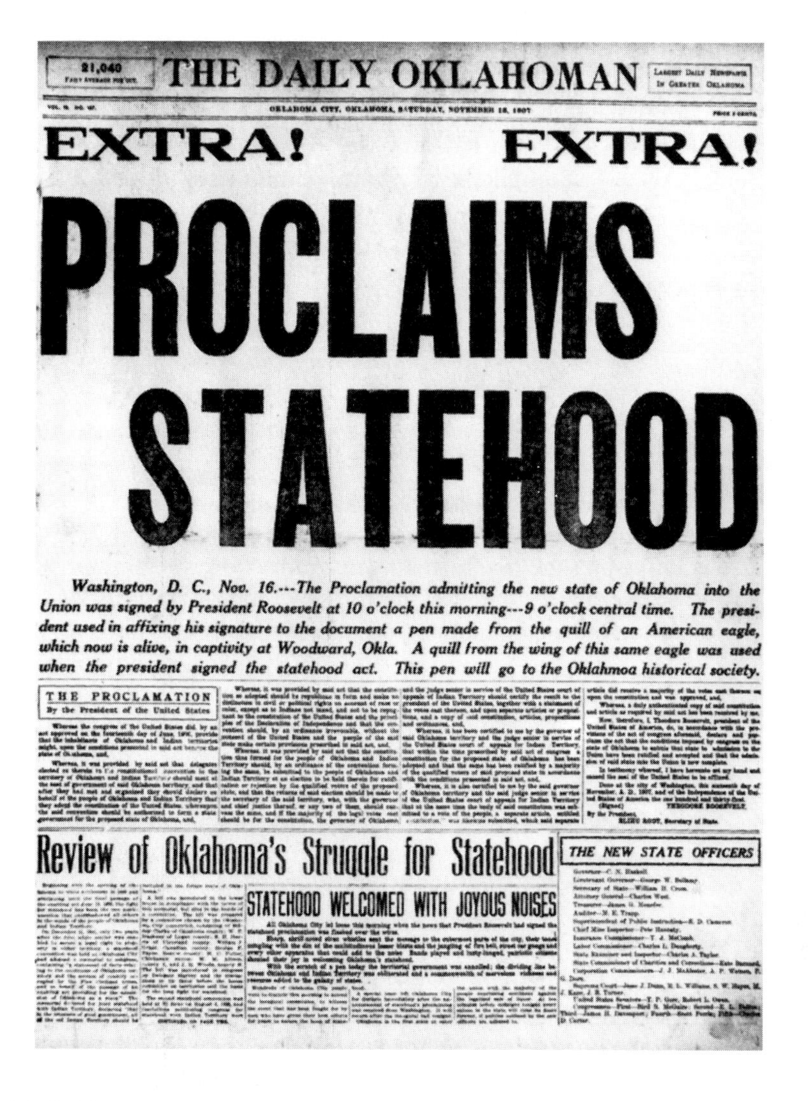

THE DAILY OKLAHOMAN

OKLAHOMA CITY, OKLAHOMA, SATURDAY, NOVEMBER 16, 1907

EXTRA! EXTRA!

PROCLAIMS STATEHOOD

Washington, D. C., Nov. 16.---The Proclamation admitting the new state of Oklahoma into the Union was signed by President Roosevelt at 10 o'clock this morning---9 o'clock central time. The president used in affixing his signature to the document a pen made from the quill of an American eagle, which now is alive, in captivity at Woodward, Okla. A quill from the wing of this same eagle was used when the president signed the statehood act. This pen will go to the Oklahmoa historical society.

THE PROCLAMATION
By the President of the United States

Review of Oklahoma's Struggle for Statehood

STATEHOOD WELCOMED WITH JOYOUS NOISES

THE NEW STATE OFFICERS

Table 4-14 Apportionment of Counselors to the Two High Schools Based on Hamilton's Method

School	Enrollment	Standard quota (Standard divisor = 100)	Apportionment
North High	1045	10.45	10
South High	8955	89.55	90
Total	10,000	100.00	100

Suppose now that a new high school (New High) is added to the district. New High has an enrollment of 525 students, so the district (using the same standard divisor of 100 students per counselor) decides to hire 5 new counselors and assign them to New High. After this is done, someone has the bright idea of having the entire apportionment recalculated (still using Hamilton's method). The surprising result is shown in Table 4-15.

Table 4-15 Apportionment of Counselors to the Three High Schools Based on Hamilton's Method

School	Enrollment	Standard quota (Standard divisor = 100.238)	Apportionment
North High	1045	10.425	11
South High	8955	89.337	89
New High	525	5.238	5
Total	10,525	105.000	105

■

Thomas Jefferson. (*The White House Collection*)

Jefferson's Method

We are now ready to study **Jefferson's method**,[7] an apportionment method of both historical and mathematical importance. Ironically, we will explain the idea behind Jefferson's method by taking one more look at Hamilton's method.

Recall that under Hamilton's method we start by dividing every state's population by a fixed number (the standard divisor). This gives us the standard quotas. Step 2 is then to round *every* state's standard quota down. Notice that up to this point Hamilton's method uses a uniform policy for all states—every state is treated in exactly the same way. If you are looking for fairness, this is obviously good! But now comes the bad part (step 3): We have some leftover seats which we need to distribute, but not enough for every state. Thus, we are forced to choose some states over others for preferential treatment. No matter how fair we try to be about it, there is no getting around the fact that some states get that extra seat, others don't. From the fairness point of view, this is the major weakness of Hamilton's method.

Now comes a new idea, based on the following thought: Wouldn't it be nice if we could eliminate step 3 in Hamilton's method? Or, to put it another way, Wouldn't it be nice if we could rig things up so that after dividing every state's population by the same number (step 1) and then rounding the resulting quotas down (step 2) we were to *end up with no surplus seats?*

How could we work such magic? In theory, the answer is simple: We need to use a different divisor (instead of the standard divisor), so that we can come up with new **modified quotas** (no longer the standard quotas) that are just right (meaning that when we round all of them down we end up with a total that is the exact number of seats to be apportioned). In essence, we have just described **Jefferson's method**. Before we give a detailed description of the method, it's time to look at an example.

Example 8. Jefferson's Method Meets Parador's Congress.

Once again we will use the Parador example. (Recall that the standard divisor in this example is 50,000.) Table 4-16 shows the calculations based on the standard

---

[7] Jefferson's method is also known as the *method of greatest divisors,* and in Europe as the *method of d'Hondt.*

Table 4-16 Republic of Parador: Calculations Using Standard Divisor

State	Population	Standard quota (Population ÷ 50,000)	Lower quota (Step 2)
A	1,646,000	32.92	32
B	6,936,000	138.72	138
C	154,000	3.08	3
D	2,091,000	41.82	41
E	685,000	13.70	13
F	988,000	19.76	19
Total	12,500,000	250.00	246

divisor. We are already familiar with these calculations—they are exactly what we used in steps 1 and 2 of Hamilton's method.

Table 4-17 shows us similar calculations based on a **modified divisor** of D = 49,500. (Let's not worry for now about where the 49,500 came from—let's just say it dropped out of the blue.) Note that using this smaller divisor, all the modified quotas are higher than the standard quotas, and the modified lower quotas add up to exactly the right total M. Thus, the last column of Table 4-17 shows an actual apportionment—the one produced by Jefferson's method.

Table 4-17 Republic of Parador: Calculations Using Modified Divisor D = 49,500

State	Population	Standard quota	Modified quota (Population ÷ 49,500)	Modified lower quota
A	1,646,000	32.92	33.25	33
B	6,936,000	138.72	140.12	140
C	154,000	3.08	3.11	3
D	2,091,000	41.82	42.24	42
E	685,000	13.70	13.84	13
F	988,000	19.76	19.96	19
Total	12,500,000	250	252.52	250

■

Before continuing with our discussion of Jefferson's method, we will make official some of the terminology we have already used: We call the number D used in step 1 the **modified divisor**, and the result of dividing the state's population by D the state's **modified quota**. We are now ready for a formal description of Jefferson's method.

Jefferson's Method

- **Step 1.** Find a number D *(modified divisor)* such that when each state's *modified quota* (state's population divided by D) is rounded *downward (modified lower quota),* the total is the exact number of seats to be apportioned.
- **Step 2.** Apportion to each state its modified lower quota.

There is one important issue still hanging over our heads: How does one go about finding this "magic" divisor D that makes Jefferson's method work? For example, how did we come up with $D = 49{,}500$ in Example 8? The answer is to use a calculator (or even better yet, a spreadsheet) and to try some educated trial and error. Let's start with the fact that the divisor we are looking for has to be a number smaller than the standard divisor (remember, we want the modified quotas to be bigger than the standard quotas, so we must divide by a smaller amount). So we pick a number D that we hope might work. We now carry out all the calculations asked for by Jefferson's method (divide the population by D; round the results downward; add up the total) and if we are lucky, the total is exactly right and we are finished. Otherwise we change our guess (make it higher if the total is too high, lower if the total is too low) and try again. In most cases, it takes at most two or three guesses before we find a divisor D that works (usually there is more than one).

Let's go through the paces using Example 8. We know we are looking for a modified divisor that is less than 50,000. Let's start with a guess that is $D = 49{,}000$. The reader is encouraged to do the calculations [Exercise 38(a)]. It turns out that this divisor doesn't work—it gives us a total of 252, which is too high. This means that we need our guess divisor to be a bit higher (thereby lowering the modified quotas), so we try $D = 49{,}500$. Bingo! Note that the divisor $D = 49{,}450$ also works, and so do many others [Exercise 38(b)].

A constitutional footnote: When presented with Jefferson's method, many people wonder, Can we get away with it? Is it legal to use divisors other than the standard divisor? The answer is yes. Nothing in the Constitution requires use of the standard divisor as the basis for apportionment. Remember that the standard divisor is the ratio of people per seat in the House of Representatives for the nation as a whole. Using the 1990 U.S. census, for example, we had a total U.S. apportionment population of $P = 248{,}102{,}973$ (for apportionment purposes the population of the District of Columbia is not included) and a House of Representatives of size $M = 435$, giving a standard divisor of approximately 570,352—give or take a limb or two. This means that for the *nation as a whole* there were about 570,352 people per seat in the House of Representatives. For individual states, however, the figures are different, because for each state the ratio of people per seat in the House of Representatives is based on the state's *actual apportionment* rather than the quota. Take, for example, California, with a population of 29,760,021 and 52 seats in the House of Representatives. If we divide these numbers out, we get 572,308 Californians for every representative—a larger ratio than the national figure.

We can see from all of the above that there is no constitutional (or mathematical) requirement to use the standard divisor as a basis for apportionment. We should think of the standard divisor as an ideal goal rather than a requirement for apportionment. At the same time, fairness considerations require that we should not stray too far from this ideal. As we will see next, this can turn out to be a problem under Jefferson's method.

Jefferson's Method and The Quota Rule

Jefferson's method suffers from one major flaw: *it violates the quota rule.* If we go back to Example 8 and look at what Jefferson's method did for state *B*, we can see it: State *B* got 140 seats. So what, you say? Now look at its standard quota (138.72). According to the quota rule, the only fair apportionments for state *B* are either 138 seats or 139 seats. No matter how one cuts it, giving state *B* a windfall of 140 seats goes against a basic principle of fairness (remember that state *B*'s gain has to be other states' loss).

The kind of quota-rule violation that took place in this example (where a state gets more than it should—in other words, more than its upper quota) is called an **upper-quota violation**. The other possible violation of the quota rule is when a state gets really shorted and gets less than its lower quota, which, of course, is called a **lower-quota violation**. It turns out that under Jefferson's method only upper-quota violations are possible [Exercise 39(a)].

When Jefferson's method was adopted in 1791, it is doubtful that anyone realized (certainly neither Jefferson nor Washington did) that it suffered from such a major flaw. It didn't take long for the problem to come up though. In the apportionment of 1832, New York, with a standard quota of 38.59, ended up with 40 seats. This horrified practically everyone except the New York delegation. Daniel Webster, among others, argued that this was actually unconstitutional:

> *The House is to consist of 240 members. Now, the precise portion of power, out of the whole mass presented by the number of 240, to which New York would be entitled according to her population, is 38.59; that is to say, she would be entitled to thirty-eight members, and would have a residuum or fraction; and even if a member were given her for that fraction, she would still have but thirty-nine. But the bill gives her forty . . . for what is such a fortieth member given? Not for her absolute numbers, for her absolute numbers do not entitle her to thirty-nine. Not for the sake of apportioning her members to her numbers as near as may be because thirty-nine is a nearer apportionment of members to numbers than forty. But it is given, say the advocates of the bill, because the process [Jefferson's method] which has been adopted gives it. The answer is, no such process is enjoined by the Constitution.*[8]

The apportionment of 1832 was to be the last apportionment of the House of Representatives based on Jefferson's method. It was clear that something new had to be tried, and the search for an apportionment method that did not violate the quota rule was on.

Adams' Method

John Quincy Adams. (The Metropolitan Museum of Art)

At about the same time that Jefferson's method was falling into disrepute because of its violation of the quota rule, John Quincy Adams was proposing a method that was a mirror image of it. It was based on exactly the same idea but instead of being based on the modified lower quotas, it was based on the *modified upper quotas.*[9]

Adams' Method

- **Step 1.** Find a modified divisor D such that when each state's *modified quota* (state's population divided by D) is rounded *upward (modified upper quota)*, the total is the exact number of seats to be apportioned.
- **Step 2.** Apportion to each state its modified upper quota.

---

[8] Daniel Webster, *The Writings and Speeches of Daniel Webster, Vol. VI* (Boston: Little, Brown and Company, 1903).

[9] Adams' method is also known as the *method of smallest divisors.*

Undoubtedly, Adams thought that by doing this he could avoid upper-quota violations, the big weakness of Jefferson's method. He was only partly right.

Example 9. Adams' Method Meets Parador's Congress

We will start by guessing a possible divisor D that we hope will work. We know that D will have to be bigger than 50,000 (we are trying to get modified quotas that are smaller than the standard quotas, so that, we hope, when we round them up, the total turns out to be 250). Remembering that 49,500 worked for Jefferson's method, we suspect a good guess might be $D = 50{,}500$.

Table 4-18 shows the calculations based upon $D = 50{,}500$.

Table 4-18 Republic of Parador: Calculations for Adams' Method Based on $D = 50{,}500$

State	Population	Modified quota (Population ÷ 50,500)	Modified upper quota	
A	1,646,000	32.59	33	
B	6,936,000	137.35	138	
C	154,000	3.05	4	
D	2,091,000	41.41	42	
E	685,000	13.56	14	
F	988,000	19.56	20	
Total	12,500,000	247.52	251	← Too much!

Since the total is too high, we need to lower the modified quotas just a little bit more, so let's try a higher divisor, say $D = 50{,}700$. Table 4-19 shows the calculations based on $D = 50{,}700$. Now it works! The last column of Table 4-19 shows the apportionment produced by Adams' method.

Table 4-19 Republic of Parador: Calculations for Adams' Method Based on $D = 50{,}700$

State	Population	Standard quota	Modified quota (Population ÷ 50,700)	Modified upper quota	
A	1,646,000	32.92	32.47	33	
B	6,936,000	138.72	136.80	137	
C	154,000	3.08	3.04	4	
D	2,091,000	41.82	41.24	42	
E	685,000	13.70	13.51	14	
F	988,000	19.76	19.49	20	
Total	12,500,000	250.00	246.55	250	← That's it!

■

Are there any problems with Adams' method? You bet! Look at state *B*'s apportionment of 137 seats and compare it with its standard quota of 138.72—a difference of 1.72 seats! This is an example of a lower-quota violation, a problem that can occur under Adams' method and which is especially frustrating to the state involved.

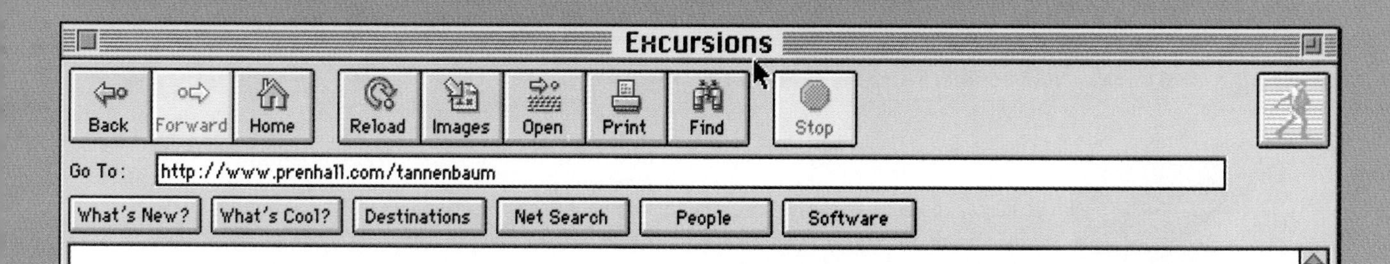

Take Your Seats!

In Europe, many countries elect representatives to congress or parliament using a type of voting called the party list method. This can be found under the Tannenbaum Website Chapter 4 section.

Like the single transferable vote method, the party list method is considered a type of proportional representation, because it gives groups of voters power in proportion to their numbers. However, the party list method leads directly to apportionment problems, since a parliament doesn't have fractional seats. One way to minimize such problems is to allow variations in the total number of seats and to use simple rounding.

Questions:

1. Why is the party list system a method of proportional representation?
2. Normally the number of seats with simple rounding will vary by only one or two seats. But in theory, a nominal 100-seat parliament could actually end up with as many as 132 seats. How?
3. A type of representation called semi-proportional can be achieved through the so-called limited vote method, also explained at the Seattle site.
4. Look back at the scenario in Get Out the Vote!, and suppose each voter gets to vote for only one candidate. How does the limited vote method improve proportional representation?
5. What are some advantages of the limited vote method compared to STV?
6. Again, consider a race among candidate *A*, *B*, *C*, and *D* for two open seats on a city council, with voters 1, 2, 3, 4, and 5. Suppose the voters' preferences among the candidates are now these:

	Voter				
Candidate	1	2	3	4	5
A	1st	1st	1st	1st	4th
B	2nd	2nd	2nd	2nd	3rd
C	3rd	3rd	3rd	3rd	2nd
D	4th	4th	4th	4th	1st

7. Who would win using a limited-vote, one-vote-per-voter method?
8. Who would win using STV? You may want to revisit the Cambridge Website to find out.
9. Which outcome in Questions 5 and 6 strikes you as more fair?

Daniel Webster. (U.S Signal Corps photo)

Webster's Method

It is clear that both Jefferson's method and Adams' method share the same philosophy: Treat all states exactly the same way. (The only difference is that whereas Jefferson's method rounds all the quotas down, Adams' method rounds all the quotas up.) For a while this sounded like a good idea, but as we now know it has serious flaws.

In 1832, Daniel Webster proposed a very basic idea. Let's round the quotas to the nearest integer, the way we round decimals in practically every other walk of life—down if the fractional part is less than 0.5, up otherwise (Webster always felt that this was the only fair way to round numbers.) But, an alert reader would argue, we have tried this idea before, and it didn't work (see Table 4-5)! There is, however, a new twist: In our first attempt, we were married to the notion that we had to use the standard quotas. Webster's idea was to use modified quotas chosen specifically so that when we are done with conventional rounding the total is exactly the number of seats to be apportioned.

Webster's Method[10]

- **Step 1.** Find a modified divisor D such that when each state's *modified quota* (state's population divided by D) is rounded the conventional way (to the nearest integer) the total is the exact number of seats to be apportioned.
- **Step 2.** Apportion to each state its modified quota rounded the conventional way.

Example 10. Webster's Method Meets Parador's Congress.

Let's apportion Parador's Congress using Webster's method. Our first decision is to make a guess at the divisor D. Should it be more than the standard divisor (50,000) or should it be less? Here we will use the standard quotas as a guideline. If we round off the standard quotas to the nearest integer (as we did in Table 4-5), we get a total of 251. This number is too high, which tells us that we should guess a divisor D larger than the standard divisor. Let us try $D = 50{,}100$. Table 4-20 shows the calculations based on $D = 50{,}100$. Rounding the modified quotas to the nearest integer works! The last column of Table 4-20 shows the apportionment under Webster's method.

Table 4-20 Republic of Parador: Calculations for Webster's Method Based on $D = 50{,}100$

State	Population	Modified quota (Population ÷ 50,100)	Rounded to
A	1,646,000	32.85	33
B	6,936,000	138.44	138
C	154,000	3.07	3
D	2,091,000	41.74	42
E	685,000	13.67	14
F	988,000	19.72	20
Total	12,500,000	249.49	250 ← It worked!

■

[10] Webster's method is sometimes known as the *Webster-Willcox method* as well as the *method of major fractions.*

Although Webster's method works in principle just like Jefferson's and Adams' methods, it is just a little bit harder to use in practice, since the modified divisor we are looking for can be smaller than, equal to, or larger than the standard divisor. (For guidelines as to how to go about making an educated guess, see Exercise 29.) On the other hand, there is something very gratifying about Webster's method—it validates what is most people's first reaction to the apportionment problem; the notion that quotas should be rounded just like ordinary numbers. The reason it didn't work for us when we first tried it is that we were doing it to the tune of standard quotas—Webster's method makes it work by modifying the quotas whenever necessary.

Webster's turns out to be a pretty good apportionment method, but as usual, there is a fly in the ointment—Webster's method also *violates the quota rule.* While Example 10 does not show this (the reader is encouraged to verify that there are no violations of the quota rule in the apportionment given in Example 10), it is possible to find examples where the quota rule is violated. Fortunately, this tends to be more of a theoretical than a practical problem, since the violations of the quota rule under Webster's method are rare and somewhat contrived. From a practical point of view, Webster's method is considered by many experts to be the best overall apportionment method available, and it could very well make a comeback as the official apportionment method for the House of Representatives, possibly in our lifetimes.

Conclusion

Balinski and Young's Impossibility Theorem

In this chapter we introduced four different *apportionment methods.* Table 4-21 summarizes the results of apportioning Parador's Congress (Example 3) under each of the four methods.

Table 4-21 Parador's Congress: A Tale of Four Methods

State	Population	Standard quota	Hamilton	Jefferson	Adams	Webster
A	1,646,000	32.92	33	33	33	33
B	6,936,000	138.72	139	140	137	138
C	154,000	3.08	3	3	4	3
D	2,091,000	41.82	42	42	42	42
E	685,000	13.70	13	13	14	14
F	988,000	19.76	20	19	20	20
Total	12,500,000	250.00	250	250	250	250

Note that, here, each of the four methods produced a different apportionment. This clearly demonstrates that the methods are indeed all different. At the same time, we should warn the reader that it is not impossible for two different methods to produce identical apportionments (see Exercises 26, 27, 28, and 42).

Of the four methods we discussed, one is based on a strict adherence to the standard quotas (Hamilton's), whereas the other three (Jefferson's, Adams', and Webster's) are based on the philosophy that quotas can be conveniently modified by the appropriate choice of divisor. While some of the methods are clearly better than others, none of them is perfect: Each either *violates the quota rule* or *produces paradoxes.* Table 4-22 summarizes the characteristics of the four methods.

Table 4-22 How the Four Methods Stack Up

	Hamilton	Jefferson	Adams	Webster
Violates quota rule	No	Yes	Yes	Yes
Alabama paradox	Yes	No	No	No
Population paradox	Yes	No	No	No
New-states paradox	Yes	No	No	No
Favoritism toward	Large states	Large states	Small states	Small states

For many years, the ultimate hope held by scholars interested in the apportionment problem, both inside and outside Congress, was that mathematicians would eventually come up with an *ideal* apportionment method—one that never violates the quota rule, does not produce any paradoxes, and treats large and small states without favoritism. ("The apportionment of Representatives to the population is a mathematical problem. Then why not use a method that will stand the test [of fairness] under a correct mathematical formula?"[11])

Indeed, why not? The answer was provided in 1980 by a surprising discovery made by two mathematicians—Michel L. Balinski and H. Peyton Young[12]—and is known as **Balinski and Young's impossibility theorem**: *There are no perfect apportionment methods. Any apportionment method that does not violate the quota rule must produce paradoxes and any apportionment method that does not produce paradoxes must violate the quota rule.*

Once again, we reach an eerily familiar conclusion in a slightly different setting: Fairness and proportional representation are inherently incompatible.

Key Concepts

Adams' method
Alabama paradox
apportionment method
apportionment problem
Balinski and Young's impossibility theorem
conventional rounding
divisor
Hamilton's method
Jefferson's method
lower quota
lower-quota violation
modified quota
new-states paradox
population paradox
quota rule
standard divisor
standard quota
upper quota
upper-quota violation
Webster's method

Exercises

Walking

Exercises 1 through 5 refer to a small country consisting of 4 states. There are 160 seats in the legislature and the populations of the states are shown as follows:

State	*A*	*B*	*C*	*D*
Population (in millions)	3.31	2.67	1.33	0.69

---

11 Congressman Ernest Gibson of Vermont, 1929 [*Congressional Record,* 70th Congress, 2d Session, 70 (1929), p. 1500].

12 Michel Balinski is a mathematician at the State University of New York at Stony Brook; H. Peyton Young is a mathematician at the University of Maryland.

1. **(a)** Find the standard divisor.
   (b) Find each state's standard quota.
   (c) Find each state's apportionment under Hamilton's method.
2. **(a)** Using the divisor 49,400, find each state's modified quota.
   (b) Find each state's apportionment under Jefferson's method.
3. **(a)** Using the divisor 50,500, find each state's modified quota.
   (b) Find each state's apportionment under Adams' method.
4. **(a)** Using the divisor 50,400, find each state's modified quota.
   (b) Using the divisor 50,350, find each state's modified quota.
   (c) Explain why the divisor 50,400 works for Adams' method but the divisor 50,350 doesn't.
   (d) Find a divisor not between 50,500 and 50,350 that will also work for Adams' method.
5. **(a)** Find a divisor that works for Webster's method. (*Hint:* You should be able to do this without using a calculator!)
   (b) Find the apportionment for each state under Webster's method.

Exercises 6 through 9 refer to a company that operates 6 bus routes (A, B, C, D, E, and F) and 130 buses. The buses are apportioned among the routes on the basis of average number of daily passengers per route, which is given in the following table.

Route	*A*	*B*	*C*	*D*	*E*	*F*
Average number of daily passengers	45,300	31,070	20,490	14,160	10,260	8,720

6. **(a)** Find the standard divisor.
   (b) Find the standard quota for each bus route.
   (c) Apportion the buses among the routes using Hamilton's method.
7. Apportion the buses among the routes using Jefferson's method.
8. Apportion the buses among the routes using Adams' method.
9. Apportion the buses among the routes using Webster's method.

Exercises 10 through 14 refer to a clinic with a nursing staff consisting of 225 nurses working in four shifts: A (7:00 A.M. to 1:00 P.M.); B (1:00 P.M. to 7:00 P.M.); C (7:00 P.M. to 1:00 A.M.); and D (1:00 A.M. to 7:00 A.M.). The number of nurses apportioned to each shift is based on the average number of patients per shift, given in the following table.

Shift	*A*	*B*	*C*	*D*
Average number of patients	871	1029	610	190

10. **(a)** Find the standard divisor. Explain what the standard divisor represents in this problem.
    (b) Find the standard quota for each shift.
11. Apportion the nurses to the shifts using Hamilton's method.
12. Apportion the nurses to the shifts using Jefferson's method. (*Hint*: Divisors don't have to be whole numbers.)
13. Apportion the nurses to the shifts using Adams' method. (*Hint*: Divisors don't have to be whole numbers.)

14. Apportion the nurses to the shifts using Webster's method. (*Hint*: Divisors don't have to be whole numbers.)

Exercises 15 through 19 refer to a small country consisting of five states. The total population of the country is 23.8 million. The standard quota of each state is given in the following table.

State	*A*	*B*	*C*	*D*	*E*
Standard quota	40.50	29.70	23.65	14.60	10.55

15. (a) Find the number of seats in the legislature.

(b) Find the standard divisor.

(c) Find the population of each state.

16. Find each state's apportionment using Hamilton's method.

17. Find each state's apportionment using Jefferson's method.

18. Find each state's apportionment using Adams' method.

19. Find each state's apportionment using Webster's method.

Exercise 20 through 23 refer to the following: Tasmania State University is made up of five different schools: Agriculture, Business, Education, Humanities, and Science. A total of 250 faculty positions must be apportioned based on the schools' respective enrollments, shown below.

School	Agriculture	Business	Education	Humanities	Science	Total
Enrollment	1646	762	2081	1066	6945	12,500

20. (a) Find the standard divisor. What does the standard divisor represent in this problem?

(b) Find each school's standard quota.

21. Find the number of faculty members apportioned to each school using Hamilton's method.

22. Find the number of faculty members apportioned to each school using Jefferson's method.

23. Find the number of faculty members apportioned to each school using Webster's method.

24. A mother wishes to distribute 10 pieces of candy among her 3 children based on the number of minutes each child spends studying, as shown in the following table.

Child	Bob	Peter	Ron
Minutes studied	54	243	703

(a) Find each child's apportionment using Hamilton's method.

(b) Suppose that, just prior to actually handing over the candy, mom finds another piece of candy and includes it in the distribution. Find each child's apportionment using Hamilton's method and 11 pieces of candy.

(c) Did anything paradoxical occur? (What's the name of this paradox?)

25. A mother wishes to distribute 11 pieces of candy among her 3 children based on the number of minutes each child spends studying, as shown in the following table.

Child	Bob	Peter	Ron
Minutes studied	54	243	703

(a) Find each child's apportionment using Hamilton's method. [Note that this is Exercise 24(b).]

(b) Suppose that before mom has time to sit down and do the actual calculations, the children decide to do a little more studying. Say Bob studies an additional 2 minutes, Peter an additional 12 minutes, and Ron an additional 86 minutes. Find each child's apportionment using Hamilton's method based on the new total time studied.

(c) Did anything paradoxical occur? Explain.

Jogging

26. Make up an apportionment problem in which Hamilton's method and Jefferson's method result in exactly the same apportionment for each state. Your example should involve at least 3 states, and the standard quotas should not be whole numbers.

27. Make up an apportionment problem in which Hamilton's method and Adams' method result in exactly the same apportionment for each state. Your example should involve at least 3 states, and the standard quotas should not be whole numbers.

28. Make up an apportionment problem in which Hamilton's method and Webster's method result in exactly the same apportionment for each state. Your example should involve at least 3 states, and the standard quotas should not be whole numbers.

29. **(a)** Consider the following situation.

State	*A*	*B*	*C*	*D*	*E*
Standard quota	11.23	24.39	7.92	36.18	20.28

You want to use Webster's method as your method of apportionment. Explain why you should look for a divisor that is *smaller* than the standard divisor.

(b) Consider the following situation.

State	*A*	*B*	*C*	*D*	*E*
Standard quota	11.73	24.89	7.92	35.68	19.78

You want to use Webster's method as your method of apportionment. Explain why you should look for a divisor that is *bigger* than the standard divisor.

(c) Under what conditions can you be assured that the standard divisor will work in Webster's method?

30. Make up an apportionment problem (different from any given in the chapter) in which Jefferson's method and Webster's method give different results.

31. Make up an apportionment problem (different from any given in the chapter) in which Adams' method and Webster's method give different results.

32. Make up an apportionment problem (different from any given in the chapter) in which Jefferson's method, Adams' method, and Webster's method all give different results.

33. Make up an apportionment problem in which Webster's method violates the quota rule. [*Hint*: Make one state much larger than the others.]

34. Consider an apportionment problem with only 2 states, *A* and *B*. Suppose that state *A* has standard quota q_1 and state *B* has standard quota q_2, neither of which is a whole number. (Of course, $q_1 + q_2 = M$ must be a whole number.) Let f_1 represent the fractional part of q_1 and f_2 the fractional part of q_2.

(a) Explain why one of the fractional parts is bigger than or equal to 0.5 and the other is smaller than or equal to 0.5.

(b) Assuming neither fractional part is equal to 0.5, explain why Hamilton's method and Webster's method must result in the same apportionment.

(c) Explain why in any apportionment problem involving only 2 states Hamilton's method can never produce the Alabama paradox or the population paradox.

(d) Explain why in the above situation Webster's method can never violate the quota rule.

35. The purpose of this example is to show that under rare circumstances the use of a modified divisor method may not work. A small country consists of 4 states with populations given as follows.

State	*A*	*B*	*C*	*D*
Population	500	1000	1500	2000

There are $M = 51$ seats in the House of Representatives.

(a) Find each state's apportionment using Jefferson's method.

(b) Attempt to apportion the seats using Adams' method with a modified divisor $D = 100$. What happens if you take $D < 100$? What happens if you take $D > 100$?

(c) Explain why Adams' method will not work for this example.

36. This exercise is based on actual data taken from the 1880 census. Here are some figures: In 1880, the population of Alabama was given at 1,262,505. With a House of Representatives consisting of $M = 300$ seats the standard quota for Alabama was 7.671.

(a) Find the 1880 census population for the United States (rounded to the nearest person).

(b) Given that the standard quota for Texas was 9.672, find the population of Texas (to the nearest person).

37. The following table shows the results of the 1790 census (the very first census of the United States taken after the Constitution was adopted).

State	**Population**
Connecticut	236,841
Delaware	55,540
Georgia	70,835
Kentucky	68,705
Maryland	278,514
Massachusetts	475,327
New Hampshire	141,822
New Jersey	179,570
New York	331,589
North Carolina	353,523
Pennsylvania	432,879
Rhode Island	68,446
South Carolina	206,236
Vermont	85,533
Virginia	630,560
Total	3,615,920

Based on the fact that the number of seats in the House of Representatives was set at $M = 105$:

(a) Find the apportionment that would have resulted under the original bill passed by Congress to use Hamilton's method.

(b) Find the apportionment that was actually used (remember that at the end it was based on Jefferson's method).

(c) Compare the answers in (a) and (b). Which state was the winner in the 1790 controversy between the two methods? Which state was the loser?

38. Suppose that you want to apportion Parador's Congress (Example 3) using Jefferson's method.

(a) Show that the divisor $D = 49{,}000$ does not work.

(b) Show that any divisor D between 49,401 and 49,542 (inclusive) works, and that there are no other whole-number divisors that work.

39. (a) Explain why, when Jefferson's method is used, any violations of the quota rule must be upper-quota violations.

(b) Explain why, when Adams' method is used, any violations of the quota rule must be lower-quota violations.

(c) Use parts (a) and (b) to justify why, in the case of an apportionment problem with just 2 states, neither Jefferson's nor Adams' method can possibly violate the quota rule.

Exercises 40 and 41 refer to a variation of Hamilton's method known as **Lowndes' method**. *(The method is also called the modified Hamilton's method.) The basic difference between Hamilton's and Lowndes' methods is that, after each state is assigned the lower quota, the surplus seats are handed out in order of relative fractional parts. (The relative fractional part of a number is the fractional part divided by the integer part. For example, the relative fractional part of 41.82 is 0.82/41 = 0.02, and the relative fractional part of 3.08 is 0.08/3 = 0.027. Notice that while 41.82 would have priority over 3.08 under Hamilton's method, 3.08 has priority over 41.82 under Lowndes' method because 0.027 is greater than 0.02.)*

40. (a) Find the apportionment of Parador's Congress (Example 3) under Lowndes' method.

(b) Verify that the resulting apportionment is different from each of the apportionments shown in Table 4-21. In particular, list which states do better under Lowndes' method than under Hamilton's method.

(c) Fill in the blank: Lowndes' method shows favoritism toward (larger, smaller) states.

41. Consider an apportionment problem with only 2 states, A and B. Suppose that state A has standard quota q_1 and state B has standard quota q_2, neither of which is a whole number. (Of course $q_1 + q_2 = M$ must be a whole number.) Let f_1 represent the fractional part of q_1 and f_2 the fractional part of q_2.

(a) Find values q_1 and q_2 such that Lowndes' method and Hamilton's method result in the same apportionment.

(b) Find values q_1 and q_2 such that Lowndes' method and Hamilton's method result in different apportionments.

(c) Write an inequality involving q_1, q_2, f_1, and f_2 that would guarantee that Lowndes' method and Hamilton's method result in different apportionments.

Running

42. Make up an apportionment problem in which all 4 methods (Hamilton's, Jefferson's, Adams', and Webster's) result in exactly the same apportionment for each state. Your example should involve at least 4 states and the standard quotas should not be whole numbers.

43. Explain why Jefferson's method cannot produce

(a) the Alabama paradox

(b) the new-states paradox.

44. Explain why Adams' method cannot produce

(a) the Alabama paradox

(b) the new-states paradox.

45. Explain why Webster's method cannot produce

(a) the Alabama paradox

(b) the new-states paradox.

Exercises 46 through 49 refer to the Huntington-Hill method described in Appendix 1. (These exercises should not be attempted without understanding Appendix 1.)

46. Use the Huntington-Hill method to find the apportionments of each state for a small country that consists of 5 states. The total population of the country is 24.8 million. The standard quotas of each state are as follows.

State	*A*	*B*	*C*	*D*	*E*
Standard quota	25.26	18.32	2.58	37.16	40.68

47. **(a)** Use the Huntington-Hill method to apportion Parador's Congress (Example 3).

(b) Compare your answer in (a) with the apportionment produced by Webster's method. What's your conclusion?

48. A country consists of 6 states with populations as follows.

State	**Population**
A	344,970
B	408,700
C	219,200
D	587,210
E	154,920
F	285,000
Total	2,000,000

There are 200 seats in the legislature.

(a) Find the apportionment under Webster's method.

(b) Find the apportionment under the Huntington-Hill method.

(c) Compare the divisors used in (a) and (b).

(d) Compare the apportionments found in (a) and (b).

49. A country consists of 6 states with populations as follows.

State	**Population**
A	344,970
B	204,950
C	515,100
D	84,860
E	154,960
F	695,160
Total	2,000,000

There are 200 seats in the legislature.

(a) Find the apportionment under Webster's method.

(b) Find the apportionment under the Huntington-Hill method.

(c) Compare the divisors used in (a) and (b).

(d) Compare the apportionments found in (a) and (b).

APPENDIX 1 The Huntington-Hill Method

The method currently used to apportion the U.S. House of Representatives is known as the **Huntington-Hill method**, and more commonly as the **method of equal proportions**.

Let's start with some historical background. The method was developed sometime around 1911, by Joseph A. Hill, chief statistician of the Bureau of Census, and Edward V. Huntington, professor of mechanics and mathematics at Harvard University. In 1929, the Huntington-Hill method was endorsed by a distinguished panel of mathematicians. The panel, commissioned by the National Academy of Sciences at the formal request of the Speaker of the House, investigated many different apportionment methods and recommended the Huntington-Hill method as the best possible one.

On November 15, 1941, President Franklin D. Roosevelt signed "An Act to Provide for Apportioning Representatives in Congress among the Several States by the equal proportions method" (Public Law 291, H.R. 2665, 55 Stat 261). Under the same Act, the size of the House of Representatives was fixed at $M = 435$. This act still stands today,[13] but political, legal, and mathematical challenges to it have come up periodically.

There are several ways to describe how the Huntington-Hill method works. For the purposes of explanation, the method is most conveniently described by comparison to Webster's method. In fact, the two methods are almost identical. Just as in Webster's method, we will find modified quotas and we will round some of them upward and some of them downward. The difference between the two methods is in the cutoff point for rounding up or down. Take, for example, a state with a modified quota of 3.48. Under Webster's method we know that we must round this quota downward, because the cutoff point for rounding is 3.5. It may seem like a bit of an overkill, but we can put it this way: The cutoff point for rounding quotas under Webster's method is exactly halfway between the modified lower quota (L) and the modified upper quota ($L + 1$).

$$\text{Cutoff for Webster's method} = \frac{L + (L + 1)}{2}$$

(Thus, for a state with a modified quota of 3.48, the cutoff point is 3.5.)

Now under the Huntington-Hill method the cutoff point for rounding quotas is computed using a different formula:[14]

$$\text{Cutoff for the Huntington-Hill method} = \sqrt{L \times (L + 1)}$$

Thus, if a state has a modified quota of 3.48, the cutoff for rounding this quota under the Huntington-Hill method would be $\sqrt{3 \times 4} = \sqrt{12} = 3.464$. Since the modified quota 3.48 is above this cutoff, under the Huntington-Hill method this state would get 4 seats.

---

[13] The current apportionment of the House of Representatives based on the Huntington-Hill method can be found in the appendix to Chapter 2 (each state's apportionment in the House of Representatives equals the number of electoral votes minus 2).

[14] There is a handy mathematical name for the Huntington-Hill cutoffs. For any two positive numbers a and b, $\sqrt{a \times b}$ is called the *geometric mean* of a and b. Thus, we can describe each Huntington-Hill cutoff as the geometric mean of the modified lower and upper quotas.

Huntington-Hill Rounding Rules

If the quota falls between L and $L + 1$, the Huntington-Hill cutoff point for rounding is $H = \sqrt{L \times (L + 1)}$. If the quota is below H, we round down, otherwise we round up.

Huntington-Hill Method

- **Step 1.** Find a number D such that when each state's modified quota (state's population divided by D) is rounded according to the Huntington-Hill rounding rules, the total is the exact number of seats to be apportioned.
- **Step 2.** Apportion to each state its modified quota, rounded using the Huntington-Hill rules.

Table 4-23 is convenient to have handy when working with the Huntington-Hill method.

Table 4-23

Modified quota between	Cutoff point for rounding under Webster's method	Cutoff point for rounding under Huntington-Hill method
1 and 2	1.5	$\sqrt{2} \approx 1.414$
2 and 3	2.5	$\sqrt{6} \approx 2.449$
3 and 4	3.5	$\sqrt{12} \approx 3.464$
4 and 5	4.5	$\sqrt{20} \approx 4.472$
5 and 6	5.5	$\sqrt{30} \approx 5.477$
6 and 7	6.5	$\sqrt{42} \approx 6.481$
7 and 8	7.5	$\sqrt{56} \approx 7.483$
8 and 9	8.5	$\sqrt{72} \approx 8.485$
9 and 10	9.5	$\sqrt{90} \approx 9.487$
10 and 11	10.5	$\sqrt{110} \approx 10.488$

We will conclude this appendix with a very simple example, which shows that the Huntington-Hill method can produce an apportionment that differs from Webster's method.

Example.

A small country consists of 3 states. We want to apportion the 100 seats in its legislature to the 3 states according to the population figures shown in Table 4-24.

Table 4-24

State	A	B	C	Total
Population	3480	46,010	50,510	100,000

We will use Webster's method first and then the Huntington-Hill method.

We start by computing the standard quotas. Since the standard divisor is 100,000/100 = 1000, this is really easy, as shown in Table 4-25.

Table 4-25

State	A	B	C	Total
Standard quota	3.48	46.01	50.51	100

It so happens that rounding the standard quotas the conventional way gives a total of 100, so the standard quotas work for Webster's method (Table 4-26).

Table 4-26

State	Population	Standard quota	Webster's apportionment
A	3,480	3.48	3
B	46,010	46.01	46
C	50,510	50.51	51
Total	100,000	100.00	100

Next, we notice that our old friend 3.48 has made an appearance. We know that under the Huntington-Hill method 3.48 is past the cutoff point of 3.464, so it has to be rounded upward (to 4). The other two standard quotas are not affected and would still be rounded as before (Table 4-27).

Table 4-27

State	Population	Standard quota (Standard divisor = 1000)	Rounded under Huntington-Hill rules to
A	3,480	3.48	4
B	46,010	46.01	46
C	50,510	50.51	51
Total	100,000	100.00	101

Since the total comes to 101, these quotas don't work—we can see they are a bit too high. But when we try a divisor just a tad bigger ($D = 1001$), the totals do come out right (Table 4-28). The last column of Table 4-28 shows the way the 100 seats would be apportioned under the Huntington-Hill method. (Note that in this example the apportionment is different from the one produced by Webster's method.)

Table 4-28

State	Population	Modified quota (Divisor = 1001)	Rounded under Huntington-Hill rules to
A	3,480	3.477	4
B	46,010	45.964	46
C	50,510	50.460	50
Total	100,000	100.00	100

APPENDIX 2 A Brief History of Apportionment in the United States

- 1787
  - Constitutional Convention meets in Philadelphia.
  - Under the "Great Compromise" the House of Representatives will be apportioned based on states' populations. Article I, Section 2, gives Congress the authority to determine the exact method of apportionment.
- 1791
  - Following the census of 1790 two methods of apportionment are proposed. Hamilton's method is supported by the Federalists, Jefferson's method by the Republicans.
  - Congress approves a bill to apportion the House of Representatives using Hamilton's method with 120 seats ($M = 120$).
  - President Washington vetoes the bill (the first exercise of a presidential veto in U.S. history!).
  - Jefferson's method is adopted using $M = 105$ seats and a divisor $D = 33{,}000$. (Jefferson's method will remain in use until 1840.)
- 1822
  - Rep. William Lowndes (South Carolina) proposes what we now call Lowndes' method. (See Exercise 40.) The proposal dies in Congress.
- 1832
  - John Quincy Adams (former president and at this time a congressman from Massachusetts) proposes what we now call Adams' method. The proposal fails.
  - Senator Daniel Webster (Massachusetts) proposes what we now call Webster's method. His proposal also fails.
  - Jefferson's method is used once again with $M = 240$.
- 1842
  - Webster's method is adopted with $M = 223$. This is one of the few times in U.S. history that M goes down. (Politicians are not inclined to legislate themselves out of work!)
- 1852
  - A bill adopting Hamilton's method as the permanent method of apportionment with $M = 233$ seats is presented in Congress.
  - Congress approves the bill with the change $M = 234$. (For this value of M, Hamilton's method agrees with Webster's method.)
- 1872
  - $M = 283$ seats is proposed, because with this number Hamilton's method and Webster's method agree.
  - The final apportionment approved by Congress ends up having 292 seats in the House and is not based on either Hamilton's or Webster's method but rather on a power grab among states. For the rest of the decade, the apportionment of the House of Representatives is in violation of the Constitution (no proper method was used).
- 1876
  - Rutherford B. Hayes becomes President of the United States based on the unconstitutional apportionment of 1872. If Hamilton's method had been used, Tilden would have had enough electoral votes to win the election.

The election of 1876. The botched apportionment of 1872 resulted in the election of Rutherford B. Hayes over Samuel Tilden.

- 1880
  - The Alabama paradox surfaces as a serious flaw of Hamilton's method.
- 1882
  - Despite serious concerns about Hamilton's method, an apportionment bill based on it with $M = 325$ seats is eventually approved. (This number is chosen so that Hamilton's method and Webster's method agree.)
- 1901
  - The Bureau of the Census submits to Congress tables showing apportionments based on Hamilton's method for all size Houses between $M = 350$ and $M = 400$.
  - For all values of M between 350 and 400 except one ($M = 357$) Colorado would get an apportionment of 3 seats. For $M = 357$ Colorado would get only 2 seats. (The Alabama paradox again!)
  - The House Committee on Apportionment proposes a bill to apportion the House of Representatives using $M = 357$ seats.
  - Congress is in an uproar; the bill is defeated, and Hamilton's method is finally abandoned for good.
  - Webster's method is adopted with $M = 386$.
- 1907
  - Oklahoma joins the Union. The new-states paradox is discovered.
- 1911
  - Webster's method is readopted with $M = 433$. (A provision is made for Arizona and New Mexico to get 1 seat each if admitted into the Union.)
  - Joseph Hill (chief statistician of the Bureau of the Census) proposes a new method, now known as the Huntington-Hill method or the method of equal proportions. (See Appendix 1.)
- 1921
  - No reapportionment is done after the 1920 census (in direct violation of the Constitution).

- 1931
  - Webster's method is used with $M = 435$.
- 1941
  - Huntington-Hill method is adopted with $M = 435$. This remains (by law) the permanent method of apportionment.
- 1992
  - Montana challenges the constitutionality of the Huntington-Hill method in a lawsuit (*Montana v. U.S. Dept. of Commerce*). The Supreme Court upholds the Huntington-Hill method as constitutional.

References and Further Readings

1. Balinski, Michel L., and H. Peyton Young, "The Apportionment of Representation," *Fair Allocation: Proceedings of Symposia on Applied Mathematics,* 33 (1985) 1–29.
2. Balinski, Michel L., and H. Peyton Young, *Fair Representation; Meeting the Ideal of One Man, One Vote.* New Haven, CT: Yale University Press, 1982.
3. Balinski, Michel L., and H. Peyton Young, "The Quota Method of Apportionment," *American Mathematical Monthly,* 82 (1975), 701–730.
4. Brams, Steven, and Philip Straffin, Sr., "The Apportionment Problem," *Science,* 217 (1982), 437–438.
5. Eisner, Milton, *Methods of Congressional Apportionment,* COMAP Module #620.
6. Hoffman, Paul, *Archimedes' Revenge: The Joys and Perils of Mathematics.* New York: W. W. Norton & Co., 1988, chap. 13.
7. Huntington, E. V., "The Apportionment of Representatives in Congress." *Transactions of the American Mathematical Society,* 30 (1928), 85–110.
8. Huntington, E. V., "The Mathematical Theory of the Apportionment of Representatives," *Proceedings of the National Academy of Sciences, U.S.A.,* 7 (1921), 123–127.
9. Meder, Albert E., Jr., *Legislative Apportionment.* Boston: Houghton Mifflin Co., 1966.
10. Saari, D. G., "Apportionment Methods and the House of Representatives," *American Mathematical Monthly,* 85 (1978), 792–802.
11. Schmeckebier, L. F., *Congressional Apportionment.* Washington, DC: The Brookings Institution, 1941.
12. Steen, Lynn A., "The Arithmetic of Apportionment," *Science News,* 121 (May 8, 1982), 317–318.
13. Webster, Daniel, *The Writings and Speeches of Daniel Webster, Vol. VI, National Edition.* Boston: Little, Brown, and Company, 1903.

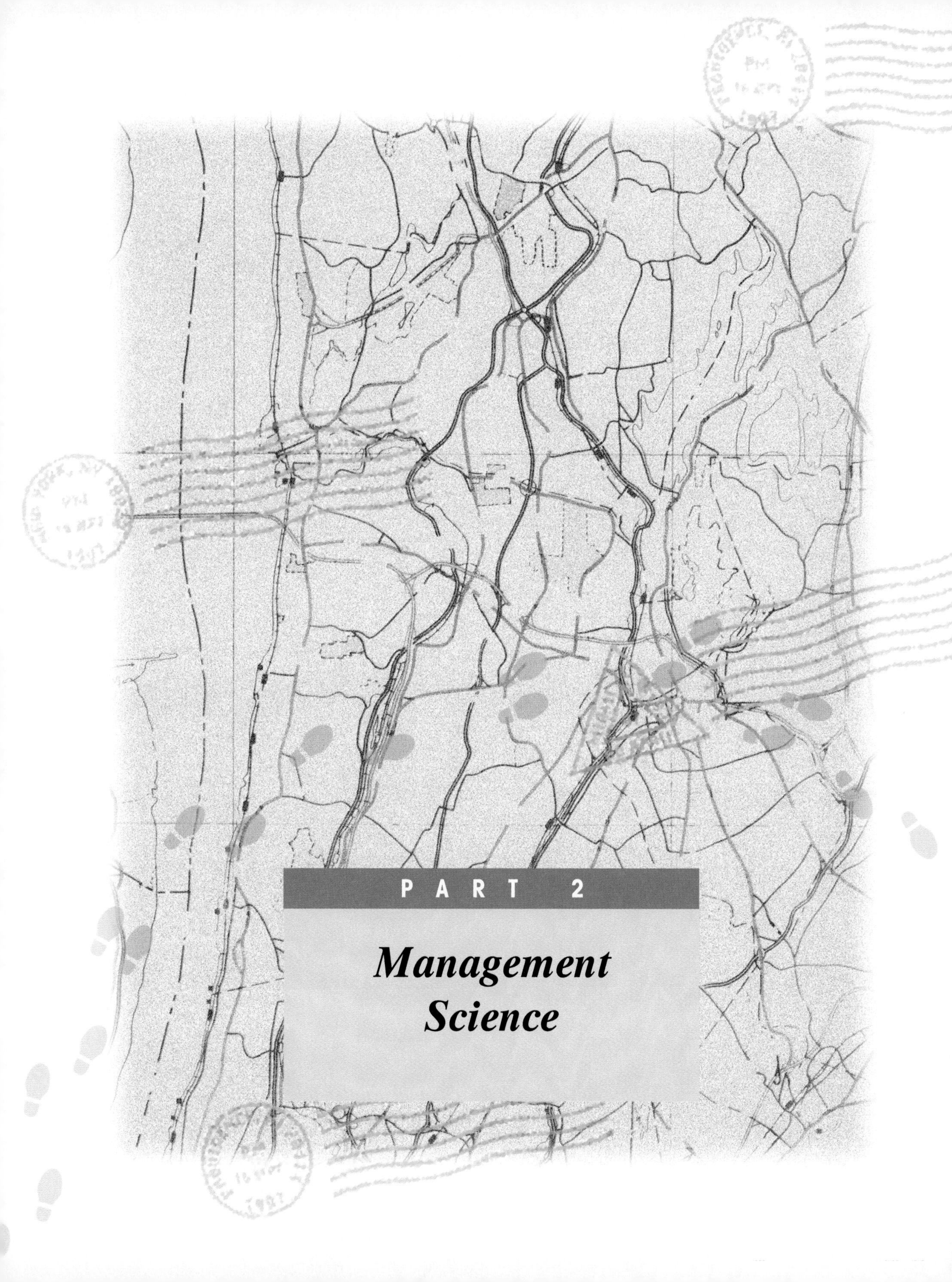

PART 2

Management Science

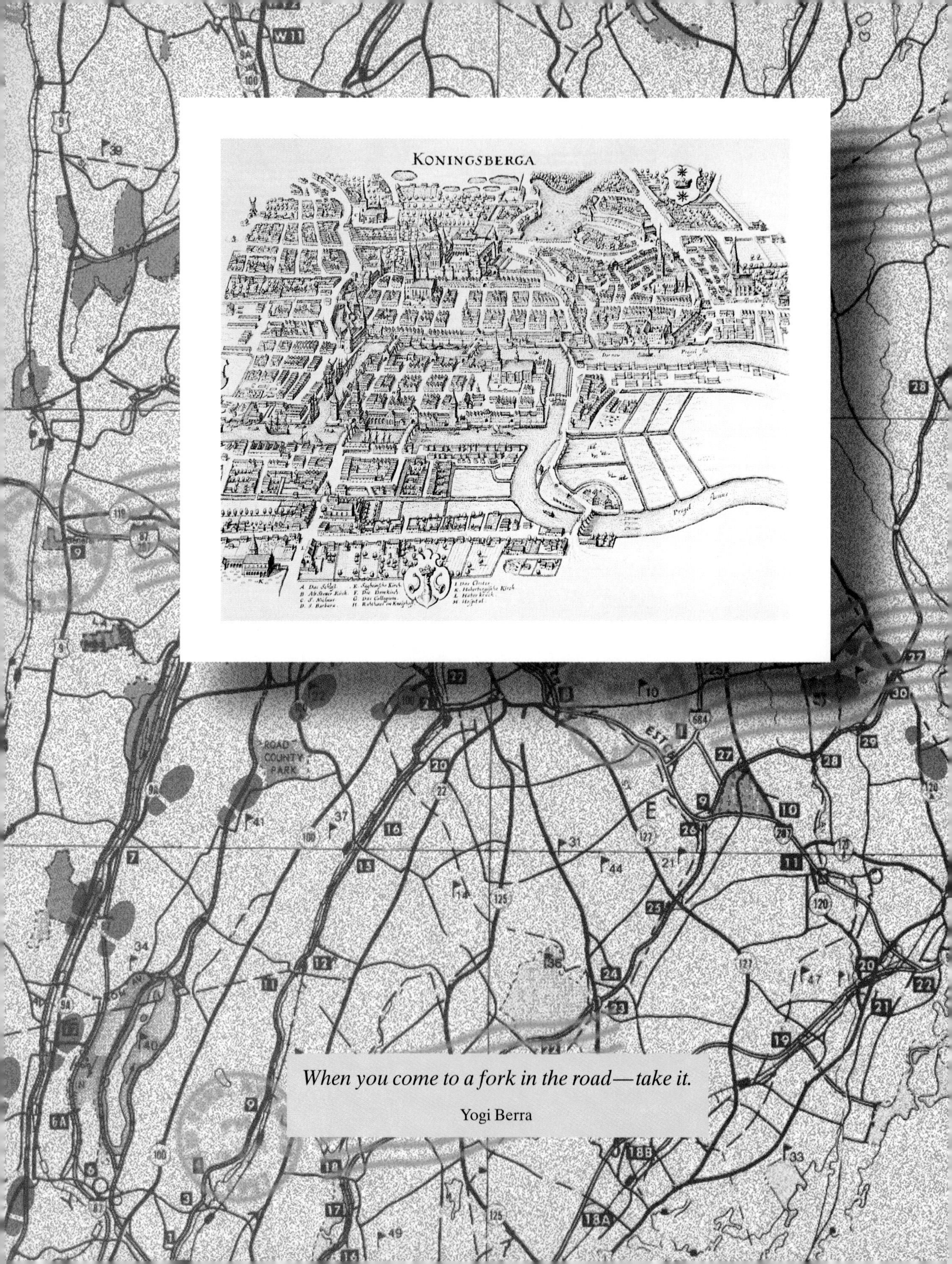

When you come to a fork in the road—take it.

Yogi Berra

Euler Circuits

The Circuit Comes to Town

Our story starts in the year 1736. A brilliant young Swiss mathematician named Leonhard Euler comes across an innocent little puzzle of disarming simplicity: The medieval town of Königsberg has a river running through it. There is an island and a fork in the river which together divide the city into four separate land areas. At the time, the four land areas were connected by seven bridges. (The map in the opposite page shows the layout of Königsberg back in the 1700s.) The puzzle asked whether it was possible for a stroller to take a walk around the town, *crossing each of the seven bridges just once.* By the time Euler heard of the puzzle, many people had tried unsuccessfully to find such a walk, and the general feeling was that such a walk was actually impossible, but of course, nobody knew for sure. Euler, perhaps sensing that something important lay behind the frivolity of this problem, immediately proceeded to solve it by demonstrating mathematically that such a walk was impossible. In so doing, he laid the foundations for what was at the time a totally new type of mathematics. He called this new type of mathematics *geometris situs* ("the geometry of location").

So what? one might ask. The story of the Königsberg bridges puzzle would be inconsequential, if it weren't that the mathematical ideas Euler set in motion when solving it have grown into one of the most important and practical branches of modern mathematics—a unique kind of geometry now known as *graph theory.* Modern applications of graph theory span practically every area of science and technology—from physics, biology, and computer science to psychology, sociology, and management science.

Over the next four chapters we will learn how graph theory is used to solve many important and unique problems in real life. For starters, in this chapter we will get acquainted with the basic concepts in graph theory and the use of *graphs* to model real-world problems. Along the way, we will also learn how Euler[1] solved the Königsberg bridge puzzle[2].

Routing Problems

How important to you is the work of your garbage collector? Hardly anyone thinks much about this except when the collectors go on strike, at which time our appreciation for their services grows significantly. Similar things can be said about the mail carrier and the policeman on the beat. What these people have in common is that they are *providers* of a service that is *delivered* to us (usually at our homes), as opposed to a service we must go out and get (haircuts, the movies, etc.). Usually, these types of services can be delivered economically only when they are delivered to many customers, and doing this properly requires planning. In this chapter we will deal, among other things, with the mathematics behind the proper planning and design of delivery routes. These kinds of mathematics problems fall under the generic title of **routing problems.**

What is a *routing problem?* To put it in the most general way, routing problems are concerned with finding ways to route the delivery of *goods* and/or *services* to an assortment of *destinations.* Examples of goods in question are packages, mail, newspaper, raw materials; examples of services are police protection, garbage collection, Internet access; examples of destinations are houses, warehouses, computer terminals, towns. In addition, *proper routes* must satisfy what we will call the *rules of the road*: (i) if there is a "direction of traffic" (as in one-way streets, pipeline flows, and communication protocols), then the direction of traffic must be followed, and (ii) if there is no direct way to get from destination X to destination Y, then a proper route cannot go directly from point X to point Y. (This seems self-evident, but in some situations it is easy to forget this rule.)

---

[1] Leonhard Euler (1707–1783) (the last name is properly pronounced "oiler", not "yuler") was born in Basel, Switzerland. From a very early age, Euler showed an incredible talent for doing enormously complicated calculations in his head. This alone is not necessarily the hallmark of a great mathematician, but Euler added to it an amazing creative talent and a tremendous work ethic. Today, Euler is acknowledged as one of the greatest mathematicians in the history of mankind as well as the most prolific (it is estimated that, when finally compiled, his collected memoirs will fill nearly 100 volumes). Euler was quite prolific in a more mundane way as well—he had 13 children of his own (it is said that he loved to work on his mathematical research with one of the younger children on his lap and other children noisily running around). In the words of one biographer, "Euler was the Shakespeare of mathematics—universal, richly detailed and inexhaustible."

[2] Königsberg, located on the southern coast of the Baltic Sea, was a medieval port, best known as the home town of the famous philosopher Immanuel Kant. In the 1700s and 1800s it was part of Prussia, but it eventually became part of Russia. After the Russian revolution its name was changed to Kaliningrad.

Two fundamental questions can come up in a routing problem:

1. Is there a proper route for the particular problem?
2. If there are many possible routes, which one is the *best* (where best is a function of some predetermined variable such as *cost, distance,* or *time*)?

Question 1 calls for a Yes/No answer, which is often (but not always) easy enough to provide. Question 2 tends to be a little more involved and in some situations (as we will see in Chapter 6) can actually be quite difficult to answer. In this chapter we will learn how to answer both types of questions for a special category of routing problems called **Euler circuit problems.**

The routing of a garbage truck, a mail truck, or a patrol car through the streets of a city is a typical example of this type of problem. Other examples might involve routing water and electric meter readers, census takers, newspaper deliverers, tour buses, etc. Whatever the case may be, the common thread in all Euler circuit problems is the need to *traverse all* the streets (roads, lanes, bridges, etc.) within a designated area—be it a whole town or a section of it.

To clarify the concept of an Euler circuit problem, we will introduce several examples of such problems (just the problems for now—their solutions will come later in the chapter).

Example 1. The Walking Patrolman

After a rash of burglaries, a private security guard is hired to patrol on foot the streets of the small neighborhood shown in Fig. 5-1. He parks his car at the corner of Baltic and Maine (*S* in Fig. 5-1). The security guard is anxious to get the job

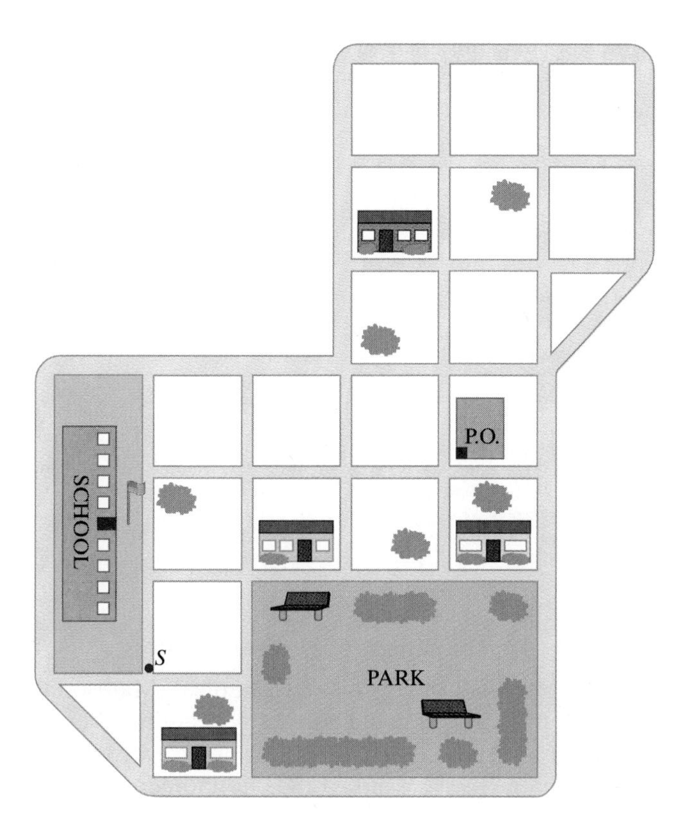

FIGURE 5-1

done with the least amount of walking (he is being paid for just one walk-through). Here are two questions he would like answered: (1) Can he walk through every block once and only once, with the walk starting and ending at the corner where he parked his car(S)? (2) If not, what is the most *efficient* possible way to walk the neighborhood, once again starting and ending at S? Here efficiency is measured in total number of blocks walked. ■

Example 2. The Walking Mail Carrier

Consider now the problem of a mail carrier, who has exactly the same neighborhood as the security guard (Fig. 5-1) as her designated mail delivery area. The big difference is that for those blocks in which there are homes on both sides of the street the mail carrier must walk through the block *twice* (she does each side of the street separately); also the mail carrier needs to start and end her trip at the local Post Office (P.O. in Fig. 5-1). The mail carrier asks two similar questions, since she is also interested in doing her route with the least amount of walking: (1) Starting at the Post Office, can she cover every sidewalk along which there are homes once and only once, ending her walk back at the Post Office? (2) If that can't be done, what is the most efficient (least number of blocks walked) way to deliver the mail throughout the neighborhood? ■

Example 3. The Seven Bridges of Königsberg

Basically, this is the true story with which we opened the chapter—with a little embellishment: A prize (7 gold coins) is offered to the first person who can find a way to walk across each one of the 7 bridges of Königsberg without recrossing any and return to the original starting point (for the reader's convenience we modernized the area map, now shown as Fig. 5-2). A smaller prize (5 gold coins) is offered for anyone who can cross each of the 7 bridges exactly once without necessarily returning to the original starting point.

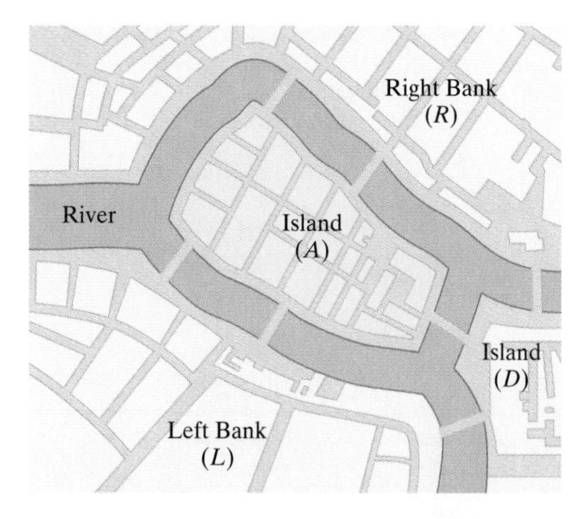

FIGURE 5-2 ■

Example 4. The Bridges of Madison County

This is a more modern version of Example 3. Madison County is a quaint old place, famous for its quaint old bridges. A beautiful river runs through the county, and there are 4 islands (A, B, C, and D) and *eleven* bridges joining the islands to both banks of the river (R and L) and one another (Fig. 5-3). A famous photogra-

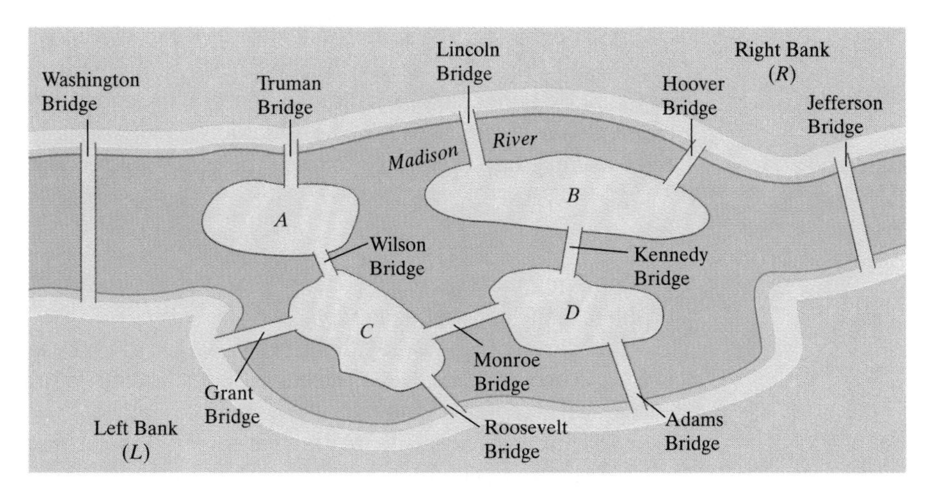

FIGURE 5-3

pher is hired to take pictures of each of the 11 bridges for a national magazine. The photographer needs to drive across each bridge once for the photo shoot. Moreover, since there is a $25 toll (the locals call it a "maintenance tax") every time an out-of-town visitor drives across a bridge, the photographer wants to minimize the total cost of his trip and to recross bridges only if it is absolutely necessary. What is the best (cheapest) route for him to follow? ■

Example 5 Child's Play?

Figure 5-4 shows some simple line drawings. For each drawing we would like to know if it can be *traced* without lifting the pencil or retracing any of the lines, starting and ending the tracing in the same place. What if we are not required to end back at the starting place?[3] Many of us played such games in our childhood (those were the good old days before video games) and may actually know (or can quickly figure out) the answers in the case of Figs. 5-4(a), (b), and (c),[4] but what about more complicated shapes, such as the one in Fig. 5-4(d)? Can this figure be traced without lifting the pencil or retracing any of the lines? If so, how? In general, any unicursal tracing problem is just an abstract example of an Euler circuit problem. ■

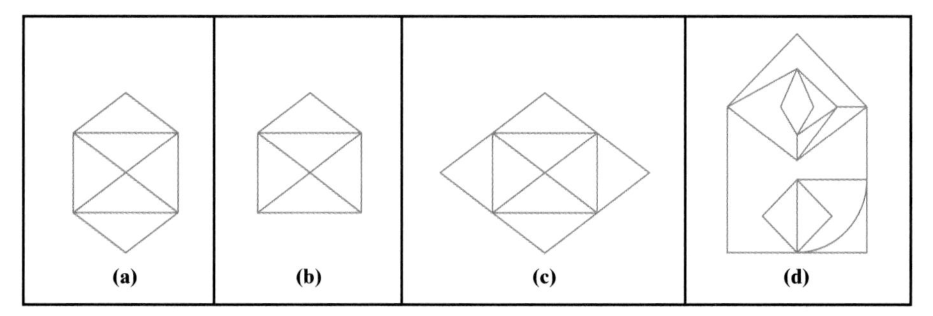

FIGURE 5-4

[3] These kinds of tracings are called **unicursal tracings**—*closed unicursal tracings* if we have to end back where we started, *open unicursal tracings* if we don't.

[4] Figure 5-4(a) has a closed unicursal tracing. Any point can be chosen as the starting and ending point. Figure 5-4(b) does not have a closed unicursal tracing but it does have an open unicursal tracing. The starting point must be one of the two bottom corners—the ending point will be the bottom corner opposite the starting point. Figure 5-4(c) has no possible unicursal tracings whatsoever.

As the preceding examples illustrate, Euler circuit problems can come in a variety of forms. Fortunately, the same basic mathematical theory is used to solve any Euler circuit problem. In the next several sections we will develop the basic elements of this theory. In the last section we will use our newly acquired knowledge to come back and solve some of these examples.

Graphs

The unifying mathematical concept that will allow us to solve any Euler circuit problem is the concept of a *graph.*

For starters, let's just say that a **graph** is a picture consisting of dots (called **vertices**) and lines (called **edges**). The edges do not have to be straight lines (curvy, wavy, etc., are all acceptable), but they always have to connect two vertices. When an edge connects a vertex back with itself (which is also allowed), then it is called a **loop**.

The foregoing is not to be taken as a precise definition of a graph, but rather as an informal description that will help us get by for the time being. To get a feel for what a graph is, let's look at a few examples.

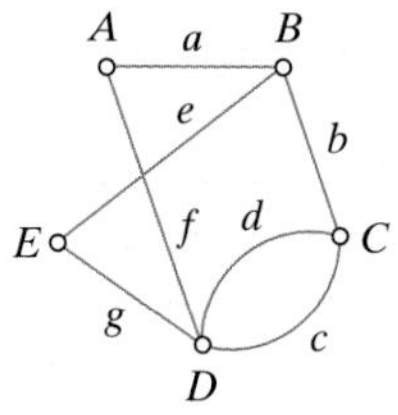

FIGURE 5-5

Example 6.

Figure 5-5 shows an example of a graph. This graph has 5 vertices called *A, B, C, D,* and *E* and 7 edges called *a, b, c, d, e, f,* and *g.* (As much as possible, we will try to be consistent and use upper-case letters for vertices and lower-case letters for edges, but this is not mandatory.) A couple of comments about this graph: First, note that the point where edges *e* and *f* cross is *not* a vertex—it is just the crossing point of two edges. One does not imply the other! Second, note that there is no rule against having more than one edge connecting the same two vertices, as is the case with vertices *D* and *C.* These are called **multiple edges.** ■

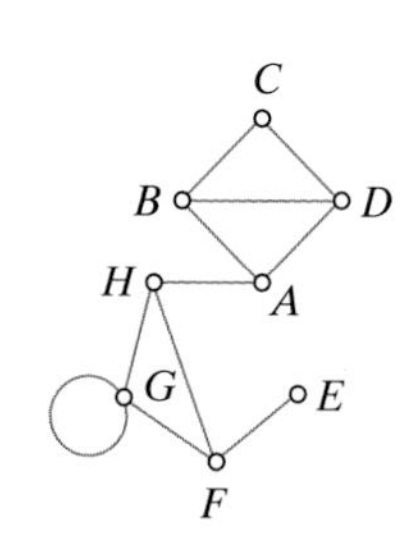

FIGURE 5-6

Example 7.

Figure 5-6 shows a graph with 8 vertices (*A, B, C, D, E, F, G,* and *H*) and 11 unlabeled edges. (There is no rule that says that we have to give names to the edges.) We can still specify an edge by naming the 2 vertices that are its end points. For example, we can talk about the edge *AH,* the edge *BD,* and so on. Note that there is a loop in this graph—it is the edge *GG.* ■

Example 8.

Does Fig. 5-7 show a graph? Yes! It is a graph with 4 vertices and no edges. While it does not make for a particularly interesting graph, there is nothing illegal about it. Graphs without any edges are permissible.[5] ■

FIGURE 5-7

Example 9.

Does Fig 5-8 show a graph? We have vertices and we have edges, so the answer is yes! The vertices have funny names, but so what? Note that the graph is made up of 2 separate, disconnected pieces. Graphs of this type are called *disconnected,* and the individual "pieces" are called the *components* of the graph. (The graph in Fig. 5-7, for example, is disconnected and has 4 components.)

[5] On the other hand, we cannot have a graph without vertices, since then there can be no edges, and without vertices or edges we have nothing!

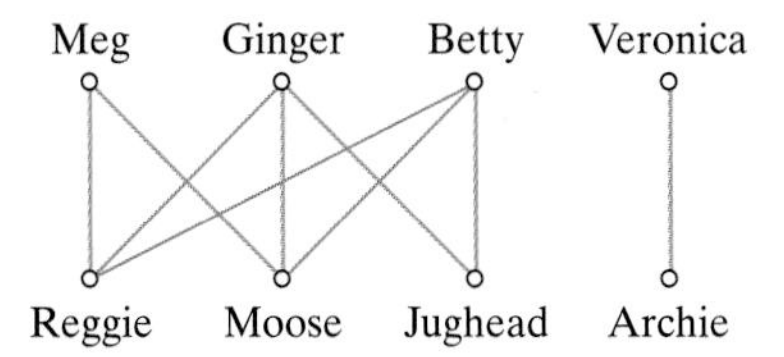

FIGURE 5-8

What might the graph in Fig. 5-8 represent? Let's suppose that Meg, Ginger, Betty, Veronica, Reggie, Moose, Jughead, and Archie are friends who went together to a party. A graph such as the one in Fig. 5-8 might be a pictorial description of who danced with whom at the party. We can learn a few things from such a picture (such as the fact that Veronica and Archie spent the evening dancing together), but most importantly, we should appreciate the fact that the picture provides such a crisp and convenient way to describe the evening's dancing arrangements. ■

Example 10.

We are now going to present a graph without a picture. This graph has 4 vertices (*A, D, L,* and *R*) and 7 edges (*AR, AR, AD, AL, AL, DR,* and *DL*). This information completely specifies the graph. The reader is encouraged to draw a picture of this graph. Where should the vertices be placed? (It doesn't matter!) What shape should the edges have—straight, curved, wiggly? (It doesn't matter either!) Figures 5-9(a) and (b) show two different representations of this same graph. ■

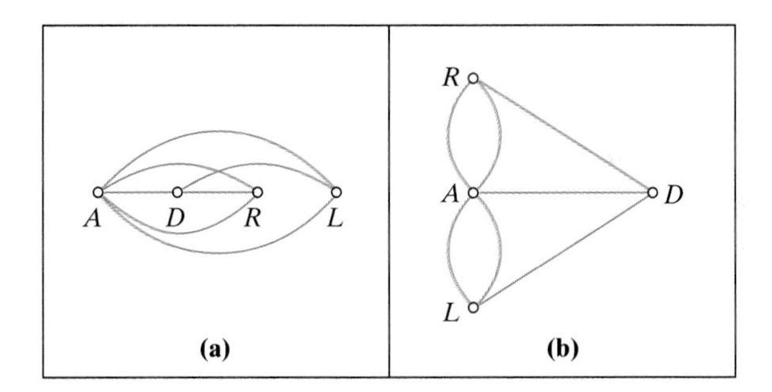

FIGURE 5-9

Example 10 illustrates an important point: A graph can be drawn in infinitely many different ways, and *it is not the shape of the graph that matters, but rather the information telling us which vertices are connected to which other vertices.*

With all of the above examples under our belt, we might be ready for a more formal definition of a graph. A graph is a *structure for describing relationships.* It tells us how a bunch of objects (the vertices) are related to each other. The story of which objects are related to which other objects is told by the edges. That is all the information we get out of a graph—it doesn't seem like much, but it is.

It follows that any time we have a relationship between objects, whatever that relationship might be (love, kinship, dance partner, etc.), *we can describe such a relationship by means of a graph.* This simple idea is the key reason for the tremendous usefulness of graph theory.

Example 11.

On any particular week of the baseball season one can look up the schedule for that week in a good newspaper or a television guide. Here is one week's schedule for the National League East exactly as it would be reported in the newspaper:

- *Monday.* Pittsburgh versus Montreal, New York versus Philadelphia, Chicago versus St. Louis
- *Tuesday.* Pittsburgh versus Montreal
- *Wednesday.* New York versus St. Louis, Philadelphia versus Chicago
- *Thursday.* Pittsburgh versus St. Louis, New York versus Montreal, Philadelphia versus Chicago.
- *Friday.* Philadelphia versus Montreal, Chicago versus Pittsburgh
- *Saturday.* Philadelphia versus Pittsburgh, New York versus Chicago, Montreal versus St. Louis.
- *Sunday.* Philadelphia versus Pittsburgh.

A different way to describe the schedule is by means of a graph. Here the vertices are the teams, and each game played during that week is described by an edge between two teams, as in Fig. 5-10. (Insofar as the description of the schedule is concerned, geography is not an issue, and note that the position of the vertices has nothing to do with the geographic location of the cities.) The main point of this example is to illustrate the convenience of the graph as a way to describe the schedule. Do you want to know how many games Pittsburgh is scheduled to play during the week? Do you want to know if New York plays Pittsburgh during the week? Where would you rather look—the list or the graph? The answer is obvious. ■

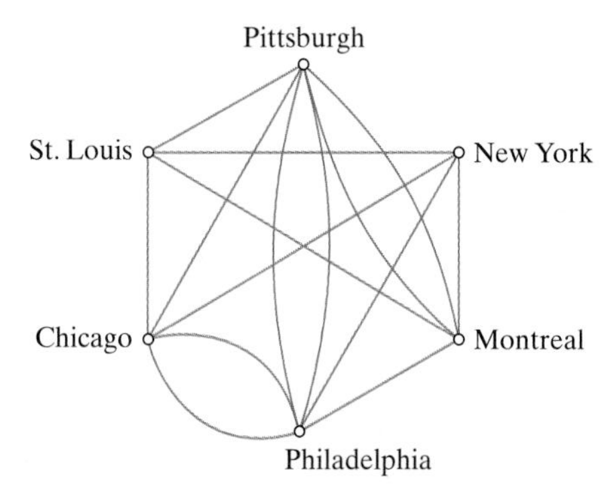

FIGURE 5-10

Graph Concepts and Terminology

Every branch of mathematics has its own peculiar jargon, and the theory of graphs has more than its share. In this section we will introduce a few essential concepts and terms that we will need in the chapter.

- **Adjacent vertices.** Two vertices are said to be **adjacent** if there is an edge joining them. (In this context, adjacent vertices do not have to be "next" to each other.) In the graph shown in Fig. 5-11 vertices E and B are adjacent; D and C are not (even though in the picture they are near each other). Also, because of the loop at E, we can say that vertex E is adjacent to itself.
- **Adjacent edges.** Two edges are **adjacent** if they share a common vertex. In Fig. 5-11, edges AB and AD are adjacent; edges AB and DE are not.
- **Degree of a vertex.** The **degree** of a vertex is the number of edges at that vertex. (A loop contributes twice toward the degree.) In the graph shown in Fig.

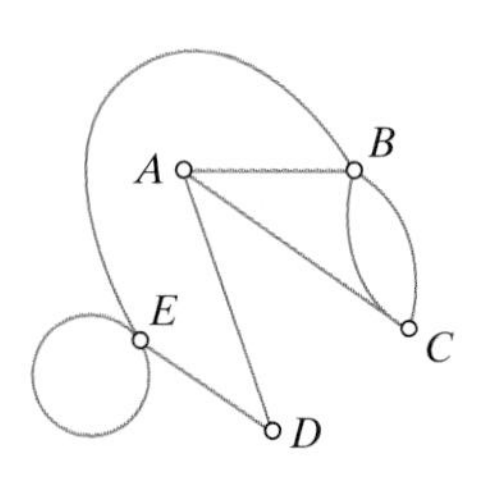

FIGURE 5-11

5-11, vertex A has degree 3 [we can write this as $\deg(A) = 3$], vertex B has degree 4 [$\deg(B) = 4$], $\deg(C) = 3$, $\deg(D) = 2$, and $\deg(E) = 4$ (because of the loop).

- **Paths.** A **path** is a sequence of vertices with the property that each vertex in the sequence is *adjacent* to the next one. Thus, a path can also be thought of describing a sequence of adjacent edges. Whereas a vertex can appear on the path more than once, an edge can be part of a path *only once.* The graph in Fig. 5-11 has many paths—here are just a few examples.

  - *A, B, E, D.* This is a path from vertex *A* to vertex *D,* consisting of edges *AB, BE,* and *ED.*
  - *A, B, C, A, D, E.* This is a path from *A* to *E.* The path "visits" vertex *A* twice, but no edge is repeated.
  - *A, B, C, B, E.* Another path from *A* to *E.* This path is possible because there are two edges connecting *B* and *C.*
  - *A, C, B, E, E, D.* This path is possible because of the loop at *E.*

  The following *are not* paths:

  - *A, C, D, E.* There is no edge connecting *C* and *D.*
  - *A, B, C, B, A, D.* The edge *AB* appears twice, so this is not a path.
  - *A, B, C, B, E, E, D, A, C, B.* In this long string of vertices, everything is OK until the very end, when the edge *CB* appears for a third time. The first two times are fine, because there are two edges connecting *B* and *C*. The third time through is one too many. One of the two edges would have to be retraveled.

- **Circuits.** A **circuit** is a path that starts and ends at the same vertex. Some of the circuits in the graph in Fig. 5-11 are:

  - *A, B, C, A.* Note that this same circuit can also be written as *B, C, A, B* or *C, A, B, C.* A circuit—like a bead necklace—has really no specified start or end. We use the words "starting vertex" and "ending vertex" only when we choose to write circuits in linear form—which is a necessity caused by the conventions of written communication.
  - *B, C, B.* A perfectly legitimate circuit. It could also be written as *C, B, C.*
  - *E, E.* Why not?

  There are several other circuits in Fig. 5-11. (The reader is encouraged to find at least a couple more.)

- **Connected graphs.** A graph is **connected** if any two of its vertices can be joined by a path. This essentially means that it is possible to travel from any vertex to any other vertex along consecutive edges of the graph. If a graph is not connected, it is said to be **disconnected**. A graph that is disconnected is made up of pieces that are by themselves connected. Such pieces are called the **components** of the graph. The graph in Fig. 5-12(a) is connected. The graphs in Figs. 5-12(b) and (c) are disconnected. The one in Fig. 5-12(b) has two components; the one in Fig. 5-12(c) has three.

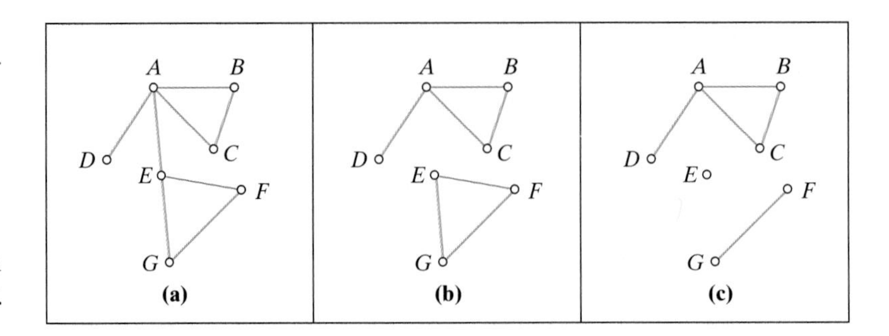

FIGURE 5-12 (a) This graph is connected; (b) graph is not connected—two components; (c) graph is not connected—three components, one of them the isolated vertex E.

- **Bridges.** Sometimes in a connected graph there is an edge such that if we were to erase it, the graph would become disconnected. For obvious reasons such an edge is called a **bridge** (burn a bridge behind you, and you'll never be able to get back to where you were). In Fig. 5-12(a), the edge AE is a bridge because when we remove it, the graph becomes disconnected. [There is another bridge in Fig. 5-12(a). We leave it to the reader to find it.]

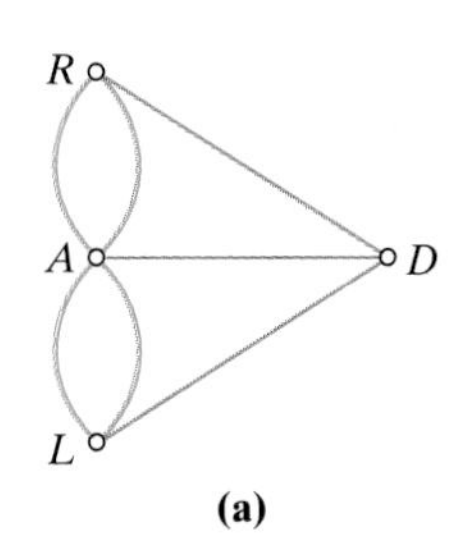

- **Euler paths.** An **Euler path** is a path that "travels" through *every* edge of a connected graph. Since it is a path, edges can only be traveled once, so the long and short of it is that an Euler path travels through every edge of the graph *once and only once*—every edge must be traveled (Euler); no edge can be retraveled (path). The definition of an Euler path should ring a bell—it sounds almost the same as the concept of a unicursal tracing. In fact, they are the same idea. The former is couched in the context of graphs, the latter in the context of ordinary line drawings. Not every graph has an Euler path—it is often the case that the graph cannot be "traced." The graph shown in Fig. 5-13(a) does not have an Euler path. On the other hand, the graph shown in Fig. 5-13(b) has several Euler paths. One of them is $L, A, R, D, A, R, D, L, A$. The reader is encouraged to find at least one more.

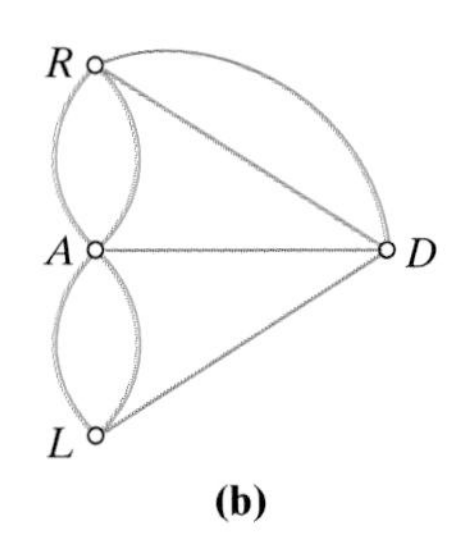

FIGURE 5-13 (a) This graph has no Euler paths; (b) this graph has several Euler paths.

- **Euler circuits.** An **Euler circuit** is a circuit that *travels* through *every* edge of a connected graph. Thus, we have the same requirements as for an Euler path, but we also ask that the starting and ending vertex be the same. (An Euler circuit is essentially the same as a closed unicursal tracing of the graph.)

Graph Models

One of Euler's most important ideas was the observation that certain types of problems can be conveniently rephrased as problems in graphs, and that, in fact, graphs offer the perfect model for describing many real-life situations. The notion of using a mathematical concept to describe and solve a real-life problem is part of one of the oldest and grandest traditions in mathematics. It is called *modeling.* Unwittingly, we have all done simple forms of modeling before, all the way back to elementary school. Every time we turn a "word problem" into an arithmetic calculation, an algebraic equation, or a geometric picture, we are modeling. We can now add to our repertoire one more tool for modeling: graph models.

Example 12. The Seven Bridges of Königsberg: Act 2

Remember that the Königsberg bridges problem as described in Example 3 was to find a walk through the city that crossed each of the bridges once and returned to

A Colorful Problem

How many colors are needed to color any map so that no two neighboring countries are the same color? This well known application of graph theory is known as the *map coloring problem*. The problem may sound simple, but it confounded mathematicians until Appel and Haken used a computer to solve it conclusively in 1976.

Chris Caldwell has written a short tutorial which will help you to understand this problem and its connection to graph theory. You will find it helpful to print out pages from the tutorial, or at least copy the problems onto paper so that you can solve them. The tutorial can be located in the Tannenbaum website under Chapter 5.

And just for fun, you may want to view this site, which uses a Java applet to color in a map of the United States.

Questions:

1. Briefly explain how to make a graph which corresponds to a map. What do the vertices represent? What do the edges represent?
2. Explain why a map in a plane will always correspond to a *planar* graph, that is, a graph which can be drawn on a plane without the edges crossing.
3. Suppose a graph has 8 vertices, each of which is connected to all of the others. How many colors are necessary to color the vertices of this graph so that adjacent vertices have different colors? Explain why your answer does not violate the four color theorem.
4. You can find a map of the counties in Vermont at http://www.lib.utexas.edu/Libs/PCL/MAP_collection/states/Vermont.gif. Sketch a graph that corresponds to the map. Then show how to color the map (or your graph) using four colors. Why is it impossible to color this map with only three colors?

the starting place (good for a prize of 7 gold coins) or doing the same without ending at the starting place (good for 5 gold coins). A stylized map of the city of Königsberg is shown once again in Fig. 5-14(a). The reader is warned that we moved the exact positions and angles of some of the bridges and in general smoothed out some of the details in the original map. Isn't this cheating? Actually not! A moment's reflection should convince us that many things on the original map are irrelevant to the problem: the shape and size of the islands and river banks, the lengths of the bridges, and even the exact location of the bridges, as long as they are still joining the same two sections of the city. Aha! We have just stumbled upon the key observation. Insofar as this problem is concerned, *the only thing that truly matters is the relationship between land masses and bridges:* which land mass is connected to which other land mass and by how many bridges [Fig. 5-14(b)]. Thus, when we strip the map of all its superfluous information, we end up with the graph shown in Fig. 5-14(c), where the vertices represent the 4 land masses and the edges represent the 7 bridges. In this new interpretation of the puzzle a stroll around the town that crosses each bridge once and goes back to the starting point can be described by an *Euler circuit* of the graph; a stroll that crosses each bridge once but does not return to the starting point corresponds to an *Euler path.* ■

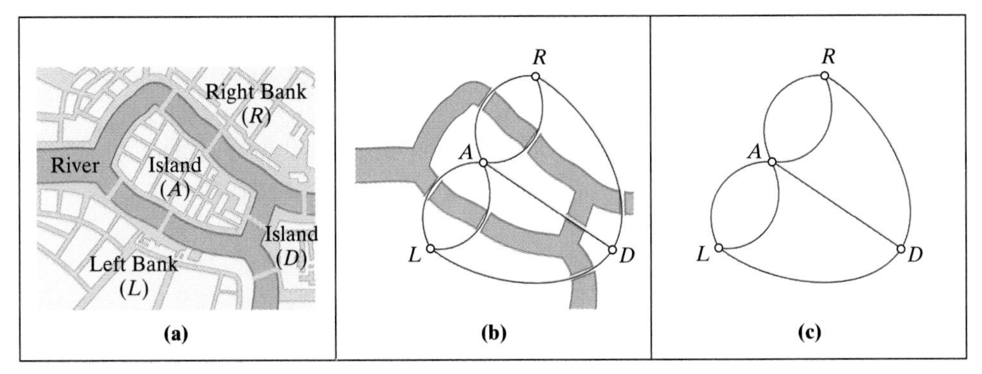

FIGURE 5-14

As big moments go, this moment may not seem like much, but the reader is encouraged to take stock of what we have accomplished in Example 12 (and, if necessary, reread the example carefully), because we have actually taken a big step—we made a connection between theory and reality.

Example 13. The Walking Patrolman: Act 2

In Example 1, we discussed the problem of a security guard who needs to walk the streets of the neighborhood shown in Fig. 5-15(a). A graph model of this problem is given in Fig. 5-15(c), with each block of the neighborhood represented by an edge in the graph, and each intersection represented by a vertex of the graph. ■

Example 14. The Walking Mail Carrier: Act 2

As the reader may recall, the mail carrier's problem of Example 2 differed from the security guard's problem in that, except for the blocks facing the school and the park, the mail carrier must deliver mail on both sides of the street which means she has to walk each side of the street separately. Consequently, an appropriate

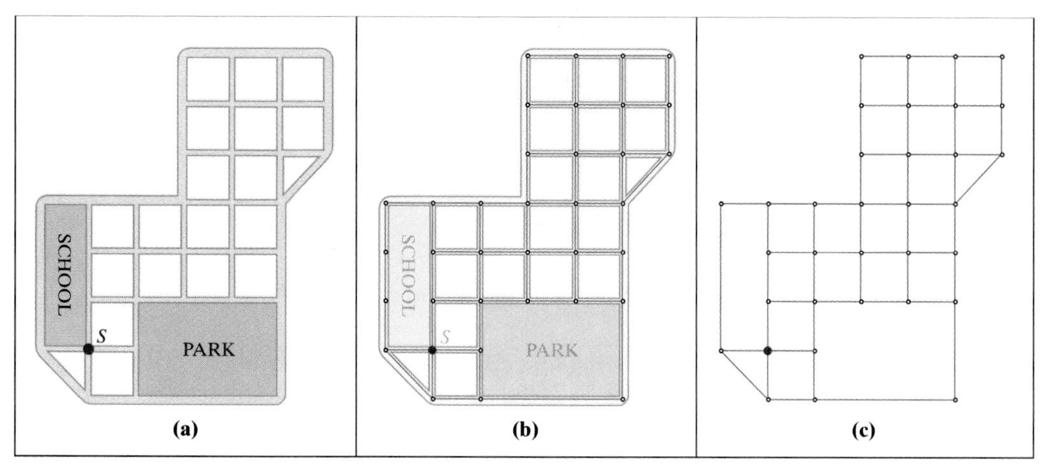

FIGURE 5-15

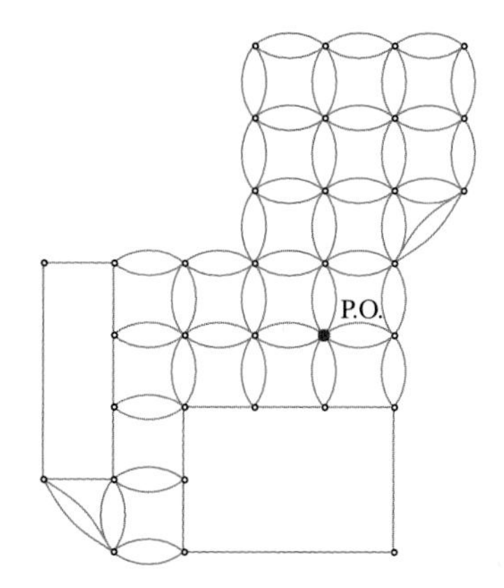

FIGURE 5-16

graph model for this problem should have one edge for every side of the street in which the mail has to be delivered, as shown in Fig. 5-16. ■

Euler's Theorems

The Königsberg bridges problem, as modeled in Example 12, was equivalent to finding an Euler circuit or an Euler path in the corresponding graph. Euler's solution to the problem was to demonstrate that neither an Euler circuit nor an Euler path is possible.

Let's start with the Euler circuit argument. Why is such a circuit impossible? Let's say for the sake of argument that the starting vertex is *L*. (Since we are looking for a round trip, it makes no difference which vertex we pick for the starting point.) Somewhere along the way we will have to go through *A*, and in fact, we will have to go through *A* more than once. Let's count exactly how many times. The first visit to *A* will use up two edges (bridges) (one getting there and a different one getting out); the second visit to *A* will use up two other edges; and the third visit to *A* will use up two more. Oops! There are only five edges to get in and out of *A*, so two visits to *A* won't do (there would be an untraveled bridge) and three visits are too many (we would have to recross one of the bridges). It follows that the walk is impossible! It's the odd number of edges at A (or anywhere else) that causes the problem. The argument can be extended and made general in a very natural way. We present it without any further ado.

Euler's Theorem 1

(a) If a graph has *any* vertices of odd degree, then it *cannot* have an Euler circuit.

(b) If a graph is *connected* and *every* vertex has even degree, then it has at least one Euler circuit (usually more).

Note that having every vertex of even degree is not enough to guarantee an Euler circuit (Fig. 5-17) unless the graph is also connected.

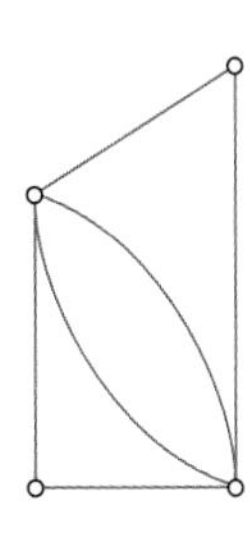
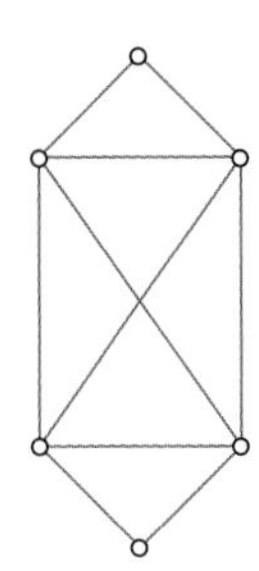

FIGURE 5-17

What do we need for a graph to have an Euler path? Similar arguments work for all the vertices (they must have even degree) except for the *starting* and *ending* vertices of the path which are special. The starting vertex requires *one edge to get out at the start* and two more for each visit through that vertex, so it must have *odd* degree. Likewise the *ending* vertex must have *odd* degree (two edges for every visit plus one more to come into the vertex at the end of the trip). Thus, we have

Euler's Theorem 2

(a) If a graph has *more than two* vertices of odd degree, then it cannot have an Euler path.

(b) If a graph is connected and has just *two* vertices of odd degree, then it has at least one Euler path (usually more). Any such path must start at one of the odd-degree vertices and end at the other one.

Example 15. The Bridges of Königsberg: Conclusion

We now know that the graph that models this problem (Fig. 5-18) has four vertices of odd degree, and thus neither an Euler circuit nor an Euler path is possible, *There is no possible way anyone can walk across all of the bridges without having to recross some of the bridges!* We leave it as an exercise to the reader to verify that a walk around old Königsberg that crosses all of the bridges is possible by *recrossing just one bridge* if we are allowed to end the walk at a place other than the starting place or, if we needed to end back at the starting point, by *recrossing just two bridges.* (See Exercise 41.) ■

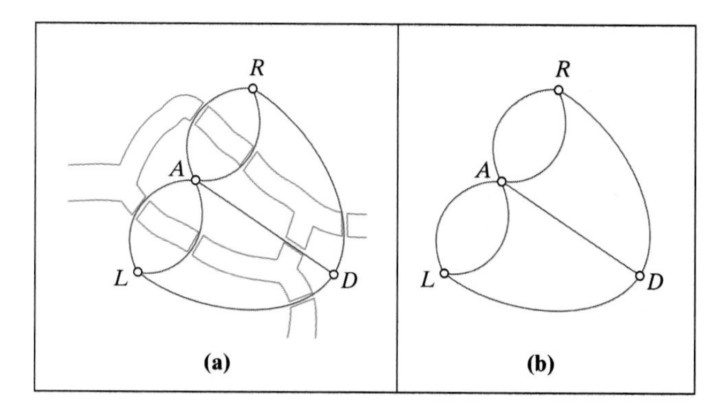

FIGURE 5-18

Example 16. Unicursal Tracings Revisited

Figure 5-19 shows four graphs. These graphs are equivalent to the line drawings in Example 5 (Fig. 5-4) (we have just put vertices at every corner and turned the drawings into graphs). We can now easily apply Euler's theorems: The graph in Fig. 5-19(a) has an Euler circuit (and thus a closed unicursal tracing), because all vertices have even degree; the graph in Fig. 5-19(b) has an Euler path (open unicursal tracing) which must start at *D* and end at *C* (or vice versa), because *D* and *C* are the only two vertices of odd degree; the graph in Fig. 5-19(c) has neither an Euler path nor an Euler circuit (there are too many vertices of odd degree); and finally the more complicated graph in Fig. 5-19(d) does have a closed unicursal tracing, since we can verify with a quick check that every vertex is of even degree. ■

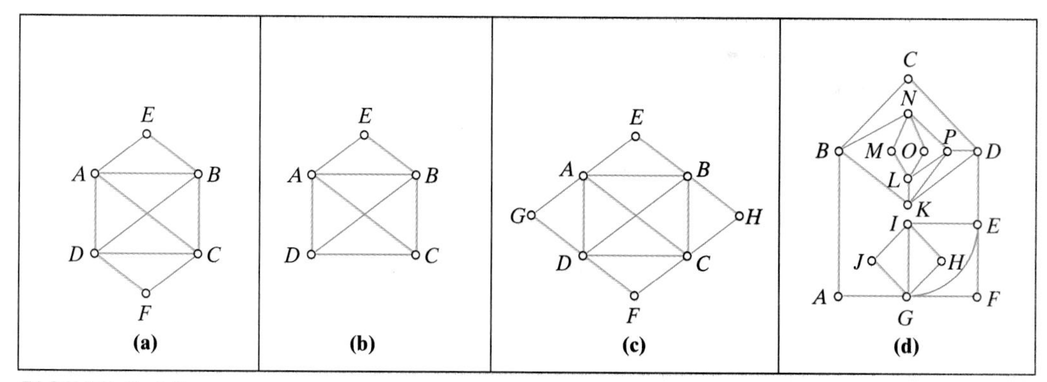

FIGURE 5-19

The careful reader may have noticed that there is an apparent gap in Euler theorems 1 and 2. The two theorems together cover the cases of graphs with *zero* vertices of odd degree (Theorem 1), *two* vertices of odd degree [Theorem 2(b)], and *more than two* vertices of odd degree [Theorem 2(a)]. What happens if a graph has *just one* vertex of odd degree? What then? Didn't Euler consider this possibility? It turns out that he did, but found that it is impossible for a graph to have *just one vertex of odd degree.*

The key observation that Euler made was that when *the degrees of fall the vertices of a graph are added up, the total is exactly twice the number of edges in the graph.* (Think about it: An edge—let's say *XY*—contributes once to the degree of vertex X and once to the degree of vertex Y, so all in all that edge contributed twice to the sum of the degrees. If still unconvinced, the reader is referred to Exercise 38.) One of the important consequences of Euler's observation is that when the degrees of all the vertices of a graph are added up, *the total must be an even number,* which means therefore that it is impossible to have only one vertex of odd degree. In fact, we can push the logic one step further, and argue just as well that it is impossible for a graph to have 3, 5, 7, . . . vertices of odd degree. We summarize the preceding observations into a theorem.

Euler's Theorem 3

(a) The sum of the degrees of all the vertices of a graph equals twice the number of edges (and therefore it must be an even number).

(b) The number of vertices of *odd* degree must be *even.*

Fleury's Algorithm

Euler's theorems give us an easy way to determine if a graph has an Euler circuit or an Euler path. Unfortunately, Euler's theorems are of no help in finding the actual Euler circuit or path, if there is one. Of course, for simple graphs such as the ones shown in Figs. 5-19(a) and (b), one can find an Euler path (or circuit) by simple trial and error. But what about the graph in Fig. 5-19(d)? Or an even more complicated graph, with hundreds of vertices and edges? Do we really want to use trial and error to find an Euler circuit or an Euler path? Of course not. A trial-and-error approach in a large graph is a crapshoot—we could get lucky and find the solution right away, or we could spend hours chasing up dead ends. What we really need here is an *algorithm.*

An **algorithm** is a set of mechanical rules that, when followed, are guaranteed to produce an answer to a specific problem. The fact that the rules making up an algorithm are mechanical means that there is no thinking involved—that's why mindless but efficient things like computers are ideally suited to carrying out algorithms. For human beings, the difficulties in carrying out algorithms (once the rules are understood) are not intellectual but rather procedural. Accuracy and fastidious attention to detail are the key virtues when carrying out the instructions in an algorithm. We offer this as a piece of friendly advice, because most of the practical things we will do in this part of the book will require the ability to correctly carry out algorithms—an ability acquired primarily through practice.

Our next major accomplishment will be to learn an algorithm for finding an Euler circuit in a connected graph in which all the vertices have even degree. The algorithm we will learn is called **Fleury's algorithm.** Because this is our first encounter with a graph algorithm, we will begin by describing it informally, work out a couple of examples, and then give the formal description.

The basic philosophy behind Fleury's algorithm is quite simple, and it can be summarized by paraphrasing an old piece of folk wisdom: *don't cross a bridge until you have to.* The only thing we have to be careful about is in our interpretation of the word *bridge.*

We know that when talking about graphs, a bridge is an edge whose removal disconnects the graph, and Fleury's algorithm specifically instructs us to travel along such edges only as a last resort. Simple enough, but there is a rub: The graph whose bridges we are supposed to avoid is not necessarily the original graph of the problem, but rather that part of the original graph which has yet to be traveled. The point is this: Each time we travel along an edge, we are done with it! We will never cross it again, and from that point on, as far as we are concerned, it is as if that edge never existed. Our concerns lie only on how we are going to get around the yet-untraveled part of the graph. Thus, when we talk about bridges that we want to leave as a last resort, we are really referring to *bridges of the untraveled part of the graph.*

Since each time we traverse an edge the *untraveled part* of the graph changes (and consequently so do the bridges), Fleury's algorithm requires some careful bookkeeping. This does not make the algorithm difficult; it just means that we must take extra pains in separating what we have already done from what we yet need to do.

While there are many different ways to accomplish this (and readers are certainly encouraged to invent their own), a fairly reliable way goes like this: We start with two separate copies of the graph, copy 1 for making decisions and copy 2 for record keeping. Every time we traverse another edge, we erase it from copy 1 but mark it (say in red) and label it with the appropriate number on copy 2. As we progress along our Euler circuit, copy 1 gets smaller and copy 2 gets redder. Copy 1 helps us decide where to go next; copy 2 helps us reconstruct our trip (just in case we are asked to demonstrate how we did it!). Let's try a couple of examples.

Example 17.

The graph in Fig. 5-20 has an Euler circuit—we know this is so because every vertex has even degree. Let's use Fleury's algorithm to find an Euler circuit. Granted, this is a bit of an overkill (this is a very simple graph which could be done easily by trial and error), but the real purpose of this example is to help us understand how the algorithm works.

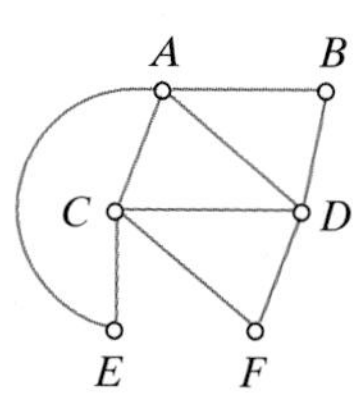

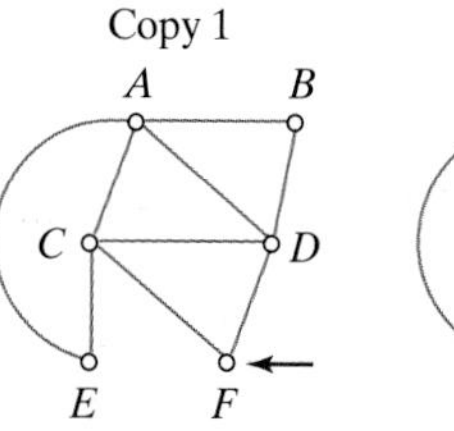

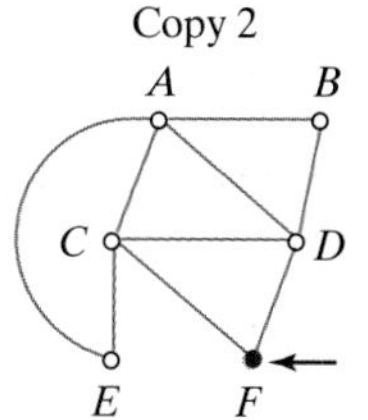

Start: We can pick any starting point we want. Let's say we start at *F*.

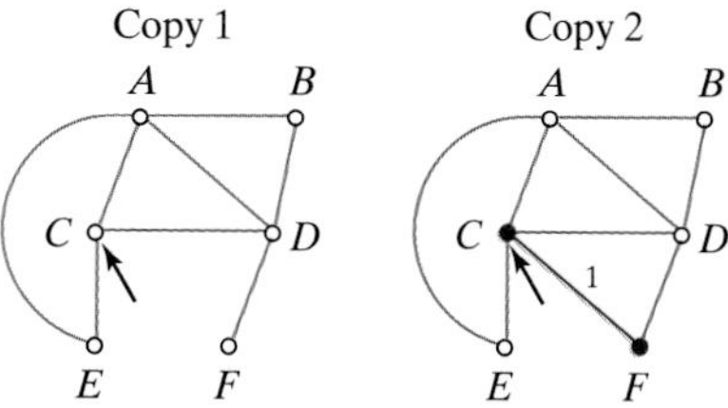

Step 1: Travel from *F* to *C*.
(Could have also gone from *F* to *D*.)

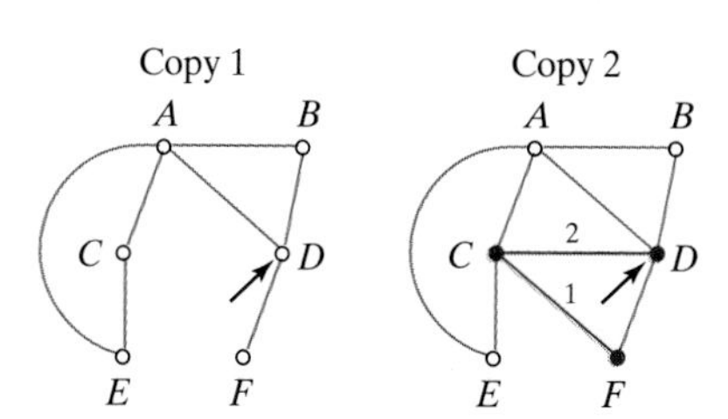

Step 2: Travel from *C* to *D*.
(Could have also gone to *A* or to *E*.)

FIGURE 5-20

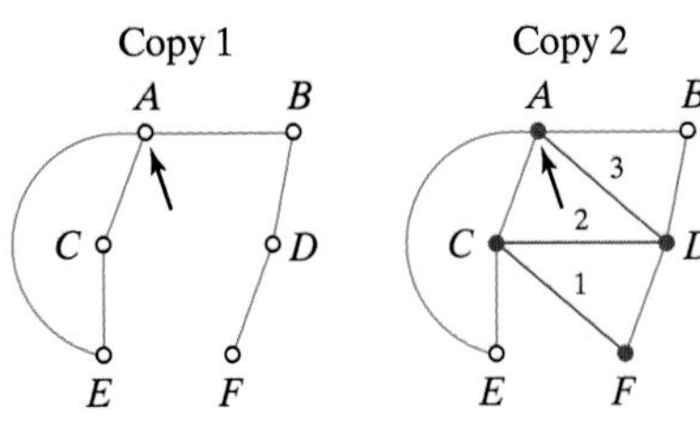

Step 3: Travel from D to A.
(Could have also gone to B but not to F—DF is a bridge!)

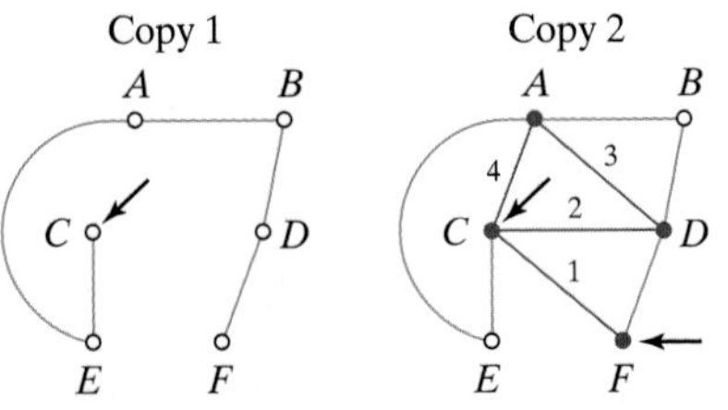

Step 4: Travel from A to C.
(Could have also gone to E but not to B—AB is a bridge!)

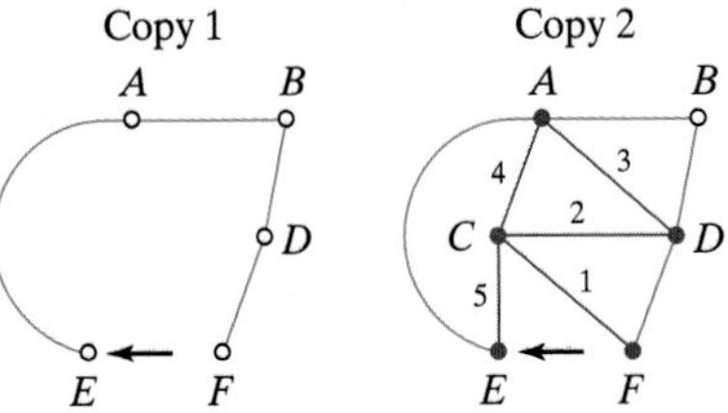
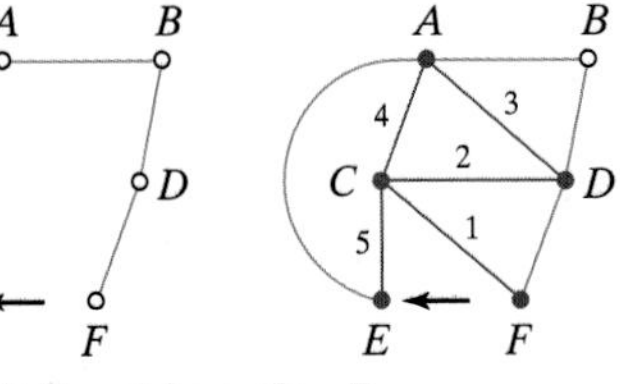

Step 5: Travel from C to E.
(There is no choice!)

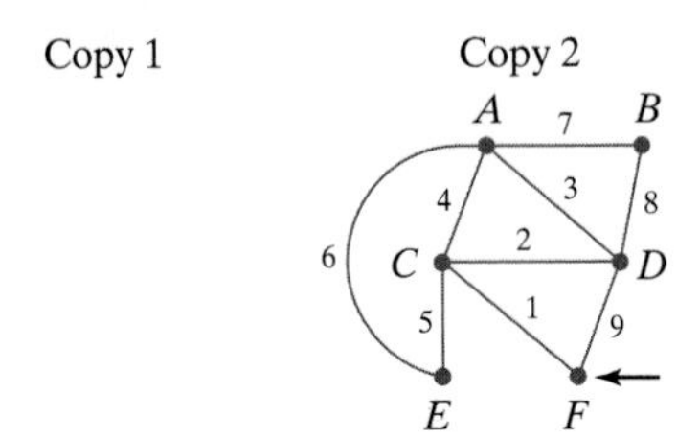

Steps 6, 7, 8, and 9: Only one way to go at each step.

FIGURE 5-20 (cont.)

■

EXAMPLE 18.

We have already confirmed that the graph in Fig. 5-21 has an Euler circuit; thus, it is possible to find a closed unicursal tracing of the picture. Let's do it! Since it would be a little impractical to show each step of Fleury's algorithm with a separate picture as we did in Example 17, we ask the reader to do some of the work. If you haven't already done so, then, get a pencil, an eraser, and some paper. Next, make two copies of the graph. Ready? Let's go!

- **Start:** Pick an arbitrary starting point, say J.
- **Step 1:** From J we can go to either I or G. Since neither JI nor JG is a bridge, we can choose either one. Say we choose JI. (We can now erase JI on copy 1; mark and label it with a 1 on copy 2.)
- **Step 2:** From I we can go to E, H, or G. Any of these choices is OK. Say we choose IH. (Now erase IH from copy 1 and mark and label it on copy 2.)
- **Step 3:** From H there is only one way to go, and that's to G. [Erase edge HG as well as vertex H (we won't be coming back to it) from copy 1 and mark and label it on copy 2.]
- **Step 4:** From G we have several choices (to A, to F, to E, to I, or to J). We should not go to J—GJ is a bridge of copy 1. Any of the other choices is OK. Say we choose GF. (Erase edge GF from copy 1, etc.)
- **Step 5:** There is only one way to go from F (to E). (Erase edge FE as well as vertex F from copy 1, etc.)
- **Step 6:** From E we have three choices, all of which are OK. Say we choose ED. (You know what to do with copy 1 and copy 2. To speed things up, from here on we will omit this part.)

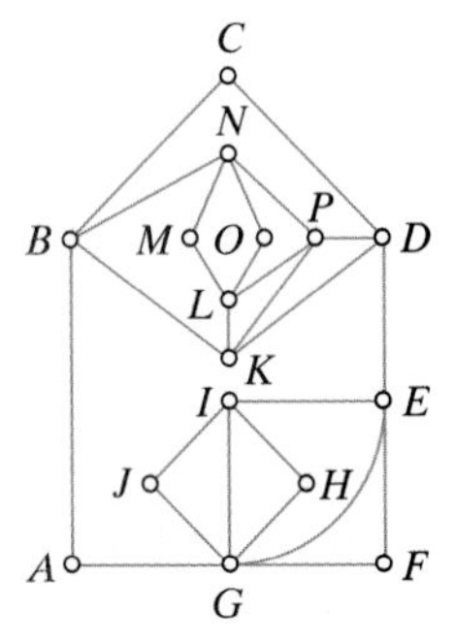

FIGURE 5-21

- **Step 7:** Three choices at *D*. All of them are OK. Say we choose *DP*.
- **Step 8:** Several choices at *P*. All of them are OK. Say we choose *PK*.
- **Step 9:** Several choices at *K*. All of them are OK. Say we choose *KB*.
- **Step 10:** Several choices at *B*. One of them is not OK (edge *BA* is a bridge in copy 1). Say we choose *BN*.
- **Step 11:** Several choices at *N*. All of them are OK. Say we choose *NO*.
- **Step 12:** Only one way to go (to *L*).
- **Step 13:** Several ways to go, but one of them (*LK*) is a bridge in copy 1. Say we choose *LP*.
- **Steps 14–22:** Only one way to go in each step. From *P* to *N* to *M*, to *L*, to *K*, to *D*, to *C*, to *B*, to *A*, to *G*.
- **Step 23:** We do have some choices at *G*, but one of them (*GJ*) is a bridge in copy 1. Say we choose *GI*.
- **Steps 24–26:** From *I* to *E* to *G* to *J*.

  We are finished! The Euler circuit we found is

  J, I, H, G, F, E, D, P, K, B, N, O, L, P, N, M, L, K, D, C, B, A, G, I, E, G, J.

  Needless to say, this is just one of many possible Euler circuits. ■

Here is a formal description of the basic rules for Fleury's algorithm.

Fleury's Algorithm for Finding an Euler Circuit

1. First make sure that the graph is connected and all the vertices have even degree.
2. Start at any vertex.
3. Travel through an edge if (a) it is not a bridge for the untraveled part, or (b) there is no other alternative.
4. Label the edges in the order in which you travel them.
5. When you can't travel any more, stop. (You are done!)

When a connected graph has exactly two vertices of odd degree, then we know that the graph does not have an Euler circuit, but it does have an Euler path, and we can find such a path using Fleury's algorithm with one minor change: In (2), *the starting point must be one of the vertices of odd degree.* Other than that, the rest of the steps [(3), (4), and (5)] are exactly the same. When they are followed properly, the trip is guaranteed to end at *the other vertex of odd degree.*

Eulerizing Graphs

We now know that when a graph has no vertices of odd degree or two vertices of odd degree, then it has an Euler circuit or an Euler path, respectively, and that when a graph has more than two vertices of odd degree, then there is no Euler circuit or Euler path. In this case there is no possible way that we can cover all the edges of the graph without having to recross some of the edges.

The question we now want to deal with is: How do we go about finding a trip that covers all the edges of the graph while recrossing the least possible number of edges? This is important because, in many real-world routing problems, there is a cost that is proportional to the amount of travel; thus, the most efficient routes are those with the least amount of wasted travel (usually called **deadhead travel**), which in this case means the least amount of duplication of edges.

The following simple rule summarizes the preceding observations:

Total cost of route = cost of traveling original edges in the graph
+ cost of deadhead travel.

Example 19.

Consider the graph in Fig. 5-22(a). Since it has 8 vertices of odd degree (*B, C, E, F, H, I, K,* and *L*—shown as red vertices), the graph has no Euler circuit or Euler path. By throwing in another copy of edges *BC, EF, HI,* and *KL* we get the graph in Fig. 5-22(b), a close cousin to the original graph The main difference between the two graphs is that 5-22(b) has an Euler circuit. Figure 5-22(c) shows one such Euler circuit (just travel the edges as numbered). The Euler circuit in Fig. 5-22(c) can be "reinterpreted" as a trip along the edges of our original graph as shown in Fig 5-22(d). In this trip we are traveling along all of the edges of the graph, but we are retracing 4 of the edges (*BC, EF, HI,* and *KL*). While this is not an Euler circuit for the original graph, it a circuit describing the most *efficient* trip (meaning a trip with the least amount of duplication) that covers all of the edges—an *optimal* such circuit, if you will. ■

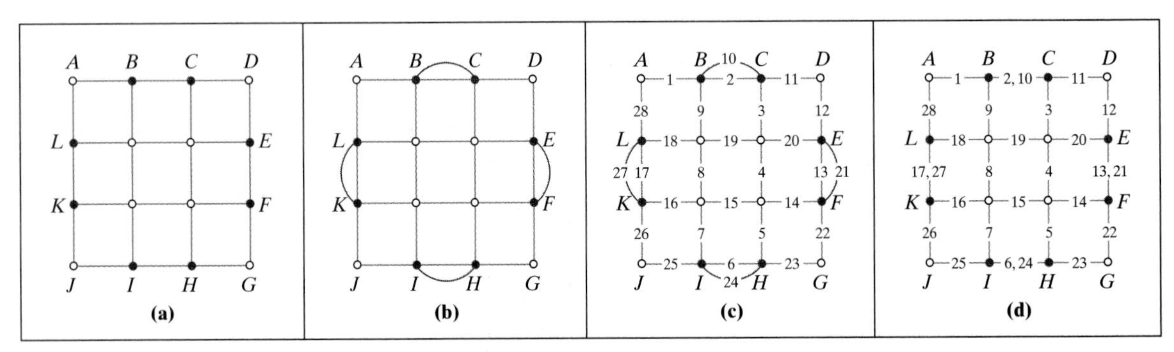

FIGURE 5-22

Before we go on to the next example, let's introduce some convenient terminology. What is the connection between the graphs in Fig. 5-22(a) and 5-22(b)?A close look shows us that Fig. 5-22(b) is the result of applying the following straightforward process to the graph in Fig. 5-22(a): Add extra edges in such a way that the vertices of odd degree become vertices of even degree (in other words, neutralize the "bad guys"). This process of changing a graph so that the vertices of odd degree are eliminated by adding additional edges is called **eulerizing** the graph. There is one thing we must be careful about: The edges that we add *must be duplicates of edges that are already in the graph.* (Remember that the point of all of this is to cover the edges of the existing graph in the best possible way without creating any new edges.) Our next example, clarifies this point.

Example 20.

Consider the graph in Fig. 5-23(a). This graph has 12 vertices of odd degree, as shown in the figure. If we want to travel along all the edges of this graph and come back to our starting point, we know we are going to have to double up on some of the edges. Which ones? The answer is provided by first eulerizing the graph, so let's discuss ways in which we can do this.

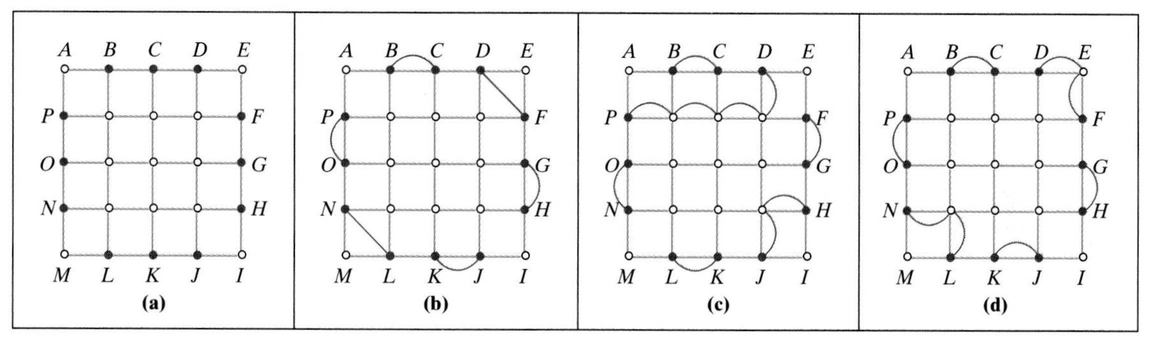

FIGURE 5-23

Figure 5-23(b) shows how *not to do it!* Adding the edges *DF* and *NL* is not allowed, since those edges were not in the original graph.

Figure 5-23(c) shows a legal, but wasteful, eulerization of the original graph. It is legal because we have eliminated all the vertices of odd degree by adding edges that duplicate already-existing edges, but it is wasteful because it is obvious that we could have accomplished the same thing by adding fewer duplicate edges. If there is a cost to traveling edges, we don't want to duplicate any more edges than is absolutely necessary!

Figure 5-23(d) shows an *optimal eulerization* of the original graph—one of several possible. This eulerization is optimal because it has the fewest possible duplicate edges (8). An optimal eulerization gives us the blueprint for an optimal round trip along the edges of the original graph. In this case we know that we are going to have to retrace exactly 8 of the edges, and in fact we know exactly which ones. Figure 5-24 shows an actual example of an optimal trip (just follow the numbers) obtained using Fleury's algorithm on Fig. 5-23(d), and we can clearly see exactly which edges are being retraced.

FIGURE 5-24 ■

Example 21.

We will now consider a simple variation of the problem in Example 20. The graph, shown in Fig. 5-25(a), is exactly the same as in Fig. 5-23(a). The name of the game is, once again, to travel the edges of this graph while duplicating the fewest possible edges. The difference is that this time we are not required to start and end in the same place. In this case we do what is called a *semi-eulerization* of the graph: We duplicate as many edges as needed to eliminate all the vertices of odd degree *except for two,* which we allow to remain odd, since we can use them as the starting and ending points of our travels. Figure 5-25(b) shows an optimal semi-eulerization of the graph in 5-25(a), with vertices *D* and *F* being the 2 vertices that remain of odd degree—all the other vertices that were originally of odd degree (*B, C, G, H, J, K, L, N, O,* and *P*) are of even degree. The semi-eulerization in Fig. 5-25(b) tells us the following: it is possible to travel all the edges of the graph in Fig. 5-25(a) by starting at *D* and ending at *F* (or vice versa) and duplicating only 6 of the edges (*BC, GH, KJ, LM, MN,* and *OP*). Of course, there are many other ways to accomplish this (but none that does it with less than 6 duplicate edges). The reader is encouraged to find a different semi-eulerization of the graph in Fig. 5-25(a) (Exercise 44).

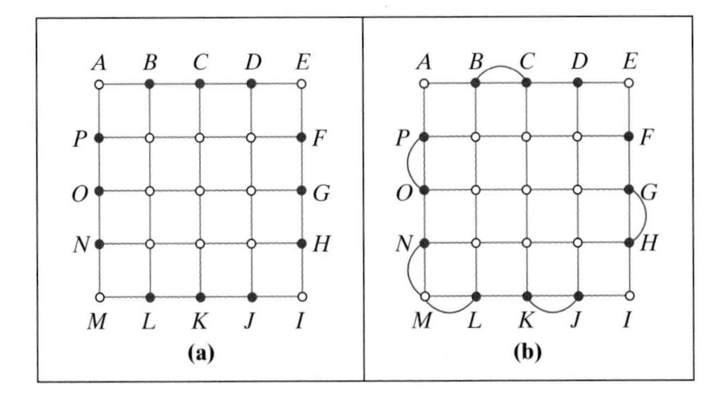

FIGURE 5-25 ■

Example 22. The Bridges of Madison County: Conclusion

A graph model for this problem is shown in Fig. 5-26, with each island and bank a vertex and each bridge an edge. The graph has 4 vertices of odd degree (*R, L, B,* and *D*), so some bridges are definitely going to have to be recrossed. The photographer plans to drive into the area, take his photographs of the bridges, and leave, so that he does not need to start and end in the same spot. Thus, the ideal route involves an optimal semi-eulerization of the graph, which leaves vertices *R* and *L* with odd degrees (they will be the starting and ending points). This can be easily accomplished by duplicating the edge *BD*. This gives us the final solution to the problem: It is possible to start a trip at *R,* cross each of the bridges once (except for the Kennedy Bridge, which will have to be crossed twice), and end the trip at *L*. The total cost of this trip (in bridge tolls) is $300 (12 bridge crossings at $25 each). ■

Example 23. The Walking Patrolman; Conclusion

We have already discussed the graph model for this problem (Example 13) Figure 5-27(a) shows the graph, with the 18 vertices of odd degree shown in red. Given that there are lots of vertices of odd degree, an Euler circuit or Euler path is out of

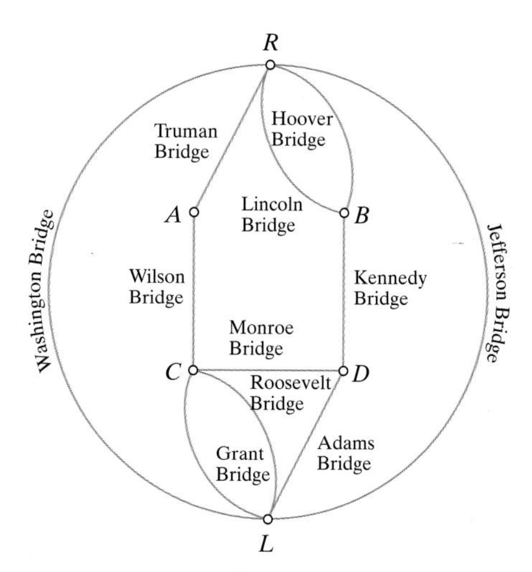

FIGURE 5-26

the question—the security guard is going to have to retrace some of his steps. We can, however, determine the optimal route that starts and ends at S by finding the optimal eulerization of the graph. This is shown in Fig. 5-27(b). The figure now tells us that the most efficient possible route will require that the patrolman double up on the 9 blocks where an extra red edge has been added. An actual optimal route (there are several) can be obtained using Fleury's algorithm or just trial and error on the graph in 5-27(b). One such route is shown in Fig. 5-27(c).

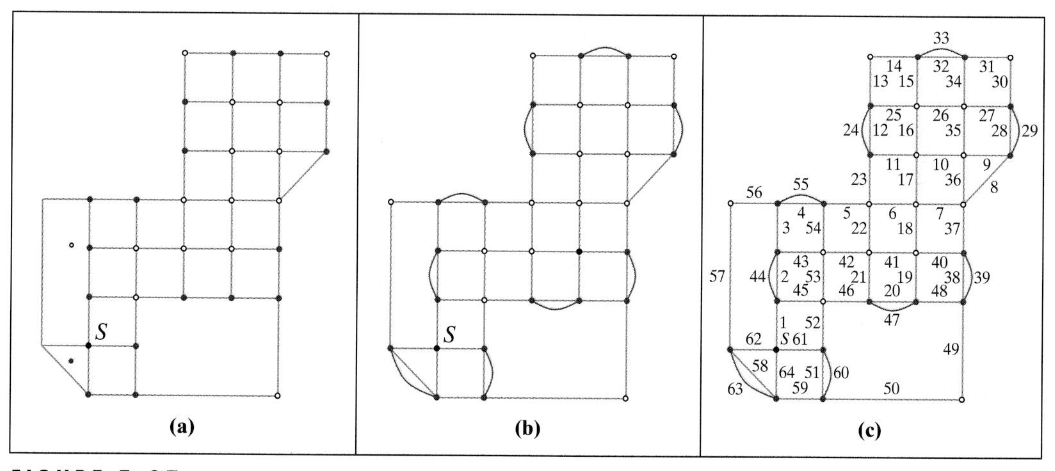

FIGURE 5-27

■

Example 24. The Walking Mail Carrier: Conclusion

We have already found that the graph model for the mail carrier's problem is the one shown in Fig. 5-28 (see Example 14). Surprisingly, this graph has all its vertices of even degree, which means that the graph has an Euler circuit and therefore that the mail carrier will not have to waste any steps if she chooses her route carefully. The actual route, which must start and end at the Post Office, can be found using Fleury's algorithm, or just common sense and some trial and error. The actual details are best left to the interested parties (Exercise 45).

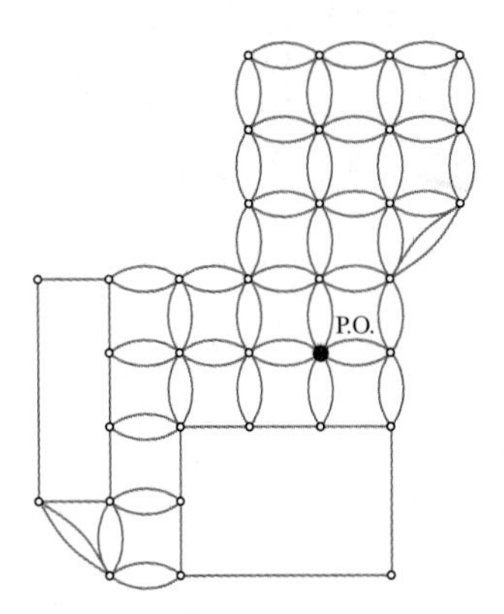

FIGURE 5-28

Conclusion

In this chapter we got our first introduction to three fundamental ideas. First, we learned about a simple but powerful concept for describing relational information between objects: the concept of a *graph.* This idea can be traced back to Euler, more than 250 years ago. Since then, the study of graphs has grown into one of the most important and useful branches of modern mathematics.

The second important idea of this chapter is the concept of a graph *model.* Every time we take a real-life problem and turn it into a mathematical problem, we are doing mathematical modeling. When our ancestors first started using their fingers to count things, they were in a sense, doing a very crude form of mathematical modeling. We have all done some form of mathematical modeling at one time or another: using arithmetic in elementary school, using equations and geometric figures (algebraic and geometric modeling) in high school. In this chapter we learned about a new type of modeling called graph modeling, in which we use graphs and the mathematical theory of graphs to solve real-life problems.

By necessity, the problems that we solved in this chapter were fairly simplistic—the Königsberg bridge problem and some Euler circuit problems, such as the patrolman and the mail-carrier routing problems—but we should not be deceived by the simplicity of our examples. In many big cities, where the efficient routing of municipal services (police patrols, garbage collection, etc.) is a significant problem, the same theory that we developed in this chapter is being used on a large scale, the only difference being that many of the more tedious details are mechanized and carried out by a computer. In New York City, for example, garbage collection, curb sweeping, snow removal, and other municipal services have been scheduled and organized using graph models since the 1970s, and the improved efficiency has yielded savings estimated in the tens of millions of dollars a year.

The third important concept we encountered in this chapter is that of an *algorithm*—a mechanical set of rules that, when followed, provide the solution to certain types of problems. Perhaps without even realizing it, we had our first exposure to algorithms in elementary school, when we learned how to add, multiply, and divide numbers following precise and exacting procedural rules. In this chapter we learned about *Fleury's algorithm,* which helps us find an Euler circuit or an Euler path in a graph. In the next few chapters we will learn many other *graph algorithms,* some quite simple, others a bit more complicated. When it comes to algorithms of any kind, be they for doing arithmetic calculations or for finding circuits in graphs, there is one standard piece of advice that always applies: *practice makes perfect.*

Key Concepts

adjacent edges
adjacent vertices
algorithm
bridge
circuit
connected graph
degree of a vertex
disconnected graph
edge
Euler circuit
Euler circuit problem
eulerizing a graph
Euler path
Euler's theorems
Fleury's algorithm
graph
graph model
loop
multiple edges
path
routing problems
semi-eulerization
vertex

Exercises

Walking

1. For each of the following graphs, list the vertices and edges and find the degree of each vertex.

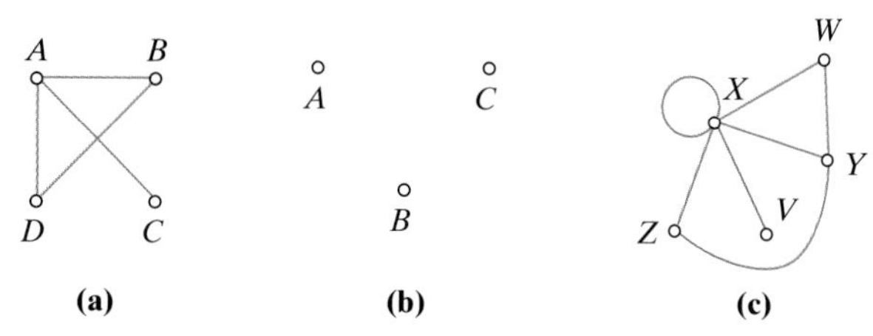

2. For each of the following graphs, list the vertices and edges and find the degree of each vertex.

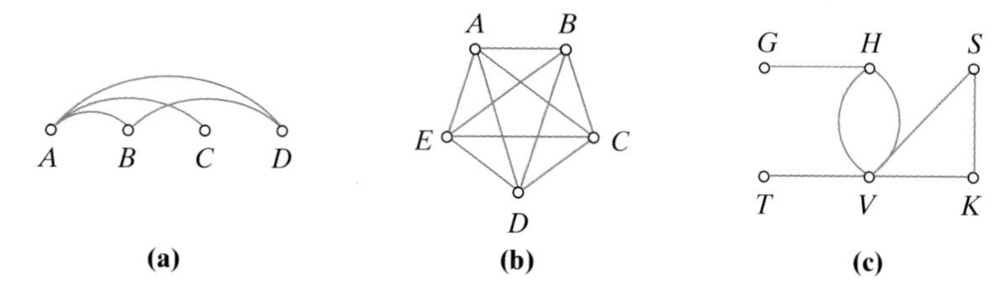

3. For each of the following, draw two different pictures of the graph.

(a) Vertices: *A, B, C, D*

Edges: *AB, BC, BD, CD*

(b) Vertices: *K, R, S, T, W*

Edges: *RS, RT, TT, TS, SW, WW, WS.*

4. For each of the following, draw two different pictures of the graph.

(a) Vertices: *L, M, N, P*

Edges: *LP, MM, PN, MN, PM*

(b) Vertices: *A, B, C, D, E*

Edges: *A* is adjacent to *C* and *E; B* is adjacent to *D* and *E; C* is adjacent to *A, D,* and *E; D* is adjacent to *B, C,* and *E; E* is adjacent to *A, B, C,* and *D.*

5. **(a)** Explain why the following figures represent the same graph.

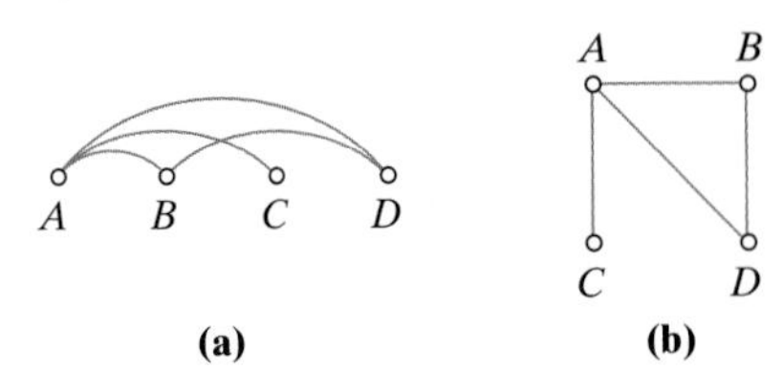

(b) Draw a third figure that represents the same graph.

6. **(a)** Explain why the following figures represent the same graph.

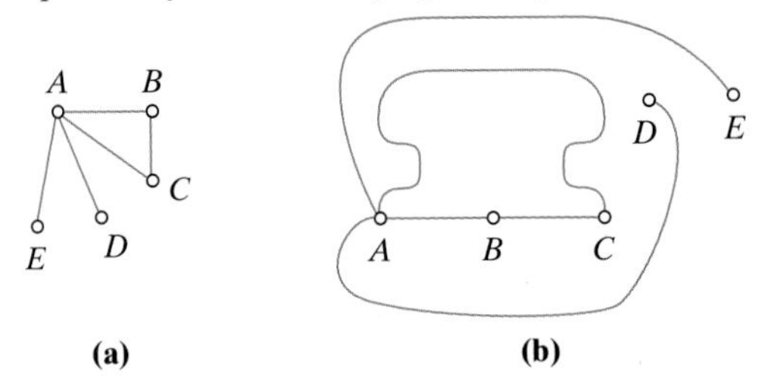

(b) Draw a third figure that represents the same graph.

7. **(a)** Draw a graph with 4 vertices and such that each vertex has degree 2.
   (b) Draw a graph with 6 vertices and such that each vertex has degree 3.

8. **(a)** Draw a graph with 4 vertices and such that each vertex has degree 1.
   (b) Draw a graph with 8 vertices and such that each vertex has degree 3.

9. Draw a graph with 4 vertices, each of degree 3 and such that
   (a) there are no loops and no multiple edges
   (b) there are loops but no multiple edges
   (c) there are multiple edges but no loops
   (d) there are both multiple edges and loops.

10. Draw a connected graph with 5 vertices, each of degree 4 and such that
   (a) there are no loops and no multiple edges
   (b) there are loops but no multiple edges
   (c) there are multiple edges but no loops
   (d) there are both multiple edges and loops.

Exercises 11 through 14 refer to the following graph.

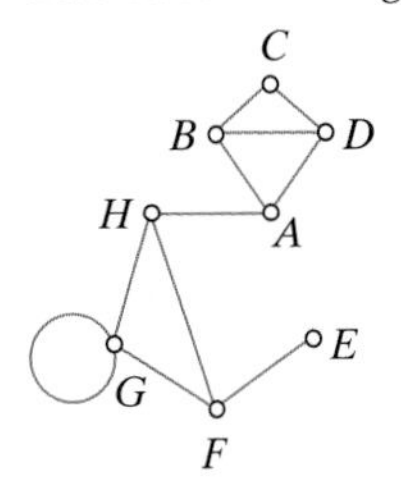

11. **(a)** Find a path from C to F passing through vertex B but not through vertex D.
   (b) Find a path from C to F passing through both vertex B and vertex D.
   (c) How many paths are there from C to A?

(d) How many paths are there from H to F?

(e) How many paths are there from C to F?

12. (a) Find a path from D to E passing through vertex G only once.

(b) Find a path from D to E passing through vertex G twice.

(c) How many paths are there from D to A?

(d) How many paths are there from H to E?

(e) How many paths are there from D to E?

13. (a) Find a circuit passing through vertex D.

(b) How many circuits start and end at vertex D?

(c) Which edges in the graph are bridges?

14. (a) Find a circuit passing through vertex H.

(b) How many circuits start and end at vertex H?

(c) Which edge can be added to this graph so that the resulting graph has no bridges?

15. (a) In the Königsberg bridge problem (Example 3), which of the real bridges are bridges in the graph-theory sense?

(b) Give an example of a connected graph with 4 vertices in which every edge is a bridge.

16. (a) In the Bridges of Madison County problem (Example 4), which of the real bridges are bridges in the graph-theory sense?

(b) Give an example of a connected graph with 6 vertices in which every edge is a bridge.

17. For each of the following, determine if the graph has an Euler circuit, an Euler path, or neither of these. Explain your answer, but do not find the actual path or circuit.

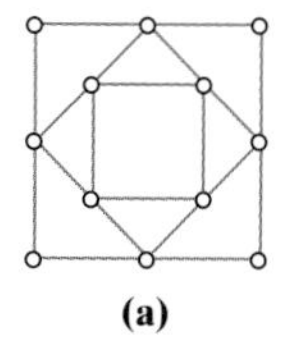
(a)

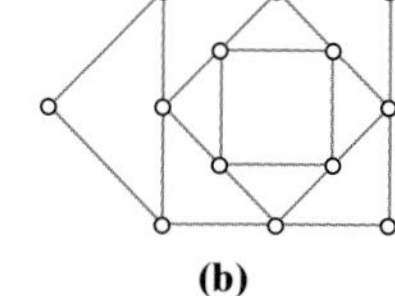
(b)

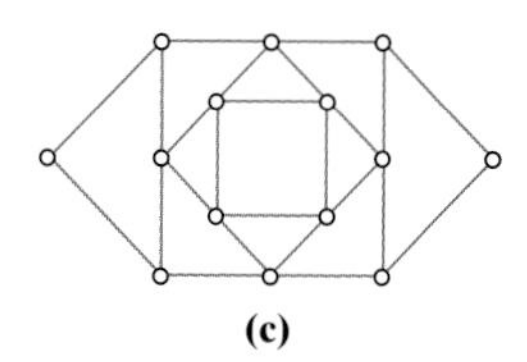
(c)

18. For each of the following, determine if the graph has an Euler circuit, an Euler path, or neither of these. Explain your answer, but do not find the actual path or circuit.

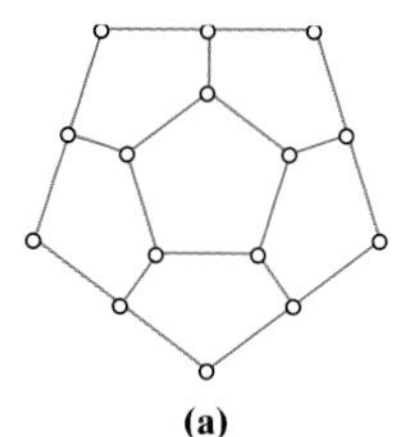
(a)

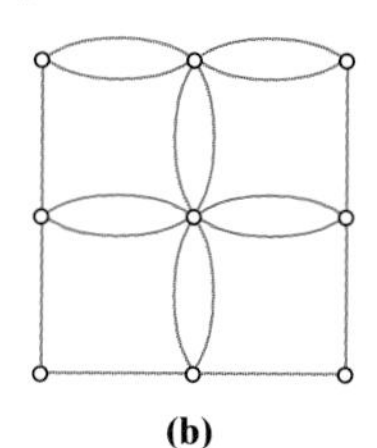
(b)

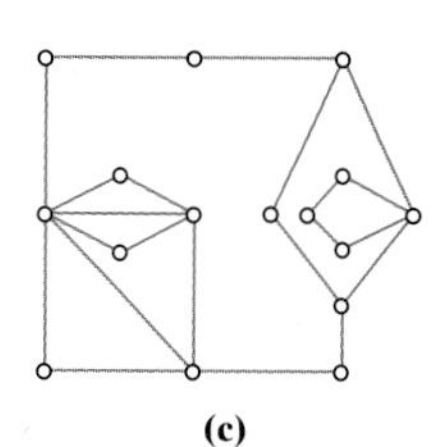
(c)

19. For each of the following, determine if the graph has an Euler circuit, an Euler path, or neither of these. Explain your answer, but do not find the actual path or circuit.

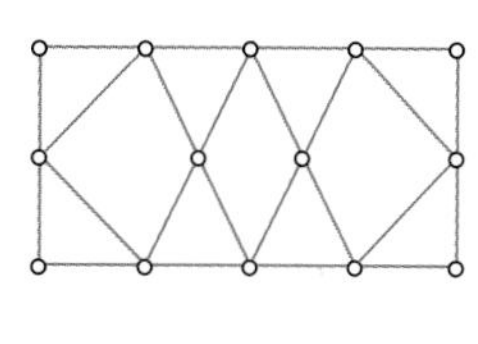
(a)

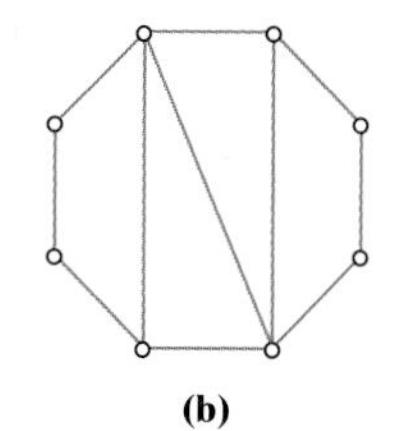
(b)

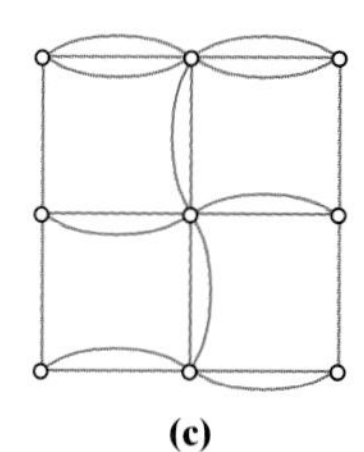
(c)

20. For each of the following line drawings, indicate whether the drawing has an open unicursal tracing, a closed unicursal tracing, or neither. (If it does have a unicursal tracing, trace it by labeling the edges 1, 2, 3, etc. in the order in which they can be traced.)

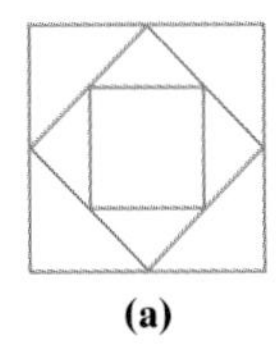

(a)

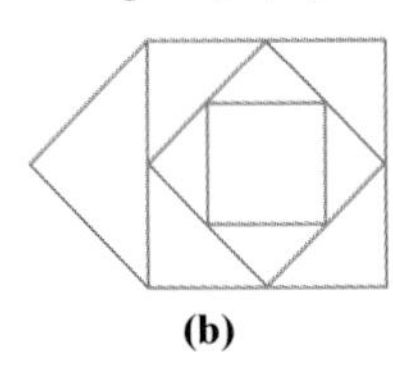

(b)

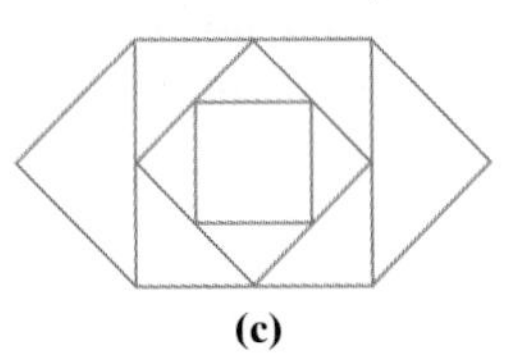

(c)

21. For each of the following line drawings, indicate whether the drawing has an open unicursal tracing, a closed unicursal tracing, or neither. (If it does have a unicursal tracing, trace it by labeling the edges 1, 2, 3, etc. in the order in which they can be traced.)

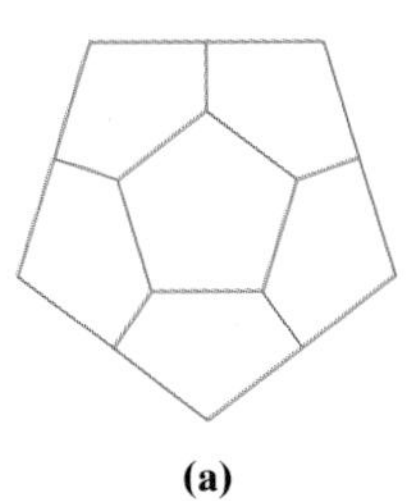

(a)

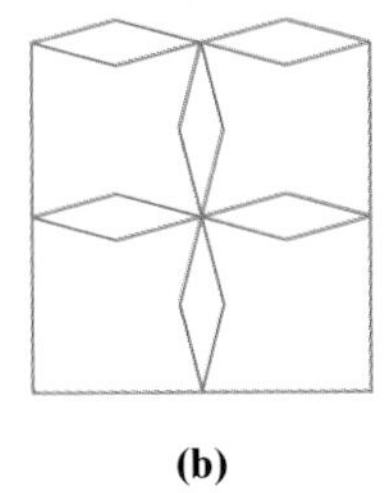

(b)

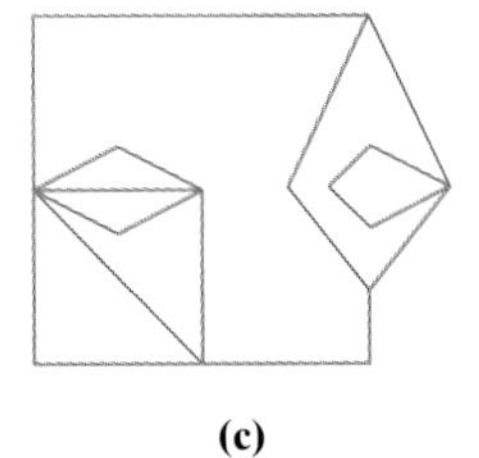

(c)

22. For each of the following line drawings, indicate whether the drawing has an open unicursal tracing, a closed unicursal tracing, or neither. (If it does have a unicursal tracing, trace it by labeling the edges 1, 2, 3, etc. in the order in which they can be traced.)

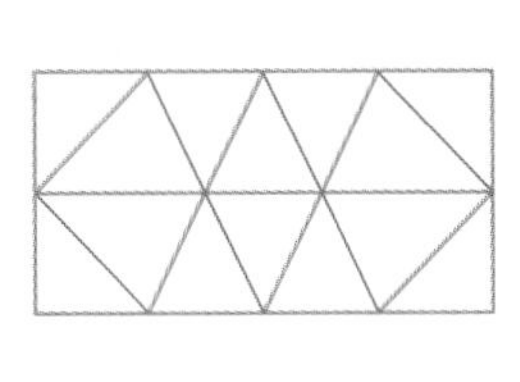

(a)

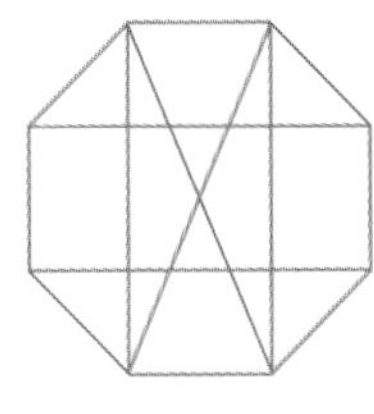

(b)

(c)

23. Find an optimal eulerization for each of the following graphs.

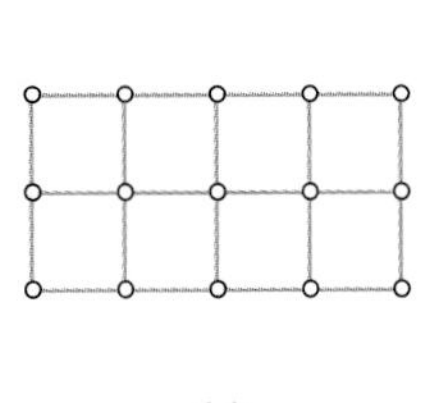

(a)

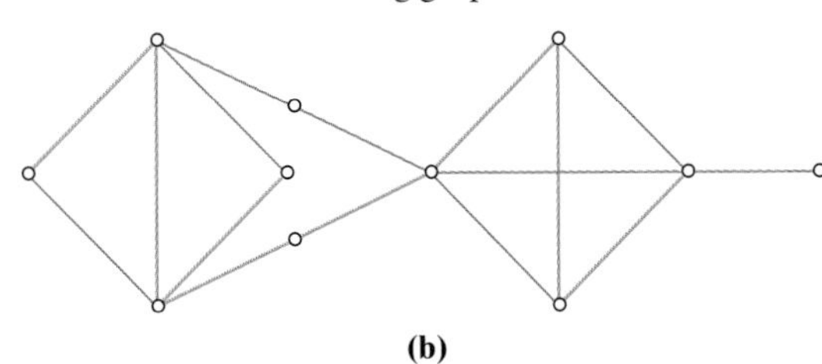

(b)

24. Find an optimal eulerization for each of the following graphs.

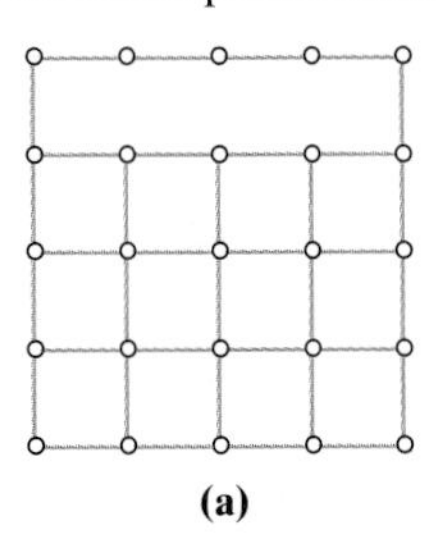

(a)

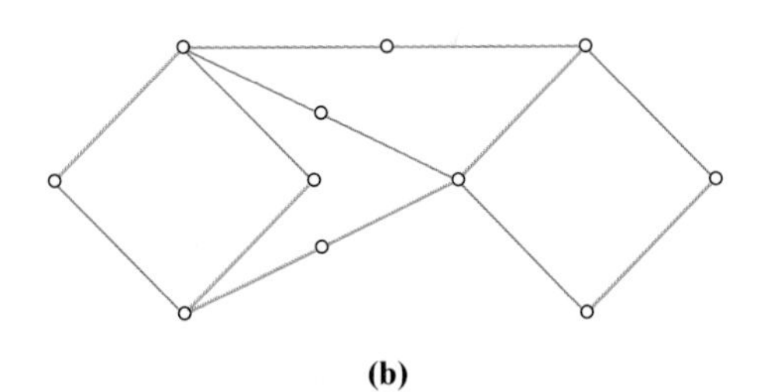

(b)

25. Find an optimal semi-eulerization for each of the following graphs.

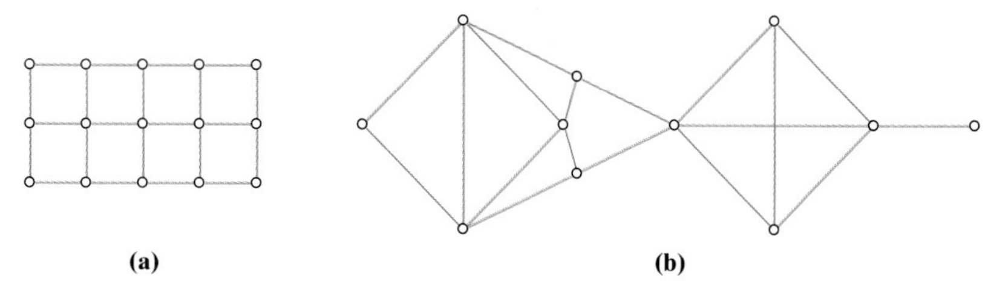

(a) (b)

26. Find an optimal semi-eulerization for each of the following graphs.

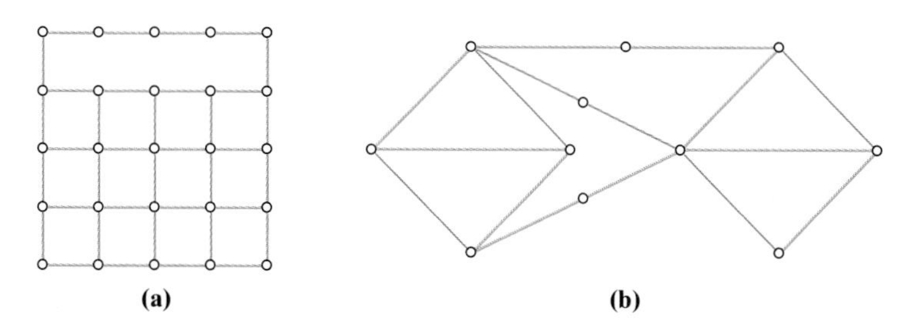

(a) (b)

27. Find an Euler circuit in the following graph.

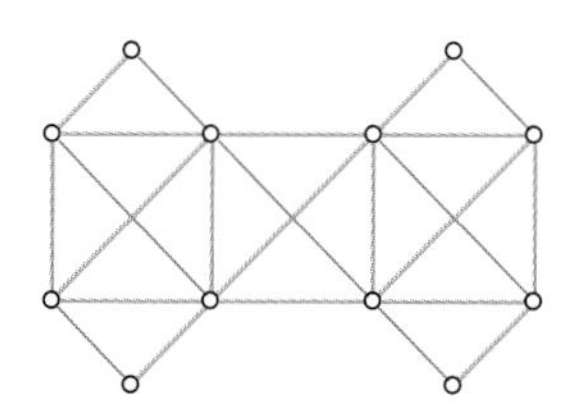

28. Find an Euler circuit in the following graph.

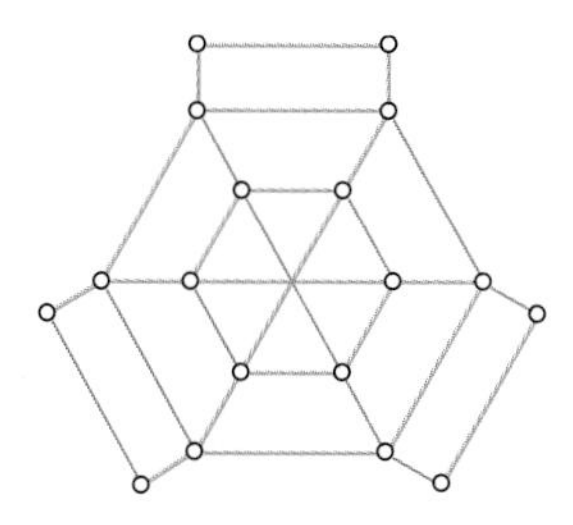

29. Use Fleury's algorithm to find an Euler circuit in the following graph.

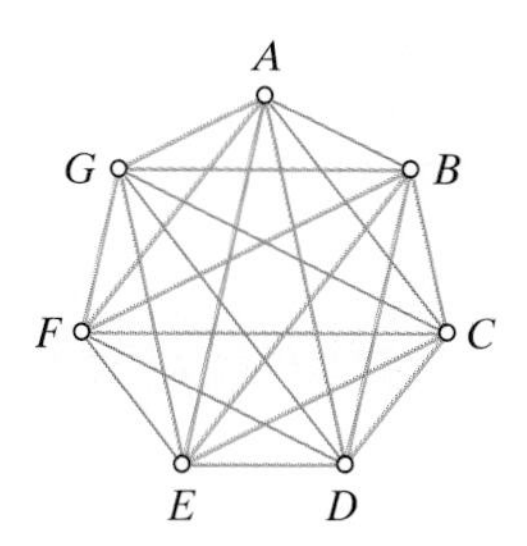

30. Use Fleury's algorithm to find an Euler circuit in the following graph.

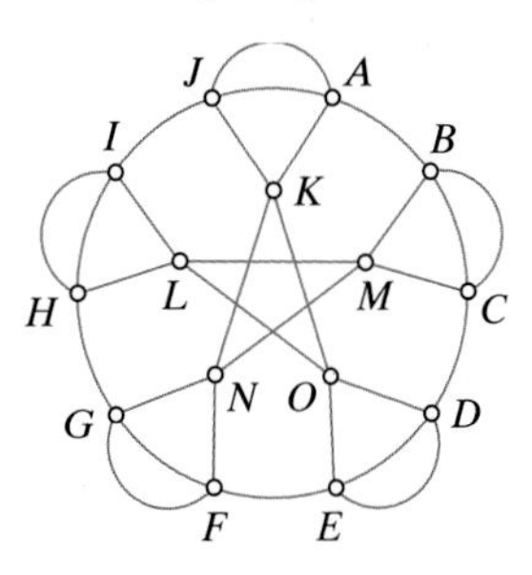

31. Use Fleury's algorithm to find an Euler path in the following graph that starts at X and ends at Y.

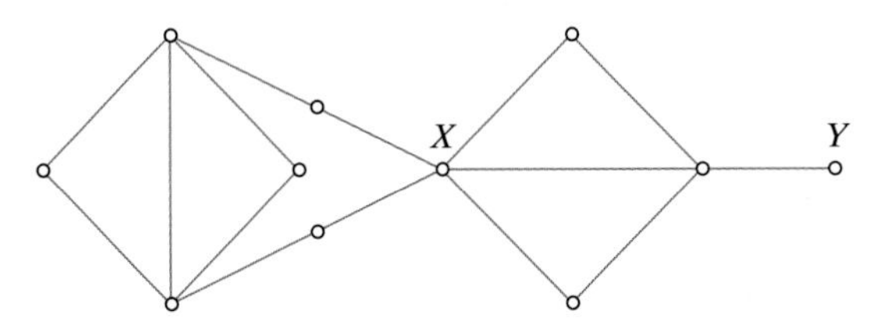

32. Use Fleury's algorithm to find an Euler path in the following graph that starts at X and ends at Y

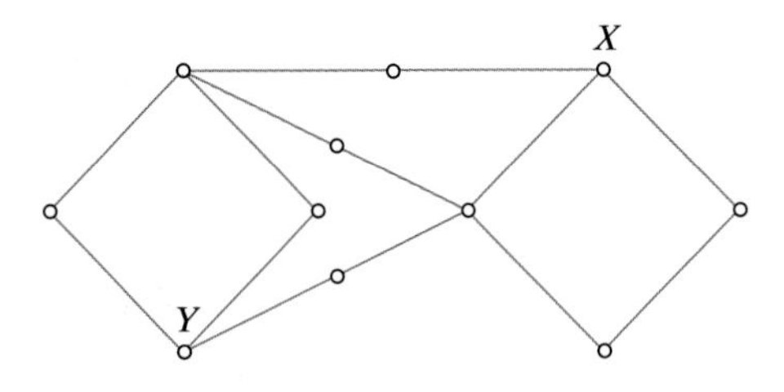

33. Suppose we want to trace the following graph with the fewest possible duplicate edges.

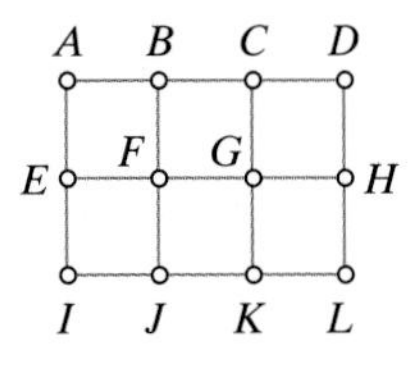

(a) Find an optimal semi-eulerization of the graph that starts at E and ends at H.

(b) Which edges of the graph will have to be retraced?

34. Suppose we want to trace the same graph as in Exercise 33, but now we want to start at B and end at K. Which edges of the graph will have to be retraced?

35. How many times would you have to lift your pencil to trace the following diagram, assuming that you do not trace over any line segment twice?

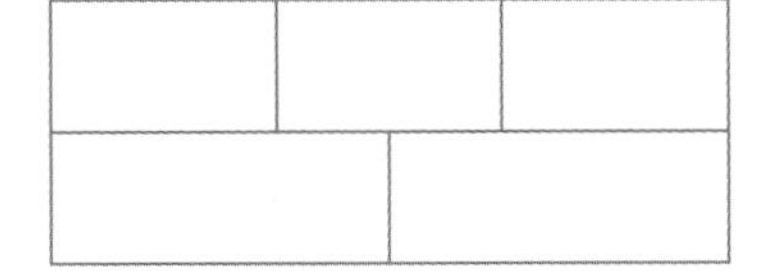

36. How many times would you have to lift your pencil to trace the following diagram, assuming that you do not trace over any line segment twice?

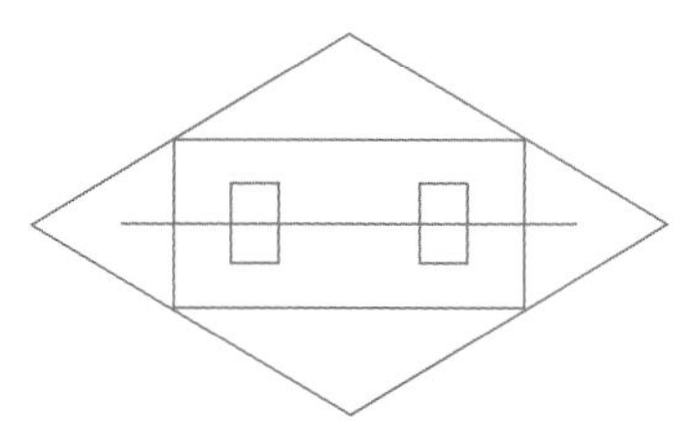

37. Draw a continuous line that **crosses (not traces)** every line segment in the diagram, but recrosses as few line segments as possible.

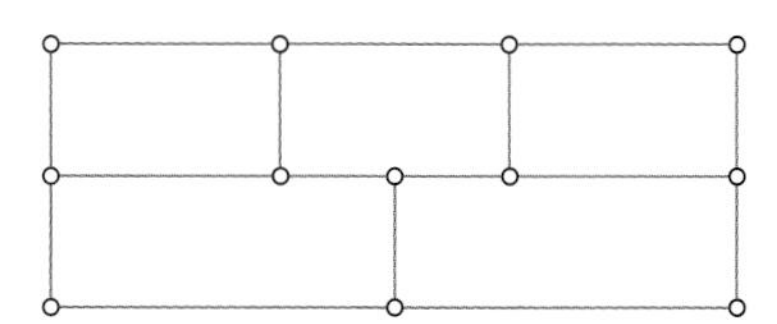

Jogging

38. (a) For each of the graphs in Exercises 1 and 2, complete the following table.

Graph	Number of Edges	Sum of the Degrees of All Vertices
Exercise 1(a)	4	$3 + 2 + 1 + 2 = 8$
Exercise 1(b)		
Exercise 1(c)		
Exercise 2(a)		
Exercise 2(b)		
Exercise 2(c)		

(b) Explain why in every graph the sum of the degrees of all the vertices equals twice the number of edges.

(c) Explain why every graph must have either zero or an even number of vertices of odd degree.

39. A garbage truck must pick up garbage along all the streets of the subdivision shown in the following figure (starting and ending at the garbage dump labeled G). All the streets are two-way streets, and garbage is picked up on both sides of the street on a single pass.

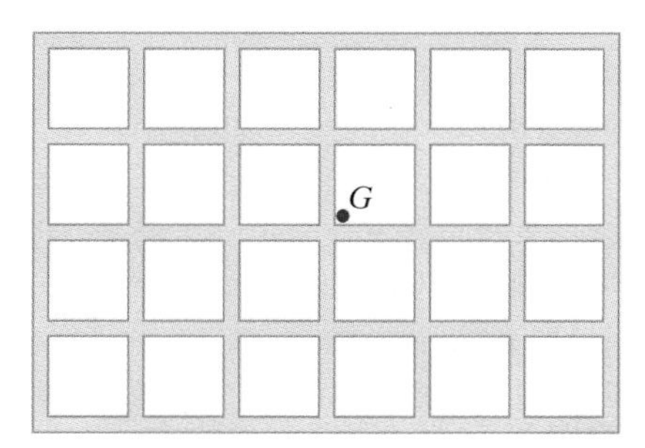

(a) Draw an appropriate graph representing this problem.

(b) Find an optimal eulerization of the graph in (a).

(c) Find an optimal route for the garbage truck. Describe the route by labeling the edges 1, 2, 3, . . . in the order in which they are traveled.

40. Consider the same subdivision as in Exercise 39, but this time assume the garbage is collected on each side of the street on separate passes.

(a) Draw an appropriate graph representing this problem.

(b) Find an optimal route for the garbage truck. Describe the route by labeling the edges 1, 2, 3, . . . in the order in which they are traveled.

41. Consider the following game: You must take a walk along the bridges of the city of Königsberg and cross each bridge at least once. It costs $1 each time you cross a bridge.

(a) Describe the cheapest possible walk you can make if you must start and end at the left bank (L).

(b) Describe the cheapest possible walk you can make if you are allowed to start and end at different places.

Exercises 42 and 43 refer to a small neighborhood described by the following street map.

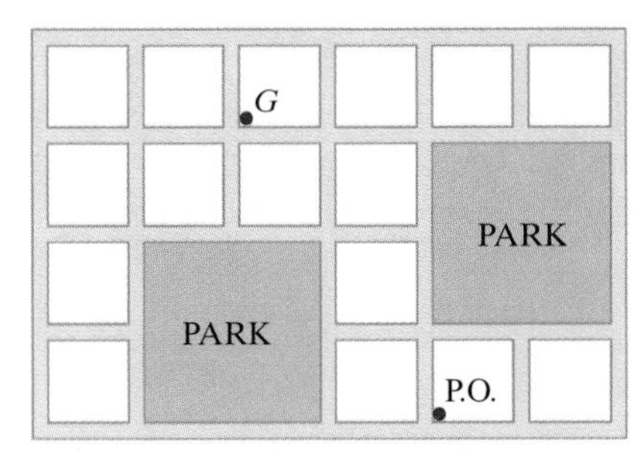

42. A night watchman must walk the streets of the neighborhood (starting and ending at G). The night watchman needs to wa!k only once along each block.

(a) Draw an appropriate graph representing this problem.

(b) Find an optimal eulerization of the graph in (a).

(c) Find an optimal route for the night watchman. Describe the route by labeling the edges 1, 2, 3, . . . in the order in which they are traveled.

43. Suppose that we want to find an optimal route for a mail carrier to deliver mail along the streets of the town. In this case the mail carrier must walk along each block twice (once for each side of the street) except for blocks facing one of the parks (here only one pass is needed).

(a) Draw an appropriate graph representing this problem.

(b) Find an optimal eulerization of the graph in (a).

(c) Describe an optimal route for the mail carrier (the starting and ending point is the Post Office, labeled P.O.).

44. Find a semi-eulerization for the graph given in Example 21 different from the one given in the text, and then exhibit an Euler path for your semi-eulerization.

45. Describe an actual optimal route for the mail carrier's problem in Examples 14 and 24 which starts and ends at the Post Office. (Label the edges 1, 2, 3, etc. in the order in which the route is traveled.)

46. The following diagram is of a hypothetical city with a river running through the middle of the city. There are 3 islands and bridges as shown in the figure. Is it possible to take a walk and cross each bridge exactly once? If so, show how; if not, explain why not.

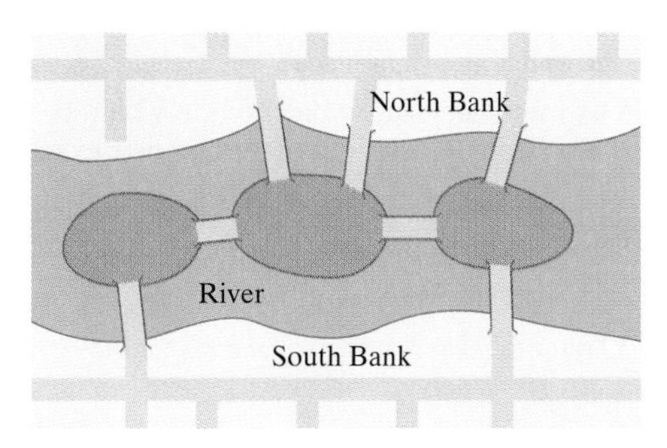

47. A policeman has to patrol along the streets of the subdivision represented by the following graph.

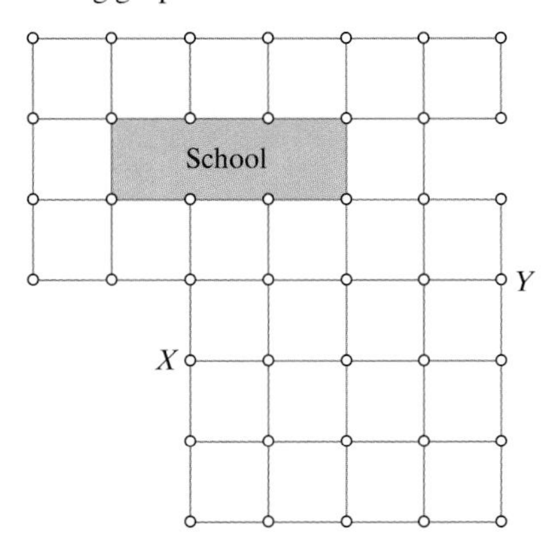

He wants to start his trip at the police station (located at X) and end the trip at his home (located at Y). He needs to cover each block of the subdivision at least once and at the same time he wants to duplicate the fewest possible number of blocks.

(a) How many blocks will he have to duplicate in an optimal trip through the subdivision?

(b) Describe an optimal trip through the subdivision. Label the edges 1, 2, 3, . . . in the order the policeman would travel them.

48. **(a)** Give an example of a graph with 15 vertices and no multiple edges that has an Euler circuit.

(b) Give an example of a graph with 15 vertices and no multiple edges that has an Euler path but no Euler circuit.

(c) Give an example of a graph with 15 vertices and no multiple edges that has neither an Euler circuit nor an Euler path.

49. The following figure is the floor plan of an office complex.

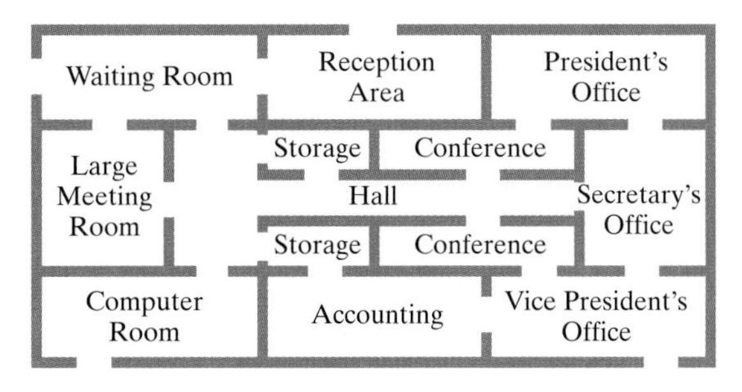

(a) Show that it is impossible to start outside the complex and walk through each door of the complex exactly once and end up outside.

(b) Show that it is possible to walk through every door of the complex exactly once (if you start and end at the right place).

(c) Show that by removing exactly one door, it would be possible to start outside the complex, walk through each door of the complex exactly once, and end up outside.

50. If a connected graph G has $2N$ vertices of odd degree what is the least number of times that you would have to lift your pencil to trace all the edges of graph, assuming that you do not trace over any edge twice? Explain.

Exercises 51 and 52 refer to the following street map for the city of Cleansburg.

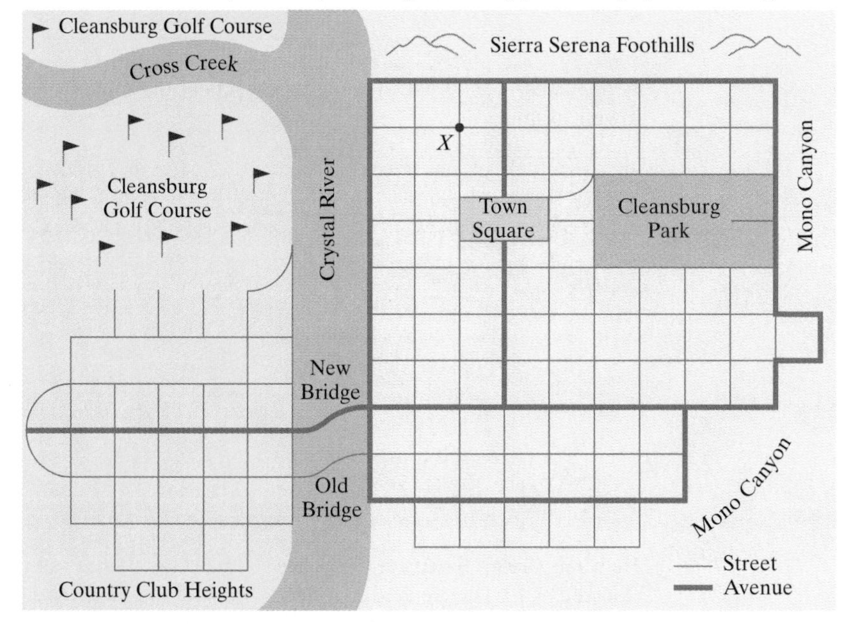

51. The city of Cleansburg wants to find an optimal route for its only garbage truck. All streets in the city of Cleansburg are two-way streets, and the garbage truck picks up garbage on both sides of the street as it passes through each block.

(a) Draw a graph model for this problem.

(b) Find an optimal eulerization for the graph.

(c) Find an optimal route through the city for the garbage truck, starting and ending at the Cleansburg Sanitation Department Equipment Yard (X in the figure).

52. The city of Cleansburg wants to find an optimal route for its only curb sweeper. All streets in the city of Cleansburg are two-way streets, and the street sweeper must pass through each block twice (once for each curb).

(a) Draw a graph model for this problem.

(b) Find an optimal eulerization for the graph.

(c) Find an optimal route through the city for the street sweeper, starting and ending at the Cleansburg Sanitation Department Equipment Yard (X in the figure).

Running

53. **(a)** Can a graph that has an Euler circuit have any bridges? If so, demonstrate it by showing an example. If not, explain why not.

(b) Can a graph that has an Euler path have any bridges? If so, how many? Explain your answer.

54. Suppose G and H are two graphs that have no common vertices and such that each graph has an Euler circuit. Let J be a (single) graph consisting of the graphs G, H, and one additional edge joining one of the vertices of G to one of the vertices of H. Explain why the graph J has no Euler circuit but does have an Euler path.

55. Explain why in any graph in which the degree of each vertex is at least 2, there must be a circuit.

56. Suppose we have a graph with 2 or more vertices and without loops or multiple edges. Explain why the graph must have at least 2 vertices with the same degree.

57. (Open-ended question) If we are given a list of N positive integers $d_1, d_2, \ldots d_N$, is there always a graph having N vertices with exactly those degrees? You should consider separately the case when the sum of the integers $d_1, d_2, \ldots, d_N$ is odd (that one is easy) and the case when the sum of the integers is even (in this case the answer is yes, and your job is to describe how to construct the graph). In this second case, what happens if neither loops nor multiple edges are allowed? Explain and give examples to illustrate your explanations.

References and Further Readings

1. Beltrami, E., *Models for Public Systems Analysis.* New York: Academic Press, Inc., 1977.
2. Beltrami, E., and L. Bodin, "Networks and Vehicle Routing for Municipal Waste Collection," *Networks,* 4 (1973), 65–94.
3. Chartrand, Gary, *Graphs as Mathematical Models.* Belmont, CA: Wadsworth Publishing Co., Inc., 1977.
4. Euler, Leonhard, "The Königsberg Bridges," trans. James Newman, *Scientific American,* 189 (1953), 66–70.
5. Minieka, E., *Optimization Algorithms for Networks and Graphs.* New York: Marcel Dekker, Inc., 1978.
6. Newman, J., ed., *Mathematics—An Introduction to Its Spirit and Its Use.* New York: W. H. Freeman & Co., 1978.
7. Roberts, Fred S., "Graph Theory and Its Applications to Problems of Society," *CBMS-NSF Monograph No. 29.* Philadelphia: Society for Industrial and Applied Mathematics, 1978, chap. 8.
8. Tucker, A. C., "Perfect Graphs and an Application to Optimizing Municipal Services," *SIAM Review,* 15 (1973), 585–590.
9. Tucker, A. C., and L. Bodin, "A Model for Municipal Street-Sweeping Operations," in *Modules in Applied Mathematics,* Vol. 3, eds. W. Lucas, F. Roberts, and R. M. Thrall. New York: Springer-Verlag, 1983, 76–111.

"Where shall I begin, please your majesty?", asked the White Rabbit.
"Begin at the beginning," the King said gravely,
"and go on till you come to the end. Then stop."
Lewis Carroll, Alice in Wonderland

The Traveling-Salesman Problem

Hamilton Joins the Circuit

CHAPTER 6

To those of us who get a big thrill out of planning for a trip (and many of us do), the next 10 to 15 years hold a truly special treat: We (mankind) will be planning for the mother of all excursions—a trip to Mars. Why are we planning to go to Mars, and what are we going to do when we get there?

The answer to the first question is easy: Mars is the next great frontier for human exploration. Its promise lies in the fact that it is the most earthlike of all the planets in the solar system; it has plenty of water, an atmosphere, and the Martian soil is rich in chemicals and minerals. Using current technology, a trip to Mars would take about six months, a long but not impossible trip (it is roughly the same time it took Europeans to travel to Australia by boat in the eighteenth century).

The ultimate attraction, however, is the hope of finding life on Mars. Of all the planets in our solar system, Mars is the most likely place to show evidence of life—probably primitive bacterial forms buried inside Martian rocks or under the Martian surface. The question is, How and where do we look for it?, which finally brings us to the point: Lurking behind the complex details of Mars exploration is a special type of graph routing problem. We will deal with problems of this type in this chapter.

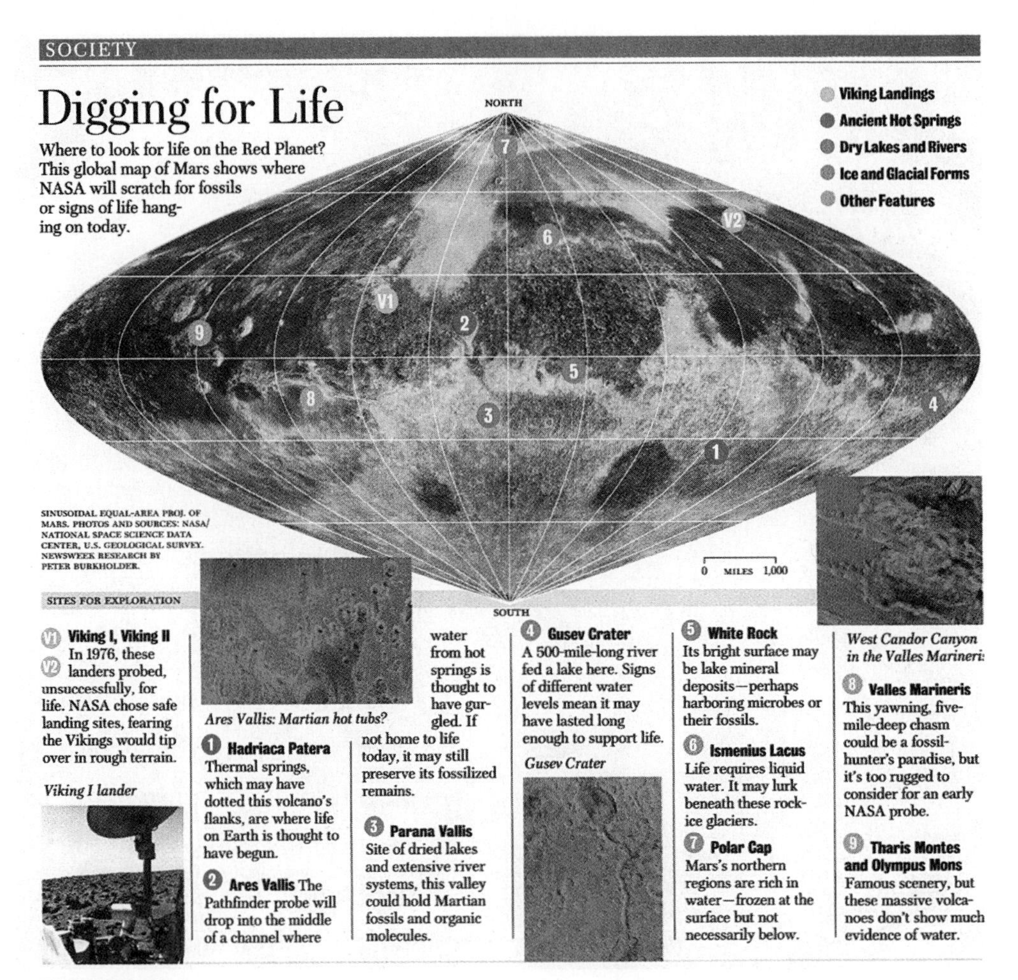

SOCIETY

Digging for Life

Where to look for life on the Red Planet? This global map of Mars shows where NASA will scratch for fossils or signs of life hanging on today.

- Viking Landings
- Ancient Hot Springs
- Dry Lakes and Rivers
- Ice and Glacial Forms
- Other Features

NORTH

SOUTH

SINUSOIDAL EQUAL-AREA PROJ. OF MARS. PHOTOS AND SOURCES: NASA/ NATIONAL SPACE SCIENCE DATA CENTER, U.S. GEOLOGICAL SURVEY. NEWSWEEK RESEARCH BY PETER BURKHOLDER.

0 MILES 1,000

SITES FOR EXPLORATION

V1 V2 Viking I, Viking II In 1976, these landers probed, unsuccessfully, for life. NASA chose safe landing sites, fearing the Vikings would tip over in rough terrain.

Viking I lander

Ares Vallis: Martian hot tubs?

1 Hadriaca Patera Thermal springs, which may have dotted this volcano's flanks, are where life on Earth is thought to have begun.

2 Ares Vallis The Pathfinder probe will drop into the middle of a channel where water from hot springs is thought to have gurgled. If not home to life today, it may still preserve its fossilized remains.

3 Parana Vallis Site of dried lakes and extensive river systems, this valley could hold Martian fossils and organic molecules.

4 Gusev Crater A 500-mile-long river fed a lake here. Signs of different water levels mean it may have lasted long enough to support life.

Gusev Crater

5 White Rock Its bright surface may be lake mineral deposits—perhaps harboring microbes or their fossils.

6 Ismenius Lacus Life requires liquid water. It may lurk beneath these rock-ice glaciers.

7 Polar Cap Mars's northern regions are rich in water—frozen at the surface but not necessarily below.

West Candor Canyon in the Valles Marineris

8 Valles Marineris This yawning, five-mile-deep chasm could be a fossil-hunter's paradise, but it's too rugged to consider for an early NASA probe.

9 Tharis Montes and Olympus Mons Famous scenery, but these massive volcanoes don't show much evidence of water.

FIGURE 6-1

Source: NASA/National Space Science Data Center, U.S. Geological Survey. *Newsweek* research by Peter Burkholder.

FIGURE 6-2

Here are the bare-bones details: Based on data collected by previous flybys, NASA has identified several places on the surface of Mars where the likelihood of finding either bacterial fossils or actual life forms is the highest (see Fig. 6-1). A main goal for the early exploration stages is to get an unmanned rover to each and every one of these locations to collect soil samples and perform experiments. One approach being proposed is a *robotic sample-return mission,* in which a lander would land at one of the designated sites (probably the Ares Valley) and release a robotic rover controlled from Earth. The rover would then travel to each of the other sites, collecting soil samples and performing experiments. After all the sites had been visited, the rover would return to the landing site, where a return rocket would bring the samples back to Earth. If things go according to plan, it is possible that the first of these sample-return missions could be launched as early as 2003.

Figure 6-3 shows one of the possible circuits which the rover might travel, as it visits each of the locations once to look for life and collect soil samples. There are, however, hundreds of other possible ways to route the rover. From an efficiency point of view, which one is the best? As a purely mathematical problem, this problem falls within a very special and important category called TSP.

The acronym TSP stands for "traveling-salesman problem," so called because one well-known variation is that of a traveling salesman who must call on customers in several cities and wants to find the most efficient route that visits each of the cities once, returning at the end to his home town. The name has stuck as a generic name applied to similar problems, even if they have nothing to do with traveling salespeople. It is best to think of the "traveling-salesman" as just a metaphor: It is a rover searching for the best route on Mars, a UPS driver trying to find the best way to deliver packages around town, or any of us, trying to plan the best route by which to run a bunch of errands on a Saturday morning.

The traveling-salesman problem is very easy to understand and, intuitively, it seems pretty easy to solve. It is always a surprise to find out, as we soon will, that TSPs represent one of the most interesting, important, and complex problems in graph theory, if not all of mathematics. Understanding TSPs and what it means to "solve" them will be the main purpose of this chapter.

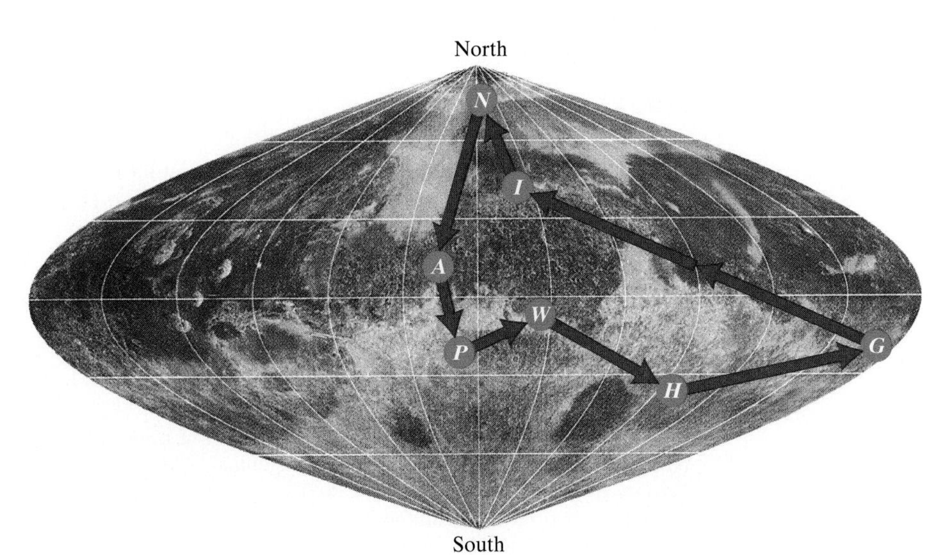

FIGURE 6-3
A: Ares Valley (starting and ending point); *G:* Gusev Crater; *H:* Hadriaca Patera; *I:* Ismenius Lacus; *P:* Parana Vallis; *N:* North Polar Cap; *W:* White Rock.

Hamilton Circuits and Hamilton Paths

From a mathematical point of view, the main concept that will concern us in this chapter is that of a *Hamilton circuit.* Let's think back to the last chapter, where we studied Euler circuits and Euler paths. In these types of circuits (or paths) the name of the game is to travel or "pass through" each *edge* of the graph once and only once. But what about a circuit that must pass through each *vertex* of the graph once and only once (except at the end, where it returns to the starting vertex)? This entirely different type of circuit is called a **Hamilton circuit.**[1] Likewise, a path that passes through every vertex of the graph is called a **Hamilton path.**

The distinction between a Hamilton circuit and an Euler circuit (or a Hamilton path and an Euler path) may appear minor (just a word of difference—substitute "vertex" for "edge"), but there are some significant differences. These can best be illustrated with an example.

Example 1.

Consider the four graphs in Fig. 6-4. The graph in Fig. 6-4(a) has many Hamilton circuits (*A, B, C, D, E, A* is one of them; *C, B, E, A, D, C* is another—why don't you try to find a couple of others?). It also has many Hamilton paths (for example, *A, B, C, D, E* and *C, B, E, A, D*)—after all, any Hamilton circuit can be shortened into a Hamilton path by just removing the last edge. On the other hand, this graph has no Euler circuits or Euler paths (it has 4 vertices of odd degree, and, according to Euler's theorems, that settles the issue).

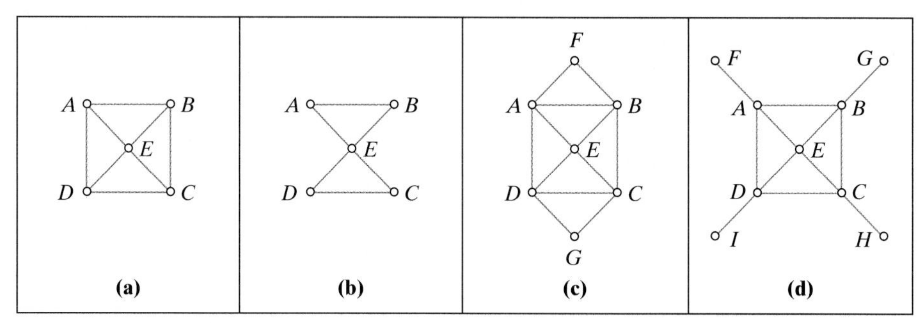

FIGURE 6-4

(a) A graph with a Hamilton circuit and a Hamilton path. (b) A graph with a Hamilton path but no Hamilton circuit. (c) A graph with a Hamilton circuit and an Euler circuit. (d) A graph with no Hamilton circuit, no Hamilton path, no Euler circuit.

The graph in Fig. 6-4(b) has Euler circuits (every vertex has even degree). At the same time, it has a Hamilton path (for example, *A, B, E, C, D*) but it has no Hamilton circuits (whatever the starting point, we are going to have to pass through vertex *E* more than once to close the circuit).

[1] Named after the great Irish mathematician and astronomer Sir William Rowan Hamilton (1805–1865). It is said that at the age of four Hamilton could read Latin, Greek, and Hebrew, as well as English. At the age of 21 he became a Professor of Astronomy at Trinity College in Dublin. Besides being a great scientist, Hamilton was an accomplished man of letters and a poet, who counted Wordsworth and Coleridge among his closest friends.

The graph in Fig. 6-4(c) has Euler circuits (every vertex has even degree). It also has several Hamilton circuits and many Hamilton paths (see Exercise 5).

Finally, the graph in Fig. 6-4(d) has no Euler circuits, no Euler paths, no Hamilton circuits, and no Hamilton paths (see Exercise 9). ■

There are a couple of observations that Example 1 brings to our attention. First, the fact that a graph has an Euler circuit says nothing about its having or not having a Hamilton circuit. Second, if a graph has a Hamilton circuit, it automatically has a Hamilton path (just skip the return to the starting vertex); on the other hand, a graph can have a Hamilton path without having a Hamilton circuit [as in Fig. 6-4(b)].

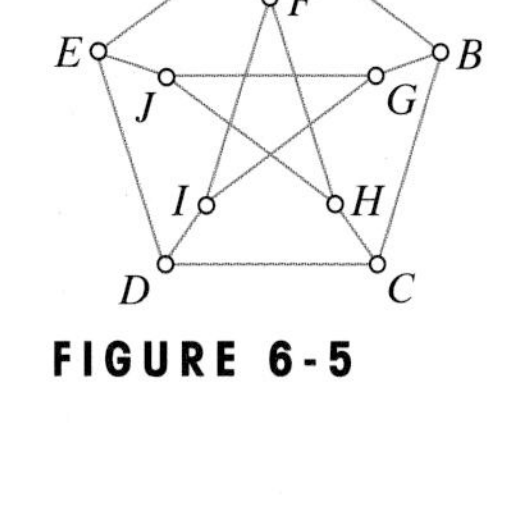

FIGURE 6-5

Given an arbitrary graph, how can we tell if it has a Hamilton circuit or a Hamilton path? How about some nice, clean-cut theorems of the kind that Euler gave us for Euler circuits and Euler paths. Unfortunately, there are no such theorems. Even for a small graph, such as the one in Fig. 6-5, it is not easy to determine whether it has a Hamilton circuit (it doesn't). For graphs with dozens or hundreds of vertices, there is often no easy way to establish conclusively whether the graph has a Hamilton circuit or a Hamilton path.

The flip side of a graph that has no Hamilton circuits is a graph in which every possible sequence of the vertices turns out to produce a Hamilton circuit. We will discuss these graphs next.

Complete Graphs

Figure 6-6 shows four graphs, having 3, 4, 5, and 6 vertices, respectively. These graphs have one characteristic in common: There is an edge connecting each pair of vertices. A graph with N vertices in which *every* pair of vertices is joined by exactly one edge is called the **complete graph** (on N vertices), and denoted by the symbol K_N. In the complete graph on N vertices, each vertex is adjacent to each of the other vertices, so each vertex has degree $N-1$. From this it follows that the total number of edges in the complete graph with N vertices is $N(N-1)/2$ (Exercise 39).

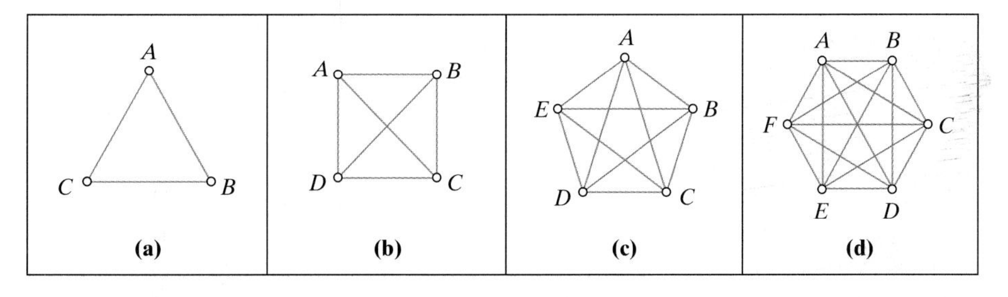

FIGURE 6-6

A complete graph has a complete repertoire of Hamilton circuits. We can write the vertices in any order we want, repeat the first vertex at the end, and presto, we have a Hamilton circuit! Take K_4, the complete graph with 4 vertices shown in Fig. 6-6(b). Let's randomly choose the 4 vertices in any order—say *C, A, D, B.* Now repeat the first vertex (*C*) at the end. The circuit *C, A, D, B, C* is indeed a Hamilton circuit of the graph. Repeat the process with a different sequence, and once again we will end up with a Hamilton circuit, although not necessarily a different one! It is important to remember that different sequences of letters can produce the same

Hamilton circuit: The circuit *C, A, D, B, C* is the same circuit as the circuit *A, D, B, C, A*—we only changed the *reference point* (in the first case, we used vertex *C* as the reference point; in the second one we used vertex *A*). There are two more sequences that give exactly the same Hamilton circuit as *C, A, D, B, C*—the ones with reference points *D* and *B* (*D, B, C, A, D* and *B, C, A, D, B*, respectively).

A complete listing of all the Hamilton circuits of K_4 is shown in Table 6-1—there are 6 altogether.

Table 6-1 The 6 Hamilton circuits of K_4. Each circuit can be written in 4 ways.

	Reference point is *A*	Reference point is *B*	Reference point is *C*	Reference point is *D*
1	*A, B, C, D, A*	*B, C, D, A, B*	*C, D, A, B, C*	*D, A, B, C, D*
2	*A, B, D, C, A*	*B, D, C, A, B*	*C, A, B, D, C*	*D, C, A, B, D*
3	*A, C, B, D, A*	*B, D, A, C, B*	*C, B, D, A, C*	*D, A, C, B, D*
4	*A, C, D, B, A*	*B, A, C, D, B*	*C, D, B, A, C*	*D, B, A, C, D*
5	*A, D, B, C, A*	*B, C, A, D, B*	*C, A, D, B, C*	*D, B, C, A, D*
6	*A, D, C, B, A*	*B, A, D, C, B*	*C, B, A, D, C*	*D, C, B, A, D*

The complete list of all the Hamilton circuits of K_5 is shown in Table 6-2. There are 24 Hamilton circuits. Notice that this time we listed them using a consistent vertex (*A*) as the reference point—this helps with the bookkeeping. We also paired each Hamilton circuit with its *mirror-image circuit* (the circuit traveled in reverse order). The reader is warned that a circuit and its mirror-image circuit are not considered equal—though they are close relatives.

Table 6-2 The 24 Hamilton circuits of K_5 (using *A* as the reference point). Each circuit is paired up with its mirror-image circuit.

1	*A, B, C, D, E, A*	13	*A, E, D, C, B, A*
2	*A, B, C, E, D, A*	14	*A, D, E, C, B, A*
3	*A, B, D, C, E, A*	15	*A, E, C, D, B, A*
4	*A, B, D, E, C, A*	16	*A, C, E, D, B, A*
5	*A, B, E, C, D, A*	17	*A, D, C, E, B, A*
6	*A, B, E, D, C, A*	18	*A, C, D, E, B, A*
7	*A, C, B, D, E, A*	19	*A, E, D, B, C, A*
8	*A, C, B, E, D, A*	20	*A, D, E, B, C, A*
9	*A, C, D, B, E, A*	21	*A, E, B, D, C, A*
10	*A, C, E, B, D, A*	22	*A, D, B, E, C, A*
11	*A, D, B, C, E, A*	23	*A, E, C, B, D, A*
12	*A, D, C, B, E, A*	24	*A, E, B, C, D, A*

There is a convenient formula that gives the number of Hamilton circuits in a complete graph. It uses the factorial, a concept we first came across in Chapter 2. Recall that if *N* is any positive integer, the number $N! = N \times (N - 1) \times \cdots \times 3 \times 2 \times 1$ is called the **factorial** of *N*. Table 6-3 shows the first 25 values of $N!$.

Table 6-3 The first 25 factorials.

1! = 1
2! = 2
3! = 6
4! = 24
5! = 120
6! = 720
7! = 5040
8! = 40,320
9! = 362,880
10! = 3,628,800
11! = 39,916,800
12! = 479,001,600
13! = 6,227,020,800
14! = 87,178,291,200
15! = 1,307,674,368,000
16! = 20,922,789,888,000
17! = 355,687,428,096,000
18! = 6,402,373,705,728,000
19! = 121,645,100,408,832,000
20! = 2,432,902,008,176,640,000
21! = 51,090,942,171,709,440,000
22! = 1,124,000,727,777,607,680,000
23! = 25,852,016,738,884,976,640,000
24! = 620,448,401,733,239,439,360,000
25! = 15,511,210,043,330,985,984,000,000

Notice that 3! = 6 (the number of Hamilton circuits of K_4) and that 4! = 24 (the number of Hamilton circuits of K_5). Coincidence? Not at all. As we now know, in a complete graph we can list the vertices in any order and get a Hamilton circuit. If we choose a specified vertex as the reference point, every possible ordering of the remaining $(N - 1)$ vertices will result in a Hamilton circuit, and the total number of ways of ordering $(N - 1)$ things is $(N - 1)!$

The complete graph with N vertices has $(N - 1)!$ Hamilton circuits.

The most important thing to notice about Table 6-3 is how quickly factorials grow and, consequently, how quickly the number of Hamilton circuits of a complete graph grows as we increase the number of vertices. A modest-size graph such as the complete graph with just a dozen vertices (K_{12}) has almost 40 million Hamilton circuits (11! = 39,916,800). Double the size of the graph to K_{24} and the number of Hamilton circuits grows to an astronomical 23!, which is more than 25 trillion billions, a number so big that it defies ordinary human comprehension.

The main lesson of this section is this: When it comes to Hamilton circuits in complete graphs, we are facing an embarassment of riches. The main question is no longer "Are there any?" but rather "How are we going to deal with so many?"

Traveling-Salesman Problems

What kind of a problem is a TSP? Here are a few examples of which only the first is self-evident.

Example 2. A Tale of Five Cities

Meet Willy Loman, a traveling salesman. Willy has customers in 5 cities, which for the sake of brevity we will call *A, B, C, D,* and *E,* and is planning an upcoming sales trip to visit each of his customers. Willy needs to start and end the trip at his home town of *A*; other than that, there are no particular restrictions as to the order in which he should visit the other 4 cities.

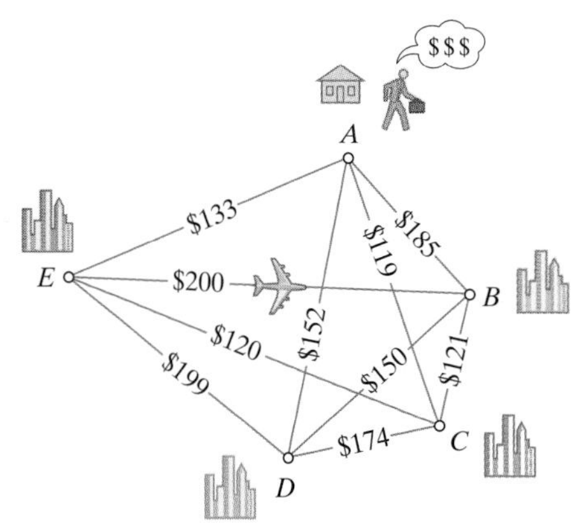

FIGURE 6-7
A 5-city TSP.

The graph in Fig. 6-7 shows the cost of a *one-way* airline ticket between each pair of cities. Naturally, Willy wants to cut down on his travel expenses as much as possible, to the point that he is willing to ask for mathematical advice. What Willy needs to know is, What is the cheapest possible sequence in which to visit the 5 cities? We will return to this question soon. ■

Example 3. Probing the Outer Reaches of Our Solar System

It is the year 2020. An expedition to explore the outer planetary moons in our solar system is about to be launched from planet Earth. The expedition is scheduled to visit Callisto, Ganymede, Titan, Hyperion, and Phoebe (the first two are moons of Jupiter; the last three of Saturn), collect rock samples at each, and then return to Earth with the loot.

Figure 6-8 shows the travel time (in years) between any two moons. What is the best way to route the spaceship so that the entire trip takes the least amount of time? ■

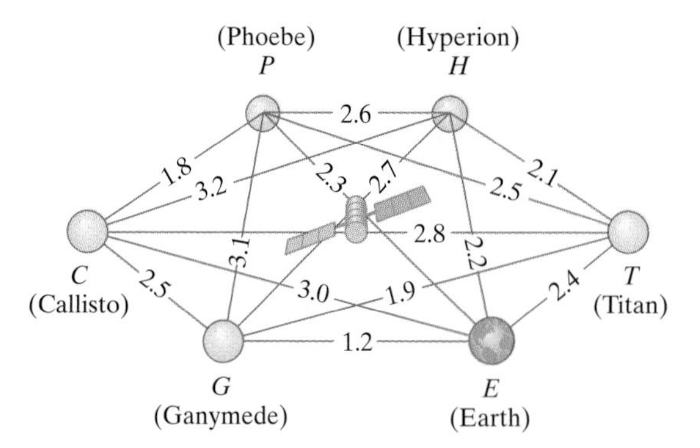

FIGURE 6-8
A 6-vertex TSP.

Example 4. Searching for Martians

Figure 6-9 shows seven locations on Mars where NASA scientists believe there is a good chance of finding evidence of life. Imagine that you are in charge of planning a *sample-return* mission, which calls for landing an unmanned rover in the Ares Valley (A) and then having the rover go to each of the other sites, collecting and analyzing soil samples at each site, finally returning to the Ares Valley landing site, where a return rocket would bring the best samples back to Earth. A trip like this would take several years and cost several billion dollars, so good planning is critical.

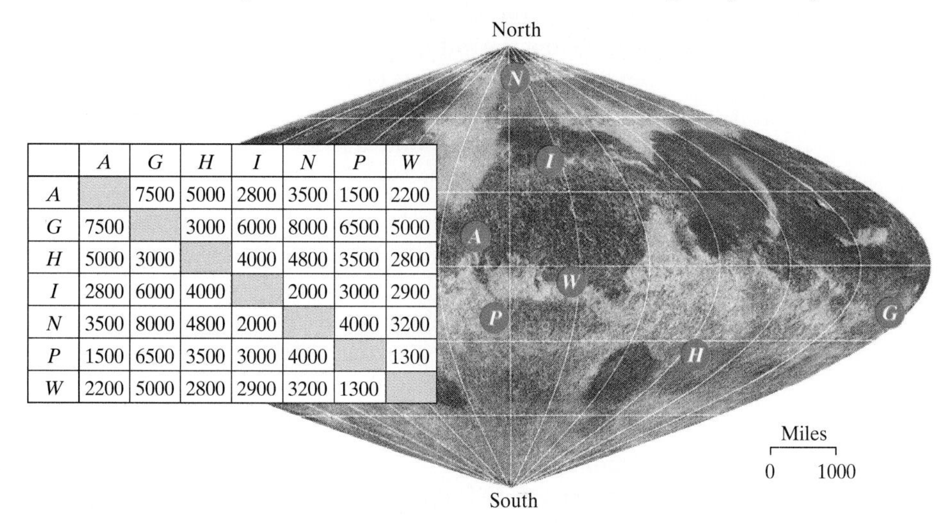

	A	G	H	I	N	P	W
A		7500	5000	2800	3500	1500	2200
G	7500		3000	6000	8000	6500	5000
H	5000	3000		4000	4800	3500	2800
I	2800	6000	4000		2000	3000	2900
N	3500	8000	4800	2000		4000	3200
P	1500	6500	3500	3000	4000		1300
W	2200	5000	2800	2900	3200	1300	

FIGURE 6-9 Approximate distances (in miles) between locations in Mars.

The table shows the estimated distances (in miles) that a rover would have to travel to get from one Martian site to another. What is the optimal sequence in which the rover should travel to the different sites so that the total distance it has to travel is minimized? ■

Examples 2, 3, and 4 are variations on a single theme. In each case, we are presented with a complete graph whose edges have numbers attached to them. (For example 4 it turns out to be easier to put the numbers in a table). Such graphs are called **complete weighted graphs** (any graph whose edges have numbers attached to them is called a **weighted graph**, and the numbers are called the **weights**[2] of the edges). In each example the weights of the graph represent a different variable: in Example 2 the weights represent *cost,* in Example 3 they represent *time,* and in Example 4 they represent *distance.* Most important of all, in each example the problem we want to solve is the same: *to find an optimal Hamilton circuit—that is, a Hamilton circuit with least total weight—for the complete weighted graph.* These kinds of problems are known generically as **TSP**s (traveling-salesman problems).

Many important real-life situations turn out to represent problems that can be formulated as TSPs. The following are just a few general examples:

- **Package Deliveries.** Companies such as United Parcel Service (UPS) and Federal Express deal with this situation daily. Each truck has packages to deliver to a list of destinations. The travel time between any two delivery locations is known or can be estimated (experienced drivers always know such

[2] This is a different usage of the word *weight* from that in Chapter 2.

things). The object is to deliver the packages to each of the delivery locations and return to the starting point in the least amount of time—clearly an example of a TSP. On a typical day, a UPS truck delivers packages to somewhere between 100 and 200 locations, so we are dealing with a TSP involving a graph with that many vertices.

- **Fabricating Circuit Boards.** In the process of fabricating integrated-circuit boards, tens of thousands of tiny holes must be drilled in each board. This is done by using a stationary laser beam and rotating the board around. Efficiency considerations require that the order in which the holes are drilled be such that the entire drilling sequence be completed in the least amount of time. This is an example of a TSP, in which the vertices of the graph represent the holes on the circuit board and the weight of the edge connecting vertices X and Y represents the time needed to rotate the board from drilling position X to drilling position Y.

- **Scheduling Jobs on a Machine.** In many industries there are machines that perform many different jobs. Think of the jobs as the vertices of the graph. After performing job X the machine needs to be set up to perform another job. The amount of time required to set up the machine to change from job X to job Y (or vice versa) is the weight of the edge connecting vertices X and Y. The problem is to schedule the machine to run through all the jobs in a cycle that requires the least total amount of time. This is another example of a TSP.

- **Running Errands around Town.** When we have a lot of errands to run around town, we like to follow the best route that will take us to each of our destinations and bring us home at the end of the day. This is an example of a TSP. (See Exercises 37 and 38.)

Simple Strategies for Solving TSPs

Example 5. A Tale of Five Cities: Part II

At the end of Example 2 we left Willy the traveling salesman pondering his upcoming sales trip, dollar signs running through his head. (For the reader's convenience the graph showing the cost of one-way travel between any two cities is given again in Fig. 6-10.) Imagine now that Willy, unwilling or unable to work out the problem himself, decides to offer a reward of $20 to anyone who can find the optional Hamilton circuit for this graph.

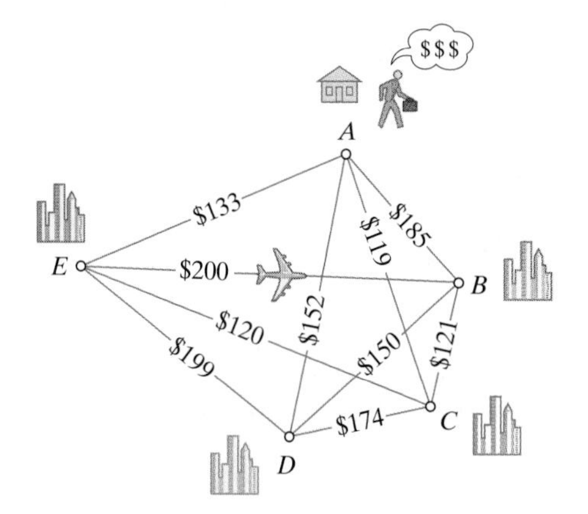

FIGURE 6-10

Would it be worth $20 to you to work out this problem? If so, How would you do it? We encourage you at this point to take a break from your reading, get a pencil and a piece of paper and try to work out the problem on your own. (Pretend you are doing it for the money.) Give yourself about 15 to 20 minutes.

. . .

If you are like most people, you probably followed one of two standard strategies in looking for an optimal Hamilton circuit:

- **Method 1.** (a) *You made a list of all possible Hamilton circuits;* (b) *you calculated the total cost for each circuit;* (c) *you selected a circuit with the least total cost for the answer.*

Table 6-4 shows the worked-out solution in all its glory: (a) There are 24 possible Hamilton circuits in a complete graph with 5 vertices. (Since the one-way airfares are the same in either direction, a circuit and its mirror-image circuit result in the same total cost and are shown on the same row of the table. This observation saves a little work.) (b) The total cost of each circuit can be easily calculated and is shown in the middle column of the table. (c) The optimal circuits with a total cost of $676 are shown in the second to last row (*A, D, B, C, E, A* and its mirror-image circuit *A, E, C, B, D, A*). We can use either one as a solution, shown graphically in Fig. 6-11.

Table 6-4 The 24 possible Hamilton circuits and their total costs.

	Hamilton Circuit	Total Cost	Mirror-Image Circuit
1	*A, B, C, D, E, A*	185 + 121 + 174 + 199 + 133 = 812	*A, E, D, C, B, A*
2	*A, B, C, E, D, A*	185 + 121 + 120 + 199 + 152 = 777	*A, D, E, C, B, A*
3	*A, B, D, C, E, A*	185 + 150 + 174 + 120 + 133 = 762	*A, E, C, D, B, A*
4	*A, B, D, E, C, A*	185 + 150 + 199 + 120 + 119 = 773	*A, C, E, D, B, A*
5	*A, B, E, C, D, A*	185 + 200 + 120 + 174 + 152 = 831	*A, D, C, E, B, A*
6	*A, B, E, D, C, A*	185 + 200 + 199 + 174 + 119 = 877	*A, C, D, E, B, A*
7	*A, C, B, D, E, A*	119 + 121 + 150 + 199 + 133 = 722	*A, E, D, B, C, A*
8	*A, C, B, E, D, A*	119 + 121 + 200 + 199 + 152 = 791	*A, D, E, B, C, A*
9	*A, C, D, B, E, A*	119 + 174 + 150 + 200 + 133 = 776	*A, E, B, D, C, A*
10	*A, C, E, B, D, A*	119 + 120 + 200 + 150 + 152 = 741	*A, D, B, E, C, A*
11	*A, D, B, C, E, A*	152 + 150 + 121 + 120 + 133 = 676	*A, E, C, B, D, A*
12	*A, D, C, B, E, A*	152 + 174 + 121 + 200 + 133 = 780	*A, E, B, C, D, A*

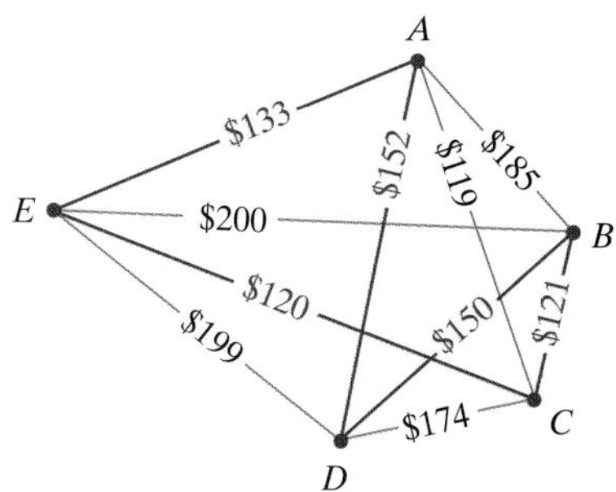

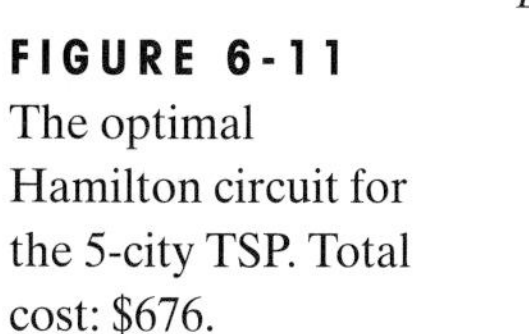
FIGURE 6-11 The optimal Hamilton circuit for the 5-city TSP. Total cost: $676.

All of the above can be reasonably done in somewhere between 10 and 20 minutes—not a bad way to earn $20!

- **Method 2.** *You started at home (A); from there you went to the city to which the cost of travel is the cheapest; from there you went to the next city to which the cost of travel is the cheapest, and so on. From the last city, you went back to A.*

Here are the worked-out details when we follow Method 2: We start at A. Looking at the graph, we see that the cheapest city to go to from A is C (cost: $119). From C, the cheapest city to go to (other than A) is E (cost: $120). From E, the cheapest remaining city to go to is D (cost $199), and from D we have little choice—the only remaining city to visit is B (cost: $150). From B we close the circuit (return home) and go back to A (cost: $185). Using this strategy, we get the circuit A, C, E, D, B, A, shown in Fig. 6-12. The total cost of this circuit is $773.

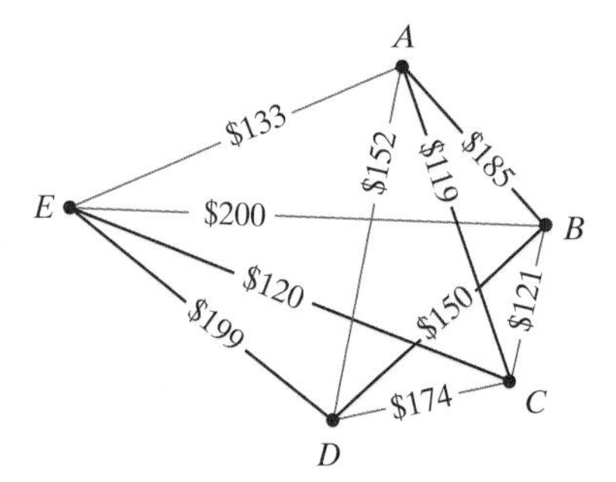

FIGURE 6-12 Circuit obtained following method 2. Total cost: $773.

Using this method, we can work out the problem in just a couple of minutes, which is nice, but there is a hitch: We are not going to collect any money from Willy. Justifiably, Willy is not pleased with this answer, which is $97 higher than the optimal answer obtained under Method 1. This idea looks like a bust! (But be patient, there is more to come.) ■

Example 6. Willy Expands His Territory

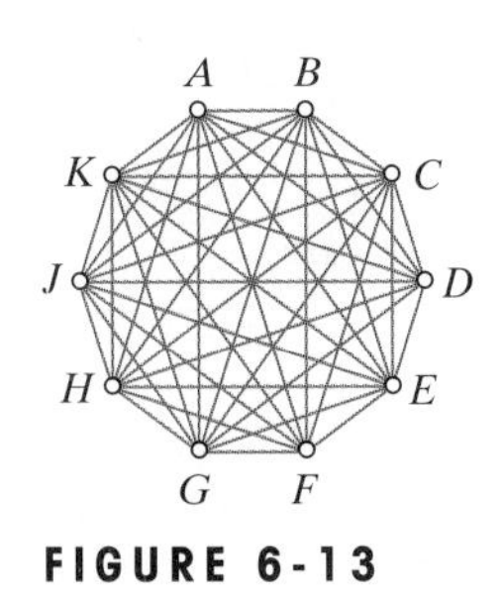

FIGURE 6-13

Let's imagine now that Willy, who has done very well with his business, has expanded his sales territory to ten cities (Fig. 6-13). Willy wants us to help him once again find an optimal Hamilton circuit that starts at A and goes to each of the other 9 cities, and flush with success and generosity, he is offering a whopping $100 as a reward for a solution to this problem. Should we accept the challenge?

The one-way cost of travel between any two cities is shown by means of Table 6-5.[3] We now know that a foolproof method for tackling this problem is Method 1. But before we plunge into it, we think of what we learned earlier in this chapter about factorials and realize that we may be biting off more than we can chew: The number of Hamilton circuits that we would have to check is 9!, which is 362,880. Still thinking about it? Here are some numbers that may help you to make up your mind: If you could do two circuits per minute—and that's working fast!—it would take about 3000 hours to do all 362,880 possible circuits. Shortcuts? Say you are

[3] With this many vertices it is a little easier to put the weights in a table like this one—the graph would get pretty cluttered if we tried to put the weights on the graph itself!

Table 6-5 Cost of travel between any two cities

	A	*B*	*C*	*D*	*E*	*F*	*G*	*H*	*J*	*K*
A	*	185	119	152	133	321	297	277	412	381
B	185	*	121	150	200	404	458	492	379	427
C	119	121	*	174	120	332	439	348	245	443
D	152	150	174	*	199	495	480	500	454	489
E	133	200	120	199	*	315	463	204	396	487
F	321	404	332	495	315	*	356	211	369	222
G	297	458	439	480	463	356	*	471	241	235
H	277	492	348	500	204	211	471	*	283	478
J	412	379	245	454	396	369	241	283	*	304
K	381	427	443	489	487	222	235	478	304	*

clever and cut the work in half—that's still 1500 hours of work (a couple of month's worth if you worked nonstop, 24 hours a day, 7 days a week).

Let's now go back to Method 2, the one that turned out to be a pretty bad idea in Example 5. This is what we get: We start at *A*. From *A*, we would travel to *C*—at a cost of $119 it is the cheapest place to go to. Continuing with our strategy of always choosing the cheapest new city available (we leave it to the reader to verify the details), we would go from *C* to *E;* from *E* to *D;* from *D* to *B;* from *B* to *J;* from *J* to *G;* from *G* to *K;* from *K* to *F;* from *F* to *H;* and then finally back to *A*. It takes just a few minutes to do this, and the Hamilton circuit obtained this way is *A, C, E, D, B, J, G, K, F, H, A* with a total cost of $2153. Well, it's something, but is it a correct answer? Willy, for one, is not convinced and refuses to pay us the $100. Are we right in asking for the money? Is this an optimal circuit? What if it isn't—is it at least close? What do you think? ■

We will return to this example and answer these questions later in the chapter. Before we do so, we will formalize some of the ideas we have just discussed.

The Brute-Force and Nearest-Neighbor Algorithms

We touched on the subject of algorithms in Chapter 5, and we now revisit the concept in a little more detail. Recall that an algorithm is a set of mechanical rules which, when properly followed, produce a specific answer to a problem. Both of the intuitive strategies for finding optimal Hamilton circuits that we discussed in the preceding section are examples of **graph algorithms**. Method 1 goes by the descriptive name of **the brute-force algorithm**; Method 2 has an equally descriptive name—**the nearest-neighbor algorithm**.

Algorithm 1: The Brute-Force Algorithm

- Make a list of all the possible Hamilton circuits of the graph.
- For each Hamilton circuit calculate its total weight by adding the weights of all the edges in the circuit.
- Find the circuits (there is always more than one) with the least total weight. Any one of these can be chosen as an optimal Hamilton circuit for the graph.

Algorithm 2: The Nearest-Neighbor Algorithm

- Pick a vertex as the starting point.
- From the starting vertex go to the vertex for which the corresponding edge has the smallest weight—we call this vertex the *nearest neighbor.* If there is more than one, choose one of them at random.
- Continue building the circuit, one vertex at a time, by always going from a vertex to the nearest neighbor of that vertex *from among the vertices that haven't been visited yet.* (Whenever there is a tie, choose at random.) Keep doing this until all the vertices have been visited.
- From the last vertex return to the starting point.

Based on what we learned in Examples 5 and 6, we know that in a general sense there are some problems with both of these algorithms.

Let's start with the brute-force algorithm. Checking through all possible Hamilton circuits to find the optimal one sounds like a great idea in theory, but, as we saw in Example 6, there is a practical difficulty in trying to use this approach. The difficulty resides in the fantastic growth of the number of Hamilton circuits that need to be checked. In fact, if we are doing things by hand, it would be quite foolhardy to try to use this algorithm except for graphs with a very small number of vertices.

A possible way to get around this, one might think, is to recruit a fast helper, such as a powerful computer. It seems that a computer would be exactly the right kind of helper, because the brute-force algorithm is essentially a mindless exercise in arithmetic with a little bookkeeping thrown in, and both of these are things that computers are very good at. Unfortunately, even the world's most powerful computer won't take us very far.

Let's imagine, for the sake of argument, that we have been given free access to the fastest supercomputer on the planet, one that can compute *ten billion* (that's 10^{10}) circuits per second. (That's more that any current computer could do, but since we are just fantasizing, let's think big!) Now there are 31,536,000 seconds in a year (which, roughly speaking, is 3×10^7). Altogether, this means that our supercomputer can compute about 3×10^{17} Hamilton circuits in one year. For graphs of up to 15 vertices, our helper can run through all the Hamilton circuits in a matter of seconds (or less). Things get more interesting when we start moving beyond 15 vertices. Table 6-6 illustrates what happens then.

Table 6-6 illustrates how extraordinarily fast the computational burden grows with the brute-force algorithm. Each time we increase the number of vertices of the graph by one, *the amount of work required to carry out the algorithm increases by a factor that is equal to the number of vertices in the graph.* For example, it takes 5 times as much work to go from 5 vertices to 6, 10 times more work to go from 10 vertices to 11, and 100 times as much work to go from 100 vertices to 101. Bad news!

The brute-force algorithm is a classic example of what is formally known as an **inefficient algorithm**—an algorithm for which the number of steps needed to carry it out grows disproportionately with the size of the problem. The trouble with inefficient algorithms is that they are of limited practical use—they can realistically be carried out only when the problem is small.

Table 6-6 Solving the TSP using the brute-force algorithm with a supercomputer.

Number of vertices N	Number of Hamilton circuits $(N-1)!$	Amount of time to check them all with a supercomputer
16	1,307,674,368,000	$\approx$ 2 minutes
17	$\approx 2.1 \times 10^{13}$	$\approx$ 35 minutes
18	$\approx 3.6 \times 10^{14}$	$\approx$ 10 hours
19	$\approx 6.4 \times 10^{15}$	$\approx 7\frac{1}{2}$ days
20	$\approx 1.2 \times 10^{17}$	$\approx$ 140 days
21	$\approx 2.4 \times 10^{18}$	$\approx 7\frac{1}{2}$ years
22	$\approx 5.1 \times 10^{19}$	$\approx$ 160 years
23	$\approx 1.1 \times 10^{21}$	$\approx$ 3,500 years
24	$\approx 2.6 \times 10^{22}$	$\approx$ 82,000 years
25	$\approx 6.2 \times 10^{23}$	$\approx$ 2 million years

Fortunately, not all algorithms are inefficient. Let's discuss now the nearest-neighbor algorithm, in which we hop from vertex to vertex using a simple criterion: Where is the next "nearest" place to go to? For a graph with 5 vertices, we have to take 5 steps.[4] What happens when we double the number of vertices to ten? We now have to take ten steps. Essentially, the amount of work doubled when the problem doubled. Could we use the nearest-neighbor algorithm in a complete graph with 100 vertices? You bet. It would take a little longer than in the case of 10 vertices (maybe an hour) but for a nice reward (say $200) it would certainly be worth our trouble.

An algorithm for which the number of steps needed to carry it out grows in proportion to the size of the input to the problem is called an **efficient algorithm**. As a practical matter, efficient algorithms are the only kind of algorithms that we can realistically use on a consistent basis to solve a graph problem.

The nearest-neighbor algorithm is an efficient algorithm, which is good! The problem with it is that, as we saw in Example 5, it doesn't give us what we are asking for—an optimal Hamilton circuit. So why should we even consider an algorithm that doesn't give us the optimal answer? As we will find out next, sometimes we have to take what we can get.

Approximate Algorithms

The ultimate goal in finding a general method for solving TSPs is to find an algorithm that is *efficient* (like the nearest-neighbor algorithm) and **optimal** (meaning that it guarantees us an optimal answer at all times, as the brute-force algorithm does). Unfortunately, nobody knows of such an algorithm. Moreover, *we don't even know why we don't know.* Is it because such an algorithm is actually a mathematical impossibility? Or is it because no one has yet been clever enough to find one?

[4] Actually, the last step (go back to the starting vertex) is automatic—but for the sake of simplicity we'll still call it a step.

Despite the efforts of some of the best mathematicians of our time, the answer to these questions has remained quite elusive. So far, no one has been able to come up with an efficient optimal algorithm for solving TSPs or, alternatively, to prove that such an algorithm is an impossibility. Because this question has profound implications in an area of computer science called complexity theory, it has become one of the most famous unsolved problems in modern mathematics.[5]

In the meantime, we are faced with a quandary: In many real-world applications, it is necessary to find some sort of solution for TSPs involving graphs with hundreds and even thousands of vertices, and to do so in real time. Since the brute-force algorithm is out of the question, and since no efficient algorithm that guarantees an optimal solution is known, the only practical strategy to fall back on is to compromise: We give up on the expectation of having an optimal solution and accept a solution that may not be optimal. In exchange, we ask for quick results. Nowadays, this is the way that TSPs are "solved."

We will use the term **approximate algorithm** to describe any algorithm that produces solutions[6] that are, most of the time, reasonably close to the optimal solution. Sounds good, but what does "most of the time" mean? And how about reasonably close? Unfortunately, to properly answer these questions would take us beyond the scope of this book. We will have to accept the fact that in the area of analyzing algorithms we will be dealing with informal ideas rather than precise definitions.

To appreciate the value of approximate algorithms, let's return to some of the questions we left hanging at the end of Example 6.

Example 6. (continued)

When we left this example a few pages back, we had (a) decided not to try the brute-force algorithm (way too much work for just a $100 reward), and (b) used the nearest-neighbor algorithm to come up with the circuit *A, C, E, D, B, J, G, K, F, H, A* (total cost: $2153). We are still hoping to collect the $100, but Willy refuses to pay. It turns out that the optimal Hamilton circuit for this problem is *A, D, B, C, J, G, K, F, H, E, A* (total cost: $1914). To find this optimal circuit using the brute-force algorithm would have required spending hundreds of hours slaving over numbers and circuits (to be truthful, we found it using a fast computer and some special software). The net savings between the optimal solution and the "quick and easy" one we got using the nearest-neighbor algorithm is $239. Are the savings worth the extra effort? This is an interesting and deep question. In this case the answer is "Probably not." In other situations the answer may be different. ■

The important point to take home from Example 6 is that approximate algorithms are not necessarily bad, and that sometimes we may be better off settling for a quick approximate answer rather than insisting on finding the optimal answer. It is often the case that approximate algorithms are the ones that give the most "bang for the buck."

[5] For an excellent account of this famous problem see Reference 5.

[6] Note that in this context the word "solution" no longer means the "best answer," but simply "an answer."

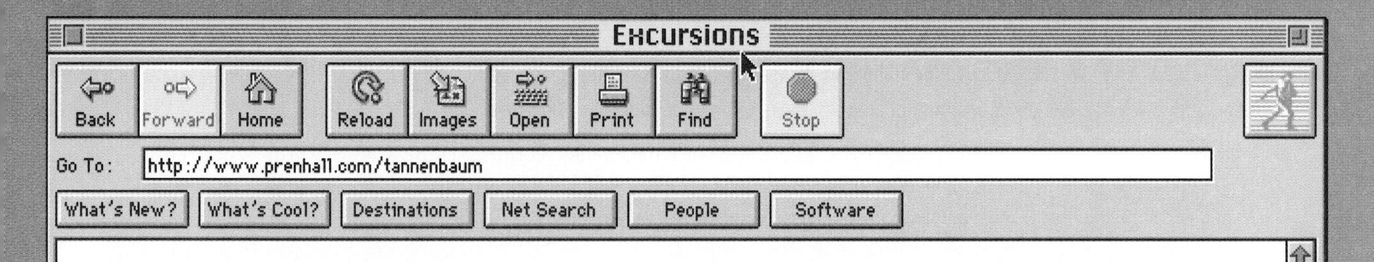

The Elastic Net Method

One ingenious approach to solving the traveling saleman problem is an algorithm created in 1987 by R. Durbin and D. Willshaw. Their approach is to start with a circular path which may miss all of the cities completely, and then gradually deform the path by pulling each point on the path both toward the nearest cities and towerd its neighbors on the path. Eventually—after deforming the path a few hundred times—a path is created which approximates a Hamilton circuit. As you can see, this algorithm requires a tremendous number of calculations, even if just 10 cities are involved, but it is efficient compared to many other methods.

The best way to understand this algorithm is to see it in action, and you can do just that by viewing the Java applet in the Tannenbaum Website and then under the Chapter 6 section.

Questions:

1. Why does "elastic net" seem like an appropriate name for the algorithm?
2. The applet performs calculations on only a limited number of points on the path. How many points does it use to define the path? What would happen if this number were too small? What would happen if it were very large?
3. What is the purpose of the "force" that pulls each point on the path toward its neighbors on the path?
4. Which "force" seems to be stronger—the one that pulls each point on the path to its neighbor, or the one that pulls it toward a city? Explain.
5. Are the "forces" used to deform the path similar to gravity? Explain.
6. The algorithm does not always produce an optimal circuit. Run the applet many times using different arrangements of cities, and see if you can describe some of the "mistakes" that it can make.

In the next two sections, we will discuss a couple of new algorithms for solving TSPs: the *repetitive nearest-neighbor algorithm* and the *cheapest-link algorithm.* Both of these are approximate algorithms.

The Repetitive Nearest-Neighbor Algorithm

As one might guess, the *repetitive nearest-neighbor algorithm* is a variation of the nearest-neighbor algorithm in which we repeat several times the entire nearest-neighbor circuit-building process. Why would we want to do this? The reason is that the Hamilton circuit one gets when applying the nearest-neighbor process depends on the choice of the starting vertex. If we change the starting vertex, it is likely that the Hamilton circuit we get will be different, and, if we are lucky, better. Since finding a Hamilton circuit using the nearest-neighbor algorithm is an efficient process, it is not an unreasonable burden to do it several times, each time using a different vertex of the graph as the starting vertex. In this way, one gets several different nearest-neighbor solutions, from which we can then pick the best.

But what do we do with a Hamilton circuit that starts somewhere other than the vertex we really want to start at? That's not a problem. Remember that once we have a circuit, we can start the circuit anywhere we want—in fact, in an abstract sense, a circuit has no starting or ending point.

To illustrate how the repetitive nearest-neighbor algorithm works, let's return to the original 5-city problem we last discussed in Example 5.

Example 7. A Tale of Five Cities: Part III

Once again, we are going to look at the TSP given by the complete graph in Fig. 6-14, representing Willy's original sales territory. We already know that the optimal Hamilton circuit is given by *A, D, B, C, E, A,* so the main point in this example is just to illustrate how the repetitive nearest-neighbor algorithm works.

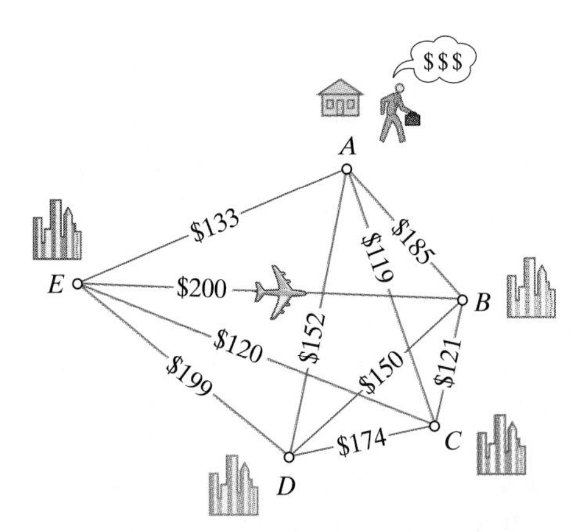

FIGURE 6-14

When we used the nearest-neighbor algorithm with *A* as the starting point, we got the Hamilton circuit *A, C, E, D, B, A* with a total cost of $773. Let's now try it with *B* as the starting point. We leave it to the reader to verify [Exercise 19(a)] that now the nearest-neighbor algorithm yields the Hamilton circuit *B, C, A, E, D, B* with a total cost of $722. Well, that is certainly an improvement! Can a person such as Willy who must start and end his trip at *A* take advantage of this

\$722 circuit? Why not? All he has to do is rewrite the circuit in the equivalent form *A, E, D, B, C, A*.

Having done so well so far, we might as well try the nearest-neighbor algorithm with *C, D,* and *E* as the starting points. We leave it to the reader [Exercise 19(b), (c), and (d)] to verify that when the starting point is *C* we get the Hamilton circuit *C, A, E, D, B, C* with a total cost of \$722; when the starting point is *D* we get the Hamilton circuit *D, B, C, A, E, D,* also with a total cost of \$722; and finally, when the starting point is *E* we get the Hamilton circuit *E, C, A, D, B, E* with a total cost of \$741. None of these improves on the circuit we found when we started at *B* (although starting at *C* and *D* actually gave us the same circuit), so there is our answer: The repetitive nearest-neighbor algorithm gives us the circuit *A, E, D, B, C, A* with a total cost of \$722 (Fig. 6-15). Not a bad improvement over the original \$773 circuit, which we got when we started at *A*.

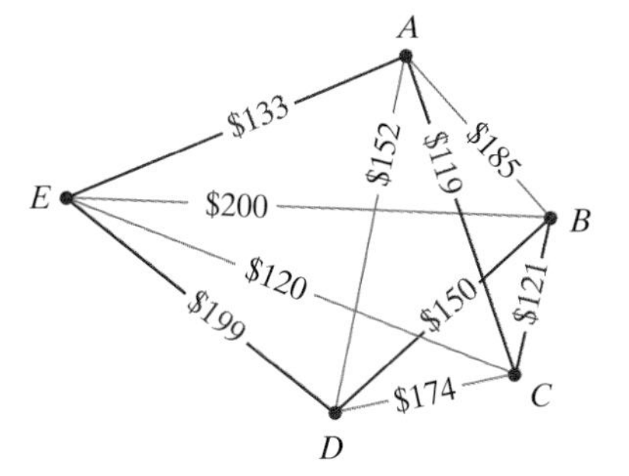

FIGURE 6-15 Hamilton circuit obtained using the repetitive nearest-neighbor algorithm. Total cost: \$722.

A formal description of the repetitive nearest-neighbor algorithm is given below.

Algorithm 3: The Repetitive Nearest-Neighbor Algorithm

- Let *X* be any vertex. Apply the nearest-neighbor algorithm using *X* as the starting vertex and calculate the total cost of the circuit obtained.
- Repeat the process using each of the other vertices of the graph as the starting vertex.
- Of the Hamilton circuits obtained, keep the best one. If there is a designated starting vertex, rewrite this circuit with that vertex as the reference point.

The Cheapest-Link Algorithm

This is the last—but not the least—of our algorithms for finding Hamilton circuits. One lesson of the repetitive nearest-neighbor algorithm is that the order in which one builds a Hamilton circuit and the order in which one actually travels the circuit do not have to be one and the same. In fact, one can build a Hamilton circuit piece by piece without requiring that the pieces be connected, so long as at the end it all comes together. People often use this strategy when putting together a large jigsaw puzzle.

The **cheapest-link algorithm** is essentially an algorithm based on this strategy. One starts by grabbing the cheapest edge of the graph, wherever it may be. Once this is done, one grabs the next cheapest edge of the graph, wherever it may be. We continue doing this, each time grabbing the cheapest edge available, subject to the following two restrictions:

(i) do not allow circuits to form (other than at the very end), and

(ii) do not allow three edges to come together into a vertex.

It is clear that if we allow either of these things to happen, it would be impossible to end up with a Hamilton circuit at the end. Fortunately, these are the only two restrictions we must worry about.

Example 8. A Tale of Five Cities: Part IV

To illustrate the cheapest-link algorithm, we will revisit one final time the 5-city TSP in Example 2. Once again, we show the cost of travel between any two cities (Fig. 6-16).

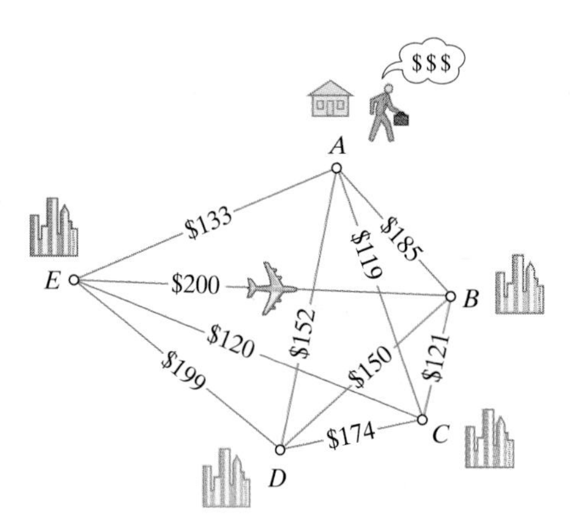

FIGURE 6-16

Our first step is to scan the graph and pick the cheapest of all possible "links," regardless of where they may be. In this case, it is the edge *AC* ($119). We will keep a record of the circuit-building process by marking the edges of our circuit in red. Figure 6-17(a) shows where we are at this point. The next step is to scan the graph again, looking for the cheapest unmarked link available, which in this case is edge *CE* ($120). We mark it in red, as shown in Fig. 6-17(b). Once again, we scan the graph looking for the cheapest unmarked link, which in this case is edge *BC* ($121). But there is a problem: This edge can't be part of the circuit, since a circuit cannot have three edges going into the same vertex—this one we have to throw away [Fig. 6-17(c)]. After *BC,* the next cheapest link is given by edge *AE* ($133). But we have to throw this one away too—the vertices *A, C,* and *E* would be linked in a "short circuit," and a Hamilton circuit can never have a smaller circuit within it [Figure 6-17(d)]!. So we persevere, scanning the graph for the next cheapest link, which is *BD* ($150). This one works, so we add it to our budding circuit [Fig. 6-17(e)]. The next cheapest link available is *AD* ($152), and it works just fine [Fig. 6-17(f)]. (*Warning:* Remember that the only cities are *A, B, C, D,* and *E,* and that other places where edges cross are not vertices, so no "short circuits" are being formed.) At this point, we have only one way to close up the Hamilton circuit (edge *BE*), as shown in Fig. 6-17(g). The Hamilton circuit in red can now be given with any vertex as reference point. Since Willy lives at *A,* we give it as *A, C, E, B, D, A* (or its mirror image). The total cost of this circuit is $741 (a little better than the nearest-neighbor solution but not as good as the repetitive nearest-neighbor solution).

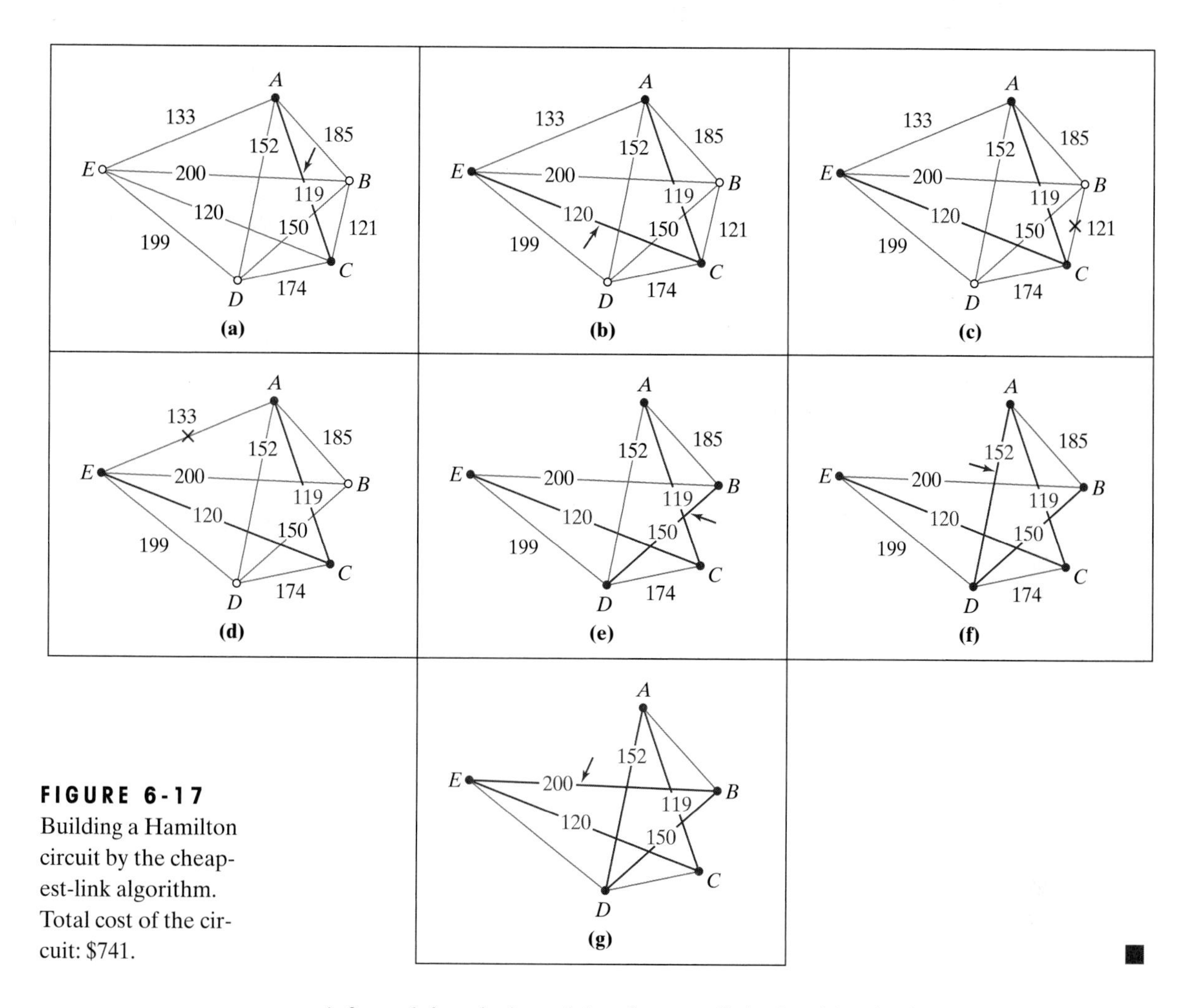

FIGURE 6-17
Building a Hamilton circuit by the cheapest-link algorithm. Total cost of the circuit: $741.

A formal description of the cheapest link algorithm is given below:

> **Algorithm 4: The Cheapest-Link Algorithm**
>
> - Pick the link with the smallest weight first (in case of a tie pick one at random). Mark the corresponding edge (say in red).
> - Pick the next cheapest link and mark the corresponding edge in red.
> - Continue picking the cheapest link available. Mark the corresponding edge in red except when
>
>   (a) it closes a circuit
>
>   (b) it results in three edges coming out of a single vertex.
> - When there are no more vertices to link, close the red circuit.

For the last example of this chapter, we will return to the problem first described in the chapter opener—that of finding an optimal route for a rover exploring Mars.

Example 9.

Figure 6-18(a) shows 7 locations on Mars identified as those where some form of bacterial life is most likely to be found. Our job is to find the shortest route for a rover that will pass through all the sites, starting and ending at A. The approximate distance (in miles) between any two sites is shown in the graph in Fig. 6-18(b)—a complete weighted graph with 7 vertices that will serve as a model for the problem.

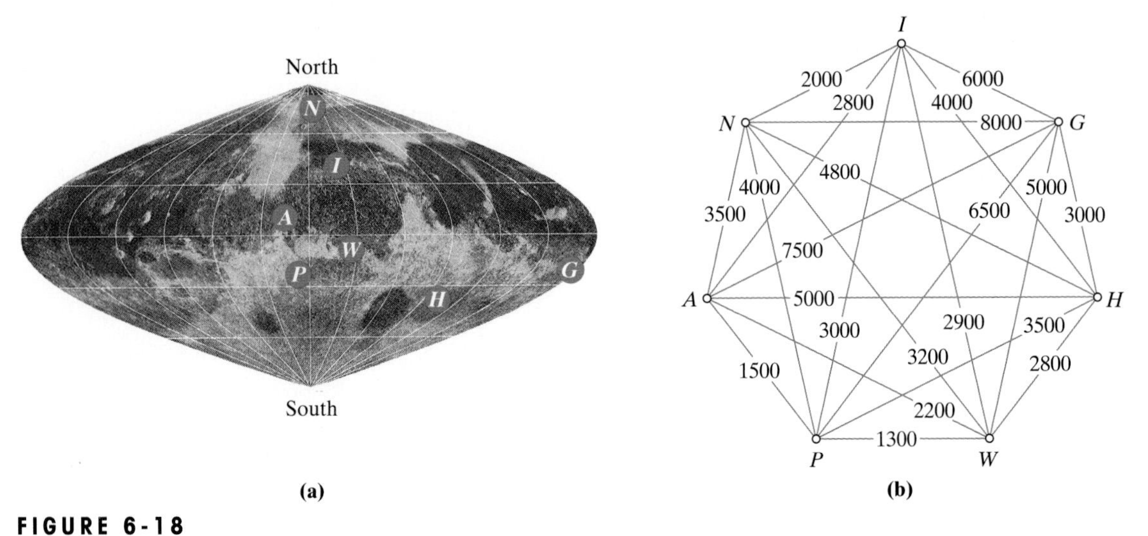

FIGURE 6-18
a) Locations most likely to show evidence of life. b)Approximate distances (in miles) between locations.

We will tackle this problem using several different approaches that we now know of.

- **The Brute-Force Approach.** This is the only method we know that is guaranteed to give us the optimal Hamilton circuit. Unfortunately, it would require us to check through 720 different Hamilton circuits (6! = 720). We will pass on that idea for now.
- **The Cheapest-Link Approach.** This is a reasonable algorithm to use—not trivial but not too hard either. A summary of the steps is shown in Table 6-7.

Table 6-7

Step	Cheapest edge available	Weight	Use in circuit?
1	*PW*	1300	yes
2	*AP*	1500	yes
3	*IN*	2000	yes
4	*AW*	2200	no
5 } tie	*HW* } tie	2800	yes
6	*AI*	2800	yes
7	*IW*	2900	no
8 } tie	*IP* } tie	3000	no
9	*GH*	3000	yes
Last	*GN* only way to close circuit	8000	yes

The circuit obtained using this algorithm is *A, P, W, H, G, N, I, A* with a total length of 21,400 miles (Fig. 6-19).

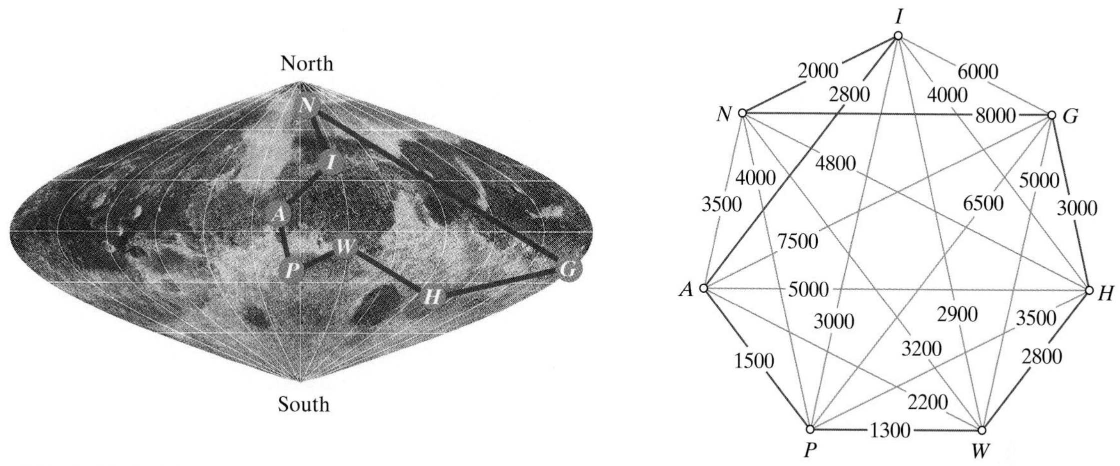

FIGURE 6-19
Hamilton circuit obtained using cheapest-link algorithm. Total length: 21,400 miles.

- **The Nearest-Neighbor Approach.** This is the simplest of all algorithms we learned. Starting from *A* we go to *P*, then to *W*, then to *H*, then to *G*, then to *I*, then to *N*, and finally back to *A*. The circuit obtained under this algorithm is *A, P, W, H, G, I, N, A* with a total length of 20,100 miles (Fig. 6-20). (We know that we can also use this approach but start it from a different location, but we won't bother with that at this time.)

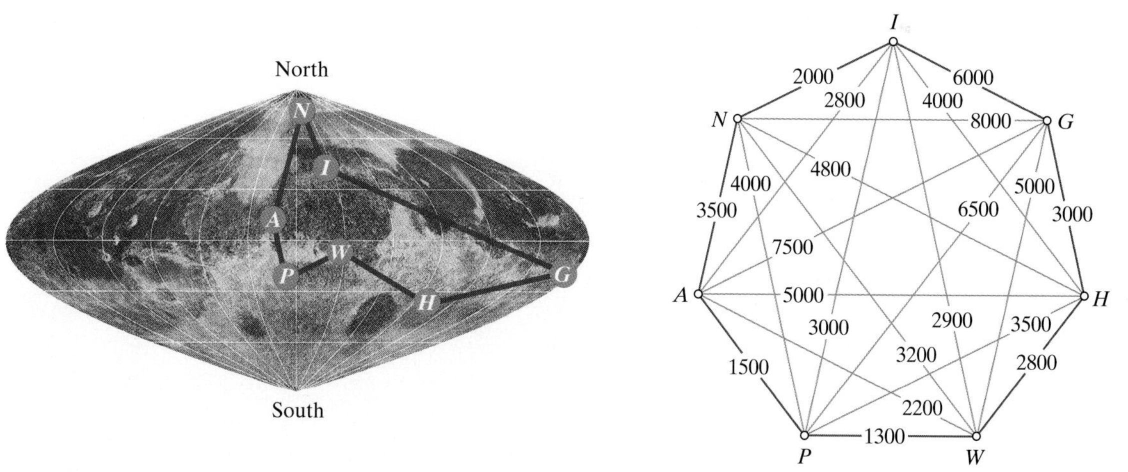

FIGURE 6-20
Hamilton circuit obtained using nearest-neighbor algorithm. Total length: 20,100 miles.

The first surprise is that the nearest-neighbor algorithm gives us a better Hamilton circuit than the cheapest-link algorithm. It happens this way about as often as it happens the other way around, so neither of the two algorithms can claim to be superior to the other one.

The second surprise is that the circuit *A, P, W, H, G, I, N, A* obtained using the nearest-neighbor algorithm turns out to be the optimal Hamilton circuit for this example. (We know this because we used a computer to find the optimal Hamilton circuit.) Essentially, this means that in this particular case, the simplest of all methods turned out to give us the optimal answer. Wouldn't it be nice if things always turned out this way? Well, maybe next chapter. ■

Conclusion

In this chapter we discussed the problem of finding optimal Hamilton circuits in complete weighted graphs, a problem usually known by the generic name of the *traveling-salesman problem (TSP)*. In many situations, finding an optimal Hamilton circuit is reasonably easy, but a completely general algorithm that would work for every TSP has eluded mathematicians who have been interested in this problem for more than 50 years. This is an extremely important and at the same time notoriously difficult problem.

The nearest-neighbor and cheapest-link algorithms, which we learned in this chapter, are two fairly simple strategies for attacking TSPs, but sophisticated variations of these strategies are in fact being used today in business and industry to solve important real-life applications involving thousands of vertices. We have seen that both algorithms are approximate algorithms, which means that they are not likely to give us an optimal solution, although, as we saw in Example 8, on a lucky day even that is possible. By the same token, on an unlucky day either of these two algorithms can give us the worst possible Hamilton circuit (see Exercises 53 and 54). In most typical problems, however, one can expect either of these algorithms to give an approximate solution that is within a reasonable margin of error. With some problems the cheapest-link algorithm gives a better solution than the nearest-neighbor algorithm; with other problems it's the other way around. Thus, while they are not the same, neither is superior to the other.

There is a great deal of interest among mathematicians and computer scientists in finding ever improving approximate algorithms for solving TSPs, and many

A 48-city TSP with a little bite: What is the shortest circuit that visits all of the state capitals in the continental United States? The *optimal solution*, shown in red, is a Hamilton circuit approximately 12,000 miles long. Even with a fast computer and special software, it might take months of calculations to come up with this answer.

For the same 48-city TSP, *approximate solutions* can be found using efficient algorithms in a matter of just minutes. This approximate solution (shown in red) was obtained by hand in less than ten minutes, using the nearest-neighbor algorithm (starting at Olympia, Washington). The total length of this trip is approximately 14,500 miles, roughly 20% longer than the optimal solution.

sophisticated algorithms are known. Some of these algorithms have *performance guarantees* certifying that the solution will never be off by more than a certain percent from the optimal solution (sort of like a certificate of accuracy on a watch). One such algorithm is given in Exercise 59. At present, the best approximate algorithms known give solutions guaranteed to be within 1% of the optimal solution for problems with up to 100,000 vertices (see Table 6-8). Constant refinements of these algorithms and improvements in technology guarantee that such levels of performance are going to get even better.

On a theoretical level, the fundamental question for TSPs remains unsolved: Is there an algorithm that is both efficient and optimal, or is such an algorithm a mathematical impossibility? This problem is still waiting for the next Euler to come along.

Table 6-8 What can be done with TSPs?

Number of vertices in the TSP	Computer time* needed for an approximate solution within 3.5% of optimal	Computer time* needed for an approximate solution within 1% of optimal	Computer time* needed for an approximate solution within 0.75% of optimal	Computer time* needed for an optimal solution
Up to 3000 vertices	A few seconds	A few minutes	Less than 1 hour	Less than 1 week
About 100,000 vertices	A few minutes	Less than 2 days	About 7 months	Can't be done
About 1 million vertices	Less than 3 hours	Hundreds of years	Can't be done	Can't be done

*All times are based on the use of state-of-the-art algorithms and a super computer or several hundred small computers working in parallel.

Key Concepts

approximate algorithm
algorithm
brute-force algorithm
cheapest-link algorithm
complete graph
complete weighted graph
efficient algorithm
factorial
Hamilton circuit
Hamilton path
inefficient algorithm
nearest-neighbor algorithm
optimal
repetitive nearest-neighbor algorithm
traveling-salesman problem (TSP)
weighted graph
weight

Exercises

Walking

1. Two of the Hamilton circuits in the graph below are *A, B, D, C, E, F, G, A* and *A, D, C, E, B, G, F, A*.

(a) Find another Hamilton circuit (not a mirror-image circuit). Use *A* as the starting point.

(b) Find a Hamilton path that starts at *A* and ends at *B*.

(c) Find a Hamilton path that starts at *D* and ends at *F*.

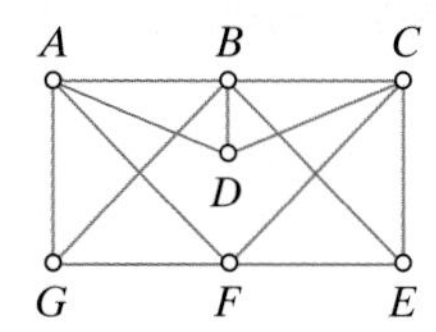

2. (a) Find two Hamilton circuits (not mirror-image circuits) in the graph below.

(b) Find a Hamilton path that starts at *A* and ends at *B*.

(c) Find a Hamilton path that starts at *F* and ends at *I*.

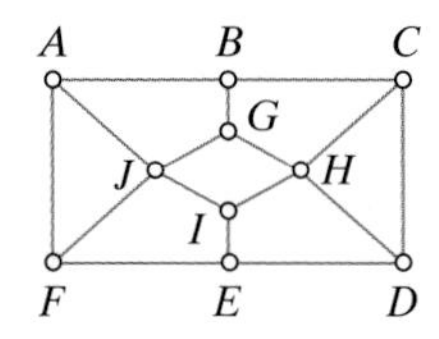

3. List all possible Hamilton circuits in the following graph.

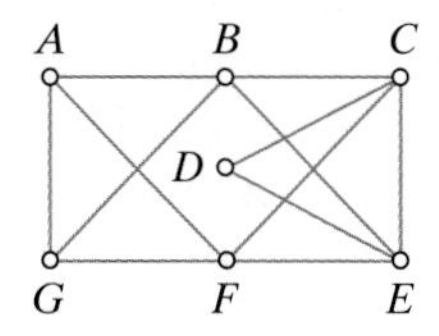

4. List all possible Hamilton circuits in the following graph.

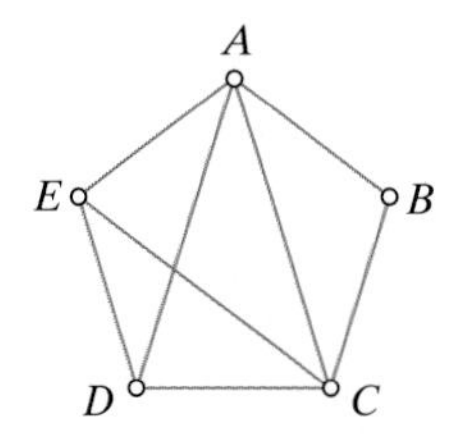

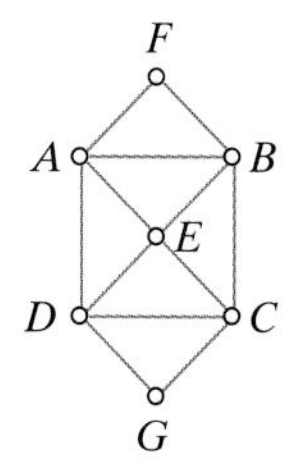

5. (a) Find two Hamilton circuits (not mirror-image circuits) in the graph shown in the margin.

(b) Find a Hamilton path that starts at *A* and ends at *E*.

(c) Find a Hamilton path that starts at *A* and ends at *C*.

(d) Find a Hamilton path that starts at *F* and ends at *G*.

6. (a) Find two Hamilton circuits (not mirror-image circuits) in the graph below.

(b) Find a Hamilton path that starts at *A* and ends at *E*.

(c) Find a Hamilton path that starts at *A* and ends at *G*.

(d) Find a Hamilton path that starts at *F* and ends at *G*.

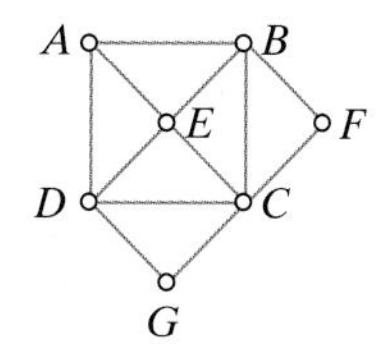

7. List all possible Hamilton circuits in the graph below.

(a) starting at vertex *A*

(b) starting at vertex *D*.

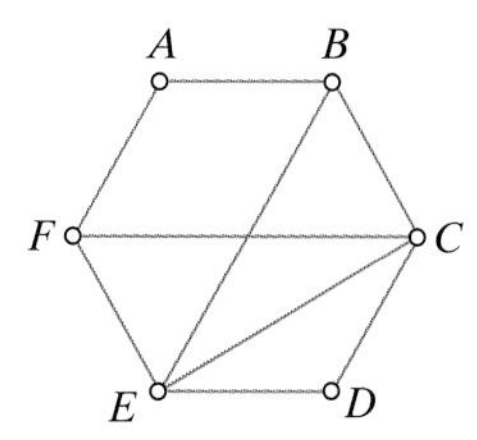

8. List all possible Hamilton circuits in the graph below.

(a) starting at vertex *A*

(b) starting at vertex *C*.

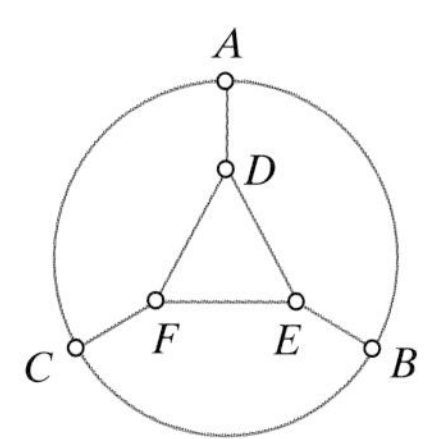

9. Explain why the following graph has no Hamilton circuit and no Hamilton paths.

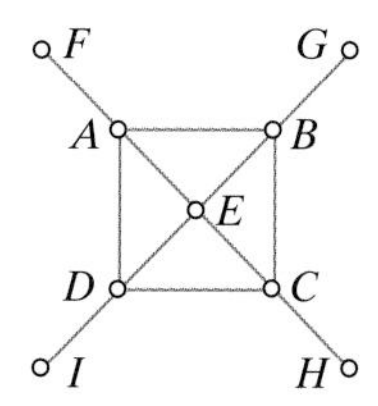

10. Explain why the following graph has no Hamilton circuit but does have a Hamilton path.

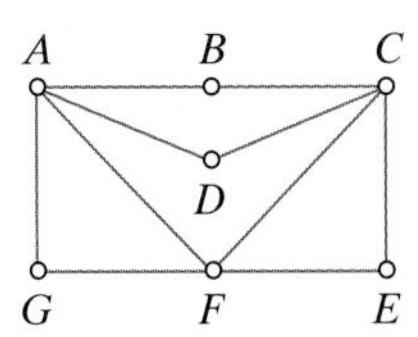

11. **(a)** In the weighted graph below, find the weight of edge BD.

(b) Find the weight of edge EC.

(c) Find a Hamilton circuit in the graph and give its weight.

(d) Find a different Hamilton circuit in the graph and give its weight.

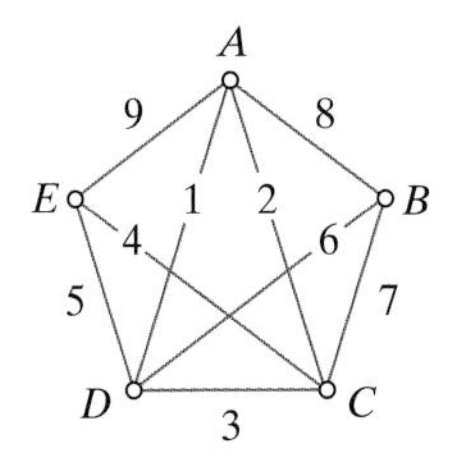

12. **(a)** In the weighted graph below, find the weight of edge AD.

(b) Find the weight of edge AC.

(c) Find a Hamilton circuit in the graph and give its weight.

(d) Find a different Hamilton circuit in the graph and give its weight.

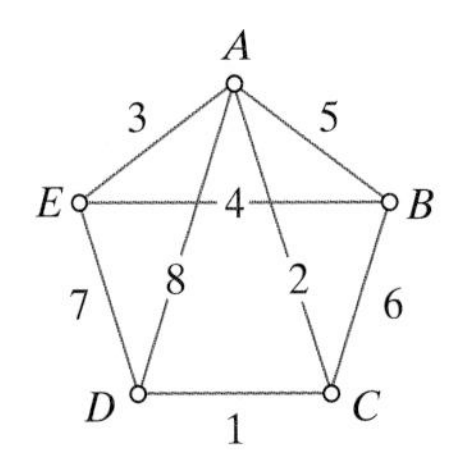

13. **(a)** In the weighted graph below, find the weight of edge BC.

(b) Find a Hamilton circuit in the graph and give its weight.

(c) Find a different Hamilton circuit in the graph and give its weight.

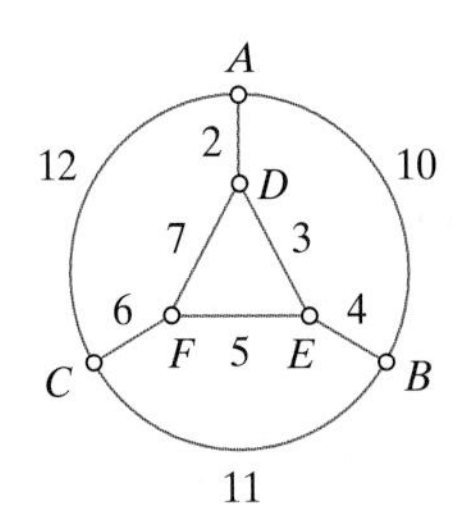

14. **(a)** In the following weighted graph, find the weight of edge AC.

(b) Find a Hamilton circuit in the graph and give its weight.

(c) Find a different Hamilton circuit in the graph and give its weight.

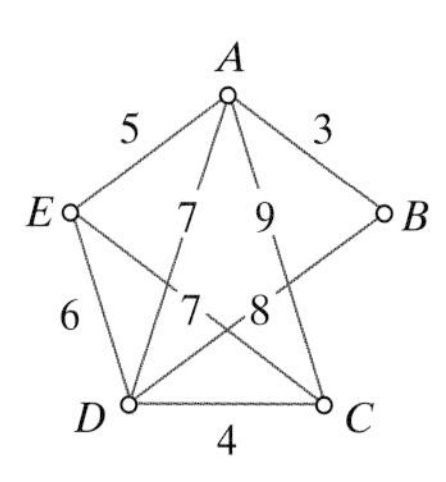

15. (a) Use the fact 5! = 120 to painlessly find 6! (No calculators, please.)

(b) Given that 10! = 3,628,800, find 9! (No calculators, please.)

(c) How many different Hamilton circuits are there in a complete graph with 10 vertices?

16. (a) Use the fact 7! = 5040 to painlessly find 8! (No calculators, please.)

(b) How many different Hamilton circuits are there in a complete graph with 9 vertices?

17. Consider the following weighted graph.

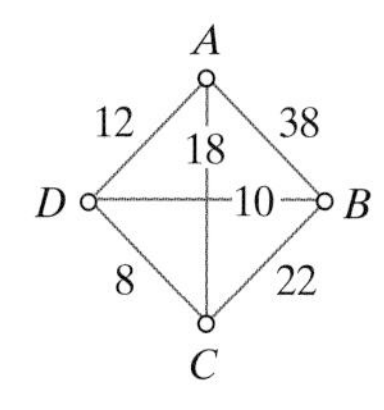

(a) Use the brute-force algorithm to find an optimal Hamilton circuit.

(b) Use the nearest-neighbor algorithm with starting vertex A to find a Hamilton circuit.

(c) Use the cheapest-link algorithm to find a Hamilton circuit.

(d) Compare the optimal solution obtained in (a) with the Hamilton circuits obtained in (b) and (c). Give the relative error in each of these solutions.

$$\text{Relative error} = \frac{(\text{cost of approximate solution}) - (\text{cost of optimal solution})}{\text{cost of optimal solution}}$$

18. Consider the following weighted graph.

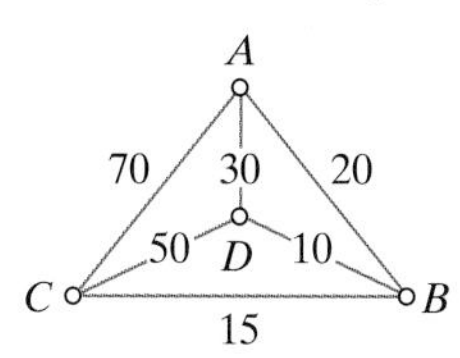

(a) Use the brute-force algorithm to find an optimal Hamilton circuit.

(b) Use the nearest-neighbor algorithm with starting vertex A to find a Hamilton circuit.

(c) Use the cheapest-link algorithm to find a Hamilton circuit.

(d) Compare the optimal solution obtained in (a) with the Hamilton circuits obtained in (b) and (c). Give the relative error in each of these solutions.

$$\text{Relative error} = \frac{(\text{cost of approximate solution}) - (\text{cost of optimal solution})}{\text{cost of optimal solution}}$$

19. The following is the weighted graph of Willy's original sales-territory problem (Examples 2, 5, 7, 8).

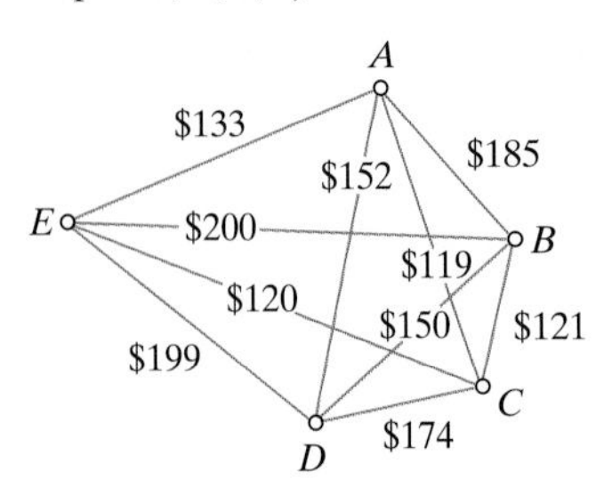

(a) Use the nearest-neighbor algorithm with starting vertex B to find a Hamilton circuit and verify that its weight is $722.

(b) Use the nearest-neighbor algorithm with starting vertex C to find a Hamilton circuit and verify that its weight is $722.

(c) Use the nearest-neighbor algorithm with starting vertex D to find a Hamilton circuit and verify that its weight is $722.

(d) Use the nearest-neighbor algorithm with starting vertex E to find a Hamilton circuit and verify that its weight is $741.

Exercises 20 through 23 refer to the following situation. Sophie, a traveling salesperson, must call on customers in five different cities (A, B, C, D, and E). Each edge of the graph shows the cost of travel between any two cities. Sophie's trip must start and end at her home town (A).

20. Apply the nearest-neighbor algorithm with starting vertex A to find a Hamilton circuit in the graph. What is Sophie's cost for this Hamilton circuit?

21. Apply the cheapest-link algorithm to find a Hamilton circuit in the graph. What is Sophie's cost for this Hamilton circuit?

22. Apply the repetitive nearest-neighbor algorithm to find a Hamilton circuit in the graph. What is Sophie's cost for this Hamilton circuit?

23. Using the brute-force algorithm, find the optimal trip for Sophie. What is Sophie's cost for the optimal trip?

Exercises 24 through 26 refer to the following situation (Example 3 in the chapter). A space expedition is scheduled to visit the moons Callisto (C), Ganymede (G), Titan (T), Hyperion (H), and Phoebe (P) to collect rock samples at each, and then return to Earth (E). The following graph summarizes the travel time (in years) between any two places.

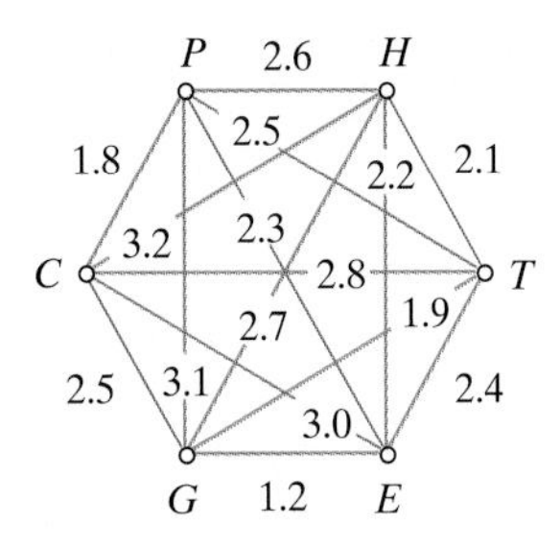

24. Apply the nearest-neighbor algorithm with starting vertex E to find a Hamilton circuit in the graph. What is the travel time for this Hamilton circuit?

25. Apply the cheapest-link algorithm to find a Hamilton circuit in the graph. What is the travel time for this Hamilton circuit?

26. Apply the repetitive nearest-neighbor algorithm to find a Hamilton circuit in the graph. What is the travel time for this Hamilton circuit?

Exercises 27 and 28 refer to the following weighted graph.

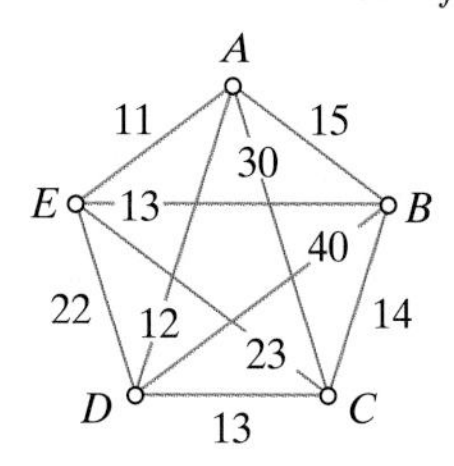

27. (a) Use the cheapest-link algorithm to find a Hamilton circuit in the weighted graph.

(b) Is there a cheaper Hamilton circuit? If so, find one. If not, explain why.

28. (a) Use the nearest-neighbor algorithm starting at vertex A to find a Hamilton circuit in the weighted graph.

(b) Is there a cheaper Hamilton circuit? If so, find one. If not, explain why.

Exercises 29 and 30 refer to the following weighted graph.

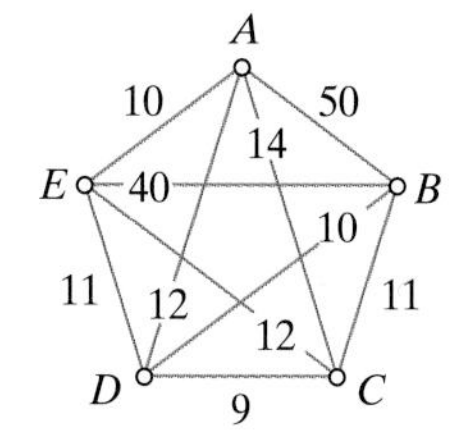

29. Use the cheapest-link algorithm to find a Hamilton circuit in the graph.

30. Use the repetitive nearest-neighbor algorithm to find a Hamilton circuit in the graph.

Exercises 31 through 33 refer to the following weighted graph.

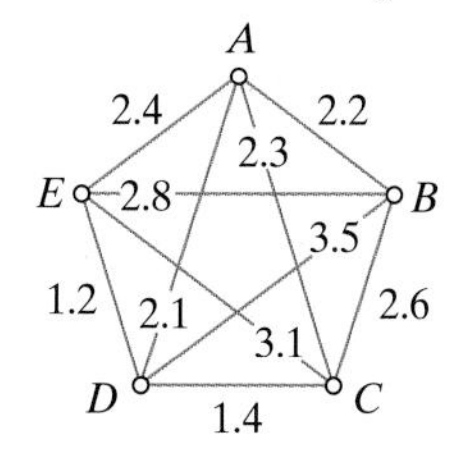

31. Use the cheapest-link algorithm to find a Hamilton circuit.

32. Use the nearest-neighbor algorithm starting at vertex A to find a Hamilton circuit.

33. Use the repetitive nearest-neighbor algorithm to find a Hamilton circuit.

Exercises 34 through 36 refer to the following weighted graph.

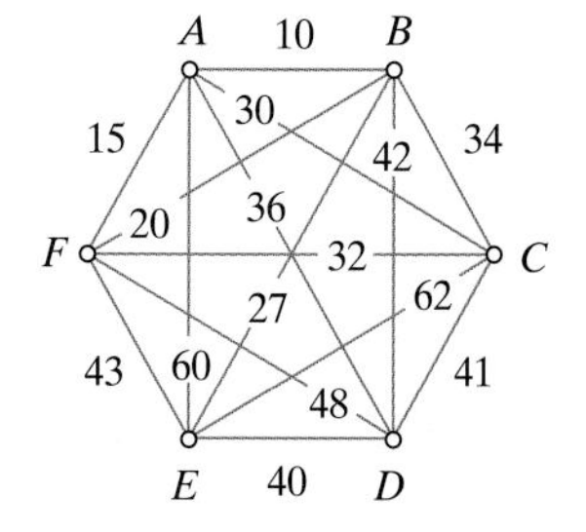

34. Use the cheapest-link algorithm to find a Hamilton circuit.

35. Use the nearest-neighbor algorithm starting at vertex A to find a Hamilton circuit.

36. Use the repetitive nearest-neighbor algorithm to find a Hamilton circuit.

37. You have a busy day ahead of you. You must run the following errands (in no particular order): go to the post office, deposit a check at the bank, pick up some French bread at the deli, visit a friend at the hospital, and get a haircut at Karl's Beauty Salon. You must start and end at home. Each block on the map is exactly 1 mile.

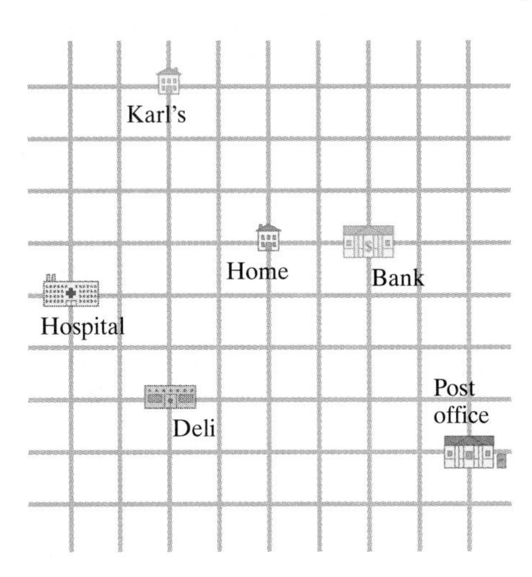

(a) Draw a weighted graph corresponding to this problem.

(b) Find the optimal (shortest) way to run all the errands. (Use any algorithm you think is appropriate.)

38. Rosa's Floral must deliver flowers to each of the five locations A, B, C, D, and E shown on the map. The trip must start and end at the flower shop, which is located at X. Each block on the map is exactly 1 mile.

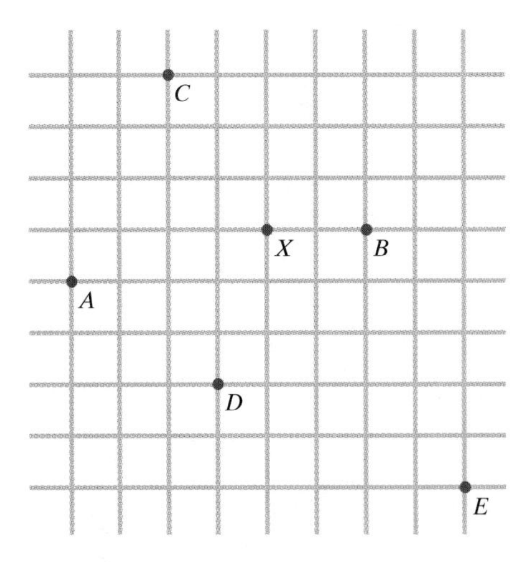

(a) Draw a weighted graph corresponding to this problem.

(b) Find the optimal (shortest) way to make all the deliveries. (Use any algorithm you think is appropriate.)

Jogging

39. Explain why the number of edges in a complete graph with N vertices is $N(N - 1)/2$.

40. Explain why the cheapest edge in any graph is always part of the nearest-neighbor algorithm solution.

41. Give an example of a complete graph with 6 vertices so that the nearest-neighbor algorithm (at every vertex) and the cheapest-link algorithm give the optimal solution. Choose the weights of the edges to be all different.

42. Hamilton's puzzle. Find a Hamilton circuit in the following graph. Indicate your solution by marking the circuit right on the graph.

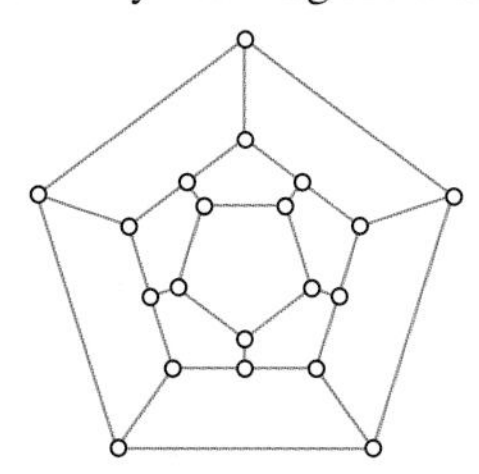

Historical footnote. Hamilton made up and marketed a game which was a three-dimensional version of this exercise, using a regular dodecahedron (see figure) in which the vertices were different cities. The purpose of the game was to find a "trip around the world" going from city to city along the edges of the dodecahedron without going back to any city (except for the return to the starting point). When a regular dodecahedron is flattened, we get the graph shown in this exercise.

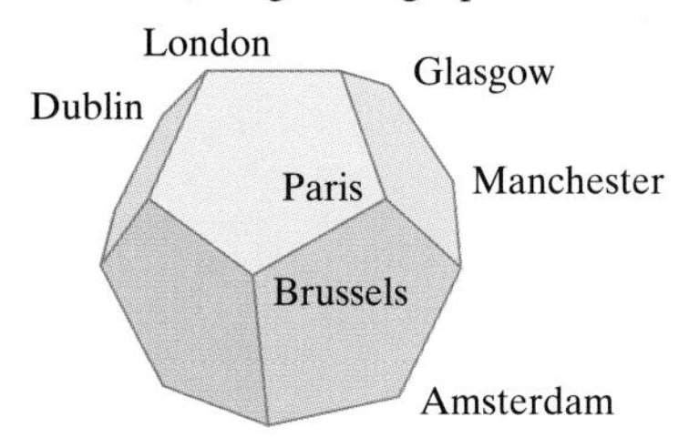

43. Find a Hamilton circuit in the following graph.

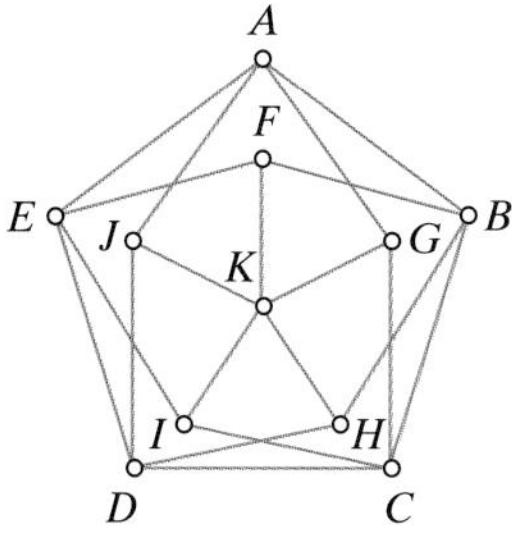

44. Petersen graph. The following graph is called the Petersen graph. Find a Hamilton path in the Petersen graph.

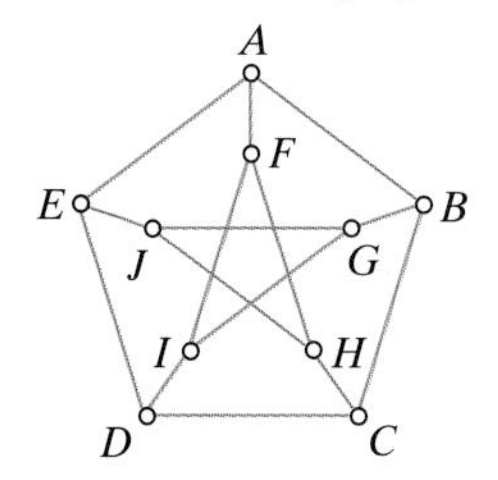

45. Find a Hamilton path in the following graph.

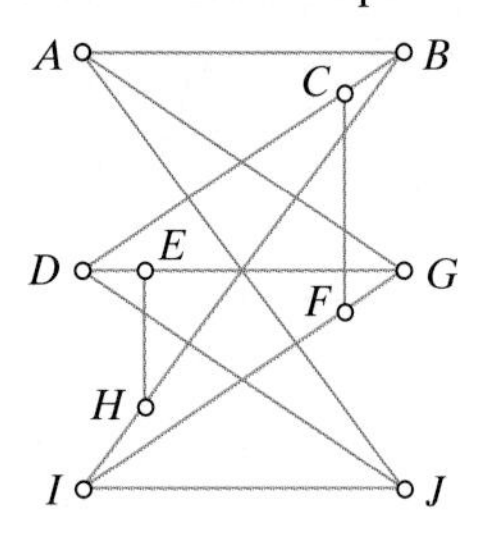

46. **(a)** Give an example of a graph with 4 vertices in which the same circuit can be both an Euler circuit and a Hamilton circuit.

(b) Give an example of a graph with N vertices in which the same circuit can be both an Euler circuit and a Hamilton circuit.

47. **(a)** Compare 2^3 with 3! Which is bigger?

(b) Compare 2^4 with 4! Which is bigger?

(c) Suppose N is more than 5. If you had a choice between 2^N and $N!$ dollars, which would you choose? Explain why the one you picked is the bigger number.

48. Explain why 21! is more than 100 billion times bigger than 10! (i.e., show that $21! > 10^{11} \times 10!$).

Exercises 49 and 50 refer to the following situation: A traveling salesperson's territory consists of the 11 cities shown on the mileage chart below. The salesperson must organize a round trip that starts and ends in Dallas (that's home) and passes through each of the other 10 cities exactly once.

49. Use the nearest-neighbor algorithm with starting city Dallas to find a Hamilton circuit for the traveling salesperson.

50. Use the cheapest-link algorithm to find a Hamilton circuit for the traveling salesperson.

Mileage Chart

	Atlanta	**Boston**	**Buffalo**	**Chicago**	**Columbus**	**Dallas**	**Denver**	**Houston**	**Kansas City**	**Louisville**	**Memphis**
Atlanta	*	1037	859	674	533	795	1398	789	798	382	371
Boston	1037	*	446	963	735	1748	1949	1804	1391	941	1293
Buffalo	859	446	*	522	326	1346	1508	1460	966	532	899
Chicago	674	963	522	*	308	917	996	1067	499	292	530
Columbus	533	735	326	308	*	1028	1229	1137	656	209	576
Dallas	795	1748	1346	917	1028	*	781	243	489	819	452
Denver	1398	1949	1508	996	1229	781	*	1019	600	1120	1040
Houston	789	1804	1460	1067	1137	243	1019	*	710	928	561
Kansas City	798	1391	966	499	656	489	600	710	*	520	451
Louisville	382	941	532	292	209	819	1120	928	520	*	367
Memphis	371	1293	899	530	576	452	1040	561	451	367	*

Running

51. Compare bipartite graphs. A complete bipartite graph is one in which the vertices can be divided into two sets A and B and each vertex in set A is adjacent to each of the vertices in set B. There are no other edges! (If there are m vertices in set A and n vertices in set B, the complete bipartite graph is written as $K_{m, n}$.

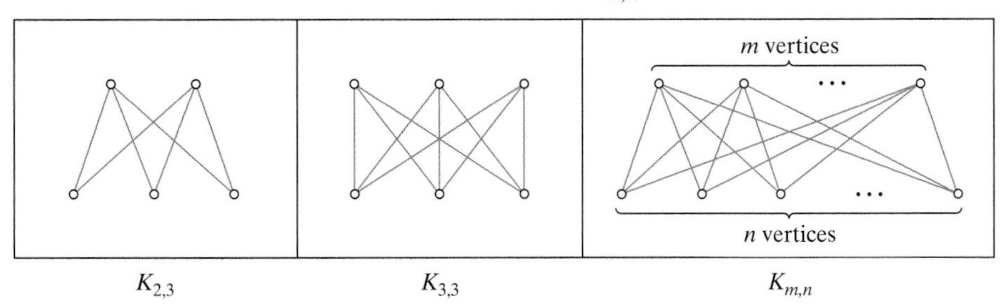

$K_{2,3}$ $K_{3,3}$ $K_{m,n}$

(a) Explain why when $m \neq n$, $K_{m, n}$ cannot have a Hamilton circuit.

(b) Explain why for $n > 1$, $K_{n, n}$ always has a Hamilton circuit.

52. Explain why the Peterson graph below (also see Exercise 44) does not have a Hamilton circuit.

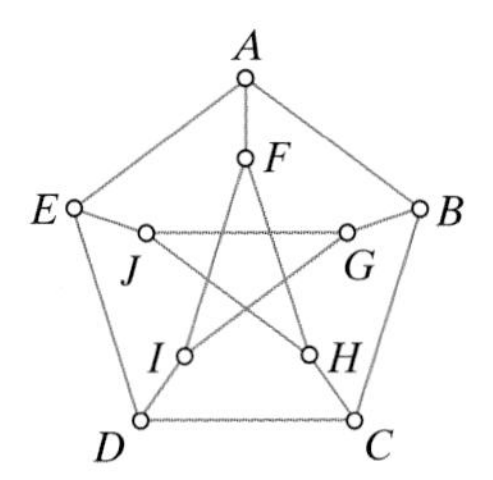

53. Make up an example of a complete weighted graph such that the Hamilton circuit produced by the nearest-neighbor algorithm (you may pick the starting vertex) gives the worst possible choice of a circuit (in other words, one whose weight is bigger than any other).

54. Make up an example of a complete weighted graph such that the Hamilton circuit produced by the cheapest-link algorithm gives the worst possible choice of a circuit.

55. Make up an example of a complete weighted graph such that the Hamilton circuit produced by the nearest-neighbor algorithm (you can pick the starting vertex) has a relative percentage error of at least 100% (in other words, the weight of the Hamilton circuit produced by the nearest-neighbor algorithm is at least twice as much as the weight of the optimal Hamilton circuit).

56. Make up an example of a complete weighted graph such that the Hamilton circuit produced by the cheapest-link algorithm has a relative percentage error of at least 100% (in other words, the weight of the Hamilton circuit produced by the cheapest-link algorithm is at least twice as great as the weight of the optimal Hamilton circuit).

57. The knight's tour. A knight is on the upper left-hand corner of a 3-by-4 "chessboard" as shown in the figure.

A	B	C	D
E	F	G	H
I	J	K	L

(a) Draw a graph with the vertices representing the squares on the board and the edges representing the allowable chess moves of the knight (e.g., an edge joining vertices *A* and *J* means that the knight is allowed to move from square *A* to square *J* or vice versa in a single move).

(b) Find a Hamilton path starting at vertex *A* in the graph drawn in (a) and thus show how to move the knight so that it visits each square of the board exactly once starting at square *A*.

(c) Show that the graph drawn in (a) does not have a Hamilton circuit and consequently that it is impossible for the knight to move so that it visits each square of the board exactly once and then returns to its starting point.

58. Using the ideas of Exercise 57, show that it is possible for a knight to visit each square of a 6-by-6 "chessboard" exactly once and return to its starting point, and that this is true regardless of which square is used as the knight's starting point.

59. The nearest-insertion algorithm. Here is a description of a different approximate algorithm for the traveling-salesman problem. The basic idea is to start with a small subcircuit of the graph and enlarge it one vertex at a time until all the vertices are included and it is a full-fledged Hamilton circuit.

- **Step 1.** Pick any vertex as a starting circuit (consisting of one vertex and zero edges). Mark it "red" ("red" is just a figure of speech for "any color").
- **Next step.** Suppose that at step k we have already built a red subcircuit with k vertices (call it C_k). We look for a black vertex in the graph that is as close as possible to some vertex of C_k. Let's call this black vertex *B*, and the vertex of C_k it is nearest to, *R*. We now create a new red circuit C_{k+1} which is the same as C_k except that *B* is inserted immediately after *R* in the sequence. Repeat until you have a Hamilton circuit.

(a) Verify that when the nearest-insertion algorithm is applied to the traveling-salesman problem in Exercises 20–23, the following sequence of circuits is produced (we use *A* as our starting vertex):

- C_1: *A*
- C_2: *A*, *D*, *A*, (*D* is the nearest vertex to C_1)
- C_3: *A*, *C*, *D*, *A*, (*C* is the nearest vertex to *A* in C_2)
- C_4: *A*, *C*, *E*, *D*, *A*, (*E* is the nearest vertex to *C* in C_3)
- C_5: *A*, *C*, *B*, *E*, *D*, *A*, (*B* is the nearest vertex to *C* in C_4).

(b) Use the nearest-insertion algorithm to find a Hamilton circuit for the graph in Exercise 31. Use *A* as the starting vertex.

(c) Use the nearest-insertion algorithm to find a Hamilton circuit for the graph in Exercise 31. Use *B* as the starting vertex.

(d) Use the nearest-insertion algorithm to find a Hamilton circuit for the graph in Exercise 31. Use *C* as the starting vertex.

60. (Open-ended question) The Great Kaliningrad Circus has been signed for an extended tour in the United States. The tour is scheduled to start and end in Miami, Florida, and visit 20 other cities in between. The cities and distances between the cities are shown in the mileage chart on p. 221. The cost of transporting an entire circus the size of the Great Kaliningrad can be estimated to be about $1000 per mile, so finding a "good" Hamilton circuit for the 21 cities is clearly an important part of the organization of the tour. Your job is to do the best you can to come up with a reasonably good tour for the circus. You should not only describe the actual tour, but also explain what strategies you used to come up with it and why you think that your answer is a reasonable one. The tools at your disposal are everything you learned in this chapter (including Exercise 59), a limited amount of time, and your own ingenuity.

Mileage Chart

	Atlanta	Boston	Buffalo	Chicago	Columbus	Dallas	Denver	Houston	Kansas City	Louisville	Memphis	Miami	Minneapolis	Nashville	New York	Omaha	Pierre	Pittsburgh	Raleigh	St. Louis	Tulsa
Atlanta	*	1037	859	674	533	795	1398	789	798	382	371	655	1068	242	841	986	1361	687	372	541	772
Boston	1037	*	446	963	735	1748	1949	1804	1391	941	1293	1504	1368	1088	206	1412	1726	561	685	1141	1537
Buffalo	859	446	*	522	326	1346	1508	1460	966	532	899	1409	927	700	372	971	1285	216	605	716	1112
Chicago	674	963	522	*	308	917	996	1067	499	292	530	1329	405	446	802	459	763	452	784	289	683
Columbus	533	735	326	308	*	1028	1229	1137	656	209	576	1160	713	377	542	750	1071	182	491	406	802
Dallas	795	1748	1346	917	1028	*	781	243	489	819	452	1300	936	660	1552	644	943	1204	1166	630	257
Denver	1398	1949	1508	996	1229	781	*	1019	600	1120	1040	2037	841	1156	1771	537	518	1411	1661	857	681
Houston	789	1804	1460	1067	1137	243	1019	*	710	928	561	1190	1157	769	1608	865	1186	1313	1160	779	478
Kansas City	798	1391	966	499	656	489	600	710	*	520	451	1448	447	556	1198	201	592	838	1061	257	248
Louisville	382	941	532	292	209	819	1120	928	520	*	367	1037	697	168	748	687	1055	388	541	263	659
Memphis	371	1293	899	530	576	452	1040	561	451	367	*	997	826	208	1100	652	1043	752	728	285	401
Miami	655	1504	1409	1329	1160	1300	2037	1190	1448	1037	997	*	1723	897	1308	1641	2016	1200	819	1196	1398
Minneapolis	1068	1368	927	405	713	936	841	1157	447	697	826	1723	*	826	1207	357	394	857	1189	552	695
Nashville	242	1088	700	446	377	660	1156	769	556	168	208	897	826	*	892	744	1119	553	521	299	609
New York	841	206	372	802	542	1552	1771	1608	1198	748	1100	1308	1207	892	*	1251	1565	368	489	948	1344
Omaha	986	1412	971	459	750	644	537	865	201	687	652	1641	357	744	1251	*	391	895	1214	449	387
Pierre	1361	1726	1285	763	1071	943	518	1186	592	1055	1043	2016	394	1119	1565	391	*	1215	1547	824	760
Pittsburgh	687	561	216	452	182	1204	1411	1313	838	388	752	1200	857	553	368	895	1215	*	445	588	984
Raleigh	372	685	605	784	491	1166	1661	1160	1061	541	728	819	1189	521	489	1214	1547	445	*	804	1129
St. Louis	541	1141	716	289	406	630	857	779	257	263	285	1196	552	299	948	449	824	588	804	*	396
Tulsa	772	1537	1112	683	802	257	681	478	248	659	401	1398	695	609	1344	387	760	984	1129	396	*

References and Further Readings

1. Bellman, R., K. L. Cooke, and J. A. Lockett, *Algorithms, Graphs and Computers.* New York: Academic Press, Inc., 1970, chap. 8.
2. Chartrand, Gary, *Graphs as Mathematical Models.* Belmont, CA: Wadsworth Publishing Co., Inc., 1977, chap. 3.
3. Knuth, Donald, "Mathematics and Computer Science: Coping with Finiteness," *Science,* 194 (December 1976), 1235–1242.
4. Kolata, Gina, "Analysis of Algorithms: Coping With Hard Problems," *Science,* 186 (November 1974), 520–521.
5. Kolata, Gina, "Math Problem, Long Baffling, Slowly Yields," *The New York Times,* March 12, 1991, B8–B10.

6. Lawler, E. L., J. K. Lenstra, A. H. G. Rinooy Kan, and D. B. Shmoys, *The Traveling Salesman Problem.* New York: John Wiley & Sons, Inc., 1985.

7. Lewis, H. R., and C. H. Papadimitriou, "The Efficiency of Algorithms," *Scientific American,* 238 (January 1978), 96–109.

8. Peterson, Ivars, *Islands of Truth.* New York: W. H. Freeman & Co., 1990, chap. 6.

9. Wilson, Robin, and John J. Watkins, *Graphs: An Introductory Approach.* New York: John Wiley & Sons, Inc., 1990.

10. Zimmer, Carl, "And one for the road," *Discover* (January 1993), 91, 92.

11. Zubrin, Robert. *The Case for Mars*. New York: Free Press, 1996.

c̲ŏn · nec̲′ tion

(1) that which connects or unites; a tie; a bond; means of joining.

(2) a line of communication from one point to another.

Webster's New 20th Century Dictionary

The Mathematics of Networks

Connections!

CHAPTER 7

As any sober person knows, the shortest distance between two points is a straight line. What, then, is the shortest distance between three points? And, what does the term *shortest distance* mean when we are dealing with three points? (With just two points the meaning is clear, but with three it is not.) We will touch upon these questions briefly in this introduction, and we will discuss them in a general context (not just three but any number of points) in greater detail in this chapter.

In 1989, a consortium of telephone companies completed the construction of a new transpacific fiber optic trunk line called TPC-3 linking Japan, Guam, and Oahu, Hawaii. (From Hawaii, the line connects with an already existing trunk line joining Hawaii to the mainland.) Since both the fiber optic cable itself and the laying of it along the ocean floor are extremely expensive propositions, it was important for the telephone companies to find a way to link the three islands using the least amount of cable. This meant, of course, finding the shortest distance connecting the three points. The solution (the red lines), is shown in the stamp on the opposite page which was issued by the Japanese post office in commemoration of the event. We will find out later in the chapter exactly what makes this network the shortest network and how such networks can be found.

The general theme of this chapter is the problem of finding efficient networks linking a set of points. In many situations besides the obvious one of telephone networks this is a problem of importance—for example, in the building of transportation networks (roads, high-speed rail systems, canals), the construction of pipelines, and the design of computer chips, etc.

The common thread in all these problems is twofold: (1) the need to link all the points so that one can go from any point to any other point (when this is accomplished, we have a *network*), (2) the desire to make the total cost of the network as small as possible. For obvious reasons, problems of this type are known as **minimum network problems**, and in this chapter we will discuss two basic variations on this theme. Once again, one of the most important tools we will use is graph theory.

Trees

Example 1. An Amazonian Telephone Network

The Amazonia Telephone Company (ATC) is the main provider of telephone services to some of the world's most inaccessible places—small towns and villages buried deep within the Amazonian jungle. Consider, for example, the following problem: Seven small villages, shown in Fig. 7-1(a), are scheduled to be connected into a small regional network using state-of-the-art fiber optic underground cable. The figure also shows the already existing network of roads connecting the 7 villages. The weighted graph in Fig. 7-1(b) is a graph model describing the situation: The edges of the graph are the existing roads and the weights of the edges represent the cost (in millions of dollars) of laying down the fiber optic cable along each road (laying down cable anywhere other than along an existing road would be prohibitively expensive in the jungle). The network that the company wants to build should link all the cities for the least money. What is the *cheapest* network?

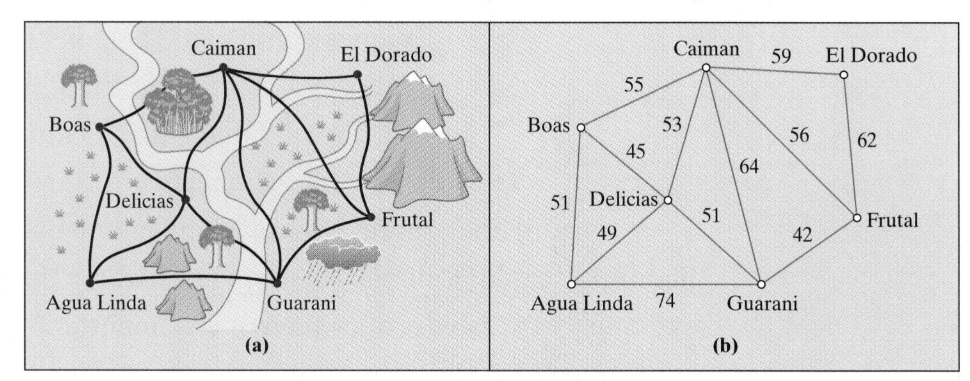

FIGURE 7-1 (a) The 7 villages in the Amazon and the roads connecting them. (b) A graph model of the possible connections. The weights of the edges are the costs (in millions) of laying fiber optic lines.

Let's start by asking a general question: What would a minimal network like the one to be built in Example 1 have to be like?

First, we observe that as a graph, the network will have to be a **subgraph** of the original graph. In other words, building the network will require choosing some of the edges in Fig. 7-1(b), but certainly not all of them. In addition, the subgraph should *reach out and touch* each and every vertex of the original graph—that's just the requirement that no village should be left out of the telephone network! In the language of graph theory, a subgraph that includes *every one of the vertices of an original graph* is called a **spanning**[1] **subgraph**. To summarize, in tech-

[1] **span,** to extend, reach, or stretch across. *Webster's New 20th Century Dictionary.*

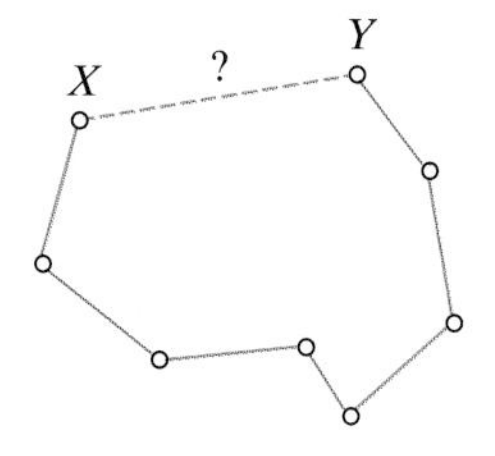

FIGURE 7-2
Is the link XY really necessary? Not when saving money is the object.

nical terms, the network we are looking for is a *spanning* (includes all of the vertices) *subgraph* (uses only some of the edges) of the graph in Fig. 7-1(b).

But that in itself is not enough. The graph should also have the following characteristics:

- It should be *connected*. This is an obvious requirement in a network where the object is to be able to reach (place a call) from any vertex (village) to any other vertex (village).
- It should *not contain any circuits.* This reflects the requirement that this network be built in the cheapest possible way. Whenever there is a circuit, there is guaranteed to be a *redundant* connection—take away the redundant connection and the network would still work. Figure 7-2 is a closeup of a hypothetical circuit. Delete any one of the edges—say XY—and the telephone calls would still go through.

These last two characteristics define a very important category of graphs called trees. A **tree** is a graph that is connected and has no circuits. Because of their importance, trees have been studied extensively, and we will take a little time to get acquainted with them ourselves. ■

Example 2.

Figure 7-3 shows examples of graphs, all of which are trees (all of them are connected and without circuits). Some of them look like a tree, but that's not a requirement.

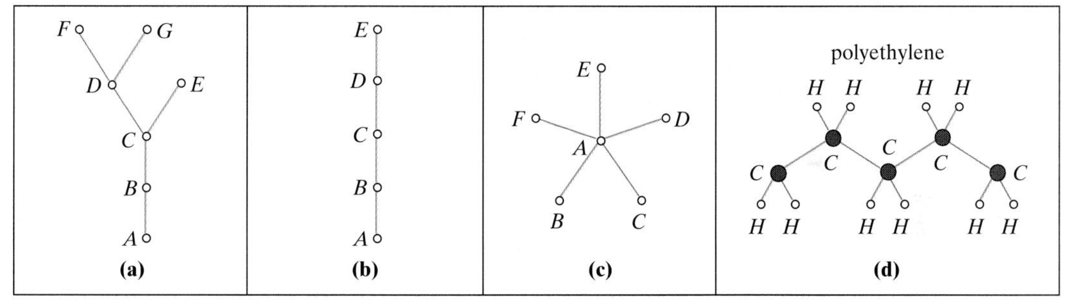

FIGURE 7-3 ■

Example 3.

Figure 7-4 shows examples of graphs that are *not* trees. Figure 7-4(a) has one circuit, Fig. 7-4(b) has several circuits. In either case, appearance notwithstanding, neither one is a tree. Figure 7-4(c) has no circuits but is not connected; and Fig. 7-4(d) fails to be a tree on both accounts—it has circuits, and it is not connected.

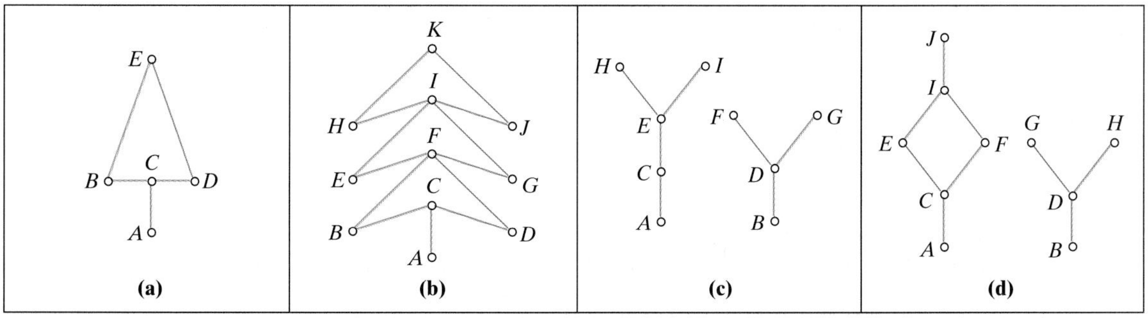

FIGURE 7-4 ■

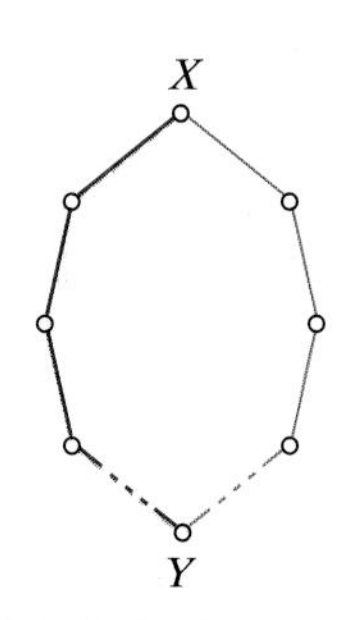

FIGURE 7-5
Two different paths joining X and Y make a circuit.

Properties of Trees

Trees are a very special type of graph, and as such have some very special properties, some of which will come in quite handy in our quest for cost-efficient networks.

Let's start with the observation that in a connected graph (be it a tree or not) there is always a path joining any one vertex to any other vertex, but if there are two or more paths joining any pair of vertices, then the graph is definitely not a tree—the two paths joining the same two vertices make a circuit (Fig. 7-5). This leads to the first important property of trees.

Property 1. If a graph is a tree, there is one and only one path joining any two vertices. Conversely, if there is one and only one path joining any two vertices of a graph, the graph must be a tree.

One practical consequence of Property 1 is that a tree is connected in a very precarious way. The removal of any edge of a tree will disconnect it (Exercise 29). We can restate this by saying that in a tree, every edge is a *bridge.*

Property 2. In a tree, every edge is a bridge. Conversely, if every edge of a connected graph is a bridge, then the graph must be a tree.

From our preceding discussion, it seems intuitively obvious that a tree cannot have too many edges. Having lots of edges is in some sense contrary to the nature of being a tree. At the same time, a tree must be connected, so a certain minimum number of edges is going to be necessary. A very important property of trees is that there is a very precise relation between the number of edges and the number of vertices: *The total number of edges is always one less than the number of vertices.*

Property 3. A tree with N vertices must have $N-1$ edges.

It would be nice to be able to turn Property 3 around and say that if a graph has N vertices and $N-1$ edges, then it must be a tree. As the graph in Fig. 7-6 shows, however, this need not be the case—it has 10 vertices and 9 edges, and yet it is not a tree. What's the problem? As you may have guessed, the problem is that the graph is not connected. Fortunately, for connected graphs the converse of Property 3 is true.

Property 4. A *connected* graph with N vertices and $N-1$ edges must be a tree.

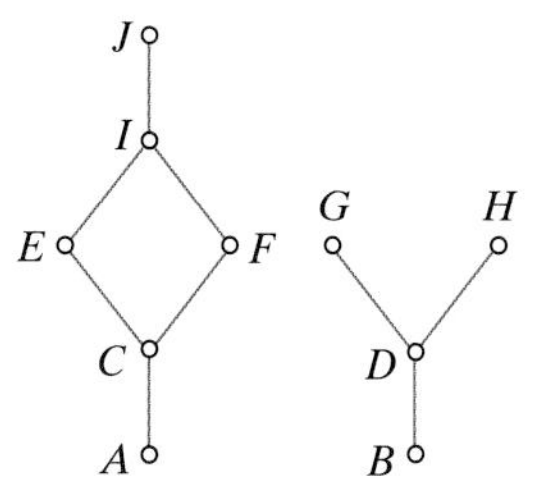

FIGURE 7-6
A graph with 10 vertices and 9 edges that is not a tree.

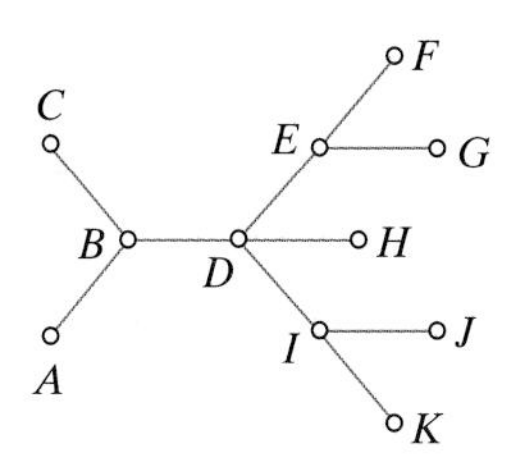

FIGURE 7-7
A connected graph with 11 vertices and 10 edges. Every edge is a bridge. A bird? A plane? No, a tree.

Example 4.

This example illustrates some of the ideas discussed so far. Let's say that we have 5 vertices with which to build a graph, and let's start putting edges on these vertices. At first, with 1, 2, or 3 edges (Fig. 7-8), we just don't have enough edges to make the graph connected. When we get to 4 edges, we can, for the first time, make the graph connected. If we do so, we have a tree (Fig. 7-9). As we add even more edges the connected graph starts picking up circuits ands stops being a tree (Fig. 7-10).

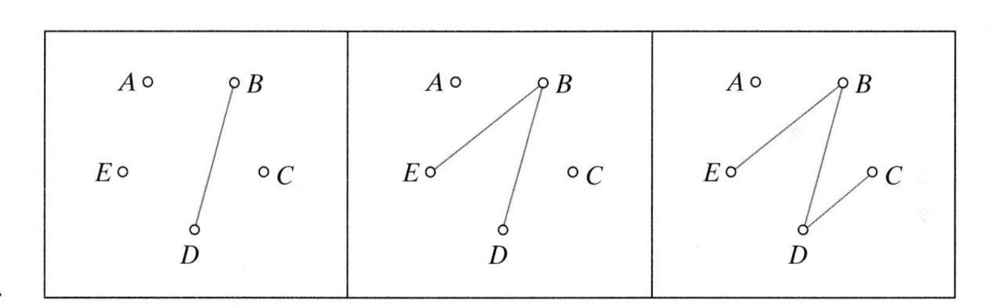

FIGURE 7-8
Five vertices, less than four edges. The graph is disconnected.

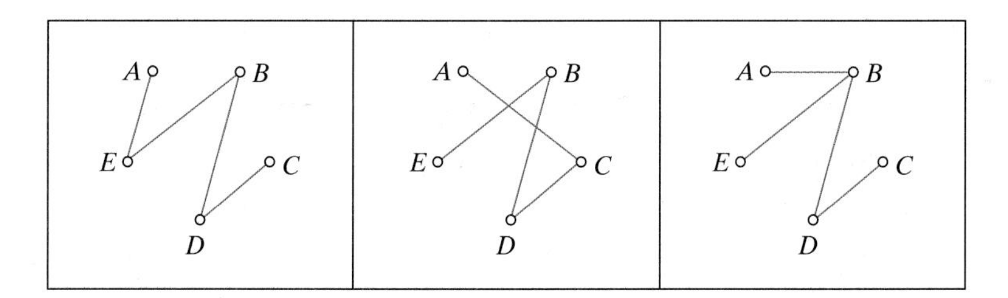

FIGURE 7-9
Five vertices, four edges—just enough to connect.

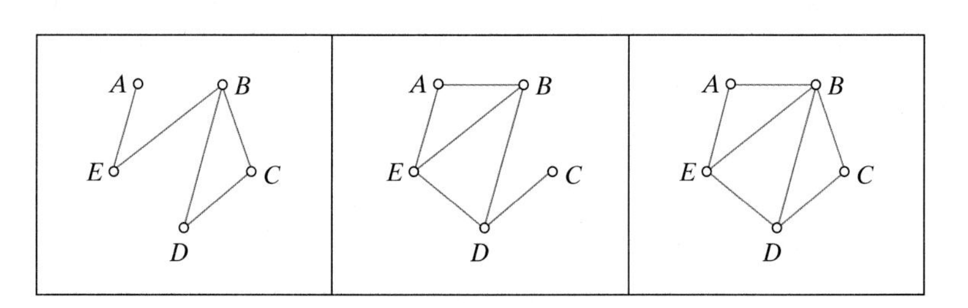

FIGURE 7-10
Five vertices, more than four edges—circuits begin to form.

■

Minimum Spanning Trees

In Fig. 7-10 we have three examples of graphs that are connected but are not trees—they have too many edges. Within such a graph we can always find a tree *spanning* the vertices of the graph—something like a bare skeleton holding up the rest of the body. We call such a tree a spanning tree of the graph. Being a tree, the spanning tree has one less edge than it has vertices.

In summary, a **spanning tree** of a connected graph G with N vertices is a connected subgraph of G having $N - 1$ of the edges of G and all N of the vertices of G.

Example 5.

Consider the weighted graph shown in Fig. 7-11(a). It is connected, it has 9 vertices and 10 edges, so we know it is not a tree. This particular graph has two separate circuits. To get a spanning tree for this graph we must "bust" each of the two circuits by removing an edge from each. Figure 7-11(b) shows one possible spanning tree, with a total weight of 43. But this is just one of 18 possible spanning trees for this graph. (Exercise 31b). Can you find one with total weight less than 43? Can you find the one with the least total weight? You should definitely give it a try.

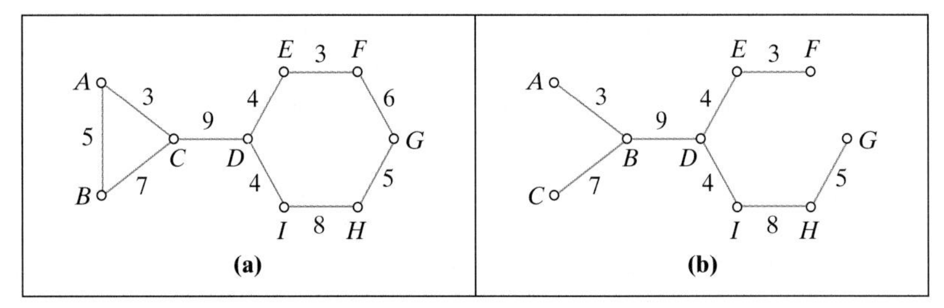

FIGURE 7-11 Busting each of the circuits in (a) gives a spanning tree (b).

■

Example 5 shows, if nothing else, that even a very modest and unpretentious graph can have lots of spanning trees. Our problem is to find a **minimum spanning tree (MST)**—that is to say, one with the least total weight. How easy is it going to be to find such an optimal solution? After our experience with Chapter 6, we are right in being skeptical. Are we in another situation where, unless we try all possibilities, we have no guarantees of an optimal answer? Not this time.

Kruskal's Algorithm

In 1956 an American mathematician, Joseph Kruskal, working at the Bell Laboratories,[2] "discovered" a very simple algorithm that will *always* find a minimum spanning tree in a weighted graph. Known as **Kruskal's algorithm**, it is a simple variation of the *cheapest-link algorithm* discussed in Chapter 6. Just as in that algorithm, the basic idea in Kruskal's algorithm is to be greedy, always picking the cheapest available link and staying away from creating any circuits. The next example illustrates how the algorithm works.

Example 6. The Amazonian Telephone Network: Part II

Let's try to find a minimum spanning tree (MST) for the graph in Fig. 7-12, which shows the costs (in millions) of connecting the various villages with telephone

[2] For many years, Bell Labs was one of the world's greatest research centers for pure and applied science, fully owned and operated by AT&T. It is now called Lucent Technologies.

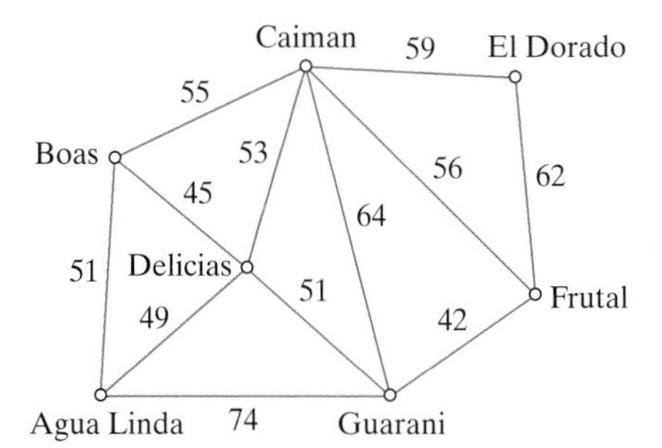

FIGURE 7-12 The graph for the Amazonian telephone-network problem.

lines. The MST that we find will represent the cheapest possible network connecting the 7 cities—exactly what we set out to do in Example 1.

- **Step 1.** Of all the possible links between a pair of villages, the cheapest one is Guarani-Frutal, at a cost of $42 million. We designate this as the first link in the network by marking it in red.[3]
- **Step 2.** The next cheapest link is Boas-Delicias, at a cost of $45 million. We also mark it in red, as it is also going to be part of the network.
- **Step 3.** The next cheapest link is Agua Linda-Delicias, costing $49 million. Same story—mark it in red.
- **Step 4.** The next cheapest link is a tie between Agua Linda-Boas and Delicias-Guarani, both at $51 million. Agua Linda-Boas, however, is now a *redundant* connection, and we do not want to use it. (For bookkeeping purposes, the best thing to do is erase it.) Delicias-Guarani, on the other hand, is just fine, so we mark that link in red.
- **Step 5.** The next cheapest link is Caiman-Delicias, at $53 million. No problems here, so we mark the link in red.
- **Step 6.** The next cheapest link is Boas-Caiman, at $55 million, but this is a redundant connection, so we discard it. The next possible choice is Caiman-Frutal at $56 million, but this is also a redundant connection (calls between Caiman and Frutal are already possible in our budding network), so we keep looking. The next possible choice is Caiman-El Dorado at $59 million, and this is OK, so we mark the link Caiman-El Dorado in red.
- **Step . . .** Wait a second—we are finished! (We can tell we are done by just looking at the red network we have built and verifying that it is a spanning tree. Or, even better, we can recognize that 6 edges—and therefore 6 steps—is exactly what it takes to build a tree with 7 vertices.)

The total cost of the telephone network we have come up with (Fig. 7-13) is $299 million, and this is, in fact, the optimal solution to the problem—there is no cheaper spanning tree!

---

[3] Note that these will not necessarily be the first two towns actually connected. We are putting the network together on paper, and the rules require that we follow a certain sequence, but in practice we can build the links in any order we want.

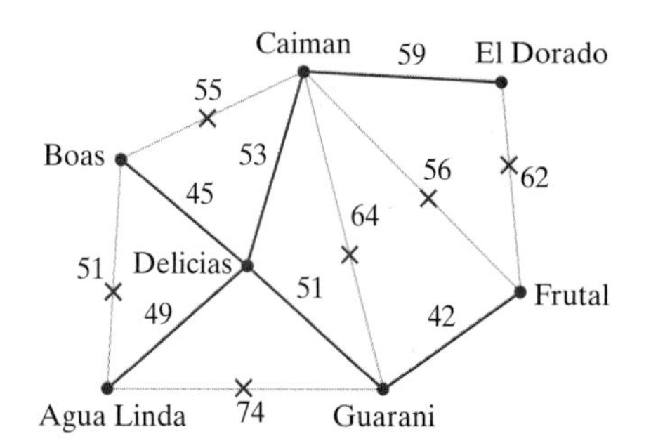

FIGURE 7-13 The MST for the Amazonian telephone-network problem (shown in red). The network has a 4-way junction at Delicias.

■

We call the reader's attention to one other fact that will become relevant later in the chapter: The network we have built has one main junction—a four-way junction at Delicias. In telephone networks, junction points are important, because these are places where switching equipment has to be installed.

Here is a formal description of Kruskal's algorithm.

Kruskal's Algorithm

- Find the cheapest link in the graph (if there is more than one, pick one at random). Mark it in red.
- Find the next cheapest link in the graph, and also mark it in red. If there is more than one, pick one at random.
- Find the next cheapest unmarked link in the graph that does not create a red circuit and mark it in red. If there is more than one, pick one at random.
- Repeat the previous step until the red edges span every vertex of the graph. The red edges are the desired MST.

The truly remarkable thing about Kruskal's algorithm is the fact that it is an *optimal algorithm*: The spanning tree that we get is guaranteed to be the cheapest possible one. In light of our experience in Chapter 6, this is a bit of a surprise (how can something so simpleminded give such great results?). Kruskal's algorithm is also an *efficient algorithm.* As we increase the number of vertices and edges in the graph, the amount of work grows more or less proportionally. For the right reward (say an "A" in a math course) it would not be unreasonable to use Kruskal's algorithm in a graph with a couple of hundred vertices.

In short, the problem of finding minimum spanning trees represents one of those rare situations where everything falls into place: We have an important real-life problem which can be solved by means of an algorithm (Kruskal's) that is easy to understand and to carry out and that is *optimal* and *efficient*—who could ask for better karma? Wouldn't it be great if things always went this well?

The Shortest Distance between Three Points

Minimum spanning trees give us optimal networks connecting a set of locations in the case in which the connections have to be along prescribed routes. Remember, for example, that in the Amazonian telephone network problem, the only possible routes along which the lines could be located were the already existing roads. But what if, in a manner of speaking, we don't have to *follow the road?* What if when we link one location to another we can make our own route? To clarify the distinction, let's look at a new type of telephone-network problem.

Example 7. An Australian telephone-network problem

This is a connection story involving three small fictional towns (Alcie Springs, Booker Creek, and Camoorea) located smack in the middle of the Australian outback, and which by sheer coincidence happen to form an equilateral triangle, 500 miles a side (Fig. 7-14). The problem, once again, is to lay telephone cable linking the three towns into a network. What is the shortest possible way to connect these towns?

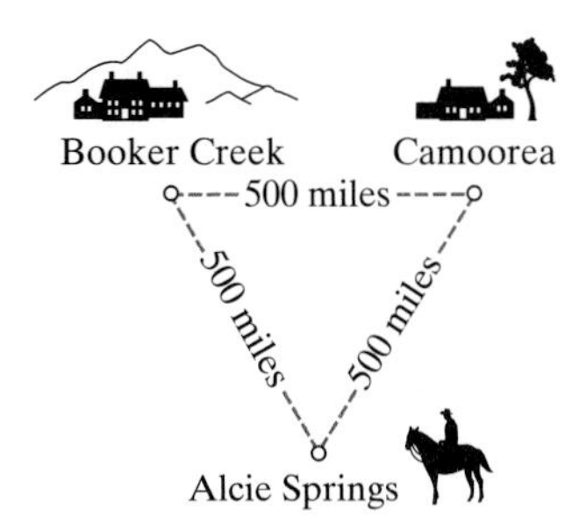

FIGURE 7-14

While this example looks like just a small-scale version of the Amazonian telephone-network problem, it is not. What makes this situation different is the nature of the terrain. The Australian outback is mostly a flat expanse of desert and, in contrast to the Amazon situation, there is little or no advantage to laying the telephone lines along roads. (In fact, let's assume that there are no roads to speak of connecting these three towns.) Because of the flat and homogeneous nature of the terrain, we can lay the telephone cable anywhere we want, and the cost per mile is always the same.

The question now is: Of all possible networks that connect the three towns, which one is the shortest? Let's start where we left off: What is a minimum spanning tree in this case? Since all three sides of the triangle are the same length, we can pick any two of them to form a minimum spanning tree, as shown in Fig. 7-15(a). The total length of the MST is 1000 miles.

It is not hard to see that the MST is not the shortest possible network connecting the three towns. Look at Fig. 7-15(b). It shows a network that is definitely shorter than 1000 miles. It has a "T" junction at a new point we call J. A little high school geometry (Pythagorean theorem) and a calculator are sufficient to verify that it is approximately 433 miles from Alcie Springs to the junction J, and that the network, therefore, is about 933 miles long (see Exercise 24).

Can we do even better? Why not? With a little extra thought and effort we might come up with the network shown in Fig. 7-15(c). Here, there is a "Y" junction at a new point called S located at the center of the triangle. This network is

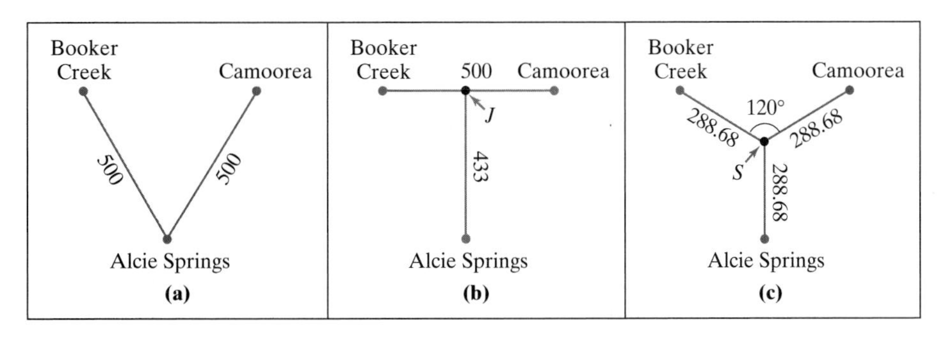

FIGURE 7-15
(a) A minimum spanning tree. (b) A shorter network with a T-junction at J. (c) The shortest network with a Y-junction at S. The three branches of the "Y" meet at equal angles.

approximately 866 miles long (more high school geometry—see Exercise 25). The most important feature of this network is that the three branches come together at the junction point S forming *equal angles,* which forces each angle to be exactly 120°. Even without a formal mathematical argument, it is not hard to buy into the fact that Fig. 7-15(c) shows the *shortest possible network* connecting the three towns. After all, mathematics should not choose sides, and this is the only network that looks the same to all three towns. ■

Let's recap what we learned from Example 7: If we do not require the junction points of our network to be chosen from among the original cities (in other words, if we are allowed to create new junction points in the network), then the minimum spanning tree may not be the best way to connect the cities. In Example 7, the optimal solution is a network having an interior junction point S in which the three branches meet at equal (120°) angles.

If we compare the length of this solution (866 miles) with the MST solution (1000 miles), we can see that creating the new junction point produced a savings of 134 miles over the original 1000-mile MST. This comes out to be a 13.4% savings. Please make a mental note of this number.

Before we go on the next example, let's introduce some new terminology.

- The shortest possible network connecting a set of points is called, not surprisingly, the **shortest network**.
- Any junction point in a network formed by three branches coming together at 120° angles is called a **Steiner point**[4] of the network.

We now return to a couple of questions first raised in the opening of this chapter—What is the *shortest distance* between Japan, Guam, and Hawaii? and, Why should anyone care?

Example 8. The TPC-3 Connection

Let's review the background for this story. In 1989, a consortium of several of the world's biggest telephone companies (among them AT&T, MCI, Sprint, and British Telephone) completed a major undertaking: the 3rd Trans Pacific Cable (TPC-3), a fiber optic trunk line linking Japan to the continental United States (via Hawaii) and to Guam. For obvious reasons, the primary consideration in designing the trunk line was its cost. By and large, laying cable along the ocean floor has a fixed cost per nautical mile, so that, unlike the Amazonian telephone network problem, here we are after the shortest network. The approximate straight-line distances (in miles) between the three endpoints of TCP-3 (Chikura, Japan; Tanguisson Point, Guam, and Oahu, Hawaii) are shown in Fig. 7-16.

By now we have a pretty good idea, that to find the shortest network, we are going to have to create a new junction point inside of the triangle Japan-Guam-Hawaii. If we don't, the best we can do is the MST, which in this example has a length of 1620 + 3820 = 5440 miles. Assuming we can do better, where should we

[4] Named after the Swiss mathematician Jakob Steiner (1796–1863).

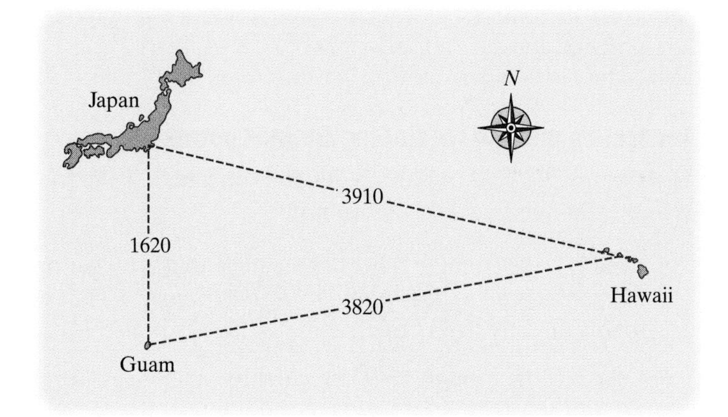

FIGURE 7-16
The distances (in miles) between the 3 vertices of the triangle Japan-Guam-Hawaii.

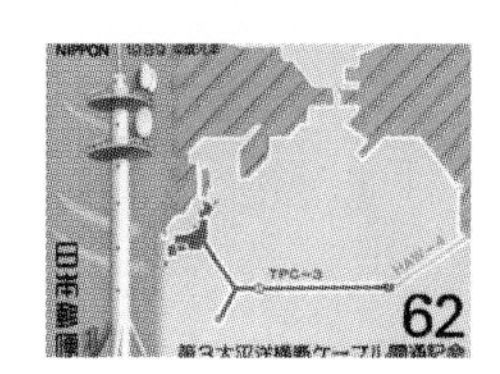

put the junction point so that we come up with the shortest possible network? A reasonable guess is that the junction point should be located so that the three branches of the network come together at (equal) 120° angles—in other words, it should be a Steiner point of the network. This turns out to be exactly the right solution, as shown in Fig. 7-17 (as well as in the Japanese stamp issued to commemorate the completion of the trunk line). (For the details of *why* this is the shortest network, the reader is referred to Exercise 55.) The total length of the network is 5180 miles.[5]

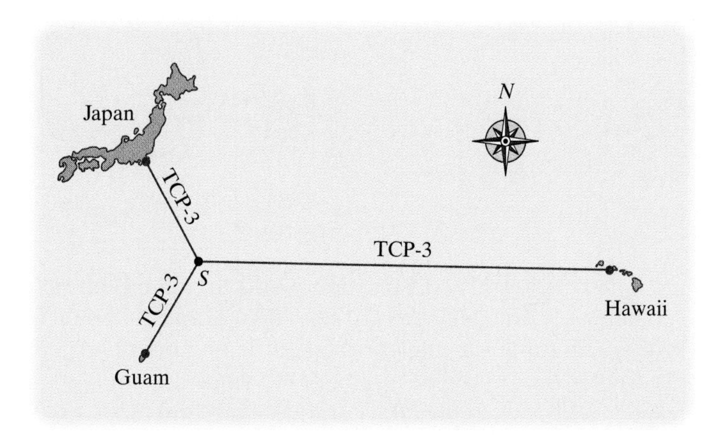

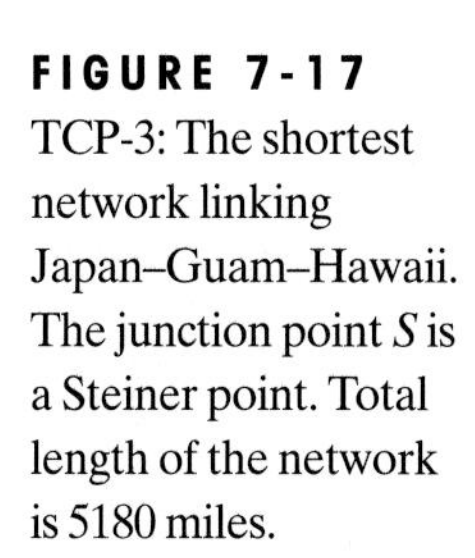

FIGURE 7-17
TCP-3: The shortest network linking Japan–Guam–Hawaii. The junction point *S* is a Steiner point. Total length of the network is 5180 miles.

One final question remains to be answered: How can we find the exact location of the Steiner point *S* using simple geometric tools? This question has a long and controversial history, going back at least 350 years. In the early 1600s the Italian Evangelista Torricelli[6] discovered a remarkably simple method for finding

[5] The theoretical length of the network is not the same as the total amount of cable used. With cable running on the ocean floor, one has to add as much as 10% to the straight-line distances. The exact length of cable used in TCP-3 is 5690 miles.

[6] Evangelista Torricelli (1608–1647) was Galileo's assistant and disciple. Although he was a brilliant mathematician, he is best known for discovering the principle of barometric pressure commonly known as Torricelli's law.

the exact location of a Steiner point inside a triangle, and all it takes is just a straightedge and a compass.

Torricelli's method for finding Steiner points

In a triangle ABC, the solution depends on whether the triangle has an angle that is bigger than or equal to 120° or not.

- **Case 1.** If the triangle ABC has an angle that is bigger or equal to 120-degrees, then *there is no Steiner point inside the triangle.* (That was easy!) The reasons follow directly from basic geometry (see Exercise 42).
- **Case 2.** If the triangle ABC has all three angles smaller than 120°, then there is a unique Steiner point inside of the triangle, which can be found as follows:

  Step 1. On the other side of BC, build the equilateral triangle BXC as shown in Fig. 7-18(b). (The new point X is on the opposite side of A.)

  Step 2. Circumscribe a circle around the equilateral triangle BXC [Fig. 7-18(c)].

  Step 3. Join X to A [Fig. 7-18(d)]. The point where the line segment XA intersects the circle *is the desired Steiner point S.* That's it!

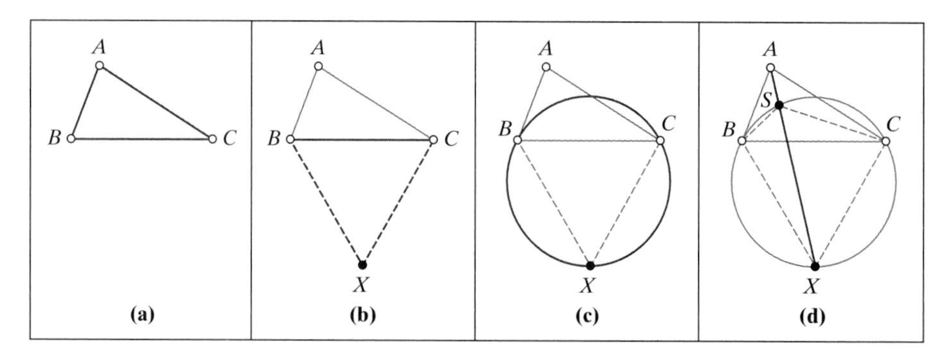

FIGURE 7-18 Finding the Steiner point S inside a triangle ABC. *Step 1.* Find point X opposite A such that BXC is an equilateral triangle. *Step 2.* Circumscribe triangle BXC in a circle. *Step 3.* Join X and A. The intersection of the circle and $\overline{AX}$ is S.

The reasons this construction works are all based on facts from high school geometry: (1) The angle BXC is 60° (BXC is an equilateral triangle). (2) The angle BSC is 120° (opposite angles of a quadrilateral inscribed in a circle add up to 180°). (3) Angles BSX and XSC are both 60° (they are equal to angles BCX and XBC, respectively). (4) Angles BSA and CSA are both 120°. (5) S has to be the desired Steiner point.

Before we conclude our discussion of finding the shortest network linking three points, we need to look at one more example.

Example 9.

Suppose that we want to build the tracks for a bullet-train network linking the cities of Los Angeles, Las Vegas, and Salt Lake City. The straight-line distances between the three cities are shown in Fig. 7-19. Once again, we want to consider the shortest network connecting the three cities.

The only substantive difference between this example and Example 8 is that here the triangle has an angle bigger than 120°, and therefore no interior Steiner point to use as a junction. Without a Steiner junction point, how do we find the shortest network? The answer turns out to be surprisingly simple: *the shortest net-*

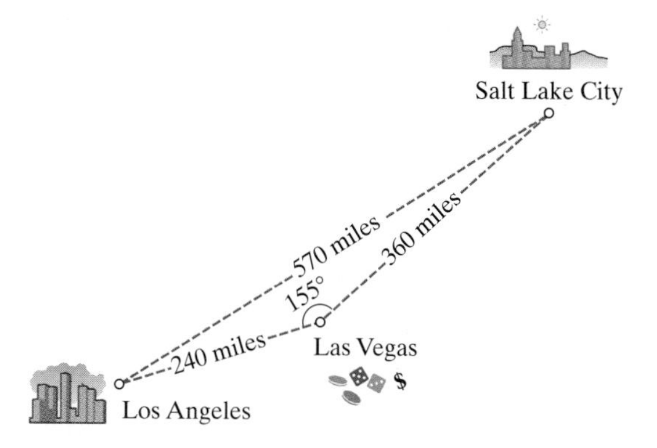

FIGURE 7-19

work is the same a the minimum spanning tree. In a triangle this simply means: pick the two shortest sides. For this example, the solution is shown in Fig. 7-20.

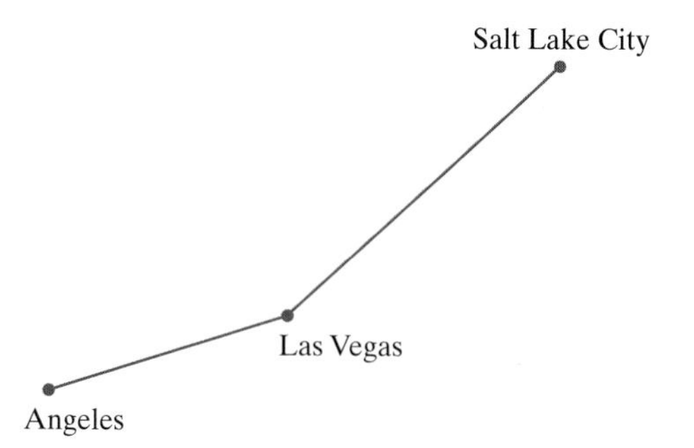

FIGURE 7-20 The shortest network connecting the three cities is the MST.

In the opening paragraph of this chapter we raised the question "What is the shortest distance between three points?" We finally have an answer. Let's summarize what we now know:

The Shortest Distance between Three Points *A, B,* and *C*

1. When one of the angles of the triangle ABC is 120° or more, then the shortest network linking the three points consists of the two shortest sides of the triangle [Fig. 7-21(a)]. In this situation the shortest network coincides with the minimum spanning tree.
2. When all the angles of the triangle are less than 120°, then the shortest network is obtained by finding a Steiner point S inside the triangle and joining S to each of the vertices A, B, and C [Fig. 7-21(b)]. The exact location of S can be found using Torricelli's construction.

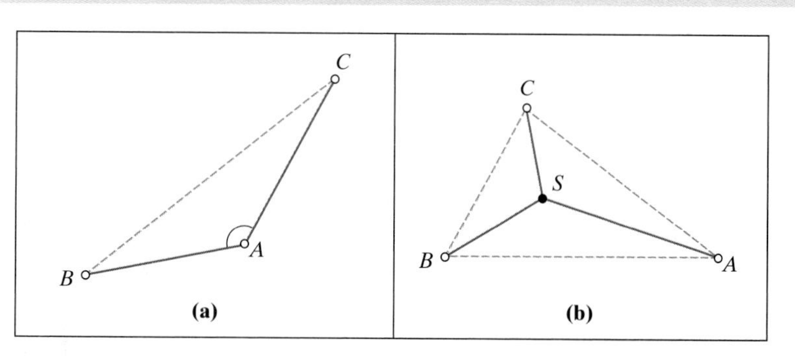

FIGURE 7-21 The shortest network connecting A, B, and C. (a) The angle at A is bigger than or equal to 120°. (b) All angles are less than 120°.

Shortest Networks Linking More than Three Points

When it comes to finding shortest networks, things get really interesting when we get past three cities.

Example 10.

Four cities (A, B, C, and D) are to be connected into a telephone network. For starters, let's imagine that the 4 cities form the vertices of a square 500 miles a side, as shown in Fig. 7-22(a). What does an optimal network connecting the 4 cities look like? It depends on the situation.

If we *don't want to introduce any new junction points in the network* (either because we don't want to venture off the prescribed paths—as in the jungle scenario—or because the cost of creating a new junction is too high), then the answer is a minimum spanning tree, such as the one shown in Fig. 7-22(b). The length of the MST is 1500 miles.

On the other hand, if interior (new) junction points are allowed, somewhat shorter networks are possible. One obvious improvement is the network shown in Fig. 7-22(c), having an "X" type of junction a O, the center of the square. The length of this network is approximately 1414 miles (see Exercise 43).

An even shorter network is possible by using two Steiner junction points S_1 and S_2 [Fig. 7-22(d)]. Using high school geometry (Exercise 44), we can find that the length of this network is approximately 1366 miles. This is it! There is no shorter network, although there is another possible network equivalent to this one [Fig. 7-22(e)].

The difference between the 1500 miles in the MST and the 1366 miles in the shortest network is 134 miles, which represents a savings of about 9% (134/1500). ■

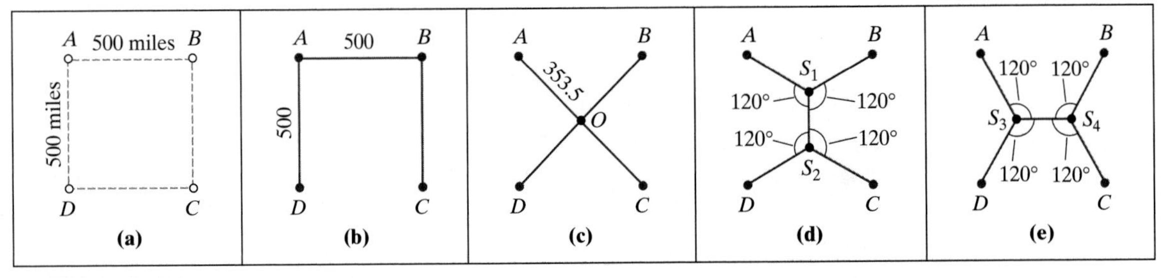

FIGURE 7-22
(a) Four cities located at the vertices of a square. (b) A minimum spanning tree network with a total length of 1500 miles. (c) A shorter network, with a total length of approximately 1414 miles, obtained by placing an interior junction point O at the center of the square. (d) A shortest network with a total length of approximately 1366 miles. The junction points S_1 and S_2 are both Steiner points. (e) A different solution, with Steiner points S_3 and S_4.

Example 11.

Let's repeat what we did in Example 10, but this time imagine that the 4 cities are located at the vertices of a rectangle, as shown in Fig. 7-23(a). By now, we have some experience on our side, so we can cut to the chase. We know that the

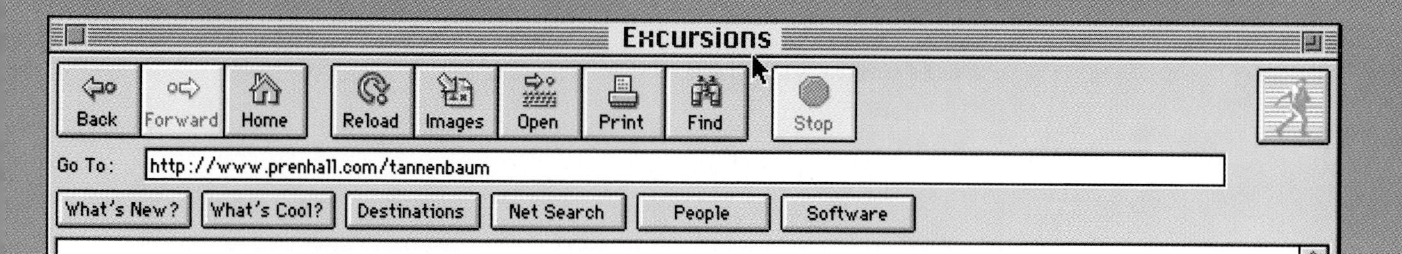

Minimum Spanning Trees in a Plane

As you have learned, a minimum spanning tree of a connected weighted graph is a spanning tree that has the least possible weight. If each vertex corresponds to a point in the plane, then each edge corresponds to a line segment, and we may define the weight of an edge as the length of its corresponding line segment. In this case, a minimum spanning tree is simply a spanning tree with the least possible total length.

A Java applet by Joe Ganley can be used to create a minimal spanning tree for an arrangement of up to 30 points (vertices) in the plane. You can click on the arrow buttons to change the number of vertices, and you can also move the vertices around by dragging them with the mouse. The applet can be found in the Tannenbaum Website under Chapter 7.

Questions:

1. Use the applet to create a graph with about eight vertices. Then experiment by dragging vertices around with the mouse. You should notice that whenever one edge disappears, another one appears. Explain why this *must* be true.
2. Use the applet to create a minimal spanning tree with 9 vertices, of which exactly 2 have degree 4. What are the degrees of the remaining vertices? Sketch a drawing of your graph.
3. Use the applet to create a minimal spanning tree with 6 vertices, of which exactly 2 have degree 1. What is the degree of each of the remaining vertices? Explain why this *must* be true.
4. a. Use the applet to create a minimal spanning tree with 4 vertices, of which 3 have degree 1. Sketch your graph.

   b. What is the largest value of N such that you can use the applet to create a minimal spanning tree with N vertices, $N - 1$ of which have degree 1? Does your answer surprise you? Sketch a drawing of your graph for this value of N.

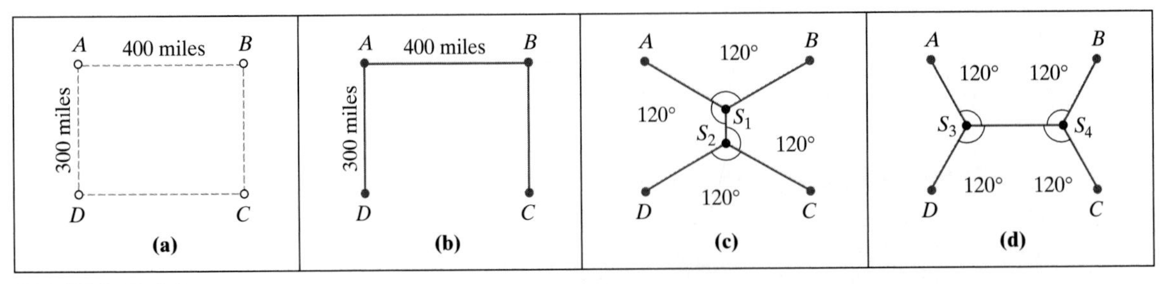

FIGURE 7-23
(a) Four cities located at the vertices of a rectangle. (b) A minimum spanning tree network (1000 miles). (c) A network with two Steiner junction points (approximately 933 miles). (d) The shortest network also has two Steiner junction points (approximately 920 miles).

minimum-spanning-tree solution is 1000 miles long [Fig. 7-23(b)]. That's the easy part.

For the shortest network solution let's think about what happened in the previous example (after all, a square is just a special case of a rectangle). An obvious candidate would be a network with two interior Steiner junction points. There are two such networks, shown in Figs. 7-23(c) and 7-23(d), but this time they are not equivalent: The network shown in Fig. 7-23(c) is approximately 993 miles [see Exercise 45(a)], while the network shown in Fig. 7-23(d) is approximately 920 miles [see Exercise 45(b)]—a pretty significant difference. Obviously, Fig. 7-23(c) cannot be the shortest network, but what about Fig. 7-23(d)? If there is any justice in this mathematical world, this one fits the pattern and ought to be it. In fact, it is! (But don't jump to any conclusions about justice just yet!) ■

Example 12.

Let's look at 4 cities once more. This time imagine the cities are located at the vertices of a skinny trapezoid as shown in Fig. 7-24(a). The minimum spanning tree is shown in Fig. 7-24(b), and it is 600 miles long.

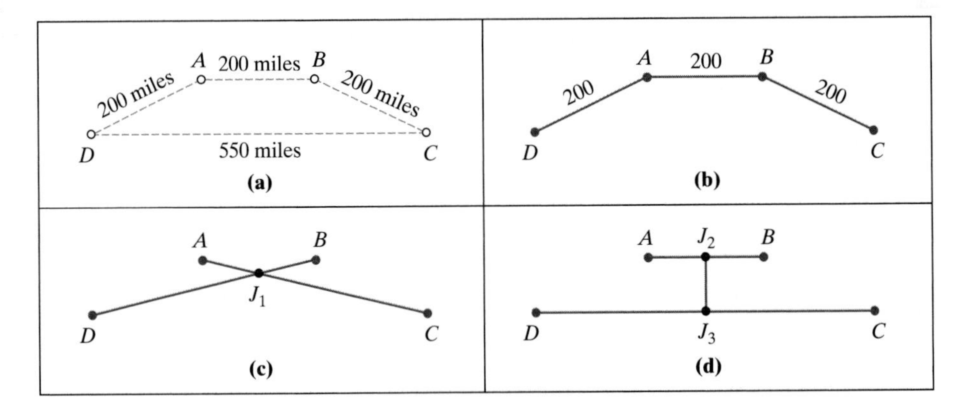

FIGURE 7-24
(a) Four cities located at the vertices of a trapezoid. (b) The minimum spanning tree. (c) A network with an X-junction at J_1 is longer than the MST (774.6 miles). (d) A network with a couple of T-junction points at J_2 and J_3 is even worse (846.8 miles).

What about the shortest network? Based on our experience with Examples 10 and 11, we are fairly certain that we should be looking for a network with a couple of interior Steiner junction points. After a little trial and error, however, we realize that such a layout is impossible! The trapezoid is too skinny, or to put it in a slightly more formal way, the angles at A and B are more than 120°. Since no Steiner points can be placed inside the trapezoid, the shortest network, whatever it is, will have to be one without Steiner junction points.

Well, we say to ourselves, if not Steiner junction points, how about other kinds of interior junction points? How about X-junctions (Fig 7-24[c]) or T-junctions (Fig. 7-24[d]), or Y-junctions where the angles are not all 120°? As reasonable as this idea sounds, a remarkable thing happens with shortest networks: *The only possible new (i.e., not original) junction points are Steiner points.* For convenience, we will call this the *interior junction rule* for shortest networks. ■

The Interior Junction Rule for Shortest Networks

In a shortest network, any new junction point has to be a Steiner point.

The interior junction rule is an important and powerful piece of information in building shortest networks, and we will come back to it soon. Meanwhile, what does it tell us in the situation of Example 12? It tells us that the shortest network cannot have any new junction points (Steiner junction points are impossible because of the geometry; other junction points are impossible because of the interior junction rule). But we also know that the shortest network without new junction points is the minimum spanning tree! Conclusion: For the four cities of Example 12, *the shortest network is the minimum spanning tree!* (Fig. 7-25).

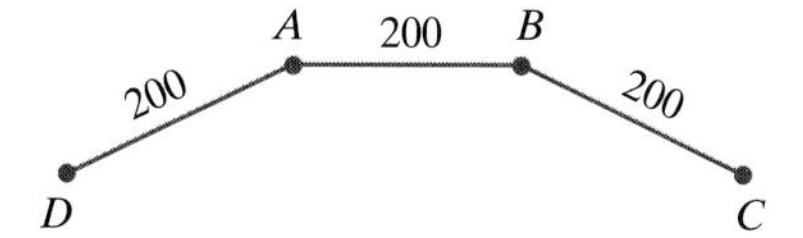

FIGURE 7-25 The shortest network and the MST are one and the same!

Example 13.

For the last time, let's look at 4 cities *A, B, C,* and *D*. This time the 4 cities sit as shown in Fig. 7-26(a). The minimum spanning tree is shown in Fig. 7-26(b), and its length is 1000 miles. Based on what happened in Example 12, we have to consider this network a serious contender for the title of shortest network. We also know that any network that is going to be shorter than this one is going to have to have some interior junction points, and if it's going to be the shortest network, then these junction points will have to be Steiner points. Because of the layout of these 4 cities, it is geometrically impossible to build a network with 2 interior Steiner points (Exercise 48). On the other hand, there are 3 possible networks with a single interior Steiner point [Figs. 7-26(c), (d), and (e)]. These are the only 3 possible challengers to the minimum spanning tree. Two of them [Figs. 7-26(c) and (d)] are the same length (approximately 1325 miles), and they are not even close. On the other hand, the network shown in Fig. 7-26(e) has a total length of approximately 982 miles. Given that there are no other possible candidates, *it must be the shortest network.* ■

The main lesson to be learned from Examples 11, 12, and 13 is twofold: First, if a shortest network is not the minimum spanning tree, then it must have interior junction points (that is to say, junction points other than one of the original points), and all of these junction points must be Steiner points. (Any network without circuits in which every interior junction is a Steiner point is called a **Steiner tree**. The

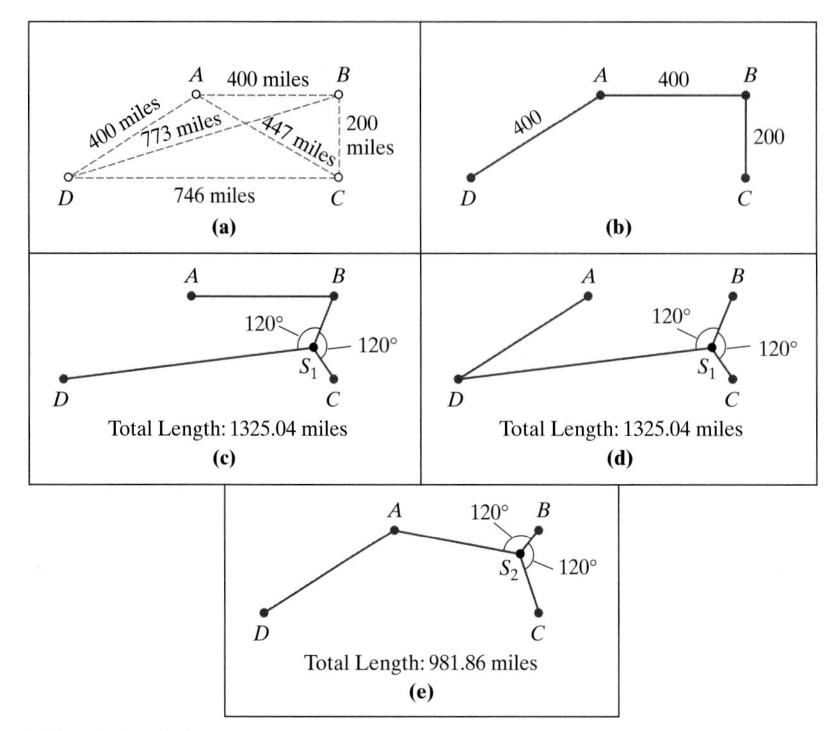

FIGURE 7-26

(a) The distances between the four cities. (b) The minimum spanning tree (total length is 1000 miles). (c) and (d) Both of these networks have one Steiner point (S_1) connecting cities *B, C,* and *D*. Both networks have the same length (1325 miles) and are much longer than the MST. (e) A different challenger with one Steiner point (S_2) and length of approximately 982 miles beats out (just barely) the MST. It turns out to be the shortest network connecting the 4 cities.

networks shown in Figs. 7-23(c) and (d) as well as Fig. 7-26(c), (d), and (e) are Steiner trees.) Second, just because a network is a Steiner tree, there is no guarantee that it is the shortest network. As Example 13 shows, there are many possible Steiner trees, and not all of them produce shortest networks.

What happens when the number of cities gets larger? How do we look for the shortest network? Here, we are in the same boat as in Chapter 6 (The Traveling Salesman Problem)—with no algorithm that is both optimal and efficient to hang our hat on. At this point, the best we can do is to take advantage of the following rule, which we informally discovered as we worked our way through the preceding examples.

The shortest network connecting a bunch of cities is either:

- the minimum spanning tree (no interior junction points), or
- a Steiner tree (some interior junction points).

This means that we can always find the shortest network by rummaging through all possible Steiner trees, finding the shortest one among them, and comparing it with the minimum spanning tree. The shorter of these two has to be the shortest network. This sounds like a good idea, but once again, the problem is the explosive growth of the number of possible Steiner trees. With as few as 10 cities, the possible number of Steiner trees we would have to rummage through is in the millions; with 20 cities it's in the billions.

What's the alternative? Just as in Chapter 6, if we are willing to settle for an approximate solution (in other words, if we are willing to accept a short network that is not necessarily *the shortest*), we can tackle the problem no matter how large the number of cities. Some excellent approximate algorithms for finding short networks are presently known, and one of them happens to be Kruskal's algorithm for finding a minimum spanning tree. The reason for this is a fairly recent discovery: in 1990 mathematicians Frank Hwang of AT&T Bell Laboratories and Ding-Zhu Du of Princeton University were able to prove that the percentage difference in length between the minimum spanning tree and the shortest network *is never more than 13.4%* (this largest possible difference can only happen with three cities located at the vertices of an equilateral triangle—see Example 7). In fact, in most real-life applications, using the minimum spanning tree to approximate the shortest network produces a relative percentage of error that is less than 5%. (Which is still a lot of money if you are building networks whose cost can run into the billions of dollars.)

Conclusion

In this chapter we discussed the problem of linking a set of points in a network in an optimal way, where "optimal" usually means least expensive or shortest. In practice, the points represent geographical locations (cities, telephone centrals, pumping stations, etc.), and the linkages can be rail lines, telephone trunk lines, pipelines, etc. Depending on the circumstances, we considered two different ways of doing this.

Version 1. In the first half of the chapter we were required to build the network in such a way that no junction points other than the original locations were allowed, in which case the optimal network is a *minimum spanning tree.* Finding a minimum spanning tree was the great success story of the chapter—Kruskal's algorithm provides the answer in an efficient and optimal way.

Version 2. In the second half of the chapter we considered what on the surface appeared to be a minor modification by removing the prohibition against new junction points. In this case the problem becomes one of finding the *shortest network* connecting the points. Here the situation gets considerably more complicated, but we do know a few things:

1. Sometimes the minimum spanning tree and the shortest network are one and the same, but most of the time they are not. When they are not the same, the shortest network is obviously shorter than the minimum spanning tree, but by how much? It took mathematicians more than two decades to completely answer this question, but the answer is now known: The difference between the minimum spanning tree and the shortest network is, in general, relatively small, and *under no circumstances can it ever be more than 13.4%.*

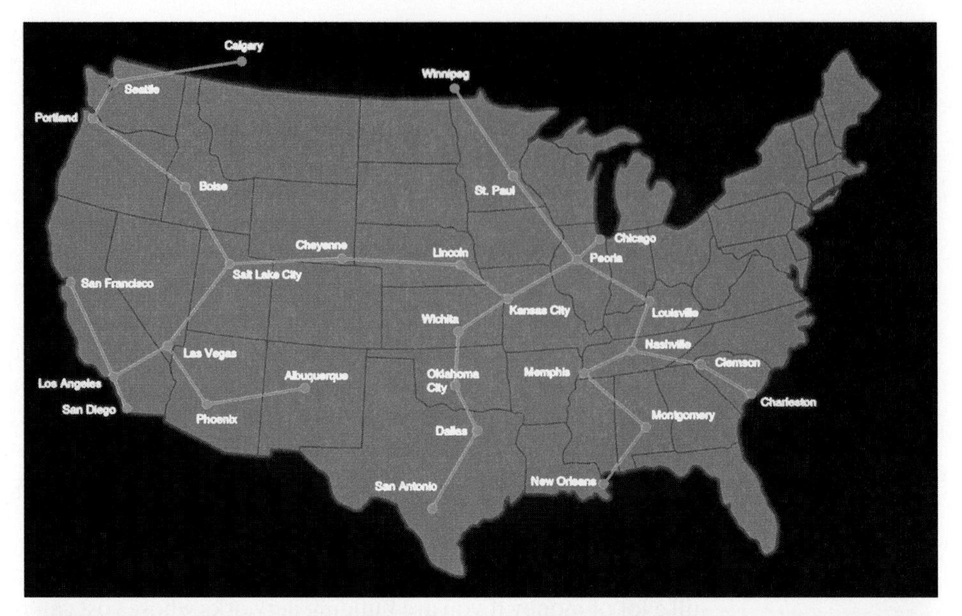

An optimal network problem with 29 cities. The top plate shows the MST, found using Kruskal's algorithm in a matter of seconds. Total length: approximately 7600 miles. The bottom plate shows the shortest network, found by a computer using a sophisticated algorithm developed by researchers at the University of Victoria (Canada). Total length: approximately 7400 miles (a savings of about 3%). This network has 13 new junction points (shown in yellow)—all of which are Steiner junction points.

2. In a shortest network, any new junction points that are created (call them *interior* junction points) must have the form of a perfect Y-junction (called a *Steiner* point). A Steiner point is always formed by three lines joining at 120° angles. Any network without circuits connecting the original points and such that all

interior junction points are Steiner points is called a *Steiner tree.* It follows that *if a shortest network has interior junction points it must be a Steiner tree, and if it doesn't, then it must be the minimum spanning tree.*

3. While in general the number of minimum spanning trees for a given set of points is small (usually just one) and easy to find (use Kruskal's algorithm), the number of possible Steiner trees for the same set of points is usually very large and difficult to find.[7] With 7 points, for example, the possible number of Steiner trees is in the thousands, and with 10 it's in the millions.

The problem of finding the shortest network connecting a set of points has a lot of similarities with the traveling-salesman problem: No efficient optimal algorithms are known, but for most real-life problems we can find approximate solutions that are very close (with margins of error less than 1%). For many applications this is good enough. In other situations it isn't, and mathematicians are constantly striving to find even better algorithms. Ultimately, an embarrassingly simple question still cannot be answered: What *is* the shortest distance between many points?

Key Concepts

Kruskal's algorithm
minimum network problem
minimum spanning tree (MST)
shortest network
spanning (subgraph)
spanning tree
Steiner point
Steiner tree
subgraph
tree

Exercises

Walking

1. For each of the following graphs determine whether the graph is a tree. If it is not a tree, give a reason why.

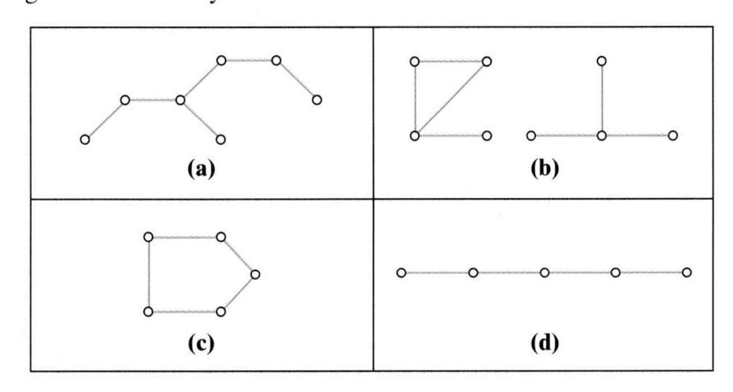

2. For each of the following graphs determine whether the graph is a tree. If it is not a tree, give a reason why.

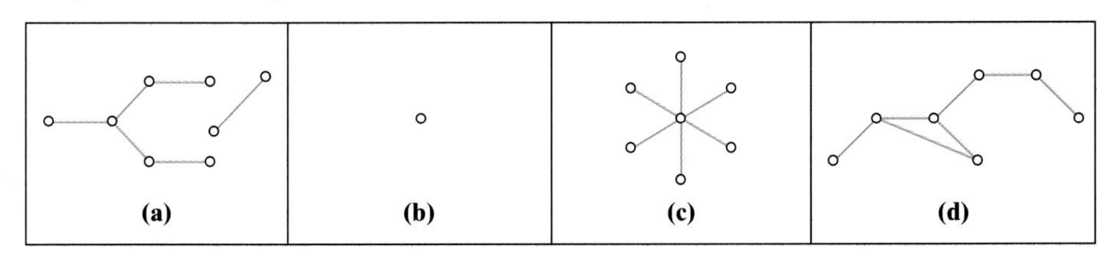

[7] There is a fun way to find some of the Steiner trees using soap-film computers (see the Appendix "The Soap-Bubble Solution" for details), but it is slippery to say the least.

3. Find three different spanning trees for the following graph.

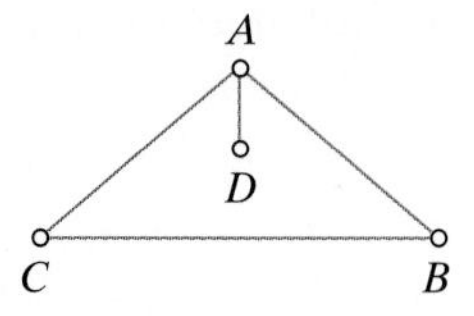

4. Find a spanning tree for each of the following graphs.

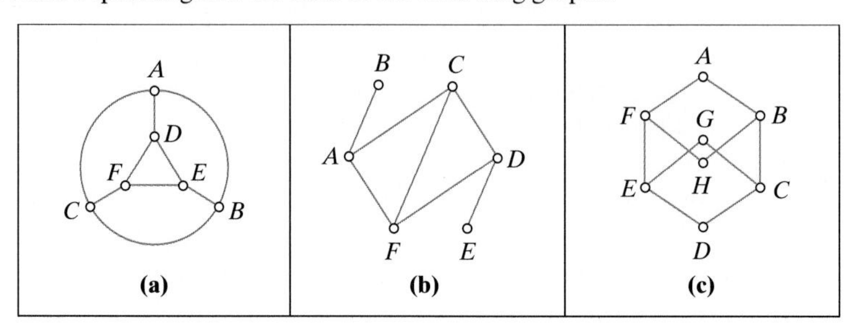

5. Find a spanning tree for each of the following graphs.

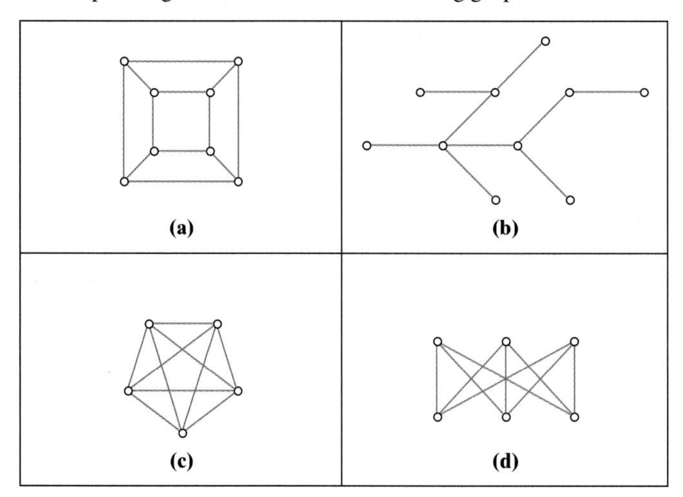

6. Find two different spanning trees for each of the following graphs.

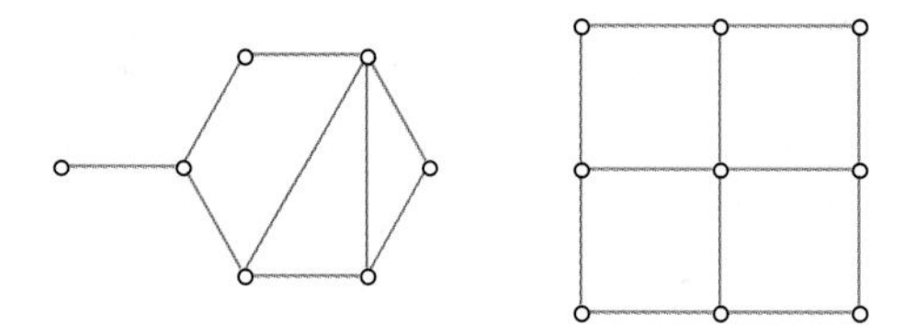

7. Find all the possible spanning trees of the following graph.

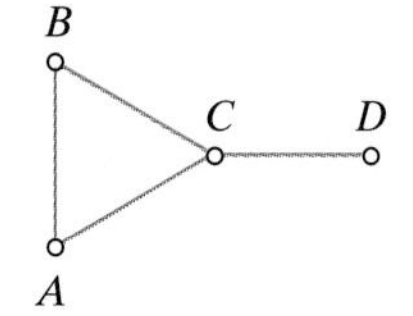

8. Find all the possible spanning trees of the following graph.

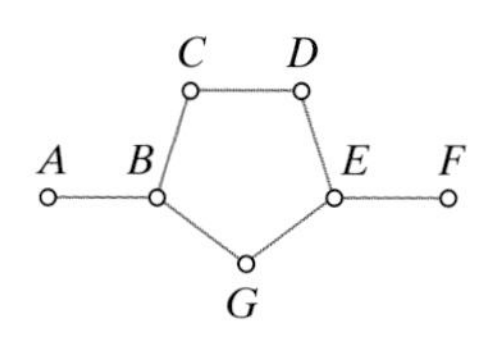

9. How many different spanning trees does each of the following graphs have?

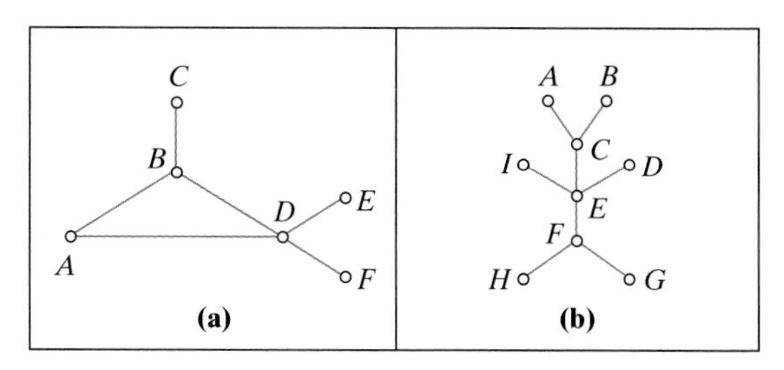

10. How many different spanning trees does each of the following graphs have?

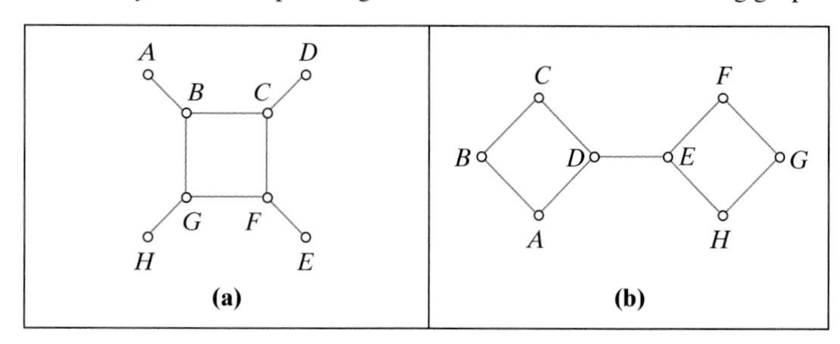

11. Use Kruskal's algorithm to find a minimum spanning tree of the following weighted graph.

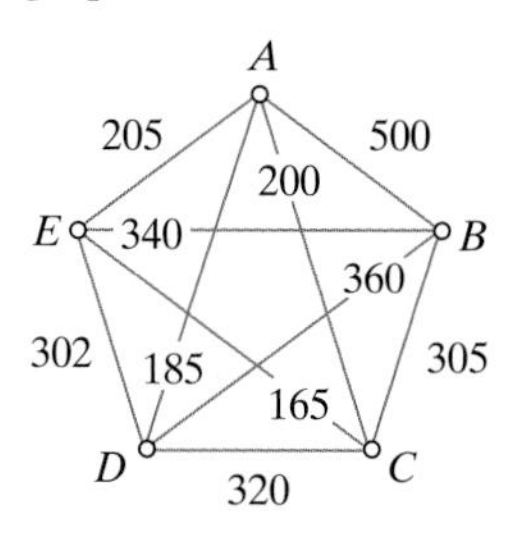

12. Use Kruskal's algorithm to find a minimum spanning tree of the following weighted graph.

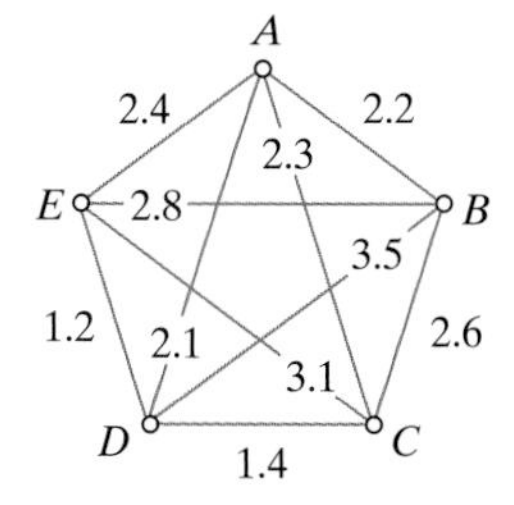

13. Use Kruskal's algorithm to find a minimum spanning tree of the following weighted graph.

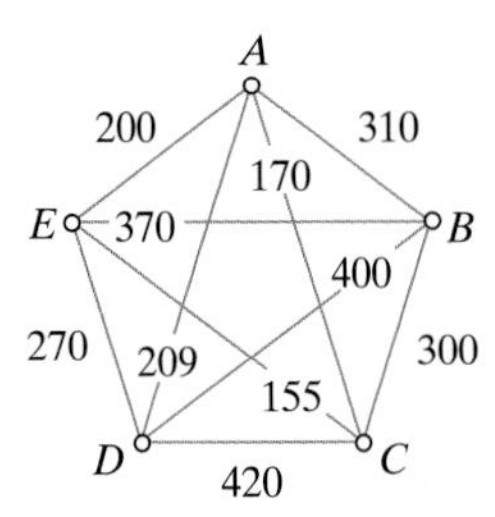

14. Use Kruskal's algorithm to find a minimum spanning tree of the following weighted graph.

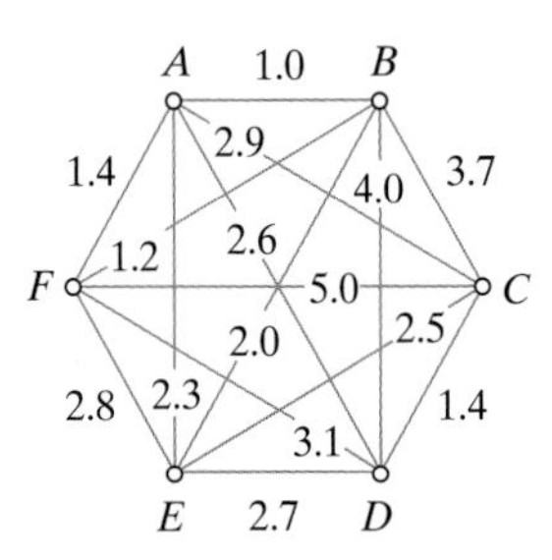

15. Use Kruskal's algorithm to find a minimum spanning tree of the following weighted graph.

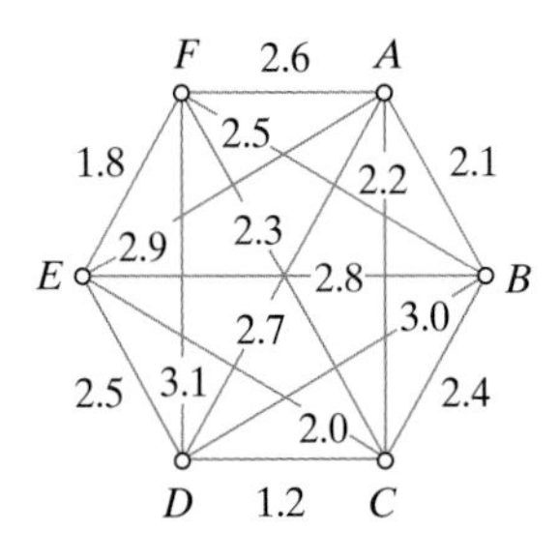

16. Use Kruskal's algorithm to find a minimum spanning tree of the following weighted graph.

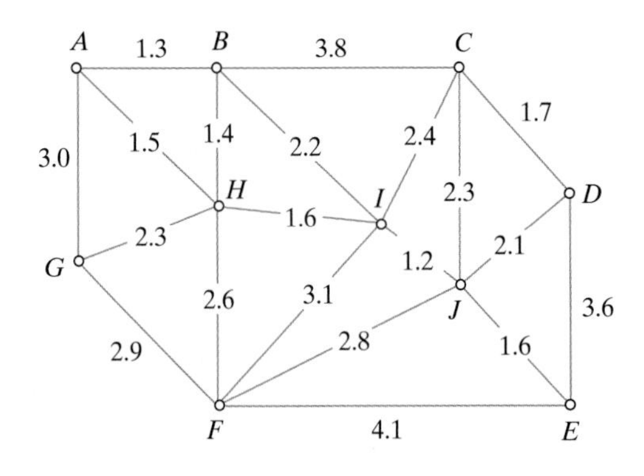

Exercises 17 and 18 refer to 5 cities (A, B, C, D, and E) located as shown in the following figure. In the figure, AC = AB, DC = DB, and angle CDB = 120°.

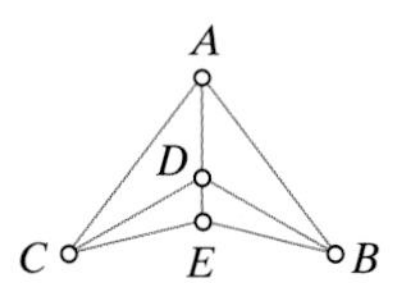

17. (a) Which is larger, $CD + DB$ or $CE + ED + EB$? Explain.

(b) What is the shortest network connecting the cities C, D, and B? Explain.

(c) What is the shortest network connecting the cities C, E, and B? Explain.

18. (a) Which is larger, $CA + AB$ or $DC + DA + DB$? Explain.

(b) Which is larger, $EC + EA + EB$ or $DC + DA + DB$? Explain.

(c) What is the shortest network connecting the cities A, B, and C? Explain.

19. Find the length of the shortest network connecting the cities A, B, and C shown in the following figure.

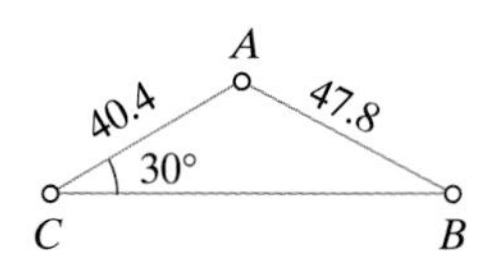

20. Find the length of the shortest network connecting the cities A, B, and C shown in the following figure.

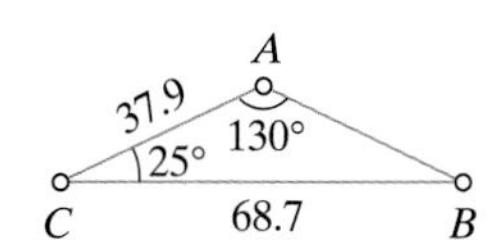

21. Find the length of the shortest network connecting the cities A, B, and C shown in the following figure.

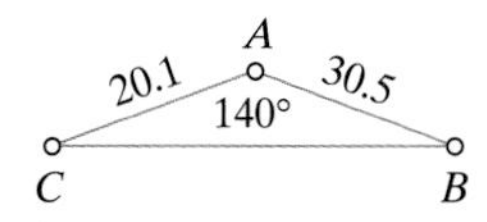

22. Find the length of the shortest network connecting the cities A, B, and C shown in the following figure.

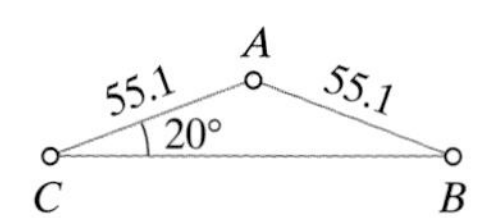

23. In a triangle ABC, angle A is 80°, angle B is 60°, and angle C is 40°.

(a) Which is the longest side of the triangle?

(b) What is the minimum spanning tree connecting A, B, and C?

(c) Is the minimum spanning tree the shortest network connecting A, B, and C? Explain.

24. Find the length $JA + JB + JC$ in the equilateral triangle ABC with altitude AJ shown in the following figure. (*Hint:* Use your knowledge of 30°-60°-90° triangles.)

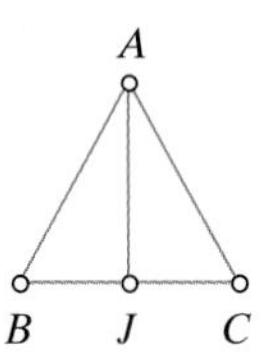

25. Find the length $SA + SB + SC$ in the equilateral triangle ABC with Steiner point S shown in the following figure. (*Hint:* Use your knowledge of 30°-60°-90° triangles.)

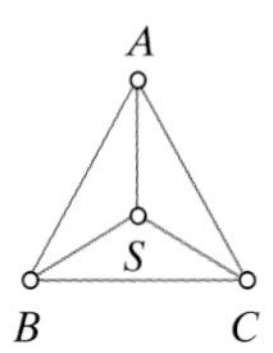

Jogging

26. Five cities (A, B, C, D, and E) are located as shown on the following figure. The 5 cities need to be connected by a railroad. and the cost of building the railroad system connecting any two cities is proportional to the distance between the two cities. Find the length of the railroad network of minimum cost (assuming that no additional junction points can be added).

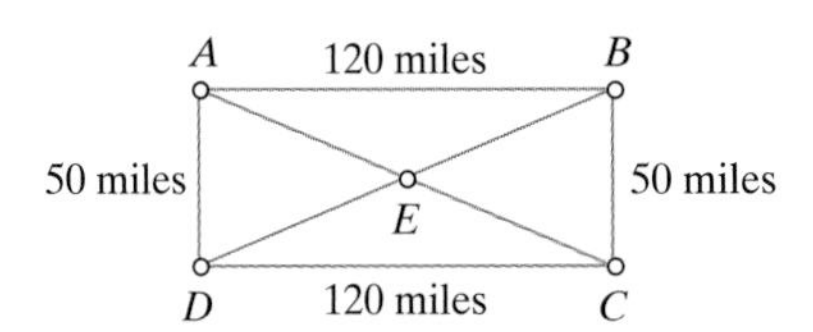

27. Four cities (A, B, C, and D), shown in the following figure, must be connected into a telephone network. The cost of laying down telephone cable connecting any two of the cities is given (in millions of dollars) by the weights of the edges in the graph. In addition, in any city that serves as a junction point of the network, expensive switching equipment must be installed. The cost of installing this equipment is given (also in millions of dollars) by the numbers inside the circles. Find the minimum-cost telephone network connecting these 4 cities.

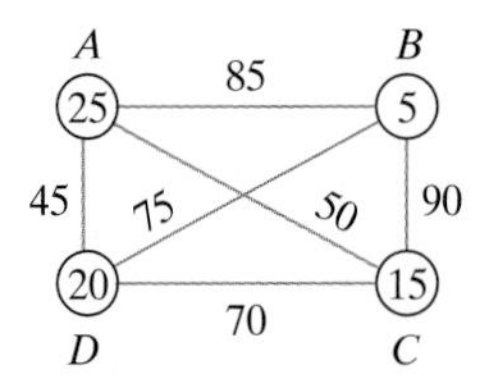

28. Explain why, if G is a connected graph with N vertices, Kruskal's algorithm will require exactly $N-1$ steps.

29. Explain why in a tree, every edge is a bridge. (Recall that a bridge is an edge whose removal disconnects the graph.)

30. Explain why in a scalene triangle (all the angles different) the minimum spanning tree consists of the two sides forming the largest angle. (See Exercise 23.)

31. (a) How many different spanning trees does the following graph have?

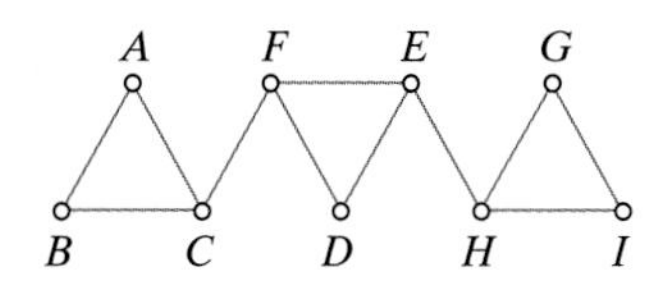

(b) How many different spanning trees does the following graph have?

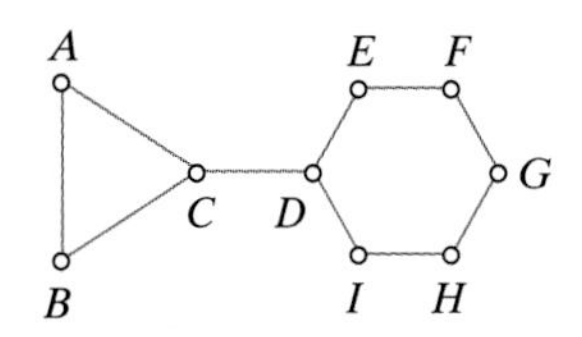

(c) How many different spanning trees does the following graph have?

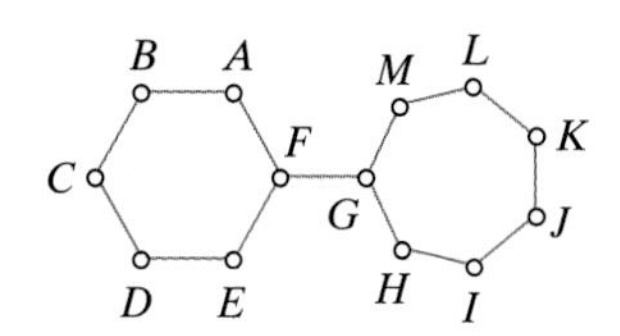

32. (a) Find all possible spanning trees of the following graph.

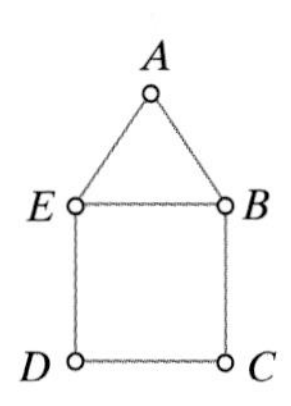

(b) How many different spanning trees does the following graph have?

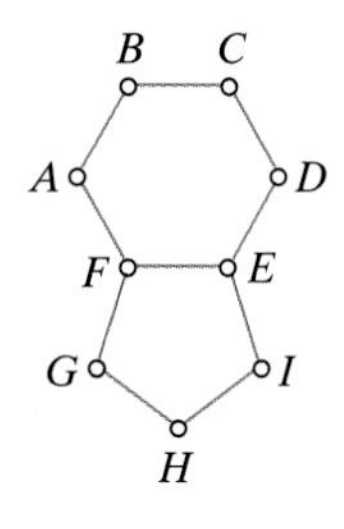

33. A theorem from graph theory (known as Cayley's theorem) states that the number of spanning trees in a complete graph with N vertices is N^{N-2}.

(a) Verify this result for the cases $N = 3$ and $N = 4$ by finding all spanning trees for complete graphs with 3 and 4 vertices.

(b) Which is larger, the number of Hamilton circuits or the number of spanning trees in a complete graph with N vertices? Explain.

34. (a) Can you give an example of a tree with 4 vertices such that the degrees of the vertices are 2, 2, 3, and 3? If yes, do so. If not, explain why not.

(b) If you have a tree with 4 vertices and you add up the degrees of all the vertices, what do you get?

(c) If you have a tree with 5 vertices and you add up the degrees of all the vertices, what do you get?

(d) If you have a tree with N vertices and you add up the degrees of all the vertices, what do you get?

35. (a) Give an example of a tree with 4 vertices such that the degrees of the vertices are 1, 1, 2, and 2.

(b) Give an example of a tree with 6 vertices such that the degrees of the vertices are 1, 1, 2, 2, 2, and 2.

(c) Give an example of a tree with N vertices such that the degrees of the vertices are $1, 1, 2, 2, 2, \ldots, 2$.

36. (a) Give an example of a tree with 4 vertices such that the degrees of the vertices are 1, 1, 1, and 3.

(b) Give an example of a tree with 5 vertices such that the degrees of the vertices are 1, 1, 1, 1, 4.

(c) Give an example of a tree with N vertices such that the degrees of the vertices are $1, 1, 1, \ldots, 1, N-1$.

37. (a) What is the smallest number of vertices of degree 1 that a tree with N ($N > 2$) vertices can have? Explain.

(b) What is the largest number of vertices of degree 1 that a tree with N ($N > 2$) vertices can have? Explain.

38. Explain why, if a single edge (but no additional vertex) is added to a tree, the resulting graph has a single circuit.

39. Suppose that T is a minimum spanning tree of a weighted graph G and suppose H is a new graph obtained by adding an additional edge (say e) of G to T. According to Exercise 38, the graph H has a single circuit. Explain why no edge in this circuit can have a weight larger than the weight of the new edge e.

40. A highway system connecting 9 cities $C_1, C_2, C_3, \ldots, C_9$ is to be built. Use Kruskal's algorithm to find a minimum spanning tree for this problem. The accompanying table shows the cost (in millions of dollars) of putting a highway between any 2 cities.

	C_1	C_2	C_3	C_4	C_5	C_6	C_7	C_8	C_9
C_1	*	1.3	3.4	6.6	2.6	3.5	5.7	1.1	3.8
C_2	1.3	*	2.4	7.9	1.7	2.3	7.0	2.4	3.9
C_3	3.4	2.4	*	9.9	3.4	1.0	9.1	4.4	6.5
C_4	6.6	7.9	9.9	*	8.2	9.7	0.9	5.5	4.9
C_5	2.6	1.7	3.4	8.2	*	4.8	7.4	3.7	3.5
C_6	3.5	2.3	1.0	9.7	4.8	*	8.9	4.4	5.8
C_7	5.7	7.0	9.1	0.9	7.4	8.9	*	4.7	3.9
C_8	1.1	2.4	4.4	5.5	3.7	4.4	4.7	*	2.8
C_9	3.8	3.9	6.5	4.9	3.5	5.8	3.9	2.8	*

41. (a) Explain why the triangle ABC cannot have an interior Steiner point.

(*Hint:* Take J to be any point inside the triangle. How does the angle BJC compare with the angle BAC?)

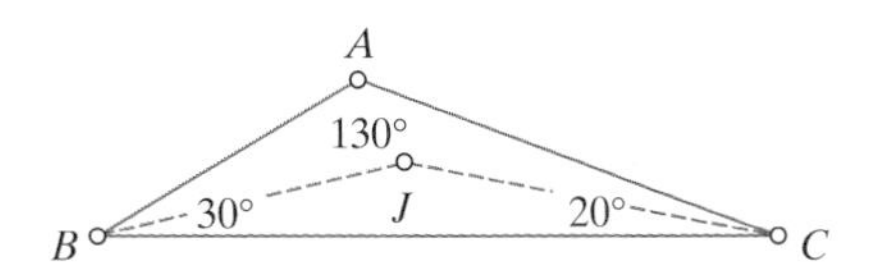

(b) Generalize your arguments in (a) to any triangle ABC with the angle BAC greater than 120°.

42. Show that if a tree has a vertex of degree M, then there are at least M vertices in the tree of degree 1.

43. Use 45°-45°-90° triangles to show that the length of the following network is $1000\sqrt{2} \approx 1414$.

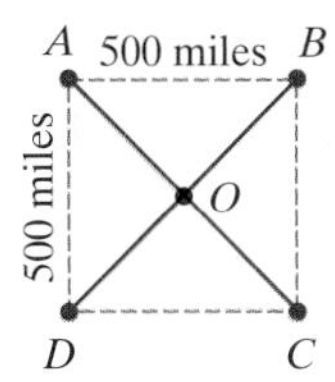

44. Use 30°-60°-90° triangles to show that the length of the following network is $500\sqrt{3} + 500 \approx 1366$.

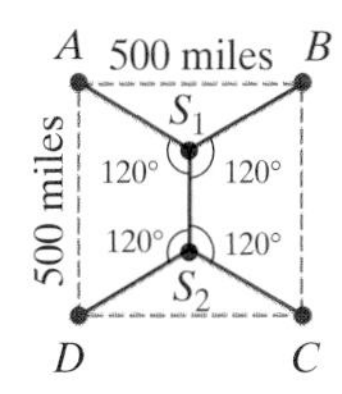

45. (a) Use 30°-60°-90° triangles to show that the length of the following network is $400\sqrt{3} + 300 \approx 993$.

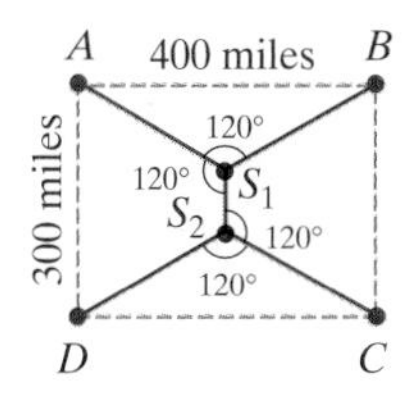

(b) Show that the length of the following network is $300\sqrt{3} + 400 \approx 919.6$.

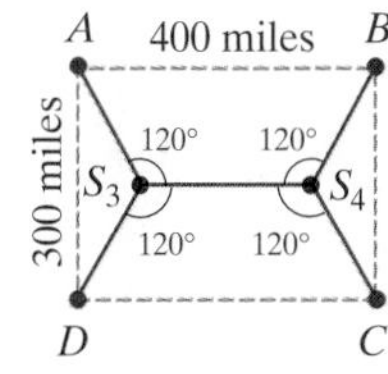

46. Consider triangle ABC with equilateral triangle EFG inside, as shown in the following figure.

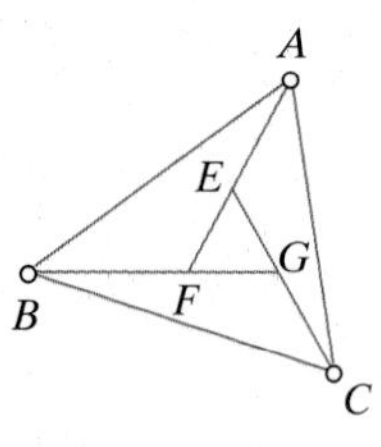

(a) Find angles BFA, AEC, and CGB.

(b) Explain why all the angles of triangle ABC are less than 120°.

(c) Explain why the Steiner point for triangle ABC lies inside triangle EFG.

47. Explain why in any tree with 3 or more vertices, it is impossible for all the vertices to have the same degree.

Running

48. Show that the following figure (see Example 13) cannot have two interior Steiner points.

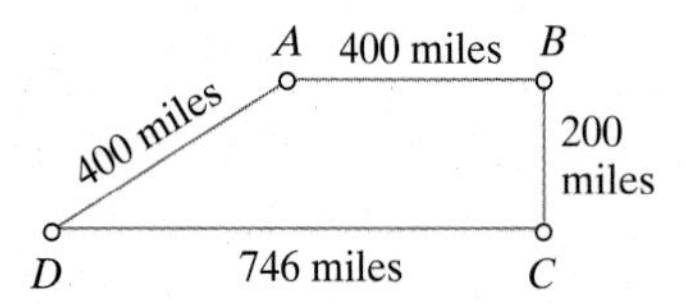

49. (a) Suppose you are asked to find a minimum spanning tree for a weighted graph that must contain a given edge. Describe a modification of Kruskal's algorithm that accomplishes this.

(b) Consider Exercise 41 again. Suppose that C_3 and C_4 are the two largest cities in the area, and the chamber of commerce insists that a section of highway directly connecting them must be built (or heads will roll). Find the minimum spanning tree that includes the section of highway between C_3 and C_4.

50. Prim's algorithm. The following algorithm for finding a minimum spanning tree is called Prim's algorithm.

- **Step 0.** Pick any vertex as a starting vertex. (Call it S.) Mark it red.
- **Step 1.** Find the nearest neighbor of S (call it P_1). Mark both P_1 and the edge SP_1 in red.
- **Step 2.** Find the nearest black neighbor to the red subgraph (i.e., the closest vertex to any red vertex). Mark it and the edge connecting the vertex to the red subgraph in red. Delete all black edges in the graph that connect red vertices.

Repeat Step 2 until all the vertices are marked red. The red subgraph is a minimum spanning tree.

Use Prim's algorithm to find a minimum spanning tree for the following graph.

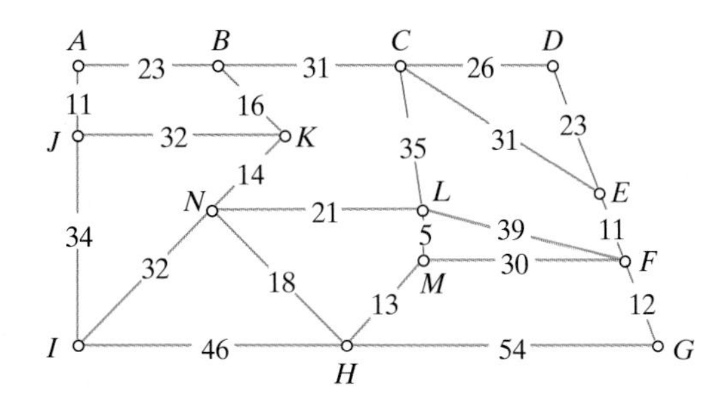

51. A graph with M components, each of which is a tree, is called a **forest** (with M trees).

(a) If G is a forest with N vertices and M trees, how many edges does G have?

(b) What is the smallest number of vertices of degree 1 that a forest with N vertices and M trees can have? (See Exercise 37a.)

(c) What is the largest number of vertices of degree 1 that a forest with N vertices and M trees can have? (See Exercise 37b.)

Exercise 52 and 53 involve the concepts of the middle of a tree. Every tree has either a ***center*** *or a* ***bicenter*** *(but not both). These can be found by carrying out the following algorithm:*

- **Step 1.** If the tree has more than 2 vertices, remove all the vertices of degree 1 along with the edges incident to these vertices.
- **Steps 2, 3, . . .** Repeat Step 1 until either

1. a single vertex remains, in which case we define this to be the **center** of the tree, or
2. two vertices joined by a single edge remain, in which case we define these two vertices along with this edge to be the **bicenter** of the tree.

52. Find the center or bicenter of each of the following trees.

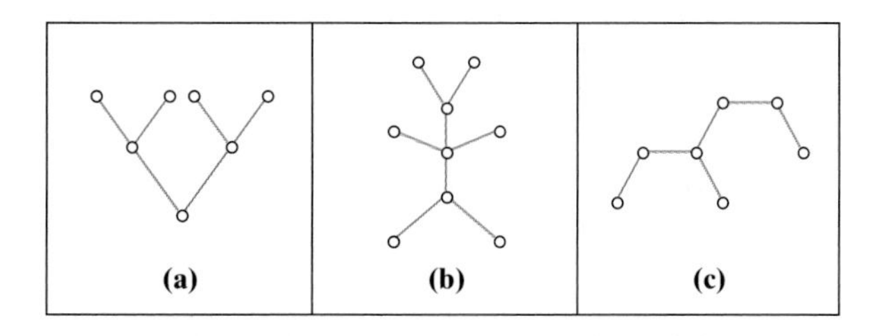

53. Find the center or bicenter of each of the following trees.

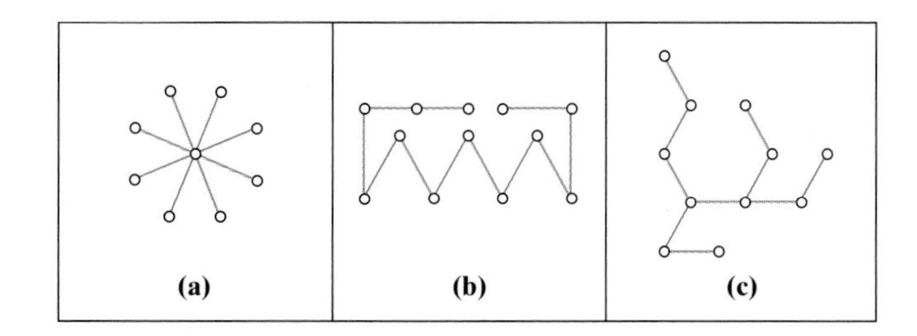

54. Consider four points (A, B, C, and D) forming the vertices of a rectangle with length b and width a as shown in the figure on p. 256 ($a < b$). Determine the conditions on a and

b so that the length of the minimum spanning tree is less than the length of one of the (two) Steiner trees.

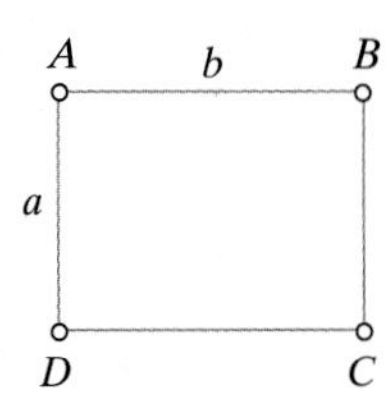

55. (a) Let P be an arbitrary point inside the equilateral triangle RQT as shown in the following figure. Suppose also that perpendiculars are drawn from P to the three sides as shown in the figure. Show that the sum of the lengths of PA, PB, and PC is equal to the length of the altitude of the triangle. (*Hint:* Compute the areas of triangle RPQ, triangle QPT, and triangle TPR. How does the sum of these three areas compare with the area of triangle RQT?)

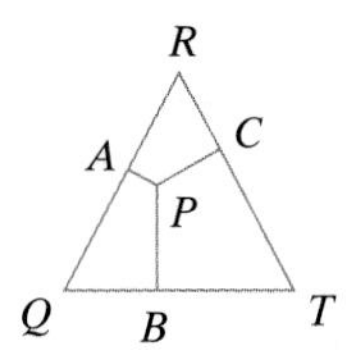

(b) Let triangle ABC be an arbitrary triangle with all angles less than 120°, an let S be a Steiner point inside the triangle. Draw lines perpendicular to SA, SB, and SC. Explain why these lines intersect to form an equilateral triangle RQT.

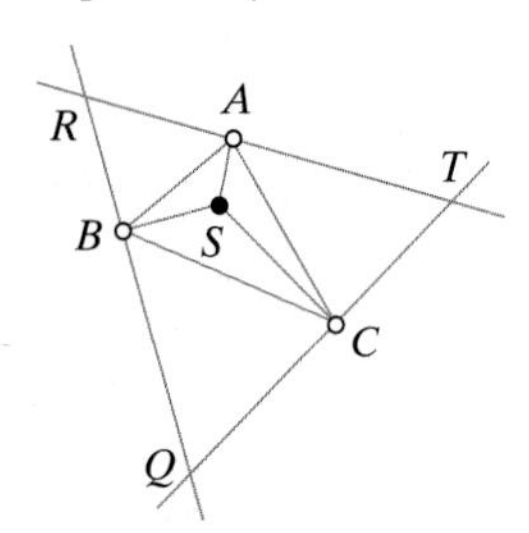

(c) Let P be any point other than S in triangle ABC, and draw perpendiculars from P to the three sides of triangle RQT as shown. Use (a) to conclude that $PA' + PB' + PC' = SA + SB + SC$, and yet $PA \geq PA'$, $PB \geq PB'$, and $PC \geq PC'$ (why?), and so $PA + PB + PC \geq SA + SB + SC$.

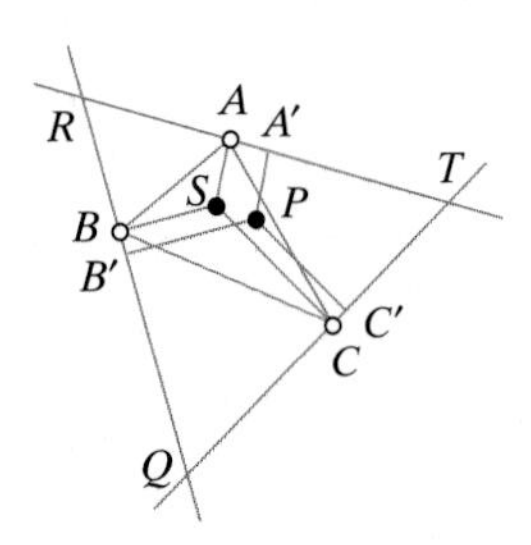

(d) Use (c) to conclude that no junction point P can give a shorter network than the Steiner point S.

56. Find the shortest network connecting the 12 vertices of the "star of David" shown in the following figure.

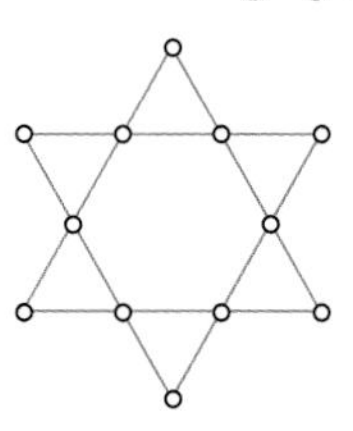

57. Eight cities (*A* through *H*) are located as in the following figure. Find the shortest network connecting the 8 cities. What is its length?

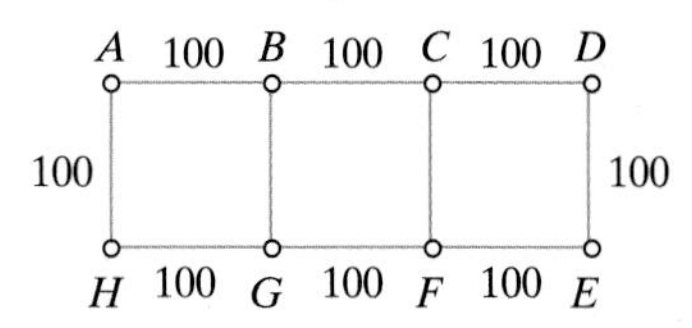

58. Find the shortest network connecting the 16 vertices of the 3-by-3 "checkerboard" shown in the following figure.

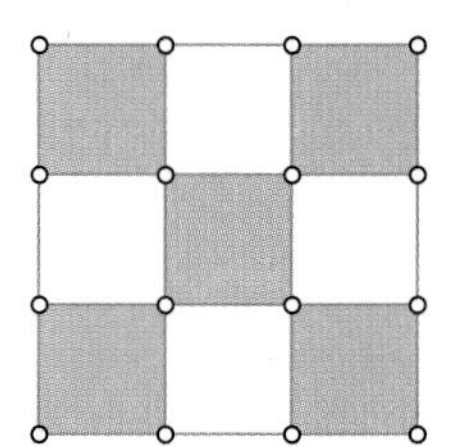

59. In the following figure, *S* is the Steiner point inside triangle *ABC* obtained using Torricelli's construction (*BXC* is an equilateral triangle). Show that the length of the shortest network ($SA + SB + SC$) equals the length of the segment *AX*.

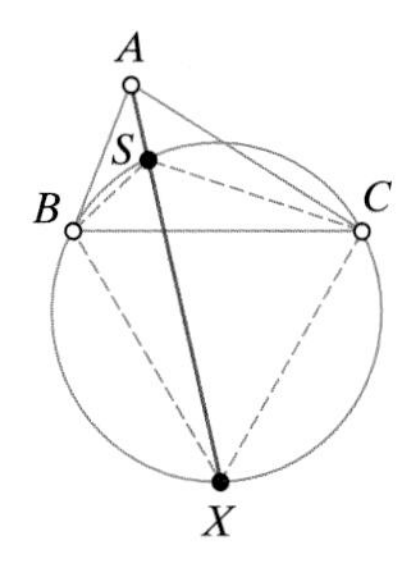

APPENDIX: THE SOAP-BUBBLE SOLUTION

Every child knows about the magic of soap bubbles: Take a wire or plastic ring, dip it into a soap-and-water solution, blow, and presto—beautiful iridescent geometric shapes magically materialize to delight and inspire our fantasy. Adults are not averse to a puff or two themselves.

What's special about soapy water that makes this happen? A very simplistic understanding of the forces of nature that create soap bubbles will help us understand how these same forces can be used to find (imagine, of all things) Steiner trees connecting a given set of points.

Take a liquid (any liquid), and put it into a container. When the liquid is at rest, there are two categories of molecules: those that are on the surface and those that are below the surface. The molecules below the surface are surrounded on all sides by other molecules like themselves and are therefore in perfect balance—the forces of attraction between molecules all cancel each other out. The molecules on the surface, however, are only partly surrounded by other molecules and are therefore unbalanced. For this reason, an additional force called **surface tension** comes into play for these molecules. As a result of this surface tension, the surface layer of any liquid behaves exactly as if it were made of a very thin, elastic material. The amount of elasticity of this surface layer depends on the structure of the molecules in the liquid. Soap or detergent molecules are particularly well suited to create an extremely elastic surface layer. (A good soap-film solution can be obtained by adding a small amount of dishwashing liquid to water, stirring gently to minimize surface bubbles, and, if necessary, adding a small amount of glycerin to make the soap film a little more stable.)

The connection between the preceding brief lesson in soapy solutions and the material in this chapter is made through one of the fundamental principles of physics: A physical system will remain in a certain configuration only if it cannot easily change to another configuration that uses less energy. Because of its extreme elasticity, the surface layer of a soapy solution has no trouble changing its shape until it feels perfectly comfortable—i.e., at a position of relatively minimal energy. When the energy is proportional to the distance, minimal energy results in minimal distance—ergo, Steiner trees.

Suppose that we have a set of points $(A_1, A_2, \ldots, A_N)$ for which we want to find a shortest network. We can find a Steiner tree that connects these points by means of an ingenious device which we will call a soap-bubble computer. To begin with, we draw the points A_1, $A_2, \ldots, A_N$ to exact scale on a piece of paper. (As much as possible, choose the scale so that points are neither too close to each other nor too far apart—somewhere from 1 to 4 inches should do just fine.) We now take two sheets of Plexiglas or Lucite and, using the paper map as a template, drill small holes on both sheets of Plexiglas at the exact locations of the points. Then we put thin metal or plastic pegs through the holes in such a way that the two sheets are held about an inch apart.

When we dip our device into a soap-and-water solution and pull it out, the soap-bubble computer goes to work. The film layer that is formed between the plates connects the various pegs. For a while it moves seeking a configuration of minimal energy. Very shortly, it settles into a Steiner tree.

It is a bit of a disappointment that the Steiner tree we get is not necessarily the shortest network. The reasons for this are beyond the scope of our discussion. (The interested reader is referred to the excellent technical discussion of soap-film computers in Reference 1.) At the same time, we should be thankful for what nature has provided: a simple device that can compute in seconds what might take hours to do with pencil and paper.

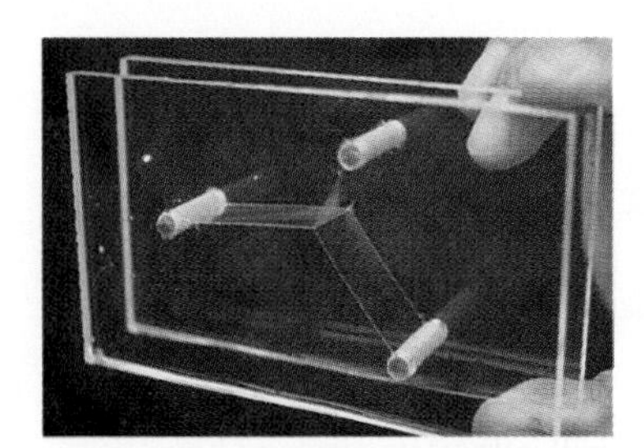

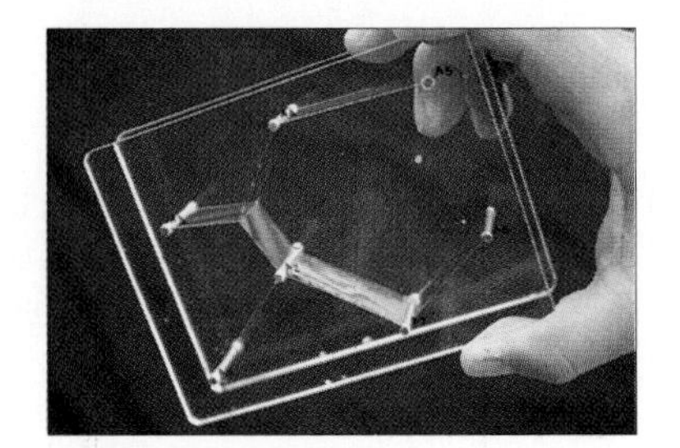

References and Further Readings

1. Almgren, Fred J., Jr., and Jean E. Taylor, "The Geometry of Soap Films and Soap Bubbles," *Scientific American,* 235 (July 1976), 82–93.
2. Bern, M., and R. L. Graham, "The Shortest Network Problem," *Scientific American,* 260 (January 1989), 84–89.
3. Chung, F., M. Gardner, and R. Graham, "Steiner Trees on a Checkerboard," *Mathematics Magazine,* 62 (April 1984) 83–96.
4. Cockayne, E. J., and D. E. Hewgill, "Exact Computation of Steiner Minimal Trees in the Plane," *Information Processing Letters,* 22 (1986), 151–156.
5. Courant, R., and H. Robbins, *What Is Mathematics*? New York: Oxford University Press, 1941.
6. Du, D.-Z., and F. K. Hwang, "The Steiner Ratio Conjecture of Gilbert and Pollack is True," *Proceedings of the National Academy of Sciences,* U.S.A., 87 (December 1990), 9464–9466.
7. Gilbert, E. N., and H. O. Pollack, "Steiner Minimal Trees," *SIAM Journal of Applied Mathematics,* 16 (1968), 1–29.
8. Gardner, Martin, "Mathematical Games: Casting a Net on a Checkerboard and Other Puzzles of the Forest," *Scientific American,* June 1986, 16–23.
9. Graham, R. L., and P. Hell, "On the History of the Minimum Spanning Tree Problem," *Annals of the History of Computing,* 7 (January 1985), 43–57.
10. Hwang, F. K., D. S. Richards, and P. Winter, *The Steiner Tree Problem.* Amsterdam: North Holland, 1992.
11. Kolata, Gina, "Solution to Old Puzzle: How Short a Shortcut?" *The New York Times,* October 30, 1990.
12. Melzak, Z. A., *Companion to Concrete Mathematics.* New York: John Wiley & Sons, Inc., 1973.
13. Pierce, A. R., "Bibliography on Algorithms for Shortest Path, Shortest Spanning Tree and Related Circuit Routing Problems (1956–1974)," *Networks,* 5 (1975), 129–149.

"Waste neither time nor money, but make the best use of both."

Ben Franklin

The Mathematics of Scheduling

Directed Graphs and Critical Paths

CHAPTER 8

How long does it take to build a house? Here is a deceptively simple question that defies an easy answer. Some of the factors involved are obvious: the size of the house, the type of construction, the number of workers, the tools and machinery used. Less obvious, but equally important, is another variable: the ability to organize and coordinate the timing of people, equipment, and work so that things get done in a timely way. For better or for worse, this last issue boils down to a graph-theory problem, just one example in a large family of problems that fall under the purview of what is known as the *mathematical theory of scheduling.* Discussing some of the basics of this theory will be the theme of this chapter.

Let's get back to our original question. According to the Building Industry Association, a national association of home builders based in Washington, DC, it takes 1092 man-hours to build the average American house. (Since we are not being picky about details, we'll just call it 1100 hours.) One way to interpret the above statement is this: Given just one worker (and assuming that this worker can do every single job required for building a house), it would take about 1100 hours of labor to finish this hypothetical average American house.

Let's now turn the question on its head: If we had 1100 equally capable workers, could we get the same house built in one hour? Of course not! In fact, we could put thousands of workers on the job and we still could not get the house built in one hour. Some inherent physical limitations to the speed with which a house can be built are outside of the builder's control. Some jobs cannot be speeded up beyond a certain point regardless of how many workers one puts on that job. Even more significantly, certain jobs can be started only after certain other jobs have been completed (roofing, for example, can be started only after framing has been completed).

A simple precedence relation in construction: framing must be finished before roofing can be started.

Given the fact, then, that it is impossible to build the house in one hour, how fast could we build it if we had as many workers as we wanted at our disposal and we cared only about speed? (To the best of our knowledge, the record is something a tad under 24 hours.) How fast could we build the house if we had 10 workers at our disposal at all times, all equally capable? What if we had only 3 workers? What if we needed to finish the entire project within a given time frame—say 3 weeks? How many workers should we hire then? All these questions could equally well be asked if we replaced "building a house" with many other types of projects—from preparing an elaborate dinner to repairing a stretch of highway. These are the kinds of questions scheduling theory is designed to deal with, and while they may sound simple, they are surprisingly difficult to answer. Whatever answers are possible, the best way to find them is by using graph models and graph algorithms, the kind of thing we have gotten used to in the last three chapters.

The Basic Elements of Scheduling

We will now introduce the principal characters in every scheduling story.

The Processors. This is the name that we give to the "workers" that carry out the work. While the word *processor* may sound a little cold and impersonal, it does underscore an important point: Processors need not be human beings. In scheduling, a processor could just as well be a robot, a computer, an automated teller machine, and so on. For the purposes of our discussion, we will use the notation $P_1, P_2, P_3, \ldots, P_N$ to represent the processors (N represents the total number of processors).

The number of processors N can range from just 1 to the tens of thousands, but when $N = 1$, the whole question of scheduling is trivial and not very interest-

Processors hard at work completing their tasks. Fine restaurants deal with sophisticated scheduling problems on a daily basis.

ing. As far as we are concerned, real scheduling problems begin when the number of processors is 2 or more.

The Tasks. In every complex project there are individual pieces of work, often called "jobs" or "tasks." We will need to be a little more precise than that, however. We will define a task as an indivisible unit of work that (either by nature or by choice) cannot be broken up into smaller units. Moreover, and most importantly, in our definition of the term, *a task will always be something that is by nature carried out by a single processor.*

To clarify the concept, let's consider a simple illustration. If a foreman assigns the wiring of a house to a single electrician, and it takes him 16 hours to do it, then we will consider wiring the house as a single 16-hour task. On the other hand, he may assign the job to 2 electricians, who together take, let's say, 7 hours. In this case, we will consider the job of wiring the house as 2 separate 7-hour tasks.

In general, we will use capital letters *A, B, C, . . .* , to represent the tasks, although, when convenient, we will also use appropriate abbreviations (such as *WE* for "wiring the electrical system," *PL* for "plumbing," etc.).

At any particular moment in time throughout the project, a task can be in one of four different states:

- *ineligible* (the task cannot be started at this time because certain other requirements have not yet been met),
- *ready* (the task could be started at this time),
- *in execution* (the task is presently being carried out by one of the processors), and
- *completed.*

The Processing Times. Associated with every task is a number called the *processing time.* It represents the amount of time, without interruption, required by one processor to carry out the task. But, one might ask, which processor? After all, how long it takes to do something often depends on who is doing it: It might take P_1 two hours to carry out a task, and it might take P_2 only one hour to carry out the same task. In general, different processors work at different rates. This surely complicates matters, so we will have to make an important concession to expediency: From now on, we will work under the assumption that *each processor can carry out each and every one of the tasks and that the processing time for a task*

The four possible states a task can be in: *ineligible* (upper left), *ready* (upper right), *in execution* (lower left), and *completed* (lower right).

is the same regardless of which processor is doing it.[1] To help things along even further, we will make a second assumption: *Once the task is started, the processor must execute it without interruptions.* A processor cannot stop in the middle of a task, be it to start another task or to take a break. With these assumptions, it now makes sense to talk about *the processing time* of a task, a single nonnegative number which we will attach (in parentheses) to the right of the task's name. Thus, when we see $A(5)$, we take this to mean that it takes 5 units of time (be it minutes, hours, or whatever) to execute the task called A, and that this is the case whether the task is done by P_1 or P_2, or any other processor.

X ○——→○ Y

FIGURE 8-1
X precedes Y.

The Precedence Relations. These are restrictions on the order in which the tasks can be executed. A typical precedence relation is of the form *task X precedes task Y,* and it means that task Y cannot be started until task X has been completed. Such a precedence relation can be conveniently abbreviated by writing $X \to Y$, or described pictorially as in Fig. 8-1. Precedence relations arise from laws that govern the order in which things are done in the real world—we just don't put our shoes on before our socks! A single scheduling problem can have hundreds or even thousands of precedence relations, each adding another restriction on the scheduler's freedom.

At the same time, it also happens fairly often that there are no restrictions as to the order of execution between two tasks in a project. When a pair of tasks X and Y have no precedence requirements between them (neither $X \to Y$ nor $Y \to X$), we say that the tasks are **independent**. When two tasks are independent,

[1] Essentially, this is the theory behind automobile repair charges: The hours charged for a given repair job (processing time) come from manuals such as *Chilton's Guide to Automobile Repairs.*

either one can be started before the other one or they can both be started at the same time—some people put their shoes on before their shirt, others put their shirt on before their shoes, and occasionally, some of us have been known to put our shoes and shirt on at the same time. Sometimes an entire project can be made up of all independent tasks (no precedence relations whatsoever to worry about). We will discuss this special situation in greater detail later in the chapter.

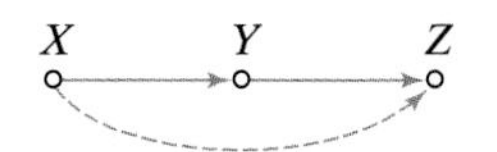

FIGURE 8-2
When $X \to Y$ and $Y \to Z$, then $X \to Z$ is implied.

Two final comments about precedence relations. First, precedence relations are *transitive:* if $X \to Y$ and $Y \to Z$, then it must be true that $X \to Z$. In a sense, the last precedence relation is implied by the first two, and it is really unnecessary to mention it (Fig. 8-2). Thus, we will make a distinction between two types of precedence relations: *basic* and *implicit.* Basic precedence relations are the ones that come with the problem and that we must follow in the process of creating a schedule. Once we do this, the implicit precedence relations will be automatically taken care of.

The second observation is that *we cannot have a set of precedence relations that form a cycle!* Imagine having to schedule the tasks in Fig. 8-3, with precedence relations as shown by the arrows Clearly, this is a logical impossibility. From here on, we will assume that there are no cycles of precedence relations among the tasks.

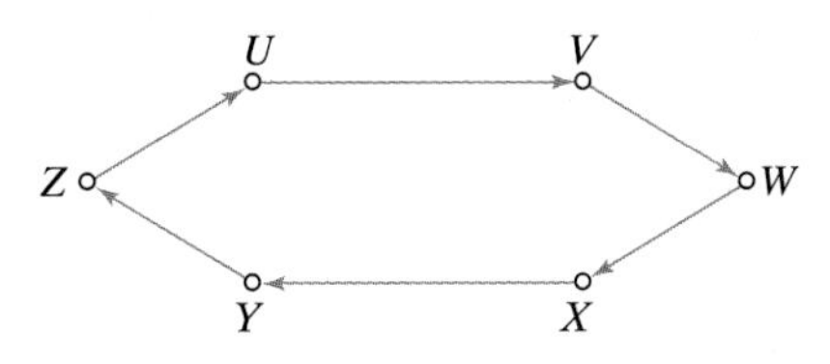

FIGURE 8-3
These tasks cannot be scheduled because of the cyclical nature of the precedence relations.

Processors, tasks, processing times, and precedence relations are the basic ingredients that make up a scheduling problem. They constitute, in a manner of speaking, the hand that is dealt to us. But how do we play such a hand? To get a small inkling of what's to come, let's look at the following (very) simple example.

Example 1. Repairing a Wreck

Imagine that you just totaled your expensive new sports car, but that (thank heavens) you are OK and the insurance company will pick up the tab. You take the car to the best garage in town, operated by the Click (P_1) and Clack (P_2) brothers. The repairs on the car can be broken into four different tasks: (A) Exterior body work (4 hours); (B) Engine repairs (5 hours); (C) painting and exterior finish work (7 hours); and (D) Repair transmission (3 hours). The only precedence relation for this set of tasks is that the painting and exterior finish work cannot be started until the exterior body work has been completed ($A \to C$). The two brothers always work together on a repair project, but each takes on a different task (they refuse to work together on any single task because they always end up in a big argument). Under these conditions, how should the different tasks be scheduled? Who should do what and when?

Even in this very simple situation, many different schedules are possible. Figure 8-4 shows several possibilities, each one illustrated by means of a timeline. Figure 8-4(a) shows a schedule that is very inefficient. All the short tasks were assigned to one mechanic (P_1) and all the long tasks to the other mechanic (P_2)—obviously not a very clever strategy. Under this schedule, the **finishing time** (the time elapsed from the beginning of the project until the last task is completed) is 12 hours. Figure

Scheduling the repairs for a wrecked car is easy—actually doing the repairs is not!

8-4(b) shows what looks like a much better schedule, but it is in violation of the precedence relation $A \rightarrow C$: We are not allowed to start task C on the third hour. In fact, this is true not only for this schedule, but for any other schedule we might think of. No matter how clever we are (and no matter how many processors we have at our disposal), task *C can never be started before the fourth hour.* Why not? Because of the requirement that $A(4)$ must be finished first. On the other hand, if we make P_2 sit idle for 1 hour, waiting for the green light so to speak, to start task C, we get a perfectly good schedule, shown in Fig. 8-4(c). The finishing time is 11 hours, and given that task $C(7)$ cannot be started until the fourth hour, this is as short as it is going to get. *No possible schedule can complete this project in less than 11 hours.* As far as finishing time is concerned, the schedule shown in Fig. 8-4(c) is *optimal.* In general, there is likely to be more than one optimal schedule, and Fig. 8-4(d) shows a different schedule with a finishing time of 11 hours. ■

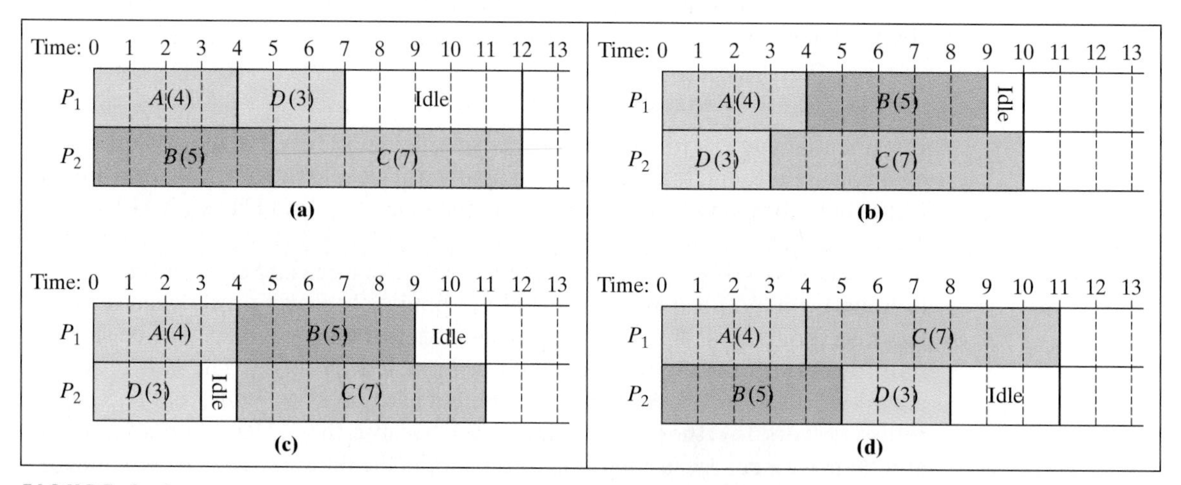

FIGURE 8-4

Some possible schedules for Example 1. (a) An *inefficient* schedule. Finishing time is 12 hours. (b) An illegal schedule. The precedence relation $A \rightarrow C$ is violated when C is started before A is completed. (c) A small adjustment in the preceding schedule makes things OK: Make P_2 sit idle for one hour before starting task C. Finishing time is 11 hours. (d) A different schedule with finishing time of 11 hours. Schedules (c) and (d) are both *optimal,* since it *is* impossible to finish the project in less than 11 hours.

As scheduling problems go, Example 1 was a fairly simple one. But even from this simple example we can draw some useful lessons. First, notice that even though we had only 4 tasks and 2 processors, we were able to create several different schedules, and the four we looked at were just a sampler—there are other possible schedules which we didn't bother to discuss. Imagine what would happen if we had hundreds of tasks and dozens of processors—the number of possible schedules to consider would boggle the mind. In looking for a good, or even optimal, schedule, we are going to need a systematic way to sort through the many possibilities—in other words, we are going to need some good *scheduling algorithms.*

The second useful thing we learned in Example 1 is that when it comes to the finishing time of a project, there is a bottom line—a minimum time barrier which no schedule can break, no matter how good an algorithm we use or how many processors we put to work. In Example 1, this minimum barrier was 11 hours, and as luck would have it, we found a schedule (actually we found two—Fig. 8-4(c) and (d)) with finishing time to match this minimum. Every project, no matter how simple or complicated, has such an absolute time barrier which depends on the processing times and precedence relations for the tasks, and not on the number of processors used. This theoretical minimum is called the **critical time** of the project, and one important thing we are going to learn later in this chapter is how to calculate the critical time of any project.

To set the stage for a more formal discussion of scheduling algorithms, we will introduce the most important example of this chapter. While couched in what seems like science fiction terms, the situation it describes is not totally far-fetched.

Example 2. Building a Dream Home On Mars

It is the year 2050, and several human colonies have already been established on Mars. Imagine that you accept a job offer to work in one of these colonies, so next thing we know you are bidding farewell to friends and family and moving to Mars.

What about housing? Like everyone else in Mars, you will be provided with a living pod called a Martian Habitat Unit (MHU). MHUs are shipped to Mars in

the form of prefabricated kits which have to be assembled on the spot, an elaborate and unpleasant job if you are going to do it yourself. A better option is to hire special "workers" that will do all of the assembly work for you. In Mars, these workers come in the form of robots called Habitat Unit Building Robots (affectionately nicknamed "Hubris"), which can be rented by the hour at the local Rent-a-Robot outlet.

Here are some questions you are going to have to deal with: How can you get your MHU built quickly? How many Hubris should you rent? How do we create a suitable work schedule that will get the job done? (A Hubri will do whatever it is told, but someone has to tell it what to do and when.)

The assembly of an MHU consists of 15 basic tasks as shown in Table 8-1, with the processing times representing Hubri-hours (i.e., the number of hours it takes one Hubri to execute the task). In addition, the tasks are constrained by 17 different precedence relations as shown in Table 8-2.

Table 8-1

Task	Symbol (Processing Time)
Assemble Pad	$AP(7)$
Assemble Flooring	$AF(5)$
Assemble Wall Units	$AW(6)$
Assemble Dome Frame	$AD(8)$
Install Floors	$IF(5)$
Install Interior Walls	$IW(7)$
Install Dome Frame	$ID(5)$
Plumbing	$PL(4)$
Install Atomic Power Plant	$IP(4)$
Install Pressurization Unit	$PU(3)$
Install Heating Units	$HU(4)$
Install Commode	$IC(1)$
Interior Finish Work	$FW(6)$
Pressurize Dome	$PD(3)$
Install Entertainment Unit (virtual reality TV, computer, music box, communication port, etc.)	$EU(2)$

Table 8-2

Precedence Relations
$AP \to IF$
$AF \to IF$
$IF \to IW$
$AW \to IW$
$AD \to ID$
$IW \to ID$
$IF \to PL$
$IW \to IP$
$IP \to PU$
$ID \to PU$
$IP \to HU$
$PL \to IC$
$HU \to IC$
$PU \to EU$
$HU \to EU$
$IC \to FW$
$HU \to PD$

We will return soon to the question of how best to assemble an MHU, but first we will develop a few helpful concepts. ■

Directed Graphs

All of the information presented in Tables 8-1 and 8-2 can be summarized in a very convenient way, shown in Fig. 8-5. The tasks are represented by vertices and the precedence relations are represented by arrows pointing from one vertex to another, so that an arrow pointing from vertex X to vertex Y indicates that task X must be completed before task Y can be started. This approach is consistent with what we already did in Figs. 8-1, 8-2, and 8-3.

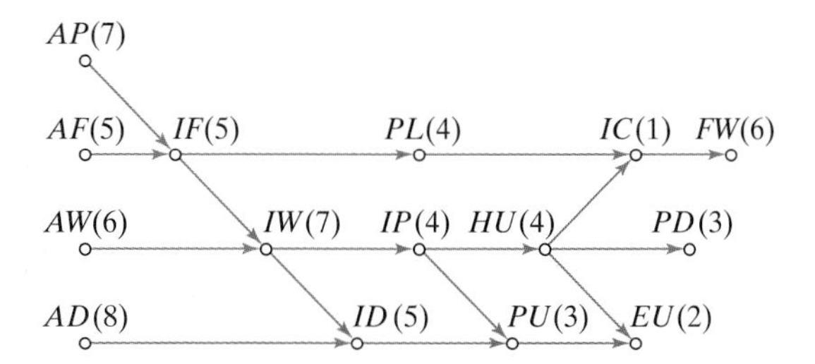

FIGURE 8-5

Figure 8-5 looks just like one of the graphs in the previous chapters except that each "edge" now has a direction associated with it. A graph in which the edges have a direction associated with them is called a **directed graph**, and more commonly a **digraph**.

Just like graphs, digraphs are used to describe relationships between objects, but in this case the nature of the relationship is such that we cannot always assume it is reciprocal. We call such relationships **asymmetric relationships**. Being in love is a good example of an asymmetric relationship: Just because X is in love with Y it does not necessarily follow that Y must be in love with X. Sometimes it happens and sometimes (sigh) it doesn't.

To distinguish digraphs from ordinary graphs, we use a slightly different terminology. Instead of *edge,* we use the word **arc** to indicate that the edge has a direction, and we describe the arc $X \rightarrow Y$ as XY, which in this case is different from YX ($Y \rightarrow X$). If there is an arc joining vertices X and Y, we can indicate its direction by saying that X is **incident to** Y if the arc is $X \rightarrow Y$, and that X is **incident from** Y if the arc is $Y \rightarrow X$; instead of the *degree* of a vertex we speak about the **indegree** and the **outdegree** of a vertex (the *indegree* is the number of arrowheads pointing toward the vertex; the *outdegree* is the number of arrowheads coming out of the vertex). In a digraph, a **path** from vertex X to vertex Y is a sequence of arcs starting at X and ending at Y, in which no arc is repeated and each vertex in the sequence is incident to the next one. In a digraph, a **cycle** is a path that starts and ends in the same place.

Example 3.

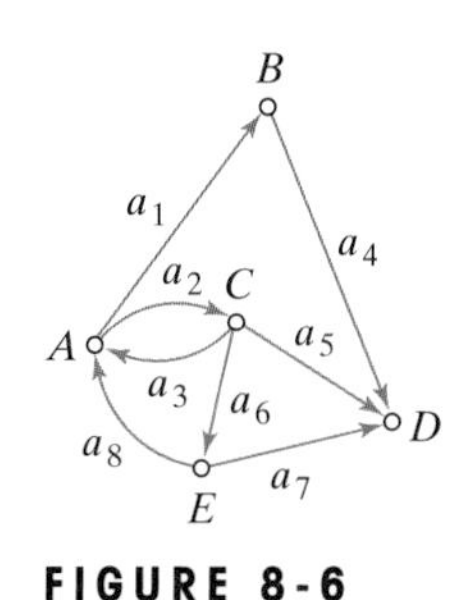

FIGURE 8-6

Consider the digraph in Fig. 8-6. This is a digraph with 5 vertices (A, B, C, D, and E) and 8 arcs (a_1, a_2, a_3, a_4, a_5, a_6, a_7, and a_8). In this digraph, for example, A is *incident to* B and C but not to E. By the same token, A is *incident from* E as well as from C. The indegree of vertex A is 2, and so is the outdegree. The indegree of vertex C is 1, and the outdegree is 3. We leave it to the reader to find the indegrees and outdegrees of each of the other vertices of the graph (Exercise 1b).

In this digraph there are several paths from A to D, such as A, C, D; A, C, E, D; A, B, D; and even A, C, A, B, D. On the other hand, A, E, D is not a path from A to D (AE is not an arc). Are there any possible paths from D to A? Why not? As for examples of cycles in this digraph, here are two: A, C, E, A; and A, C, A. Can you find any others? ■

Many real-life situations can be represented by digraphs. Here are a few examples:

- **Transportation.** Here the *vertices* might represent locations within a city, and the *arcs* might represent one-way streets.

- **The Internet.** Here the *vertices* represent sources of information, and the *arcs* the possible flows of information.
- **Pipelines.** Here the *vertices* might represent pumping stations, and the *arcs* the direction of flow in the pipeline.
- **Chain of command.** In a corporation or in the military we can use a digraph to describe the chain of command. The *vertices* are individuals, and an *arc* from X to Y indicates that X can give orders (is a superior) to Y.

Example 4. Building a Dream Home on Mars: Part II

Let's return now to the problem of assembling a Martian Habitat Unit. Hubris will do the labor, we will do the thinking. We now know that the main elements of the problem can be conveniently described by the directed graph shown in Fig. 8-7(a) (it's just Fig. 8-5 shown once again for the reader's convenience). Figure 8-7(b) shows a slight modification of Fig. 8-7(a), where we have added two fictitious tasks, START and END. These two "tasks" are not real—we just added them for convenience. We can now visualize the entire project as a flow that begins at START and concludes at END. By giving these fictitious tasks zero processing time we avoid affecting the time calculations for the project. The digraph shown in Fig. 8-7(b) is called the **project digraph.**

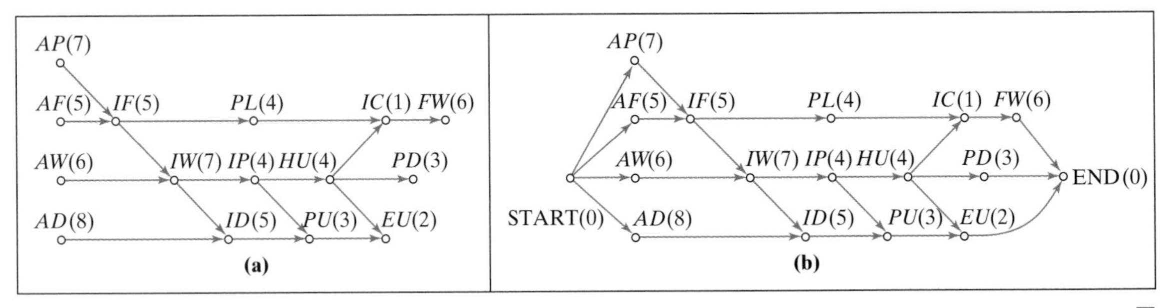

FIGURE 8-7

The Priority List Model for Scheduling

The project digraph is the basic graph model used to conveniently describe all the information in a scheduling problem, but there is nothing in the project digraph itself that specifically tells us how to create a schedule. We are going to need something else, some set of instructions as to the order in which tasks should be executed. The basic idea we will use to accomplish this is that of a **priority list**. A priority list is nothing more than a list of all the tasks in a particular order. In principle, the order of the tasks in a priority list is arbitrary (it doesn't have to make any special sense and it is unrelated to the precedence relations). We should think of the priority list as the order in which the scheduler would prefer to see the tasks executed. The only reason not to follow that exact order would be that some precedence relation prohibits one from doing so. Precedence relations override the priority list, but other than that we assign tasks to processors according to the priority list.

Since each time we change the order of the tasks we get a different priority list, there are as many priority lists as there are ways to order the tasks. (For 3 tasks there are 6 possible priority lists; for 4 tasks there are 24 priority lists; for 10 tasks there are more than three million priority lists; and for 100 tasks there are more priority lists than there are molecules in the universe.) Clearly, a shortage of priority lists is not going to be our problem. Sounds familiar? By now a familiar concept to us (Chapter 6 as well as Chapter 2)—the *factorial*—is once again entering the scene.

For a project consisting of M tasks, the number of possible priority lists is

$$M! = 1 \times 2 \times 3 \times \cdots \times M.$$

Before we proceed, we will illustrate how the priority-list model for scheduling works with a couple of small examples.

Example 5. Preparing for Launch

Before the launching of a satellite into space, five different system checks need to be performed by the computers on board. For simplicity, we will call the system checks $A(6)$, $B(5)$, $C(7)$, $D(2)$, and $E(5)$, with the numbers in parentheses representing the hours it takes one computer to perform that system check. In addition, there are precedence relations: D cannot be started until both A and B have been finished, and E cannot be started until C has been finished. All of the above information can be summarized by the project digraph shown in Fig. 8-8.

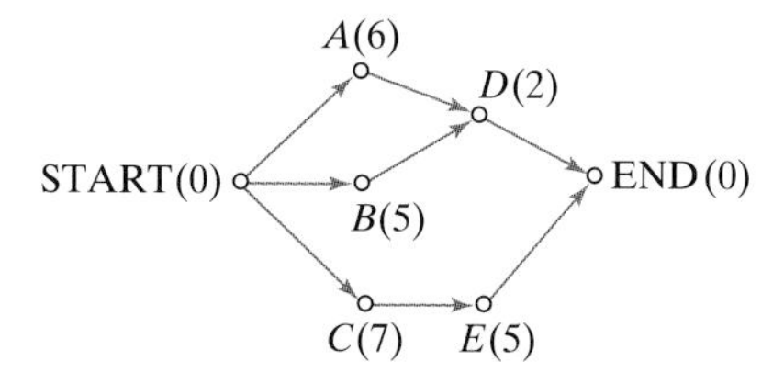

FIGURE 8-8

Let's suppose that two computers (P_1 and P_2) are available to carry out the system checks and that each individual system check can only be carried out by one of the computers.

How does one create a schedule to get all five system checks done? To start, we will need a priority list. Let's say that we are given a priority list in which the tasks are simply listed alphabetically:

Priority List: $A(6)$, $B(5)$, $C(7)$, $D(2)$, $E(5)$

- **Time $T = 0$ hr (start of project).** $A(6)$, $B(5)$, and $C(7)$ are the only *ready* tasks. Following the priority list, we assign $A(6)$ to P_1 and $B(5)$ to P_2.
- **Time $T = 5$ hr.** P_1 is still *busy* with $A(6)$; P_2 has just *completed* $B(5)$. $C(7)$ is the only available *ready* task. We assign $C(7)$ to P_2.
- **Time $T = 6$ hr.** P_1 has just *completed* $A(6)$; P_2 is *busy* with $C(7)$. $D(2)$ has just become a *ready* task (A and B have been completed). We assign $D(2)$ to P_1.
- **Time $T = 8$ hr.** P_1 has just *completed* $D(2)$; P_2 is still *busy* with $C(7)$. There are no *ready* tasks at this time for P_1, so P_1 has to sit *idle.*
- **Time $T = 12$ hr.** P_1 *idle, P_2* has just *completed* $C(7)$. Both processors *ready* for work. $E(5)$ is the only ready task, so we assign $E(5)$ to P_1. P_2 sits *idle.*

- **Time T = 17 hr**. P_1 has just *completed* $E(5)$. Project is completed. Finishing time is 17 hours.

The final schedule can be seen in Fig. 8-9. Is this a good schedule? All we have to do is look at all the idle time to see that it is a very bad schedule. When it comes to the finishing time, we would be hard put to come up with a worse schedule. Maybe we should try again! We can try a different strategy by simply changing the priority list.

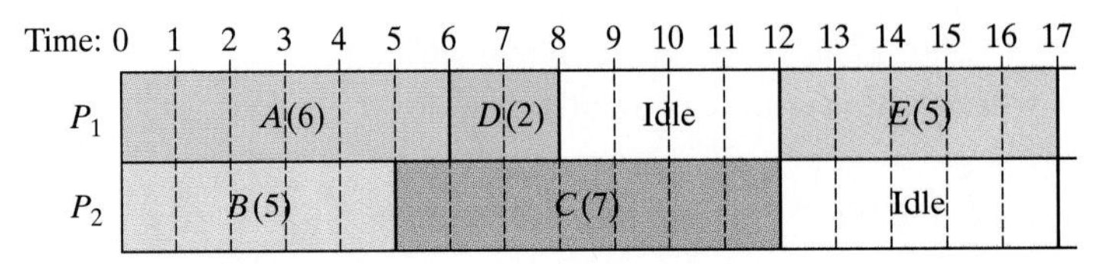

FIGURE 8-9 The final schedule for Example 5. ■

Example 6. Preparing for Launch: Part II

We are going to schedule the same project and with the same processors, but with a different priority list. (The project digraph is shown again in Fig. 8-10—exactly the same digraph as Fig. 8-8.)

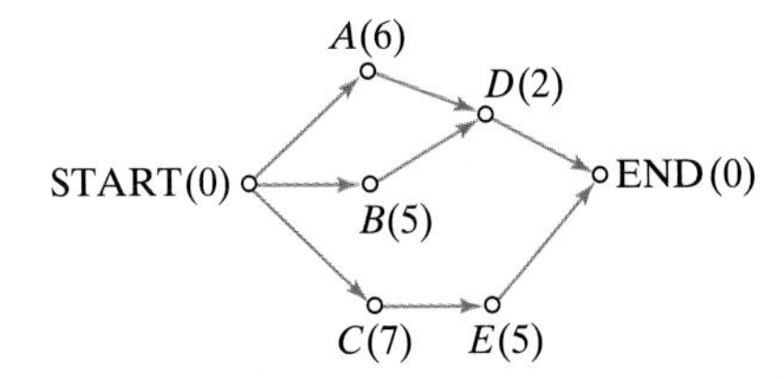

FIGURE 8-10

This time we will try to go in reverse alphabetical order (why not?).

Priority List: $E(5)$, $D(2)$, $C(7)$, $B(5)$, $A(6)$

- **Time T = 0 hr (start of project)**. $C(7)$, $B(5)$, and $A(6)$ are the only *ready* tasks. Following the priority list, we assign $C(7)$ to P_1 and $B(5)$ to P_2.
- **Time T = 5 hr**. P_1 is still *busy* with $C(7)$; P_2 has just *completed* $B(5)$. $A(6)$ is the only available ready task. We assign $A(6)$ to P_2.
- **Time T = 7 hr**. P_1 has just *completed* $C(7)$; P_2 is *busy* with $A(6)$. $E(5)$ has just become a *ready* task, and we assign it to P_1.
- **Time T = 11 hr**. P_2 has just *completed* $A(6)$; P_1 is *busy* with $E(5)$. $D(2)$ has just become a *ready* task, and we assign it to P_2.
- **Time T = 12 hr**. P_1 has just *completed* $E(5)$; P_2 is *busy* with $D(2)$. No tasks left, so P_1 sits idle.
- **Time T = 13 hr**. P_2 has just *completed* the last task, $D(2)$. Project is completed. Finishing time is 13 hours. The actual schedule is shown in Fig. 8-11.

Time: 0 1 2 3 4 5 6 7 8 9 10 11 12 13

P_1: C(7) | E(5) | Idle

P_2: B(5) | A(6) | D(2)

FIGURE 8-11 The final schedule for Example 6.

It's easy to see that this schedule is a lot better than the one in Fig. 8-9. In fact, as long as we have two processors to do the work, this is an optimal schedule—the finishing time of 13 hours cannot be improved! (See Exercise 29.) ■

What would happen if we kept the same priority list but added a third computer to the team? One would think that this would speed things up. Let's check it out.

Example 7. Preparing for Launch: Part III

This example is the same as the previous example (same project digraph and same priority list), but we will now create the schedule based on the fact that there are three processors (P_1, P_2, and P_3) to do all the work. For the reader's convenience the project digraph is shown again in Fig. 8-12. (It's hard to schedule without the project digraph in front of you!)

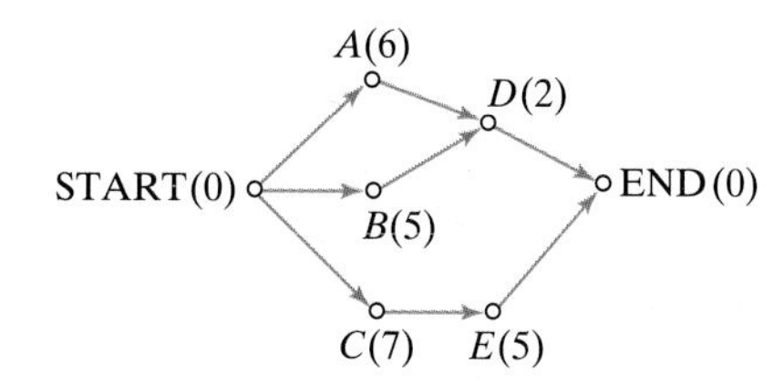

FIGURE 8-12

Priority List: $E(5), D(2), C(7), B(5), A(6)$

- **Time $T = 0$ hr (start of project).** $C(7)$, $B(5)$, and $A(6)$ are the *ready* tasks. We assign $C(7)$ to P_1, $B(5)$ to P_2, and $A(6)$ to P_3.
- **Time $T = 5$ hr.** P_1 is *busy* with $C(7)$; P_2 has just *completed* $B(5)$; and P_3 is *busy* with $A(6)$. There are no available ready tasks for P_2 [$E(5)$ can't be started until $C(7)$ is done and $D(2)$ can't be started until $A(6)$ is done], so P_2 sits idle.
- **Time $T = 6$ hr.** P_3 has just *completed* $A(6)$; P_2 is *idle;* and P_1 is still *busy* with $C(7)$. $D(2)$ has just become a *ready* task. We assign $D(2)$ to P_2.
- **Time $T = 7$ hr.** P_1 has just *completed* $C(7)$ and $E(5)$ has just become a *ready* task, so without any further ado we assign it to P_1. There are no other tasks to assign.
- **Time $T = 8$ hr.** P_2 has just *completed* $D(2)$. There are no other tasks to assign, so P_2 sits *idle.*
- **Time $T = 12$ hr.** P_1 has just *completed* the last task, $E(5)$, so the project is completed. Finishing time is 12 hours.

The actual schedule is shown in Fig. 8-13. The surprise, if we could call it that, is in the fact that adding a third processor didn't really help all that much. We'll come back to this point later in the chapter.

Time: 0 1 2 3 4 5 6 7 8 9 10 11 12

	0–5	5–6	6–7	7–8	8–12
P_1	C(7)			E(5)	
P_2	B(5)	Idle	D(2)		Idle
P_3	A(6)		Idle		

FIGURE 8-13 The final schedule for Example 7. ■

The *priority-list model* is a set of ground rules telling us how to assign tasks to processors for a given priority list. We have seen in the previous three examples how this works: At any particular moment in time throughout a project, a processor can be either *busy* or *idle,* and a task can be *ineligible, ready, in execution,* or *completed.* Depending on the various combinations of these, there are three different scenarios to consider:

- *All processors are busy.* There is nothing we can do but wait.
- *One processor is free.* We scan the priority list from left to right, looking for the first *ready* task in the priority list, which we assign to that processor. (Remember that for a task to be *ready,* all the tasks that are incident to it in the project digraph must have been completed.) If there are no ready tasks at that moment, the processor must stay idle until things change.
- *More than one processor is free.* In this case the first ready task on the priority list is given to the first free processor, the second ready task is given to the second free processor, and so on. If there are more free processors than ready tasks, some of the processors will remain idle. [Since the processors are identical and (at least in theory) have no say in the matter, the choice of which free processor gets which ready task is totally arbitrary (we could just as well toss a coin). It makes things easier for us if we choose among the processors in order.]

It's fair to say that the basic idea behind the priority-list model is not difficult, but there is a lot of bookkeeping involved, and that becomes critical when the number of tasks is large. At each stage of the schedule the scheduler must keep track of the status of each task—which tasks are *ready* for processing, which tasks are *in execution,* which tasks have been *completed,* which tasks are still *ineligible.* One convenient record-keeping strategy goes like this: On the priority list itself, circle all the *ready* tasks in red [Fig. 8-14(a)]. When a ready task is picked up by a processor and goes into *execution,* put a single red slash through the red circle [Fig. 8-14(b)]. When a task that has been in execution is completed, put a second red slash through the circle [Fig. 8-14(c)]. At this point, it is also important to check the project digraph to see if any new tasks have all of a sudden become eligible. Tasks that are *ineligible* remain unmarked [Fig. 8-14(d)].

We will now show how to implement this strategy to assemble a Martian Habitat Unit.

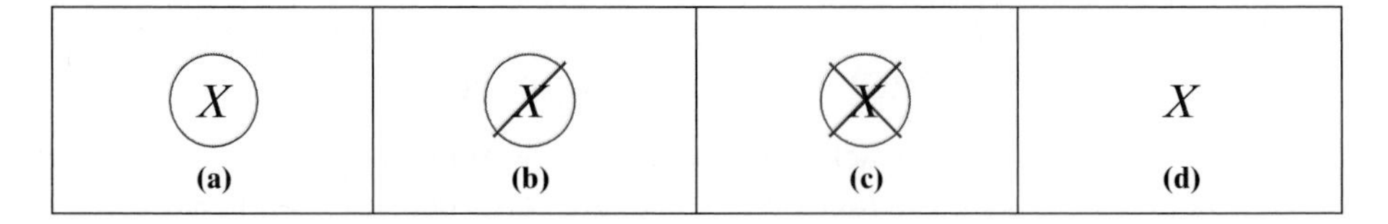

FIGURE 8-14 "Road" signs on a priority list. (a) Task X is ready. (b) Task X is in execution. (c) Task X is completed. (d) Task X is ineligible.

Example 8. Assembling an MHU: Part III

We are finally ready to get going with the project of assembling our MHU, and, like any good scheduler, we will first work the entire schedule out with pencil and paper. Let's start with the assumption that maybe we can get by with just two robots (P_1 and P_2). For the reader's convenience, we show the project digraph again (Fig. 8-15).

The Hubble Space Telescope

The Hubble Space Telescope (HST), named after astronomer Edwin Hubble, was launched into orbit on April 25, 1990. By sending a telescope above the earth's atmosphere, modern-day astronomers have obtained a much clearer view of the planets, the stars, and other galaxies that is about ten times clearer than achievable from the surface of earth.

Naturally, the tremendous demand for the telescope has led to a horrendously complex scheduling problem. Many of the projects can only be accommodated at a certain time of the day or a certain time of the year, and last-minute adjustments are frequently necessary when the instruments behave in unexpected ways. A computer program called Spike has been developed to use artificial intelligence (AI) techniques to schedule the telescope as efficiently as possible.

Questions:

1. Spike is designed to create a schedule that "satisfies all hard constraints and as many soft constraints as possible." What do you think is meant by "soft constraints" and "hard constraints?"
2. Briefly explain Spike's approach to solving the optimization problem.
3. What do you think is meant by hillclimbing and backtracking repair procedures?
4. The Spike developers learned that change can have a tremendous impact on telescope scheduling. Why might this also be true for a more down-to-earth application, such as scheduling automobile repairs?

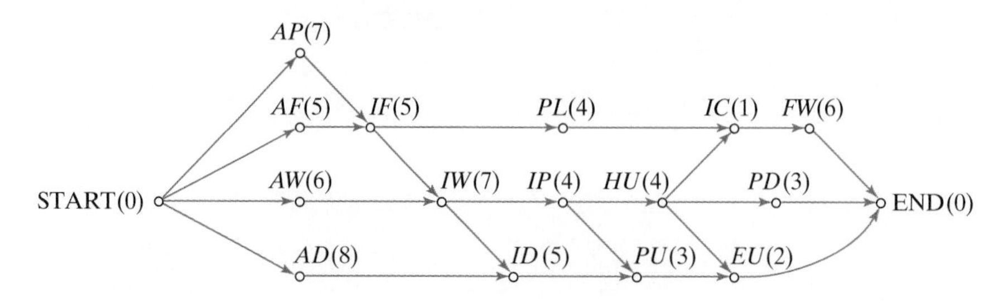

FIGURE 8-15

Suppose that we decide to use the following priority list:

Starting Priority List: *AD*(8), *AW*(6), *AF*(5), *IF*(5), *AP*(7), *IW*(7), *ID*(5), *IP*(4), *PL*(4), *PU*(3), *HU*(4), *IC*(1), *PD*(3), *EU*(2), *FW*(6) (ready tasks are circled).

- **Time: $T = 0$**

  Status of Processors: P_1 starts *AD;* P_2 starts *AW.*

  Priority List (Updated Status): *AD*, *AW*, *AF*, *IF*, *AP*, *IW, ID, IP, PL, PU, HU, IC, PD, EU, FW.*

- **Time: $T = 6$**

  Status of Processors: P_1 busy (executing *AD*); P_2 completed *AW,* starts *AF.*

  Priority List (Updated Status): *AD*, *AW*, *AF*, *IF*, *AP*, *IW, ID, IP, PL, PU, HU, IC, PD, EU, FW.*

- **Time: $T = 8$**

  Status of Processors: P_1 completed *AD,* starts *AP;* P_2 is busy (executing *AF*).

  Priority List (Updated Status): *AD*, *AW*, *AF*, *IF*, *AP*, *IW, ID, IP, PL, PU, HU, IC, PD, EU, FW.*

- **Time: $T = 11$**

  Status of Processors: P_1 busy (executing *AP*); P_2 completed *AF,* but since there are no ready tasks to take on, it must remain idle.

  Priority List (Updated Status): *AD*, *AW*, *AF*, *IF*, *AP*, *IW, ID, IP, PL, PU, HU, IC, PD, EU, FW.*

- **Time: $T = 15$**

  Status of Processors: P_1 just completed *AP.* Now *IF* becomes a ready task and is given to P_1; P_2 stays idle.

  Priority List (Updated Status): *AD*, *AW*, *AF*, *IF*, *AP*, *IW, ID, IP, PL, PU, HU, IC, PD, EU, FW.*

At this point we will let the reader take over and finish the schedule, hopefully following the above footsteps (see Exercise 30). Remember, the main point here is to learn how to keep track of the status of each task, and the only way to learn how to do it is with practice. (Besides, explaining the same thing over and over can get monotonous to both the explainer and the explainee.) After a fair amount of work, one obtains the final schedule shown in Fig. 8-16. The finishing time for the project is 44 hours.

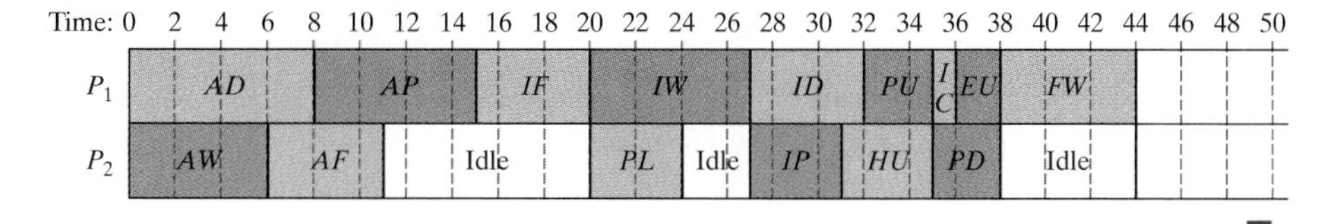

FIGURE 8-16

Scheduling under the priority-list model can be thought of as a two-part process: (1) choose a priority list, and (2) use the priority list and follow the rules of the model to come up with a schedule. As we have seen in the last example, the second part is long and tedious, but purely mechanical—it can be done by anyone (or anything) that is able to properly follow a set of instructions, be it a meticulous student or a properly programmed computer. We will use the term "scheduler" to describe the entity (be it student or machine) that takes a priority list as input and produces the schedule as output (Fig. 8-17).

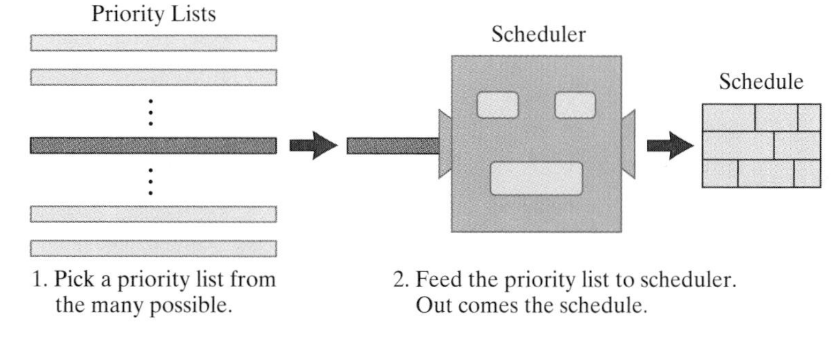

FIGURE 8-17 The scheduling process.

Ironically, it is the seemingly easiest part of this process—choosing a priority list—that is actually the most interesting. How do we know which of the many possible priority lists will give us an optimal schedule? (We will call such a priority list an **optimal priority list**.) How do we even pick a priority list that gives us a decent schedule?

In the rest of this chapter we will try to find some answers to these questions.

The Decreasing-Time Algorithm

Our first attempt in trying to find a good priority list is to formalize what is a commonly used and seemingly sensible strategy: *do the longer jobs first, leave the shorter jobs for last.* Formally, this translates into writing the priority list by listing the tasks in decreasing order of processing times, with longest first, second longest next, and so on. (When there are two or more tasks with equal processing times we will break the tie randomly.) We will call this the **decreasing-time list**, and we will call the process of creating a schedule using the decreasing-time list combined with the priority list model the **decreasing-time algorithm (DTA)**.

Example 9. Building an MHU Using the DTA

A decreasing-time list for the 15 tasks required to assemble an MHU is:

Decreasing-Time List: $AD(8)$, $AP(7)$, $IW(7)$, $AW(6)$, $FW(6)$, $AF(5)$, $IF(5)$, $ID(5)$, $IP(4)$, $PL(4)$, $HU(4)$, $PU(3)$, $PD(3)$, $EU(2)$, $IC(1)$.

Using two processors and the decreasing-time algorithm, we get the schedule shown in Fig. 8-18, with finishing time 42 hours. In the interest of fairness, we show a summary (no explanations) of the step-by-step details in Table 8-3. The reader is advised to carefully check the table—or, better yet, work out the details independently.

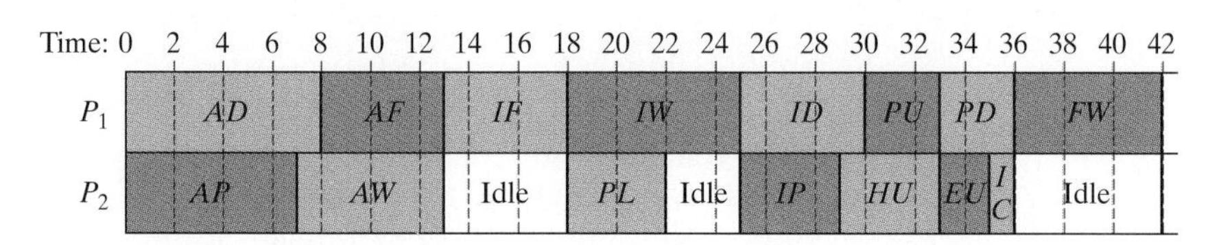

FIGURE 8-18 The schedule for assembling an MHU using the decreasing-time algorithm.

When looking at the finishing time under the decreasing-time algorithm, one can't help but feel disappointed. This promising idea of doing the longer jobs first and the shorter jobs later turned out to be a bit of a dud—at least in this example! What went wrong? If we work our way backward from the end, we can see that we made a bad choice at $T = 33$ hours. At this point there were three ready tasks [$PD(3)$, $EU(2)$, and $IC(1)$] and both processors were available. Following the decreasing-time priority list, we chose the two longest tasks, $PD(3)$ and $EU(2)$. Bad strategy! If we had looked at what was down the road, we would have seen that $IC(1)$ is a much more "critical" task than the other two because we can't start $FW(6)$ until we finish $IC(1)$. In short, we were shortsighted: We made our choices based on the immediate rather than the long-term benefits, and we ended up paying the price.

An even more blatant example of how the decreasing-time algorithm can lead to bad choices in scheduling occurs at the very start of the Example 9 schedule: We failed to notice that it is critical to start $AP(7)$ and $AF(5)$ as early as possible. Until we finish *AP* and *AF*, we cannot start $IF(5)$; and unless we finish *IF*, we cannot start $IW(7)$; and until we finish *IW*, we cannot start $IP(4)$ and $ID(5)$; and so on down the line. ■

The lesson to be learned from what happened in Example 9 is that a task should not be prioritized by how long it is but rather by the total length of all tasks that lie ahead of it. Simply put, the greater the *total amount of work lying ahead of a task,* the sooner that task should be started.

Critical Paths

To formalize the notion of *total amount of work lying ahead of a task,* we introduce the concept of *critical path.* For a given vertex *X* of a project digraph, the **critical path for *X*** is the *path from X to END that has the longest total sum of processing times.* The actual total time in the critical path for *X* is called the **critical time for *X*.**

Table 8-3 The Decreasing Time Algorithm applied to the MHU project

Step	Time	Priority-List Status	Schedule Status
1	$T = 0$	$AD(8)$ $AP(7)$ $IW(7)$ $AW(6)$ $FW(6)$ $AF(5)$ $IF(5)$ $ID(5)$ $IP(4)$ $PL(4)$ $HU(4)$ $PU(3)$ $PD(3)$ $EU(2)$ $IC(1)$	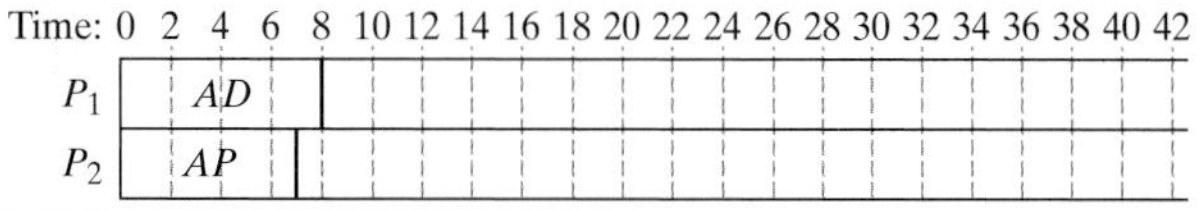 Time: 0 2 4 6 8 10 12 14 16 18 20 22 24 26 28 30 32 34 36 38 40 42 P_1: AD P_2: AP
2	$T = 7$	$AD(8)$ $AP(7)$ $IW(7)$ $AW(6)$ $FW(6)$ $AF(5)$ $IF(5)$ $ID(5)$ $IP(4)$ $PL(4)$ $HU(4)$ $PU(3)$ $PD(3)$ $EU(2)$ $IC(1)$	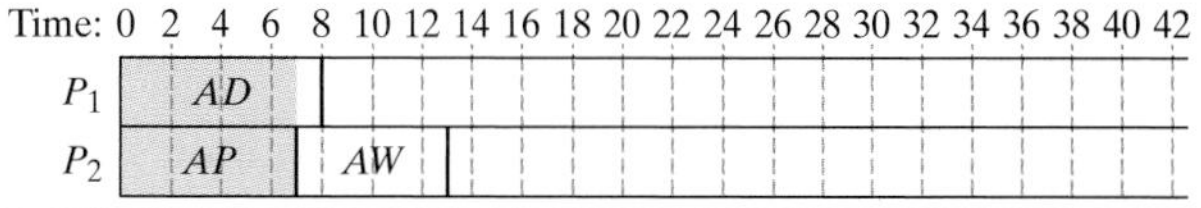 Time: 0 2 4 6 8 10 12 14 16 18 20 22 24 26 28 30 32 34 36 38 40 42 P_1: AD P_2: AP, AW
3	$T = 8$	$AD(8)$ $AP(7)$ $IW(7)$ $AW(6)$ $FW(6)$ $AF(5)$ $IF(5)$ $ID(5)$ $IP(4)$ $PL(4)$ $HU(4)$ $PU(3)$ $PD(3)$ $EU(2)$ $IC(1)$	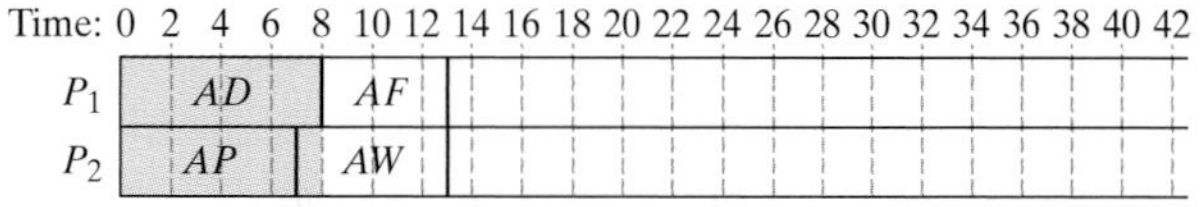 Time: 0 2 4 6 8 10 12 14 16 18 20 22 24 26 28 30 32 34 36 38 40 42 P_1: AD, AF P_2: AP, AW
4	$T = 13$	$AD(8)$ $AP(7)$ $IW(7)$ $AW(6)$ $FW(6)$ $AF(5)$ $IF(5)$ $ID(5)$ $IP(4)$ $PL(4)$ $HU(4)$ $PU(3)$ $PD(3)$ $EU(2)$ $IC(1)$	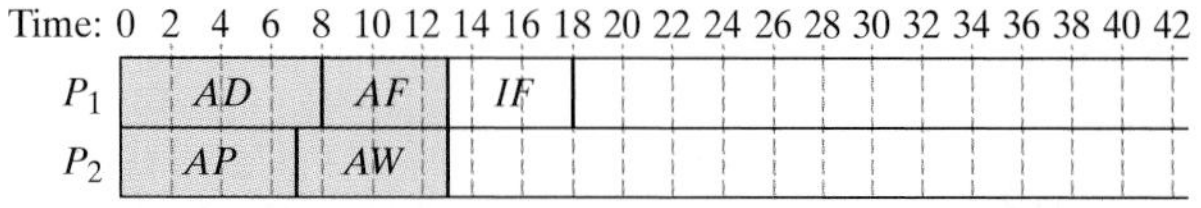 Time: 0 2 4 6 8 10 12 14 16 18 20 22 24 26 28 30 32 34 36 38 40 42 P_1: AD, AF, IF P_2: AP, AW
5	$T = 18$	$AD(8)$ $AP(7)$ $IW(7)$ $AW(6)$ $FW(6)$ $AF(5)$ $IF(5)$ $ID(5)$ $IP(4)$ $PL(4)$ $HU(4)$ $PU(3)$ $PD(3)$ $EU(2)$ $IC(1)$	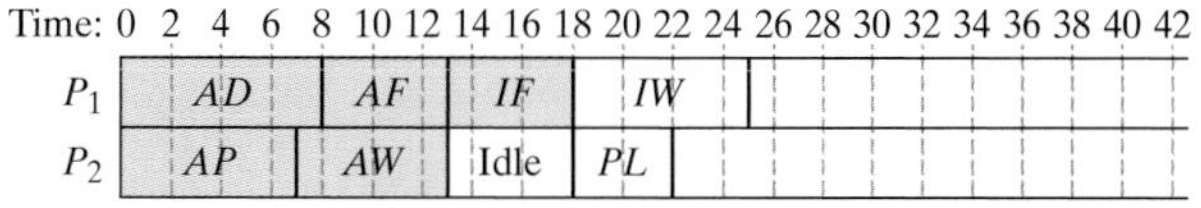 Time: 0 2 4 6 8 10 12 14 16 18 20 22 24 26 28 30 32 34 36 38 40 42 P_1: AD, AF, IF, IW P_2: AP, AW, Idle, PL
6	$T = 22$	$AD(8)$ $AP(7)$ $IW(7)$ $AW(6)$ $FW(6)$ $AF(5)$ $IF(5)$ $ID(5)$ $IP(4)$ $PL(4)$ $HU(4)$ $PU(3)$ $PD(3)$ $EU(2)$ $IC(1)$	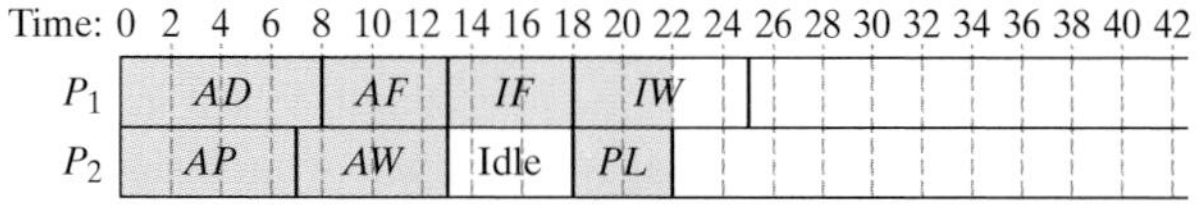 Time: 0 2 4 6 8 10 12 14 16 18 20 22 24 26 28 30 32 34 36 38 40 42 P_1: AD, AF, IF, IW P_2: AP, AW, Idle, PL
7	$T = 25$	$AD(8)$ $AP(7)$ $IW(7)$ $AW(6)$ $FW(6)$ $AF(5)$ $IF(5)$ $ID(5)$ $IP(4)$ $PL(4)$ $HU(4)$ $PU(3)$ $PD(3)$ $EU(2)$ $IC(1)$	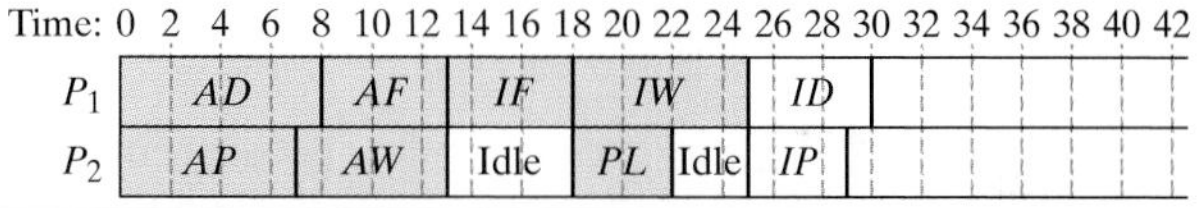 Time: 0 2 4 6 8 10 12 14 16 18 20 22 24 26 28 30 32 34 36 38 40 42 P_1: AD, AF, IF, IW, ID P_2: AP, AW, Idle, PL, Idle, IP
8	$T = 29$	$AD(8)$ $AP(7)$ $IW(7)$ $AW(6)$ $FW(6)$ $AF(5)$ $IF(5)$ $ID(5)$ $IP(4)$ $PL(4)$ $HU(4)$ $PU(3)$ $PD(3)$ $EU(2)$ $IC(1)$	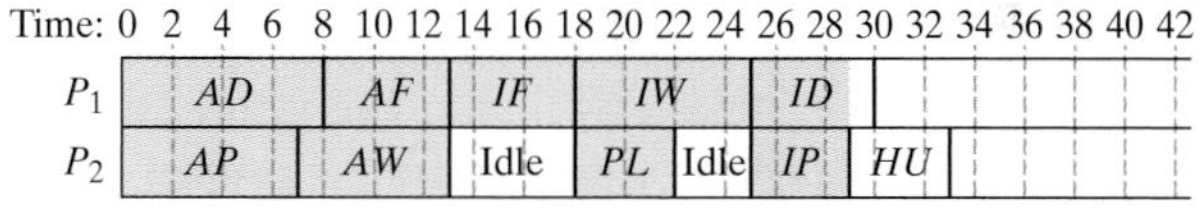 Time: 0 2 4 6 8 10 12 14 16 18 20 22 24 26 28 30 32 34 36 38 40 42 P_1: AD, AF, IF, IW, ID P_2: AP, AW, Idle, PL, Idle, IP, HU
9	$T = 30$	$AD(8)$ $AP(7)$ $IW(7)$ $AW(6)$ $FW(6)$ $AF(5)$ $IF(5)$ $ID(5)$ $IP(4)$ $PL(4)$ $HU(4)$ $PU(3)$ $PD(3)$ $EU(2)$ $IC(1)$	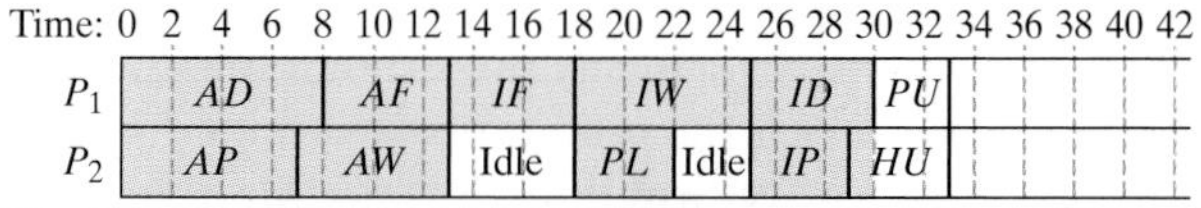 Time: 0 2 4 6 8 10 12 14 16 18 20 22 24 26 28 30 32 34 36 38 40 42 P_1: AD, AF, IF, IW, ID, PU P_2: AP, AW, Idle, PL, Idle, IP, HU
10	$T = 33$	$AD(8)$ $AP(7)$ $IW(7)$ $AW(6)$ $FW(6)$ $AF(5)$ $IF(5)$ $ID(5)$ $IP(4)$ $PL(4)$ $HU(4)$ $PU(3)$ $PD(3)$ $EU(2)$ $IC(1)$	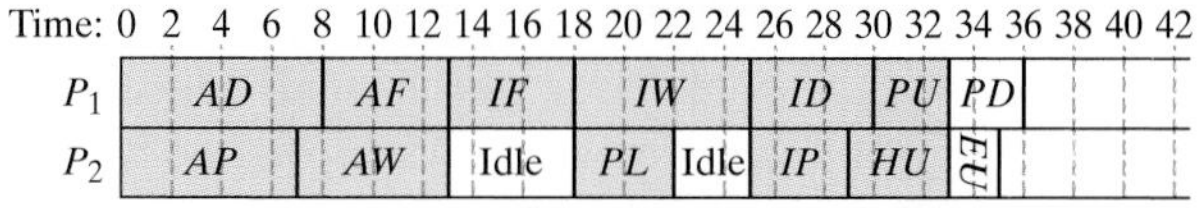 Time: 0 2 4 6 8 10 12 14 16 18 20 22 24 26 28 30 32 34 36 38 40 42 P_1: AD, AF, IF, IW, ID, PU, PD P_2: AP, AW, Idle, PL, Idle, IP, HU, EU
11	$T = 35$	$AD(8)$ $AP(7)$ $IW(7)$ $AW(6)$ $FW(6)$ $AF(5)$ $IF(5)$ $ID(5)$ $IP(4)$ $PL(4)$ $HU(4)$ $PU(3)$ $PD(3)$ $EU(2)$ $IC(1)$	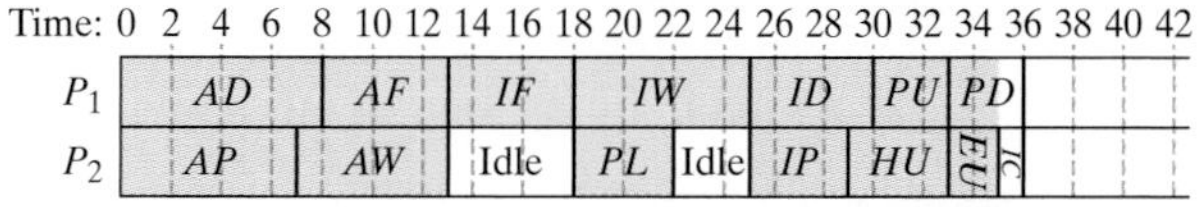 Time: 0 2 4 6 8 10 12 14 16 18 20 22 24 26 28 30 32 34 36 38 40 42 P_1: AD, AF, IF, IW, ID, PU, PD P_2: AP, AW, Idle, PL, Idle, IP, HU, EU, IC
12	$T = 36$	$AD(8)$ $AP(7)$ $IW(7)$ $AW(6)$ $FW(6)$ $AF(5)$ $IF(5)$ $ID(5)$ $IP(4)$ $PL(4)$ $HU(4)$ $PU(3)$ $PD(3)$ $EU(2)$ $IC(1)$	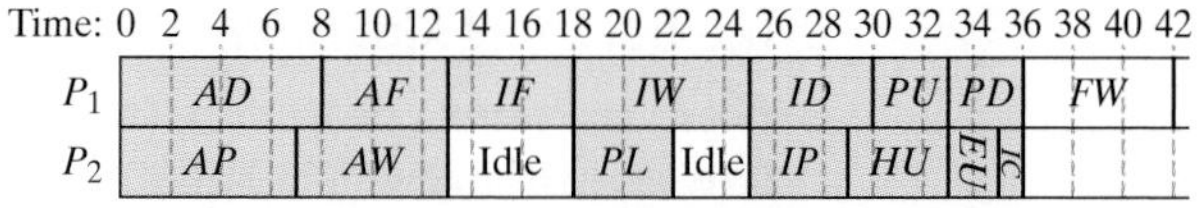 Time: 0 2 4 6 8 10 12 14 16 18 20 22 24 26 28 30 32 34 36 38 40 42 P_1: AD, AF, IF, IW, ID, PU, PD, FW P_2: AP, AW, Idle, PL, Idle, IP, HU, EU, IC
13	$T = 42$	$AD(8)$ $AP(7)$ $IW(7)$ $AW(6)$ $FW(6)$ $AF(5)$ $IF(5)$ $ID(5)$ $IP(4)$ $PL(4)$ $HU(4)$ $PU(3)$ $PD(3)$ $EU(2)$ $IC(1)$	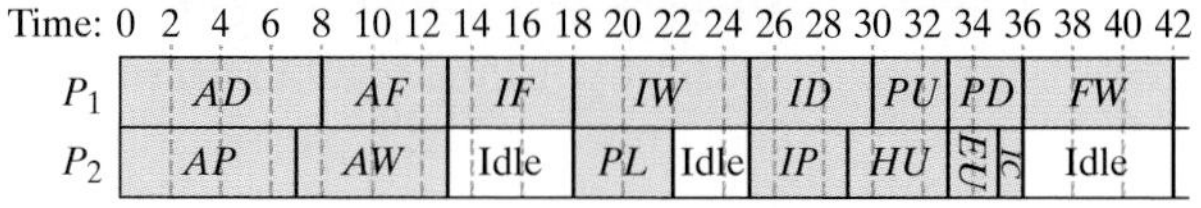 Time: 0 2 4 6 8 10 12 14 16 18 20 22 24 26 28 30 32 34 36 38 40 42 P_1: AD, AF, IF, IW, ID, PU, PD, FW P_2: AP, AW, Idle, PL, Idle, IP, HU, EU, IC, Idle

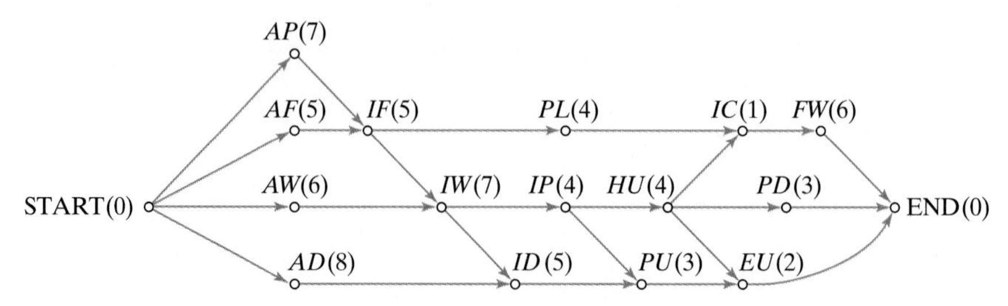

Here are some examples of critical paths and critical times for tasks in the MHU project digraph:

Example 10.

Let's try to find the critical path for vertex *HU*. There are three paths from *HU* to END. They are (1) *HU, IC, FW,* END; (2) *HU, PD,* END; and (3) *HU, EU,* END. The sum of the processing times in (1) is $4 + 1 + 6 = 11$; the sum of the processing times in (2) is $4 + 3 = 7$; and the sum of the processing times in (3) is $4 + 2 = 6$. Of the three, path (1) has the largest sum, so it is the *critical path for HU,* and the critical time for *HU* is 11. ■

Example 11.

When we try to find the critical path for vertex *AD,* we notice that there is only one path from *AD* to END, namely *AD, ID, PU, EU,* END. Since this is the only path, it is the *critical path for AD.* The critical time for *AD* is 18. ■

Example 12.

There are quite a few paths from START to END. After looking at the project digraph for a little while, however, we can pretty much "see" that the one with the longest total processing time seems to be START, *AP, IF, IW, IP, HU, IC, FW,* END with a total time of 34 hours. This is indeed the *critical path for the vertex* START, and it is called **the critical path for the project** or, more briefly, just **"the" critical path.** ■

In any project digraph, the critical path is of fundamental importance, and so is the total processing time for all the tasks in the critical path, which is called the **critical time**. For the Martian Habitat Unit assembly project, the critical path is START, *AP, IF, IW, IP, HU, IC, FW,* END and the critical time is 34 hours. We will come back to the significance of the critical time and the critical path soon, but before we do so, let's discuss how to find critical paths from any vertex of a project digraph. After all, we can hardly be expected to find critical paths in large-project digraphs the way we did it in Examples 10, 11, and 12, where we were pretty much flying by the seat of our pants. We need an efficient algorithm!

The Backflow Algorithm

There is a simple procedure (which for lack of a better name we will call the **backflow algorithm**) which will allow us to find the critical time and the critical path for each and every vertex of a project digraph. The basic idea is to build the critical path by working backward from the END to the START. Once we know the critical times for all the vertices "ahead" of a given vertex *X,* we choose among these the one with the *largest critical time* (call it *C*). The critical time of *X* is then obtained by adding the *processing time* of X to the *critical time* of C (see Fig. 8-19). To keep the record keeping straight, it is suggested that you write the critical time of the vertex in [square brackets] to distinguish it from the processing time in (parentheses). Once we have the critical times, the critical path for a vertex is obtained by always following the forward rule of moving to the vertex with largest critical time.

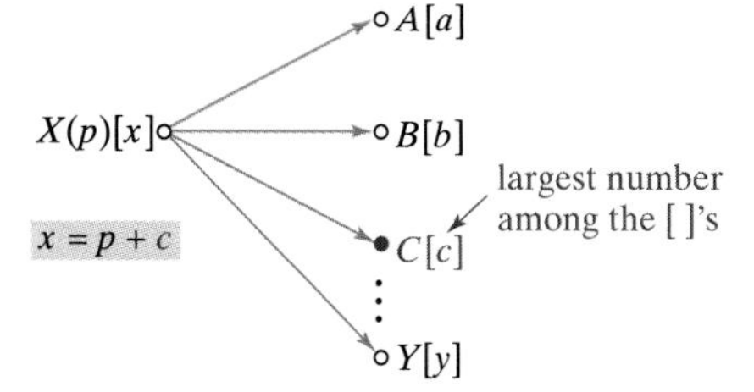

FIGURE 8-19 Critical time for X = processing time for X + critical time for C.

While it sounds a little complicated in words, the backflow algorithm is actually pretty easy to do, as we will show in the next example.

Example 13. The Critical Path for the MHU Project

We will use the backflow algorithm to find the critical times for every task in the MHU assembly project. The project digraph is shown in Fig. 8-20.

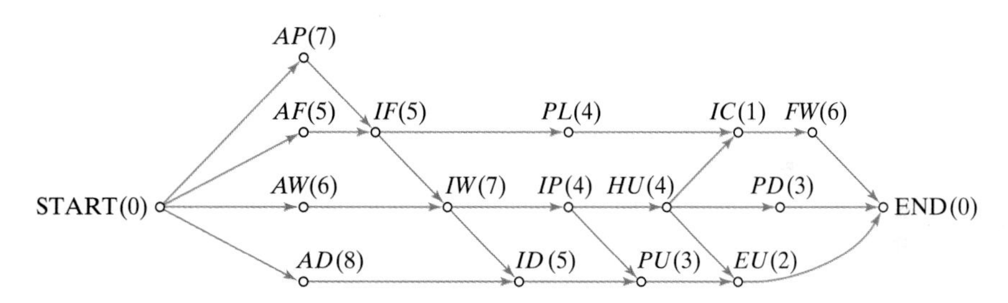

FIGURE 8-20

- **Step 1.** We start at END and arbitrarily assign to it a critical time of zero. (It's the only value that makes sense!)
- **Step 2.** We move backward to the three vertices that are incident to END, namely *FW*(6), *PD*(3), and *EU*(2). For each of them, the critical time is their processing time plus zero, so the critical times are *FW*[6], *PD*[3], and *EU*[2].
- **Step 3.** From *FW*[6] we move backward to *IC*(1). The only vertex incident from *IC*(1) is *FW*[6], so the critical time for *IC* is [1 + 6 = 7]. We record a [7] next to *IC* in the graph.
- **Step 4.** We move backward to *HU*(4). There are three vertices incident from it (*IC*[7], *PD*[3], and *EU*[2]), and the one with the largest critical time is *IC*[7]. It follows that the critical time for *HU* is [4 + 7 = 11]. At this stage we can also find the critical times of *PL*(4) and *PU*(3). For *PL*(4) the only vertex incident from it is *IC*[7], so its critical time is [4 + 7 = 11]. Likewise, for *PU*(3) the only vertex incident from it is *EU*[2], so its critical time is [3 + 2 = 5].
- **Step 5.** We move backward to *IP*(4). There are two vertices incident from it (*HU*[11] and *PU*[5]). The critical time for *IP* is [4 + 11 = 15]. We can also move backward to *ID*(5) and find its critical time, which is [10]. (Right?)

- **Step 6.** We can now move backward to *IW*(7). We leave it to the reader to verify that its critical time is [7 + 15= 22].
- **Step 7.** We can now move backward to *IF*(5). We leave it to the reader to verify that its critical time is [5 + 22 = 27].
- **Step 8.** We now move backward to *AP*(7), *AF*(5), *AW*(6), and *AD*(8). Their respective critical times are [7 + 27 = 34]; [5 + 27 = 32]; [6 + 22 = 28]; and [8 + 10 = 18].
- **Step 9.** Finally, we move backward to *START.* We still follow the same rule: the critical time is [0 + 34 = 34]. This is the critical time for the entire project!

The critical time for every vertex of the project digraph is shown (in [red]) in Fig. 8-21. To find the critical path we just go from vertex to vertex following the scent of largest critical times: START, *AP, IF, IW, IP, HU, IC, FW,* END—just as we suspected!

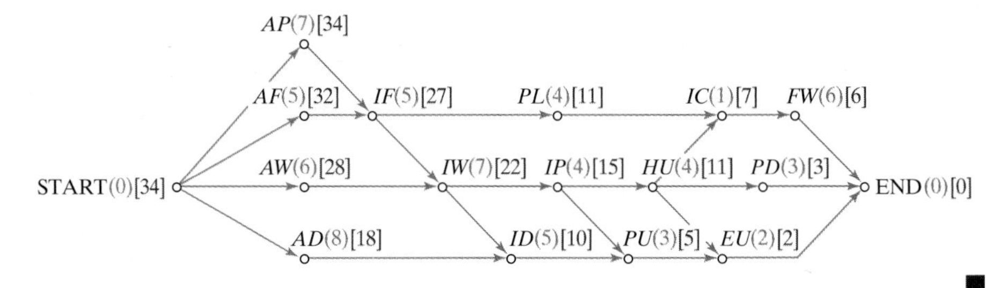

FIGURE 8-21 Processing times in blue; critical times in red.

■

Why are the critical path and critical time of a project of special significance? There are two reasons: (1) As we discussed at the start of this chapter, in every project there is a theoretical time barrier for the finishing time—a certain minimum amount of time below which the project can never be finished, regardless of how clever the scheduler is or how many processors are used. *This theoretical barrier turns out to be the project's critical time.* (2) If a project is going to be finished in the absolute minimum time (i.e., in the critical time), it is absolutely necessary that all the tasks in the critical path be done at the earliest possible time. Any delay in starting up one of the tasks in the critical path will necessarily delay the finishing time of the entire project. (By the way, this is why it is called the *critical path.*)

Unfortunately, it is not always possible to schedule the tasks on the critical path one after the other, bang, bang, bang, without delay. For one thing, processors are not always free when we need them (and remember one of our rules—a processor cannot stop in the middle of one task, start a new task, and leave the other one for later). Another reason is the problem of uncompleted predecessor tasks. We cannot concern ourselves only with tasks along the critical path and disregard other tasks that might affect them through precedence relations. There is a whole web of interrelationships that we need to worry about. Optimal scheduling is extremely complex.

The Critical-Path Algorithm

It is possible to use the concept of critical paths to generate very good (although not necessarily optimal) schedules. The idea is the same as the one we used with the decreasing-time algorithm, but at a higher level of sophistication: Instead of prioritizing the tasks in decreasing order of processing times, we will prioritize them in decreasing order of critical times. (Think of it as a mathematical version of strategic planning.) The priority list obtained by writing the tasks in decreasing order of critical times (ties are broken randomly) is called the **critical-path list**. The process of creating a schedule using the priority-list model applied to the critical-path list is called the **critical-path algorithm (CPA)**.

Example 14. Building an MHU Using the CPA

We will apply the critical-path algorithm to the MHU problem. We already know the critical times for each vertex (Fig. 8-22).

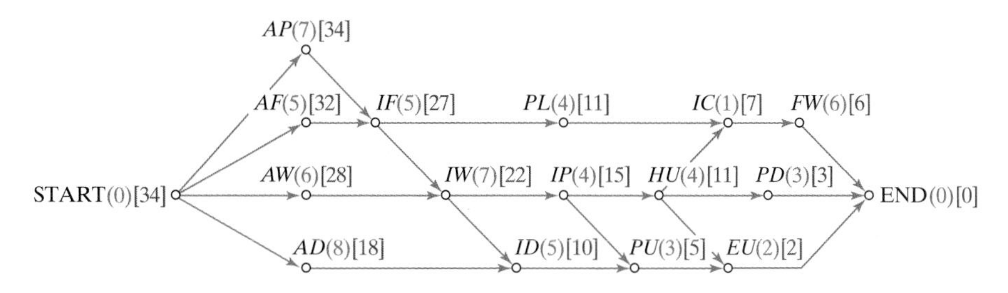

FIGURE 8-22

One possible critical-path list is given below:

Critical-Path List: *AP*[34], *AF*[32], *AW*[28], *IF*[27], *IW*[22], *AD*[18], *IP*[15], *PL*[11], *HU*[11], *ID*[10], *IC*[7], *FW*[6], *PU*[5], *PD*[3], *EU*[2].

With two processors, the schedule that results from this priority list is shown in Fig. 8-23. This time we leave the details to the reader (Exercise 31).

FIGURE 8-23 Schedule for the MHU project obtained using the critical-path algorithm.

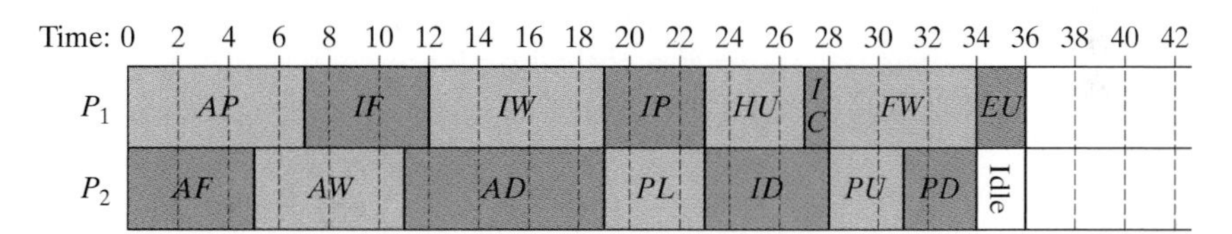

In the MHU problem, the finishing time using the critical-path algorithm is 36 hours, a big improvement over the 42 hours produced by the decreasing-time algorithm. Is this an optimal solution? Figure 8-24 shows a schedule with a finishing time of 35 hours, so our 36-hour schedule, while good, is obviously not optimal.

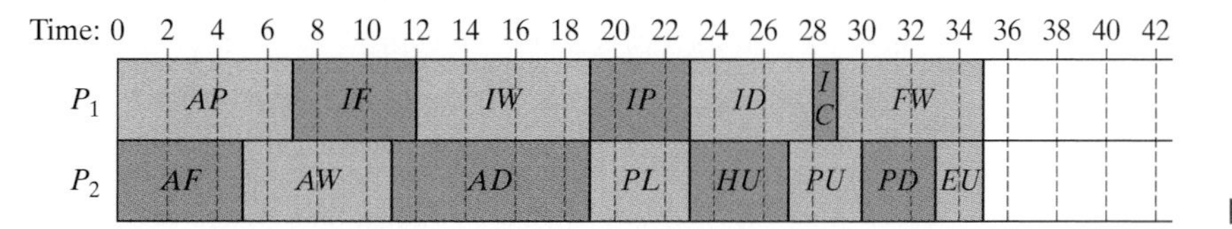

FIGURE 8-24 Optimal schedule for the MHU project using two processors.

■

By and large, the critical-path algorithm is an excellent approximate algorithm for scheduling a project (in most cases far superior to the deceasing-time algorithm), but, in general, it will not produce an optimal schedule. As it turns out, no efficient scheduling algorithm is presently known that always gives an optimal schedule. In this regard, scheduling problems are a lot like TSPs (Chapter 6) and shortest-network problems (Chapter 7)—there are efficient algorithms that can produce good schedules, but there are no efficient optimal algorithms. Of the standard scheduling algorithms, the critical-path algorithm is by far the best known and most commonly used. Other, more sophisticated algorithms have been developed in the last 20 years, and under specialized circumstances they can outperform the critical-path algorithm, but as an all-purpose algorithm for scheduling, the critical-path algorithm is hard to beat.

Scheduling with Independent Tasks

In this section we will briefly discuss what happens to scheduling problems in the special case when there are no precedence relations to worry about. This situation arises whenever we are scheduling tasks that are all independent—for example, scheduling a group of typists in a steno pool to type a bunch of reports of various lengths.

It is tempting to think that without precedence relations hanging over one's head, scheduling becomes a simple problem, and one should be able to find optimal schedules without difficulty, but appearances are deceiving. *There are no optimal and efficient algorithms known for scheduling, even when the tasks are all independent.*

While, in a theoretical sense, we are not much better able to schedule independent tasks than to schedule tasks with precedence relations, from a purely practical point of view there are a few differences. For one thing, there is no getting around the fact that the nuts-and-bolts details of creating a schedule using a priority list become tremendously simplified when there are no precedence relations to mess with. In this case we just assign the tasks to the processors as they become free in exactly the order given by the priority list. Second, without precedence relations, the critical-path time of a task equals its processing time, which means that the *critical-path list* and *decreasing-time list* are exactly the same list, and therefore the critical-path algorithm is the same as the decreasing-time algorithm. Before we go on, let's look at a couple of examples of scheduling with independent tasks.

Example 15. Cooking Up a Storm

Imagine that you and your two best friends are cooking a 9-course meal as part of a charity event. Each of the 9 courses is an independent task, to be done by just one of the 3 cooks (we'll call you P_1, P_2, and P_3). The 9 courses are $A(70)$, $B(90)$ $C(100)$ $D(70)$, $E(80)$, $F(20)$, $G(20)$, $H(80)$, and $I(10)$, with their processing times given in minutes. Let's do it first using an alphabetical priority list.

Priority List: $A(70)$, $B(90)$, $C(100)$, $D(70)$, $E(80)$, $F(20)$, $G(20)$, $H(80)$, $I(10)$.

Since there are no precedence relations, there are no ineligible tasks, and all tasks start out as ready tasks. As soon as a processor is free, it picks up the next available task in the priority list. From the bookkeeping point of view, this is a piece of cake. We leave it to the reader to verify that the resulting schedule is the one in Fig. 8-25, with finishing time of 220 minutes. It is obvious from the figure that this is not a very good schedule.

Time: 0 10 20 30 40 50 60 70 80 90 100 110 120 130 140 150 160 170 180 190 200 210 220 230 240 250

P_1: $A(70)$ | $D(70)$ | $H(80)$

P_2: $B(90)$ | $E(80)$ | Idle

P_3: $C(100)$ | $F(20)$ | $G(20)$ | $I(10)$ | Idle

FIGURE 8-25 Schedule for Example 15 using an alphabetical priority list.

If we use the critical-path algorithm, the priority list is the decreasing-time list:

Decreasing-Time List: $C(100)$, $B(90)$, $E(80)$, $H(80)$, $A(70)$, $D(70)$, $F(20)$, $G(20)$, $I(10)$.

The resulting schedule is shown in Fig. 8-26 with a finishing time of 180 minutes. Clearly, this schedule is optimal, since all three processors are working for the entire time.

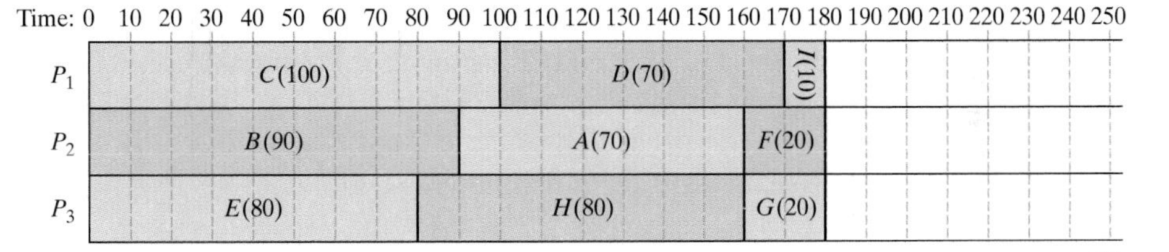

FIGURE 8-26 Schedule for Example 15 using a decreasing-time list.

In Example 15, the critical-path algorithm gave us the optimal schedule, but unfortunately this need not always be the case.

Example 16.

After the success of your last banquet, you and your two friends are asked to prepare another banquet. This time it will be a 7-course meal. The courses are all independent tasks, and their processing times (in minutes) are: $A(50)$, $B(30)$, $C(40)$, $D(30)$, $E(50)$, $F(30)$, and $G(40)$.

Using the critical-path algorithm the priority list is: $A(50)$, $E(50)$, $C(40)$, $G(40)$, $B(30)$, $D(30)$, and $F(30)$. The schedule one gets is shown in Fig. 8-27 with finishing time of 110 minutes. With a little trial and error, we can do better than this. The schedule shown in Fig. 8-28 is an optimal schedule, with finishing time of 90 minutes.

For Example 16 we can precisely measure how "well" the critical-path algorithm performed by computing the **relative percentage of error** = (computed fin-

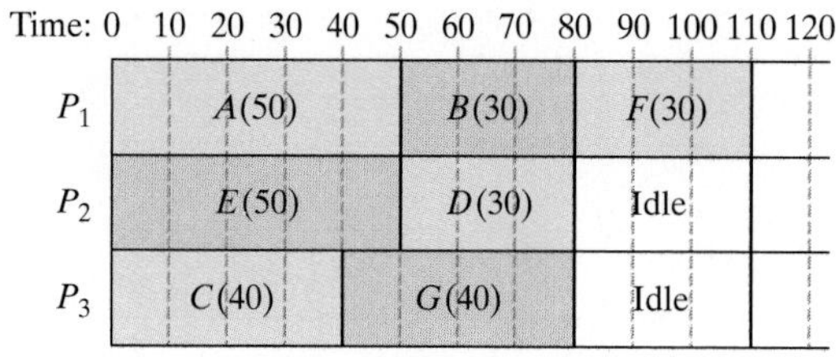

Time: 0 10 20 30 40 50 60 70 80 90 100 110 120

P_1: $A(50)$ | $B(30)$ | $F(30)$

P_2: $E(50)$ | $D(30)$ | Idle

P_3: $C(40)$ | $G(40)$ | Idle

FIGURE 8-27 Schedule for Example 16 using a decreasing-time list.

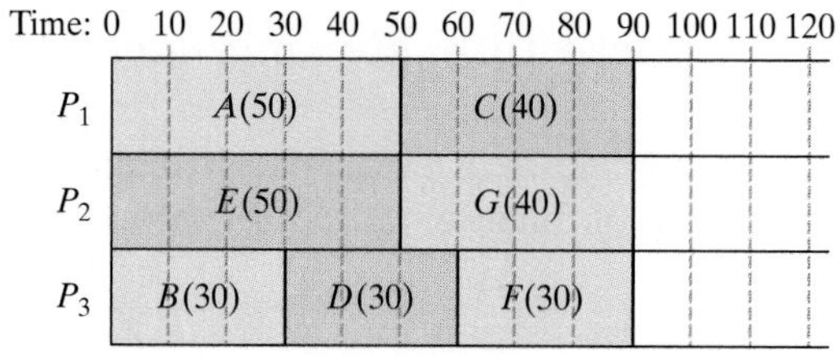

Time: 0 10 20 30 40 50 60 70 80 90 100 110 120

P_1: $A(50)$ | $C(40)$

P_2: $E(50)$ | $G(40)$

P_3: $B(30)$ | $D(30)$ | $F(30)$

FIGURE 8-28 Optimal schedule for Example 16.

ishing time − optimal finishing time)/optimal finishing time. In this case the relative percentage of error is $(110 - 90)/90 = 20/90 \approx 0.2222 = 22.22\%$. This tells us that in this particular example the critical-path algorithm gave us a schedule that is 22.22% longer than the optimal schedule. ■

In 1969, American mathematician Ronald L. Graham of AT&T Bell Laboratories showed that for independent tasks, the critical-path algorithm will always produce schedules with finishing times that cannot be off by more than a fixed percentage from the optimal finishing time. Specifically, Graham showed that for independent tasks, when the number of processors is *M*, the relative per-

Table 8-4 Max Error (*CP*) Represents the Largest Possible Percentage Error Under the CPA When the Tasks Are Independent.

Number of Processors (M)	Max Error (CP) $\left(\frac{M-1}{3M}\right)$
2	$\frac{2-1}{3\times 2} = \frac{1}{6} \approx 16.66\%$
3	$\frac{3-1}{3\times 3} = \frac{2}{9} \approx 22.22\%$
4	$\frac{4-1}{3\times 4} = \frac{3}{12} = 25\%$
5	$\frac{5-1}{3\times 5} = \frac{4}{15} \approx 26.66\%$
Ú	Ú
100	$\frac{100-1}{3\times 100} = \frac{99}{300} = 33\%$

centage of error using the critical-path algorithm is at most $(M - 1)/3M$ (see Table 8-4). This maximum value for the relative percentage of error increases slowly as the number of processors increases, but is always less than $33\frac{1}{3}\%$ (see Exercise 45). Graham's discovery essentially reassures us that when the tasks are independent, the finishing time we get from the critical-path algorithm can't be too far off from the optimal finishing time—no matter how many tasks need to be scheduled or how many processors are available to carry them out.

Conclusion

In one form or another, the scheduling of human (and nonhuman) activity is a pervasive and fundamental problem of modern life. At its most informal, it is part and parcel of the way we organize our everyday living (so much so that we are often scheduling things without realizing we are doing so). In its more formal incarnation, the systematic scheduling of a set of activities for the purposes of saving either time or money is a critical issue in industry, science, government, and so on.

By now it should not surprise us that at their very core, scheduling problems are mathematical in nature, and that the mathematics of scheduling can be both simple and profound. By necessity, we focused on the simple side, but it is important to realize that there is a great deal more to scheduling than what we learned in this chapter.

In this chapter we discussed scheduling problems where we are given a set of *tasks,* a set of *precedence relations* among the tasks, and a set of identical *processors.* The objective is to schedule the tasks by properly assigning tasks to processors so that the *finishing time* for all the tasks is as small as possible.

To systematically tackle these scheduling problems we first developed a graph model of such a problem, called the *project digraph,* and a general framework by means of which we can create, compare, and analyze schedules, called the *priority-list model.* Within the priority-list model, many strategies can be followed (each strategy leading to the creation of a specific priority list). In the chapter we considered two basic strategies for creating schedules. The first was the *decreasing-time algorithm,* a strategy that intuitively makes a lot of sense but which in practice often results in inefficient schedules. The second strategy, called the *critical-path algorithm,* is generally a big improvement over the decreasing-time algorithm, but it falls short of the ideal goal: an efficient optimal algorithm. The critical-path algorithm is by far the best known and most widely used algorithm for scheduling in business and industry.

When scheduling with independent tasks, the decreasing-time algorithm and the critical-path algorithm become one and the same, and the finishing times they generate are never off by much from the optimal finishing times.

Although several other, more sophisticated strategies for scheduling have been discovered by mathematicians in the last 30 years, no optimal, efficient scheduling algorithm is presently known, and the general feeling among the experts is that there is little likelihood that such an algorithm actually exists. Good scheduling, nonetheless, will always remain a significant human goal—a task, if you will, in the grand cosmic schedule of humanity.

Key Concepts

arc
backflow algorithm
critical path
critical-path algorithm
critical-path list
critical time
cycle (in a digraph)
decreasing-time algorithm
decreasing-time list
digraph
factorial
finishing time
incident (to and from)
indegree (outdegree)
independent tasks
path (in a digraph)
precedence relation
priority list
priority-list model
processing time
processor
project digraph
relative percentage of error
task (ineligible, ready, in execution, completed)

Exercises

Walking

1. For each of the following digraphs make and complete a table similar to the one shown here.

Vertex	Degree	Indegree	Outdegree	Vertex is incident to	Vertex is incident from
A					
B					
⋮					

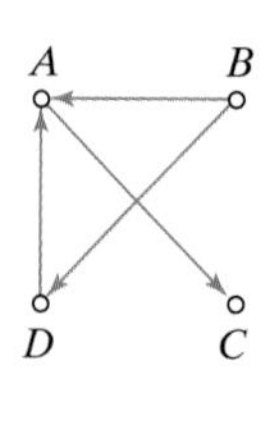

(a)

(b)

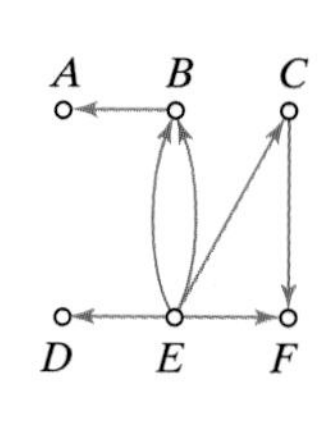

(c)

2. For each of the following digraphs list the vertices and arcs. (Use *XY* to represent an arc from *X* to *Y*.) Give the indegree and outdegree of each vertex, and list the vertices incident to and incident from each vertex.

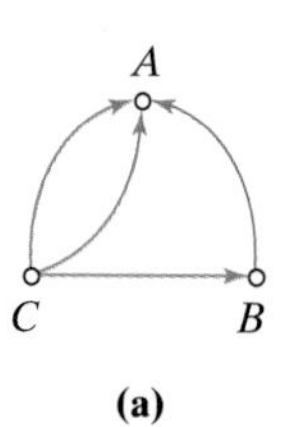

(a)

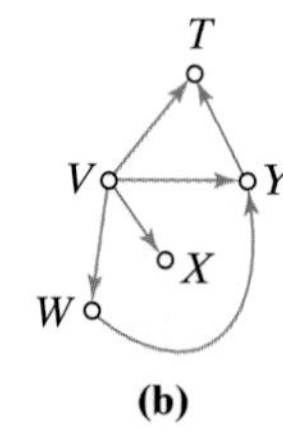

(b)

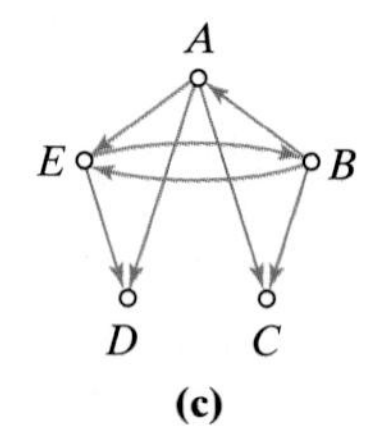

(c)

3. For each of the following, draw a picture of the digraph.

(a) Vertices: *A, B, C, D*.

Arcs: *A* is incident to *B* and *C; D* is incident from *A* and *B*.

(b) Vertices: A, B, C, D, E.

Arcs: A is incident to C and E; B is incident to D and E; C is incident from D and E; D is incident from C and E.

4. For each of the following, draw a picture of the digraph.

(a) Vertices: A, B, C, D.

Arcs: A is incident to B, C, and D; C is incident from B and D.

(b) Vertices: V, W, X, Y, Z.

Arcs: X is incident to V, Z, and Y; W is incident from V, Y, and Z; Z is incident to Y and incident from W and V.

5. A city has several one-way streets as well as two-way streets. The White Pine neighborhood is a rectangular area 6 blocks long and 2 blocks wide. Blocks alternate one-way, two-way, as shown in the following figure. Draw a digraph that represents the traffic flow in this neighborhood.

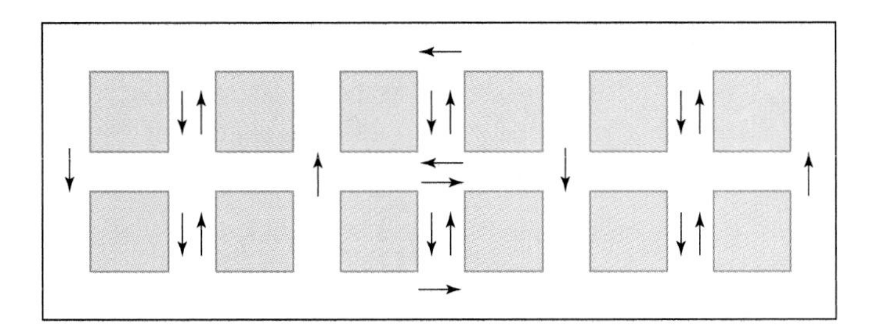

6. A mathematics textbook for liberal arts students consists of 10 chapters. While many of the chapters are independent of the others, some chapters require that previous chapters be covered first. The accompanying diagram illustrates the dependence. Draw a digraph that represents the dependence/independence relation among the chapters in the book.

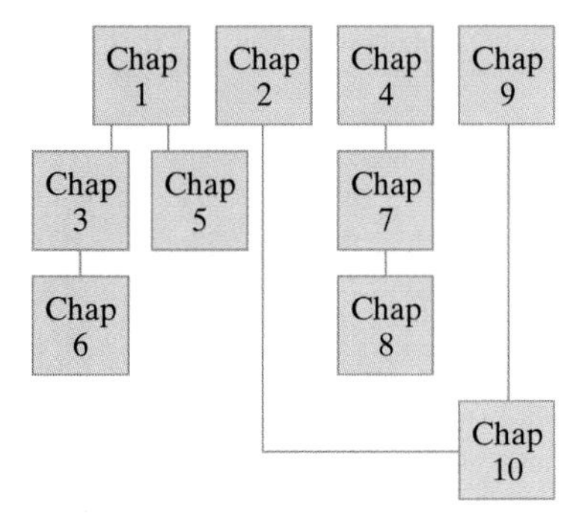

7. Give an example of a directed graph with 4 vertices, no loops or multiple arcs, and

(a) with each vertex having an indegree different than its outdegree.

(b) with 1 vertex of outdegree 3 and indegree 0, and the remaining 3 vertices each having indegree 2 and outdegree 1.

8. Give an example of a directed graph with 7 vertices, no loops or multiple arcs, and 3 vertices with indegree of 1 and outdegree of 1, 3 vertices of indegree 2 and outdegree 2, and 1 vertex of indegree 3 and outdegree 3.

Exercises 9 through 13 refer to an apartment maintenance organization that refurbishes apartments before new tenants move in. The following tables show the tasks performed, the average time required for each task (measured in 15-minute increments), and the precedence relations between tasks.

Job	Symbol/Time
Bathrooms (clean)	$B(8)$
Carpets (shampoo)	$C(4)$
Filters (replace)	$F(1)$
General cleaning	$G(8)$
Kitchen (clean)	$K(12)$
Lights (replace bulbs)	$L(1)$
Paint	$P(32)$
Smoke detectors (battery)	$S(1)$
Windows (wash)	$W(4)$

Precedence relations
$L \to P$
$P \to K$
$P \to B$
$K \to G$
$B \to G$
$F \to G$
$G \to W$
$G \to S$
$W \to C$
$S \to C$

9. Using the priority list *B, C, F, G, K, L, P, S, W*:

(a) Make a schedule for refurbishing an apartment using a single worker.

(b) Make a schedule for refurbishing an apartment using 2 workers.

10. Using the priority list *W, C, G, S, K, B, L, P, F*:

(a) Make a schedule for refurbishing an apartment using a single worker.

(b) Make a schedule for refurbishing an apartment using 2 workers.

11. Explain what is illegal about the following schedule for refurbishing an apartment with 1 worker.

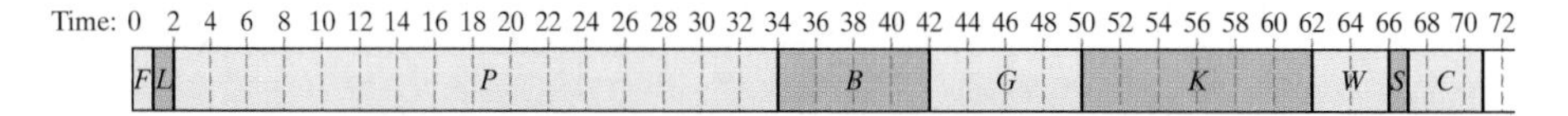

12. Explain what is illegal about the following schedule for refurbishing an apartment with 2 workers.

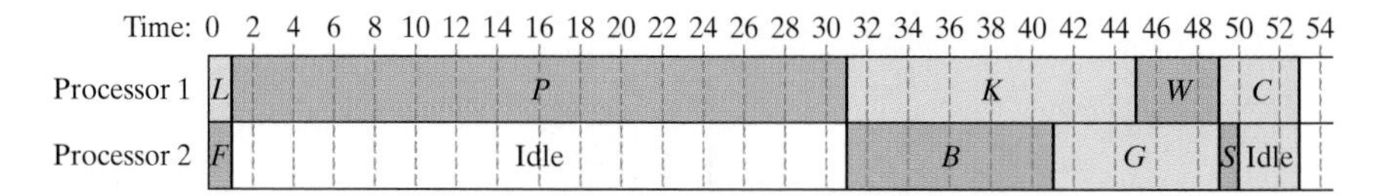

13. Explain what is illegal about the following schedule for refurbishing an apartment with 3 workers.

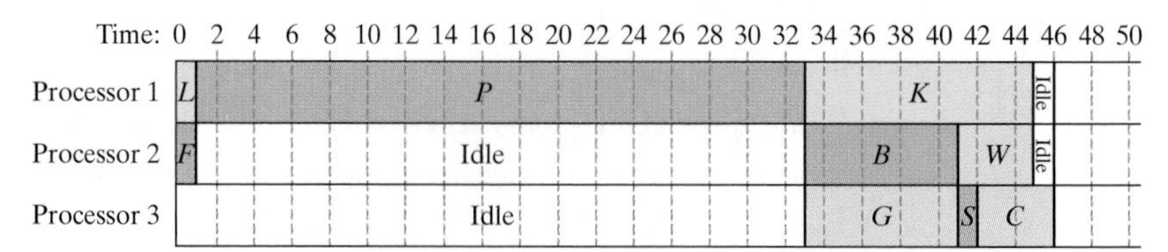

14. A schedule for a certain project using 5 processors has a completion time of 19 hours. The project consists of 11 tasks with the following processing times: $A(10)$, $B(7)$, $C(11)$, $D(8)$, $E(9)$, $F(5)$, $G(3)$, $H(6)$, $I(4)$, $J(7)$, $K(5)$. What is the total idle time in the schedule?

Exercises 15 through 21 refer to a copy center that must copy 13 research proposals. The times required (in minutes) for the 13 jobs in increasing order are: 3, 3, 4, 4, 5, 5, 5, 5, 6, 6, 7, 7, 12.

15. Schedule the copying jobs with no idle time using 4 copiers.

16. Schedule the copying jobs with a completion time of 27 minutes using 4 copiers.

17. Schedule the copying jobs using the decreasing-time algorithm using 4 copiers.

18. Schedule the copying jobs with no idle time using 6 copiers.

19. Schedule the copying jobs with a completion time of 22 minutes using 6 copiers.

20. Schedule the copying jobs using the decreasing-time algorithm using 6 copiers.

21. Is it possible to schedule the copying jobs with no idle time using 5 copiers? Explain.

Exercises 22 through 24 refer to the problem of scheduling 7 tasks (A, B, C, D, E, F, and G) in accordance with the following project digraph.

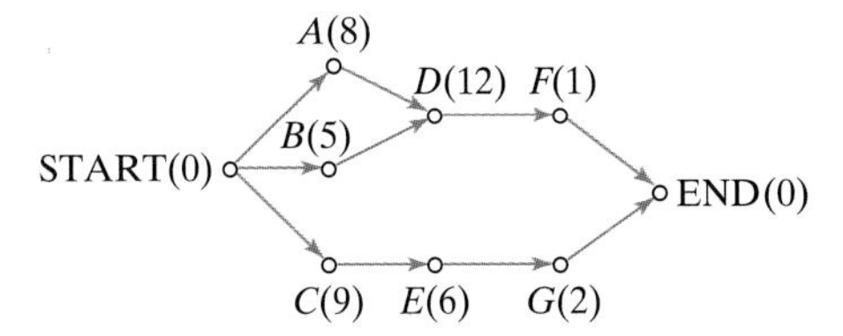

22. Using the priority list $G(2)$, $F(1)$, $E(6)$, $D(12)$, $C(9)$, $B(5)$, $A(8)$, schedule the project using 2 processors.

23. Use the decreasing-time algorithm to schedule the project using 2 processors.

24. **(a)** Find the length of the critical path from each vertex.

(b) What is the length of the critical path of the project digraph?

(c) Use the critical-path algorithm to schedule the project using two processors.

(d) Explain why the schedule obtained in this case is optimal.

Exercises 25 through 28 refer to the problem of scheduling 11 tasks in accordance with the following project digraph.

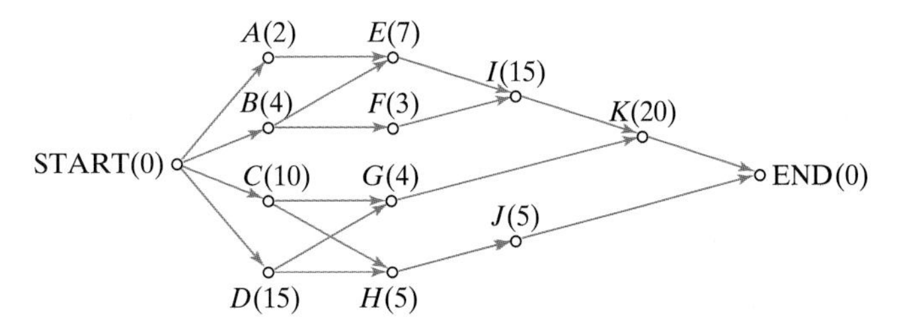

25. Use the decreasing-time algorithm to schedule the project using 2 processors.

26. **(a)** Find the length of the critical path from each vertex.

(b) What is the length of the critical path of the project digraph?

(c) Use the critical-path algorithm to schedule the project using 2 processors.

(d) Explain why the schedule obtained in this case is not optimal.

27. Use the critical-path algorithm to schedule the project using 3 processors.

28. Use the decreasing-time algorithm to schedule the project using 3 processors.

29. Consider the schedule obtained in Example 6 in this chapter. Explain why this is an optimal schedule for 2 processors.

30. Find a schedule for building an MHU with 2 processors using the priority list: $AD(8)$, $AW(6)$, $AF(5)$, $IF(5)$, $AP(7)$, $IW(7)$, $ID(5)$, $IP(4)$, $PL(4)$, $PU(3)$, $HU(4)$, $IC(1)$, $PD(3)$, $EU(2)$, $FW(6)$. (See Example 8 in this chapter.)

31. Find a schedule for building an MHU with 2 processors using the critical-path algorithm. (See Example 14 in this chapter.)

Exercises 32 through 34 refer to the following scheduling problem: Ten computer programs need to be executed. Three of the programs require 4 minutes each to complete, 3 more require 7 minutes each to complete, and 4 of the programs require 15 minutes each to complete. Moreover, none of the 15-minute programs can be started until all of the 4-minute programs have been completed.

32. Draw a project digraph for this scheduling problem.

33. Use the decreasing-time algorithm to schedule the programs on

(a) 2 computers.

(b) 3 computers.

(c) 4 computers.

34. Use the critical-path algorithm to schedule the programs on

(a) 2 computers.

(b) 3 computers.

(c) 4 computers.

Exercises 35 through 37 refer to the following scheduling problem: Eight computer programs need to be executed. One of the programs requires 10 minutes to complete, 2 programs require 7 minutes each to complete, 2 more require 12 minutes each to complete, and 3 of the programs require 20 minutes each to complete. Moreover, none of the 20-minute programs can be started until both of the 7-minute programs have been completed, and the 10-minute program cannot be started until both of the 12-minute programs have been completed.

35. Draw a project digraph for this scheduling problem.

36. Use the decreasing-time algorithm to schedule the programs on

(a) 2 computers.

(b) 3 computers.

37. Use the critical-path algorithm to schedule the programs on

(a) 2 computers.

(b) 3 computers.

38. (a) Draw a project digraph for a project consisting of the 8 tasks described by the following table.

Task	Length of task	Tasks that must be completed before the task can start
A	5	C
B	5	C, D
C	5	
D	2	G
E	15	A, B
F	6	D
G	2	
H	2	G

(b) Use the critical-path algorithm to schedule this project using 2 processors.

39. **(a)** Draw a project digraph for a project consisting of the 8 tasks described by the following table.

Task	Length of task	Tasks that must be completed before the task can start
A	3	
B	10	*C, F, G*
C	2	*A*
D	4	*G*
E	5	*C*
F	8	*A, H*
G	7	*H*
H	5	

(b) Use the critical-path algorithm to schedule this project using 2 processors.

40. A toy store is having a contest among its employees to find a team of two employees who can assemble a new toy on the market the quickest. The assembling of the toy involves 7 tasks (*A, B, C, D, E, F,* and *G*). Two teams enter the contest: The Red Team (Joey and Sue) and the Green Team (Sharon and Jose). The rules of the contest specify that each task must be done by a single member of the team and that no team member can remain idle if there is a task to be done. The precedence relations for the tasks are shown in the following project digraph.

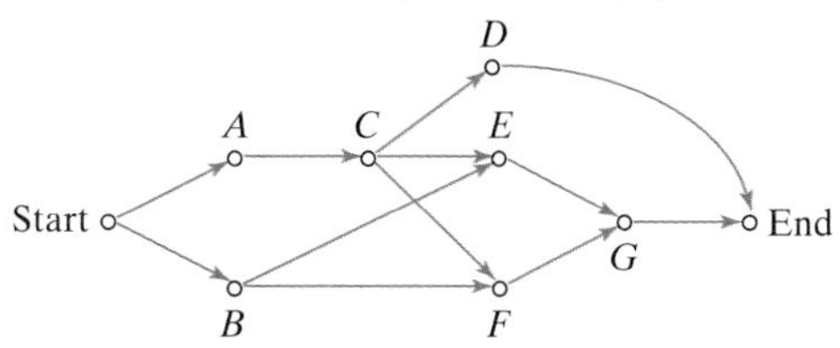

The Red Team practiced a lot, and both Joey and Sue are able to complete each task in the following time (in minutes): $A(1)$, $B(3)$, $C(1)$, $D(9)$, $E(4)$, $F(4)$, and $G(9)$.

The Green Team did not have as much time to practice, but both Sharon and Jose are able to complete each task in the following time: $A(2)$, $B(4)$, $C(2)$, $D(10)$, $E(5)$, $F(5)$, and $G(10)$.

(a) Find an optimal schedule for the Red Team.

(b) Find an optimal schedule for the green Team.

(c) Which team will win the contest?

(d) What would happen if the Red team slowed their work a little on task *C*, each taking 2 minutes rather than 1 minute?

Jogging

41. Explain why, in any digraph, the sum of all the indegrees must equal the sum of all the outdegrees.

42. **Symmetric and totally asymmetric digraphs**. A digraph is called **symmetric** if, whenever there is an arc from vertex *X* to vertex *Y*, there is *also* an arc from vertex *Y* to vertex *X*. A digraph is called **totally asymmetric** if, whenever there is an arc from vertex *X* to vertex *Y*, there *is not* an arc from vertex *Y* to vertex *X*. For each of the following, state whether the digraph is symmetric, totally asymmetric, or neither.

(a) A digraph representing the streets of a town in which all streets are one-way streets.

(b) A digraph representing the streets of a town in which all streets are two-way streets.

(c) A digraph representing the streets of a town in which there are both one-way and two-way streets.

(d) A digraph in which the vertices represent a bunch of men, and there is an arc from vertex X to vertex Y if X is a brother of Y.

(e) A digraph in which the vertices represent a bunch of men, and there is an arc from vertex X to vertex Y if X is the father of Y.

43. True or false? (If true, explain. If false, show with an example.)

(a) A schedule in which none of the processors is idle must be an optimal schedule.

(b) In an optimal schedule, none of the processors is idle.

44. Let W represent the sum of all the processing times of the tasks, M be the number of processors, and F be the finishing time for the project.

(a) Explain the meaning of the inequality

$$F \geq \frac{W}{M}$$

and why it is true for any schedule.

(b) Under what circumstances is $F = W/M$?

(c) What does the value $MF - W$ represent?

45. For independent tasks, the critical-path algorithm is never off by more than Max Error = $(M - 1)/3M$, where M is the number of processors.

(a) Calculate the value of Max Error = $(M - 1)/3M$ for $M = 5, 6, 7, 8, 9$, and 10.

(b) Explain why $(M - 1)/3M \leq 1/3$ for any value of M.

46. In 1961, T. C. Hu of the University of California showed that in any scheduling problem in which all the tasks have equal processing times and in which the original project digraph (without the START and END vertices) is a tree, the critical-path algorithm will give an optimal schedule. Using this result, find an optimal schedule for the scheduling problem with the following project digraph, using 3 processors. Assume each task takes 3 days. (Notice that we have omitted the START and END vertices.)

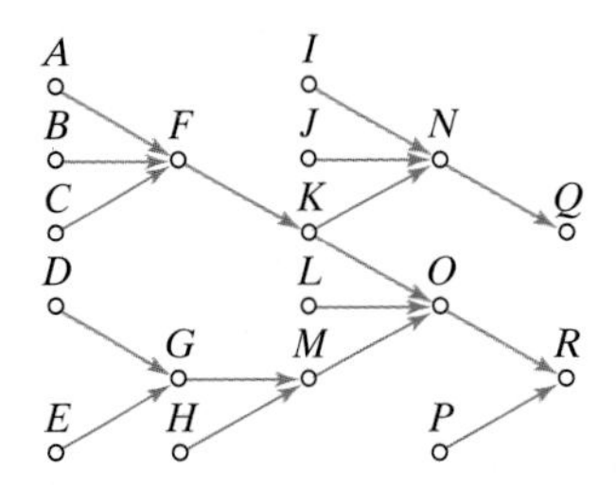

47. The following 9 tasks are all independent: $A(4)$, $B(4)$, $C(5)$, $D(6)$, $E(7)$, $F(4)$, $G(5)$, $H(6)$, $I(7)$. Four processors are available to carry out these tasks.

(a) Find a schedule using the critical-path algorithm.

(b) Find an optimal schedule.

48. The following 7 tasks are all independent: $A(4)$, $B(3)$, $C(2)$, $D(8)$, $E(5)$, $F(3)$, $G(5)$. Three processors are available to carry out these tasks.

(a) Find a schedule using the critical-path algorithm.

(b) Find an optimal schedule.

49. Use the critical-path algorithm to schedule independent tasks of length 1, 1, 2, 2, 5, 7, 9, 13, 14, 16, 18, and 20 using 3 processors. Is this schedule optimal? Explain.

Exercises 50 through 52 illustrate how it is sometimes possible to schedule independent tasks in such a way as to almost double the optimal completion time. The solution to these problems can be modeled after this example. The following schedule using 4 processors is optimal, having completion time of 8 hours (twice the number of processors).

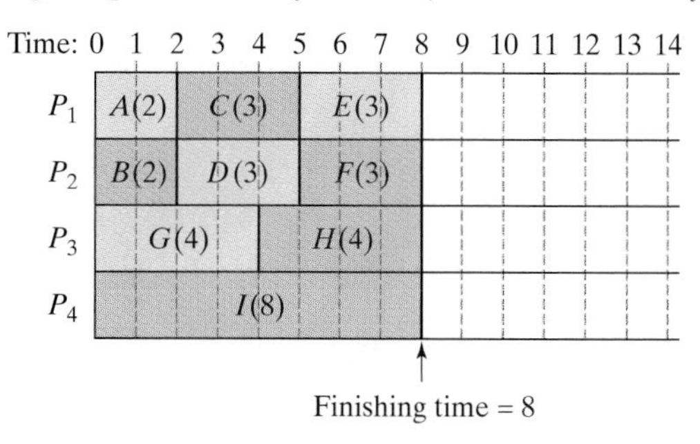

The same independent tasks are scheduled using 4 processors again, but this time the completion time is 14 hours (2 hours less than twice the optimal completion time).

Time: 0 1 2 3 4 5 6 7 8 9 10 11 12 13 14

P_1	$A(2)$	$G(4)$	$I(8)$
P_2	$B(2)$	$H(4)$	Idle
P_3	$C(3)$	$D(3)$	Idle
P_4	$E(3)$	$F(3)$	Idle

Finishing time = 14

50. Find two schedules, using 5 processors, of a project of independent tasks with

(a) optimal completion time of 10 hours (twice the number of processors), and

(b) completion time 18 hours (2 hours less than twice the optimal completion time).

51. Find two schedules, using 6 processors, of a project of independent tasks with

(a) optimal completion time of 12 hours (twice the number of processors), and

(b) completion time 22 hours (2 hours less than twice the optimal completion time).

52. Find two schedules, using 7 processors, of a project of independent tasks with

(a) optimal completion time of 14 hours (twice the number of processors), and

(b) completion time 26 hours (2 hours less than twice the optimal completion time).

53. Consider the problem of scheduling 9 tasks in accordance with the following project digraph.

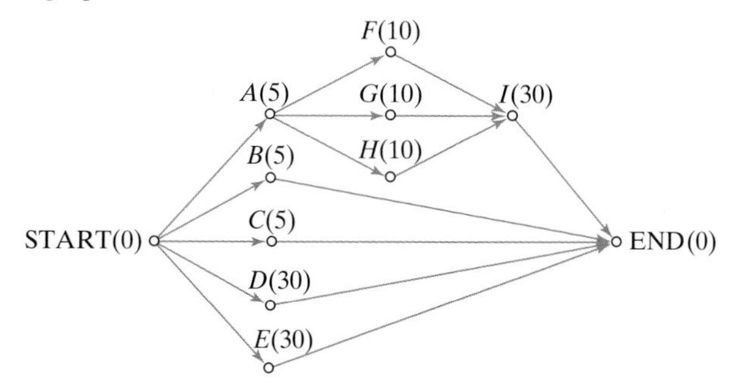

(a) Use the critical-path algorithm to schedule the project with 3 processors.

(b) Find an optimal schedule for the project using 3 processors.

54. The speed-up paradox.

(a) Use the critical-path algorithm to schedule a project with 9 tasks using 2 processors according to the following project digraph.

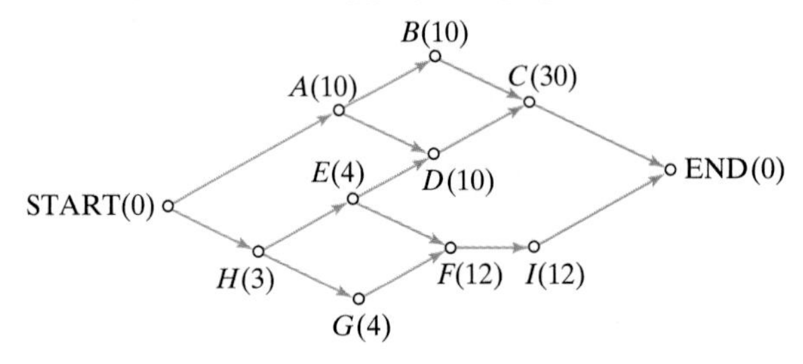

(b) Now suppose that the processing times for each of the tasks are decreased by 1 (a faster model of processor is used), giving the following project digraph. Use the critical-path algorithm to reschedule this project using 2 processors.

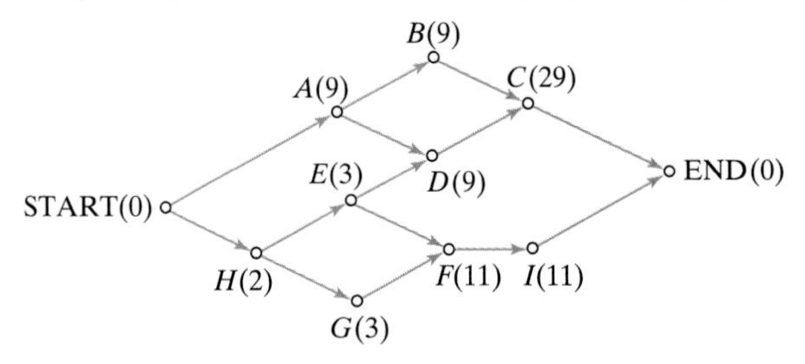

(c) Compare the times obtained in (a) and (b). Explain how this can happen.

55. The more-is-less paradox.

(a) Use the critical-path algorithm to schedule a project with 8 tasks using 2 processors according to the following project digraph.

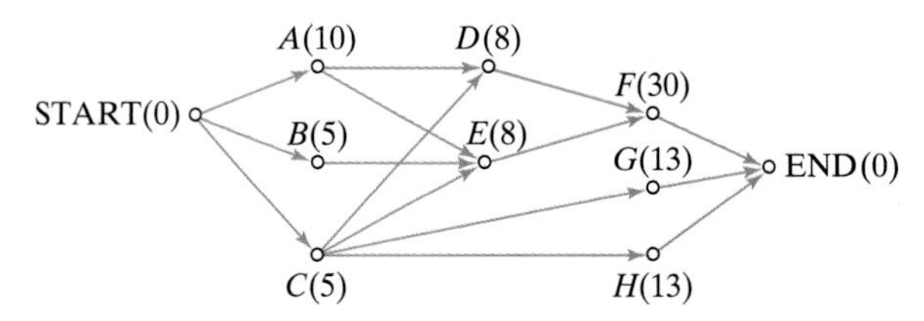

(b) Now suppose that the number of processors is increased to 3. Use the critical-path algorithm to reschedule the project.

(c) Compare the finishing times obtained in (a) and (b). Explain how this can happen.

Running

Exercises 56 through 58 refer to the following fact: In 1966, Ronald L. Graham of AT&T Bell Laboratories showed that if T_{OPT} is the optimal finishing time for a given scheduling problem, then the finishing time T for any other schedule for the problem must satisfy the *inequality* $T \leq [2 - (1/M)]T_{OPT}$, *where M is the number of processors. For example, with two processors* ($M = 2$) *the finishing time T for any schedule must satisfy the inequality* $T \leq \frac{3}{2}T_{OPT}$, *so that no scheduling will result in a time longer than* $1\frac{1}{2}$ *times the optimal time.*

56. Show that if $T_1 = 21$ hours and $T_2 = 12$ hours are finishing times of two different schedules for the same scheduling problem with 4 processors, then T_2 is the optimal finishing time for the scheduling problem and T_1 is the longest possible finishing time for the problem.

57. Suppose we have a scheduling problem with 2 processors and we come up with a schedule with finishing time $T_1 = 9$ hours. Explain why the optimal finishing time for this scheduling problem cannot be less than 6 hours.

58. Suppose we have a scheduling problem with 3 processors and we come up with two different schedules with finishing times $T_1 = 12$ hours and $T_2 = 15$ hours. Explain why the optimal finishing time for this scheduling problem has to be somewhere between 9 and 12 hours.

Exercises 59 and 61 refer to the following: It has been shown that when the number of processors is M, and all of the tasks are independent, the maximum relative percentage error using the critical path algorithm is $(M - 1)/3M$.

59. Show that the statement above is equivalent to the statement: If T_{OPT} is the optimal finishing time for a given scheduling problem in which all the tasks are independent, then the finishing time T for any schedule for the problem obtained by using the critical-path algorithm must satisfy the inequality $T \leq [4/3 - 1/(3M)]T_{OPT}$, where M is the number of processors.

60. Give an example of a scheduling problem using 3 processors in which all the tasks are independent and such that the finishing time using the critical-path algorithm is 11/9 of the optimal finishing time.

61. Give an example of a scheduling problem using 5 processors in which all the tasks are independent and such that the finishing time using the critical-path algorithm is 19/15 of the optimal finishing time.

References and Further Readings

1. Baker, K. R., *Introduction to Sequencing and Scheduling.* New York: John Wiley & Sons, Inc., 1974.
2. Coffman, E. G., *Computer and Jobshop Scheduling Theory.* New York: John Wiley & Sons, Inc., 1976, chaps. 2 and 5.
3. Conway, R. W., W. L. Maxwell, and L. W. Miller, *Theory of Scheduling.* Reading, MA: Addison-Wesley Publishing Co., Inc., 1967.
4. Dieffenbach, R. M., "Combinatorial Scheduling," *Mathematics Teacher,* 83 (1990), 269–273.
5. Garey, M. R., R. L. Graham, and D. S. Johnson, "Performance Guarantees for Scheduling Algorithms," *Operations Research,* 26 (1978), 3–21.
6. Graham, R. L., "The Combinatorial Mathematics of Scheduling," *Scientific American,* 238 (1978), 124–132.
7. Graham, R. L., "Combinatorial Scheduling Theory," in *Mathematics Today,* ed. L. Steen. New York: Springer-Verlag, Inc., 1978, 183–211.

8. Graham, R. L., E. L. Lawler, J. K. Lenstra, and A. H. G. Rinnooy Kan, "Optimization and Approximation in Deterministic Sequencing and Scheduling: A Survey," *Annals of Discrete Mathematics,* 5 (1979), 287–326.
9. Hillier, F. S., and G. J. Lieberman, *Introduction to Operations Research,* 3d ed. San Francisco: Holden-Day, Inc., 1980, chap. 6.
10. Roberts, Fred S., *Graph Theory and Its Applications to Problems of Society,* CBMS/NSF Monograph No. 29. Philadelphia: Society for Industrial and Applied Mathematics, 1978.
11. Zubrin, Robert, *The Case for Mars.* New York: Free Press, 1996.

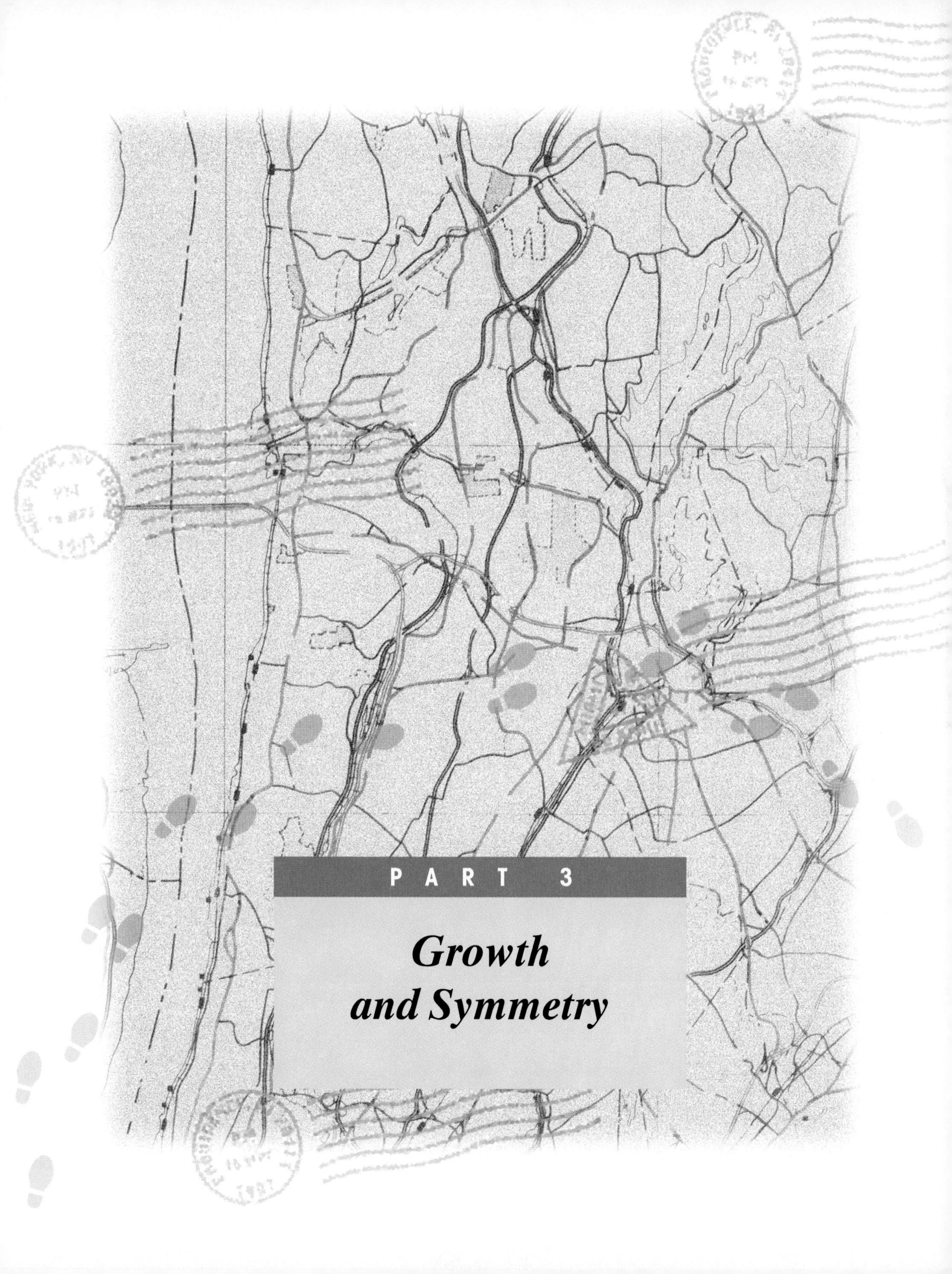

PART 3

Growth and Symmetry

Come forth into the light of things,
let Nature be your teacher.
William Wordsworth

Spiral Growth in Nature

Fibonacci Numbers and the Golden Ratio

CHAPTER 9

In nature's portfolio of architectural works, the magnificent shell of the chambered nautilus holds a place of special distinction. The spiral-shaped shell, with its revolving interior stairwell of ever-growing chambers, is more than a splendid piece of natural architecture—it is also a work of remarkable mathematical creativity.

Humans have imitated nature's wondrous spiral designs in their own architecture for centuries, all the while trying to understand how the magic works. What are the physical laws that govern spiral growth in nature? Why do so many different and unusual mathematical concepts come into play? How do the mathematical concepts and physical laws mesh together? Trying to give a partial answer to these questions is the goal of this chapter.

More than 2000 years ago, the ancient Greeks got us off to a great head start in our quest to understand nature with two great contributions—Euclidean geometry and irrational numbers. The next important mathematical connection came 800 years ago, with the serendipitous discovery by a medieval scholar named Fibonacci, of an amazing group of numbers now called Fibonacci numbers. Next came the discovery, by the French philosopher-mathematician René Descartes in 1638, that an equation he had studied for purely theoretical reasons ($r = ae^{\Theta}$) is the very same equation that describes the spirals generated by seashells. Since then, other surprising connections between spiral growing organisms and seemingly abstract mathematical concepts have been discovered, some as recently as 20 years ago.

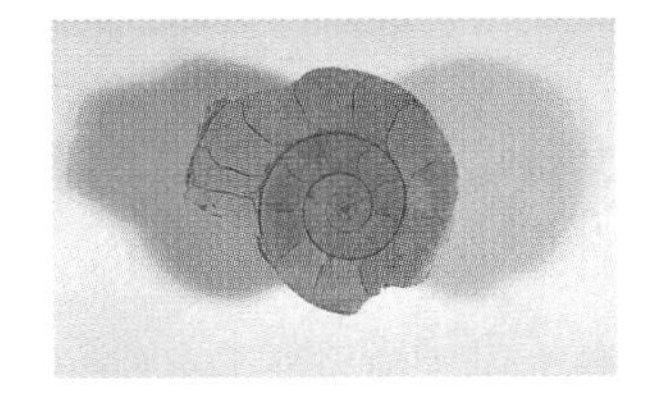

Exactly why and how these concepts play such a crucial role in the development of natural forms is not yet fully understood—not by humans, that is—a humbling reminder that nature still is the oldest and wisest of teachers.

Fibonacci Numbers

Listed in Table 9-1 is a widely known and disarmingly simple group of numbers called the **Fibonacci numbers**. They are named after the Italian Leonardo de Pisa, better known by the nickname Fibonacci.[1] It doesn't take long to see the pattern these numbers follow. After the first two, which seem to stand on their own, each subsequent number is the sum of the two numbers before it: $2 = 1 + 1$, $3 = 2 + 1$, $5 = 3 + 2, \ldots, 144 = 89 + 55$, and so on.

Table 9-1

1, 1, 2, 3, 5, 8, 13, 21, 34, 55, 89, 144, . . .

When does the list of Fibonacci numbers stop? Actually, we can keep applying the rule, each time adding the last two to get the next one. If we are willing, we can keep doing this forever. Mathematicians prefer to say the same thing by saying that the Fibonacci numbers are an **infinite sequence** of numbers. Not surprisingly, this particular sequence is called the **Fibonacci sequence**.

As with any other sequence, being infinite is no excuse for being disorganized, and there is a definite order to the Fibonacci numbers: a first Fibonacci number (1), a second (1), a third (2), . . . , a seventh (13), . . . , a tenth (55), an eleventh (89), and so on. Each Fibonacci number has its *place* in the Fibonacci sequence, and we are well advised not to mess it up. The standard mathematical notation to describe a Fibonacci number is an F followed by a subscript indicating its place in the sequence. For example, F_8 stands for the *eighth* Fibonacci number, which is $21 (F_8 = 21)$; $F_{12} = 144$, and so on. A generic Fibonacci number can be written as F_N (its place in the sequence being the generic position N). If we want to describe the Fibonacci number that comes after F_N, we write F_{N+1} (which

[1] Fibonacci (c. 1170–1250) was the son of a merchant and as a young man traveled extensively with his father throughout northern Africa. There he learned the Arabic system of numeration and algebra, which he introduced to Christian Europe in his book *Liber Abaci* (*The Book of the Abacus*), published in 1202. Although he is best remembered for the discovery of Fibonacci numbers, they were only a minor part of his book and of his contributions to Western civilization.

is not the same as $F_N + 1$!); if we want to describe the Fibonacci number that comes before F_N, we write F_{N-1}, and so on.

With a good understanding of the notation in hand (you should try Exercises 3 and 4 before going on), we can conviently describe the rule for Fibonacci numbers as follows:

$$\underbrace{F_N}_{\text{a generic Fibonacci number}} = \underbrace{F_{N-1}}_{\text{the Fibonacci number right before it}} + \underbrace{F_{N-2}}_{\text{the Fibonacci number two positions before it}}$$

Of course, the above rule cannot be applied to the first two Fibonacci numbers F_1 (which has no predecessors) and F_2 (which has only one predecessor, so for a complete description we must also give the values of the starting two Fibonacci numbers: $F_1 = 1$ and $F_2 = 1$. The combination of facts $F_N = F_{N-1} + F_{N-2}$ (which applies when $N > 2$), together with $F_1 = 1$ and $F_2 = 1$, gives a complete description of the Fibonacci numbers. It is, in essence, their definition.

Fibonacci Numbers

$F_1 = 1$
$F_2 = 1$
$F_N = F_{N-1} + F_{N-2}, \quad (N > 2)$

Using the above definition one could, in principle, compute any Fibonacci number, but this is easier said than done. Could we find, for example, F_{100}? You bet. How? It would be easy if we knew F_{99} and F_{98}, which we don't. In fact, there are plenty of Fibonacci numbers between $F_{12} = 144$ (the last one in Table 9-1) and F_{100} that presumably we don't know. At the same time, it is clear that if we set our minds to it, we could slowly but surely march up the Fibonacci ladder one rung at a time: $F_{13} = 144 + 89 = 233$, $F_{14} = 233 + 144 = 377$, and so on. Let's cheat a little bit and say that we got to $F_{97} = 83621143489848422977$ and then to $F_{98} =$ 135301852344706746049. Next comes

$$\begin{aligned} F_{99} &= 135301852344706746049 + 83621143489848422977 \\ &= 218922995834555169026 \end{aligned}$$

and finally

$$\begin{aligned} F_{100} &= 218922995834555169026 + 135301852344706746049 \\ &= 354224848179261915075. \end{aligned}$$

A definition of a sequence in which each new number is defined in terms of other (earlier) numbers is called a **recursive**[2] definition. Our definition of the Fibonacci numbers is a recursive definition.

---

[2] We have already encountered the idea of a recursive definition in Chapter 1, where we discussed recursive ranking methods in elections.

While recursive definitions have an elegant theoretical simplicity, they do have practical limitations. We saw that to calculate a Fibonacci number like F_{100} we first have to calculate all the preceding Fibonacci numbers $(\ldots, F_{96}, F_{97}, F_{98}, F_{99})$. Each can be calculated by a simple addition, but the numbers get big fast, and the process, when done by hand, can be excruciatingly long and boring. Imagine, if you will, calculating $F_{10,000}$ this way. Just thinking about it is painful. Is there another way?

A much more complicated-looking but direct definition for Fibonacci numbers was discovered by Leonhard Euler (remember him from Chapter 5?) about 250 years ago. It is generally known as **Binet's formula**.[3]

Binet's Formula for Fibonacci Numbers

$$F_N = \frac{\left(\frac{1+\sqrt{5}}{2}\right)^N - \left(\frac{1-\sqrt{5}}{2}\right)^N}{\sqrt{5}}$$

In spite of its rather nasty appearance, this formula has one advantage over a recursive definition—it gives us an explicit recipe for calculating any Fibonacci number without having to first calculate other Fibonacci numbers. For this reason, Binet's formula is called an **explicit definition** of the Fibonacci numbers. Three constants appear in Binet's formula (all three are irrational numbers, so we can only give decimal approximations to their exact values):

$$\sqrt{5} = 2.236067977\ldots;$$

$$\frac{(1-\sqrt{5})}{2} = -0.6180339887\ldots;$$

$$\frac{(1+\sqrt{5})}{2} = 1.6180339887\ldots;$$

The last of these numbers will be especially important to us in this chapter.

Until the advent of computers, Binet's formula was of limited practical use. Raising irrational numbers to high powers is hardly the kind of thing one would want to do by hand. With a good calculator or, better yet, with a good computer one can use Binet's formula to directly calculate the values of fairly large Fibonacci numbers.

Fibonacci Numbers in Nature

Our interest in Fibonacci numbers stems from their frequent occurrence in nature. Take flowers, for example. Consistently, the number of petals in a daisy is a Fibonacci number, which depends on the variety: 13 for Blue daisies; 21 for English daisies; 34 for Oxeye daisies; 55 for African daisies, and so on. What's true for daisies is also true for many other types of flowers (geraniums, chrysanthemums, lilies, etc.).

Fibonacci numbers also appear consistently in conifers and seeds: The bracts in a pine cone spiral in two different directions in 8 and 13 rows; the scales in a

---

[3] The formula was actually discovered first by Leonhard Euler, and then rediscovered (almost 100 years later) by the Frenchman Jacques Binet, who somehow ended up getting the credit.

pineapple spiral in three different directions in 8, 13, and 21 rows; the seeds in the center of a sunflower spiral in 55 and 89 rows.

Exactly why and how this unusual connection between a purely mathematical concept (Fibonacci numbers) and natural objects occurs is still not fully understood (there are several theories—the latest one fully detailed in Reference 9), but it is clear that it has something to do with the spiral nature of their growth.

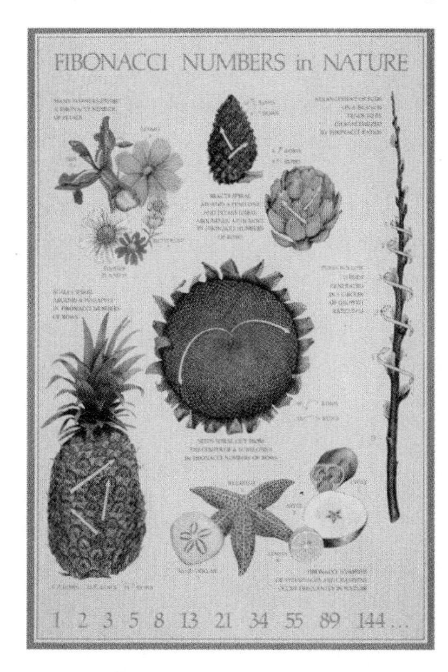

The Equation $x^2 = x + 1$ and the Golden Ratio

We next discuss a simple quadratic equation. At first, this looks like another one of those typical high school algebra questions: *Solve the quadratic equation* $x^2 = x + 1$.

You probably know how to do this without help, but here it is anyway: First, move all terms to the left-hand side to get $x^2 - x - 1 = 0$. Next, use the *quadratic formula*[4] of Algebra II fame. When the dust settles, the two solutions we get are

$$\left(\frac{1 + \sqrt{5}}{2}\right) \text{ and } \left(\frac{1 - \sqrt{5}}{2}\right).$$

[4] Just in case you forgot the quadratic formula: The solutions of $ax^2 + bx + c = 0$ are given by $x = (-b \pm \sqrt{b^2 - 4ac})/2a$. For $x^2 - x - 1 = 0$ we get $x = (1 \pm \sqrt{1 + 4})/2$. (For a brief review of the quadratic formula, see Exercises 18 through 21.)

We already saw these two numbers as part of Binet's formula for the Fibonacci numbers. (*Reminder:* These two numbers are irrational and therefore have infinite, nonrepeating decimal expansions.) For working purposes, we will use an approximation to three decimal places:

$$\left(\frac{1+\sqrt{5}}{2}\right) \approx 1.618 \text{ and } \left(\frac{1-\sqrt{5}}{2}\right) \approx -0.618.$$

(The "≈" reminds us that these are approximate values.) Notice that, of the two solutions, one is positive and the other is negative, and their decimal expansions are identical, which is not a coincidence (Exercise 35).

The Golden Ratio

Let's focus on just the positive solution $(1 + \sqrt{5})/2$. This number is important enough to have its own symbol and name: It is called the **golden ratio** and represented by the Greek letter Φ (phi). Thus, from now on, $\Phi = (1 + \sqrt{5})/2$.

The fact that the golden ratio Φ is a solution of the equation $x^2 = x + 1$ means that $\Phi^2 = \Phi + 1$. Using this fact repeatedly, we can calculate other powers of Φ. (Powers of Φ show up, for example, in Binet's formula.) Multiplying both sides of the equation $\Phi^2 = \Phi + 1$ by Φ, we get $\Phi^3 = \Phi^2 + \Phi$. We can now substitute $\Phi + 1$ for Φ^2 and get $\Phi^3 = (\Phi + 1) + \Phi$. After all is said and done, we get $\Phi^3 = 2\Phi + 1$. To get Φ^4 we multiply both sides of our last equation ($\Phi^3 = 2\Phi + 1$) by Φ and get $\Phi^4 = 2\Phi^2 + \Phi$. Substituting $\Phi + 1$ for Φ^2 gives us $\Phi^4 = 2(\Phi + 1) + \Phi = 3\Phi + 2$. Continuing this way, we get

$$\Phi^5 = 3\Phi^2 + 2\Phi = 3(\Phi + 1) + 2\Phi = 5\Phi + 3,$$

$$\Phi^6 = 5\Phi^2 + 3\Phi = 5(\Phi + 1) + 3\Phi = 8\Phi + 5,$$

$$\Phi^7 = 8\Phi^2 + 5\Phi = 8(\Phi + 1) + 5\Phi = 13\Phi + 8,$$

and so on.

Once again, we see a connection between the Fibonacci numbers and the powers of Φ, which can be described by the formula

$$\Phi^N = F_N\Phi + F_{N-1}.$$

In a sense, this formula is the mirror image of Binet's. In Binet's formula we use powers of Φ to calculate Fibonacci numbers; here we use Fibonacci numbers to calculate powers of Φ.

The third connection between the Fibonacci numbers and the golden ratio is possibly the most surprising one. What do we get when we divide two consecutive Fibonacci numbers?

Table 9-2 Ratios of consecutive Fibonacci numbers

F_N	**1**	**1**	**2**	**3**	**5**	**8**	**13**	**21**	**34**	**55**	**89**	**144**	**233**	**377**	**610**	**987**
$\frac{F_N}{F_{N-1}}$		$\frac{1}{1} = 1.0$	$\frac{2}{1} = 2.0$	$\frac{3}{2} = 1.5$	$\frac{5}{3} = 1.\overline{6}$	$\frac{8}{5} = 1.6$	$\frac{13}{8} = 1.625$	$\frac{21}{13} = 1.\overline{615384}$	$\frac{34}{21} = 1.\overline{619047}$	$\frac{55}{34} = 1.617647\ldots$	$\frac{89}{55} = 1.6\overline{18}$	$\frac{144}{89} = 1.617977\ldots$	$\frac{233}{144} = 1.6180\overline{5}$	$\frac{377}{233} = 1.618025\ldots$	$\frac{610}{377} = 1.618037\ldots$	$\frac{987}{610} = 1.618032\ldots$

Table 9-2 shows the first 15 values of the ratio F_N/F_{N-1}. What's going on? It appears that, after some early hesitation, the ratio "settles down" at a value of approximately 1.61803 . . . . If we play this game with bigger Fibonacci numbers and write the decimals to many more decimal places, the pattern becomes even more apparent. Using a computer, we have calculated to 41 decimal places the ratios

$$\frac{F_{99}}{F_{98}} = \frac{218922995834555169026}{135301852344706746049} \approx 1.61803398874989484820458683436563811772033,$$

$$\frac{F_{100}}{F_{99}} = \frac{354224848179261915075}{218922995834555169026} \approx 1.61803398874989484820458683436563811772030.$$

These two numbers, while not identical, match up everywhere except in the last digit, so that the difference between them is truly insignificant.

But there is more to it than that. The magic number that the ratios settle into is $\Phi = (1 + \sqrt{5})/2$. Essentially this means that *each Fibonacci number is approximately equal to Φ times the preceding one, and the approximation gets better as the numbers get larger.*

The Golden Ratio in Art, Architecture, and Design

As a special number, the golden ratio Φ is considered, right along with π (the ratio between the circumference and the diameter of a circle) among the great mathematical discoveries of antiquity. The golden ratio was known to the ancient Greeks, who ascribed to it mystical and religious meaning and called it the *divine proportion.* The Greeks used the golden ratio as a benchmark for proportion and scale in art and architecture. The famous Greek sculptor Phidias consistently used the golden ratio in the proportions for his sculptures, as well as in his design of the Parthenon, perhaps the best-known building of ancient Greece.[5]

During the Renaissance, famous artists such as Leonardo da Vinci and Botticelli know about the golden ratio and used it in their paintings. In 1509 the friar Luca Pacioli wrote a book called *De Divina Proportione* (*The Divine Proportion*). The book was the first mathematical treatise on the golden ratio and

[5] In fact, the choice of Φ to represent the golden ratio comes from Phidias's name.

Luca Pacioli, by Jacopo de Barbari. (1440–1515), Fra Luca Pacioli and a young man, probably Duke Guidobaldo da Montefeltro (1472–1509), (1495). Oil on wood, 99 × 120 cm. Museo Nazionale di Capodimonte, Naples, Italy. Erich Lessing/Art Resource, NY.

its fame is due in part to its illustrations, drawn by none other than Leonardo da Vinci. Since Luca Pacioli's first book, literally hundreds of books and articles have been written about the golden ratio and its role in geometry, art, architecture, music and nature. An extensive bibliography can be found in Reference 8 at the end of the chapter.

Gnomons

Time for a little geometry. In geometry, a **gnomon** to a figure A is a *connected* figure (i.e., it cannot have separate parts), which, when suitably *attached*[6] to A produces a new figure which is similar (in the geometric sense) to A. Informally, we will describe it this way: G is a gnomon to A if *G&A is similar to A.* (Here the symbol & should be taken to mean "attached in some suitable way.") Gnomons are not standard fare in high school geometry, but they do play an important role in spiral growth. The study of gnomons goes back to the Greeks, presumably to Aristotle and his disciples, more than 2300 years ago.

Before we go into gnomons, let's quickly review the concept of geometric **similarity**. We know from high school geometry that *two objects are similar if one is a scaled version of the other.* When a slide projector takes the image in a slide and blows it up onto a screen, it creates a *similar,* but larger, image. When a photocopy machine reduces the image on a sheet of paper it creates a *similar,* but smaller, image.

Here are some very basic facts about similarity that we will use in this section:

---

[6] By *attached,* we mean that they are joined together without overlapping anywhere.

- Two triangles are similar if their sides are proportional. Alternatively, two triangles are similar if the sizes of their respective angles are the same.
- Two squares are *always* similar.
- Two rectangles are similar if their sides are proportional; that is,

$$\frac{\textit{long side } 1}{\textit{long side } 2} = \frac{\textit{short side } 1}{\textit{short side } 2}.$$

- Two circles are *always* similar.
- Two circular rings are similar if their inner and outer radii are proportional; that is,

$$\frac{\textit{outer radius } 1}{\textit{outer radius } 2} = \frac{\textit{inner radius } 1}{\textit{inner radius } 2}.$$

We are now ready to take on gnomons.

Example 1.

The square S in Fig. 9-1(a) has the L-shaped figure G in Fig. 9-1(b) as a gnomon, because when G is attached to S as in Fig. 9-1(c), we get a square S'.

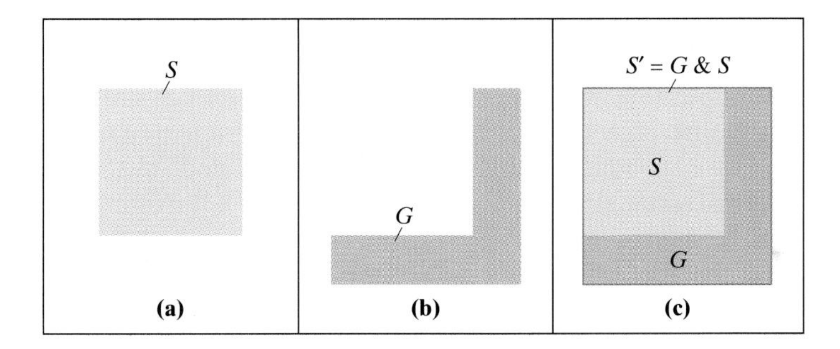

FIGURE 9-1
(a) A square S.
(b) Its gnomon G.

Note that the wording is *not* reversible: The square S *is not* a gnomon to the L-shaped figure G, since *there is no way to attach the two to form an L-shaped figure that is similar to G* (Exercise 36). On the other hand, there are other gnomons for the square S besides an L-shaped figure. Can you think of any? ■

Example 2.

The circle C in Fig. 9-2(a) has as a gnomon an O-ring G like the one in Fig. 9-2(b) (the inner radius has to be r; the outer radius R can be any number bigger than r). We can see that we can attach the O-ring G to C and get a new circle C' [Fig. 9-2(c)], and, as we know, circles are always similar.

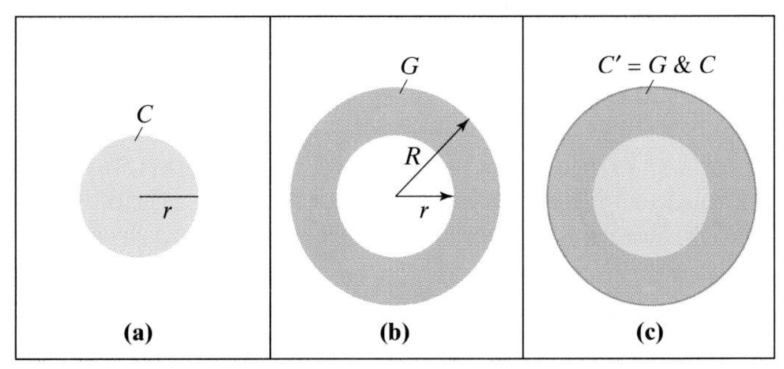

FIGURE 9-2
(a) A circle C.
(b) Its gnomon G.

■

Example 3.

Consider now an O-ring O with outer radius r [Fig. 9-3(a)], and the O-ring H with inner radius r and outer radius R [Fig. 9-3(b)]. Is H a gnomon to the original O-ring O?

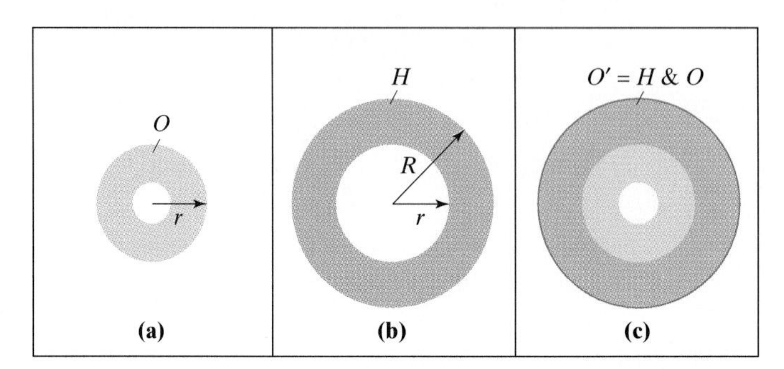

FIGURE 9-3 (a) An O-ring O. (b) Another O-ring H. (c) The combined figure is not similar to O—H is not a gnomon to O.

One is tempted to think that with the right choice of the outer radius R it might work, but it never does. No matter how we choose the outer radius of the O-ring H, when we attach the two O-rings together [Fig. 9-3(c)], O' cannot be made similar to O (Exercise 38). ■

Example 4.

Suppose that we have a rectangle R of height h and base b [Fig. 9-4(a)].The L-shaped object G shown in Fig. 9-4(b) is a gnomon to rectangle R if the ratios b/h and y/x are equal. In this case, G can be "cuddled" next to R so that together they form a rectangle R' similar to R (see Exercise 43). A simple geometric way to build the L-shaped gnomon G is by noticing that the line through the diagonal of the original rectange R must also be the diagonal through the L-corner in G.

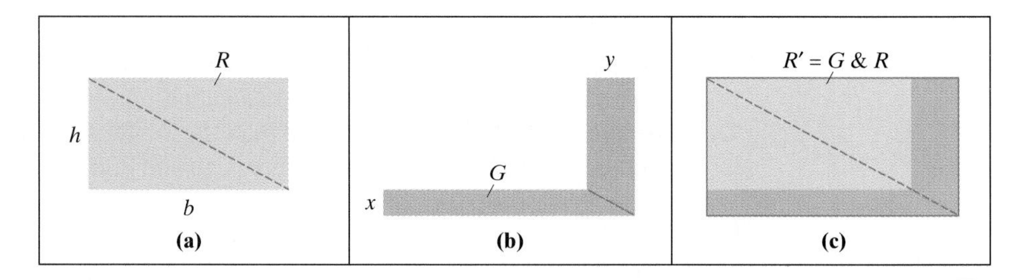

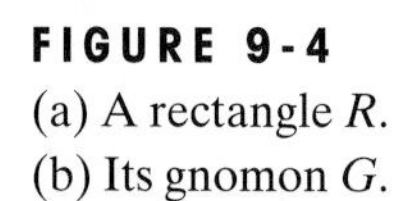

FIGURE 9-4 (a) A rectangle R. (b) Its gnomon G.

■

Example 5.

In this example we are going to do things a little bit backward. Let's start with an isosceles triangle T, with vertices B, C, and D and angles of 72°, 72°, and 36°, respectively. [Fig. 9-5(a)]. We now mark off on side DC of T a point A so that BA is congruent to BC [Fig. 9-5(b)]. (This can easily be done by centering a compass at B and drawing an arc of a circle of radius BC.) The triangle T' with vertices at C, B, and A is isosceles, with equal angles at C and A, and therefore with angles of 72°, 36° and 72°. This makes T' similar to the original triangle T. So what? you may ask. Where is the gnomon to triangle T? We don't have one yet! But we *do* have a gnomon to triangle T', and it is triangle G' [Fig. 9-5(c)]. After all, when triangle G' is attached to triangle T', we get triangle T. Note that gnomon G' is also an isosceles triangle: its angles are 36°, 36°, and 108°.

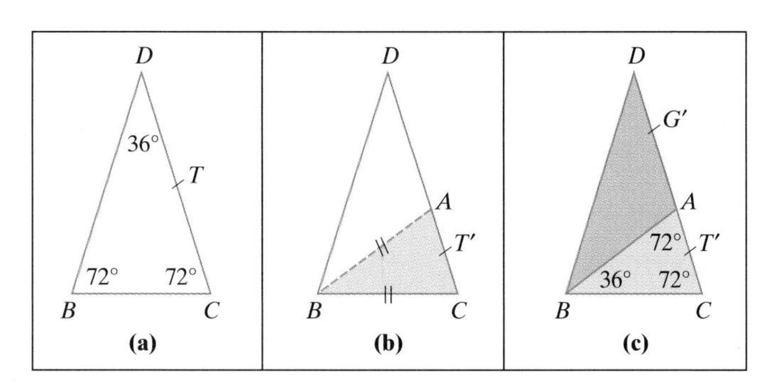

FIGURE 9-5 (a) A 72°-72°-36° isosceles triange T. (b) An isosceles triangle T' is constructed inside of T. (c) G' is a gnomon to T'.

■

We now know how to find a gnomon not only to triangle T' but to any 72°-72°-36° triangle, including the original triangle T: attach a 36°-36°-108° triangle to one of the longer sides [Fig. 9-6(a)]. We can repeat this process indefinitely, and when we do, we get a spiraling series of ever-increasing 72°-72°-36° triangles [Fig. 9-6(b)]. It's not too farfetched to use a family analogy: triangles T and G are the *parents*, with T having the *dominant* genes: The *offspring* of their union looks just like T (but bigger). The offspring then has offspring of its own (looking exactly like their grandfather), and so on ad infinitum.

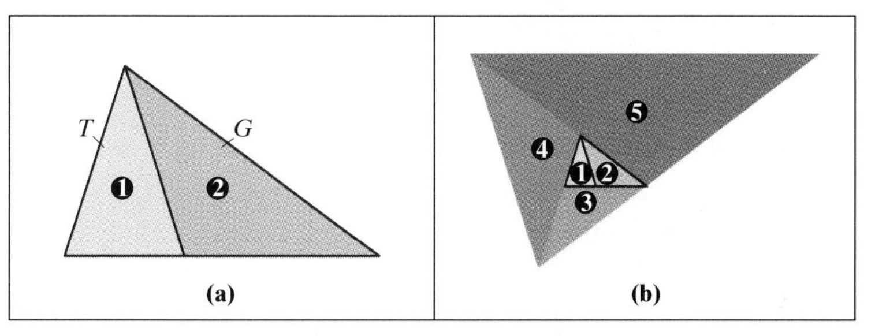

FIGURE 9-6 The process of adding a 36°-36°-108° gnomon to a 72°-72°-36° triangle can be repeated indefinitely, producing a spiraling chain of increasing triangles.

Example 5 is of special interest to us for two reasons. First, this is the first time we have an example where the figure and its gnomon are of the same type (isosceles triangles). Second, the isosceles triangles in this story (72°-72°-36° and 36°-36°-108°) have a property that makes them unique: in both cases, the ratio of their sides (longer side over shorter side) is the golden ratio (Exercise 45). These are the only two isosceles triangles with this property, and for this reason they are called **golden triangles**.

Example 6.

In this example we start with a rectangle R whose shorter side has a length 1 and whose longer side has some unspecified length x [Fig. 9-7(a)]. We would like to find out if it's possible for this rectangle to have a square gnomon [Fig. 9-7(b)], and if so, what should x be? The reason we are interested in this question is that

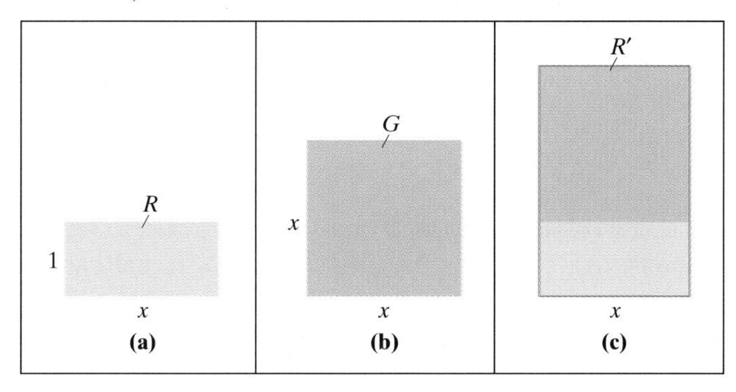

FIGURE 9-7 Can G be a gnomon to R?

squares are one of the fundamental building blocks in nature, and make for particularly nice gnomons.

For R' to be similar to R, we must have

$$\frac{\textit{long side of } R'}{\textit{long side of } R} = \frac{\textit{short side of } R'}{\textit{short side of } R};$$

in other words,

$$\frac{x+1}{x} = \frac{x}{1},$$

which turns out to be equivalent to the equation $x^2 = x + 1$, an old friend of ours. Since x is the length of the side of rectangle R, it must be positive, and the positive solution to the equation $x^2 = x + 1$ is *the golden ratio* $\Phi = \left(\frac{1+\sqrt{5}}{2}\right)$! ■

Our choice for the dimensions of rectangle R (x and 1) was dictated by convenience (having the short side equal 1 simplifies the computations), but what is true about R is certainly true for any other rectangle similar to R: *any rectangle whose sides are in the proportion of the golden ratio has a square gnomon, and vice versa.*

Rectangles whose sides are in the proportion of the golden ratio are called **golden rectangles**. We will discuss these next.

Golden Rectangles

Figure 9-8 shows an assortment of rectangles. In Figs. 9-8(a) and (b) we have *exact* golden rectangles: the ratios between the longer and shorter sides are $\Phi/1 = \Phi$, and $1/(1/\Phi) = \Phi$, respectively. In Figs. 9-8(c) and (d) we have rectangles whose sides are consecutive Fibonacci numbers.[7] The ratios between their longer and shorter sides are $55/34 = 1.617647\ldots$, and $89/55 = 1.61818\ldots$. These numbers are so close to the golden ratio that we might as well call these rectangles golden, too. The last rectangle shows the front of a box of "Corn Pops." The dimensions are 12 in. by $7\frac{1}{2}$ in., with a ratio of $12/7.5 = 1.6$, also very close to the golden ratio. It is safe to say that, at least to the naked eye, all of these rectangles are golden.

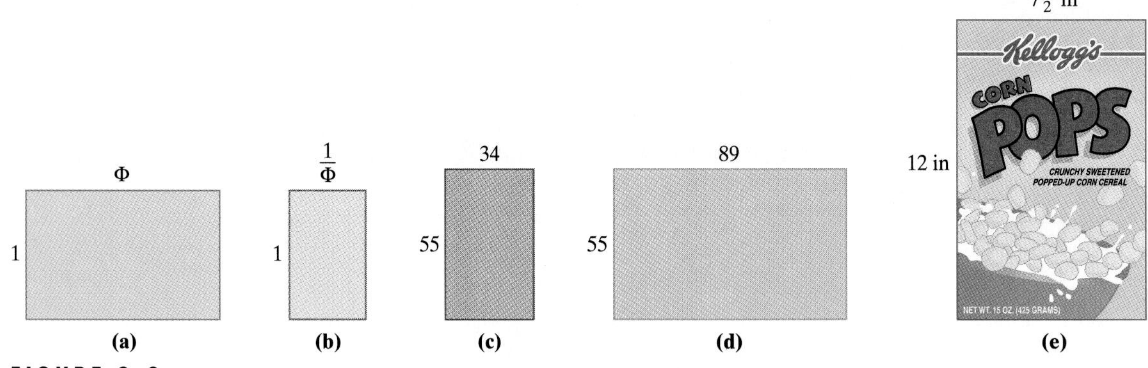

FIGURE 9-8
An assortment of golden and almost golden rectangles. (a) and (b) are exact golden rectangles; (c) and (d) are Fibonacci rectangles; (e) has proportions in a ratio of 1.6. To the naked eye, they all have the same proportions.

[7] Such rectangles are usually called **Fibonacci rectangles.**

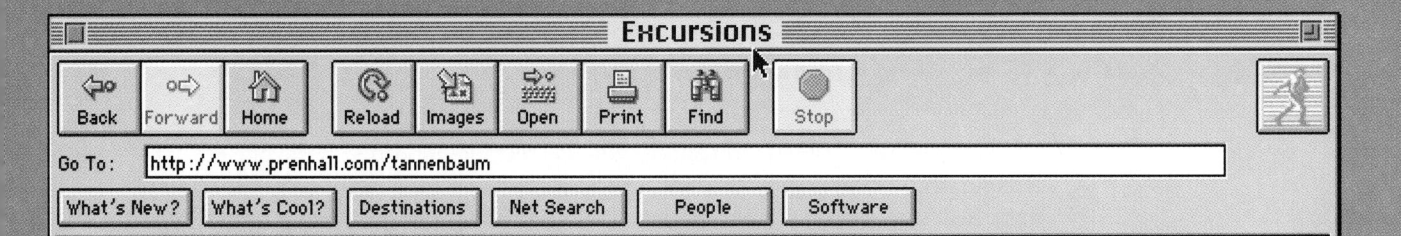

The Fibonacci Numbers and the Golden Ratio in Nature

One of the most impressive things about the Fibonacci numbers is the frequency with which they occur in nature. Fibonacci numbers appear in flowers, in shells, in vegetables, in flowers, in pine cones, and even in models of the reproduction of rabbits and bees.

An introduction to Fibonacci numbers and the golden ratio in nature can be found at the web address below. Be sure to view the Quick Time animations involving the positions of seeds on flowerheads.

Questions:

1. Suppose each female rabbit took three months to begin reproducing rabbits instead of two. How would you calculate each term in the new sequence? Write down the first ten terms of the new sequence.
2. Now suppose that, after the usual two months, each female produces *two* new pairs per month, instead of one pair. How would you calculate each term in the new sequence? Write down the first ten terms of the new sequence.
3. Briefly explain how the Fibonacci numbers occur in populations of bees.
4. Why is the pattern produced using a turn of Φ called an optimal packing? Describe the difference between the seed patterns produced using a turn of Φ between successive seeds and a turn of 0.61.
5. Predict what the seed pattern would look like using a turn of 0.25 between successive seeds. Then predict what it would look like using a turn of 0.35.

Aesthetically, golden (and almost golden) rectangles represent the perfect balance in proportion—they are neither too *skinny and long* nor too *square*. Their natural aesthetic appeal was confirmed by a well-known psychology experiment.

In 1876 the German psychologist Gustav Fechner decided to investigate what sorts of proportions were most appealing to the human eye, a question of more than passing interest to artists and designers. To do so, he performed several experiments. In one of them he showed the subjects an assortment of rectangles of different proportions, from squares to some in which one side was much longer than the other one. The subjects were asked to choose the rectangle which they found the most aesthetically pleasing. The overwhelming favorites were golden and almost golden rectangles: 75% of the subjects chose rectangles with long-side to short-side rations between 1.49 and 1.75, with 35% choosing the ratio 1.618.

In nature, where form usually follows function, the perfect balance of a golden rectangle shows up in spiral-growing organisms, often in the form of consecutive Fibonacci numbers. To see how this connection works, consider the following example which serves as a model for certain natural growth processes.

Example 7. Spiraling Fibonacci Rectangles

Start with a 1-by-1 square [marked ❶ in Fig. 9-9(a)]. Tack onto it another 1-by-1 square [marked ❷ in Fig. 9-9(b)]. Squares ❶ and ❷ together form a 2-by-1 rectangle, as shown in Fig. 9-9(b). We will call this the "second-generation" shape. For the third generation, tack on a 2-by-2 square ❸ as shown in Fig. 9-9(c). The "third-generation" shape (❶, ❷, and ❸ together) is the 2-by-3 rectangle in Fig. 9-9(c). Next tack onto it a 3-by-3 square ❹ as shown in Fig. 9-9(d), giving a 5-by-3 rectangle. Then tack on a 5-by-5 square ❺ as shown in Fig. 9-9(e), resulting in an 8-by-5 rectangle.

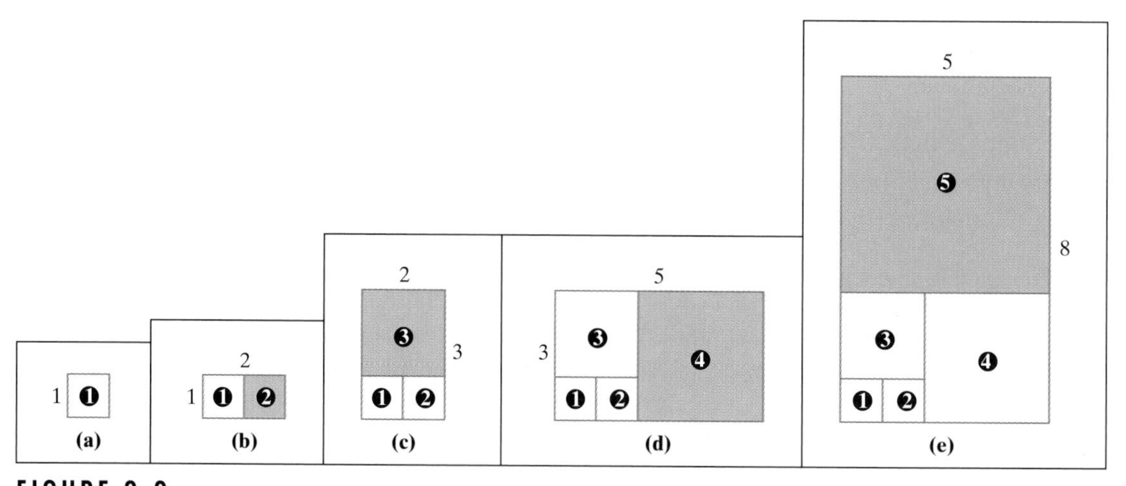

FIGURE 9-9
Fibonacci rectangles beget Fibonacci rectangles.

We can continue this process indefinitely, at each generation getting a bigger rectangle. The rectangles are all rectangles whose sides are consecutive Fibonacci numbers and, by the fifth generation, almost golden rectangles. By the time this process reaches the tenth generation, we have a 55-by-89 rectangle, with a long-side to short-side ratio of 1.61818—for all practical purposes, a golden rectangle. As the rectangles become for all practical purposes golden, they also become, for all practical purposes similar to each other, a fundamental prerequisite in natural growth. ■

The next example is a simple variation of Example 7.

Example 8. The "Chambered" Fibonacci Rectangle

Let's repeat the process in the previous example, except that now we have added to each square an interior "chamber" in the form of a quarter-circle. We now need to be a little more careful about where we attach the "chambered" square in each successive generation, but other than that we can repeat the sequence of steps in Example 7 to get the sequence of shapes shown in Fig. 9-10, showing consecutive generations in the evolution of the *chambered Fibonacci rectangle.* ■

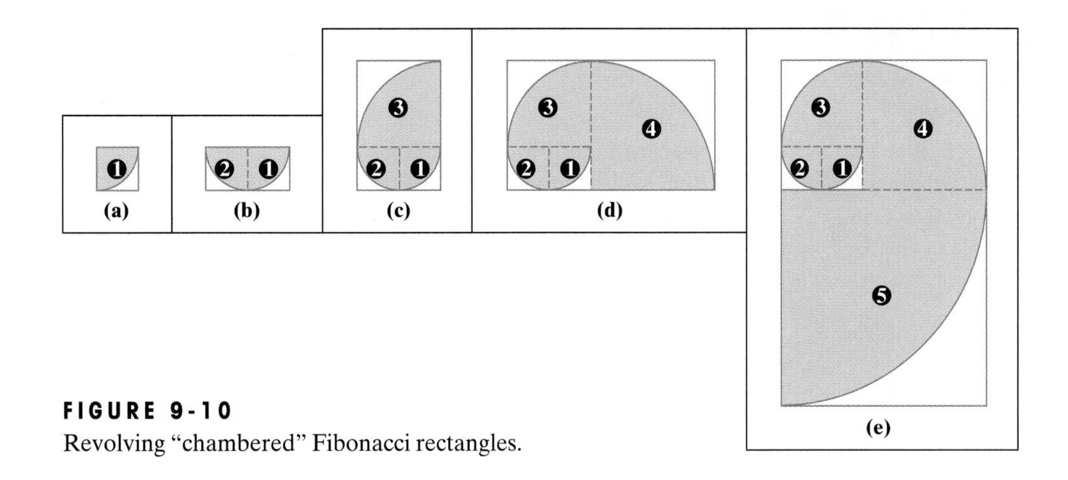

FIGURE 9-10
Revolving "chambered" Fibonacci rectangles.

Gnomonic Growth

As soon as humans realized that nature is a gifted builder, architect, and designer, from which they could learn much about the form and function of things, understanding the laws that govern the growth and form of natural organisms became an important part of natural science. (Just for plants alone, for example, there is a discipline called **phyllotaxis**, whose primary concern is the study of the patterns of growth and distribution of *lateral organs:* leaves, petals, stalks, scales, and so on.)

Natural organisms grow in essentially two different ways. The more common type of growth (and the one we are most familiar with) is the growth exhibited by humans, animals, and many plants. This can be called *all-around growth,* in which all living parts of the organism grow simultaneously (although not necessarily at the same rate). One characteristic of this type of growth is that there is no obvious way to distinguish between the newer and the older parts of the organism. In fact, the distinction between new and old parts does not make much sense. The historical record (so to speak) of the organism's growth is lost. By the time the child becomes an adult, no identifiable traces of the child (as an organism) remain—that's why we need photographs!

Contrast this with the kind of growth exemplified by the shell of the chambered nautilus, or a ram's horn, or the trunk of a redwood tree. This we may informally call *growth at one end* or *asymmetric growth.* In this type of growth the organism has a part added to it (either by its own or outside forces) in such a way that the old organism together with the added part form the new organism. At any stage of the growth process we can see not only the present but the organism's entire past. All the previous stages of growth are the building blocks that make up the present structure.

The second relevant fact is that most such organisms grow in a way that preserves their overall shape; in other words, they remain similar to themselves. This is where gnomons come into the picture: Regardless of how the new growth comes about, its shape is a gnomon of the entire organism. We will call this kind of growth process **gnomonic growth**.

We have already seen abstract mathematical examples of gnomonic growth (Examples 7 and 8). Here are a pair of more realistic examples.

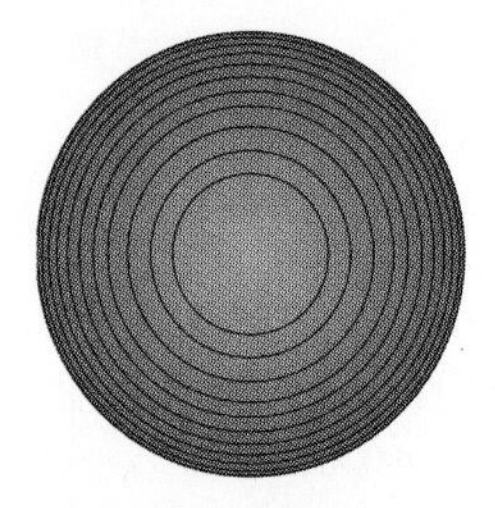

FIGURE 9-11 The growth rings in a redwood tree—an example of circular gnomonic growth.

Example 9.

We know from Example 2 that the gnomon to a circle is an O-ring with an inner radius equal to the radius of the circle. We can thus have circular growth (Fig. 9-11). Rings added one layer at a time to a starting circular structure preserve the circular shape throughout the structure's growth. When carried to three dimensions, this is a good model for the way the trunk of a redwood tree grows. ■

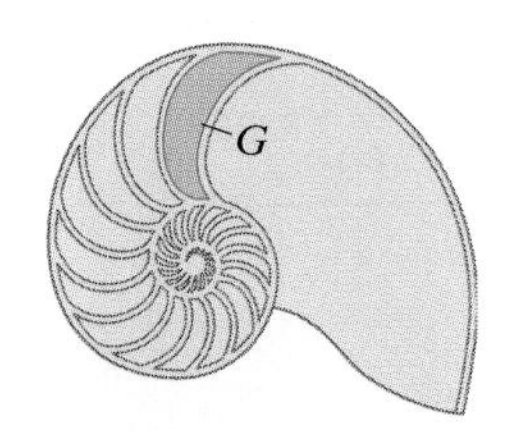

FIGURE 9-12 Gnomonic growth in the chambered nautilus.

Example 10.

Figure 9-12 shows a diagram of a cross section of the chambered nautilus, the example we used to open this chapter. The chambered nautilus builds its shell in stages, each stage adding a chamber to the already existing shell. At every stage of its growth, the shape of the chambered nautilus shell remains the same—the beautiful and distinctive spiral shown in the photograph. This as a classic example of gnomonic growth—each new chamber added to the shell is a gnomon of the entire shell. The gnomonic growth of the shell proceeds, in essence, as follows: Starting with the

shell of a baby chambered nautilus (which is a tiny spiral similar in all respects to the adult spiral shape), the animal builds a chamber (by producing a special secretion around its body that calcifies and hardens). The resulting, slightly enlarged spiral shell is similar to the original one. The processs then repeats itself over many stages, each one a season in the growth of the animal. Each new chamber is a gnomon to the shell, creating an enlarged shell that is larger but in all other respects similar to the younger shell. This process continues until the animal building the shell dies. ■

More complex examples of gnomonic growth occur in sunflowers, daisies, pineapples, pine cones, and so on. Here, the rules that govern growth are somewhat more involved and not fully understood. The most recent theories are based on the dynamics of efficient packings (for details, see References 3 and 9).

Conclusion

Some of the most beautiful shapes in nature arise from a basic principle of design: *form follows function.* The beauty of natural shapes is a result of their inherent elegance and efficiency, and imitating nature's designs has helped humans design and build beautiful and efficient structures of their own.

In this chapter we examined a special type of growth—gnomonic growth—where an organism grows by the addition of gnomons, thereby preserving its basic shape even as it grows. Many beautiful spiral-shaped organisms, from sea shells to flowers, exhibit this type of growth.

To us, understanding the basic principles behind spiral growth was relevant because it introduced us to some important mathematical concepts which have been known and studied in their own right for centuries: Fibonacci numbers, the golden ratio, gnomons, golden triangles, and golden rectangles.

To humans, these abstract mathematical concepts have been, by and large, intellectual curiosities. To nature—the consumate artist and builder—they are the building tools for some if its most beautiful creations. Whatever lesson one wishes to draw from that, it should include something about the inherent value of good mathematics.

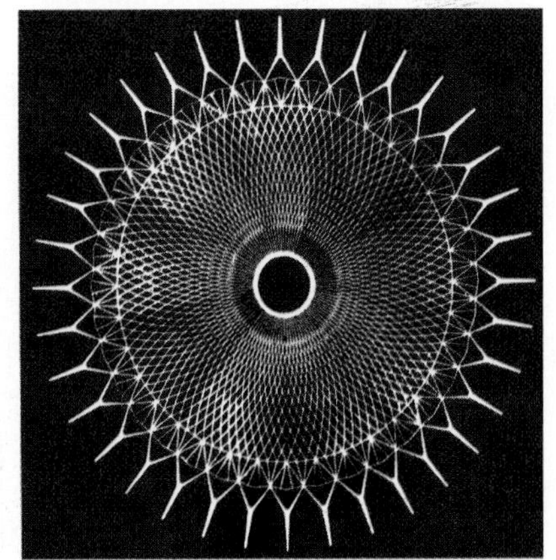

Sunflowers, by Vincent Van Gogh. 1888. Neue Pinakothek, Munich, Germany. Scala/Art Resource, NY; Sunflower; Rome Sports Palace dome design (Pier Paolo Nervi, architect).

Key Concepts

Binet's formula
Fibonacci number
Fibonacci sequence
gnomon
gnomonic growth
golden ratio
golden rectangle
golden triangle
similarity

Exercises

Walking

1. Find F_{15}, F_{16}, F_{17}, and F_{18}.

2. Find F_{19}, F_{20}, F_{21}, F_{22}, and F_{23}.

3. Find the numerical value for each of the following.

(a) F_{10}.

(b) $F_{10} + 2$.

(c) F_{10+2}.

(d) $F_{10} - 8$.

(e) F_{10-8}.

(f) $3F_4$.

(g) $F_{3\times4}$.

4. Describe in words what each of the expressions stands for.

(a) F_{N-2}.

(b) $F_N - 2$.

(c) $F_N + F_{N+1}$.

(d) $2F_N$.

(e) F_{2N}.

5. Given that $F_{36} = 14{,}930{,}352$ and $F_{37} = 24{,}157{,}817$:

(a) Find F_{38}.

(b) Find F_{35}.

6. Given that $F_{31} = 1{,}346{,}269$ and $F_{33} = 3{,}524{,}578$:

(a) Find F_{32}.

(b) Find F_{34}.

7. (a) What is the difference between the following two equations?

$$F_N = F_{N-1} + F_{N-2} \quad \text{and} \quad F_{N+2} = F_{N+1} + F_N.$$

Are they saying the same thing? different things? Explain.

(b) Give two equivalent recursive definitions for the Fibonacci numbers.

Exercise 8 requires the use of a calculator with an exponent key. (On most calculators the exponent key looks something like [y^x]*. To calculate an exponent, say for example $(2.3)^7$, enter first 2.3, then enter* [y^x]*, and finally enter 7 followed by* [=]*.)*

8. (a) Calculate:

(1) $\left(\frac{1+\sqrt{5}}{2}\right)^{10}$.

(2) $\left(\frac{1-\sqrt{5}}{2}\right)^{10}$.

(3) $F_{10} = \frac{\left(\frac{1+\sqrt{5}}{2}\right)^{10} - \left(\frac{1-\sqrt{5}}{2}\right)^{10}}{\sqrt{5}}$.

(b) Compare your answer obtained in part (3) of (a) with the known value $F_{10} = 55$, and explain any discrepancy.

9. Find integers N and M (other than $N = 1, M = 1$) satisfying the equation $F_N = M^2$.

10. Find integers N and M (other than $N = 1, M = 1$) satisfying the equation $F_N = F_M^3$.

11. Write each of the following integers as the sum of *distinct* Fibonacci numbers.

(a) 47.

(b) 48.

(c) 207.

(d) 210.

12. Write each of the following integers as the sum of *distinct* Fibonacci numbers.

(a) 52.

(b) 53.

(c) 107.

(d) 112.

13. Fact: $(F_1 + F_2 + F_3 + \cdots + F_N) + 1 = F_{N+2}$. Verify this fact for:

(a) $N = 4$.

(b) $N = 5$.

(c) $N = 10$.

(d) $N = 11$.

14. Fact: *If we make a list of any 10 consecutive Fibonacci numbers, the sum of all these numbers divided by 11 is always equal to the seventh number in the list.*

(a) Using F_N as the first Fibonacci number on the list, write the above fact as a mathematical equation.

(b) Verify this fact for $N = 5$.

(e) Verify this fact for $N = 6$.

15. Fact: *If we make a list of any four consecutive Fibonacci numbers, twice the third one minus the fourth one is always equal to the first one.*

(a) Using F_N as the first Fibonacci number in the list, write the above fact as a mathematical equation.

(b) Verify this fact for $N = 1$.

(c) Verify this fact for $N = 4$.

(d) Verify this fact for $N = 8$.

16. Fact: *If we make a list of any four consecutive Fibonacci numbers, the first one times the fourth one is always equal to the third one squared minus the second one squared.*

(a) Using F_N as the first Fibonacci number in the list, write the above fact as a mathematical equation.

(b) Verify this fact for $N = 1$.

(c) Verify this fact for $N = 4$.

(d) Verify this fact for $N = 8$.

Exercises 17 through 20 are intended for readers who need a brief review of the quadratic formula. (For a more extensive review readers are encouraged to look at any intermediate algebra textbook.) Any quadratic equation can be solved by first putting it in the form $ax^2 + bx + c = 0$ and then using the ***quadratic formula*** $x = (-b \pm \sqrt{b^2 - 4ac})/2a$. *A calculator with a square-root key is often needed to carry out this calculation.*

17. Use the quadratic formula to find the two solutions of $x^2 = 2x + 1$. Use a calculator to approximate the solutions to three decimal places.

18. Use the quadratic formula to find the two solutions of $x^2 = 3x + 2$. Use a calculator to approximate the solutions to three decimal places.

19. Use the quadratic formula to find the two solutions of $3x^2 = 5x + 8$. Use a calculator to approximate the solutions to five decimal places.

20. Use the quadratic formula to find the two solutions of $5x^2 = 8x + 13$. Use a calculator to approximate the solutions to five decimal places.

21. Use your calculator to find F_{14}/F_{13}, F_{16}/F_{15}, and F_{18}/F_{17}. What are your conclusions?

22. Use your calculator to find F_{20}/F_{19}, F_{21}/F_{20}, F_{22}/F_{21}, and F_{23}/F_{22}. What are your conclusions?

23. True or false? If true, explain. If false, give a counterexample.

(a) The corresponding angles of two similar polygons are equal.

(b) If the angles of one polygon are equal to the angles of another polygon, the polygons are similar.

24. P and P' are similar polygons.

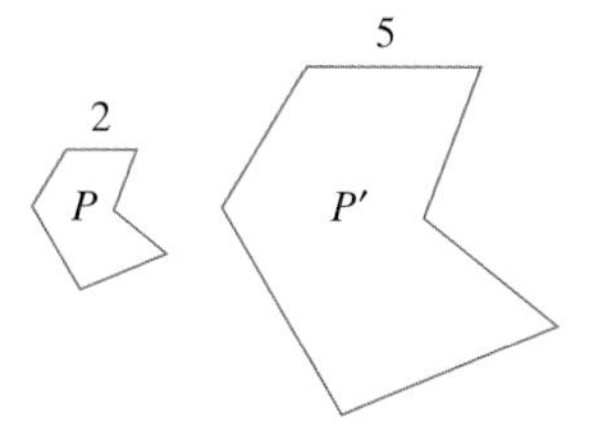

(a) If the perimeter of P is 10, what is the perimeter of P'?

(b) If the area of P is 30, what is the area of P'?

25. Rectangle A is 10 by 20. Rectangle B is a gnomon to rectangle A. What are the dimensions of rectangle B?

26. Rectangle A is 2 by 3. Rectangle B is a gnomon to rectangle A. What are the dimensions of rectangle B?

27. Find the length c of the shaded rectangle so that it is a gnomon to the white rectangle with sides 3 and 9.

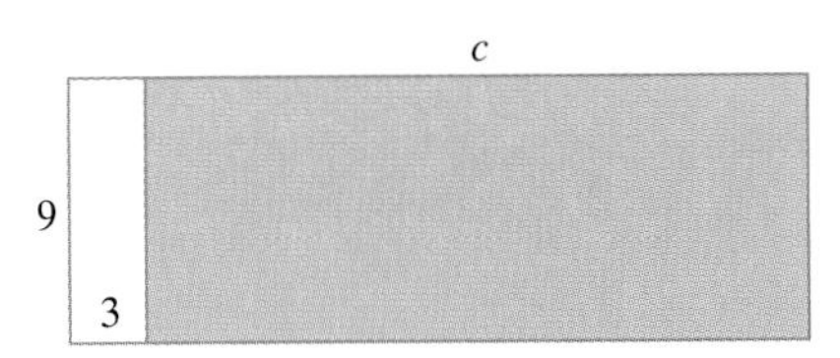

28. Find the value of x so that the shaded figure is a gnomon to the white rectangle.

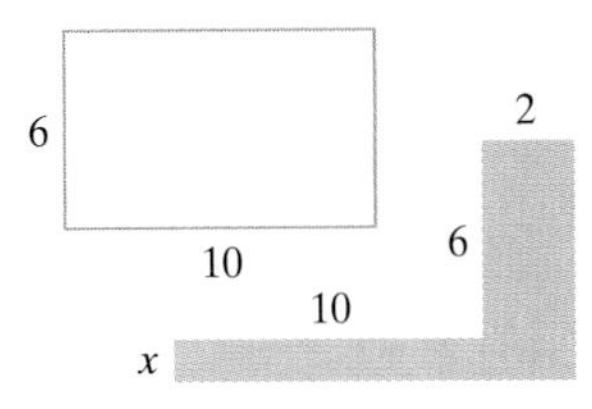

29. Find the value of x so that the shaded "rectangular ring" is a gnomon to the white rectangle.

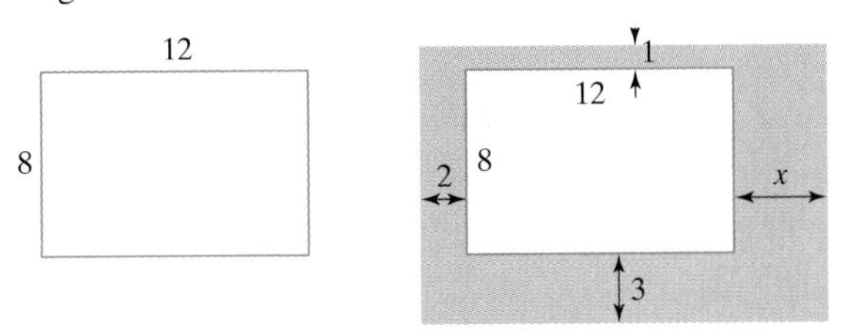

30. Find the value of x so that the shaded "rectangular ring" is a gnomon to the white rectangle.

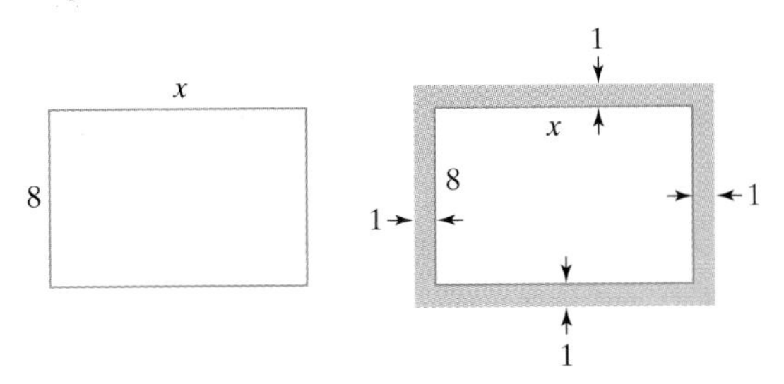

31. Find the values of x and y so that the shaded figure is a gnomon to the white triangle.

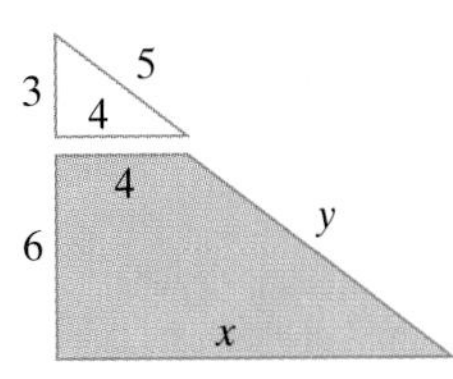

32. Find the values of x and y so that the shaded triangle is a gnomon to the white triangle ABC.

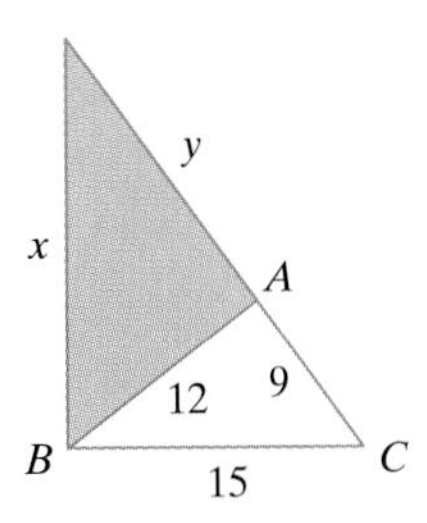

33. A rectangle has a 10 by 10 square gnomon. What are the dimensions of the rectangle?

34. What are the dimensions of a rectangle that is a gnomon to itself. (*Hint:* Label the sides of the rectangle x and 1, as shown in the figure.)

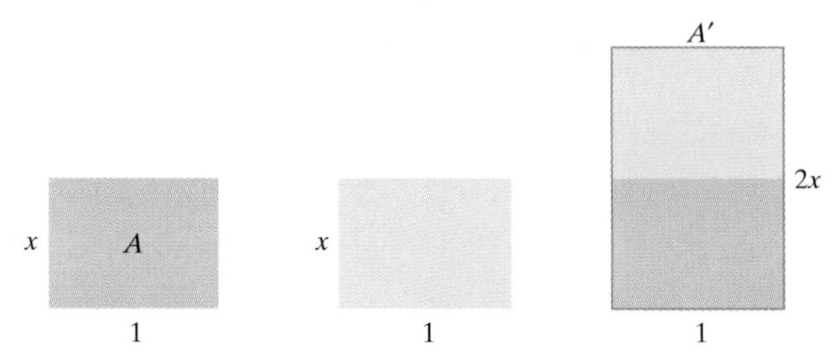

Jogging

35. Explain why the two solutions of the quadratic equation $x^2 - x - 1 = 0$ have exactly the same decimal expansions. (*Hint:* If s_1 and s_2 are the two solutions of the quadratic equation $x^2 + bx + c = 0$, then $s_1 + s_2 = -b$.)

36. Explain why a square (regardless of size) cannot be a gnomon to the L-shaped figure.

37. Find the values of x, y, and z so that the shaded figure has an area eight times the area of the white triangle and at the same time is a gnomon to the white triangle.

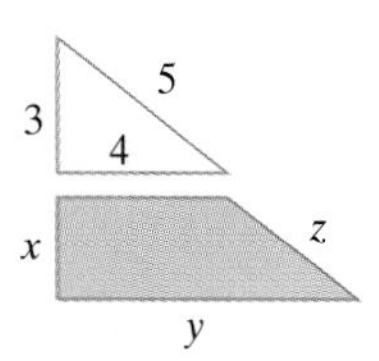

38. **(a)** Which of the following O-rings is similar to I? Explain your answer.

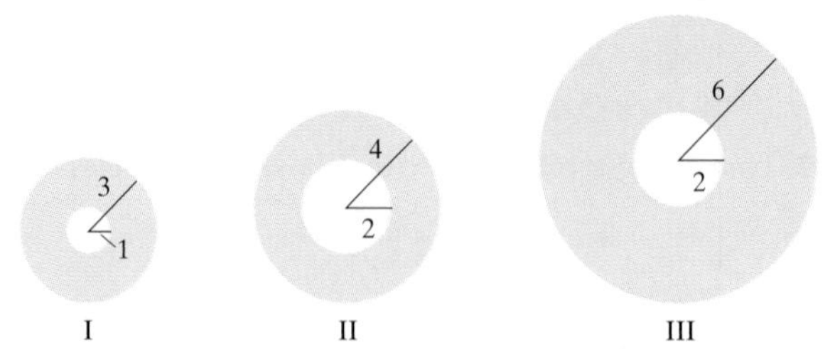

(b) Explain why an O-ring cannot have a gnomon.

39. Find the values of x and y so that the shaded figure has an area of 75 and at the same time is a gnomon to the white rectangle.

6
10
10
6
y
x

40. Under what conditions is a triangle its own gnomon?

41. Suppose you are given that $\Phi^N = a\Phi + b$. Show that $\Phi^{N+1} = (a + b)\,\Phi + a$.

42. In the following figure $ABCD$ is a square and the three triangles I, II, and III have equal areas. Show that x/y is the golden ratio.

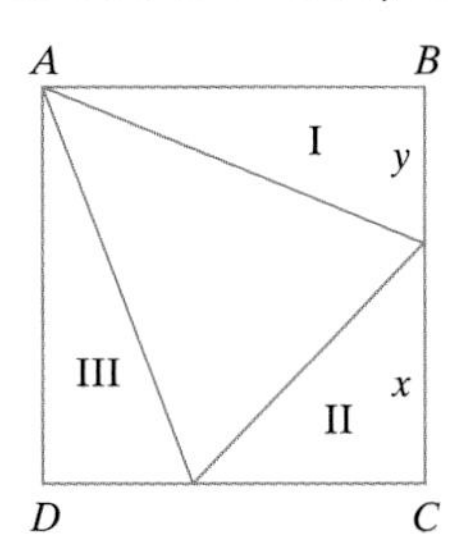

43. Show that the L-shaped object in the following figure is a gnomon for rectangle A as long as the ratios b/h and y/x are equal.

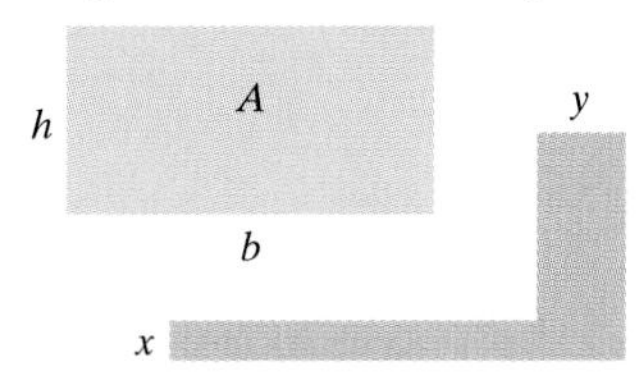

44. A rectangle has a square gnomon. The new rectangle obtained by attaching the square gnomon to the original rectangle has longer leg 20. What are the dimensions of the original rectangle?

45. In the figure, triangle BCD is a 72°-72°-36° triangle with base of length 1 and the longer sides of length x. (Using this choice of values, the ratio of the longer side to the shorter side is $x/1 = x$.)

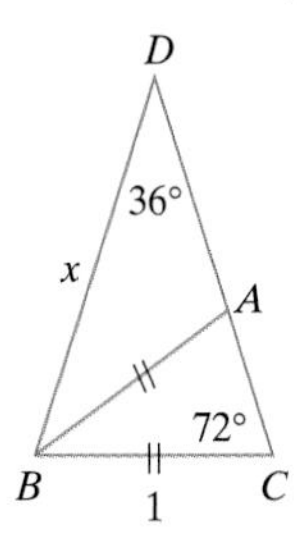

(a) Show that $x = \Phi = \left(\frac{1 + \sqrt{5}}{2}\right)$. (*Hint:* Use the fact that triangle ACB is similar to triangle BCD.)

(b) What are the interior angles of triangle DAB?

(c) Show that in the isosceles triangle DAB, the ratio of the longer to the shorter side is also Φ.

46. Let

$$a = \frac{1+\sqrt{5}}{2} \text{ and } b = \frac{1-\sqrt{5}}{2}.$$

Without using a calculator, expand and simplify:

(a) $a + b$.

(b) ab.

(c) $a^2 + b^2$.

(d) $a^3 + b^3$.

47. Consider the following sequence of numbers: 5, 5, 10, 15, 25, 40, 65, ... If A_N is the Nth term of this sequence, write A_N in terms of F_N.

48. Consider the sequence 1, 3, 4, 7, 11, 18, 29, ... (These are known as Lucas numbers.) If L_N represents the Nth term of this sequence, write L_N in terms of the Fibonacci numbers F_{N+1} and F_N.

49. Suppose that $T_N = aF_{N+1} + bF_N$, where a and b are fixed numbers.

(a) What is T_1?

(b) What is T_2?

(c) Show that $T_N = T_{N-1} + T_{N-2}$.

50. Let $ABCD$ be an arbitrary rectangle as shown in the following figure. Let AE be perpendicular to the diagonal BD and EF perpendicular to AB as shown. Show that the rectangle $BCEF$ is a gnomon to rectangle $ADEF$. (*Hint:* Show that the rectangle $ADEF$ is similar to the rectangle $ABCD$.)

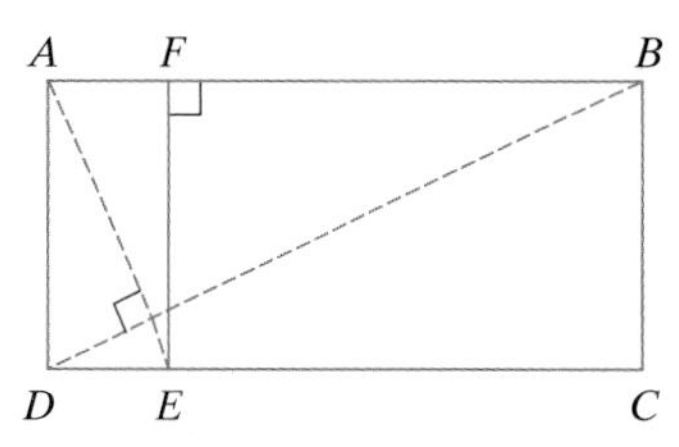

51. The regular pentagon in the following figure has sides of length 1. Show that the length of any one of its diagonals is Φ.

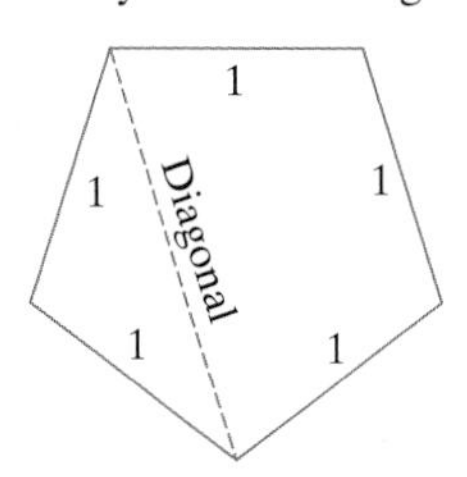

52. (a) A regular decagon (10 sides) is inscribed in a circle of radius 1. Find the perimeter in terms of Φ.

(b) Same as (a) except the radius is r. Find the perimeter in terms of Φ and r.

53. A regular decagon (10 sides) can be inscribed in a circle of radius r using a ruler and a compass as follows:

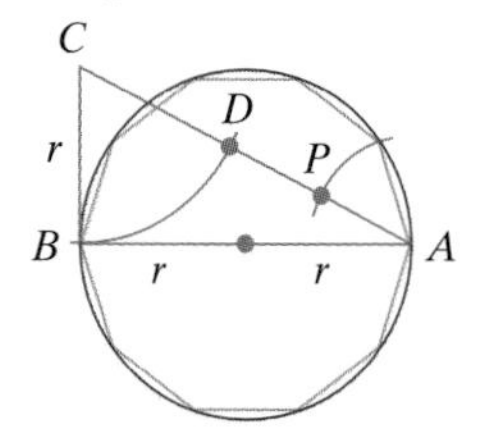

- Construct a right triangle ABC with longer leg AB a diameter of the given circle and shorter leg BC having length equal to the radius of the circle. (See figure.)
- Locate point D on segment CA so that $CD = r$.
- Locate the midpoint P of segment AD.
- The segment AP is the length of a side of the decagon.

Explain why AP is the length of a side of the decagon.

Running

54. Find the values of x, y, and z so that the shaded "triangular ring" is a gnomon to the white triangle.

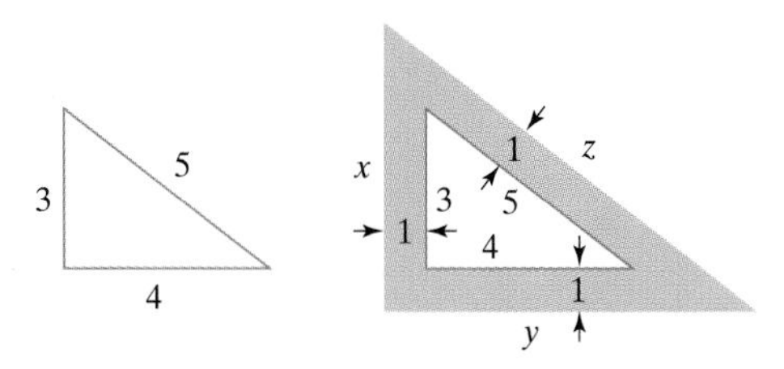

55. Show that $F_1 + F_2 + F_3 + \cdots + F_N = F_{N+2} - 1$.

56. Show that $F_1 + F_3 + F_5 + \cdots + F_N$ (here we are adding the Fibonacci numbers with the odd subscripts up to N) $= F_{N+1}$.

57. Show that every positive integer greater than 2 can be written as the sum of distinct Fibonacci numbers.

58. Consider the following equation relating various terms of the Fibonacci sequence.

$$F_{N+2}^2 - F_{N+1}^2 = F_N \cdot F_{N+3}.$$

Using the algebraic identitiy $A^2 - B^2 = (A - B)(A + B)$, show that the equation is true for every positive integer N.

59. Show that the sum of any 10 consecutive Fibonacci numbers is a multiple of 11.

60. Suppose that T is a "Fibonacci-type" sequence; that is, $T_N = T_{N-1} + T_{N-2}$ but $T_1 = c$ and $T_2 = d$.

(a) Show that there are constants a and b such that $T_N = aF_{N+1} + bF_N$ (where F is the Fibonacci sequence).

(b) Show that $T_{N+1}/T_N \approx \Phi$ when N is large.

61. During the time of the Greeks the star pentagram was a symbol of the Brotherhood of Pythagoras. A typical diagonal of the large outside regular pentagon is broken up into three segments of lengths x, y, and z, as shown in the following figure.

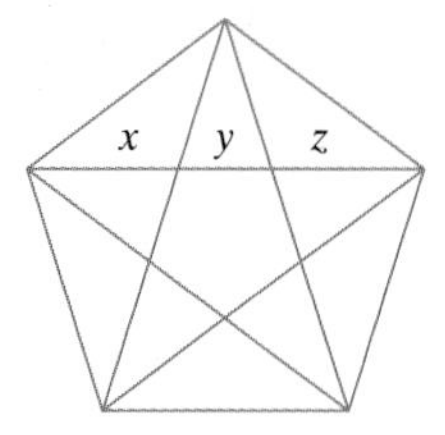

(a) Show that $\dfrac{x}{y} = \Phi$, $\dfrac{x+y}{z} = \Phi$ and $\dfrac{x+y+z}{x+y} = \Phi$.

(b) Show that if $y = 1$, then $x = \Phi$, $x + y = \Phi^2$, and $x + y + z = \Phi^3$.

62. **(a)** A regular pentagon is inscribed in a circle of radius 1. Find the perimeter in terms of Φ.
    (b) Same as (a) except the radius is r. Find the perimeter in terms of Φ and r.
63. **The puzzle of the missing area.** Consider a square 8 units on a side and cut into four pieces as shown in the accompanying figure.

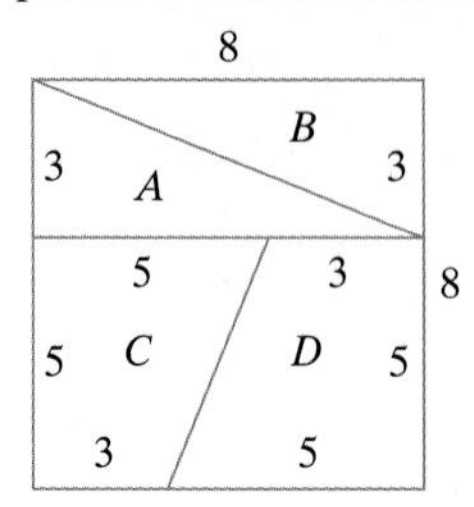

If we rearrange the pieces into a rectangle, as shown in the next figure, we see that although the square has area $8 \times 8 = 64$, the rectangle has area $13 \times 5 = 65$.

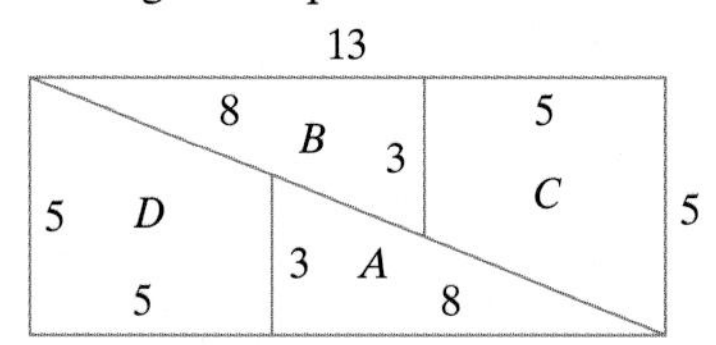

(a) Draw similar figures using other (larger) Fibonacci numbers. How does the area of the square compare with that of the rectangle?

(b) Explain the discrepancies in the areas.

(c) Consider the following figures. What are the conditions on a and b so that this puzzle is not a puzzle—that is, the areas are the same?

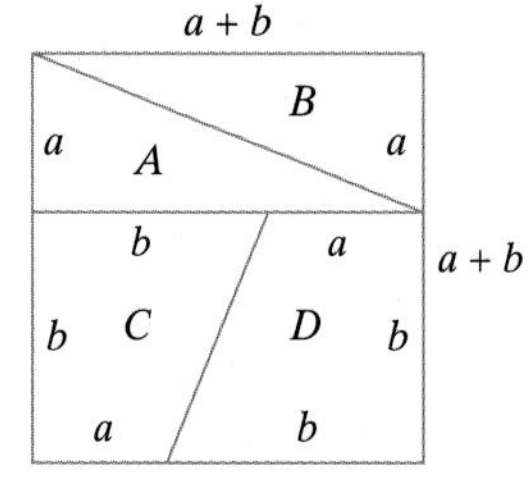

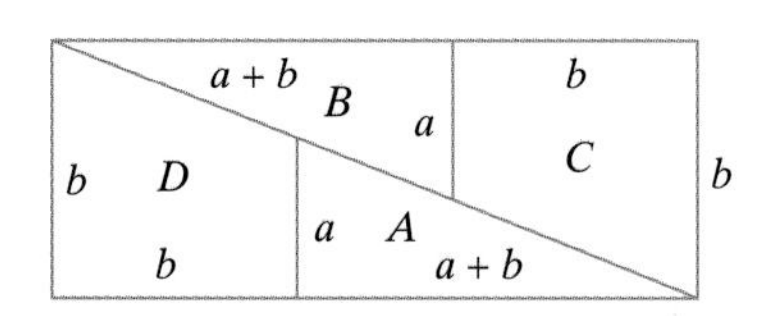

References and Further Readings

1. Coxeter, H. S. M., "The Golden Section, Phyllotaxis, and Wythoff's Game, *Scripta Mathematica,* 19, 1953, 129–133.
2. Coxeter, H. S. M., *Introduction to Geometry.* New York: John Wiley & Sons, Inc., 1961, chap. 11.
3. Douady, S., and Y. Couder, "Phyllotaxis as a Self-Organized Growth Process," in *Growth Patterns in Physical Sciences and Biology,* eds. J. M. Garcia-Ruiz et al. New York: Plenum Press, 1983.
4. Erickson, R. O., "The Geometry of Phyllotaxis," in *The Growth and Functioning of Leaves,* eds. J. E. Dale and F. L. Milthrope. New York: Cambridge University Press, 1983.
5. Gardner, Martin, "About Phi, an Irrational Number That Has Some Remarkable Geometrical Expressions," *Scientific American,* 201 (August 1959), 128–134.
6. Gardner, Martin, "The Multiple Fascinations of the Fibonacci Sequence," *Scientific American,* 220 (March 1969), 116–120.

7. Jean, R. V., *Mathematical Approach to Pattern Form in Plant Growth.* New York: John Wiley & Sons, Inc., 1984.
8. Kappraff, J., *Connections: The Geometric Bridge Between Art and Science.* New York: McGraw-Hill Book Company, 1991.
9. Prusinkiewicz, P., and A. Lindenmayer, *The Algorithmic Beauty of Plants.* New York: Springer-Verlag, Inc., 1990, chap. 4.
10. Stewart, Ian, "Daisy, Daisy, Give Me Your Answer, Do," *Scientific American,* January 1995, 96–99.
11. Thompson, D'Arcy, *On Growth and Form.* New York: Macmillan Publishing Co., Inc., 1942, chaps. 11, 13, and 14.
12. Zusne, Leonard, *Visual Perception of Form.* New York: Academic Press, 1970.

. . . and you, be ye
fruitful and multiply.
Genesis 9:7

The Mathematics of Population Growth

There Is Strength in Numbers

CHAPTER 10

In Chapter 9 we discussed the concept of *growth* when applied to a single organism. In this chapter we will discuss the concept of *growth* as it applies to entire populations. The connection between the study of populations and mathematics goes back to the very beginnings of civilization. One of the main reasons humans invented a numbering system was their need to handle the rudiments of counting populations—how many sheep in the flock, how many people in the tribe, and so on. By biblical times, simple models of population growth were being used to measure crop production and even to estimate the yields of future crops. Today, mathematical models of population growth are a fundamental tool in our efforts to understand the rise and fall of endangered wildlife populations, fishery stocks, agricultural pests (such as locusts, cicadas, boll weevils), infectious diseases, radioactive waste, ordinary trash, and so on. Entire modern disciplines, such as *mathematical ecology, population biology,* and *biostatistics* are built around the mathematics of population growth.

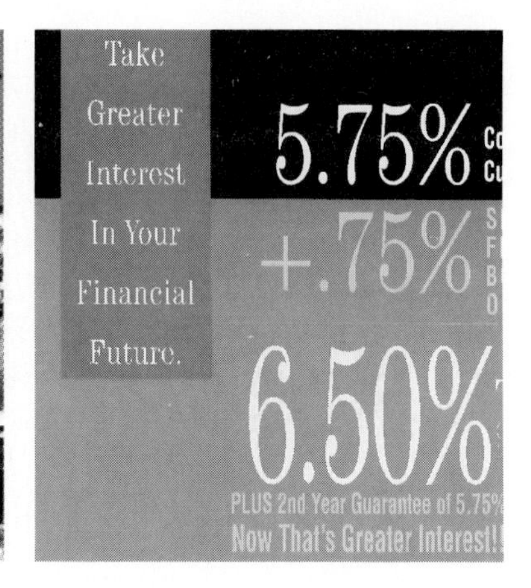

As the role of mathematics in the study of population growth as expanded, so has its complexity, and the mathematical tools in use today to study populations can be quite sophisticated. The overall mathematical principles involved, however, are reasonably simple (the devil is always in the details!) This chapter touches upon the basic principles behind the mathematics of population growth and presents some of the simpler models that can be used in studying its *dynamics*.

Before we proceed, let us clarify some terminology. In its modern usage, the term "population growth" has become very broad and all encompassing, owing primarily to the broad meaning given nowadays to both "population" and "growth." The Latin root of *population* is *populus* (which means "people"), so that in its original interpretation the word refers to human populations. Over time, this scope has been expanded to include any collection of objects (animate or inanimate) about which we want to make a numerical or quantitative statement. Thus, we can speak of a population of penguins, of tires, of bacteria, of dollars and cents, and, needless to say, of people.

Second, we normally think of the word "growth" as being applied to things that get bigger, but in this chapter we will ascribe a slightly more technical meaning to it: "Growth" can mean *negative growth or decay* (i.e., getting smaller) as well as *positive growth* (i.e., getting bigger). This is convenient, because often we don't know ahead of time which way a population is going to go: Is it going to go up or down? By allowing "growth" to mean either, we need not concern ourselves with making the distinction.

The Dynamics of Population Growth

The growth of a population is a **dynamical process**, meaning that it represents a situation that changes over time. Mathematicians distinguish two kinds of situations: continuous growth and discrete growth. In **continuous growth** the dynamics of change are in effect all the time—every hour, every minute, every second, there is change. The classic example of this kind of growth is represented by money left in an account that is drawing interest on a continuous basis (yes, there are banks that offer such accounts). We will not study continuous growth in this chapter—the mathematics involved (calculus) would take us beyond the scope of this book.

The second type of growth, **discrete growth**, is the most common and natural way by which populations change. We can think of it as a *stop-and-go* type of situation: For a while nothing happens, then there is a sudden change in the population (we will call this a **transition**), then for a while nothing happens again, then another transition takes place, and so on. Of course the "for a while" (i.e., the period between transitions) can be 100 years, an hour, a second, or a nanosecond. To us the length of time between transitions will not make a difference. The human population of our planet is an example of what we mean: Nothing happens until someone is born or someone dies, at which point there is a change (+ 1 or − 1); then, again there is no change until the next birth or death. Since, however, someone is born every fraction of a second and someone dies slightly less often, it is somewhat tempting to think that the world's human population is for all practical purposes changing in a continuous way. On the other hand, the laws of growth affecting the world's population may be only quantitatively different from the laws affecting the population of Hinsdale County, Colorado (population 408[1]), where a change in the population may not come about for months or even years.

The basic problem of population growth is to figure out what happens to a given population over time. Sometimes we talk about a specific period of time ("The Hispanic population of the United States will grow by 30% before the end of the century"[2]), and sometimes we may talk about the long-term behavior of the population ("The black rhino population is heading for extinction")[3]. In either case, the most basic way to deal with the question of growth of a particular population is to find the rules that govern the transitions. We will call these the **transition rules**. After all, if we have a way to figure out how the population changes each time there is a transition, then (with a little help from mathematics) we can usually figure out how the population changes after many transitions. In this sense, the ebb and flow of a particular population over time can be conveniently thought of as an unending list of numbers called the **population sequence**. Figure 10-1 is a schematic illustration of how a population sequence can be generated.

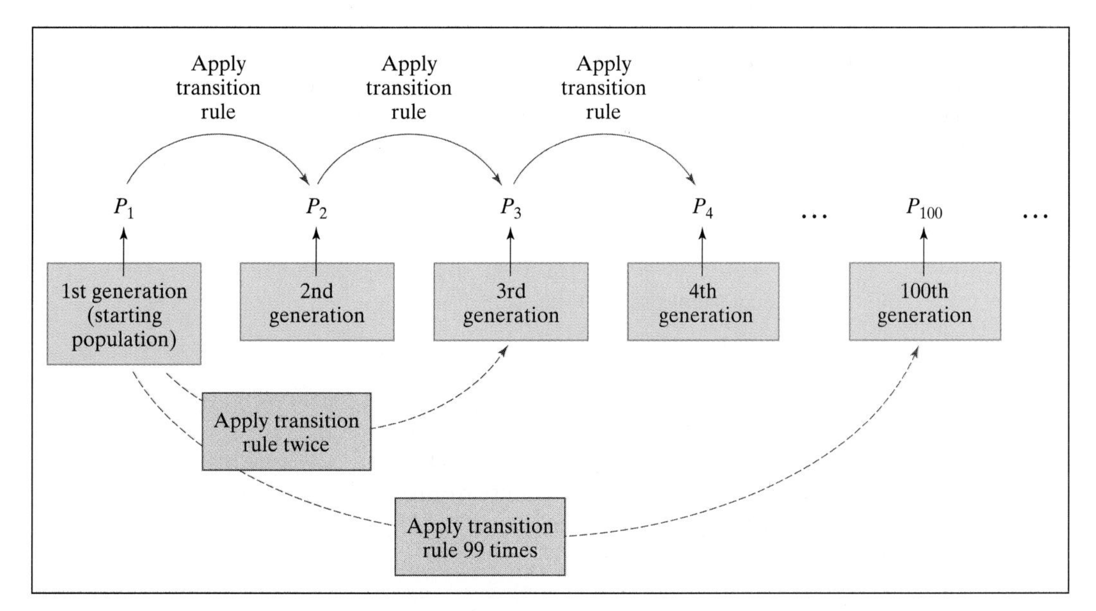

FIGURE 10-1
A generic population sequence. P_N is the population size in the Nth generation.

[1] *Source: The World Almanac and Book of Facts,* 1990.
[2] U.S. Bureau of the Census, *Statistical Abstract of the United States: 1990,* 110th edition, Washington DC, 1990, 14.
[3] *New York Times,* May 7, 1991, B5.

Just to get our feet wet, we will start with one of the oldest and best-known examples of a problem in population growth.

Example 1. Fibonacci Rabbits

A Fibonacci rabbit is a very unusual breed of bunny. Like clockwork, *at the end of each month a mature male-female couple produces as offspring a male-female pair of babies.* It takes one month for a newborn baby rabbit to become a mature rabbit and start producing offspring of its own.

In his famous book *Liber Abaci* (see footnote 1, Chapter 9), Fibonacci raised the following question: If you start with a single pair of male-female newborn Fibonacci rabbits, how many rabbits will there be at the end of the year?

For the sake of convenience, we will count rabbits in male-female pairs, and, following the notation we have just adopted, we will let P_1, P_2, P_3, etc. represent the number of pairs in the first, second, third, etc. generations. Figure 10-2 illustrates the pattern of growth for the first six months.

	First generation	Second generation	Third generation	Fourth generation	Fifth generation	Sixth generation	Seventh generation
Elapsed time	0	1 month	2 months	3 months	4 months	5 months	6 months
Number of baby pairs	1	0	1	1	2	3	5
Number of mature pairs	0	1	1	2	3	5	8
Total number of pairs	$P_1 = 1$	$P_2 = 1$	$P_3 = 2$	$P_4 = 3$	$P_5 = 5$	$P_6 = 8$	$P_7 = 13$

FIGURE 10-2
A simple model of population growth: Fibonacci rabbits. P_N is the number of male-female couples in the Nth generation.

We can see from Fig. 10-2 that $P_1 = 1$; $P_2 = 1$; $P_3 = P_2 + P_1$; $P_4 = P_3 + P_2$; and so on. Thus, in each generation, the total number of pairs is the corresponding Fibonacci number (see Chapter 9), i.e.,

$$P_N = F_N.$$

It follows that at the end of the year (13th generation) we will have $P_{13} = F_{13} = 233$ pairs of rabbits—a grand total of 466 rabbits!

As long as all the rabbits stay alive and continue breeding according to our rules, we can describe the generic transition rule for passing from one generation to the next by the equation

$$\underbrace{P_N}_{\text{number of pairs in the } N\text{th generation}} = \underbrace{P_{N-1}}_{\text{total number of pairs in the preceding generation; same as the number of mature pairs in the } N\text{th generation}} + \underbrace{P_{N-2}}_{\text{total number of pairs in the } (N-2)\text{nd generation; same as the number of baby pairs in the } N\text{th generation}}$$

■

Because Fibonacci bunnies are so methodical and precise, and because they keep breeding and breeding and breeding, people tend to confuse them with the Energizer bunnies, but one must keep in mind that there were Fibonacci bunnies long before there were Energizer bunnies. And before Fibonacci rabbits there were real rabbits, which are not as accommodating as Fibonacci rabbits—they live and breed by considerably more complicated rules, which we could never hope to capture in a simple equation. And what's true about rabbits is generally true about most other types of populations.

Is there any use then, for simplistic mathematical models of how populations grow? The answer is yes! We can make excellent predictions about the growth of a population over time, even when we don't have a completely realistic set of transition rules. The secret is to capture the variables that are really influential in determining how the population grows, put them into a few transition rules that describe how the variables interact, and forget about the small things. This, of course, is easier said than done. In essence, it is what population biologists and mathematical ecologists do for a living, and it is as much an art as it is a science.

In the rest of this chapter we will discuss three of the most basic models of population growth: the *linear growth model,* the *exponential growth model,* and the *logistic growth model.*

The Linear Growth Model

The linear growth model is the simplest of all models of population growth. In this model, in each generation *the population increases (or decreases) by a fixed amount.* The easiest way to see how the model works is with an example.

Example 2.

The city of Cleansburg is considering a new law that would restrict the monthly amount of garbage allowed to be dumped in the local landfill to a maximum of 120 tons a month, which is way below what the landfill has been taking in. There is concern among local officials that, unless restrictions on dumping are imposed, the landfill will reach its maximum capacity of 20,000 tons in a few years. Currently, there are 8000 tons of garbage already in the landfill. Assuming that the law is passed and the landfill collects exactly 120 tons of garbage each month, how much garbage will there be in the landfill 5 years from now? How long before the landfill reaches its 20,000-ton capacity?

While the circumstances are fictitious, the questions we raise are realistic and important. The *population* in this example is the garbage in the landfill, and since we only care about monthly totals, we define the transitions as happening once a month. (This is a convenient way to look at it on paper; in reality, a transition occurs every time a garbage truck dumps a load, but for our purposes that's just a stinking detail.) The essential fact about this population-growth problem is that the monthly garbage at the landfill grows by a constant amount of 120 tons a month.

Our starting population (P_1) is 8000 tons. Thus, we have the following population sequence:

$$P_1 = 8000;\ P_2 = 8120;\ P_3 = 8240;\ P_4 = 8360;\ \ldots.$$

In 5 years we will have had 60 transitions, each representing an increase of 120 tons. The population after 5 years is given by the 61st term in the population sequence, which is obtained by adding 60 transitions of 120 tons each to the existing 8000 tons in the landfill—in other words,

$$P_{61} = 8000 + 60(120) = 15{,}200 \text{ tons}.$$

To find out how many months it would take for the landfill to reach its 20,000-ton maximum, we set up the equation

$$8000 + 120x = 20{,}000 \text{ tons},$$

which has as a solution $x = 100$. This means that it will take 100 months (8 years and 4 months) for the landfill to reach its maximum capacity. Based on this information, local officials should start making plans soon for a new landfill. ■

Example 2 is a typical example of the general linear growth model, whose basic characteristic is that in each transition a constant amount—call it d—is added to the previous population. Mathematically, the general linear growth model can be described as follows:

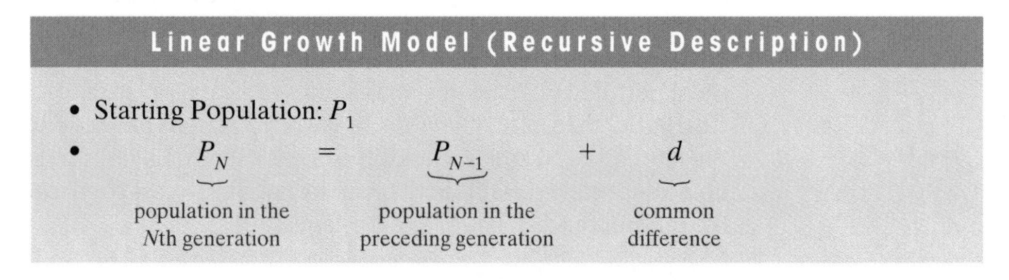

The population sequence that results from a linear growth model is commonly known as an **arithmetic sequence**. Technically speaking, the arithmetic sequence is just the numerical description of a population which is growing according to a linear growth model—informally, linear growth and arithmetic sequences can be considered synonymous. The number d is called the **common difference** for the arithmetic sequence, because any two consecutive values of the arithmetic sequence will always differ by the amount d.

The equation $P_N = P_{N-1} + d$ gives a **recursive description** of the population sequence because it calculates values of the population sequence using earlier values of that sequence. While recursive descriptions tend to be nice and tidy, they have one major drawback: To calculate one value in the population sequence we essentially have to first calculate all the earlier values. As we learned in Chapter 9 with regard to the Fibonacci numbers, this can be quite an inconvenience.

Fortunately, in the present case there is a very convenient way to describe the population sequence which does not require the use of other values in the sequence:

Linear Growth Model (Explicit Description)

$$P_N = P_1 + (N - 1) \times d.$$

The equation follows from the fact that to get to the Nth term in the sequence we need to go through $N-1$ transitions, each of which consists of adding d (Fig. 10-3).

The equation $P_N = P_1 + (N - 1) \times d$ gives an **explicit description** of the population sequence, because it allows one to calculate any value of the sequence explicitly, without having to know any of the preceding values except for the starting population P_1.

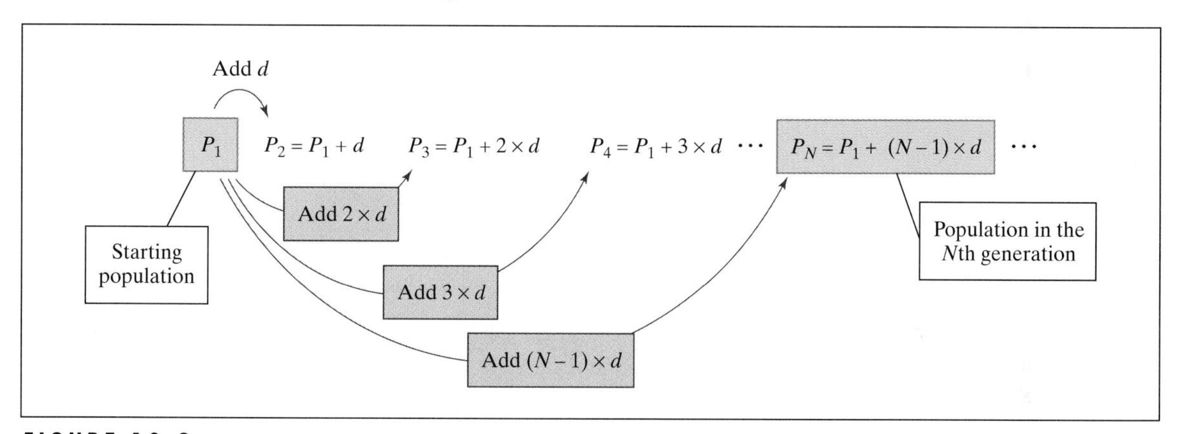

FIGURE 10-3
The linear growth model.

Example 3.

A population grows according to a linear growth model. The starting population is $P_1 = 37$ and the common difference is $d = 6$. (a) What is the population in the 16th generation? (b) What is the population after 25 transitions?

Question (a) asks for P_{16}. Using the explicit description of linear growth, we immediately find

$$\underbrace{P_{16}}_{N} = \underbrace{37}_{P_1} + \underbrace{15}_{N-1} \times \underbrace{6}_{d} = 127$$

Question (b) asks for P_{26}, the population after 25 transitions:

$$P_{26} = 37 + 25 \times 6 = 187.$$

■

Plotting Population Growth

A very convenient way to describe population growth is by means of a **plot** or **graph**. The horizontal axis usually represents time (with the tick marks generally corresponding to the transitions), and the vertical axis usually represents the size of the population. Because we have complete freedom in choosing both the horizontal and vertical scales, plots can be misleading. Consider Figs. 10-4(a) and (b). Which population sequence is growing faster?

Actually, both plots represent the growth of the same population: the garbage problem in Example 2. These plots illustrate why linear growth is called linear growth—no matter how we plot it, the values of the population line up in a straight line.

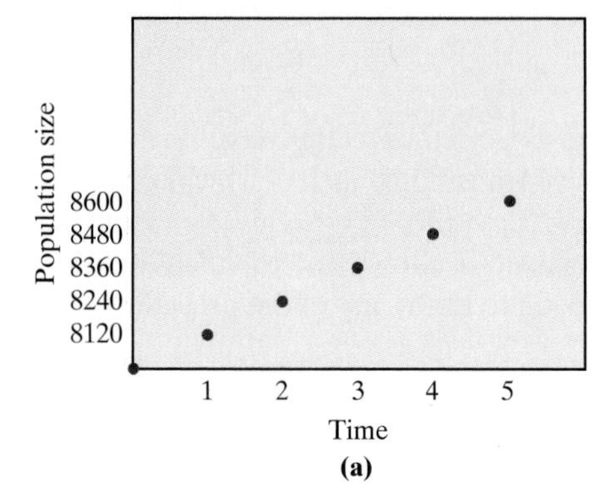

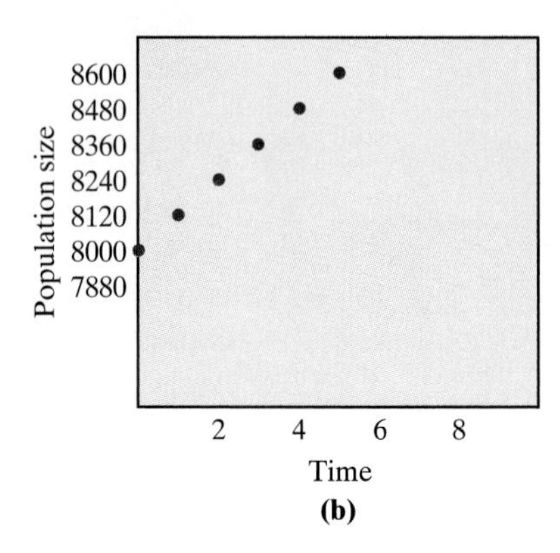

FIGURE 10-4 Which population sequence is growing faster?

Adding Terms of an Arithmetic Sequence

Example 4.

Jane Doe is a company that manufactures tractors. The company has decided to start up a new plant. On the first of each month (for a period of 6 years) a module of equipment that will produce 3 tractors per month will be installed in this plant. What is the total number of tractors that the company will produce over the 6 years the program is in effect?

In this problem, the total production of each module conforms to a linear growth pattern, but the number of months each module works is different. Let's make a list:

- Module installed the 1st month works for 72 months, producing $3 \times 72 = 216$ tractors.
- Module installed the 2nd month works for 71 months, producing $3 \times 71 = 213$ tractors.
- Module installed the 3rd month works for 70 months, producing $3 \times 70 = 210$ tractors.

.

.

.

- Module installed the 72nd month works for 1 month, producing 3 tractors.

The total number of tractors produced at the end of 72 months is

$$216 + 213 + 210 + \cdots + 3.$$

This sum is the sum of consecutive terms of the arithmetic sequence 3, 6, 9, . . . , 213, 216. We could, of course, add these numbers up, with or without a calculator, but that's dull. Let's take a slightly more elegant tack. Let's write our total twice, once forward and once backward:

$$\text{Total} = 216 + 213 + 210 + \cdots + 6 + 3$$

$$\text{Total} = 3 + 6 + 9 + \cdots + 213 + 216$$

If we add each term in the first row to each term in the second row, we get

$$2 \times \text{Total} = 219 + 219 + 219 + \cdots + 219 + 219.$$

Since there are 72 such terms, we end up with

$$2 \times \text{Total} = 219 \times 72,$$

and therefore

$$\text{Total} = \frac{219 \times 72}{2} = 7884.$$ ■

The approach in Example 4 works with *any* arithmetic sequence: We can add up any number of consecutive terms in a very convenient way. The formula is:

Adding *N* Consecutive Terms of an Arithmetic Sequence

$$\text{First term} + \cdots + \text{last term} = \frac{(\text{first term} + \text{last term}) \times N}{2}.$$

Example 5.

$$\underbrace{5 + 12 + 19 + 26 + 33 + \cdots}_{132 \text{ terms}} = ?$$

Here we are adding 132 consecutive terms of an arithmetic sequence. The first term is $P_1 = 5$; the common difference is $d = 7$. We need to find the 132nd term P_{132}. We already know how to do this: $P_{132} = 5 + 131 \times 7 = 922$. We can now apply the formula:

$$5 + 12 + 19 + 26 + 33 + \cdots + 922 = \frac{(5 + 922) \times 132}{2} = 61{,}182.$$ ■

Example 6.

$$4 + 13 + 22 + 31 + 40 + \cdots + 922 = ?$$

Here we are adding the terms of an arithmetic sequence with $P_1 = 4$ and common difference $d = 9$. To apply the formula we need to first find the number of terms N. To find N we set up an equation: $922 = 4 + 9(N-1)$. From it we get $9(N-1) = 918$, and therefore $N-1 = 102$ and $N = 103$. It follows that

$$4 + 13 + 22 + 31 + 40 + \cdots + 922 = \frac{(4 + 922) \times 103}{2} = 47{,}689.$$ ■

The Exponential Growth Model

The exponential growth model is another basic model of population growth. The main characteristic of this model is that in each transition, the population changes by a *fixed proportion.*

Before we start our discussion of exponential growth in earnest, let's develop some background. The next two examples have to do with the use of percentages to calculate increases and decreases.

Example 7.

A firm manufactures an item at a cost of C dollars. The item is marked up 10% and sold to a distributor. The distributor then marks the item up 20% (based on the price he paid) and sells the item to a retailer. The retailer marks the price up 50% and sells the item to the public. By what percent has the item been marked up over its original cost?

- Original cost of item: C.
- Price to distributor after 10% markup (D): $D = 110\%$ of $C = (1.1)C$.
- Price to retailer after 20% markup (R): $R = 120\%$ of $D = (1.2)D = (1.2)(1.1)C = (1.32)C$.
- Price to the public after 50% markup (P): $P = 150\%$ of $R = (1.5)R = (1.5)(1.32)C = (1.98)C$.

Therefore, the markup over the original cost is 98%. ■

Example 8.

A retailer buys an item for C dollars and marks it up 80%. He then puts the item on sale for 40% off the marked price. What is the net percentage markup on this item?

- Original cost of item: C.
- Price after 80% markup (P): $P = 180\%$ of $C = (1.8)C$.
- Sale price after 40% discount (S): $S = 60\%$ of $P = (0.6)P = (0.6)(1.8)C = (1.08)C$.

The net markup is 8%. ■

The main point of Examples 7 and 8 is the following: Increasing a number C by $x\%$ is equivalent to multiplying C by the quantity $(1 + x/100)$—the $x/100$ represents $x\%$ in decimal form; the 1 represents the fact that we are increasing the original number C.

Let's return now to the exponential growth model.

Example 9.

The sum of \$1000 is deposited in a retirement account that pays 10% *annual* interest (i.e., interest is paid once a year at the end of the year). How much money is there in the account after 25 years, if the interest is left in the account?

Table 10-1 will help us get started.

Table 10-1

	Account Balance at Beginning of Year	Interest Earned for the Year	Account Balance at End of Year
Year 1	\$1000	\$100	\$1100
Year 2	1100	110	1210
Year 3	1210	121	1331
.	.	.	.
.	.	.	.
.	.	.	.
Year 24	?	?	?
Year 25	?	?	?

The critical observation is that the account balance at the end of the year 1 is obtained by adding the *principal* (\$1000) and the interest earned for the year (10% of \$1000), which is the same as taking 110% of \$1000—in other words, $\$1000 \times 1.1$. Repeating the argument for year 2, the account balance at the end of year 2 is

$$\underbrace{(\text{Account balance at beginning of year 2})}_{\$1000 \times (1.1)} \times (1.1) = \$1000 \times (1.1)^2.$$

Likewise, the account balance after 25 years (in other words, at the start of year 26) is

$$\$1000 \times (1.1)^{25} = \$10{,}834.71.$$

It isn't hard to see what's happening: Each transition (which occurs at the end of the year) corresponds to taking 110% of the balance at the start of that year, which is the same as multiplying the balance at the start of the year by 1.1. ■

We can now give a general rule describing the balance in the account in Example 9: at the beginning of the $(N + 1)$st year the balance in the account is

$$P_{N+1} = \$1000 \times (1.1)^N.$$

Figure 10-5 plots the growth of the money in the account for the first 8 years.

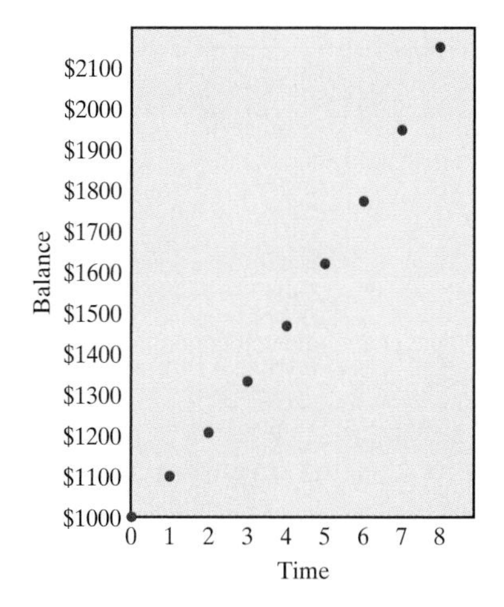

FIGURE 10-5 Plot for Example 9: Cumulative growth of \$1000 at 10% interest compounded annually.

Example 9 is a classic example of exponential growth: The money draws interest; then the money plus the interest draw interest; and so on. While the most familiar examples of exponential growth have to do with the growth of money, the exponential growth model is useful in the study of biological populations as well. The essence of exponential growth is *repeated multiplication:* each transition consists of multiplying the size of the population by a constant factor. In Example 9, the constant factor is 1.1.

A sequence defined by this property—that every term in the sequence after the first is obtained by multiplying the preceding term by a fixed amount r—is called a **geometric sequence**. The constant factor r is called the **common ratio** of the geometric sequence (it is the ratio of two successive terms in the sequence). (To insure that the population sequence does not have negative numbers, we will restrict the values of the common ratio r to positive numbers, although no such restriction is necessary when dealing with geometric sequences in general.)

The general exponential growth model can be described recursively by

Exponential Growth Model (recursive description)

$$P_N = P_{N-1} \times r \quad (r > 0).$$

As in the case of arithmetic sequences, we can also define the terms of the sequence using an *explicit description:*

Exponential Growth Model (explicit description)

$$P_N = P_1 \times r^{N-1}.$$

A common misconception is that exponential growth implies that the population always gets bigger. This need not be the case.

Example 10.

A population grows according to an exponential growth model with common ratio $r = 0.3$, starting population $P_1 = 1{,}000{,}000$, and transition periods of 1 year. What is the size of the population at the end of 6 years? (Remember, this means we want to find P_7!)

In this case, we have $P_7 = 1{,}000{,}000 \times (0.3)^6 = 726$. Figure 10-6 plots the

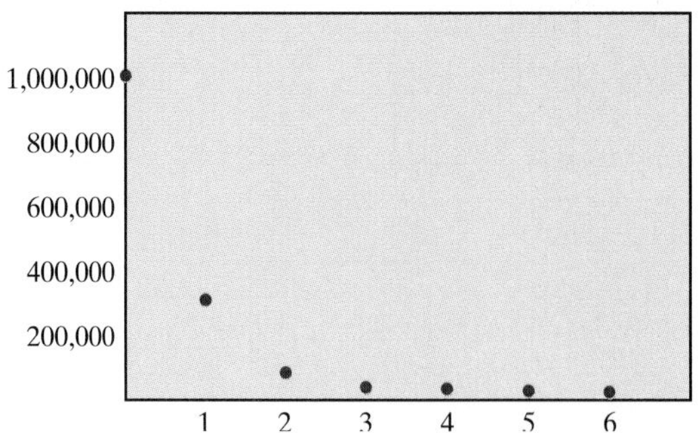

FIGURE 10-6 Plot for Example 10. Exponential "growth" with $r = 0.3$.

"growth" of this population for the first 6 years, and we can clearly see that it is heading toward extinction. ■

It is often convenient to distinguish between exponential growth situations in which populations get bigger (as in Example 9) and those in which populations actually decrease (as in Example 10). The latter situation is commonly referred to as **exponential decay**. The difference between growth and decay is in the value of the common difference r: *For* $r < 1$*, we have decay; for* $r > 1$*, we have actual growth (for* $r = 1$ *we have a constant population).*

Putting Your Money Where Your Math Is

Let's discuss now a general version of Example 9: A certain sum of money P_1 (called the *principal*) is deposited in an account that draws interest at an *annual* interest rate i (the interest is paid once a year at the end of the year). How much money is in the account at the end of N years, assuming that principal and interest are left in the account to accumulate?

We know now that we are dealing with a geometric population sequence whose terms are given explicitly by the formula

$$P_{N+1} = P_1 \times r^N.$$

How do we find the common ratio r? When the annual interest was 10% (Example 9), we got the common ratio 1.1 (110%). If the annual interest had been 12%, the common ratio would have been 1.12 (112%), and if the annual interest had been $6\frac{3}{4}$%, the common ratio would have been 1.0675 (106.75%). In general, if we write the annual interest rate as a decimal i (rather than a percent) then the common ratio r will be $1 + i$. Replacing r with $(1 + i)$ in the previous formula gives the general rule for money that is compounded annually.

principal ↘ annual interest ↘ number of years ↙

$$P_{N+1} = P_1 \times (1 + i)^N.$$

Example 11.

Suppose you deposit \$367.51 in a savings account yielding an annual interest of $9\frac{1}{2}$% a year, and you leave both the principal and the interest in the account for a full 7 years. How much money will there be in the account at the end of the 7 years? Here $P_1 = 367.51$; $i = 0.095$; and $r = 1.095$. The answer is

$$P_8 = 367.51 \times (1.095)^7 = 693.69.$$

■

Example 12.

Let's now consider a variation of Example 9. Suppose we find a bank that pays 10% *annual interest with the interest compounded monthly.* If we deposit \$1000 (and leave the interest in the account), how much money will there be in the account at the end of 5 years?

This problem is still one in exponential growth. The big difference now is that the period between transitions is a month (instead of a year). At the end of 5 years we will have gone through 60 transitions, so in this example we are after P_{61}. Since the population sequence is a geometric sequence, we have

$$P_{61} = \$1000 \times r^{60},$$

and, just as before, it all boils down to finding the value of r. Since the interest rate of 10% is *annual* but the transitions occur *monthly,* we must divide the annual 10% interest rate by 12, which gives the **periodic interest rate *p***

$$p = \frac{10\%}{12} = 0.0083333\ldots$$

The common ratio is then $r = 1.0083333\ldots$, and therefore

$$P_{61} = \$1000 \times (1.0083333\ldots)^{60} \approx \$1645.31.$$

Just out of curiosity, what would happen if we left the money in the account for 25 years? Everything is the same as in our last computation, except that the number of transitions is $25 \times 12 = 300$. It follows that after 25 years the amount of money in the account will be

$$P_{301} = \$1000 \times (1.0083333\ldots)^{300} \approx \$12{,}056.95.$$ ■

Example 13.

Now suppose we find a bank that pays 10% annual interest compounded *daily.* If we deposit \$1000 for 5 years (just as in Example 12), how much will we have at the end of the 5 years?

In this case the period between transitions is 1 day. The total number of transitions in 5 years is $365 \times 5 = 1825$, and the periodic interest rate is $p = 0.10/365 \approx 0.00027397$. The value of r for this exponential growth problem is

$$r = \left(1 + \frac{0.10}{365}\right) \approx 1.00027397,$$

so that the final answer is

$$1000\left(1 + \frac{0.10}{365}\right)^{365 \times 5} \approx 1000(1.00027397)^{1825} \approx 1648.61.$$ ■

Examples 12 and 13 illustrate the general rule for growth under compound interest, which is shown in Fig. 10-7.

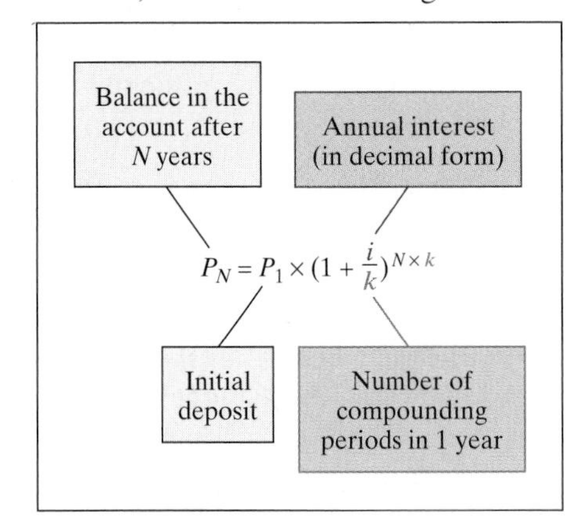

FIGURE 10.7 Periodic compounding rule.

Example 14. Shopping for a bank

You have an undisclosed amount of money to invest. Bank A offers savings accounts that pay 10% annual interest *compounded yearly.* Bank B offers accounts that pay 9.75% annual interest *compounded monthly.* Bank C offers accounts that pay 9.5% annual interest *compounded daily.* Which bank offers the best deal?

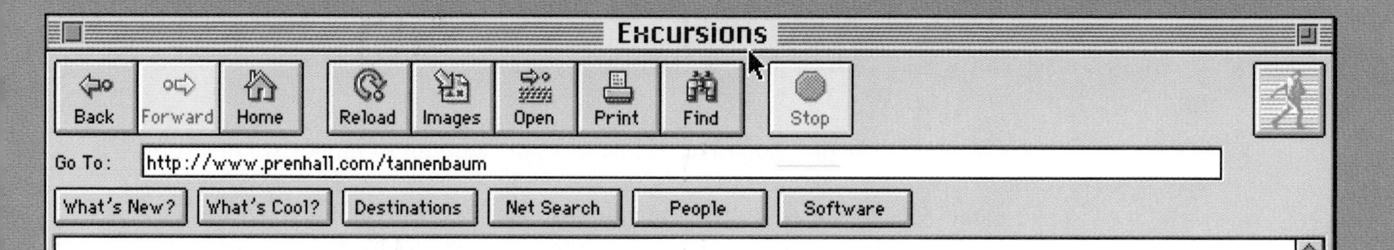

To Bee or Not To Bee

Researchers at the Carl Hayden Bee Research Center in Tucson, Arizona, have developed a mathematical model of the population of a beehive over the course of a year. The model shows how the population of worker bees is affected by climate and by events such as a swarm, a brood disease, resource depletion, a new queen, or pesticide use.

The worker bees shown in the model are divided into three categories—brood (larvae), house bees, and foragers. These three categories actually refer to the same bees at different stages in life. A young worker bee stays in the protected environment of the hive as a house bee, taking care of the storage cells and of the next generation of larvae. Later, she becomes a forager, collecting nectar, pollen, and water from the outside world.

The population model, as well as a wealth of information about honeybees, can be found at the site above. You will need to run the simulation several times and read the above related information at the Tannenbaum site in order to answer the questions.

Questions:

1. Run the basic simulation for a city of your choice. How are the curves for the brood and the house bees related? Why do you think this occurs? About how many days do you think a bee remains in the larva stage?
2. Run the basic simulations for Tucson and Buffalo, starting with an initial colony size of your choice. Compare the graphs for the number of foragers in the two locations. See if you can explain any patterns you see.
3. Run the simulation for Atlanta, with an initial colony size of 5000. Compare the results with and without "chalkbrood" occurring on July 1. About what percent of the brood is killed by the disease? How can you tell?
4. Run the simulation for Kansas City, with an initial colony size of 10,000. Explain how each of the following events are reflected in the graph:
   a. A swarm
   b. Supercedure of the queen
5. Run the simulation for Portland, with an initial colony size of 5000. Explain how each of the following events are reflected in the graph:
   a. Pesticide use
   b. Resource depletion

Note that the problem does not indicate the amount of money we invest or the length of time we plan to leave the money in the account. The answer to the problem depends only on the annual interest and the compounding period. The way to compare these different accounts is to use a common yardstick—for example, How much does \$1 grow in 1 year?

With bank A, in 1 year \$1 becomes \$1.10.

With bank B, in one year \$1 becomes

$$\$\left(1+\frac{0.0975}{12}\right)^{12} \approx \$1.102.$$

And with bank C, in one year \$1 becomes

$$\$\left(1+\frac{0.095}{365}\right)^{365} \approx \$1.0996.$$

We can now see that bank B offers the best deal. ■

These same calculations are described by banks in a slightly different form called the *annual yield.* The **annual yield** is the percentage increase that the account will produce in 1 year. In Example 14, the annual yield for bank A is 10%, for bank B, 10.2%, and for bank C, 9.96%. These numbers can be read off directly from the preceding calculations.

Adding Terms in a Geometric Sequence

We learned in Example 4 that a straightforward formula allows us to add up the beginning terms of any arithmetic sequence. We would like to be able to do something similar with a geometric sequence. The basic fact that we need is given by the following formula:

Adding N Consecutive Terms of a Geometric Sequence

$$a + ar + ar^2 + \dots + ar^{N-1} = \frac{a(r^N - 1)}{r - 1}.$$

Example 15.

$$8 + 8 \times 3 + 8 \times 3^2 + 8 \times 3^3 + \cdots + 8 \times 3^{13} = ?$$

Here $a = 8$, $r = 3$, and $N - 1 = 13$. Plugging these values into the formula gives

$$\frac{8 \times (3^{14} - 1)}{3 - 1} = 19{,}131{,}872.$$ ■

(The reader is encouraged to try Exercises 32 and 33 at this point.)

Example 16.

A mother decides to set up a college trust for her newborn child. The plan is to deposit \$100 a month for the next 18 years (i.e., 216 months) in a savings account that pays 6% annual interest compounded monthly. How much money will there be in the account at the end of 18 years.

This is a problem of exponential growth with a twist: Each \$100 deposit grows at the same monthly rate $[r = 1 + (0.06/12) = 1.005]$, but the number of periods it compounds is different for each deposit. Let's make a list:

- First deposit of \$100 draws interest compounded for 216 months, producing $\$100(1.005)^{216}$.
- Second deposit of \$100 draws interest compounded for 215 months, producing $\$100(1.005)^{215}$.
- Third deposit of \$100 draws interest compounded for 214 months, producing $\$100(1.005)^{214}$.

$$\vdots$$

- Two-hundred-sixteenth deposit of \$100 draws interest for 1 month, producing $\$100(1.005)$.

The total amount in the account at the end of 18 years will be

$$100(1.005)^{216} + 100(1.005)^{215} + \cdots + 100(1.005).$$

This is the sum of the terms of a geometric sequence. Using the formula for adding consecutive terms of a geometric sequence, we get (see Exercise 48)

$$\$\left\{\frac{100(1.005)[(1.005)^{216} - 1]}{0.005}\right\} \approx \$38{,}929.$$ ■

The Logistic Growth Model

When dealing with animal populations, the two models we have studied so far are mostly inadequate. As we now know, *linear growth* represents the case in which there is a fixed amount of growth during each period between transitions. This model might work with inanimate objects (garbage, production goods, sales figures, and so on) but fails completely when there is some form of breeding which must be taken into account. *Exponential growth,* on the other hand, represents the case in which there is unrestrained breeding (as with money left to compound in a bank account, and sometimes in the early stages of an actual animal population). In population biology, however, it is generally the case that the rate of growth of an animal population is not always the same. It depends on the relative sizes of other interacting populations (predators, prey, and so on) and, even more importantly, on the relative size of the population itself. When the relative size of the population is small (we will define more precisely what we mean by this soon) and there is plenty of room to grow, then the rate of growth is high. As the population gets larger, there is less room to grow, and the growth rate starts to taper off. Sometimes the population gets too large for its own good, leading to decay and possibly to extinction.

A well-known experiment with rats in a cage illustrates some of these ideas. Put a few rats in a cage with plenty of food. If the cage is big enough, the rats will start breeding in an unrestrained fashion, and for a while the growth of the rat population will follow an exponential growth model. As the cage gets more crowded, the rate of growth will slow down dramatically. The force that regulates this slowdown is competition for the resources that are essential for growth: food, sex, and space. Eventually, the competition gets so keen that the rats start killing each other off—it is their own quick fix for dealing with the overcrowding problem. Often when the population gets back down to an acceptable level, the killing frenzy stops. Sometimes nature's growth-regulating mechanism may get out of kilter—the killing frenzy may not stop quite in time, and the rats will wipe each other out in a total rodent holocaust.

The above scenario applies (with variations) to almost every situation in which there is a limited environment for a population. Population biologists call such an environment the **habitat**. The habitat might be a cage (as in the example of the rats), a lake (as for a population of fish), a garden (as for a population of snails), and, of course, the planet itself, which is everyone's habitat.

Of the many mathematical models that attempt to deal with a variable growth rate in a fixed habitat, the simplest is the **logistic growth model**. To put it very informally, the key idea is that the rate of growth of the population is directly proportional to the amount of "elbow room" available in the population's habitat. Thus, lots of "elbow room" means a high growth rate; little "elbow room" means a low growth rate (possibly less than 1, which, as we know, means that the population is actually going down); and finally, if the habitat ever gets to be completely saturated, the population will die out.

There are two equivalent ways we can describe the situation mathematically. If C is some constant that describes the total saturation point of the habitat (population biologists call C the **carrying capacity** of the habitat), then for a population of size P_N we can say that the amount of "elbow room" is the difference between the carrying capacity and the population size, namely $(C - P_N)$. Then, if the growth rate is proportional to the amount of elbow room (as described above), we have

$$\text{growth rate for period } N = R(C - P_N)$$

(where R is a constant of proportionality that depends only on the particular population we are studying). Using the fact that (population at period N) $\times$ (growth rate for period N) $=$ population for period $(N + 1)$, we get the following transition rule for the logistic growth model:

$$P_{N+1} = R(C - P_N)P_N.$$

There are two constants in the above transition rule: R, which depends on the population we are studying, and C, which depends on the habitat.

A slightly more convenient way to describe the same thing is to put everything in relative terms: The maximum of the population is 1 (i.e., 100% of the habitat is taken up by the population); the minimum is 0 (i.e., the population is extinct); and every possible population size P_N is represented by some fraction between 0 and 1, which we will denote by p_N (to distinguish it from P_N). The relative amount of elbow room is then $(1 - p_N)$, and the transition rules for the logistic model can be rewritten in the form of the following equation, called the **logistic equation**:[4]

Logistic Equation

$$p_{N+1} = r(1 - p_N)p_N.$$

[4] This is sometimes known as the *Verhulst equation* after the Belgian Pierre Francois Verhulst, who proposed it in the late nineteenth century.

In this equation the value p_N represents the fraction of the habitat's carrying capacity taken up by the actual population P_N ($p_N = P_N/C$), and the constant r depends on both the original growth rate R and the habitat's carrying capacity C. We will call r the **growth parameter**.

Because it looks at population growth using a single common yardstick (the fraction of its habitat's carrying capacity taken up by the population), the second description is particularly convenient when making growth comparisons between populations and is preferred by ecologists and population biologists. We will stick to it ourselves. In the examples that follow we will look at the growth pattern of an imaginary population under the logistic growth model. In each case, all we need to get started is the original population p_1 (given as a fraction of the habitat's carrying capacity) and the value of the growth parameter r (p_1 should always be between 0 and 1 and, for mathematical reasons, we will restrict r to be between 0 and 4). The logistic equation and a good calculator will do the rest. Be forewarned, however, that the calculations shown in the examples that follow were done with a computer and carried to 16 decimal places before being rounded off to 3 or 4 decimal places—they may not match exactly with identical calculations done with a hand calculator.

Example 17.

Suppose we are planning to go into the business of fish farming, which when properly done can be quite profitable. We have a pond in which we plan to raise a special and expensive variety of trout. Let's say that the growth parameter for this type of trout is $r = 2.5$.

We decide to start the business by stocking the pond with 20% of its carrying capacity. In the language of the logistic growth model, this is the same as saying $p_1 = 0.2$. Now let's see what the logistic growth model predicts for our future business.

After the first breeding season[5] we have

$$p_2 = 2.5 \times (1 - 0.2) \times (0.2) = 0.4.$$

The population of the pond has doubled and things are looking good! Since the fish are small, we decide to continue with the program. After the second breeding season we have

$$p_3 = 2.5 \times (1 - 0.4) \times (0.4) = 0.6. \quad \text{(Not too bad!)}$$

After the third breeding season, we have

$$p_4 = 2.5 \times (1 - 0.6) \times (0.6) = 0.6. \quad \text{(A surprise!)}$$

Stubbornly, we try one more breeding season:

$$p_5 = 2.5 \times (1 - 0.6) \times (0.6) = 0.6.$$

It is quite clear that by the third generation the trout population has stabilized at 60% of the pond's carrying capacity, and unless some external change is made, it will remain at the same level for all future generations. It's time to start selling some of the fish. ■

---

[5] In animal populations, the transitions usually correspond to breeding seasons.

Example 18.

Suppose that we have the same pond and the same variety of trout as in Example 17 (in other words, we still have $r = 2.5$), but we wonder what would happen if we stocked the pond differently—let's say we started with $p_1 = 0.3$. We now have

$$p_2 = 2.5 \times (1 - 0.3) \times (0.3) = 0.525,$$

$$p_3 = 2.5 \times (1 - 0.525) \times (0.525) \approx 0.6234,$$

$$p_4 = 2.5 \times (1 - 0.6234) \times (0.6234) \approx 0.5869,$$

$$p_5 = 2.5 \times (1 - 0.5869) \times (0.5869) \approx 0.6061,$$

$$p_6 = 2.5 \times (1 - 0.6061) \times (0.6061) \approx 0.5968,$$

$$p_7 = 2.5 \times (1 - 0.5968) \times (0.5968) \approx 0.6016.$$

Something different is happening now, or is it? After the second breeding season the population of the pond starts fluctuating—up, down, up again, back down—but in a rather special way. We leave it to the reader to verify that as one continues with the population sequence, the p-values inch closer and closer to 0.6 in an oscillating (up, down, up, down, . . .) manner. ■

Example 19.

What happens in Example 18 if $p_1 = 0.7$? After the first generation the population behaves identically with that in Example 18. This follows from the fact that in both cases we get the same value for p_2:

$$p_2 = 2.5 \times 0.3 \times 0.7 = 2.5 \times 0.7 \times 0.3 = 0.525.$$

A useful general rule about logistic growth can be spotted here: If we replace p_1 with its complement $(1 - p_1)$, then after the first generation the populations will behave identically. ■

Example 20.

Let's say that based on what we learned from the previous examples we decide to try to raise a different population of fish—a special variety of catfish for which the growth parameter is $r = 3.1$.

What happens if we start with $p_1 = 0.2$? For the sake of brevity we will write the values of the population in sequence form and leave the calculations to the reader.

$$\begin{array}{llll} p_1 = 0.2, & p_2 = 0.496, & p_3 \approx 0.775, & p_4 \approx 0.541, \\ p_5 \approx 0.770, & p_6 \approx 0.549, & p_7 \approx 0.767, & p_8 \approx 0.553, \\ p_9 \approx 0.766, & p_{10} \approx 0.555, & p_{11} \approx 0.766, & p_{12} \approx 0.556, \\ p_{13} \approx 0.765, & p_{14} \approx 0.557, & p_{15} \approx 0.765, & p_{16} \approx 0.557, \quad \ldots. \end{array}$$

An interesting pattern emerges here: After a few breeding seasons the population settles into a two-period cycle, alternating between a high-population period at 0.765 and a low-population period at 0.557. ■

There are many animal populations whose behavior parallels that of the fish population in Example 20—a lean season followed by a boom season followed by a lean season, and so on.

Example 21.

We are now out of the fish-farming business and have acquired an interest in beetles. We are going to study the behavior of a special type of beetle for which the growth parameter is $r = 3.5$.

Let's suppose that the starting population is given by $p_1 = 0.56$ and use the logistic equation to predict the growth of this beetle population. We leave it to the reader to verify these numbers and fill in the missing details (a calculator is all that is needed).

$p_1 = 0.560$, $p_2 \approx 0.862$, $p_3 \approx 0.415$, $p_4 \approx 0.850$,
$p_5 \approx 0.446$, $p_6 \approx 0.865$, ... $p_{21} \approx 0.497$,
$p_{22} \approx 0.875$, $p_{23} \approx 0.383$, $p_{24} \approx 0.827$, $p_{25} \approx 0.501$,
$p_{26} \approx 0.875$, ...

It took a while, but we can now see a pattern: Since $p_{26} = p_{22}$, the population will repeat itself in a four-period cycle ($p_{27} = p_{23}$, $p_{28} = p_{24}$, $p_{29} = p_{25}$, $p_{30} = p_{26} = p_{22}$, etc.), an interesting and surprising turn of events. ■

Many insect populations follow cyclical patterns of various lengths—7-year cycles (locusts), 17-year cycles (cicadas), and so on.

Example 22.

Our last and most remarkable example is a population sequence determined by the logistic growth model with a growth parameter of $r = 4$. Let's start with $p_1 = 0.2$. The first 20 values of the population sequence are given by

$p_1 = 0.2000$, $p_2 = 0.640$, $p_3 \approx 0.9216$, $p_4 \approx 0.2890$,

$p_5 \approx 0.8219$, $p_6 \approx 0.5854$, $p_7 \approx 0.9708$, $p_8 \approx 0.1133$,

$p_9 \approx 0.4020$, $p_{10} \approx 0.9616$, $p_{11} \approx 0.1478$, $p_{12} \approx 0.5039$,

$p_{13} \approx 0.9999$, $p_{14} \approx 0.0002$, $p_{15} \approx 0.0010$, $p_{16} \approx 0.0039$,

$p_{17} \approx 0.0157$, $p_{18} \approx 0.0617$, $p_{19} \approx 0.2317$, $p_{20} \approx 0.7121$.

Figure 10-8 plots the behavior of the population for the first 20 generations. The reader is encouraged to chart this population for a few additional generations. The surprise here is the absence of any predictable pattern. Even though the population sequence is governed by a very precise rule (the logistic equation), to an outside observer the pattern of growth appears to be quite erratic and seemingly random.

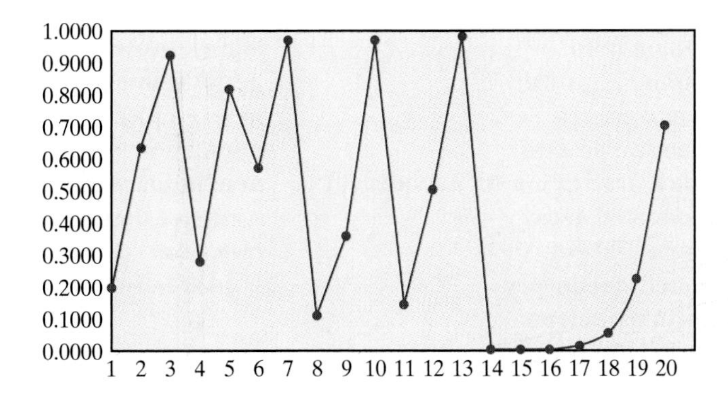

FIGURE 10-8 First twenty generations of a population under the logistic growth model ($r = 4.0$).

■

The behavior of populations under the logistic growth model exhibits many interesting surprises. In addition to doing Exercises 34 through 38 at the end of this chapter, the reader is encouraged to experiment on his or her own in a manner similar to the work we did in the preceding examples. (Choose a p_1 between 0 and 1, choose an r between 0 and 4, and fire up both your calculator and your imagination!) An excellent nontechnical account of the surprising patterns produced by the logistic growth model can be found in reference 2. More technical accounts of the logistic equation can be found in references 7 and 8.

Conclusion

In this chapter we studied three simple models that describe the way that populations grow.

In the *linear model* of population growth, the population sequence is described by an arithmetic sequence, and at each transition period the population grows by a constant amount called the *common difference.* Linear growth is most common with populations consisting of inanimate objects.

In the *exponential model* of population growth, the population is described by a geometric sequence. Here in each transition period the population is multiplied by a constant amount called the *common ratio.* Exponential growth is typical of situations in which there is unrestrained breeding. Money drawing interest in a bank account is one such example.

The *logistic model* of population growth represents situations in which the rate of growth of the population varies from one season to the next, depending on the amount of space available in the population's habitat. Many animal populations are governed by the logistic model or simple variations of it.

Most serious studies of population growth involve models with much more complicated mathematical descriptions, but to us that is neither here nor there. Ultimately, the details are not as important as the overall picture: a realization that mathematics can be useful even in its most simplistic forms to describe and predict the rise and fall of populations in many fields—from the human realm of industry and finance to the natural world of population biology and animal ecology.

Key Concepts

annual yield
arithmetic sequence
carrying capacity
common difference
common ratio
continuous growth
discrete growth
dynamical process
explicit description (of a sequence)
exponential decay
exponential growth
geometric sequence
growth parameter
habitat
linear growth
logistic equation
logistic growth
logistic growth model
periodic interest
plot (graph)
population growth
population sequence
recursive description (of a sequence)
transition
transition rule

Exercises

Walking

1. A population of laboratory rats grows according to the following transition rule: $P_N = P_{N-1} + P_{N-2}$. The starting population is $P_1 = 6$ and the population in the second generation is $P_2 = 10$.
   (a) Find the population in the fourth generation.
   (b) Find the population after 5 transitions.
   (c) Can the size of the rat population ever be an odd number? Explain.

2. A population of guinea pigs grows according to the following transition rule: $P_N = 2P_{N-1} + P_{N-2}$. The starting population is $P_1 = 3$ and the population in the second generation is $P_2 = 5$.
   (a) Find the population in the third generation.
   (b) Find the population after 4 transitions.
   (c) Can the size of the guinea pig population ever be an even number? Explain.

3. Mr. GQ is a snappy dresser and has an incredible collection of neckties. At the end of each month, he buys himself 5 new neckties. Let P_N represent the number of neckties in his collection during the Nth month. Assume he started out with just 3 neckties, and that he never throws anything away.
   (a) Give a recursive description for P_N.
   (b) Give an explicit description for P_N.
   (c) Find P_{300}.

4. A nuclear power plant produces 12 lb of radioactive waste every month. The radioactive waste must be stored in a special storage tank. At present, the tank holds 25 lb of radioactive waste. Let P_N represent the amount of radioactive waste (in pounds) in the storage tank in the Nth month.
   (a) Give a recursive description for P_N.
   (b) Give an explicit description for P_N.
   (c) If the maximum capacity of the storage tank is 500 lb, how long will it take before the tank has reached its maximum capacity?

5. Consider a population that grows according to the linear growth model $P_N = P_{N-1} + 125$, with starting population $P_1 = 80$.
   (a) Find P_2, P_3, and P_4.
   (b) Find P_{100}.
   (c) Give an explicit description of the population sequence.

6. Consider a population that grows according to the linear growth model $P_N = P_{N-1} + 23$, with starting population $P_1 = 57$.
   (a) Find P_2, P_3, and P_4.
   (b) Find P_{201}.
   (c) Give an explicit description of the population sequence.

7. Consider a population that grows according to a linear growth model. The starting population is $P_1 = 8$, and the population in the tenth generation is $P_{10} = 35$.
   (a) Find the common difference d.
   (b) Find P_{51}.
   (c) Give an explicit description of the population sequence.

8. Consider a population that grows according to a linear growth model. The population in the sixth generation is $P_6 = 37$, and the population in the seventh generation is $P_7 = 42$.

(a) Find the common difference d.

(b) Find the starting population P_1.

(c) Give an explicit description of the population sequence.

9. Find $\underbrace{2 + 7 + 12 + \cdots}_{100 \text{ terms}}$.

10. Find $\underbrace{21 + 28 + 35 + \cdots}_{57 \text{ terms}}$.

11. Find $12 + 15 + 18 + \cdots + 309$.

12. Find $1 + 10 + 19 + \cdots + 2701$.

13. Consider a population that grows according to a linear growth model. The starting population is $P_1 = 23$ and the common difference is $d = 7$.

(a) Find $P_1 + P_2 + \cdots + P_{1000}$.

(b) Find $P_{101} + P_{102} + \cdots + P_{1000}$.

14. Consider the arithmetic sequence with the first four terms 7, 11, 15, 19.

(a) Find P_{100}.

(b) Find P_N.

(c) Find $P_1 + P_2 + \cdots + P_{1000}$.

(d) Find $P_{101} + P_{102} + \cdots + P_{1000}$.

15. The city of Lightsville currently has 137 street lights. As part of an urban renewal program the city council has decided to install and have operational 2 additional street lights at the end of each week for the next 52 weeks. Each street light cost $1 to operate for 1 week.

(a) How many street lights will the city have at the end of 38 weeks?

(b) How many street lights will the city have at the end of N weeks? ($N \leq 52$.)

(c) What is the cost of operating the original 137 lights for 52 weeks?

(d) What is the additional cost for operating the newly installed lights for the 52-week period during which they are being installed?

16. A manufacturer currently has on hand 387 widgets. During the next 2 years, the manufacturer will be increasing his inventory by 37 widgets per week. Each widget costs 10 cents a week to store.

(a) How many widgets will the manufacturer have on hand 21 weeks from today?

(b) How many widgets will the manufacturer have on hand N weeks from today? (Assume $N \leq 104$.)

(c) What is the cost of storing the original 387 widgets for 2 years?

(d) What is the additional cost of storing the increased inventory of widgets for the next 2 years?

17. You have a coupon worth 15% off any item in the store (including sale items). The particular item you want is on sale at 30% off the marked price of $100. The store policy allows you to use your coupon before the 30% discount or after the 30% discount (i.e., you can take 15% off the marked price first and then take 30% off the resulting price, or you can take 30% off the marked price first and then take 15% off the resulting price).

(a) What is the dollar amount of the discount in each case?

(b) What is the total percentage discount in each case?

(c) Suppose the article cost P dollars (instead of \$100). What is the percentage discount in each case?

18. You have \$1000 to invest in one of two competing banks (bank A or bank B), both of which are paying 10% annual interest on deposits left for 1 year. Bank A is offering a 5% bonus credited to your account at the time of the initial deposit, provided the funds are left in the account for a year. Bank B is offering a 5% bonus paid on your account balance at the end of the year after the interest has been credited to your account.

(a) How much money would you have a the end of the year if you invested in bank A? In bank B?

(b) What is the total percentage gain (interest plus bonus) at the end of the year for each of the two banks?

(c) Suppose you invested P dollars (instead of \$1000). What is the total percentage gain (interest plus bonus) at the end of the year for each bank?

19. At Tasmania State University, during a 3-year period, the tuition increased by 10%, 15%, and 10%, respectively, each year. What was the total percentage increase overall during the 3-year period? (*Hint:* The answer is not 35%!)

20. A membership store gives a 10% discount on all purchases to its members. If the store marks each item up 50% (based on its cost), what is the markup actually realized by the store when an item is sold to a member?

21. The amount of \$3250 is deposited in a savings account that draws 9% annual interest, with interest credited to the account at the end of each year. Assuming no withdrawals are made, how much money will be in the account after 4 years?

22. The amount of \$1237.50 is deposited in a savings account that draws 8.25% annual interest, with interest credited to the account at the end of each year. Assuming no withdrawals are made, how much money will be in the account after 3 years?

23. (a) The amount of \$5000 is deposited in a savings account that pays 12% annual interest compounded monthly. Assuming no withdrawals are made, how much money will be in the account after 5 years?

(b) What is the annual yield on this account?

24. (a) The amount of \$874.83 is deposited in savings a account that pays $7\frac{3}{4}\%$ annual interest compounded daily. Assuming no withdrawals are made, how much money will be in the account after 2 years?

(b) What is the annual yield on this account?

25. You have some money to invest. The Great Bulldog Bank offers accounts that pay 6% annual interest. The First Northern Bank offers accounts that pay 5.75% annual interest compounded monthly. The Bank of Wonderland offers 5.5% annual interest compounded daily. What is the annual yield for each bank?

26. Complete the following table:

Annual interest rate	Compounded	Annual yield
12%	Yearly	12%
12%	Semiannually	?
12%	Quarterly	?
12%	Monthly	?
12%	Daily	?

27. You decide to open a Christmas Club account at a bank that pays 6% annual interest compounded monthly. You deposit \$100 on the first of January and on the first of each succeeding month through November. How much will you have in your account on the first of December?

28. You decide to save money to buy a car by opening a special account at a bank that pays 8% annual interest compounded monthly. You deposit \$300 on the first of each month for 36 months. How much will you have in your account at the end of the 36th month?

29. You are interested in buying a car 5 years from now, and you estimate the future cost will be \$10,000. You decide to deposit money today in an account that pays interest, so that 5 years hence you have the \$10,000 necessary to purchase your "dream" car. How much money do you need to deposit if the account you deposit your money in:

(a) Has an interest rate of 10% compounded annually?

(b) Has an interest rate of 10% compounded quarterly?

(c) Has an interest rate of 10% compounded monthly?

30. A population grows according to an exponential growth model. The starting population is $P_1 = 8$ and the common ratio is $r = 1.5$.

(a) Find P_2.

(b) Find P_{10}.

(c) Give an explicit description for the population sequence.

31. A population grows according to an exponential growth model. The starting population is $P_1 = 11$ and the common ratio is $r = 1.25$.

(a) Find P_2.

(b) Find P_{10}.

(c) Give an explicit description for the population sequence.

32. Consider the geometric sequence with first four terms 1, 3, 9, 27.

(a) Find P_{100}.

(b) Find P_N.

(c) Find $P_1 + P_2 + \cdots + P_{100}$.

(d) Find $P_{50} + P_{51} + \cdots + P_{100}$.

33. Consider the geometric sequence with first term $P_1 = 3$ and common ratio $r = 2$.

(a) Find P_{100}.

(b) Find P_N.

(c) Find $P_1 + P_2 + \cdots + P_{100}$.

(d) Find $P_{50} + P_{51} + \cdots + P_{100}$.

Exercises 34 through 38 refer to the logistic growth model $p_{N+1} = r(1 - p_N)p_N$. *For most of these exercises, a calculator with a memory register is suggested.*

34. A population grows according to the logistic growth model, with growth parameter $r = 1.5$ and $p_1 = 0.8$.

(a) Find p_2.

(b) Find p_3.

(c) Find the population after four transitions.

35. A population grows according to the logistic growth model, with growth parameter $r = 2.8$ and $p_1 = 0.15$.

(a) Find p_2.

(b) Find p_3.

(c) Find the population after four transitions.

36. Suppose we know that under the logistic growth model the growth parameter for a particular colony of birds is $r = 2.5$. Suppose also that the starting population is $p_1 = 0.8$.

(a) What is p_2?

(b) What is the population after the second transition?

(c) What does the logistic growth model predict in the long term for this population?

(d) What other starting population would result in the same population sequence after the first transition?

37. A population grows according to the logistic growth model, with growth parameter $r = 3.0$ and $p_1 = 0.15$. Find the values of p_2 through p_{10}.

38. A population grows according to the logistic growth model, with growth parameter $r = 3.8$ and $p_1 = 0.23$. Find the values of p_2 through p_{10}.

Jogging

39. How much should a retailer mark up her goods so that when she has a 25%-off sale, the resulting prices will still reflect a 50% markup (on her cost)?

40. What annual interest rate compounded semiannually gives an annual yield of 21%?

41. Before Annie set off for college, Daddy Warbucks offered her a choice between the following two incentive programs:

- *Option* 1. A \$100 reward for every A she gets in a college course.
- *Option* 2. One cent for her first A, 2 cents for the second A, 4 cents for the third A, 8 cents for the fourth A, and so on.

Annie chose option 1. After getting a total of 30 A's in her college career, Annie is happy with her reward of $\$100 \times 30 = \$3000$. Unfortunately, Annie did not get an A in math. Help her figure out how much she would have made had she chosen option 2.

42. Consider a population that grows according to the logistic growth model with starting population $p_1 = 0.7$. What growth parameter r would keep the population constant?

43. Suppose that you are in charge of stocking a lake with a certain type of alligator with a growth parameter $r = 0.8$. Assuming that the population of alligators grows according to the logistic growth model, is it possible for you to stock the lake so that the alligator population is constant? Explain.

44. Consider a population that grows according to the logistic growth model with growth parameter r ($r > 1$). What should the starting population p_1 be so that the population is constant?

45. Suppose the habitat of a population of snails has a carrying capacity of $C = 20{,}000$ and the current population is 5000. Suppose also that the growth parameter for this particular type of snail is $r = 3.0$. What does the logistic growth model predict for this population after four transition periods?

Exercises 46 and 47 refer to the following situation. If B dollars are borrowed at a periodic interest rate I and N equal periodic payments are to be made, then the periodic payment p is given by the formula

$$p = \frac{BI(1 + I)^N}{(1 + I)^N - 1}.$$

46. (a) You buy a house for \$120,000 with \$20,000 down and finance the balance over 30 years at 9% annual interest (with equal monthly payments). What is your monthly payment?

(b) What is the monthly payment if the loan described in (a) is financed over 40 years instead of 30 years?

47. You decide that you can afford a \$1000-per-month house payment. The current going interest rate on 30-year home loans is 11%. How much money can you borrow at this rate so that your payment will not exceed \$1000 per month?

48. Find $100(1.005)^{216} + 100(1.005)^{215} + \cdots + 100(1.005)$. Use the formula for adding the terms of a geometric sequence and a calculator. (*Hint:* Read the sum from right to left. What is a? What is r?)

49. $1 + 5 + 3 + 8 + 5 + 11 + 7 + 14 + \cdots + 99 + 152 = ?$

50. $1 + 1 + 2 + \frac{1}{2} + 4 + \frac{1}{4} + 8 + \frac{1}{8} + \cdots + 4096 + \frac{1}{4096} = ?$

Running

51. (a) Show that

$$1 + r + r^2 + \cdots + r^{100} = \left(\frac{r^{101} - 1}{r - 1}\right).$$

[*Hint:* Do the multiplication

$$(1 + r + r^2 + \cdots + r^{100})(r - 1)$$

and see what you get!]

(b) Show that

$$1 + r + r^2 + \cdots + r^N = \left(\frac{r^{N+1} - 1}{r - 1}\right).$$

(c) Show that

$$a + ar + ar^2 + \cdots + ar^N = a\left(\frac{r^{N+1} - 1}{r - 1}\right).$$

52. Show that the sum of the first N terms of an arithmetic sequence with first term c and common difference d is

$$\frac{N}{2}[2c + (N - 1)d].$$

53. You are purchasing a home for \$120,000 and are shopping for a loan. You have a total of \$31,000 to put down, including the closing costs of \$1000 and any loan fee that might be charged. Bank A offers a 10%-annual-interest loan amortized over 30 years with 360 equal monthly payments. There is no loan fee. Bank B offers a 9.5%-annual-interest loan amortized over 30 years with 360 equal monthly payments. There is a 3% loan fee (i.e., a one-time up-front charge of 3% of the loan). Which loan is better?

54. A friend of yours sells his car to a college student and takes a personal note (cosigned by the student's rich uncle) for \$1200 with no interest payable at \$100 per month for 12 months. Your friend immediately approaches you and offers to sell you this note. How much should you pay for the note if you want an annual yield of 12% on your investment?

55. The purpose of this exercise is to understand why we assume that, under the logistic growth model, the growth parameter r is between 0 and 4.

(a) What does the logistic equation give for p_{N+1} if $p_N = 0.5$ and $r > 4$. Is this a problem?

(b) What does the logistic equation predict for future generations if $p_N = 0.5$ and $r = 4$?

(c) If $0 \leq p \leq 1$, what is the largest possible value of $(1 - p)p$?

(d) Explain why, if $0 < p_1 < 1$ and $0 < r < 4$, then $0 < p_N < 1$ for every positive integer N.

56. Suppose $r > 3$. Using the logistic growth model, find a population p_1 such that $p_1 = p_3 = p_5 \cdots$, but $p_1 \neq p_2$. (*Hint:* See Exercise 44.)

References and Further Readings

1. Clark, C. W., *Bioeconomics Modeling and Fishery Management.* New York: John Wiley & Sons, 1985.
2. Gleick, James, *Chaos: Making a New Science.* New York: Viking Penguin, Inc., 1987, chap. 3.
3. Hoppensteadt, Frank, *Mathematical Methods of Population Biology.* Cambridge: Cambridge University Press, 1982.
4. Hoppensteadt Frank, *Mathematical Theories of Population: Demographics, Genetics and Epidemics.* Philadelphia: Society for Industrial and Applied Mathematics, 1975.
5. Hoppensteadt Frank, and Charles Peskin, *Mathematics in Medicine and the Life Sciences.* New York: Springer-Verlag, 1992.
6. Kingsland Sharon E., *Modeling Nature: Episodes in the History of Population Ecology.* Chicago: University of Chicago Press, 1985.
7. May, Robert M., "Biological Populations with Nonoverlapping Generations: Stable Points, Stable Cycles and Chaos," *Science,* 186 (1974), 645–647.
8. May, Robert M., "Simple Mathematical Models with Very Complicated Dynamics," *Nature,* 261 (1976) 459–467.
9. May, Robert M., and George F. Oster, "Bifurcations and Dynamic Complexity in Simple Ecological Models," *American Naturalist,* 110 (1976) 573–599.
10. Smith, J. Maynard, *Mathematical Ideas in Biology.* Cambridge: Cambridge University Press, 1968.

Symmetry, as wide or as narrow as you may define its meaning, is one idea by which man through the ages has tried to comprehend and create order, beauty and perfection

Hermann Weyl

Symmetry

Mirror, Mirror, Off The Wall . . .

CHAPTER 11

It is said that Eskimos have dozens of different words for ice. Ice is, after all, a universal theme in the Eskimo's world. Along the same lines, we would expect science and mathematics to have dozens of different words to describe the notion of symmetry, since symmetry is a recurrent theme in the world around us. Surprisingly, just the opposite is the case: We use a single word—*symmetry*—to cover an incredibly diverse set of situations and ideas.

Exactly what is symmetry? The answer depends very much on the context of the question. In everyday language, *symmetry* is most often taken to mean *mirror symmetry* (also called *bilateral* or *left-right symmetry*) such as the (almost perfect but not quite) left-right symmetry exhibited externally[1] by the human body or the perfect left-right symmetry exhibited by a snowflake. In everyday language, *symmetry* is also used to describe an aesthetic value—people think of something *symmetric* as well balanced, pleasing to the eye, well proportioned. This is often how the word is used in art and architecture. Along the same lines, *symmetry* is used in musical

---

[1] Internally, the human body is not even close to having left-right symmetry—the heart and stomach, for example, are essentially on the left side; the liver on the right.

composition to describe special melodic effects. (The music of Bach, for example is often described and analyzed in terms of its symmetry.)[2] Even in poetry and literature, symmetry is used as an important element of literary form.[3]

In this chapter we will take a geometric perspective of *symmetry,* focusing on its application to both real-world physical objects and abstract geometric shapes. Simply put, this chapter is about how to *see, read,* and *make sense of* the symmetry of the world around us.

Geometric Symmetry

When applied to physical objects and geometric shapes, symmetry is often referred to as **geometric symmetry**. What does this mean? The famous Russian crystallographer E. S. Fedorov defined geometric symmetry as "the property of figures to repeat their parts, or more precisely, their property of coinciding with their original position when in different positions."[4] Sounds confusing. A check in Webster's[5] shows several entries under **sym′me·try**; the one most appropriate to our discussion gives: "the correspondence of parts or relations; similarity of arrangement." Not much help there, either. Geometric symmetry is one of those concepts that is almost harder to define than to understand.

Let's start with a small example to get a feel for this hard-to-put-in-a-few-words concept.

Example 1.

Figure 11-1 shows three triangles. Triangle I is equilateral; triangle II is isosceles; triangle II is scalene (all three sides are different). Imagine a very tiny, almost microscopic observer standing at one of the vertices of triangle I and looking inward. The observer would see exactly the same thing whether standing at vertex *A, B,* or *C.* In fact, if the vertices were not labeled and there were no other frames of reference, the observer would be unable to distinguish one position from the other. In triangle II, the observer would see the same thing when standing at *B* or *C* but not when standing at *A.* In triangle III, the observer would see a different thing at each vertex. Informally, we might say that triangle I has more symmetry than triangle II, which in turn has more symmetry than triangle III.

Well, this is still a little vague, but a start nonetheless. We will start by saying that an object has **symmetry** if it looks exactly the same when seen from two or

[2] See, for example, Douglas Hofstadter's book *Gödel, Escher, Bach: An Eternal Golden Braid.*

[3] A trip to a medium-sized university library produced approximately 150 books with the word *Symmetry* somewhere in the title. Of these, roughly 10% were in poetry, literature, or music; another 10% in art and/or architecture; approximately 50% in chemistry, physics, or engineering; and 30% in pure or applied mathematics.

[4] Quoted in I. Hargittai and M. Hargittai, *Symmetry through the Eyes of a Chemist,* 2d ed. (New York: Plenum Press, 1995).

[5] *Webster's New Twentieth Century Dictionary,* 2d ed. (New York: Simon and Schuster, 1979).

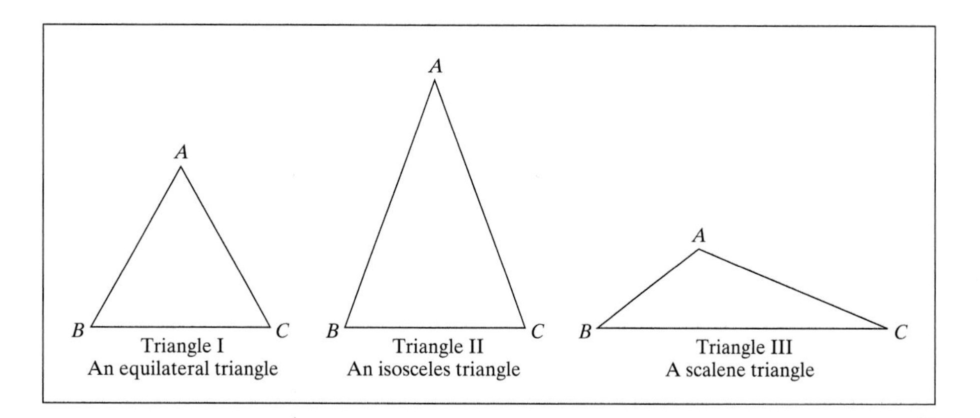

FIGURE 11-1

more different vantage points. A different, but equivalent, way to think about it is this: Rather than moving the observer, why don't we move the object itself? To say, for example, that triangle II in Fig. 11-1 looks exactly the same to an observer standing at vertex B as it does to the same observer standing at vertex C is the same as saying that there must be some way to *move* triangle II so that vertex B is where vertex C used to be and yet the triangle as a whole is in exactly the same position as before.

Informally, an object's symmetry is somehow related to the fact that we can move the object in such a way that when all the moving is done, the object sits exactly as it did before. Thus, to fully understand symmetry, we need to understand the different ways in which we can "move" an object.

Rigid Motions

The act of taking an object and moving it from some starting position to some ending position *without altering its shape or size* is called a **rigid motion**. (When, in the process of moving the object, we stretch it, tear it, or generally alter its shape or size, that's *not* a rigid motion.) Since in a rigid motion the size and shape of an object are not altered, distances between points are preserved: *The distance between any two points X and Y in the starting position is the same as the distance between the same two points in the ending position* (Fig. 11-2).

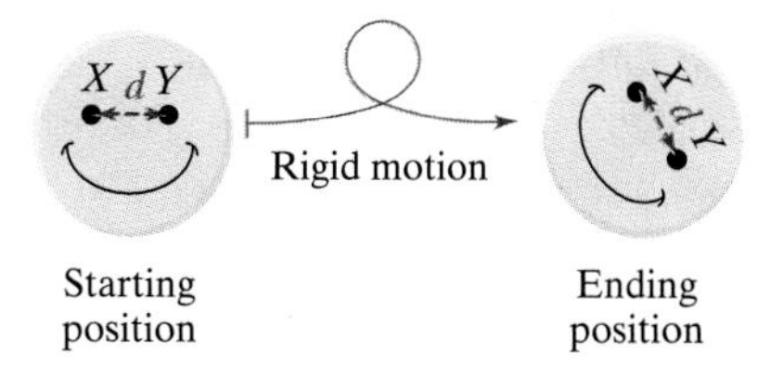

FIGURE 11-2 A rigid motion always preserves distances between points.

In studying rigid motions, *the only thing that we will care about are the starting and ending positions, and not what happens in between.* To illustrate this point, consider the adventures of a humble quarter sitting on top of a dresser. In the morning we might pick it up, put it in a pocket, drive around town with it, take it out of the pocket, flip it in the air, put it back in a different pocket, go home, take it out of the pocket, and finally put it back on top of the dresser again. While the actual trip taken by the quarter was long and eventful, the end result certainly wasn't: The quarter started somewhere on top of the dresser and ended somewhere else on the dresser. From the quarter's perspective, we could have accomplished the whole thing in a much simpler way—possibly a little slide along the top of the dresser, possibly a single flip over (if the starting and final positions had opposite faces up).

When two rigid motions accomplish the same net effect, they are said to be **equivalent** rigid motions. From our point of view, to say that two rigid motions are equivalent essentially means that they are the same. It is a remarkable fact that every rigid motion, no matter how complicated, is always equivalent to something very basic.

Let's focus our attention, momentarily, on *two-dimensional* objects and shapes. (Although we live in a three-dimensional world, it is going to be a lot easier to understand symmetry in the two-dimensional world of the page than in the three-dimensional world of space.) In the case of two-dimensional objects in a plane, every rigid motion is equivalent to a rigid motion *of one of only four possible kinds:* it's either a **reflection**, a **rotation**, a **translation**, or a **glide reflection**. We will call these four types of rigid motions the **basic rigid motions in the plane**.[6]

A rigid motion (let's call it M) in the plane moves each point in the plane from its starting position P to an ending position P', also in the plane. We will call the point P' the **image** of the point P under the rigid motion M and describe this informally by saying that *M moves P to P'*. (Throughout the chapter we will stick to the convention that the image point has the same label as the original point but with a prime symbol added.) It is possible for a point P to end up back where it started under M ($P' = P$), in which case we call P a **fixed point** of the rigid motion.

It would appear that to "completely know" a rigid motion, one would need to know where it moves every point of the plane, but fortunately, we will see that we need know where the rigid motion moves just a very few points (three at the most). Because the motion is rigid, the behavior of those few points forces all the rest of the points to follow their lead. (We can think of this as a sort of "Pied Piper effect.")

We will now discuss each of the basic rigid motions in the plane in a little more detail.

Reflections

A **reflection** in the plane is a rigid motion that moves an object into a new position that is a mirror image of the starting position. In two dimensions, the "mirror" is just a line, called the **axis** of the reflection. For the flat image of a cat and its reflection on the water, the axis of reflection is the water line.

A reflection is completely described by its axis. (In other words, if we know the axis of the reflection, we know everything we need to know about the reflection.) Figures 11-3 and 11-4 show examples of reflections.

An important characteristic of a reflection is that it reverses all the traditional frames of reference one uses for orientation. As illustrated in Fig. 11-4, in a reflection, left is interchanged with right, and clockwise with counterclockwise. We will

---

[6] For three-dimensional objects in space, there is a similar, but slightly more complicated fact: *Every rigid motion is equivalent to a rigid motion of one of only 6 possible types—reflection, rotation, translation, glide reflection, rotary reflection, and screw displacement.* These are called the basic rigid motions in space.

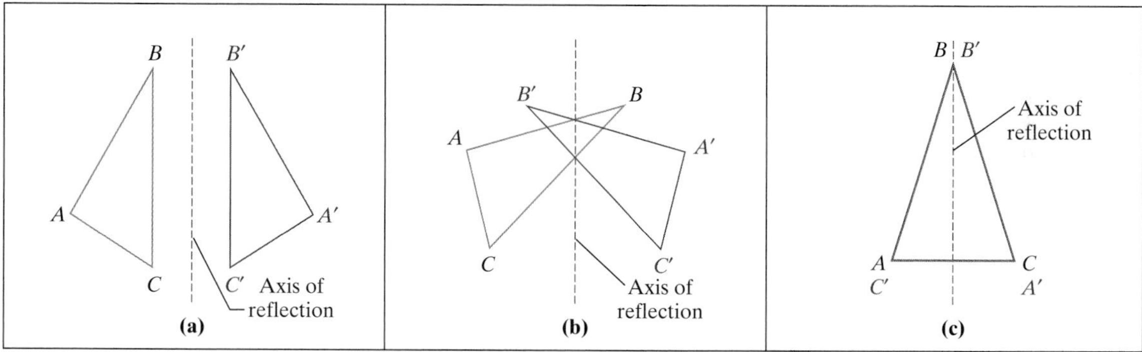

FIGURE 11-3
Original figure in blue; reflected figure in red.

say that reflection is an **improper** rigid motion to indicate the fact that it reverses the left-right and clockwise-counterclockwise orientations.

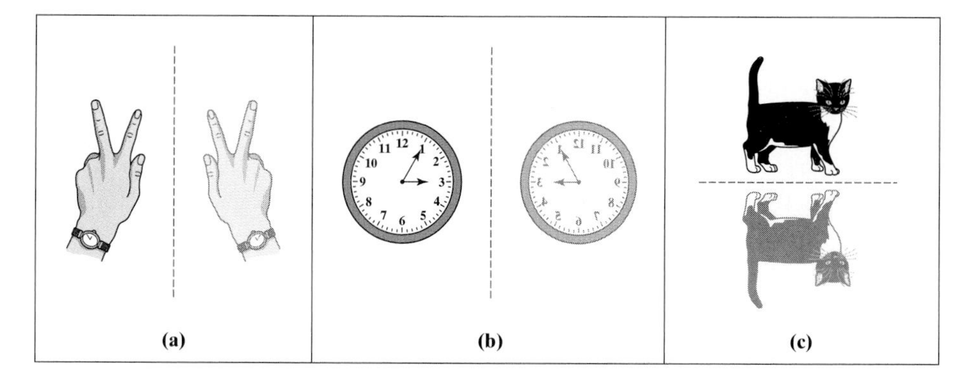

FIGURE 11-4

From a purely geometric point of view, a reflection can best be described by showing how it moves a generic point P: Given the axis of the reflection, the image of a point P is found by drawing a line through P perpendicular to the axis and finding the point P' that is on this line and at the same distance as P from the axis (Fig. 11-5). If P is on the axis itself, it is a fixed point of the reflection. Conversely, if we know any point P and its image P' under the reflection, we can find the axis: It is the perpendicular bisector of the segment joining the two points

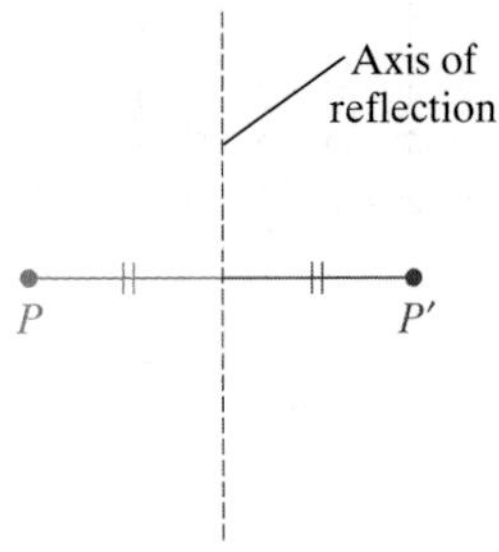

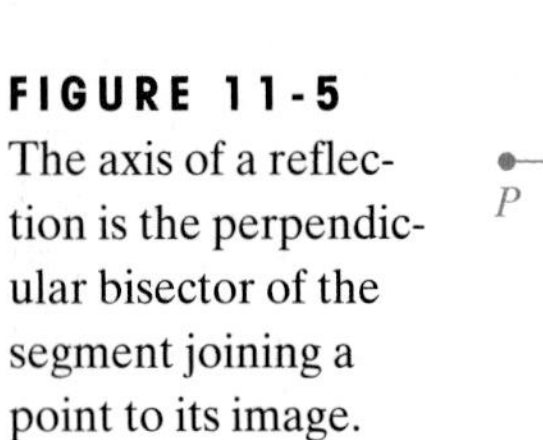
FIGURE 11-5
The axis of a reflection is the perpendicular bisector of the segment joining a point to its image.

(Fig. 11-5). A useful consequence of this is that a point P and its image P' under the reflection completely specify where the axis is and, thus, completely specify the reflection.

Another important fact about reflections is the following: If we apply the same reflection twice, every point ends up exactly where it started. In other words, *the net effect of applying the same reflection twice is the same as not having moved the object at all.* This leads us to an interesting semantic question; Should not moving an object at all be considered in itself a rigid motion? On the one hand, it seems rather absurd to say yes. If we are talking about motion, then there should be some kind of movement, however small. On the other hand, we are equally compelled to argue that the result of combining two (or more) consecutive rigid motions should itself be a rigid motion. If this is the case, then combining two consecutive reflections with the same axis (which produces the same result as no motion at all) should be a rigid motion. We will opt for the latter alternative, because it is the mathematically correct way to look at things. We will formally agree therefore that not moving an object at all is itself a very special kind of rigid motion of the object, which we will call the **identity motion**.

Rotations

The second type of rigid motion we will discuss is **rotation**. For two-dimensional figures, a rotation is described by specifying a point (called the **center** of the rotation or **rotocenter**) and an angle (called the *angle* or *amount* of the rotation). Figures 11-6 and 11-7 show examples of rotations.

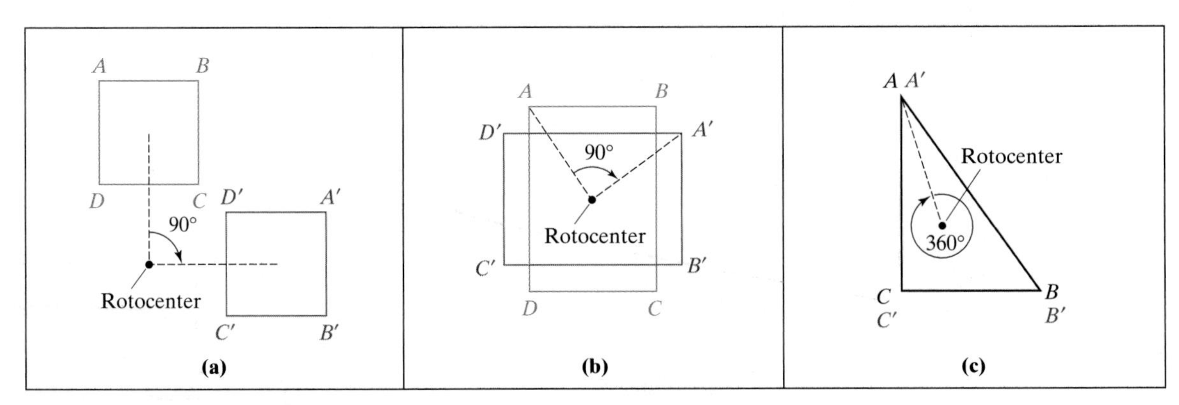

FIGURE 11-6
Original figure in blue, rotated figure in red.

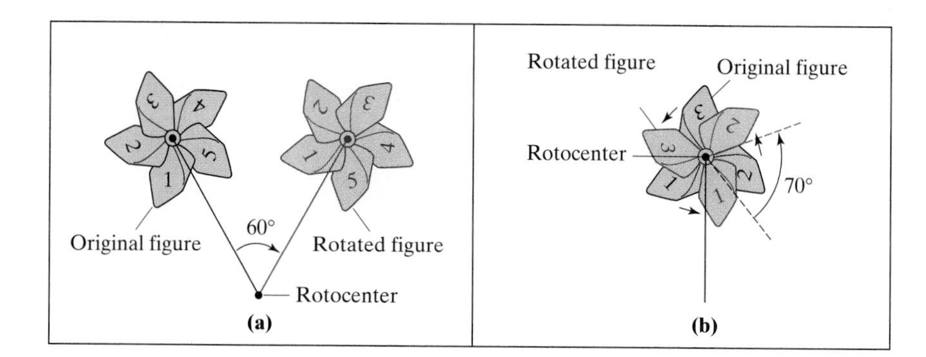

FIGURE 11-7 (a) 60° clockwise rotation; (b) 70° counterclockwise rotation.

A few comments about the examples in the figures are in order. In each example we have specified the angle of rotation in degrees. This is strictly a matter of personal choice—some people prefer degrees, others radians.[7]

Our second observation starts with the well-known fact that a rotation by 360° leaves the figure unchanged—it is the identity motion. This has several useful consequences. First, any rotation by an angle that is more than 360° is equivalent to another rotation with the same center by an angle that is between 0° and 360°. All we have to do is divide the angle by 360 and take the remainder. For example, as a rigid motion, a clockwise rotation by 759° is the same as a clockwise rotation by an angle of 39°, because 759 divided by 360 gives a quotient of 2 and a remainder of 39. Second, any rotation that is specified in a clockwise orientation can just as well be specified in a counterclockwise orientation. In Fig. 11-7(a), for example, the angle of rotation was given as 60° clockwise, but it could just as well be given as 300° counterclockwise. In the special case when the rotation is by an angle of 180°, clockwise and counterclockwise turn out to be the same.

Can a rotation ever be equivalent to a reflection? The answer is no. A rotation, regardless of its center and angle, always leaves the original orientations (left, right, clockwise, counterclockwise) unchanged. Any rigid motion that does this is called a **proper** rigid motion. We have already observed that a reflection is an *improper* rigid motion.

Our final comment about rotations is that, unlike reflections, they cannot be completely described by giving a single point P and its image P'. There are infinitely many rotations that send P to P'. Any point on the perpendicular bisector of the segment PP' can be the center of a possible rotation sending P to P' [Fig. 11-8(a)]. A second pair Q, Q' will allow us to nail down the rotation: The center is

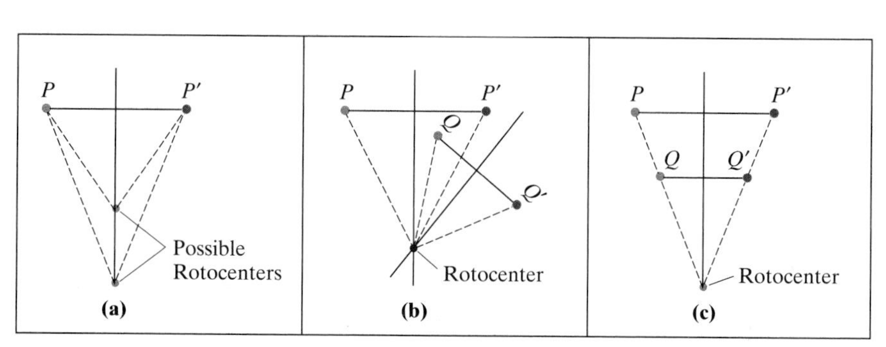

FIGURE 11-8 Finding the rotocenter requires at least two pairs of points P, P' and Q, Q'.

[7] Throughout the chapter we will stick with degrees but the reader is reminded that one can always change degrees to radians using the equation: radians = $(\pi/180) \times$ degrees.

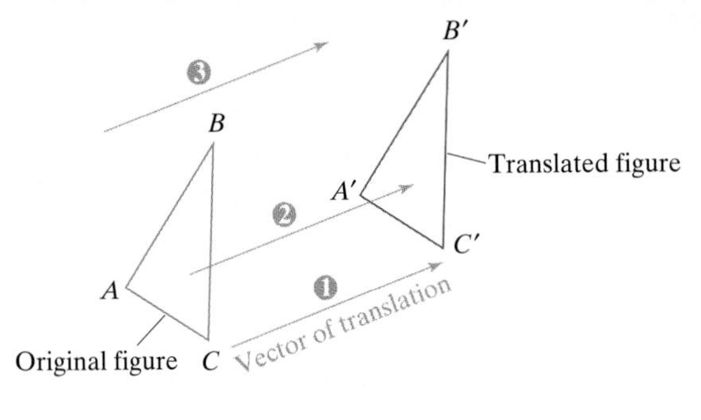

FIGURE 11-9 Any one of the arrows can be used to indicate the vector of transition.

the point where the perpendicular bisectors of PP' and QQ' meet [Fig. 11-8(b)]. In the special case where PP' and QQ' are parallel, the center of rotation is the intersection of PQ and $P'Q'$ [Fig. 11-8(c)].

Translations

A **translation** is essentially a slide of an object in the plane. It is completely specified by the direction and amount of the slide. These two pieces of information are combined in the form of a **vector**. A vector can be represented by an arrow giving its direction and length. As long as the arrow points in the proper direction and has the right length, its actual placement is immaterial, as shown in Fig. 11-9.

Translations, like rotations, are *proper* rigid motions of the plane: They do not change the left-right or clockwise-counterclockwise orientations. On the other hand, translations are like reflections in the sense that they can be completely described by giving a point P and its image P'. The arrow joining P to P' gives us the vector of the translation. Once we have the vector, we know where the translation sends any other point.

Glide Reflections

A **glide reflection**, as the name suggests, is a rigid motion consisting of a translation (the glide part) followed by a reflection. The axis of the reflection *must* be parallel to the direction of the translation. The wording "translation followed by a reflection" is somewhat misleading: We can just as well do the reflection first and the translation second, and the end result will be the same. Figure 11-10 shows a glide reflection broken down into stages.

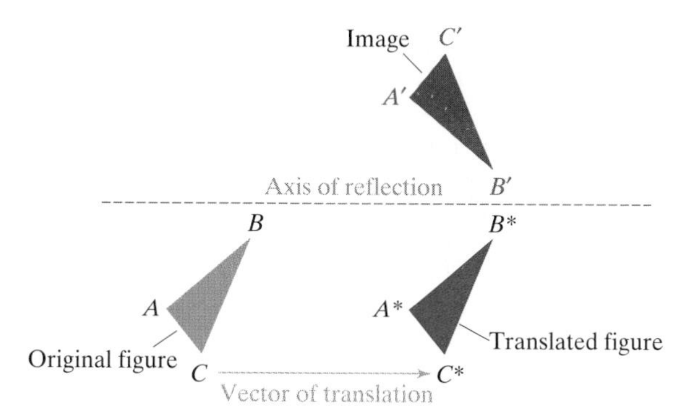

FIGURE 11-10

A glide reflection is an *improper* rigid motion—it changes left-right and clockwise-counterclockwise orientations. We can thank the reflection part of the glide reflection for that.

A glide reflection cannot be determined by just one point P and its image P'. As with a rotation, another point Q and its image Q' are needed. Given the two pairs P, P' and Q, Q', the axis of the reflection can be found by joining the midpoints of the segments PP' and QQ' [Fig. 11-11(a)]. (This follows from the fact that in any glide reflection the midpoint between a point and its image belongs to the axis of the reflection.) Once the axis of reflection is known, the vector of the translation can be determined by locating the intermediate point P^* which is the image of P' under the reflection [Fig. 11-11(b)]. In the unlikely event that the midpoints of PP' and QQ' are the same point M, then the line passing through P and Q must be perpendicular to the axis of reflection [Fig. 11-11(c)], so the axis of reflection is obtained by taking a line perpendicular to the line PQ and passing through the common midpoint M.

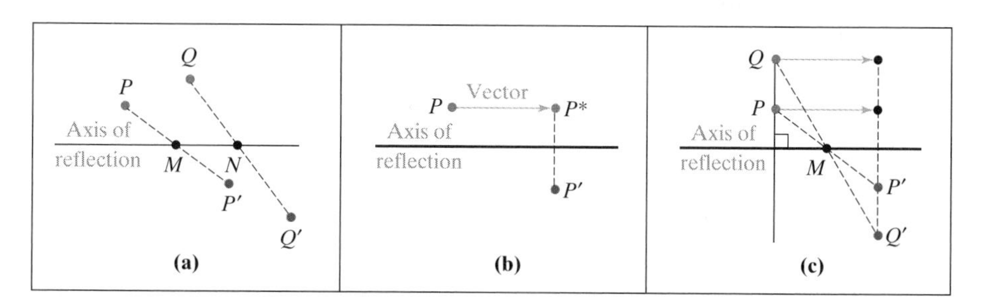

FIGURE 11-11 A glide reflection is determined by two pairs P, P' and Q, Q'.

Of the four basic rigid motions of the plane, the glide reflection is unique in that it is defined as a combination of two other rigid motions—a translation and a reflection. Simply put, there is no simpler way to describe that particular combination of motions. Surprisingly, any other combination of motions, no matter how complex, is guaranteed to be equivalent to one (that's it—just one!) of the four basic rigid motions. And what about the identity motion? Which of the basic rigid motions is it equivalent to? The best answer to this question is to think of the identity motion as a *rotation* of 0° (or 360° if you prefer).

Symmetry Revisited

With an understanding of rigid motions and their classification, we will be able to consider the concept of geometric symmetry in a much more precise way. Here, finally, is a good definition of geometric symmetry, one that probably would not have made much sense at the start of this chapter: *A symmetry of an object or shape is a rigid motion that moves the object back onto itself.* In other words, in a symmetry one cannot tell, at the end of the motion, that the object has been moved. It is important to note that this does not necessarily force the rigid motion to be the identity motion. Individual parts of the object may be moved to different starting and ending positions, even while the whole object is moved back into itself. And, of course, the identity motion is itself a symmetry, one possessed by every object and which from now on we will call simply the **identity**.

Since symmetries are themselves rigid motions, they can be classified accordingly. For two-dimensional objects there are only four possible types of symmetry: *reflection symmetry, rotation symmetry, translation symmetry,* and *glide reflection symmetry.*

Example 2. The Symmetries of a Square

What are the possible rigid motions that move the square in Fig. 11-12(a) onto

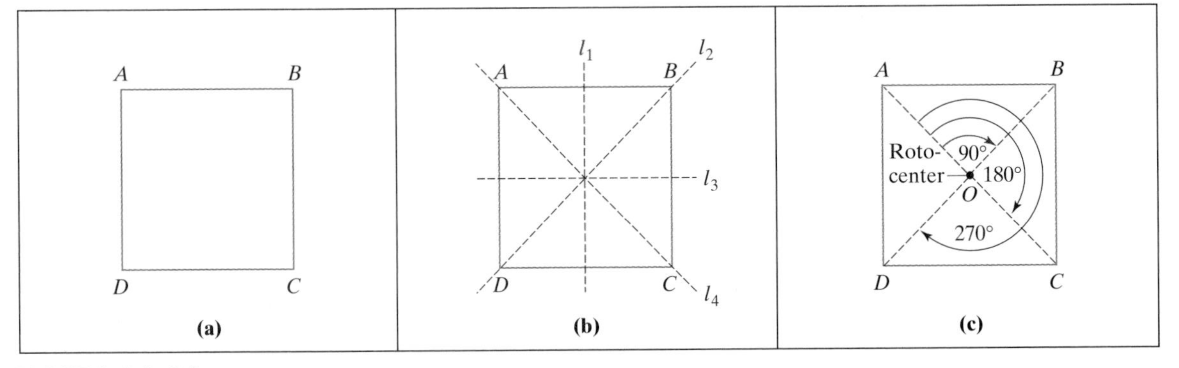

FIGURE 11-12
The symmetries of a square. (b) Four reflection symmetries (axes are l_1, l_2, l_3, and l_4), (c) four rotation symmetries with rotocenter O (90°, 180°, 270°, and the identity).

itself? First, there are *reflection symmetries.* For example, if we use the line l_1 in Fig. 11-12(b) as the axis of reflection, the square falls back into itself; points A and B interchange places and so do C and D. It is not hard to think of three other reflection symmetries, with axes l_2, l_3, and l_4 shown in Fig. 11-12(b). Are there any other symmetries? Yes. There are rotation symmetries with rotocenter O, the center of the square. The angles of rotation are 90°, 180°, 270°, and 360°—this last one being none other than the identity.

All in all, we have easily found 8 symmetries for the square in Fig. 11-12(a): 4 of them are reflections, and 4 are rotations. Could there be more? What if we combined one of the reflections together with one of the rotations? A symmetry combined with another symmetry, after all, has to be itself a symmetry. It turns out that the 8 symmetries we listed are all there are—no matter how we combine them, we always end up with one of the 8 (see Exercises 26 and 54). ■

Since what is true about the square in Fig. 11-12(a) is true for any other square, we can now confidently make a very general statement: A square has 8 symmetries—4 reflections and 4 rotations.

Example 3. The Symmetries of a Propeller

Consider the 4-bladed propeller shown in Fig. 11-13(a). What can we say about its symmetries? It's not hard to see that, once again, there are 4 reflection symmetries [Fig. 11-13(b)] as well as 4 rotations: the identity, 90°, 180°, and 270°. And there are no other possible symmetries.

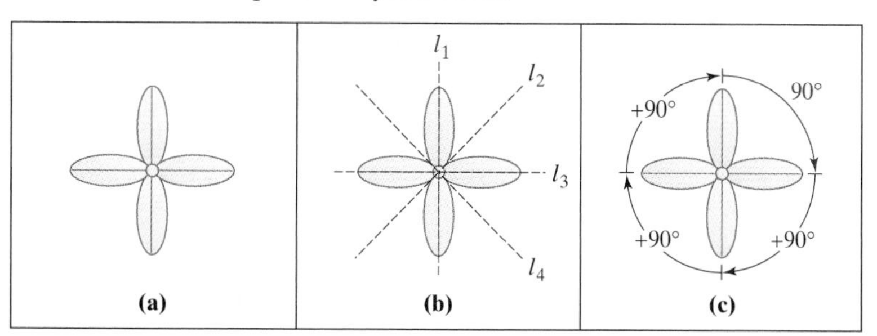

FIGURE 11-13 (a) The symmetries of a propeller, (b) four reflections, (c) four rotations (including the identity). ■

An important lesson lurks behind Examples 2 and 3: *Two different-looking objects can have exactly the same set of symmetries.* A good way to think about this is that the square and the propeller, while certainly different objects, are blood relatives—both members of the same "symmetry family."

Formally, we will say that two objects or shapes are of the same **symmetry type** if they have exactly the same set of symmetries. The symmetry type for the square (as well as the propeller) is called D_4—which is shorthand for 4 reflections and 4 rotations. Figure 11-14 shows several objects with symmetry type D_4.

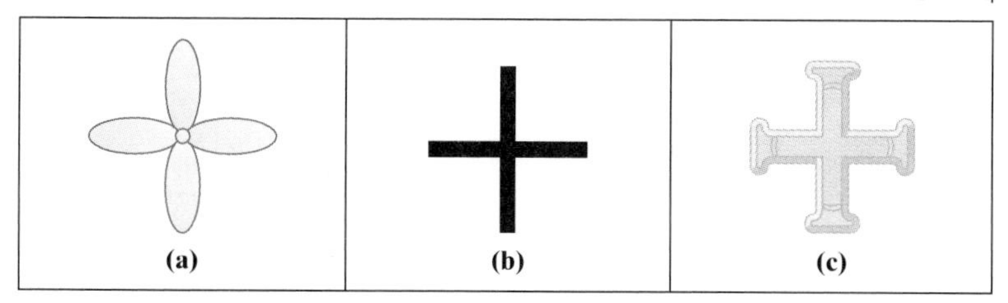

FIGURE 11-14 Symmetry type D_4. (a) Propeller, (b) "plus" sign, (c) Crusader cross.

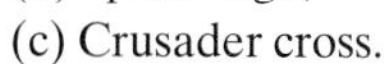

Example 4. The Symmetries of a Propeller: Part II

Let's consider now the propeller shown in Fig. 11-15(a), a slightly different propeller from the one in Example 3. The difference is subtle, but from the symmetry point of view significant. Does this figure have 4 reflection symmetries? Certainly not! A vertical reflection, for example, would not give us an identical propeller [Fig. 11-15(b)], and for that matter, neither would a horizontal or any other kind of reflection. This propeller has no reflection symmetries at all! On the other hand, it still has the 4 rotations (identity, 90°, 180°, and 270°). This propeller has *only* the four rotation symmetries (Exercise 27) and belongs therefore to a new symmetry family called Z_4 (which is shorthand for the symmetry type of objects having 4 rotations only).

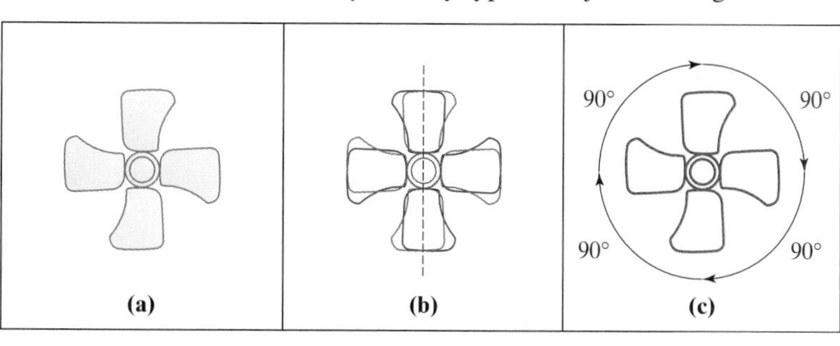

FIGURE 11-15 This propeller has four rotation symmetries only. Reflections don't work. (Symmetry type: Z_4.)

Example 5. The Symmetries of a Propeller: Part III

Here is one last propeller example. Every once in a while a propeller looks like the one in Fig. 11-16(a), which is kind of a cross between Figs. 11-15(a) and 11-14(a): only opposite blades are the same. This figure has no reflection symmetries (try it!), and a 90° rotation won't work either [Fig. 11-16(b)]. Only the identity and a 180° rotation are possible as symmetries of this propeller. Any object having only these symmetries is of symmetry type Z_2. Figure 11-17 shows several additional examples of shapes of symmetry type Z_2.

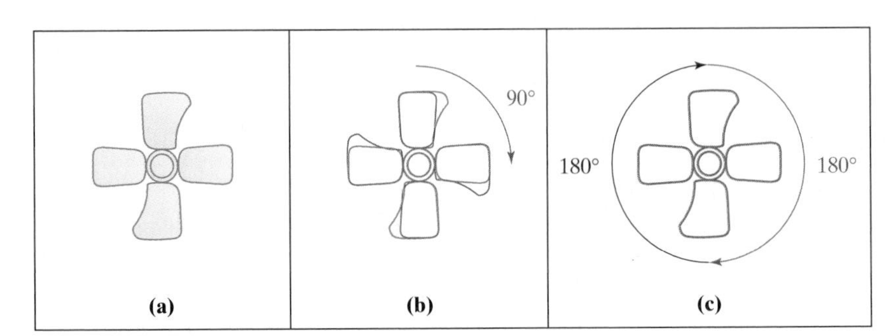

FIGURE 11-16 A propeller with only two rotation symmetries and no reflection symmetries. (Symmetry type Z_2.)

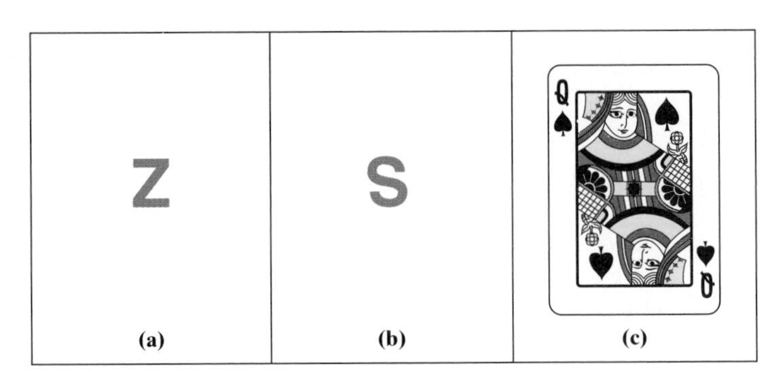

FIGURE 11-17 Objects with symmetry type Z_2. (a) The letter Z, (b) the letter S (in some fonts but not in others), (c) the Queen of Spades (and many other cards in the deck).

■

Example 6. The Symmetries of a Butterfly, etc.

One of the most common symmetry types occurring in nature is that of objects having only 1 reflection symmetry and 1 rotation symmetry (the identity). This symmetry type is called D_1. Figure 11-18 shows several examples of shapes and objects having symmetry type D_1. Notice that it doesn't matter if the axis of reflection is vertical, horizontal, or anywhere in between: If the figure has only 1 reflection symmetry, it is guaranteed to be of symmetry type D_1 (the identity is automatic).

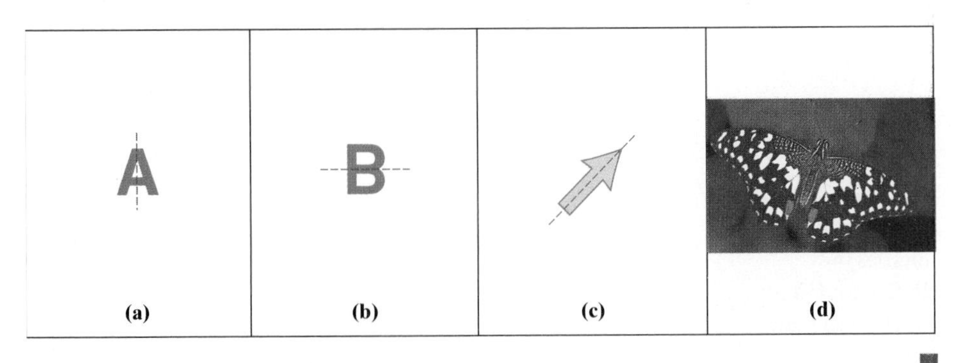

FIGURE 11-18 Shapes and objects with only 1 reflection symmetry (Symmetry type D_1). The axis of the reflection is in red.

■

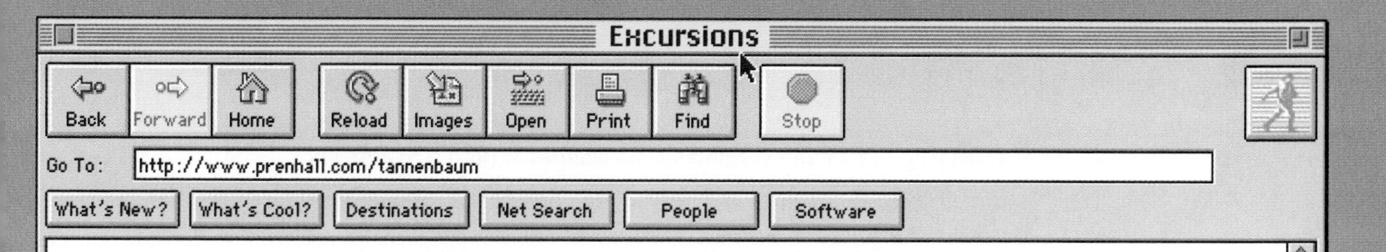

If You Start Repeating Yourself, Frieze

In architecture, a horizontal decorated strip across the side of a building is called a frieze. Friezes often use repeating border patterns like those described in this chapter. Some frieze patterns, with different symmetry types, are shown under Chapter 11 in the Tannenbaum Website.

Questions:

1. Using the standard notation, classify the frieze labeled "Pattern 2: Glide reflection" in the first Website.
2. Classify the frieze labeled "Pattern 4: Two half turns" by the same method.
3. Why is there no such thing as type *m*2?
4. Why is there no type 1*g*2?
5. Why is there no type *mg*2?
6. How are the two basic translation directions of the wallpaper patterns at the third site related to each other? How many other translation directions are generated by the two basic ones?

Example 7. The Symmetries of "Shapes With No Symmetry"

Many objects and shapes are informally considered to have no symmetry at all, but this is a misnomer, since *every object has at least the identity symmetry.* Objects whose only symmetry is the identity are said to have symmetry type Z_1. Figure 11-19 shows a few examples of objects of symmetry type Z_1—there are plenty of such objects around. ■

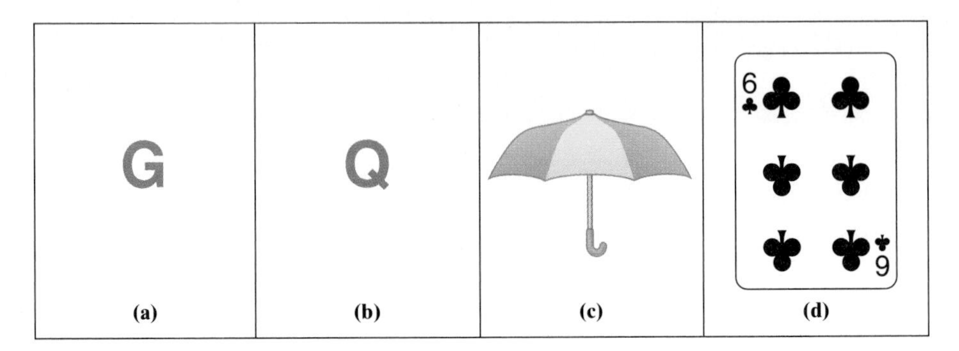

FIGURE 11-19 Shapes and objects with only the identity symmetry (Symmetry type Z_1). Why doesn't the six of clubs have a 180° rotation symmetry?

Example 8. Shapes with Many Rotations and Many Reflections

In everyday language, certain objects and shapes are said to be "highly symmetric" when they have lots of rotation and reflection symmetries. Figure 11-20(a) shows a snowflake, with six reflection symmetries (can you find all six axes of symmetry?) and six rotation symmetries (the rotocenter is the center of the snowflake and the angles are 60°, 120°, 180°, 240°, 300°, and 0°). The snowflake, like all other snowflakes, has symmetry type D_6 (short for 6 reflections and 6 rotations).[8] Figure 11-20(b) shows a ceramic plate—it has 18 reflections and 18 rotation symmetries, and its symmetry type (not surprisingly) is called D_{18}. Finally, in Fig. 11-20(c) we have a picture of a daisy with 21 petals. When perfect, it has 21 reflections and 21 rotations and is of symmetry type D_{21}. ■

FIGURE 11-20 (a) Snowflake (Symmetry type D_6); (b) ceramic plate, (Symmetry type D_{18}), (c) daisy (Symmetry type D_{21}).

In each case illustrated in Example 8, the number of reflections matches the number of rotations. This was also true in Examples 2, 3, and 6. Coincidence? Not at all. When an object or shape has *both* reflection and rotation symmetries, the number of rotation symmetries (remember, this includes the identity) has to match the number of reflection symmetries! Any finite object or shape with exactly N

[8] This symmetry type occurs often in nature; it is commonly known as *hexagonal symmetry* because it is the symmetry type of the regular hexagon.

reflection symmetries and N rotation symmetries is said to have symmetry type D_N. The standard example for a shape with symmetry type D_N is the regular polygon with N sides, commonly known as the *regular N-gon.*

Example 9. Shapes with Infinitely Many Rotations and Reflections

If we are looking for a two-dimensional shape that has *as much symmetry as possible,* we don't have to look past the wheels of a car to get the picture. The wheel works so wonderfully well as a means of locomotion because of the infinitely many rotation and reflection symmetries of the circle. In a circle, a rotation with center the center of the circle by any angle whatsoever is a symmetry, and any line passing through the center of the circle can be used as an axis of reflection symmetry. We will call the symmetry type of the circle D_{infinity}. ■

Example 10. Shapes With Rotations But No Reflections

We now know that if a finite two-dimensional shape has rotations *and* reflections, then it must have exactly the same number of each and it has symmetry type which we will informally denote as $D_{\text{something}}$.[9] But we also saw examples (Examples 4, 5, and 7) of shapes that have rotations *but no* reflections. In this case we used the letter Z to describe the symmetry type, with a subscript indicating the actual number of rotations. Figure 11-21 shows examples of shapes having symmetry types of the Z-something variety.

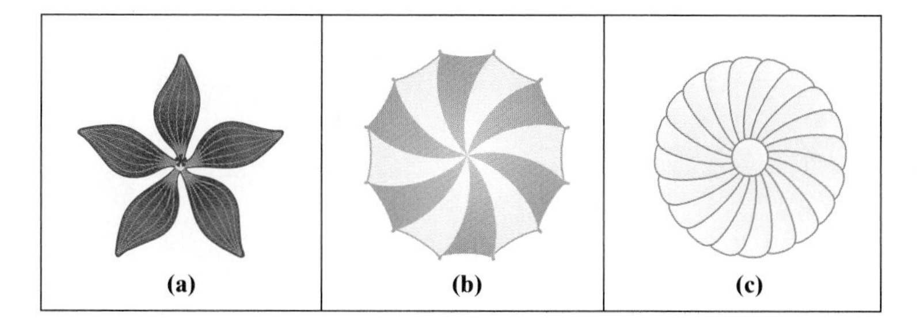

FIGURE 11-21 (a) Hibiscus (Symmetry type Z_5), (b) Parasol (top view) (Symmetry type Z_6), (c) Turbine (Symmetry type Z_{20}). ■

Patterns

Well, we've come a long way, but we have yet to see examples of shapes having translation and/or glide reflection symmetry. In fact, if we think of objects and shapes as finite, then translation symmetry is impossible (there is no way that a finite object can be slid a certain distance and still be exactly where it was before!). On the other hand, if we broaden our interpretation of a "shape" and allow infinitely repeating patterns, then translation symmetry is not only possible, it is necessary!

An infinite shape made up of one or more infinitely repeating themes is called a **pattern**. A pattern is really an abstraction—in the real world there are no infinite objects as such, although the idea of an infinitely repeating pattern is all too familiar to us from wallpaper, textiles, carpets, ribbons, and so on.

---

[9] The formal mathematical name for this class of symmetry types is *dihedral symmetry.*

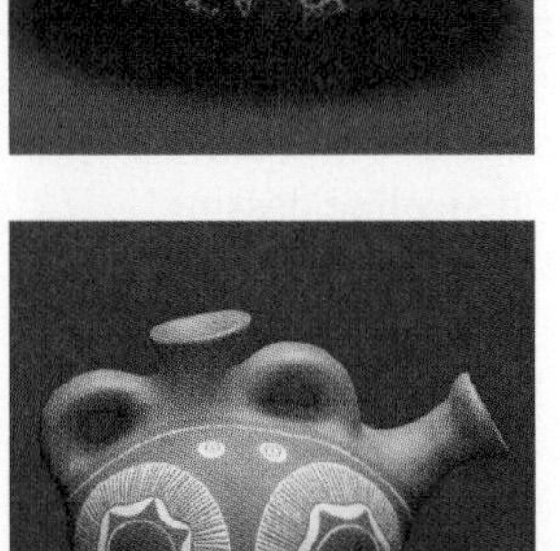

Just like finite shapes, patterns fall into symmetry types. The classification of patterns according to their symmetry type is of fundamental importance in the study of molecular and crystal organization in chemistry, so it is not surprising that some of the first people to seriously investigate the symmetry types of patterns were crystallographers. Archeologists and anthropologists have also found that the symmetry types characteristic of a particular culture (in their textile and pottery) can be used as a means to gain a better understanding of that culture.

We will briefly discuss the symmetry types of one-dimensional and two-dimensional patterns. A comprehensive study of patterns is beyond the scope of this book, so we will not go into as much detail as we did with finite shapes.

Border (One-Dimensional) Patterns

In a one-dimensional pattern the infinitely repeating theme repeats itself in only one direction. One-dimensional patterns are commonly known as **border patterns**, and they are found in ribbons, borders, baskets, pottery, and so on.

The most common direction in a border pattern (what we will call the *direction of the pattern*) is horizontal, but in general a border pattern can be in any direction. (For typesetting in a book, it is more efficient to display a border pattern horizontally, and we will do so from now on. Thus, when we say "horizontal direction" we really mean *the direction of the pattern,* and it follows that when we say "vertical direction" we really mean *the direction perpendicular to the direction of the pattern.*)

In a border pattern, a theme repeats itself in one direction, be it in a straight line or wrapping itself into a circle. (*Center:* Enamelled brick frieze from the Palace of Artaxerxes II, Susa. Achaemenid period, 5th c. BCE. Louvre, Paris, France. Giraudon/Art Resource, NY.)

We will now discuss the possible symmetries of a border pattern. At first, one might think that there are more possibilities for symmetry in a border pattern than in a finite shape, but in fact the opposite is true. The possibilities for symmetry in a border pattern are fairly limited.

- **Translations.** Every border pattern has translation symmetry, and the translation is always in the direction of the pattern (Fig. 11-22).

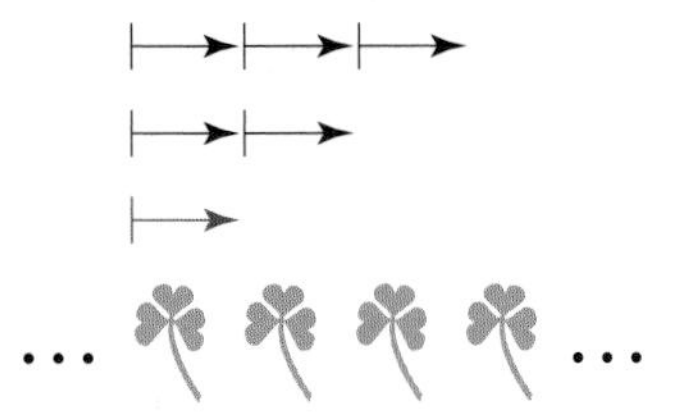

FIGURE 11-22 The basic unit of translation is shown in red. Any multiples of the red translation are also translation symmetries.

- **Reflections.** A border pattern can have (a) no reflection symmetry (Fig. 11-22); (b) only horizontal (i.e., in the direction of the pattern) reflection symmetry [Fig. 11-23(a)]; (c) only vertical reflection symmetries [Fig. 11-23(b)]; (d) both horizontal and vertical reflection symmetries [Fig. 11-23(c)]. No other reflection symmetries are possible (Exercise 33).

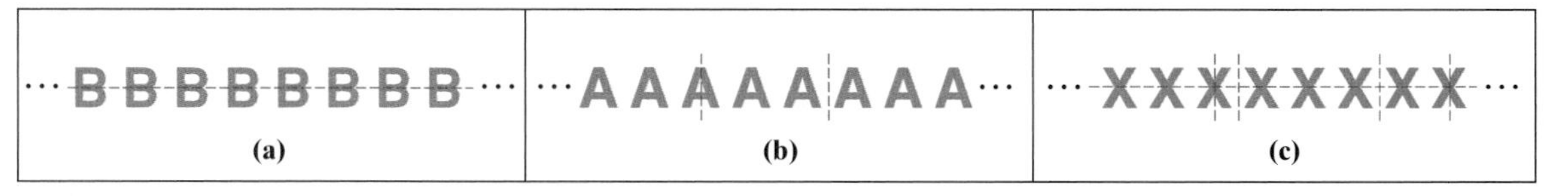

FIGURE 11-23
Border patterns with (a) horizontal symmetry only, (b) vertical symmetries only (many axes are possible) and, (c) both.

- **Rotations.** A border pattern can have (a) only one rotation symmetry—the identity [Fig. 11-24(a)]; or (b) two rotation symmetries—the identity and 180° rotation [Fig. 11-24(b)]. Remarkably, no other rotation symmetries are possible (Exercise 34).

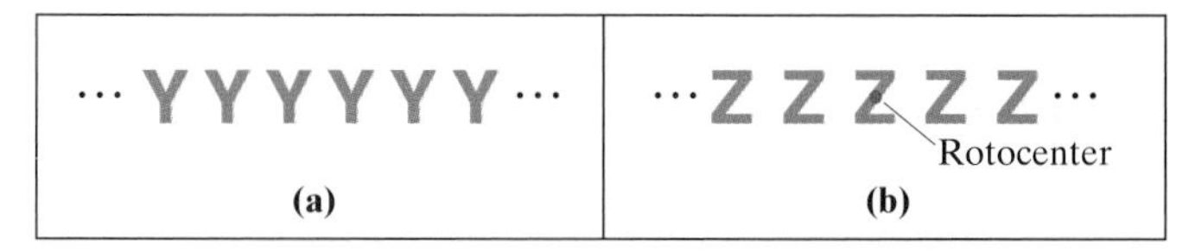

FIGURE 11-24
Border patterns with (a) one rotation symmetry (the identity);
(b) two rotation symmetries (identity and 180°).

- **Glide reflections.** A border pattern can have (a) no glide reflection symmetry [Fig. 11-25(a)]; or (b) glide reflection symmetries. The latter can happen only under fairly restrictive conditions: the axis of reflection *has* to be a line along the center of the pattern, and the reflection in the glide reflection cannot itself be a symmetry of the pattern.[10]

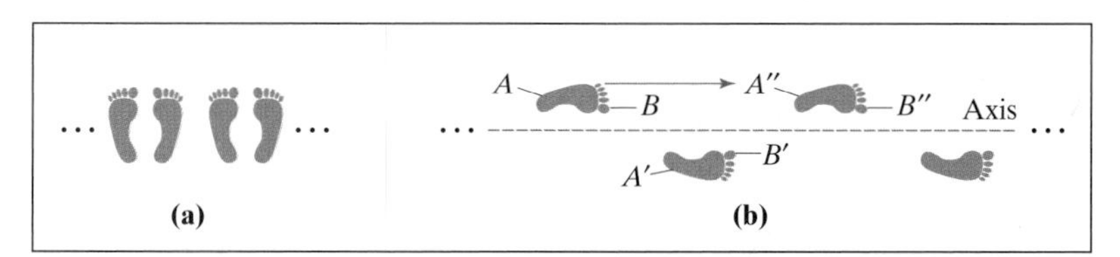

FIGURE 11-25
(a) No glide refletion symmetry; (b) glide reflection symmetry. The vector of translation and the axis of reflection shown in red. Neither the glide alone nor the reflection alone are a symmetry.

In how many different ways can the different possible symmetries of a border pattern be combined into symmetry types? Surprisingly, the answer is very few: Every border pattern falls into one of *only 7 possible symmetry types.* These, together with their odd names, are illustrated in Table 11-1.

---

[10] Thus, patterns such as those in Fig. 11-23(a) and (c), which have both translation and horizontal reflection symmetry, are *not* considered to have glide reflection symmetry.

Table 11.1 The seven symmetry types for border patterns.

Symmetry Type*	Translations	Horizontal Reflections	Vertical Reflections	180° Rotations	Glide Reflection	Example
1. **11**	Yes	No	No	No	No	
2. **1*m***	Yes	Yes	No	No	No	
3. ***m*1**	Yes	No	Yes	No	No	
4. **12**	Yes	No	No	Yes	No	
5. **1*g***	Yes	No	No	No	Yes	
6. ***mg***	Yes	No	Yes	Yes	Yes	
7. ***mm***	Yes	Yes	Yes	Yes	No	

(*Standard notation used in crystallography.)

Wallpaper (Two-Dimensional) Patterns

Two-dimensional patterns are patterns that repeat themselves in at least two different (nonparallel) directions in the plane. Two-dimensional patterns are commonly called **wallpaper patterns**. Typical examples of can be found in wallpaper (of course), carpets, textiles, and so on.

With wallpaper patterns things get a bit more complicated, so we will skip the details.

- **Translations.** Every wallpaper pattern has translation symmetry in at least two different (nonparallel) directions (Fig. 11-26).

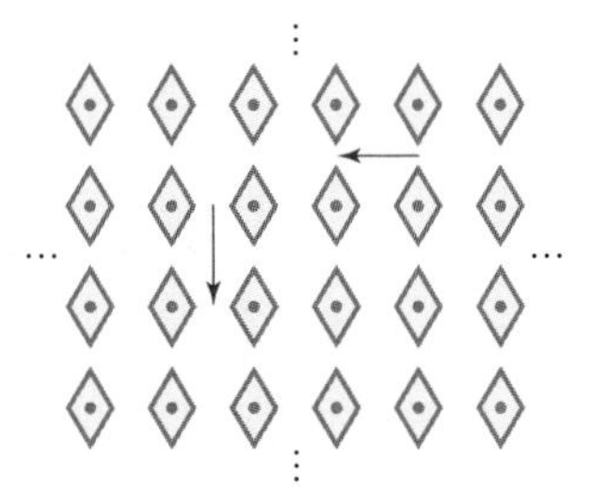

FIGURE 11-26

- **Reflections.** A wallpaper pattern can have (a) no reflections; (b) reflections in only 1 direction; (c) reflections in 2 nonparallel directions, (d) reflections in 3 nonparallel directions; (e) reflections in 4 nonparallel directions; (f) reflections in 6 nonparallel directions. There are no other possibilities. Note that particularly conspicuous in its absence is the case of reflections in exactly 5 different directions. (Examples are shown in the chapter appendix.)
- **Rotations.** In terms of rotation symmetries, a wallpaper pattern can have (a) the identity only; (b) 2 rotations (identity and 180°); (c) 3 rotations (identity,

120°, and 240°) (d) 4 rotations (identity, 90°, 180°, and 270°); and (e) 6 rotations (identity, 60°, 120°, 180°, 240°, and 300°). There are no other possibilities. Once again, note that a wallpaper pattern cannot have exactly 5 different rotations. (Examples are shown in the chapter appendix.)

- **Glide Reflections.** Just like reflections, a wallpaper pattern can have (a) no glide reflections; (b) glide reflections in only 1 direction; (c) glide reflections in 2 nonparallel directions; (d) glide reflections in 3 nonparallel directions; (e) glide reflections in 4 nonparallel directions; (f) glide reflections in 6 nonparallel directions. There are no other possibilities. (Examples are shown in the chapter appendix.)

It is a truly remarkable fact that in spite of all these possibilities, the symmetries of a wallpaper pattern can be combined into only *17 distinct symmetry types.* The hundreds and thousands of wallpapers one can find at a decorating store all fall into just 17 different symmetry families. They are listed and illustrated in the chapter appendix.

Conclusion

Real-life tangible physical objects as well as abstract shapes from geometry, art, and ornamental design are often judged and measured by a yardstick that can be both mathematical and aesthetic: *How much symmetry and what kinds of symmetry does it have?*

The possibilities, while limitless, fall into a small and well-defined set of categories. For two-dimensional objects and shapes that are finite, there are really only two possible scenarios: The object has rotation symmetries only (a *Z-something* kind of shape), or it has both rotation and reflection symmetries in equal amounts (a *D-something* kind of shape). It is quite remarkable that there are no other possibilities. Nowhere in the universe of two-dimensional shapes does there exist, for example, a shape with three reflection symmetries and five rotation symmetries—it just can't happen.

Patterns—that is, shapes with an infinitely repeating theme—are even more surprising in their symmetry pedigrees. One-dimensional patterns, commonly known as *border patterns,* fall into just *seven* different symmetry types, whereas two-dimensional patterns, such as those found in wallpapers and textiles, fall into just *seventeen* different symmetry types. It wasn't until 1924 that a rigorous mathematical proof of the latter fact was given by the Hungarian mathematician George Polya.

In this chapter we learned that there is a lot more to symmetry than a reflection in a mirror, and that the key to unlocking its mysteries can be found in mathematics. We conclude with a brief quote from the great mathematician Hermann Weyl:

> *Symmetry is a vast subject, significant in art and nature. Mathematics lies at its root, and it would be hard to find a better one on which to demonstrate the working of the mathematical intellect.*

Key Concepts

angle (of rotation)
axis (of reflection, of symmetry)
basic rigid motions of the plane
bilateral symmetry
border pattern (one-dimensional pattern)
equivalent rigid motion
fixed point
glide reflection (rigid motion)
glide reflection symmetry
identity (rigid motion)
image
improper (rigid motion)
pattern
proper (rigid motion)
reflection (rigid motion)
reflection symmetry
rigid motion
rotation (rigid motion)
rotation symmetry
rotocenter
symmetry (geometric symmetry)
symmetry type
translation (rigid motion)
translation symmetry
vector (of translation)
wallpaper pattern (two-dimensional pattern)

Exercises

Walking

1. Given a reflection that sends the point P to the point P' as shown in the figure, find:

(a) The axis of reflection,

(b) Q' (the image of Q) under the reflection.

(c) The image of triangle PQR under the reflection.

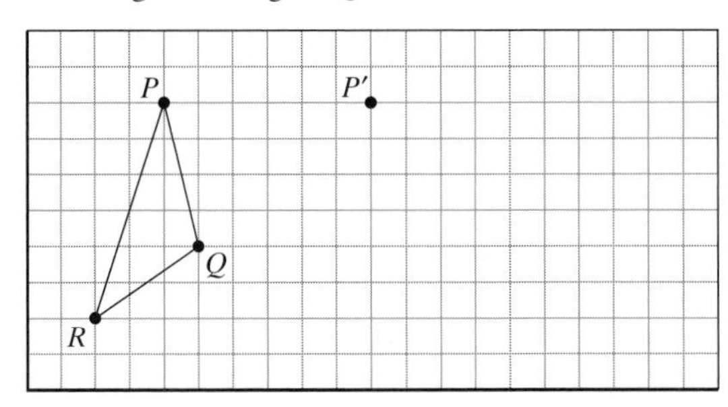

2. Given a reflection with the axis of reflection as shown in the figure, find:

(a) P' (the image of P) under the reflection.

(b) The image of triangle PQR under the reflection.

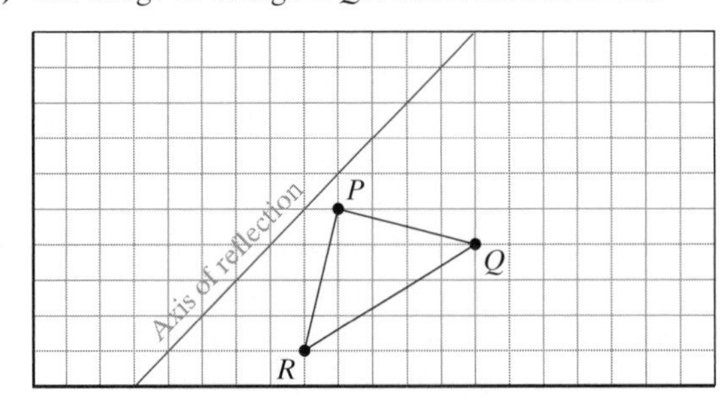

3. In each of the following give an answer between 0° and 360°.

(a) A clockwise rotation by an angle of 500° is equivalent to a clockwise rotation by an angle of ________.

(b) A clockwise rotation by an angle of 3681° is equivalent to a counterclockwise rotation by an angle of ________.

4. In each of the following give an answer between 0° and 360°.

   (a) A clockwise rotation by an angle of 500° is equivalent to a counterclockwise rotation by an angle of ________.

   (b) A clockwise rotation by an angle of 3681° is equivalent to a clockwise rotation by an angle of ________.

5. Given a rotation that sends the point B to the point B' and the point C to the point C' as shown in the figure, find:

   (a) The center of the rotation.

   (b) The image of triangle ABC under the rotation.

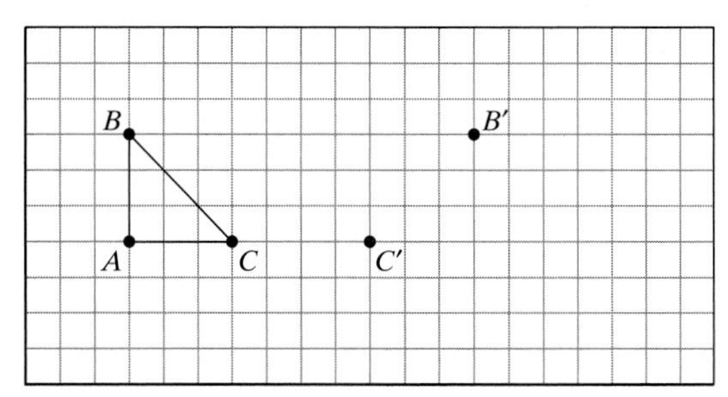

6. Given a rotation that sends the point B to the point B' and the point C to the point C' as shown in the figure, find:

   (a) The center of the rotation.

   (b) The image of triangle ABC under the rotation.

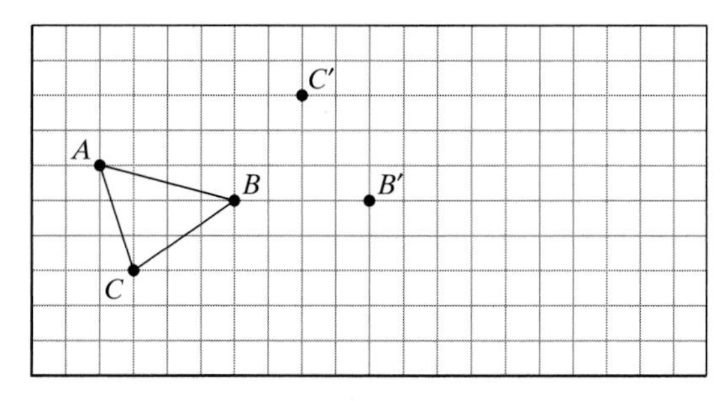

7. Given a translation that sends the point E to the point E' as shown in the figure, find:

   (a) The image A' of A under the translation.

   (b) The image of figure $ABCDE$ under the translation.

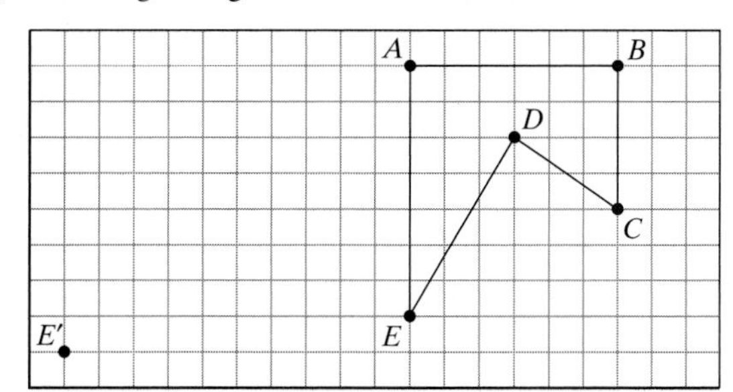

8. Given a translation that sends the point Q to the point Q' as shown in the figure, find:

   (a) The image P' of P under the translation.

   (b) The image of figure $PQRS$ under the translation.

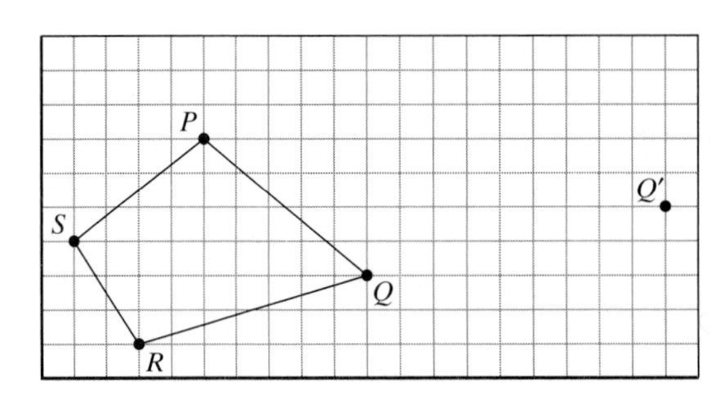

9. Given a glide reflection that sends the point B to the point B' and the point D to the point D' as shown in the figure, find:

(a) The axis of the glide reflection.

(b) A' (the image of A) under the glide reflection.

(c) The image of figure $ABCDE$ under the glide reflection.

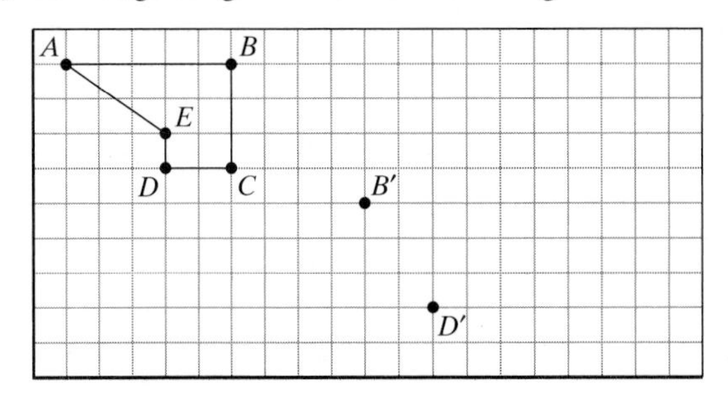

10. Given a glide reflection that sends the point A to the point A' and the point C to the point C' as shown in the figure, find:

(a) The axis of the glide reflection.

(b) B' (the image of B) under the glide reflection.

(c) The image of figure $ABCD$ under the glide reflection.

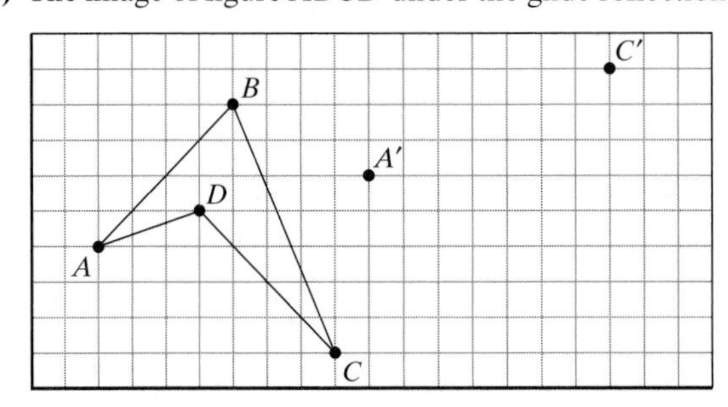

11. What "word" do you get if you reflect each of the given "words" about a vertical axis?

(a) TAd

(b) TIM

(c) bAd

(d) Yod

12. What "word" do you get if you rotate each of the given "words" 180° about the center of the "O"?

(a) SOS

(b) WOW

(c) MON

(d) NOS

13. Find the symmetry type for each of the following symbols.

(a) A

(b) D

(c) +

(d) Z

(e) Q

14. Find the symmetry type for each of the following figures.

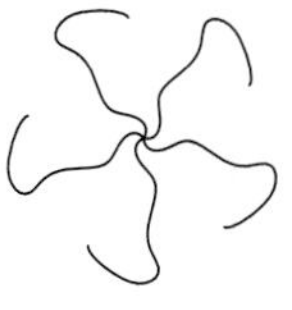

(a) **(b)** **(c)** **(d)**

15. Give an example of a capital letter of the alphabet that has symmetry type:

(a) Z_1.

(b) D_1.

(c) Z_2.

(d) D_2.

(e) D_{infinity}.

16. Give an example of a numeral that has symmetry type:

(a) Z_1.

(b) D_1.

(c) Z_2.

(d) D_2.

(e) D_{infinity}.

17. (a) Give an example of a figure that has symmetry type D_3.

(b) Give an example of a figure that has symmetry type Z_3.

18. **(a)** Give an example of a figure that has symmetry type D_8.

(b) Give an example of a figure that has symmetry type Z_8.

19. Find the symmetry type for each of the following figures.

(a) ⊠

(b) ✥

(c) ✾

(d) ✻

20. Find the symmetry type for each of the following figures.

(a) ❧

(b) ♣

(c) ➤

(d)  “”

21. Describe all the symmetries of each of the following border patterns.

(a) · · · A A A A A A A A · · ·

(b) · · · D D D D D D D D · · ·

(c) · · · S S S S S S S S · · ·

(d) · · · L L L L L L L L · · ·

22. Describe all the symmetries of each of the following border patterns.

(a) · · · ⊠ ⊠ ⊠ ⊠ ⊠ ⊠ ⊠ · · ·
(b) · · · ✤ ✤ ✤ ✤ ✤ ✤ · · ·
(c) · · · ✾ ✾ ✾ ✾ ✾ ✾ ✾ ✾ · · ·
(d) · · · ✱ ✱✱✱ ✱ ✱ · · ·

23. Describe all the symmetries of each of the following border patterns.

(a) · · · ❧ ❧ ❧ ❧ ❧ ❧ · · ·
(b) · · · ♣ ♣ ♣ ♣ ♣ ♣ · · ·
(c) · · · ➤ ➤ ➤ ➤ ➤ · · ·
(d) · · · ❝❞ ❝❞ ❝❞ ❝❞ · · ·

24. Explain why any proper rigid motion that has a fixed point must be equivalent to a rotation.

25. Explain why any rigid motion other than the identity that has two or more fixed points must be equivalent to a reflection.

Jogging

26. Consider a square with axes of reflection l_1, l_2, l_3, and l_4 as shown in the figure. Let r_1, r_2, r_3, r_4 denote reflections with axis l_1, l_1, l_3, l_4, respectively. Also, let R_{90}, R_{180}, R_{270} denote clockwise rotations of 90°, 180°, 270°, respectively, about the center of the square and I denote the identity motion.

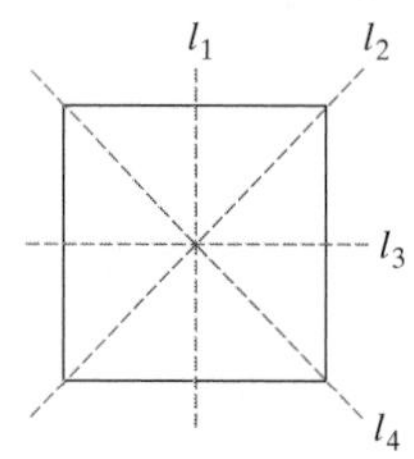

In each of the following, specify the single rigid motion (as described above) that is equivalent to the given rigid motions.

(a) r_1 followed by r_2.

(b) r_2 followed by r_1.

(c) r_1 followed by R_{90}.

(d) R_{90} followed by r_1.

(e) R_{270} followed by R_{270}.

(f) R_{270} followed by r_3 followed by R_{90}.

27. Explain why the propeller shown in the figure cannot have any reflection symmetries.

28. **(a)** Rotation 1 has center C and a clockwise angle of 30°, and rotation 2 has the same center C and a clockwise angle of 50°. Show that the result of applying rotation 1 followed by rotation 2 is equivalent to a rotation with center C and a clockwise angle of 80°.

(b) Show that if rotation 1 has center C and clockwise angle α and rotation 2 has center C and clockwise angle β, then the result of applying rotation 1 followed by rotation 2 is equivalent to a rotation with center C and clockwise angle $\alpha + \beta$.

(c) Show that for the two rotations described in (b) the result of applying rotation 1 followed by rotation 2 is equivalent to applying rotation 2 followed by rotation 1.

29. **(a)** Given a glide reflection with axis and vector as shown, find the image Q'' of the point Q when the glide reflection is applied twice.

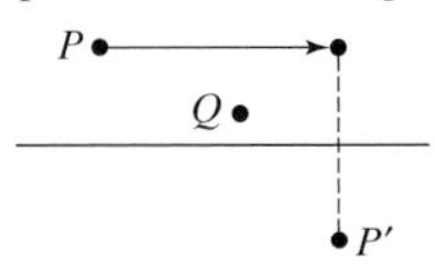

(b) Show that the result of applying the same glide reflection twice is equivalent to a translation. Describe the direction and amount of the translation in terms of the direction and amount of the original glide.

30. Reflection 1 has axis l_1; reflection 2 has axis l_2; l_1 and l_2 are parallel; and the distance between them is d.

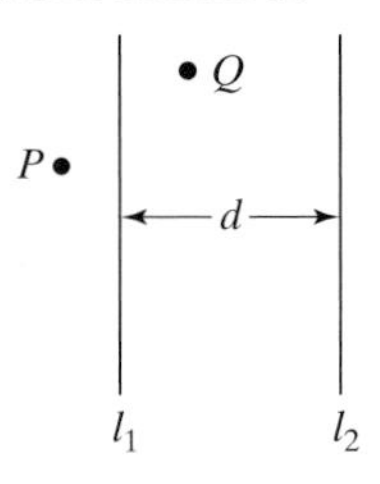

(a) Find the images of P and Q when we apply reflection 1 followed by reflection 2.

(b) Show that the result of applying reflection 1 followed by reflection 2 is a translation. Describe the direction and amount of the translation.

(c) Show that the result of applying reflection 1 followed by reflection 2 is not the same as the result of applying reflection 2 first followed by reflection 1. Describe the difference.

31. Reflection 1 has axis l_1; reflection 2 has axis l_2; l_1 and l_2 intersect at C. The angle between l_1 and l_2 as shown in the figure is α.

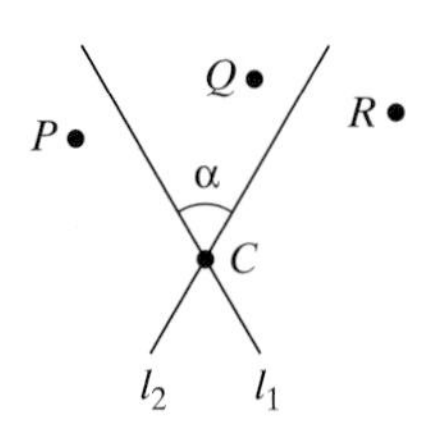

(a) Find the images of P, Q, and R when we apply reflection 1 followed by reflection 2.

(b) Show that the result of applying reflection 1 followed by reflection 2 is a rotation with center C. Give the clockwise angle of rotation.

(c) Show that the result of applying reflection 2 first followed by reflection 1 is a different rotation from the one found in (b). Describe the difference.

32. Translation 1 moves point P to point P'; translation 2 moves point Q to point Q', as shown in the figure.

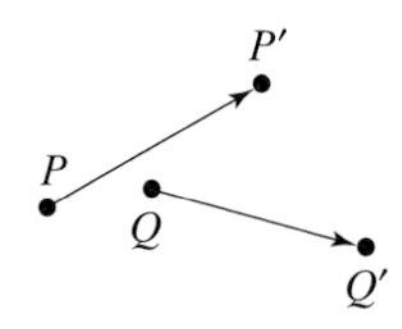

(a) Find the images of P and Q when we apply translation 1 followed by translation 2.

(b) Find the images of P and Q when we apply translation 2 followed by translation 1.

(c) Show that the result of applying translation 1 followed by translation 2 is a translation. Give a geometric description of the vector of the translation.

33. (a) Explain why a border pattern cannot have a reflection symmetry along an axis forming 45° with the direction of the pattern.

(b) Explain why a border pattern can have only horizontal and/or vertical reflection symmetry.

34. (a) Explain why a border pattern cannot have a rotation symmetry of 90°.

(b) Explain why a border pattern can have only the identity or a 180° rotation symmetry.

35. Describe all the symmetries of each of the following border patterns.

(a) ··· MOWMOWMO ···

(b) ··· p d p d p d ···

(c) ··· p d b q p d b q ···

36. For each of the following sets of symmetries, give an example of a border pattern that has exactly those symmetries (no more and no less). Do not use any of the patterns in Exercise 35. You can use letters of the alphabet, numbers, or symbols to create the patterns.

(a) Translations only.

(b) Translations and vertical reflections.

(c) Translations and horizontal reflections.

(d) Translations and 180° rotations.

(e) Translations and glide reflections.

(f) Translations, vertical reflections, glide reflections, and 180° rotations.

(g) Translations, vertical reflections, horizontal reflections, and 180° rotations.

37. A rigid motion moves the triangle PQR into the triangle $P'Q'R'$ as shown in the figure.

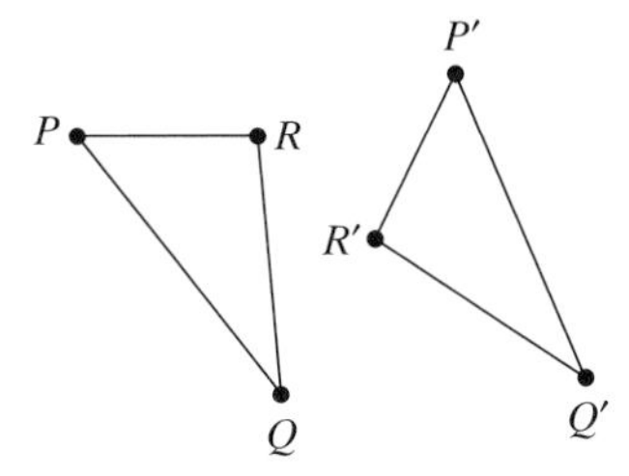

(a) Explain why the rigid motion cannot possibly be a rotation or a translation. (*Hint:* Is the rigid motion proper or improper?)

(b) Explain why the rigid motion cannot possibly be a reflection.

(c) What kind of rigid motion is it, then?

38. What single rigid motion is the result of applying two consecutive reflections with parallel axes followed by a translation?

39. What single rigid motion is the result of applying two consecutive reflections with intersecting axes followed by a rotation about the point of intersection of the axes?

40. What single rigid motion is the result of applying two consecutive reflections with parallel axes followed by two more consecutive reflections with parallel axes? (*Note:* The second pair of parallel axes need not be parallel to the first pair of parallel axes.)

41. What single rigid motion is the result of applying the same glide reflection three times?

42. A *palindrome* is a word that is the same when read forward or backward. **MOM** is a palindrome, and so is **ANNA**. (For simplicity we will assume all letters are capitals.)

(a) Explain why if a word has vertical reflection symmetry, then it must be a palindrome.

(b) Give an example of a palindrome (other than **ANNA**) that doesn't have vertical reflection symmetry.

(c) If a palindrome has vertical reflection symmetry, what can you say about the symmetries of the individual letters in the word?

(d) Find a palindrome with 180° rotational symmetry.

43. Find examples from the real world (ribbons, borders, baskets, etc.) of each of the 7 border-pattern symmetry types.

Running

44. Suppose that a rigid motion moves points P to P', Q to Q', and R to R' (as in the figure). We do not know what kind of rigid motion it is.

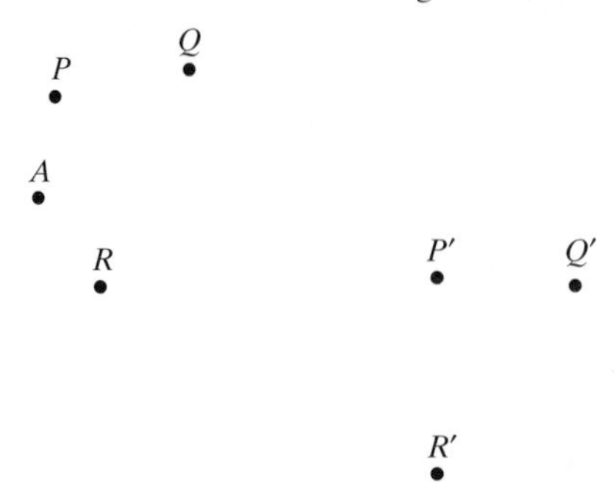

(a) For an arbitrary point A in the plane, find the image A'. (*Hint:* $AP = A'P'$, $AQ = A'Q'$, and $AR = A'R'$.)

(b) Describe a general procedure for finding the image of a point A under some rigid motion when we know three points P, Q, and R (not on a straight line) and their images P', Q', and R'.

45. Rotation 1 is a rotation of 90° clockwise about center A. Rotation 2 is a rotation of 60° clockwise about center B.

(a) Show that the result of applying rotation 1 followed by rotation 2 is another rotation, and find the center and angle of this rotation.

(b) Show that the result of applying rotation 2 followed by rotation 1 is another rotation, and find the center and angle of this rotation.

(c) Generalize the results of (a) and (b) to the case where the angles of rotation are any angles between 0° and 180°.

46. Find all the symmetries of the following wallpaper pattern.

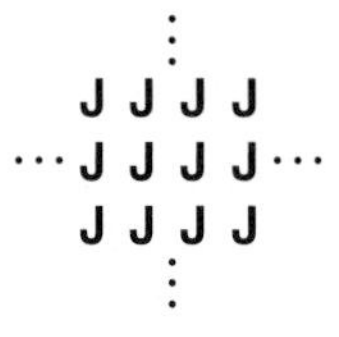

47. Find all the symmetries of the following wallpaper pattern.

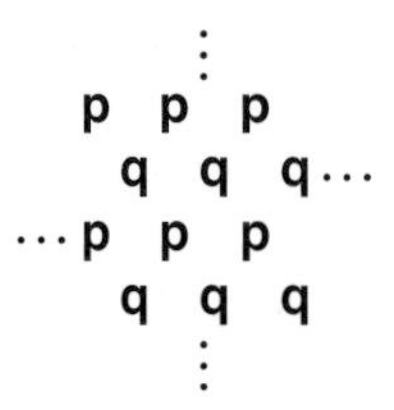

48. Find all the symmetries of the following wallpaper pattern.

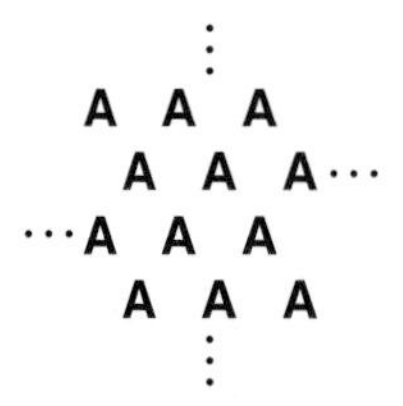

49. Find all the symmetries of the following wallpaper pattern.

```
       ⋮
 S   S   S
   S   S   S ⋯
⋯S   S   S
   S   S   S
       ⋮
```

50. Find all the symmetries of the following wallpaper pattern.

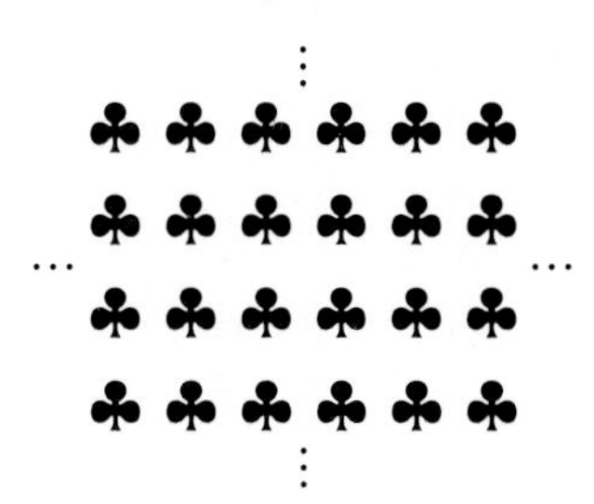

51. Find all the symmetries of the following wallpaper pattern.

52. Find all the symmetries of the following wallpaper pattern.

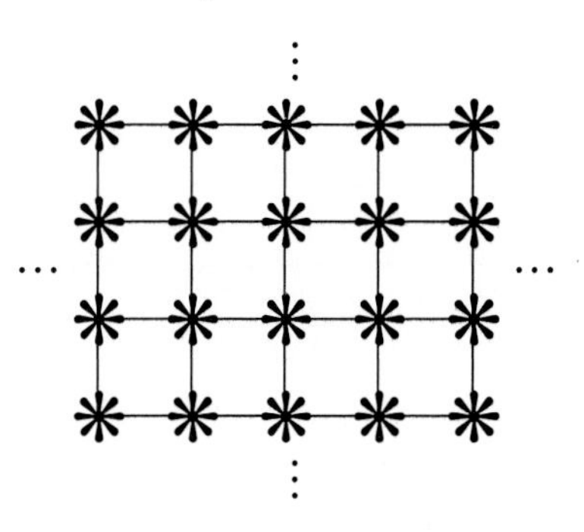

53. Find all the symmetries of the following wallpaper pattern.

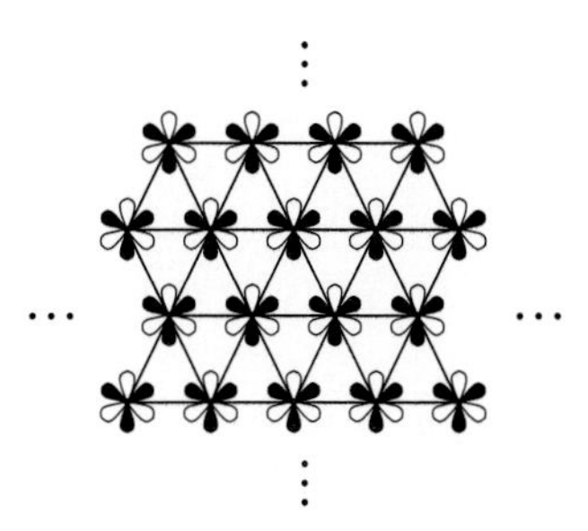

54. Show that the *only* symmetries of a square are the 8 symmetries given in Example 2.

55. (Open-ended problem) Find examples from the real world (wallpaper, wrapping paper, etc.) of each of the 17 wallpaper-pattern symmetry types.

APPENDIX The 17 Wallpaper Pattern Types

Symmetry Type*	Translation	Rotations (given by smallest angle) 60°	90°	120°	180°	Reflections (Number of Directions) 1	2	3	4	6	Glide Reflections (Number of Directions) 1	2	3	4	6	Example
pmg	✓				✓	✓					✓					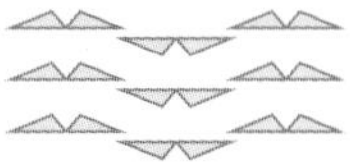
pgg	✓				✓							✓				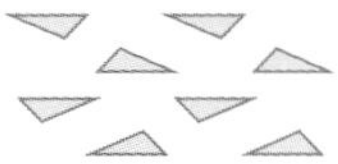
p2	✓				✓											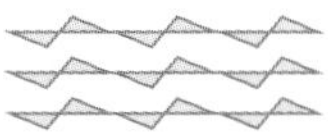
p4m	✓		✓[a]		✓[b]				✓			✓				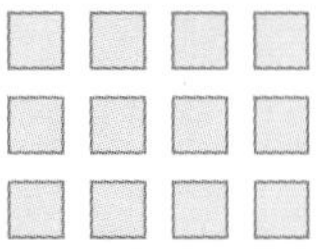
p4g	✓		✓[a]		✓[b]		✓							✓		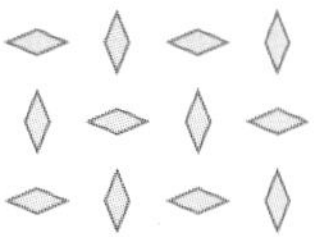
p4	✓		✓[a]		✓[b]											

*Notation adopted by the International Union of Crystallography (1952).
[a], [b]: Different rotocenters.

Symmetry Type*	Translation	Rotations (given by smallest angle)				Reflections (Number of Directions)					Glide Reflections (Number of Directions)					Example
		60°	90°	120°	180°	1	2	3	4	6	1	2	3	4	6	
cm	✓					✓					✓					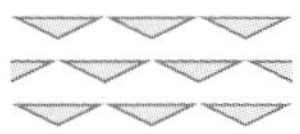
pm	✓					✓										
pg	✓										✓					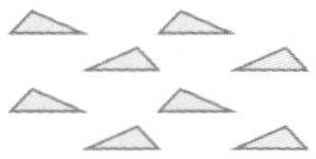
*p*1	✓															
pmm	✓				✓		✓									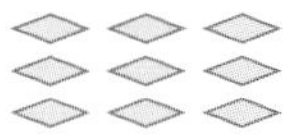
cmm	✓				✓		✓					✓				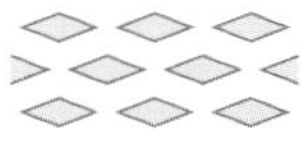

Symmetry Type*	Translation	Rotations (given by smallest angle)				Reflections (Number of Directions)					Glide Reflections (Number of Directions)					Example
		60°	90°	120°	180°	1	2	3	4	6	1	2	3	4	6	
*p*3*m*1	✓			✓[c]				✓					✓			
*p*31*m*	✓			✓[d]				✓					✓			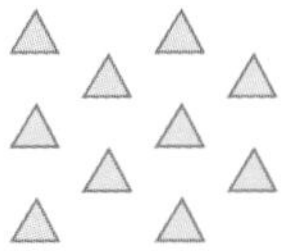
*p*3	✓			✓												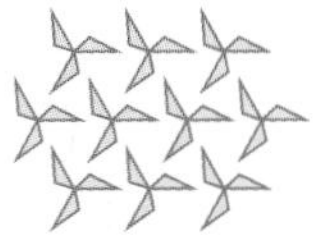
*p*6*m*	✓	✓		✓	✓					✓					✓	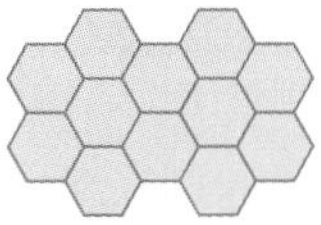
*p*6	✓	✓		✓	✓											

[c]: All rotocenters on axes of reflection.
[d]: Not all rotocenters on axes of reflection.

References and Further Readings

1. Bunch, Bryan, *Reality's Mirror: Exploring the Mathematics of Symmetry.* New York: John Wiley & Sons, Inc., 1989.
2. Coxeter, H. S. M., *Introduction to Geometry,* 2d ed. New York: John Wiley & Sons, Inc., 1967.
3. Crowe, Donald W., "Symmetry, Rigid Motions and Patterns," *UMAP Journal,* 8 (1987), 206–236.
4. Field, M., and M. Golubitsky, *Symmetry in Chaos.* New York: Oxford University Press, 1992.
5. Gardner, Martin, *The New Ambidextrous Universe: Symmetry and Asymmetry from Mirror Reflections to Superstrings,* 3d ed. New York: W. H. Freeman & Co., 1990.
6. Grunbaum, Branko, and G. C. Shephard, *Tilings and Patterns: An Introduction.* New York: W. H. Freeman & Co., 1989.
7. Hofstadter, Douglas R., *Gödel, Escher, Bach: An Eternal Golden Braid.* New York: Vintage Books, 1980.
8. Martin, George E., *Transformation Geometry: An Introduction to Symmetry.* New York: Springer Publishing Co., Inc., 1982.
9. Rose, Bruce, and Robert D. Stafford, "An Elementary Course in Mathematical Symmetry," *American Mathematical Monthly,* 88 (1981), 59–64.
10. Schattsneider, Doris, *Visions of Symmetry: Notebooks, Periodic Drawings, and Related Work of M. C. Escher.* New York: W. H. Freeman & Co., 1990.
11. Shubnikov, A. V., and V. A. Kopstik, *Symmetry in Science and Art.* New York; Plenum Publishing Corp., 1974.
12. Stewart, I., and M. Golubitsky, *Fearful Symmetry.* Cambridge, MA: Blackwell Publishers, 1992.
13. Weyl, Hermann, *Symmetry.* Princeton, NJ: Princeton University Press, 1952.
14. Wigner, Eugene, *Symmetries and Reflections.* Bloomington, IN: Indiana University Press, 1967.

Nature is a mutable cloud
Which is always and never the same
Ralph Waldo Emerson

CHAPTER 12

Fractal Geometry

Fractally Speaking

There is something unique and distinctive about many of nature's most beautiful creations. Mountains always look like mountains, even as they differ in their details. There is an undefinable but unmistakable "mountain look." And we can always tell a fake mountain from a real mountain, can't we? It is practically impossible to capture, outside of a photograph, that true "mountain look"—isn't it? And what's true for mountains is true for clouds, rivers, trees, and so on. So, which of the images on the facing page are real photographs and which are fakes? (*Hint:* Not all are real.)

Over the last 20 years an entirely new type of geometry has allowed humans to understand and reconstruct many of nature's most complex "looks," ranging from the everyday world of mountains and clouds to the microscopic world of the body's vascular system to the otherworldly look of planets and galaxies.

In this chapter we will introduce the basic ideas behind this new geometry of natural shapes, called **fractal geometry.** The conceptual building blocks of fractal geometry are the notions of *recursive replacement rules* and *self-similarity,* and we will illustrate both of these concepts by means of several important examples.

The Koch Snowflake

The Koch snowflake is a remarkable geometric shape "discovered" by the Swedish mathematician Helge von Koch in the early 1900s. The construction of the Koch snowflake proceeds as follows:

- **Start.** Start with a solid *equilateral* triangle of arbitrary size [Fig. 12-1(a)]. (For simplicity we will assume that the sides of the triangle are of length 1 and call the area of the triangle A.)
- **Step 1.** **Procedure KS:** *Attach in the middle of each side a smaller equilateral triangle, with sides of length one-third of the previous side* [Fig. 12-1(b)]. When we are done, the result is a "star of David" with 12 sides, each of length 1/3 [Fig. 12-1(c)].

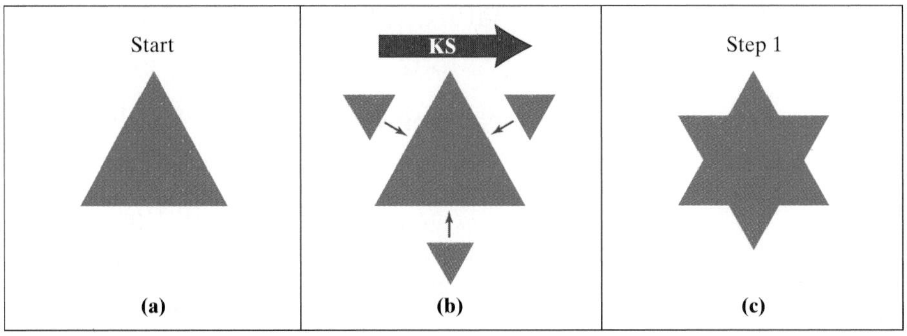

FIGURE 12-1 (a) A solid blue equilateral triangle. (b) Smaller copies of the original are added on each side. (c) A solid blue star.

- **Step 2.** For each of the 12 sides of the star of David in Step 1, repeat *procedure KS* : In the middle of each side attach a small equilateral triangle (with dimensions one-third of the dimensions of the side). The resulting shape has 48 sides, each of length 1/9 [Fig. 12-2(a)].
- **Steps 3, 4, etc.** Continue repeating *procedure KS* ad infinitum [Fig. 12-2(b), (c), etc.].

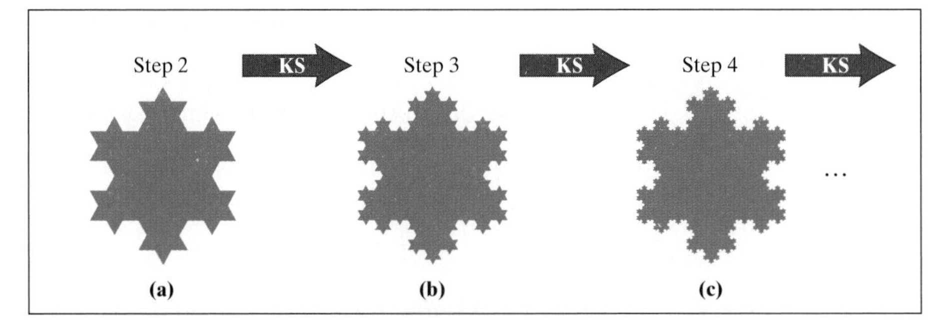

FIGURE 12-2 Successive steps in the recursive process leading toward the Koch snowflake.

At each step of this construction the figure changes a little, but after a while the changes are less and less noticeable. By the seventh or eight step the process has become *visually stable* (we really can't see the difference between the seventh and eighth step with the naked eye). For all practical purposes, what we are seeing is the ultimate shape we want: the **Koch snowflake** itself (Fig. 12-3).

FIGURE 12-3
A rendering of the Koch snowflake.

It is clear that, because the process of building the Koch snowflake is infinite, a perfect picture of it is impossible. However, this should not deter us from rendering good versions of it (as in Fig.12-3) or from using such renderings to study its mathematical properties. (This is very similar to the situation in high school geometry where we learned a lot about squares, triangles, and circles, even when our drawings of them were far from perfect.)

Recursive Processes

The construction of the Koch snowflake is an example of a *recursive process,* a process in which the same set of rules is applied over and over, with the end product at each step becoming the starting point for the next step. The concept of a recursive process is not new: It appeared in Chapter 1 (recursive ranking methods), Chapter 9 (Fibonacci numbers), and Chapter 10 (transition rules for population growth). In the case of the Koch snowflake the objects of the recursive process are shapes rather than numbers; other than that, the basic principles are quite similar.

One main advantage of recursive processes is that they allow for very simple and efficient descriptions of objects, even when the objects themselves are quite complicated. The Koch snowflake, for example, is a fairly complicated geometric shape, but we could describe it in two lines using a form of shorthand we will call a **recursive replacement rule.**

Recursive Replacement Rule for the Koch Snowflake

- Start with a solid equilateral triangle ▲.
- Whenever you see a line segment ______, replace it with__▲__.

If we look at the Koch snowflake from the perspective of traditional geometry, we will find it has some very unusual properties. Let's start by discussing two typical questions that always come up in traditional geometry: *perimeter* and *area.*

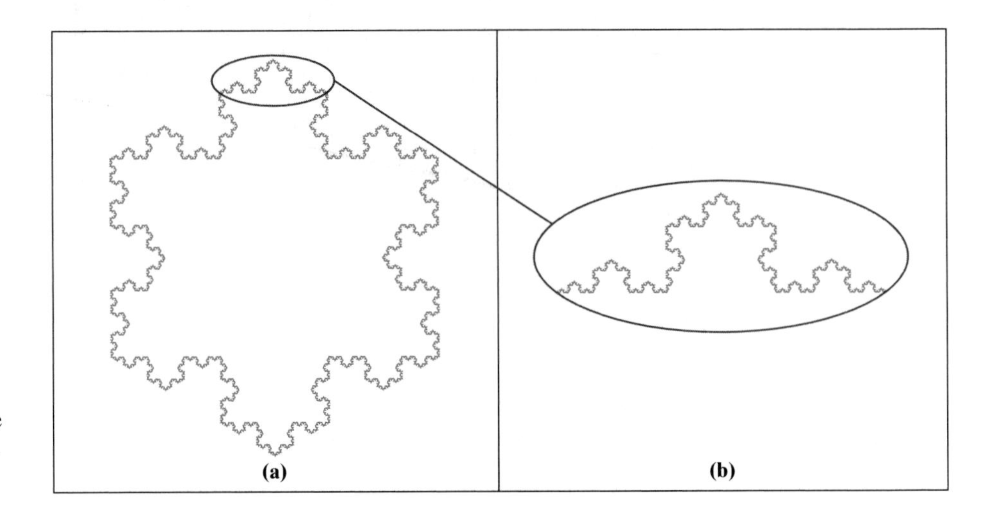

FIGURE 12-4
(a) The Koch curve
(b) detail (magnification: ×3)

Perimeter of the Koch Snowflake

The most interesting part of the Koch snowflake is its boundary. If we forget about the solid interior and look at just the boundary, we get a "curve" (Fig. 12-4) commonly known as the **Koch curve** (or **snowflake curve**).

How long is the Koch curve? Figure 12-5 shows the perimeter for the first few generations of the Koch curve. In each step, the perimeter grows by a factor of $\frac{4}{3}$ (in other words, in each step the curve is $33\frac{1}{3}\%$ longer than in the previous step). After infinitely many such steps, the length of the curve is infinite. This is our first important fact:

The *boundary* of the Koch snowflake has infinite length.

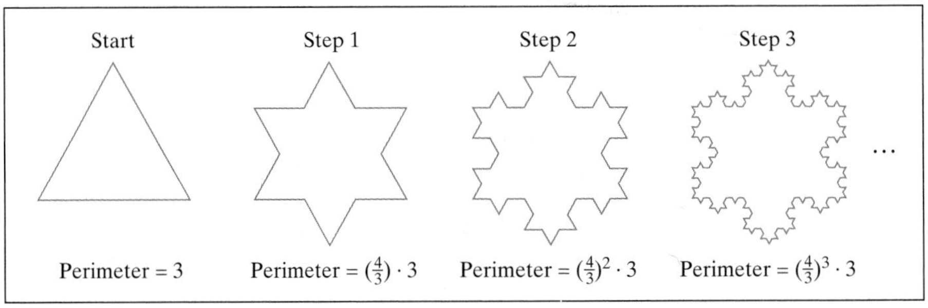

Start — Perimeter = 3

Step 1 — Perimeter = $(\frac{4}{3}) \cdot 3$

Step 2 — Perimeter = $(\frac{4}{3})^2 \cdot 3$

Step 3 — Perimeter = $(\frac{4}{3})^3 \cdot 3$ ...

FIGURE 12-5
At each step of the recursive process, the length of the curve is multiplied by 4/3.

Area of the Koch Snowflake

Here we will start with the facts:

The *area* of the Koch snowflake is 1.6 times the area of the starting equilateral triangle.

This fact is at first, very surprising. The Koch snowflake represents a shape with a finite area enclosed within an infinite boundary—something that seems contrary to our geometric intuition, but which is close to the way many important shapes are in nature. The vascular system of veins and arteries in the human body, for example, occupies a very small fraction of the body's volume, yet its' length is practically infinite: Laid end to end, the veins, arteries, and capillaries of a single human being would reach over 40,000 miles.

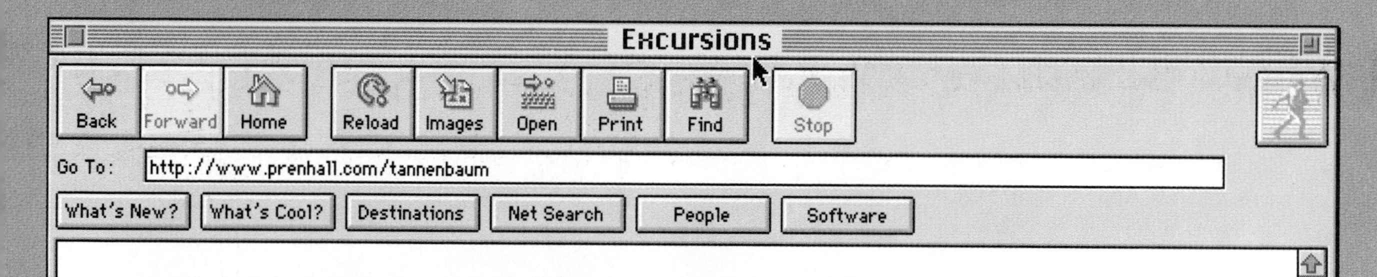

On a Magic Carpet Ride

As in the case of the Sierpinski gasket, fractal patterns can be generated in more than one way. In the case of the Sierpinski gasket we were able to create it using the SG procedure or by playing the Chaos Game. As different as the procedures seemed, they produced the same image.

Another fractal pattern that we can generate is Sierpinski's carpet. To see two ways to generate this go to the Tannenbaum site at: http://www.prenhall.com/tannenbaum.

Questions:

1. Look at method 1 and determine the pattern used to generate the carpet. How would you describe this to another student?

2. In method 1 the middle third of a line segment is replaced by a rectangle. Suppose that the rectangle is *added* to the line segment. What difference (if any) do you believe this will make to the pattern? (To see what does happen, use the link to the Peano Curve.)

3. Look at method 2 and determine the pattern used to generate the carpet. How would you describe this to another student?

4. If the area of the original square is X, then what is the area of the second figure? The third figure?

5. Determine a pattern to find the area for the nth figure.

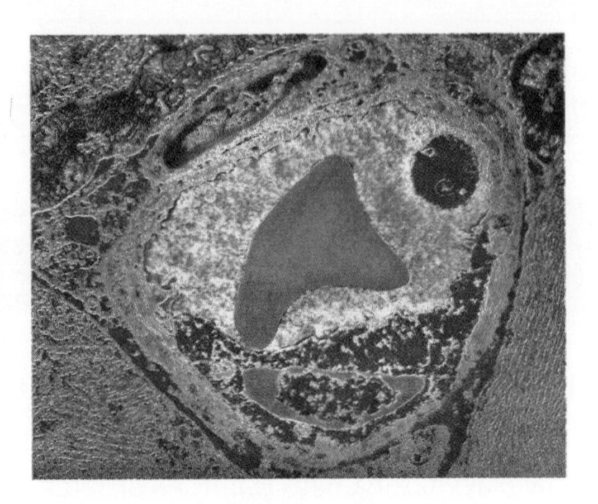

Left The network of veins, arteries, and capillaries in the human circulatory system.
Right Cross section of a blood capillary, with a single red blood cell in the center (Magnification: 7070 times).

In what follows, we will give an outline of the argument showing that the area of the Koch snowflake is 1.6 times the area of the original equilateral triangle, leaving the technical details as exercises for the reader. In fact, the reader who wishes to do so may skip the forthcoming explanation without prejudice.

The key to calculating the area of the Koch snowflake can be found by studying Fig. 12-6 carefully. At each step, we can compute how many new triangles are being added and how much is the area of each. From Fig. 12-6 we can see that in the Nth step we are adding a total of $3(4^{N-1})$ new triangles, each having an area of $(1/9)^N A$, which altogether gives an added area of $(4/9)^{N-1}(1/3)A$. The total area at the Nth step is the sum of the original equilateral triangle's area plus the areas added at each step of the way:

$$A + \left(\frac{1}{3}\right)A + \left(\frac{4}{9}\right)\left(\frac{1}{3}\right)A + \left(\frac{4}{9}\right)^2\left(\frac{1}{3}\right)A + \cdots + \left(\frac{4}{9}\right)^{N-1}\left(\frac{1}{3}\right)A.$$

Except for the first term, we are looking at the sum of terms of a geometric sequence (we discussed these kinds of sums in Chapter 10). Using the formula given in Chapter 10 for adding the consecutive terms of a geometric sequence, we can simplify the preceding expression to

$$A + \left(\frac{3}{5}\right)A\left[1 - \left(\frac{4}{9}\right)^N\right].$$

[We leave the technical details to the reader (Exercise 34).]

We are now ready to wrap this up. We need only figure out what happens to the expression $(4/9)^N$ as N gets bigger and bigger. In fact, what happens to any positive number less than 1 when we raise it to higher and higher powers? If you know the answer, then you are finished. If you don't, take a calculator, enter a number between 0 and 1, and multiply it by itself repeatedly. (You will readily convince yourself that the result gets closer and closer to 0.) The bottom line is that as N gets bigger and bigger, the expression inside the square brackets gets closer and closer to 1, and therefore the area gets closer and closer to $A + (3/5)A = (1.6)A$.

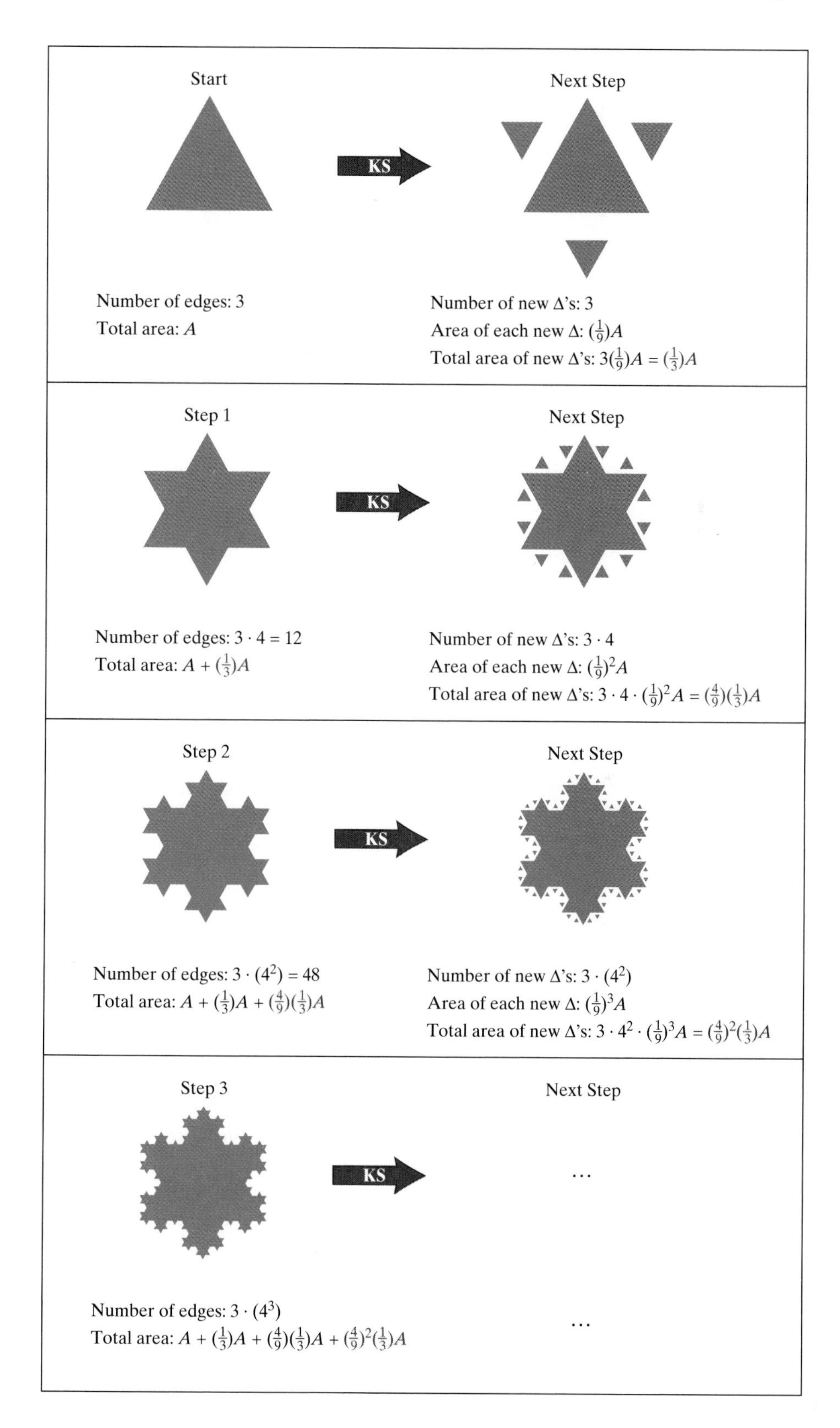

FIGURE 12-6 Caluculating the area of the Koch snowflake.

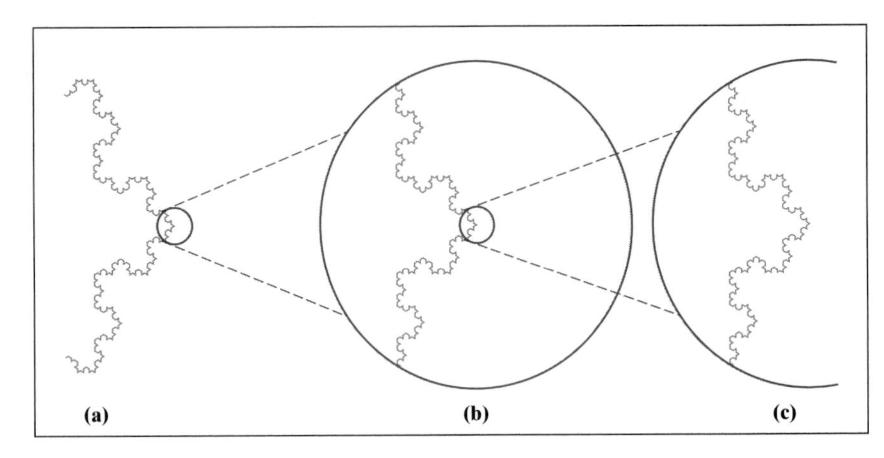

FIGURE 12-7 (a) A section of the Koch curve. (b) Detail of small section of the Koch curve (magnification: × 9). (c) Detail of a tiny section of the Koch curve (magnification: × 81).

Self-Similarity

What does the fine detail of the Koch curve look like? If we magnify a small section of the curve [Fig. 12-7(a)], we get the image in Fig. 12-7(b). Further magnification is not much help—Fig. 12-7(c) shows a detail of the Koch curve after magnification by a factor of almost 100.

We can see from Fig. 12-7 that something very surprising is happening: the fine detail of the Koch curve can look exactly the same as the rough detail! This remarkable characteristic of the Koch curve is called *self-similarity* (or *symmetry of scale*). As the name suggests, it is a symmetry that carries itself across different scales—a symmetry between the large-scale structure and the small-scale structure of an object.

We will say that a shape has **self-similarity (symmetry of scale)** if parts of the shape appear at infinitely many different scales. In the case of the Koch curve, there is a specific pattern[1] which shows up at every scale.

The Sierpinski Gasket

This is another interesting shape exhibiting self-similarity. It was first studied (in a slightly modified form) by the Polish mathematician Waclaw Sierpinski around 1915.

The construction starts with an arbitrary triangle ABC [Fig. 12-8(a)] but this time, instead of *adding* smaller copies of the original triangle, we will *remove* smaller copies of the original triangle according to the following procedure:

- **Start.** Start with an arbitrary solid triangle ABC.
- **Step 1. Procedure (SG):** *Remove the triangle whose vertices are the midpoints of each of the sides of the triangle* (we'll call this triangle the *middle* triangle) [Fig. 12-8(b)]. This leaves a white triangular hole in the original solid triangle, and three solid triangles, each of which is a half-scale version of the original.
- **Step 2.** For each of the solid triangles in the previous step, repeat *Procedure SG* (i.e., remove its *middle* triangle). This leaves us with 9 solid triangles (all similar to the original triangle ABC) and 4 triangular white holes [Fig. 12-8(c)].

---

[1] _/_

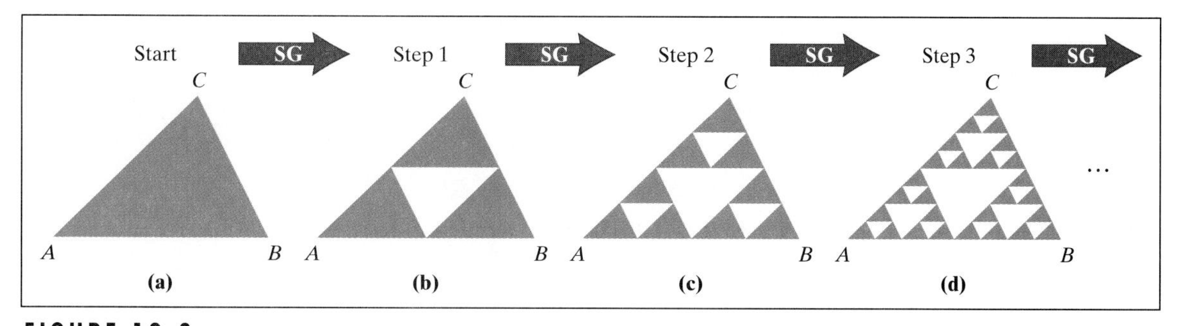

FIGURE 12-8
First three steps in the construction of the Sierpinski gasket.

- **Steps 3, 4, etc.** Continue repeating *Procedure SG* on every solid triangle ad infinitum.

After seven or eight steps the figure becomes visually stable, looking like Fig. 12-9, an unusual looking gasket called the **Sierpinski gasket**.

FIGURE 12-9
A rendering of the Sierpinski gasket.

Looking at Fig. 12-9, it would appear that the Sierpinski gasket is made of a huge number of tiny solid triangles, but these are the result of poor eyesight and the inadequacies of printing. The Sierpinski gasket has no solid triangles! If we were to magnify any one of those small solid specks, we would see another Sierpinski gasket (Fig. 12-10). This, of course, is another example of self-similarity.

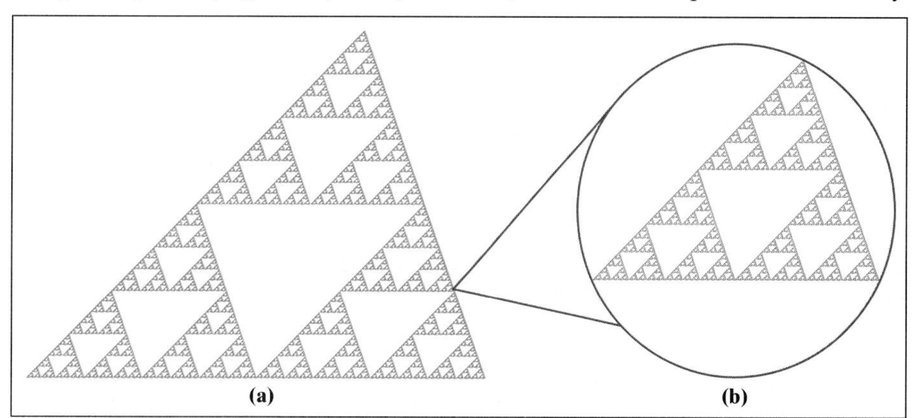

FIGURE 12-10
Detail of a small section of the Sierpinski gasket (magnification: × 256).

The Sierpinski gasket can be described in a very convenient way by a recursive replacement rule.

Recursive Replacement Rule for the Sierpinski Gasket

- Start with an arbitrary solid triangle ▲.
- Whenever you see a ▲, replace it with a ⁂.

We leave the following two facts as exercises to be verified by the reader:

- The Sierpinski gasket has zero area (Exercise 3).
- The Sierpinski gasket has an infinitely long boundary (Exercise 4).

The Chaos Game

This example involves the laws of chance. We start with an arbitrary triangle with vertices *A, B,* and *C* and an honest die [Fig. 12-11(a)]. To each of the vertices of the triangle we assign two of the six possible outcomes of rolling the die. For example, *A* is the "winner" if we roll a 1 or a 2; *B* is the "winner" if we roll a 3 or a 4; and *C* is the "winner" if we roll a 5 or 6. We are now ready to play the game.

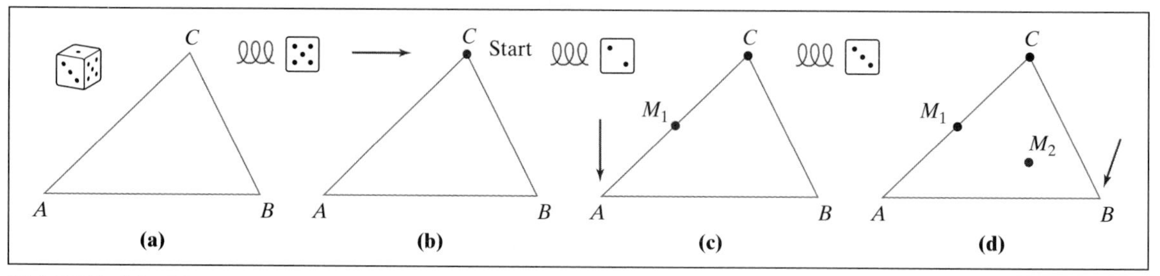

FIGURE 12-11
The Chaos Game: Always move from previous position towards the chosen vertex and stop halfway.

- **Start**. Roll the die. Start at the "winning" vertex. Say we roll a 5—we then start at vertex *C* [Fig. 12-11(b)].
- **Step 1**. Roll the die again. Say we roll a 2, so the winner is vertex *A*. *We now move straight from our previous position toward the winning vertex, but stop halfway.* Mark the new position (M_1) [Fig. 12-11(c)].
- **Step 2**. *Roll the die again, and move straight from the last position toward the winning vertex, but stop halfway.* (If the roll is 3, for example, stop at M_2 halfway between M_1 and *B* [Fig. 12-11(d)]. Mark your new position.
- **Steps 3, 4, etc**. Continue rolling the die, each time moving to a point halfway from the last position to the winning vertex.

Figure 12-12(a) shows the trail of points after 50 rolls of the die—just a bunch of scattered dots. Figure 12-12(b) shows the trail of points after 500 rolls. Figure 12-12(c) shows the trail of points after 5000 rolls—the pattern is unmistakable: a

Sierpinski gasket! After 10,000 rolls, it would be impossible to tell the difference between the trail of points and the Sierpinski gasket as shown in Fig. 12-9. This is a truly surprising turn of events. After all, the pattern of points created by the chaos game is ruled by the laws of chance, and one would not expect that any predictable pattern would appear. Instead, we get a very predictable pattern: There are no ifs or buts about it: the longer you play the chaos game, the closer you get to the Sierpinski gasket!

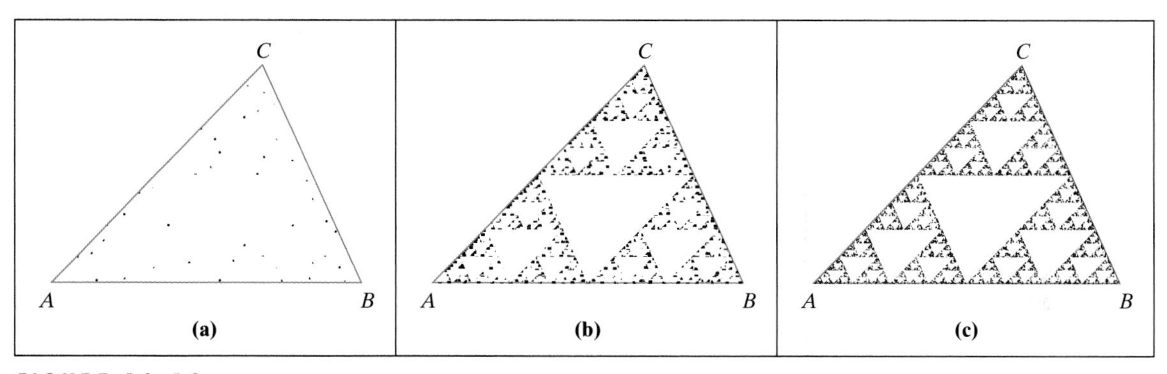

FIGURE 12-12
The "footprint" of the chaos game after (a) 50 rolls of the dice, (b) 500 rolls of the dice, (c) 5000 rolls of the dice.

The Twisted Sierpinski Gasket

Our next example is a simple variation of the original Sierpinski gasket. For lack of a better name, we will call it the **twisted Sierpinski gasket**.

The construction starts out exactly like the one for the regular Sierpinski gasket, with a solid triangle ABC [Fig. 12-13(a)] from which we remove the "middle" triangle, whose vertices we will call M, N, and L [Fig. 12-13(b)].The next move is new (we will call it the "twist"): Each of the points M, N, and L is moved a small amount in a random direction (as if an earthquake had randomly displaced them to new positions M', N', and L'). The resulting shape is shown in Fig. 12-13(c).

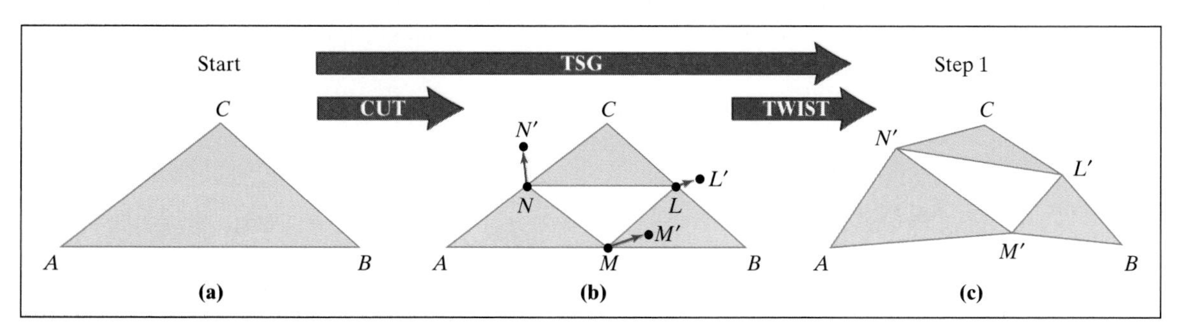

FIGURE 12-13
The two moves in **Procedure TSG:** The **cut** and the **twist**.

When the process of cutting and twisting is repeated again and again ad infinitum, we get the twisted Sierpinski gasket.

The following is a formal description of the process for building a twisted Sierpinski gasket.

- **Start.** Start with an arbitrary solid triangle *ABC*.
- **Procedure (TSG):**

(a) Cut. *Remove the middle triangle from a solid triangle* [Fig. 12-13(b)].

(b) Twist. *Move each of the midpoints of the triangle in an arbitrary direction and by a random amount that is small*[2] *in relation to the length of the corresponding side.*

- **Step 1.** Apply *Procedure TSG* to the starting solid triangle. [After step 1 is complete, we end up with 3 twisted solid triangles and 1 twisted hole in the middle, as shown in Fig. 12-14(b)].

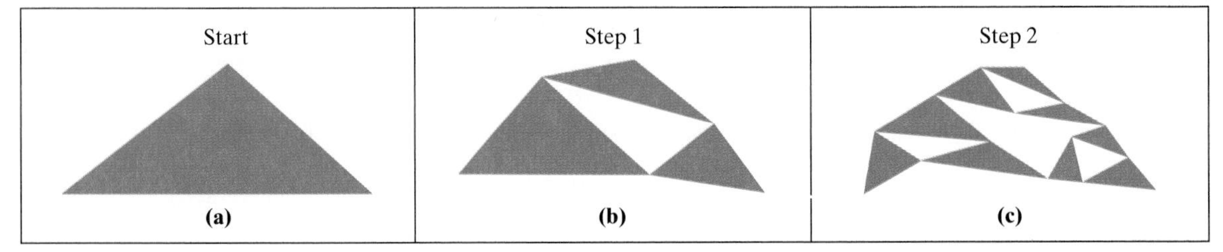

FIGURE 12-14
The first two steps in generating a twisted Sierpinski gasket.

- **Step 2.** For each of the solid triangles in the previous step, repeat *Procedure TSG.* This leaves us with 9 twisted solid triangles and 4 twisted white holes [Fig. 12-14(c)].
- **Steps 3, 4, etc.** Continue repeating *procedure TSG* on each solid triangle.

Figure 12-15 shows an example of a *twisted Sierpinski gasket* after 8 steps. Probably the most surprising thing about the figure is that there is an unmistakable "mountain look" to it. By changing the shape of the starting triangle, we can change the *shape* of the mountain, and by changing the rules for how far we allow the random displacements to go, we get a different "texture" for the mountain, but just as in nature's true mountains, we will always get that unmistakable "mountain look."

---

[2] In Fig. 12-13, we did not allow the displacement to be more than 25% of the length of the corresponding side.

FIGURE 12-15
A twisted Sierpinski gasket (step 8).

The most remarkable thing of all is that these complicated-looking geometric shapes can be described in just a few lines by means of a simple recursive replacement rule:

Recursive Replacement Rule for the Twisted Sierpinski Gasket

- Start with an arbitrary solid triangle.
- Wherever you see a black triangle, apply *Procedure TSG* to it.

What about self-similarity? Does the twisted Sierpinski gasket have it? Well, not exactly. Whenever we magnify a part of it, we don't see exactly the same identical things, but we do see variations of a single theme: that special "mountain look" is going to show up at every scale.

When an object or shape has the kind of symmetry of scale where approximate (but not identical) versions of a common theme appear at every scale, we will say it has **approximate self-similarity** (which from now on will stand in contrast to **exact self-similarity**, such as the one exhibited by the Koch snowflake and the ordinary Sierpinski gasket).

Approximate self-similarity is a common property of many natural objects and shapes: mountains, trees, plants, clouds, lightning, the human vascular system,

Two closeups, at different levels of magnification, of a head of cauliflower. One of these photos is a closeup of the other, magnified by a factor of 2. Which is which?

and so on. In fact, it is what gives these things their distinctive natural look, and only by understanding the mathematical details of this type of symmetry can we hope to realistically imitate and understand the objects themselves.

Symmetry of Scale in Art and Literature

The notion of self-similarity is not unique to formal geometry—it pops up in various forms in art, poetry, and literature.

In art, a special kind of self-similarity has been around for a long time. It is the concept of an *infinite regress*—an infinite sequence of repeatingly smaller versions of the same image. Figure 12-16 shows two examples of "infinite" regress in art. As in any work of the human hand, the infinite regress in these pictures is only illusory: At some point, the detail has to stop.

Figure 12-16(a) is a woodcut by the famous Dutch artist M. C. Escher, undoubtedly the best-known "symmetry-inspired" artist of our century. Explaining the details of how he created the woodcut, Escher wrote:

> *In this woodcut I have consistently and almost maniacally continued the reduction down to the limit of practical execution. I was dependent on four factors: the quality of my wood material, the sharpness of my tools, the steadiness of my hand, and especially my keen-sightedness . . . .*[3]

Figure 12-16(b) shows a simpler example of an infinite regress. In this cover of *TV Week* a man is sitting in an armchair holding a remote control in his left hand and an issue of the very same *TV Week* in his right hand, with himself on it holding another *TV Week* and so on ad infinitum, or at least that's the idea. In reality the regress only goes four steps deep. Interestingly enough, this piece of art can be described by means of a fairly simple recursive replacement rule:

- **Start**. Start with the big picture of the man with the TV, the remote, the armchair, and so on (everything except the issue of *TV Week* in his right hand).
- **Procedure (TVG)**. (a) Reduce the picture down to 35% of its original size; (b) translate and rotate the reduced picture so that it "slips" into the man's right hand in the old picture; (c) bring the man's fingers into the foreground.

[3] M. C. Escher, *Escher on Escher* (New York: H. N. Abrams, 1989).

(a)

(b)

FIGURE 12-16

(a) *Smaller and Smaller* by M.C. Escher (© 1956 M.C. Escher/Cordon Art, Baarn, Holland) (b) *Couch potato* (Reproduced by permission of the *Fresno Bee*).

Procedure TVG could be repeated indefinitely, creating an infinite regress that converges toward a single point. If we kept zooming in at that point, we would continue seeing little men with remote controls and *TV Week*s in their hands.

In literature, infinite regresses of various types can be found in the works of such diverse writers as E. E. Cummings (*Him*), Aldous Huxley (*Point Counter Point*), and Norman Mailer (*The Notebook*). We conclude this section with an infinite regress in an often-quoted poem by Jonathan Swift:

> So Nat'ralists observe, A Flea
> Hath Smaller Fleas that on him prey
> and these have smaller Fleas to bite 'em
> And so proceed, ad infinitum.

The Mandelbrot Set

We now return to a much more mathematical example. In fact, the mathematics in this example goes a bit beyond the level of this book, so we will describe the overall idea in general terms. The actual purpose here is not to get bogged down in the mathematical details but rather to illustrate one of the most interesting and beautiful geometric objects ever created by the human "hand". The object is called the *Mandelbrot set* (and sometimes simply the *M-set*) after the Polish-born mathematician Benoit Mandelbrot.[4] Mandelbrot was the first person to extensively study

[4] Mandelbrot is a research mathematician at IBM, an IBM Fellow, and a professor at Yale University.

and fully appreciate the importance of this beautiful and complex mathematical object.

Before we do anything else, let's take a brief look at the Mandelbrot set, first in black and white [Fig. 12-17(a)] and then over a colored background [Fig. 12-17(b)]. In both cases, the Mandelbrot set itself is the black region which, in the minds of many people, resembles some sort of bug—a flea from some exotic planet, maybe? The overall structure of this "flea" can be described as consisting of a body, a head, and an antenna coming out of the middle of the head. Both the head and the body are full of smaller "fleas." We can even see in Fig. 12-17 that these smaller "fleas" have "fleas" of their own. Maybe Jonathan Swift was onto something.

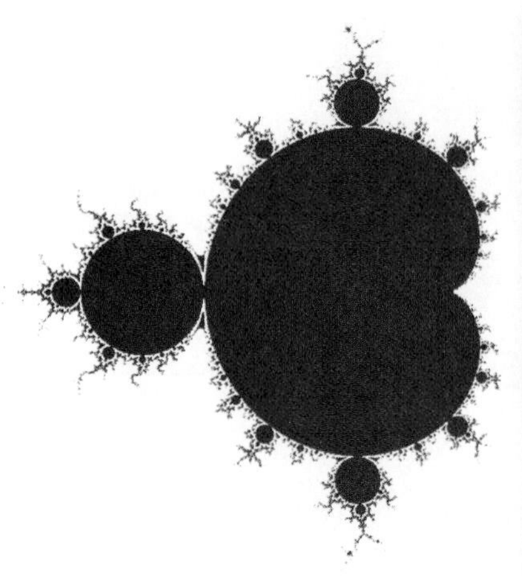

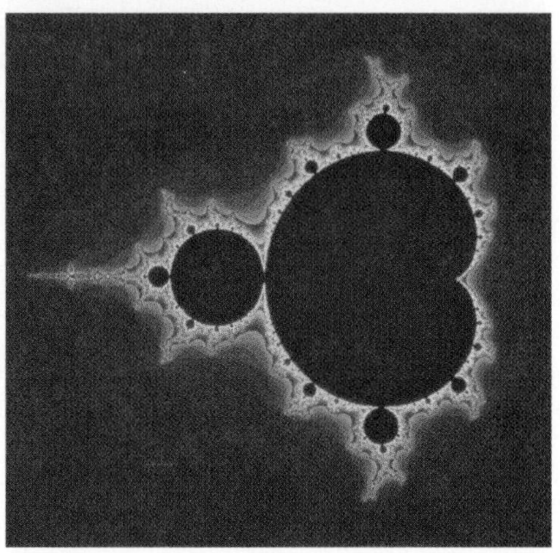

FIGURE 12-17 The black shape is the Mandelbrot set itself: over a white background (*left*), over a colored background (*right*). (Computer generated images by Rollo Silver.)

Notice that one thing is different in the main "flea" that distinguishes it from the smaller "fleas" on it: it has "buttocks." There are hundreds of "fleas" of various sizes (some pretty big others so small we can hardly see them) surrounding the big "flea." When we magnify the view around two of these (Fig. 12-18), we can see the common theme ("fleas" of many sizes crawling all over the place), but we can also see some new and surprising shapes, such as swirling green clusters and "seahorse tails" in Fig. 12-18(a), and tendrils and snowflakes in Fig. 12-18(b). If we look carefully, we can also see, floating in different parts of Figs. 12-18(a) and (b), tiny replicas of the original "flea" (buttocks included).

FIGURE 12-18 A closeup of two different regions around the Mandelbrot set. In both cases, replicas of the original Mandelbrot set can be seen. (Approximate magnification: $\times$ 10.) [Computer images by Rollo Silver.]

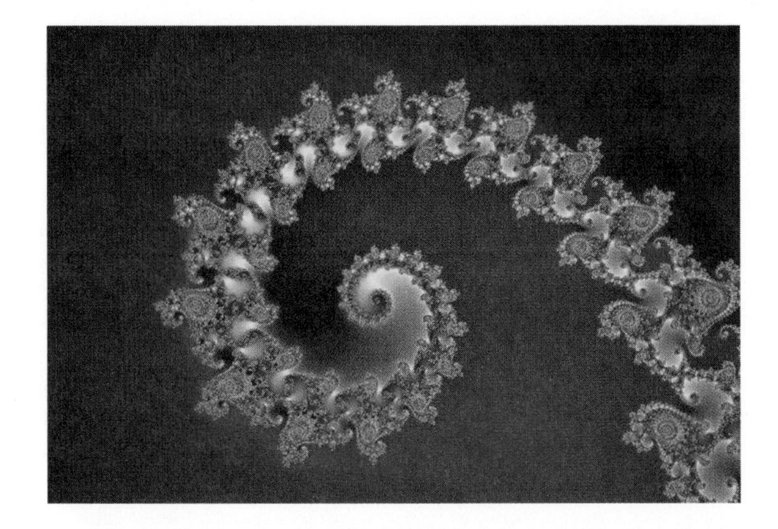

FIGURE 12-19
(a) A closeup of a "seahorse tail" from Fig. 12-18(a). (b) Detail. (c) Further detail with small Mandelbrot set. (Approximate magnification: $\times$ 10,000). [Computer images by Rollo Silver.]

Figure 12-19(a) now takes us to a closeup of one of the seahorse tails in Fig. 12-18(a). A further closeup of a section of Fig. 12-19(a) is shown in Fig. 12-19(b), and an even further magnification in Fig. 12-19(c), revealing a small replica of the original Mandelbrot set surrounded by a beautiful arrangement of swirls, spirals, and seahorse tails. The Mandelbrot set in this picture is about 50,000 times smaller than the original. Anywhere we choose to look in this picture (or any of the others) we will find (if we magnify enough) replicas of the original Mandelbrot set, always surrounded by an infinitely changing but always stunning background. The Mandelbrot set has a very exotic and complex form of approximate self-similarity—infinite repetition and infinite variety mingle together at every scale in a landscape as diverse as nature itself.

The Mandelbrot set has been rightfully described as 'the most complex object ever devised by man," even though it wasn't until the advent of powerful computers in the last 20 years that images such as these could be generated.

Constructing the Mandelbrot Set: Mandelbrot Sequences

How does this delicate mix of beauty and complexity called the Mandelbrot set come about? Incredibly, the Mandelbrot set itself can be described mathematically by a very simple process involving just numbers. The only rub is that the numbers are *complex numbers.*

Complex Numbers

You may recall having seen such numbers before (probably in Algebra II). These are the ones that allow us to take square roots of negative numbers, solve quadratic equations of any kind, and so on. The basic building block for complex numbers is the number $\sqrt{-1} = i$. Using i, we can build all other complex numbers such as $(3 + 2i)$, $(\frac{5}{3} - \frac{4}{3}i)$, and the generic complex number $(a + bi)$.

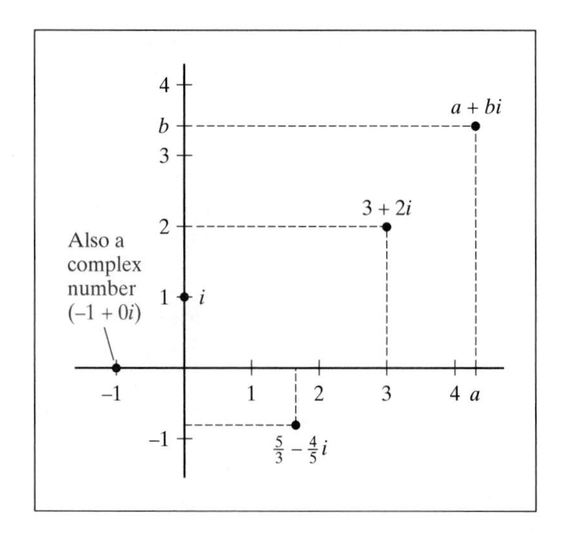

FIGURE 12-20 Every complex number "is" a point in the plane.

For our purposes, the most important fact about complex numbers is that each of them can be identified with a unique point in a coordinate plane. The basic idea is to identify the complex number $(a + bi)$ with the point (a, b) in a Cartesian coordinate system (Fig. 12-20). Once we realize this, we can talk about complex numbers and points in the plane as being one and the same. (For a review of the basic facts about complex numbers the reader is encouraged to look at any standard Algebra II text.)

Mandelbrot Sequences

Our basic construction will be to start with a complex number (point in the plane) and from it create an infinite sequence of numbers (points) that depend on the starting number. The sequence of numbers we will call a *Mandelbrot sequence,* and the starting point we will call the *seed* of the sequence. The basic recursive rule for a Mandelbrot sequence is that *each number in the sequence equals the preceding number in the sequence squared plus the seed.* The general description of a Mandelbrot sequence with seed s is shown in Fig. 12-21.

Seed
Step 1
Step 2
Step 3
. . .
s
M
s² + s
=
s₁
M
s₁² + s
=
s₂
M
s₂² + s
=
s₃
M
. . .

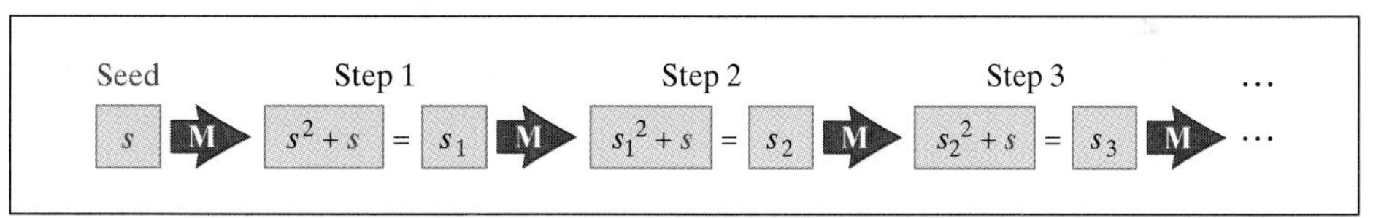

FIGURE 12-21
A Mandelbrot sequence with seed s.

A Mandelbrot sequence can also be easily described by means of a recursive rule (just as we did with the Koch snowflake and Sierpinski gasket).

Recursive Rule for a Mandelbrot Sequence

- **Start:** Seed (s).
- **Procedure:** If x is a term in the sequence, the next term is $x^2 + s$.

The choice of the name "seed" for the starting value of each sequence gives us a convenient metaphor: Each seed, when planted, produces a different sequence of numbers (the tree). The recursive replacement rule is like a rule telling the tree how to grow from one season to the next.

Let's look at some examples of Mandelbrot sequences. (*Reminder:* Integers and decimals are also complex numbers, so they make perfectly acceptable seeds.)

Example 1.

Figure 12-22 shows the first four steps in the Mandelbrot sequence with seed $s = 1$.

Seed
Step 1
Step 2
Step 3
Step 4
s = 1
M
s₁ = 1² + 1
=
2
M
s₂ = 2² + 1
=
5
M
s₃ = 5² + 1
=
26
M
s₄ = 26² + 1
=
677
. . .

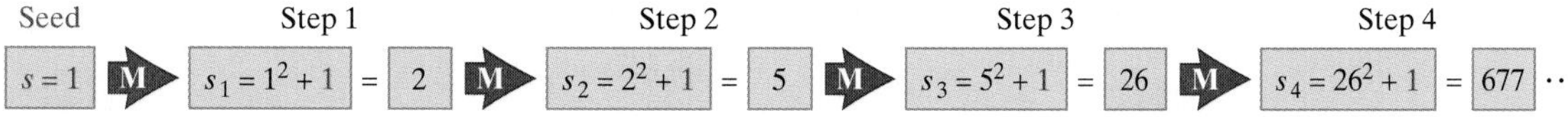

FIGURE 12-22
Mandelbrot sequence with seed $s = 1$ (*escaping*).

The pattern that emerges is clear: the numbers are getting bigger and bigger. The corresponding points are getting further and further away from the origin. We will call this kind of sequence an **escaping** Mandelbrot sequence. For such a sequence the point in the plane identified with the seed will be a *nonblack point* (it's a funny way to put it, but we are noncommittal—the only thing certain is that it will *not* be a black point!). ■

Example 2.

Figure 12-23 shows the first three steps in the Mandelbrot sequence with seed $s = -1$.

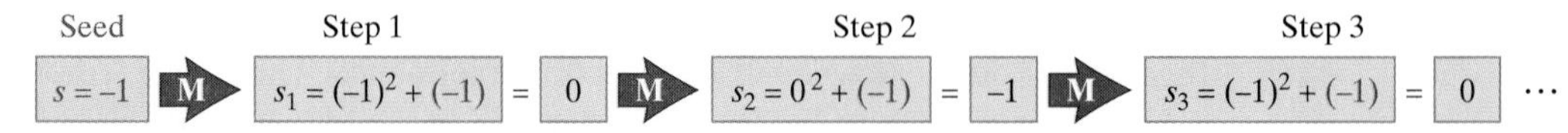

FIGURE 12-23
Mandelbrot sequence with seed $s = -1$ (*periodic*).

The pattern that emerges here is also clear: the numbers jump back and forth between 0 and -1. We say in this case that the Mandelbrot sequence is **periodic**. For such a sequence the point in the plane identified with the seed *will definitely be a black point.* ■

Example 3.

Figure 12-24 shows the first four steps in the Mandelbrot sequence with seed $s = -0.75$. Here a calculator will probably come in handy.

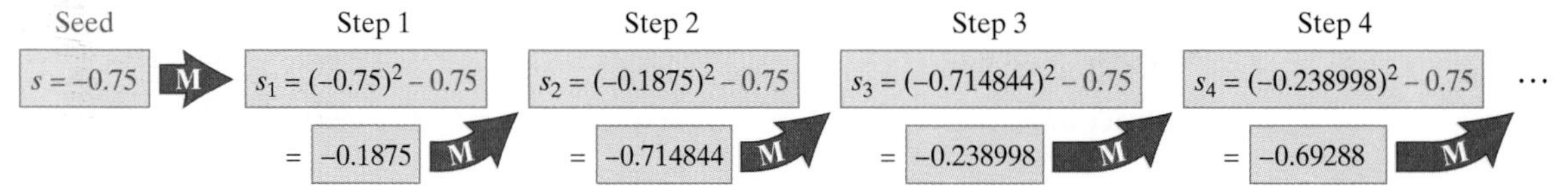

FIGURE 12-24
Mandelbrot sequence with seed $s = -0.75$ (*attracted*).

In this case, the pattern is not obvious, and additional terms of the sequence are needed. As an exercise (Exercise 35), the reader should carry this example out for another 20 steps and verify that the values of the Mandelbrot sequence get closer and closer to -0.5. In this case, we will say that the Mandelbrot sequence is **attracted** to -0.5. When the sequence is attracted toward a number, no matter what number it is, the point in the plane identified with the seed *will always be a black point* (just as for periodic sequences). ■

Example 4.

Figure 12-25 shows the first four steps in the Mandelbrot sequence with seed $s = i$. Here for the first time we are dealing with a true complex number. (Reminder: $i^2 = -1$.)

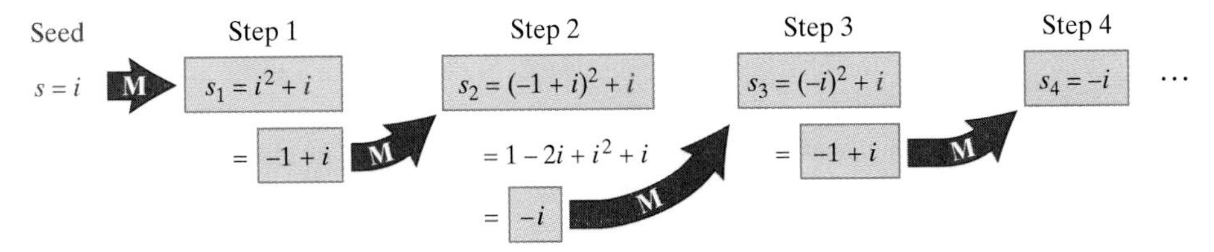

FIGURE 12-25
Mandelbrot sequence with seed $s = i$ (*periodic*).

Just as in Example 2, the Mandelbrot sequence here is periodic; thus, in the Cartesian plane the seed will be a black point.

The Mandelbrot Set (Definition)

We are ready (finally) to explain how the Mandelbrot set comes about. By now the definition will sound simple: The Mandelbrot set consists of all the points in the plane (complex numbers) that are black seeds of Mandelbrot sequences. Thus, our entire discussion can be summarized by the following logical sequence:

- Each point in the Cartesian plane is a complex number and can be used as a seed for a Mandelbrot sequence.
- If the Mandelbrot sequence is *periodic* or *attracted,* the point is part of the Mandelbrot set. If the sequence is *escaping,* the point is *not* in the Mandelbrot set. [In this case, the point can be given different colors, depending on the speed of escape (for example, hot colors—red, yellow, orange—if it escapes slowly; cool colors—blues, purples, etc.—if it escapes quickly). The coloring of the escaping points is what livens up the amazing pictures that we saw in Figs. 12-17, 12-18, and 12-19.]

Because the Mandelbrot set provides a bounty of aesthetic returns for a relatively small mathematical investment, it has become one of the most popular mathematical playthings of our time. There are now literally hundreds of software programs available (many of them shareware) that allow one to explore the beautiful landscapes surrounding the Mandelbrot set.

Conclusion: Fractals

The word **fractal** (from the Latin *fractus,* "broken up, fragmented") was coined by the mathematician Benoit Mandelbrot in the mid-1970s to describe objects as diverse as the Koch curve, the Sierpinski gasket, the twisted Sierpinski gasket, and the Mandelbrot set, as well as many shapes in nature such as clouds, trees, mountains, lightning, the vascular system in the human body, and so on.

These objects share one key characteristic: They all have some form of self-similarity. (This is not the only defining characteristic of a fractal—others, such as *fractional dimension,* would take us beyond the scope of this chapter.)

---

[5] Benoit Mandelbrot, *The Fractal Geometry of Nature* (New York: W. H. Freeman & Co., 1983).

The discovery and study of fractals and their geometric structure has become one of the hottest mathematical topics of the last 20 years. It is a part of mathematics that combines complex and interesting theories, beautiful graphics, and extreme relevance to the real world. In his classic book, *The Fractal Geometry of Nature,* Mandelbrot[5] wrote:

> *Why is [standard] geometry often described as "cold" and "dry"? One reason lies in its inability to describe the shape of a cloud, a mountain, a coastline, or a tree. Clouds are not spheres, mountains are not cones, coastlines are not circles, and bark is not smooth nor does lightning travel in a straight line. . . . Many patterns of Nature are so irregular and fragmented, that compared with [standard geometry] Nature exhibits not only a higher degree but an altogether different level of complexity. The number of distinct scales of length of natural patterns is for all practical purposes infinite.*

There is a striking visual difference between the kinds of shapes we discussed in this chapter and the shapes of traditional geometry. It is difficult to mistake one for the other. The shapes of traditional geometry (squares, circles, cones, etc.) and the objects we build based on them (bridges, machines, buildings, etc.) have a distinct man-made look. Many of the shapes of nature (mountains, trees, clouds, etc.) have a completely different kind of look, one that man has always had difficulty re-creating. Mandelbrot, who is the father of *fractal geometry,* was the first to realize that the foundation of this natural look is some form of self-similarity, and that geometric objects

There is a distinctive "feel" to man-made geometry, very different from . . .

. . . the fractured "feel" of natural geometry.

Earth, as viewed from Apollo 11 (courtesy of NASA) and *Umbra*, a computer generated fractal (© F. Kenton Musgrave). Which is which?

built on the principles of self-similarity can be used to model many shapes and pattens in nature. Today, the principles of fractal geometry are used to study the patterns of clouds and how they affect the weather; the pattern of contractions of a human heart; the behavior of the stock market; and to create the truly incredible computer graphics that animate many of the latest science fiction movies.

Geometry as we have known it in the past was developed by the Greeks about 2000 years ago and passed on to us essentially unchanged. It was (and still is) a great triumph of the human mind, and it has allowed us to develop much of our technology, engineering, architecture, and so on. As a tool and language for modeling and representing nature, Greek geometry has by and large been a failure. The discovery of fractal geometry seems to have given science the right mathematical language to remedy this failure, and as such it promises to be one of the great achievements of twentieth-century mathematics.

Key Concepts

approximate self-similarity
attracted (sequence)
chaos game
escaping (sequence)
exact self-similarity
fractal
fractal geometry
Koch curve (snowflake curve)
Koch snowflake
Mandelbrot set
Mandelbrot sequence
periodic (sequence)
recursive replacement rule
seed (of a Mandelbrot sequence)
Sierpinski gasket
self-similarity (symmetry of scale)
twisted Sierpinski gasket

Exercises

Walking

The Sierpinski Gasket. *Exercises 1 through 4 refer to the regular Sierpinski gasket. The figure shows the first few steps of the construction.*

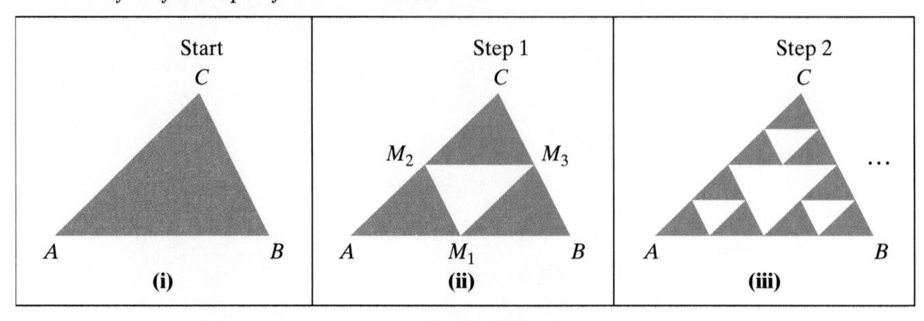

1. Suppose the area of the starting blue triangle ABC is X.

(a) Find the area of the blue triangle AM_1M_2 in (ii). Explain.

(b) Find the area of the blue triangle M_1BM_3 in (ii). Explain.

(c) Find the area of the white triangle $M_1M_2M_3$ in (ii). Explain.

(d) Find the area of the entire blue region in (ii).

2. Suppose the perimeter of the starting blue triangle ABC is P.

(a) Find the perimeter of the blue triangle AM_1M_2 in (ii).

(b) Find the perimeter of the blue triangle M_1BM_3 in (ii).

(c) Find the perimeter of the white triangle $M_1M_2M_3$ in (ii).

(d) Find the perimeter of the entire blue region in (ii).

3. Suppose the area of the starting blue triangle ABC is X.

(a) Find the area of the blue region obtained at step 2 in the construction of the Sierpinski gasket.

(b) Find the area of the blue region obtained at step 3 in the construction of the Sierpinski gasket.

(c) Find the area of the blue region obtained at step N (a generic step) in the construction of the Sierpinski gasket.

(d) Explain why the area of the Sierpinski gasket is zero. [Use your answer in (c).]

4. Suppose the perimeter of the starting blue triangle ABC is P.

(a) Find the perimeter of the blue figure obtained at step 2 in the construction of the Sierpinski gasket.

(b) Find the perimeter of the blue figure obtained at step 3 in the construction of the Sierpinski gasket.

(c) Find the perimeter of the blue figure obtained at step N in the construction of the Sierpinski gasket.

(d) Explain why the Sierpinski gasket has an infinitely long boundary.

The Mitsubishi Gasket. *Exercises 5 through 8 refer to the Mitsubishi gasket, a fractal defined by the following recursive procedure:*

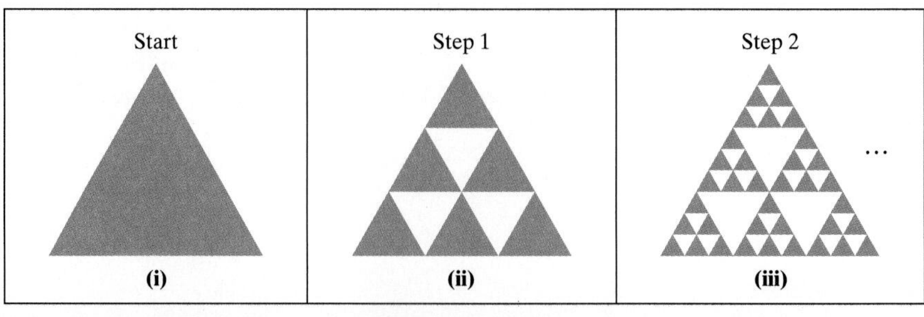

- ***Start.*** *Start with a blue equilateral triangle as shown in (i).*
- ***Step 1. Procedure MG:*** *Subdivide the blue triangle into 9 equal subtriangles and remove the 3 "interior" subtriangles, as shown in (ii).*
- ***Step 2.*** *For each of the remaining blue triangles repeat* Procedure MG. *The result is shown in (iii).*
- ***Steps 3, 4, etc.*** *Repeat* Procedure MG *ad infinitum.*

5. How many solid blue triangles are there at the start of step 3 in the construction? How about at step 10? How about at step N?

6. Suppose the area of the starting blue triangle is X.

(a) Find the area of one of the white triangles in (ii).

(b) Find the area of the blue region in (ii).

(c) Find the area of the blue region in (iii).

7. Suppose the perimeter of the starting blue triangle is P.

(a) Find the perimeter of one of the white triangles in (ii).

(b) Find the perimeter of the blue region in (ii).

(c) Find the perimeter of the blue region in (iii).

8. Suppose the area of the starting blue triangle is X.

(a) Find the area of the blue region at step N in the construction of the Mitsubishi gasket.

(b) Explain why the area of the Mitsubishi gasket is zero. [Use your answer in (a).]

The Sierpinski Carpet. *Exercises 9 through 12 refer to the Sierpinski carpet, a fractal defined by the following recursive procedure:*

- ***Start.*** *Start with a solid square.*
- ***Step 1. Procedure SC:*** *Subdivide the square into 9 equal subsquares and remove the central subsquare.*
- ***Steps 2, 3, etc.*** *On every remaining solid square, repeat* Procedure SC *ad infinitum.*

9. Using graph paper, carefully draw the figures that result at steps 1, 2, and 3 of the Sierpinski carpet.

10. Describe a recursive replacement rule for the Sierpinski carpet. (It has the following form: Start with a ________. Whenever you see a ________, replace it with a ________.)

11. Suppose the area of the starting square is X.

(a) How many solid squares are there in the figure obtained at step 3?

(b) What is the area of the figure obtained at step 3?

(c) How many solid squares are there in the figure obtained at step N?

(d) What is the area of the figure obtained at step N?

(e) Explain why the area of the Sierpinski carpet is zero.

12. Suppose that the starting square has sides of length 1.

(a) Find the length of the boundary of the figure at steps 1 and 2.

(b) Find the length of the boundary of the figure at step N.

(c) Explain why the length of the boundary of the Sierpinski carpet is infinite.

The Square Snowflake. *Exercises 13 through 16 refer to the "square snowflake," a fractal defined by the following recursive procedure:*

- ***Start.*** *Start with a solid square.*
- ***Step 1. Procedure SS:*** *Divide each side of the square into 3 equal segments. Attach to the middle segment of each side a small solid square with sides one-third the sides of the starting square.*
- ***Steps 2, 3, etc.*** *Repeat* Procedure SS *ad infinitum.*

13. Describe a recursive replacement rule for the *square snowflake.* (It has the following form: Start with a ________. Wherever you see a ________, replace it with a ________.)

14. Using graph paper, carefully draw the figures that result at steps 1, 2, and 3 of the *square snowflake.*

15. Suppose that the starting square has sides of length 1.

(a) Find the perimeter of the figure at steps 1 and 2.

(b) Find the perimeter of the figure at step N.

16. Suppose the area of the starting square is X.

(a) What is the area of the figure obtained at step 3?

(b) What is the area of the figure obtained at step N?

The Checkered Flag. *Exercises 17 through 20 refer to the "checkered flag," a fractal defined by the following recursive replacement rule:*

- *Start with a white rectangle* 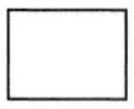 .
- *Whenever you see a white rectangle* , *replace it with a*  .

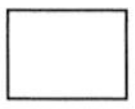

17. Using graph paper, carefully draw the figures that result at steps 1, 2, and 3 of a "checkered flag."

18. Describe in words the procedure for construction of a "checkered flag."

19. Suppose the area of the starting rectangle is X. Find the area of the blue region at step N in the construction of the "checkered flag."

20. Suppose the perimeter of the starting rectangle is P. Find the length of the boundary of the blue region at step N in the construction of the "checkered flag."

The Quadratic Koch Island. *Exercises 21 through 24 refer to the "quadratic Koch island," a fractal defined by the following recursive replacement rule:*

- *Start with a square:*
- *Whenever you see a horizontal line segment*

replace it by

- *Whenever you see a vertical line segment*

replace it by

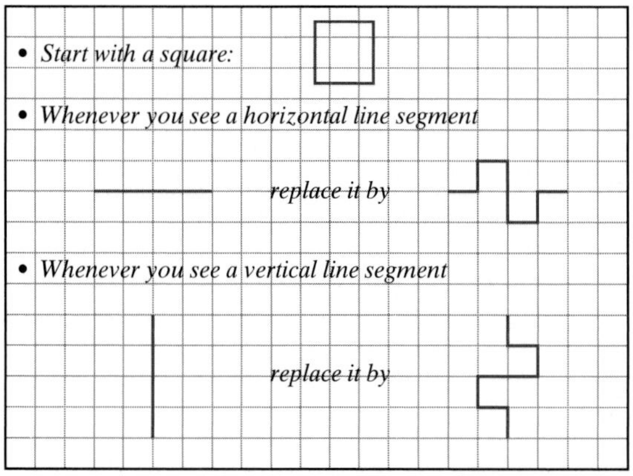

21. Using graph paper, draw the figures that result at steps 1 and 2 in the construction of the quadratic Koch island.

22. The figure obtained at step N in the construction of the quadratic Koch island is a polygon. How many sides does it have?

23. Suppose the perimeter of the starting square is P.

(a) Find the perimeter of each of the polygons obtained at steps 1 and 2 in the construction of the quadratic Koch island.

(b) Find the perimeter of the polygon obtained at step N in the construction of the quadratic Koch island.

24. Suppose that the starting square encloses an area X.

(a) Find the areas enclosed by the figures at steps 1 and 2 in the construction of the quadratic Koch island.

(b) Find the area enclosed by the figure at step N in the construction of the quadratic Koch island.

Exercises 25 through 27 refer to the chaos game as described in the chapter. Assume that we start with an arbitrary triangle ABC and that we will roll an honest die. Vertex A is assigned numbers 1 and 2; vertex B is assigned numbers 3 and 4; and vertex C is assigned numbers 5 and 6.

25. Suppose the die is rolled 16 times and the outcomes are 3, 4, 2, 3, 6, 1, 6, 5, 5, 3, 1, 4, 2, 2, 2, 3. Draw the points P_1 through P_{16}, corresponding to these outcomes.

26. Suppose the die is rolled 12 times and the outcomes are 5, 5, 1, 2, 4, 1, 6, 3, 3, 6, 2, 5. Draw the points P_1 through P_{12} corresponding to these outcomes.

27. Suppose that the die is rolled 6 times and that by sheer coincidence the outcomes are 1, 2, 3, 4, 5, 6. Show the trajectory of the trip, as one travels from point to point.

Exercises 28 through 30 refer to Mandelbrot sequences. (You may need a calculator to do the computations.)

28. **(a)** Find the first 5 terms of the Mandelbrot sequence with seed $s = 2$.

(b) Is this Mandelbrot sequence *escaping, periodic,* or *attracted?*

29. **(a)** Find the first 5 terms of the Mandelbrot sequence with seed $s = -2$.

(b) Is this Mandelbrot sequence *escaping, periodic,* or *attracted?*

30. **(a)** Find the first 10 terms of the Mandelbrot sequence with seed $s = -0.25$.

(b) Is this Mandelbrot sequence *escaping, periodic,* or *attracted?*

Jogging

31. This exercise refers to the construction of the *Sierpinski gasket.* Explain why there are $(3^N - 1)/2$ white triangles in the figure obtained at step N in the construction.

32. This exercise refers to the construction of the Sierpinski carpet (see Exercises 9 through 12). How many white squares does the figure obtained at step N in the construction have?

33. Explain why the construction of the Sierpinski gasket does not end up in an all-white triangle.

34. Use the formula for adding consecutive terms of a geometric sequence (see Chapter 10) to show that:

(a) $1 + \left(\frac{4}{9}\right) + \left(\frac{4}{9}\right)^2 + \cdots + \left(\frac{4}{9}\right)^{N-1} = \frac{9}{5}\left[1 - \left(\frac{4}{9}\right)^N\right]$

(b) $\left(\frac{1}{3}\right)A + \left(\frac{4}{9}\right)\left(\frac{1}{3}\right)A + \left(\frac{4}{9}\right)^2\left(\frac{1}{3}\right)A + \cdots + \left(\frac{4}{9}\right)^{N-1}\left(\frac{1}{3}\right)A = \frac{3}{5}A\left[1 - \left(\frac{4}{9}\right)^N\right].$

Exercises 35 through 40 are about Mandelbrot sequences. (You will need a calculator.)

35. Find the first 20 terms of the Mandelbrot sequence with seed $s = -0.75$. Toward what number is this Mandelbrot sequence attracted?

36. Consider the Mandelbrot sequence with seed $s = 0.2$. Is this Mandelbrot sequence *escaping, periodic,* or *attracted?* If attracted, to what number?

37. Consider the Mandelbrot sequence with seed $s = 0.25$. Is this Mandelbrot sequence *escaping, periodic,* or *attracted?* If attracted, to what number?

38. Consider the Mandelbrot sequence with seed $s = -1.25$. Is this Mandelbrot sequence *escaping, periodic,* or *attracted?* If attracted, to what number?

39. Consider the Mandelbrot sequence with seed $s = \sqrt{2}$. Is this Mandelbrot sequence *escaping, periodic,* or *attracted?* If attracted, to what number?

40. Consider the Mandelbrot sequence with seed $s = -\sqrt{2}$. Is this Mandelbrot sequence *escaping, periodic,* or *attracted?* If attracted, to what number?

Running

41. The Koch antisnowflake. This fractal shape is obtained by essentially reversing the process for constructing the Koch snowflake. At each stage, instead of adding a blue equilateral triangle to the outside on the middle of each side, we remove a blue equilateral triangle from the inside on the middle of each side. The following figure shows the first three steps in the construction of the *Koch antisnowflake.* If the area of the starting equilateral triangle is X, what is the area of the Koch antisnowflake?

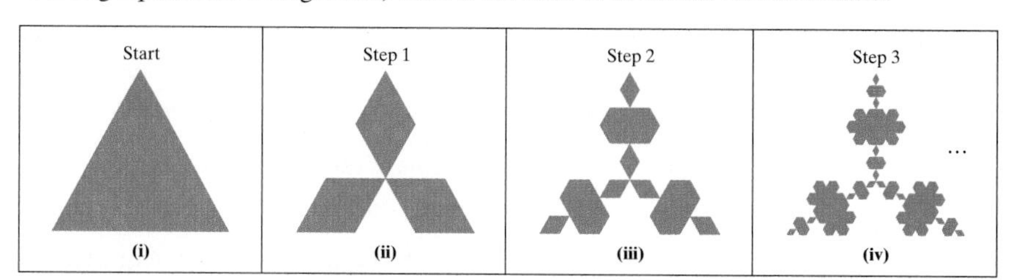

42. Find the area of the *square snowflake* defined in Exercises 13 through 16.

43. Suppose that we play the chaos game using triangle ABC, and M_1, M_2, M_3 are the midpoints of the three sides of the triangle. Explain why it is impossible at any time during the game to land inside triangle $M_1M_2M_3$.

44. Find the first few terms (as many as necessary) of the Mandelbrot sequence for each of the following seeds:

(a) $s = 1 + i$. Is s in the Mandelbrot set or not?

(b) $s = 1 - i$. Is s in the Mandelbrot set or not?

45. Find the first few terms (as many as necessary) of the Mandelbrot sequence for each of the following seeds:

(a) $s = -0.25 + 0.25i$. Is s in the Mandelbrot set or not?

(b) $s = -0.25 - 0.25i$. Is s in the Mandelbrot set or not?

46. Show that the Mandelbrot set has a reflection symmetry. (*Hint:* Review Exercises 44 and 45.)

References and Further Readings

1. Briggs, John, *Fractals: The Patterns of Chaos.* New York: Simon and Schuster, 1992.
2. Dewdney, A. K., "Computer Recreations: A computer microscope zooms in for a look at the most complex object in mathematics," *Scientific American,* 253 (August 1985), 16–24.
3. Dewdney, A. K., "Computer Recreations: A tour of the Mandelbrot set aboard the Mandelbus," *Scientific American,* 260 (February 1989), 108–111.
4. Dewdney, A. K., "Computer Recreations: Beauty and profundity. The Mandelbrot set and a flock of its cousins called Julia," *Scientific American,* 257 (November 1987), 140–145.
5. Field, M., and M. Golubitsky, *Symmetry in Chaos.* New York: Oxford University Press, 1992.
6. Gleick, James, *Chaos: Making a New Science.* New York: Viking Penguin, Inc., 1987, chap. 4.
7. Gleick, James, "The Man Who Reshaped Geometry," *New York Times Magazine,* 135 (December 8, 1985), 64.
8. Goldberger, A. L., D. R. Rigney, and B. J. West, "Chaos and Fractals in Human Physiology," *Scientific American,* 262 (February 1990), 44–49.
9. Jurgens, H., H. O. Peitgen, and D. Saupe, "The Language of Fractals," *Scientific American,* 263 (August 1990), 60–67.
10. Mandelbrot, Benoit, *The Fractal Geometry of Nature.* New York: W. H. Freeman & Co., 1983.
11. Peitgen, H. O., H. Jurgens, and D. Saupe, *Chaos and Fractals: New Frontiers of Science.* New York: Springer-Verlag, Inc., 1992.
12. Peitgen, H. O., H. Jurgens, and D. Saupe, *Fractals for the Classroom.* New York: Springer-Verlag, Inc., 1992.
13. Peitgen, H. O., and P. H. Richter, *The Beauty of Fractals.* New York: Springer-Verlag, Inc., 1986.
14. Peterson, Ivars, *The Mathematical Tourist.* New York: W. H. Freeman & Co., 1988, chap. 5.
15. Schechter, Bruce, "A New Geometry of Nature," *Discover,* 3 (June 1982), 66–68.
16. Schroeder, Manfred, *Fractals, Chaos, Power Laws: Minutes from an Infinite Paradise.* New York: W. H. Freeman & Co., 1991.

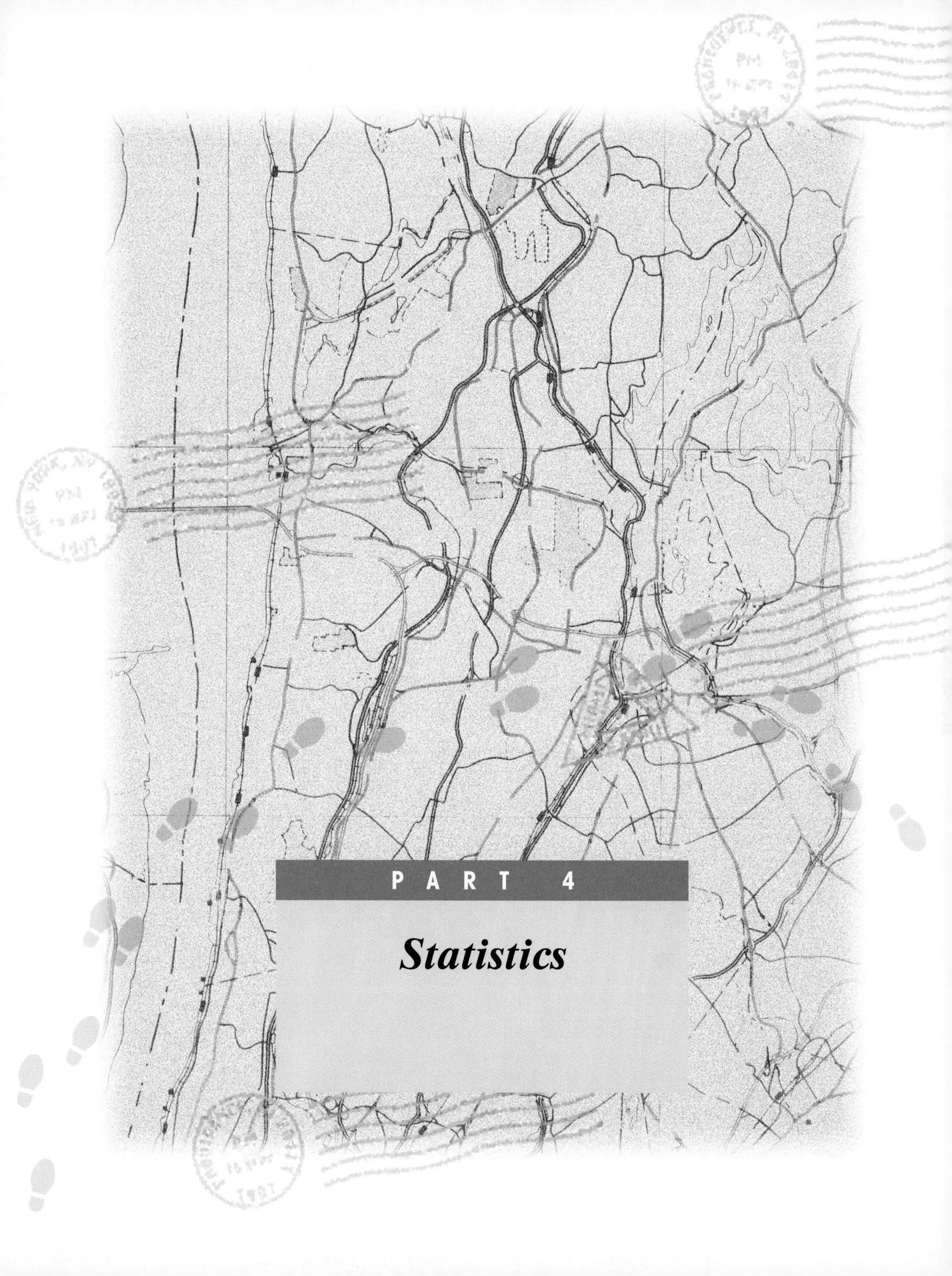

PART 4

Statistics

2 1 3 1
Cal Ripken Breaks Lou Gehrig's 2130 Consecutive Games Record.
New York Times, September 7, 1995.
U.S. Traffic Deaths Increase by 4.9 Percent in Last 2 Years.
National Safety Council, October 1995.
Weather forecasts 'wrong half the time.'
London's Daily Telegraph, Dec. 9, 1996.

Collecting Statistical Data

Censuses, Surveys, and Studies

CHAPTER 13

If there are areas of modern life that statistics doesn't touch, they are few and far between. Open today's paper and look at the sports page—plenty of stats there, no question about that. Don't like sports? Check the health section, or the business section, or the weather news. All of them are spiked with statistics. Some 60 years ago, the famous novelist H. G. Wells predicted: "Statistical reasoning will some day be as important for efficient citizenship as the ability to read or write." That day seems to have arrived with a vengeance.

Statistically speaking, today's world is a jungle. Our goal in this part of the book is to introduce and discuss the basic tools of statistical literacy needed to safely and efficiently move through this jungle—what H. G. Wells called "statistical reasoning . . . for efficient citizenship."

What is *statistics?* Statistics is, in a sense, the combination of two fundamental skills we all learned separately in school: handling information and manipulating numbers. When one combines communication skills (speaking, reading and writing) with numerical skills (arithmetic and algebra), one ends up with statistics. Here is a slightly more formal description: When information is packaged in numerical form it is called *data,* and to put it in a nutshell, *statistics* is the science of dealing with data. This includes gathering data, organizing data, interpreting data, and understanding data. We will discuss all of these things in the next four chapters.

Behind every statistical statement there is a story, and like any story it has a beginning, a middle, an end, and a moral. In this first statistics chapter we begin with the beginning, which in statistics typically means the process of gathering or collecting data. Data are the raw material of which statistical information is made, and in order to get good statistical information one needs good data.

There is a deceptive simplicity to the idea of collecting data. However, collecting good data in an efficient and timely manner is often the most difficult part of the statistical story. This chapter illustrates the do's and don'ts of gathering data. We will do this in two ways: simple examples and in-depth case studies.

The Population

Every statistical statement refers directly or indirectly, to some group of individuals or objects. In statistical terminology, this collection of individuals or objects is called the **population**. The first question we should ask ourselves in trying to make sense of a statistical statement is, What is the population to which the statement applies? If we are lucky, the population is clearly defined and we are off to a good start. Most often, this happens with statistical statements that are very specific and direct.

Example 1.

> **Snapshots to Improve the Tallies of Tigers**
>
> In a recent count . . . the Indian Government [gave] a current country-wide population estimate of 4,200 tigers.
>
> *New York Times,* July 30, 1996

Here is a very specific statistical statement about a well-defined population: *What?* Tigers; *Where?* Throughout India; *When?* Around 1995–96.

Once one has a clear understanding of the population of this story, the rest of the story is pretty straightforward: Nobody has a good handle on how to accurately estimate the size of a tiger population, and the merits of various strategies used by wildlife biologists, game wardens and government officials for getting a good head count are discussed. (By the way, this in itself is an important issue, and we will come back to it later in the chapter.) ■

Often, though no fault of our own, it can be difficult if not impossible to determine the exact definition of a population, especially when facts are given broadly and details are left uncovered.

Example 2.

> **Census Bureau Places Population at 249.6 Million**
>
> After one of the most controversial head counts ever, the Census Bureau today put the population of the United States at 249,632,692 . . . .
>
> *New York Times,* December 27, 1990

It's clear that this story is about the *population* of the United States. But exactly who is included in this definition? Are resident aliens included? What about illegal aliens? What about Americans stationed overseas? Clearly, some digging is needed to fully understand who the 249.6 million people are. We will discuss the United States Census in greater detail later in this chapter. ■

The *N*-value

If one were able to make an accurate head count of every member of a population, one would get a whole number N, sometimes informally called the *N-value* for the population. It's important to remember that the N-value is just one specific measurement of a population, not to be confused with the population itself.

Example 3.

A child has a coin jar full of nickels, dimes, and quarters. He is hoping there is enough money to buy a new baseball glove. Dad says to go count them, and if there isn't enough he will make up the difference. The child comes back with 54 nickels, 42 dimes, and 38 quarters. Here, if we define the population as the coins in the jar, the N-value is $N = 134$. ■

Knowing the size of a population is often an important part of drawing reliable statistical conclusions, but it is not always easy to get a good handle on the N-value of a population. When populations are small and accessible, one can actually get an exact N-value by simply counting "heads" the way the child counted the coins in the jar. In many situations, however, even a small population is too elusive and changeable to be accurately counted.

Example 4.

> **Siberian Tigers**
>
> It is not known exactly how many Siberian tigers roam free; like most large, secretive carnivores, they are hard to count. The latest estimate is around 430. Our work is aimed at increasing that number through better protection of the animals.
>
> *National Geographic,* February 1997

From this statement the only thing certain is that the population of Siberian tigers in the wild is small and endangered, and that the N-value $N = 430$ is only a rough estimate. ■

For large populations, an accurate N-value is usually expensive and difficult, sometimes impossible.

Example 5.

> **Census Experts Can't Count on Numbers: The Complexity of Counting a Nation Confounds Statisticians**
>
> Most experts agree that the 1990 census failed to ferret out about four million to five million Americans . . . .
>
> *New York Times,* August 6, 1991

This happened even though over 500,000 people were hired as census takers to do the counting, and over 2.5 *billion* dollars were spent to carry out the head count. The details are given in our first case study. ■

Case Study 1: The 1990 U.S. Census

Article 1, Section 2, of the Constitution of the United States mandates that a national census be conducted every 10 years. The original intent of the census was to "count heads" for a twofold purpose: taxes and political representation. Like everything else in the Constitution, Article 1, Section 2, was a compromise of many competing interests: The count was to exclude "Indians not taxed" and to count slaves as "three-fifths of a free Person." Since then, the scope and purpose of the U.S. Census has been modified and expanded by the Fourteenth Amendment and the courts in many ways:

- For the purposes of the census, the United States population is defined as consisting of "all persons *physically present* and *permanently residing* in the United States." Citizens, legal resident aliens, and even illegal aliens are meant to be included. Starting with the 1990 Census, military personnel and other federal workers stationed overseas or on American ships are also included.
- Besides counting heads, the U.S. Census Bureau now collects additional information about the population: sex, age, race, and ethnicity, marital status, housing, income, and employment data. Some of this information is updated on a regular basis, not just every ten years.
- Census data is now used for many important purposes beyond its original ones of *taxation* and *representation:* the allocation of billions of federal dollars to states, counties, cities, and municipalities; the collection of other important government statistics such as the Consumer Price Index and the Current Population Survey; the redrawing of legislative districts within each state; and the strategic planning of production and services by business and industry.

Given the critical importance of the U.S. Census and given the tremendous resources put behind the effort by the federal government, why was the head count off (according to some experts) by as much as 5 million people? How can the best intentions and tremendous resources of our government fail so miserably in an activity that on a smaller scale can be carried out by a single child?

Nowadays the notion that, if we put enough money and effort into it, all individuals living in the United States can be counted like coins in a jar is totally out of tune with the times. In 1790, when the first U.S. Census was carried out, the population was smaller and relatively homogeneous, people tended to stay in one place, and by and large they felt comfortable in their dealings with the government. Under these conditions it might have been possible for census takers to accurately count heads. Today's conditions are completely different: People are constantly on the move, many distrust the government, in large urban areas many people are homeless or don't want to be counted,[1] and then

[1] In some large cities the response rate for the 1990 U.S. Census was under 50%.

there is the apathy of many people who think of a census form as another piece of junk mail.

To make matters worse, the inability of the census to accurately count the population and measure the nation's demographics is biased against the urban poor, who were undercounted by as much as 20% in some large cities. After the 1990 Census, the cities of New York, Los Angeles, Chicago, Houston, the states of New York and California, the Leagues of Cities, the U.S. Conference of Mayors, the N.A.A.C.P., and the League of United Latin American Citizens were all parties to various lawsuits against the Census Bureau and its parent agency, the Commerce Department, claiming that the undercount of some ethnic minorities was depriving some cities and states of billions of dollars in federal funds.

In response to the problems and controversies generated by the 1990 Census, the United States Census Bureau has acknowledged what population biologists have known for a long time: it is impossible to get an exact head count of a large animal population. This applies to human beings as well. Starting with the 2000 Census, the United States population will be counted using a statistical technique which biologists call the *capture-recapture* method. We will discuss this method later in the chapter.

Surveys

A much more economical alternative to collecting data from each and every member of a population is to collect data only from a selected subgroup and then use this data to draw conclusions and make statistical inferences about the entire population. Statisticians call this approach a **survey** and the subgroup of the population from which the data is collected a **sample**.

The basic idea behind a survey is simple and well understood: If we have a sample that is "representative" of the entire population, then whatever we want to know about a population can be found out by getting the information from the sample.

Implementing a survey is far from simple, however. The critical issues are: (a) How do we find a sample that is "representative" of the population? and (b) How big should the sample be? These two questions go hand in hand, and we will discuss them next.

Sometimes a very small sample can be used to get reliable information about a population, no matter how large. This is the case when the population is highly homogeneous. An extreme example would be a population of identical individuals (clones). A completely reliable sample could be just one individual—in this case that single individual truly represents the entire population. Any information we want about the population we can get from that individual. A less extreme, but more realistic example is a blood sample that a doctor draws for a lab test. Because a person's blood is essentially the same everywhere in the body, a very small sample can yield reliable information on all of that person's blood.

The more heterogeneous a population gets, the more difficult it is to find a representative sample. The perils and difficulties of surveys of large, heterogeneous populations can be illustrated by examples from the history of *public opinion polls.*

Public Opinion Polls

We are all familiar with public opinion polls, such as the Gallup poll, the Harris poll, and many others. A public opinion poll is a special kind of survey in which the members of the sample provide information by answering specific questions from an "interviewer." The question/answer exchange can be done through a questionnaire, a personal telephone interview, or a direct face-to-face interview.

Nowadays, public opinion polls are used regularly to measure "the pulse of the nation." They give us statistical information ranging from voters' preferences before an election to opinions on issues such as the environment, abortion, and the economy.

Given their widespread use and the influence they exert, it is important to ask: How much can we trust the information that we get from public opinion polls? This is a complex question which goes to the very heart of mathematical statistics. We'll start our exploration of it with some historical examples.

Case Study 2: The 1936 Literary Digest Poll

The U.S. presidential election of 1936 pitted Alfred Landon, the Republican governor of Kansas, against the incumbent President, Franklin D. Roosevelt. At the time of the election, the nation had not yet emerged from the Great Depression, and economic issues such as unemployment and government spending were the dominant themes of the campaign.

The *Literary Digest,* one of the most respected magazines of the time, conducted a poll a couple of weeks before the election. The magazine had been polling the electorate since 1916, always accurately predicting the results of the election. Based on its 1936 poll, the *Literary Digest* predicted that Landon would get 57% of the vote against Roosevelt's 43%. The actual results of the election were 62% for Roosevelt against 38% for Landon. The difference between the poll's prediction and the actual election results was a whopping 19%, the largest error ever in a major public opinion poll.

Ironically, the 1936 *Literary Digest* poll was one of the largest and most expensive ever conducted, based on a huge sample of approximately 2.4 million people. For the same election, a Gallup poll based on a much smaller sample of approximately 50,000 people was able accurately to predict a victory for Roosevelt. What happened?

The sample for the *Literary Digest* poll was chosen by putting together, in one enormous list, the names of every person listed in a telephone directory anywhere in the United States, as well as the names of people on magazine subscription lists and rosters of clubs and professional associations. Altogether, a mailing list of about 10 million names was created. Every name on this list was mailed a mock ballot and asked to mark it and return it to the magazine.

One cannot help but be impressed by the sheer ambition of such a project. And it is not surprising that the magazine's confidence in the results was in direct proportion to the magnitude of the effort. In its issue of August 22, 1936, the *Literary Digest* crowed:

Once again, [we are] asking more than ten million voters—one out of four, representing every county in the United States—to settle November's election in October.

Next week, the first answers from these ten million will begin the incoming tide of marked ballots, to be triple-checked, verified, five-times cross-classified and totaled. When the last figure has been totted and checked, if past experience is a criterion, the country will know to within a fraction of 1 percent the actual popular vote of forty million [voters].

When reality hit, it hit hard. Soon after the election, with its credibility badly damaged and its sales drying out, the *Literary Digest* magazine went out of business, the victim of a statistical *faux-pas.*

The first thing seriously wrong with the *Literary Digest* poll was in the selection process for the names on the mailing list. Names were taken from telephone directories, rosters of club members, lists of magazine subscribers, and the like. Such a list was inherently slanted toward members of the middle and upper classes. Telephones in 1936 were something of a luxury. So, too, were club memberships and magazine subscriptions, at a time when 9 million people were unemployed. At least with regard to *economic status,* the *Literary Digest* mailing list was far from being a representative cross section of the population. This was a critical problem, because voters often vote on economic issues, and given the economic conditions of the time, this was especially true in 1936.

When the choice of the sample has a built-in tendency (whether intentional or not) to exclude a particular group or characteristic within the population, we say that a survey suffers from **selection bias**. It is obvious that selection bias must be avoided, but it is not always easy to detect it ahead of time. Even the most scrupulous attempts to eliminate selection bias can fall short (as will become apparent in our next case study).

The cover of the *Literary Digest* the week after the election.

The second problem with the *Literary Digest* poll was that out of the 10 million people whose names were on the original mailing list, only about 2.4 million responded to the survey. Thus, the size of the sample was about one-fourth of what was originally intended. When the proportion of respondents to the total number of people intended to be in the sample,[2] called the **response rate**, is low, a survey is said to suffer from **nonresponse bias**. For the *Literary Digest* poll the response rate was 24%, which is extremely low.

It is well known that people who respond to surveys are different from people who don't, not only in the obvious way (their attitude toward the usefulness of surveys) but also in more subtle ways: They tend to be better educated and in higher economic brackets (and are in fact more likely to vote Republican). Thus, nonresponse bias is a special type of selection bias—it excludes from the sample reluctant and disinterested people. But don't we want them represented?

Eliminating nonresponse bias from a survey is difficult. In a free country we cannot force people to participate, and paying them is hardly ever a solution, since it can introduce other forms of bias. Some ways of minimizing nonresponse bias are known, however. The *Literary Digest* survey was conducted by mail. This approach is the most likely to magnify nonresponse bias, because people often consider a mailed questionnaire just another form of junk mail. Of course, considering the size of the sample, the *Literary Digest* really had no other choice but to use mailed questionnaires. Here we see how a big sample size can be more of a liability than an asset.

Nowadays, almost all legitimate public opinion polls are conducted either by telephone or by personal interviews. Telephone polling is subject to slightly more nonresponse bias than personal interviews, but it is considerably cheaper. In some special situations, however, telephone polls can be so biased as to be useless.[3]

The *Literary Digest* story has two morals: (1) *you'll do better with a well-chosen small sample than with a badly chosen big one,* and (2) *watch out for selection bias and nonresponse bias.*

Our next case study illustrates how difficult it can be, even with the very best intentions, to get rid of selection bias.

Case Study 3: The 1948 Presidential Election

Despite the fiasco of 1936, and possibly because of the lessons learned from it, by 1948 the use of public opinion polls to measure the American electorate was thriving. Three major polls competed for the prize of correctly predicting the outcome of the national elections: the Gallup poll, the Roper poll, and the Crossley poll.

---

[2] Recall that by its own admission the *Literary Digest* was expecting about 10 million respondents.

[3] A blatant example of selection bias occurs when the sample is self-selected—you are in the sample because you volunteer to be in it. The worst instances of this are Area Code 900 telephone polls, where a person actually has to pay (sometimes as much as $2) to be part of the sample. It goes without saying that people who are willing to pay to express their opinions are hardly representative of the general public and that information collected from such polls should be considered totally unreliable.

In 1948 these three polls were using a much more "scientific" method for choosing their samples—**quota sampling**. George Gallup had introduced quota sampling as early as 1935 and had successfully used it to predict the winner of the 1936, 1940, and 1944 presidential elections. Quota sampling is a systematic effort to force the sample to fit a certain national profile by using quotas: The sample should have so many women, so many men, so many blacks, so many whites, so many under 40, so many over 40, and so on. The proportions in each category in the sample should be the same as those in the electorate at large.

If we assume that every important characteristic of the population is taken into account when the quotas are set up, it is reasonable to expect that quota sampling will produce a good cross section of the population and therefore lead to accurate predictions.

For the 1948 election between Thomas Dewey and Harry Truman, Gallup conducted a poll with a sample size of approximately 3250. Each individual in the sample was interviewed in person by a professional interviewer to minimize nonresponse bias, and each interviewer was given a very detailed set of quotas to meet—for example, 7 white males under 40 living in a rural area, 5 black males over 40 living in a rural area, 6 white females under 40 living in a rural area, and so on. By the time all the interviewers met their quotas, the entire sample was expected to accurately represent the entire population in every respect: gender, race, age, and so on.

Based on his sample, Gallup predicted a victory for Dewey, the Republican candidate. The predicted breakdown of the vote was 50% for Dewey, 44% for Truman, and 6% for third-party candidates Strom Thurmond and Henry Wallace. The other two polls made similar predictions. The actual results of the election turned out to be almost exactly reversed: 50% for Truman, 45% for Dewey, and 5% for the third-party candidates.

Truman's victory was a great surprise to the nation as a whole. So convinced was the *Chicago Daily Tribune* of Dewey's victory that it went to press on its early edition for November 4, 1948, with the headline "Dewey defeats Truman"—a blunder that led to Truman's famous retort, "Ain't the way I heard it." The picture of Truman holding aloft a copy of the *Tribune* (see photo) has become part of our national folklore. To pollsters and statisticians, the erroneous predictions of the 1948 election showed that quota sampling is intrinsically flawed.

"Ain't the way I heard it." Truman gloats while holding an early edition of the *Chicago Daily Tribune* in which the headline erroneously claimed a Dewey victory based on the predictions of all the polls. (UPI/Bettmann)

The basic idea of quota sampling appears to be a good one: Force the sample to be a representative cross section of the population by having each important characteristic of the population proportionally represented in the sample. Since income is an important factor in determining how people vote, the sample should have all income groups represented in the same proportion as the population at large. Ditto for sex, race, age, and so on. Right away we can see a potential problem; Where do we stop? No matter how careful we might be, we might miss some criterion that would affect the way people vote, and the sample could be deficient in this regard.

An even more serious flaw in quota sampling is that, other than meeting the quotas, the interviewers are free to choose whom they interview. This opens the door to selection bias.

Looking back over the history of quota sampling, one can see a clear tendency to overestimate the Republican vote. In 1936, using quota sampling, Gallup predicted the Republican candidate would get 44% of the vote, but the actual number was 38%. In 1940 the prediction was 48%, and the actual vote was 45%; in 1944 the prediction was 48%, and the actual vote was 46%. Nonetheless, Gallup was able to predict the winner correctly in each of these elections, mostly because the spread between the candidates was large enough to cover the error. In 1948 Gallup (and all the other pollsters) simply ran out of luck. It was time to ditch quota sampling.

The failure of quota sampling as a method for getting representative samples has a simple moral: *Even with the most carefully laid plans, human intervention in choosing the sample can result in selection bias.*

Random Sampling

If human intervention in choosing the sample is always subject to bias, what are the alternatives? The answer is to let the laws of chance decide who is in the sample—to draw the names, as it were, out of a hat. Some people find this method hard to believe in. Isn't it possible, they wonder, to get by sheer chance a sample that is very biased (say, for example, all Republican?) In theory such an outcome is possible, but in practice, when the sample is large enough, the odds of its happening are so low that we can pretty much rule it out. Most present-day methods of product quality control in industry, corporate audits in business, and public opinion polling are based on **random sampling** methods—that is, methods for choosing the sample in which chance intervenes in one form or another. The reliability of data collected by random sampling methods is supported by both practical experience and mathematical theory. We will discuss some of the details of this theory in Chapter 16.

Simple Random Sampling

The most basic form of random sampling is called **simple random sampling**. It is based on the same principle as a lottery is: Any set of numbers of a given size has an equal chance of being chosen as any other set of numbers of that size. Thus, if a lottery ticket consists of 6 winning numbers, a fair lottery is one in which any combination of 6 numbers has the same chance of winning as any other combination of 6 numbers. In sampling, this means that any group of members of the population should have the same chance of being the sample as any other group of the same size.

In theory, simple random sampling is easy to implement: We put the name of each individual in the population in "a hat," mix the names well, and then draw as

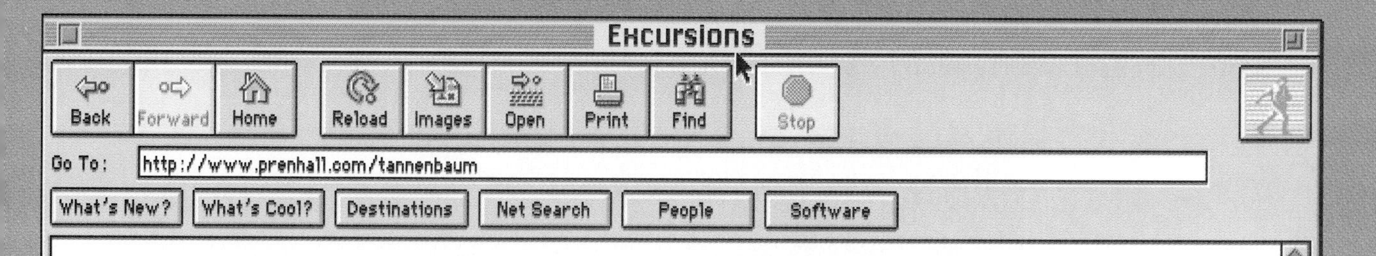

The Census

As required by the Constitution, the Census Bureau reports on the population every ten years. The population totals are used for apportioning seats in the House of Representatives. Other data gathered include information about racial and ethnic origin, population structure, and household structure.

The census data are used to influence decisions on government funding in a variety of areas such as education, health care, and energy. The data also help present a better view of the world around us and can be helpful in making decisions. Data of your county of residence from the 1990 census can be found in the Tannenbaum website under Chapter 13.

Use the map to locate your state, region of the state, and then county. From the county page select the STF1A data table. Within the STF1A data table there are many categories to choose from.

Questions:

1. What is the *N*-value, that is the number of persons for your county in 1990? Which category did you use to find this?
2. There are many categories listed. Do you believe this to be necessary?
3. Look in the category of Race (25) . Use these numbers to calculate the racial makeup of your county by percent. Does this seem to be accurate by your estimation?
4. Look over the twenty-five subcategories under Race (25). Do you feel this effectively categorizes the racial makeup of the population?
5. Suppose you were to decide on whether or not to start a day care service. Which categories would you look at to determine if there is a need?
6. Compare data for your county in the categories from question 5 with other counties near yours. Use the data to determine which county would be better. What other information would be helpful when comparing among the counties?

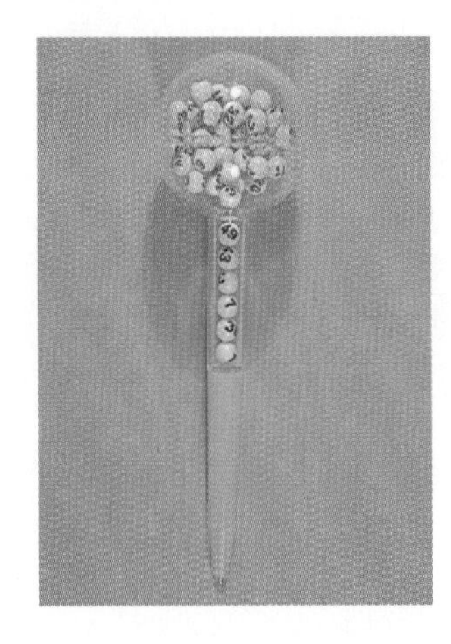

A simple example of *simple random sampling*. A gadget costing 89¢ which randomly generates numbers for lottery play.

many names as we need for our sample. Of course "a hat" is just a metaphor. If our population is 100 million voters and we want to choose a simple random sample of 2000, we will not be putting all the names in a real hat and then drawing 2000 names one by one. The modern way to do any serious simple random sampling is by computer: Make a list of members of the population, enter it into the computer, and then let the computer randomly select the names.

While simple random sampling works well in many cases, for national surveys and public opinion polls it presents some serious practical difficulties. First, it requires us to have a list of all the members of the population. The population, however, may not be clearly defined, or, even if it is, a complete list of the members may not be available. Second, implementing simple random sampling in national public opinion polls raises problems of expediency and cost. Interviewing several thousand people chosen by simple random sampling means chasing people all over the country. This requires an inordinate amount of time and money. For most public opinion polls, especially those done on a regular basis, the time and money needed to do this are simply not available.

Our next case study describes the sampling method currently used in most public opinion polls.

Case Study 4: Modern Public Opinion Polls: Stratified Samples

Present-day methods for conducting public opinion polls need to take into account two sets of considerations: (1) minimizing sample bias, and (2) choosing a sample that is accessible in a cost-efficient, timely manner. A random sampling method that deals in a satisfactory way with both these issues is **stratified sampling**. The basic idea of stratified sampling is to break the population into categories called **strata** and then randomly choose a sample from these strata. The chosen strata are then further divided into categories called substrata, and a random sample is taken from these substrata. The selected substrata are further subdivided, a random sample is taken from them, and so on. The process goes on for a predetermined number of layers.

In public opinion polls, the strata and substrata are usually defined by criteria that involve a combination of geographic and demographic elements. For example, at the first level, the nation is divided into "size of community" strata (big cities, medium cities, small cities, villages, rural areas, etc.). Each of these strata is then subdivided by geographical region (New England, Middle Atlantic, East Central, etc.). Within each geographical region and within each

The Gallup Poll

Design of the Sample

The design of the sample used by the Gallup Poll for its standard surveys of public opinion is that of a replicated area probability stratified sample down to the block level in the case of urban areas and to segments of townships in the case of rural areas.

After stratifying the nation geographically and by size of community in order to insure conformity of the sample with the 1990 Census distribution of the population, over 360 different sampling locations or areas are selected on a mathematically random basis from within cities, towns, and counties which have in turn been selected on a mathematically random basis. The interviewers have no choice whatsoever concerning the part of the city, town, or county in which they conduct their interviews.

Approximately five interviews are conducted in each randomly selected sampling point. Interviewers are given maps of the area to which they are assigned and are required to follow a specific travel pattern on contacting households. At each occupied dwelling unit, interviewers are instructed to select respondents by following a prescribed systematic method. This procedure is followed until the assigned number of interviews with male and female adults have been completed . . . .

Since this sampling procedure is designed to produce a sample which approximates the adult civilian population (18 and older) living in private households (that is, excluding those in prisons and hospitals, hotels, religious and educational institutions, and on military reservations) the survey results can be applied to this population for the purpose of projecting percentages into numbers of people. The manner in which the sample is drawn also produces a sample which approximates the population of private households in the United States. Therefore, survey results also can be projected in terms of numbers of households.

Sampling Error

In interpreting survey results, it should be borne in mind that all sample surveys are subject to sampling error, that is, the extent to which the results may differ from those that would be obtained if the whole population surveyed had been interviewed.

Source: The Gallup Report. Princeton, NJ: American Institute of Public Opinion, 1991.

size of community stratum, some communities are selected by simple random sampling. The selected communities (called *sampling locations*) are the only places where interviews will be conducted. To further randomize things, each of the selected sampling locations is subdivided into geographical units called *wards,* and within each sampling location some of its wards are once again selected by simple random sampling. The selected wards are then divided into smaller units called *precincts,* and within each ward some of its precincts are selected by simple random sampling. At the last stage, *households* are selected for interviewing by simple random sampling within each precinct. The interviewers are then given specific instructions as to which households in their assigned area they must conduct interviews in, and the order which they must follow.

The efficiency of stratified sampling compared to simple random sampling in terms of cost and time is clear. The members of the sample are clustered in well-defined and easily manageable areas significantly reducing the cost of conducting interviews as well as the response time needed to get the data together.

For a large, heterogeneous nation like the United States, stratified sampling has generally proved to be a reliable way to collect national data, and most modern public opinion polls are based on stratified samples.

What about the size of the sample? Surprisingly, it does not have to be very large. (A typical Gallup poll is based on samples consisting of less than 1500 individuals.) Even more surprisingly, about the same size sample is used to poll the population of a small city as the population of the United States. *The size of the sample does not have to be proportional to the size of the population.* How can this be? George Gallup, one of the fathers of modern public opinion polling, explained it this way[4]:

Whether you poll the United States or New York State or Baton Rouge (Louisiana) . . . you need . . . the same number of interviews or samples. It's no mystery really—if a cook has two pots of soup on the stove, one far larger than the other, and thoroughly stirs them both, he doesn't have to take more spoonfuls from one than the other to sample the taste accurately.

Before we continue with our examples and case studies, let's review some of the key concepts in sampling and introduce some new terminology.

Sampling: Terminology and Key Concepts

As we now know, except for a *census,* the common way to collect statistical information about a population is by means of a *survey.* In a survey, we use a subset of the population called a *sample* as the source of our information, and from it we try to generalize and draw conclusions about the entire population. Statisticians use the term **statistic** to describe any kind of numerical information drawn from a sample. A statistic is always an estimate for some unknown measure of the population called a **parameter**. Let's put it this way: A *parameter* is the numerical information we would like to have—the pot of gold at the end of the statistical rainbow, so to speak. Calculating a parameter is difficult, often impossible, since the only way to get the exact value for a parameter is to use a census. If we use a sample, then we can get only an estimate for the parameter, and this estimate is called a *statistic.*

We will use the term **sampling error** to describe the difference between a parameter and a statistic used to estimate that parameter. Sampling error can be attributed to two factors: *chance error* and *sampling bias.*

- **Chance error** is the result of a basic fact about surveys: A statistic cannot give exact information about the population because it is by definition based on partial information (the sample). In surveys, chance error is the result of **sampling variability**: the fact that two different samples are likely to give two different statistics, even when the samples are chosen using the same sampling method. While sampling variability, and thus chance error, are unavoidable, with careful selection of the sample and the right choice of sample size they can be kept to a minimum.
- **Sampling bias** is the result of having a poorly chosen sample. Even with the best intentions, getting a sample that is representative of the entire population can be very difficult, and many subtle factors can affect the "representativeness" of the sample. Sample bias is the result. As opposed to chance error, sampling bias can be eliminated by using proper methods of sample selection.

---

[4] As quoted in "The Man Who Knows How We Think," *Modern Maturity,* 17, no. 2 (April–May 1974).

Last, a few comments about the size of the sample, usually denoted by the letter n to contrast with N, the size of the population. The ratio n/N is called the **sampling rate**. When expressed as a percentage, a sampling rate of $x\%$ tells us that the sample is $x\%$ of the population. We now know that in sampling, it is not the sampling rate that matters but rather choosing a good sample of a reasonable size. In public opinion polls, for example, whether $N = 100{,}000$ or $N = 250$ million, a well-chosen sample of $n = 1500$ is sufficient to get reliable statistics.

For our last example of sampling and its uses we return to a problem that we discussed earlier: how to determine the N-value of a moving, elusive population. This is a common problem in population biology, and it is exactly the problem confronted by the United States Census.

Example 6.

(2000) Census Is to Use Sampling

Officials say that in the 2000 census, mail-in forms and enumerators will be used until 90 percent of the households in a county have been counted. Then a statistical sample of 10 percent of the remaining households will be selected, and enumerators will be dispatched, repeatedly if necessary, to count them. The results will be used to estimate the total number of those who were originally missed.

New York Times, Feb. 29, 1996

The statistical method that will be used in the 2000 Census to estimate the number of people originally missed by the traditional count is called the *capture-recapture method.* The method consists of two steps, which we will describe in its most frequently used setting of population biology:

- **Step 1. "The Capture":** Capture (choose) a sample of size n_1, *tag* (mark, identify) the animals (objects, people), and release them back into the general population.

- **Step 2. "The Recapture":** After a certain period of time, capture (choose) a new sample of size n_2, and take an exact head count of the *tagged* individuals (those that were also in the first sample). Let's call this number k.

If we can assume that the recaptured sample is representative of the entire population, then the proportion of tagged individuals in it equals the proportion of the tagged individuals in the population. In other words, the ratio k/n_2 is equal to the ratio n_1/N. From this we can solve for N and get that $N = n_1 n_2/k$.

For example, suppose that we want to know how many fish there are in a pond. In the first step we capture, say, 200 fish ($n_1 = 200$), tag them, and release them unharmed back in the pond. After giving enough time for the released fish to mingle and disperse throughout the pond, we recapture 300 fish ($n_2 = 300$), of which 60 (20%) are tagged fish. We can now estimate the original population of the pond to be $N = 1000$ fish (since 200 is 20% of 1000).

The Census Bureau plans to use the *capture-recapture method* in the 2000 Census to adjust the count it will get from its traditional head-counting methods. Minority groups and Republicans in Congress are both disputing the validity of this approach, and once again, a "simple" question of statistical methodology is likely to be settled in court. ■

Clinical Studies

So far, we have focused our attention on the issue of sample selection. Once we got past that point, all our examples pretty much assumed that the data itself was available to the observer in a direct and objective manner. *If the election were held today, would you vote for candidate X or candidate Y? How many teenagers live in this household? Which TV programs did you watch last night?*

A very different and important type of data collection involves questions for which there is no clear, immediate answer. *Does smoking increase your chances of lung disease? Will taking aspirin reduce your chances of having a heart attack? Does listening to classical music while taking a test improve your test score?* These kinds of questions have two things in common: (1) They involve a cause and an effect, and (2) the answers require observation over an extended period of time.

The standard approach for answering questions of this sort is to set up a *study.* When one wants to know if a certain cause X produces a certain effect Y, one sets up a study in which cause X is produced and its effects are observed. If the effect Y is observed, then it is possible that X was indeed the cause of Y. We have established an *association* between the cause X and the effect Y. The problem, however, is the nagging possibility that some other cause Z different from X sneaked in and produced the effect Y and that X had nothing to do with it. Just because we established an association, we have not established a cause-effect relation between the variables. Statisticians like to explain this by a simple saying: *Association is not causation.*

Let's illustrate with an example This one is fictitious but not all that farfetched. Suppose we want to find out if too much chocolate in one's diet can increase one's chance of becoming diabetic. Here the cause X is eating too much chocolate, and the effect Y is diabetes We set up an experiment in which 100 rats are fed a pound of chocolate a day for a period of six months. At the end of a six-month period, 15 of the 100 rats have diabetes. Since in the general rat population only 3% are diabetic, we are tempted to conclude that the diabetes in the rats is indeed caused by the excessive chocolate in the diet. The problem is that there is no certainty that the chocolate diet was the cause. Could there be another unknown reason for the observed effect?

For most cause-and-effect situations, especially those complicated by the involvement of human beings, a single effect can have many possible and actual causes What causes heart attacks? Unfortunately, there is no single cause—diet, lifestyle, stress, and heredity are all known to be contributory causes. The extent to which each of these causes contributes individually and the extent to which they interact with each other are extremely difficult questions that can be answered only by means of carefully designed statistical studies.

For the remainder of this chapter we will illustrate an important type of study called a **clinical study**. Generally, clinical studies are concerned with determining whether a single variable or treatment (usually a vaccine, drug, therapy, etc.) can cause a certain effect (a disease, a symptom, a cure, etc.). The importance of such clinical studies is self-evident: Every new vaccine, drug, or treatment must "prove" itself by means of clinical study before it is officially approved for public use.

Likewise, almost everything that is bad for us (cigarettes, caffeine, cholesterol, etc.) gets its "official" certification of badness by means of a clinical study.

Properly designing a clinical study can be both difficult and controversial, and as a result we are often bombarded with conflicting information produced by different studies examining the same cause-and-effect question. The basic principles guiding a clinical study, however, are pretty much established by statistical practice and are almost always followed. We will discuss them next.

The first and most important issue in any clinical study is to isolate the cause (treatment, drug, vaccine, therapy, etc.) that is under investigation from all other possible contributing causes (called **confounding variables**) that could produce the same effect. This is accomplished by performing the experiment on two different groups: a **treatment group** and a **control group**. The treatment group receives the treatment, and the control group should not. The control group is there for *comparison* purposes only: If a cause-and-effect relationship exists, then the treatment group should show the effects of the treatment and the control group should not. The comparison is most effective when the treatment and control groups are identical to each other in all other respects (except that one group is receiving the treatment and the other one isn't). If this is accomplished and there are significant differences between the groups in the effects of the treatment, then these differences can be safely attributed to the treatment. We have established a cause-and-effect relationship between the treatment and the results.

Any study in which a cause-and-effect relationship is established by comparing the results in a treatment group with the results in a control group is called a **controlled study**. Our example here is a famous clinical study carried out in 1954 to determine the effectiveness of a new vaccine against polio.

Case Study 5. The 1954 Salk Polio Vaccine Field Trials

Polio (infantile paralysis) has been practically eradicated in the western world. In the first half of the twentieth century, however, it was a major public health problem. Over one-half million cases of polio were reported between 1930 and 1950, and the actual number may have been considerably higher.

Because polio attacks mostly children and because its effects can be so serious (paralysis or death), eradication of the disease became a top public health priority in the United States. By the late 1940s it was known that polio is a virus and as such can best be treated by a vaccine which is itself made up of a virus. The vaccine virus can be a closely related virus that does not have the same harmful effects, or it can be the actual virus that produces the disease but which has been killed by a special treatment. The former is known as a *live-virus vaccine,* the latter as a *killed-virus vaccine.* In response to either vaccine the body is known to produce *antibodies* which remain in the system and give the individual immunity against an attack by the real virus.

Both the live-virus and the killed-virus approaches have their advantages and disadvantages. The live virus approach produces a stronger reaction and better immunity, but at the same time it is also more likely to cause a harmful reaction and in some cases even to produce the very disease it is supposed to prevent. The killed-virus approach is safer in terms of the likelihood of producing a harmful reaction, but it is also less effective in providing the desired level of immunity.

These facts are important because they help us understand the extraordinary amount of caution that went into the design of the study that tested the effectiveness of the polio vaccine. By 1953, several potential vaccines had been developed, one of the more promising of which was a killed-virus vaccine developed by Jonas Salk at the University of Pittsburgh. The killed-virus approach was chosen because there was a great potential risk in testing a live-virus vaccine in a large-scale study, and a large-scale study was needed to collect enough information on polio (which in the 1950s had a rate of incidence among children of about 1 in 2000).

The testing of any new vaccine or drug creates many ethical dilemmas which have to be taken into account in the design of the study. With a killed-virus vaccine the risk of harmful consequences produced by the vaccine itself is small, so one possible approach could have been to distribute the vaccine widely among the population (ideally giving it to every child, but this was not possible because supplies were limited) and then follow up on whether there was a decline in the national incidence of polio in subsequent years. This is called the *vital statistics* approach and is the simplest way to test a vaccine. This is essentially the way the smallpox vaccine was determined to be effective. The problem with such an approach for polio is that polio is an epidemic type of disease, which means that there is a great variation in the incidence of the disease from one year to the next. In 1951, there were close to 60,000 reported cases of polio in the United States, but in 1952 the number of reported cases had dropped to almost half (about 35,000). Since no vaccine or treatment was used, the cause of the drop was the natural variability that is typical of epidemic diseases. So, if a totally ineffective polio vaccine had been tested in 1951 without a control group, the observed effect of a large drop in the incidence of polio in 1952 could have been interpreted as a proof that the vaccine worked.

The final decision on how best to test the effectiveness of the Salk vaccine was left to an advisory committee of doctors, public officials, and statisticians convened by the National Foundation for Infantile Paralysis and the Public Health Service. In order to isolate the cause under investigation (the Salk vaccine) from other possible causes of the desired effect (a reduction in the incidence of polio), it was decided that the study would be controlled, with a treatment group (those receiving the actual vaccine) and a control group (those receiving a **placebo**—in this case a shot of harmless salt solution). A study of this kind is called a **controlled placebo study**.

The reasons for using placebos in controlled studies go back to our desire that the treatment and control groups be as equal as possible in all respects, except of course that one group is receiving the vaccine and the other one isn't. It is a well-known fact that the mere *thinking* that one is getting a helpful vaccine or pill can actually produce positive results. Placebos are the standard way of eliminating this confounding variable known as the *placebo effect.*

It goes without saying that in a controlled placebo study it is essential that neither the members of the treatment group nor the members of the control group know to which of the two groups they belong. When this is the case the study is said to be a **blind study**. It is also desirable that the scientists conducting the study not know which subjects are taking the actual treatment and which are taking the placebo. The purpose is to make the observation, analysis, and interpretation of the results of the study as impartial as possible. A controlled placebo study in which neither the subjects or the scientists conducting the study know which subjects are in the treatment group and which are in the control group is called a **double-blind study**.

Making the Salk vaccine study double-blind was particularly important because polio is not an easy disease to diagnose—it comes in many different forms and degrees. Sometimes it can be a borderline call, and if the doctor collecting the data had prior knowledge of whether the subject had received the real vaccine or the placebo, the diagnosis could have been subjectively tipped one way or the other.

With all this background we can now describe the actual details of the experiment. Approximately 750,000 children were randomly selected to participate in the study. Of these, about 340,000 declined to participate and another 8500 dropped out in the middle of the experiment. The remaining children were divided into two groups—a treatment group and a control group—with approximately 200,000 children in each group. The choice of which children were selected for the treatment group and which for the control group was made by *random selection.* (Any study in which the treatment group and control group are chosen by random selection is called a **randomized controlled study**.) Some of the figures and results of the study are shown in Table 13-1.

Table 13-1 Results of the Salk Vaccine Field Trials

	Number of Children	Number of Reported Cases of Polio	Number of Paralytic Cases of Polio	Number of Fatal Cases of Polio
Treatment group	200,745	82	33	0
Control group	201,229	162	115	4
Declined to participate in the study	338,778	182*	121*	0*
Dropped out in the middle	8,484	2*	1*	0*
Total	749,236	428	270	4

*These figures are not a reliable indicator of the actual number of cases—they are only self-reported cases. (Adapted from Thomas Francis, Jr., et al., "An Evaluation of the 1954 Poliomyelitis Vaccine Trials—Summary Report." *American Journal of Public Health,* 45 (1955) 25.)

While Table 13-1 shows only a small part of the data collected by the Salk vaccine field trials, it can be readily seen that the difference between the treatment and control groups was significant and could rightly be interpreted as a clear indication that the vaccine was indeed effective.

Based on the data collected by the 1954 field trials, a massive inoculation campaign was put into effect. Today, all children are routinely inoculated against polio,[5] and polio has essentially been eradicated in the United States. Statistics played a key role in this important public health breakthrough.

---

[5] The Salk vaccine, which had been used for many years, was replaced some years ago by the Sabin vaccine, an oral vaccine based on the live-virus approach.

Conclusion

In this chapter we have discussed different methods for collecting data. In principle, the most accurate method is a *census,* a method which relies on collecting data from each member of the population. In most cases, because of considerations of cost and time, a census is a completely unrealistic strategy. When data are collected from only a subset of the population (called a *sample*), the data collection method is called a *survey.* The most important rule in designing good surveys is to eliminate or minimize *sample bias.* Today, almost all strategies for collecting data are based on surveys in which the laws of chance are used to determine how the sample is selected, and these methods for collecting data are called *random sampling* methods. Random sampling is the best way known to minimize or eliminate sample bias. Two of the most common random sampling methods are *simple random sampling* and *stratified sampling.* In some special situations other, more complicated types of random sampling can be used.

Sometimes identifying the sample is not enough. In cases in which cause-and-effect questions are involved, the data may come to the surface only after an extensive study has been carried out. In these cases isolating the cause variable under consideration from other possible causes (called *confounding variables*) is an essential prerequisite for getting reliable data. The standard strategy for doing this is a *controlled study* in which the sample is broken up into a *treatment group* and a *control group.* Controlled studies are now used (and sometimes abused) to settle issues affecting every aspect of our lives We can thank this area of statistics for many breakthroughs in social science, medicine, and public health, as well as for the constant and dire warnings about our health, our diet, and practically anything that is fun.

Key Concepts

blind study
census
chance error
clinical study
confounding variable
control group
controlled study
controlled placebo study
data
double-blind study
nonresponse bias
parameter
placebo
population
quota sampling
randomized controlled study
random sampling
sample
sample bias
sampling variability
sampling error
sampling rate
selection bias
simple random sampling
statistic
strata
stratified sampling
survey
treatment group

Exercises

Walking

Exercises 1 through 4 refer to the following survey. In 1988 "Dear Abby" asked her readers to let her know whether they had cheated on their spouses or not. The readers' responses are summarized in the accompanying table.

Status	Women	Men
Faithful	127,318	44,807
Unfaithful	22,468	15,743
Total	149,786	60,550

Based on the results of this survey, Dear Abby concluded that the amount of cheating among married couples is much less than people believe. (In her words, "The results were astonishing. There are far more faithfully wed couples than I had surmised.")

1. (a) Describe as specifically as you can the population for this survey.

(b) What was the size of the sample?

(c) How was the sample chosen?

(d) Eighty-five percent of the women who responded to this survey claimed to be faithful. Is the number 85% a parameter? A statistic? Neither? Explain your answer.

2. (a) Explain why this survey was subject to selection bias.

(b) Explain why this survey was subject to nonresponse bias.

3. (a) Based on the Dear Abby data, estimate the percentage of married men who are faithful to their spouses.

(b) Based on the Dear Abby data, estimate the percentage of married people who are faithful to their spouses.

(c) How accurate do you think these estimates are? Explain.

4. If money were no object, could you devise a survey that might give more reliable results than the Dear Abby survey? Describe briefly what you would do.

Exercises 5 through 8 refer to the following hypothetical situation. The Cleansburg Planning Department is trying to determine what percent of the people in the city want to spend public funds to revitalize the downtown mall. In order to do so, they decide to conduct the following survey: Five professional interviewers (A, B, C, D, and E) are hired, and each is asked to pick a street corner of their choice within the city limits. Everyday between 4:00 and 6:00 P.M. the interviewers are to ask each passerby if he or she wishes to respond to a survey sponsored by Cleansburg City Hall and to make a record of their response. If the response is yes, the person is asked to respond to the next question: Are you in favor of spending public funds to revitalize the downtown mall? Yes or no? The interviewers are asked to return to the same street corner as many days as are necessary until each one has conducted a total of 100 interviews. The data collected are seen in Table 13-2.

Table 13-2

Interviewer	Yes[a]	No[b]	Nonrespondents[c]
A	35	65	321
B	21	79	208
C	58	42	103
D	78	22	87
E[d]	12	63	594

[a]In favor of spending public funds to revitalize the downtown mall.

[b]Opposed to spending public funds to revitalize the downtown mall.

[c]Declined to be interviewed.

[d]Got frustrated and quit.

5. (a) Describe as specifically as you can the population for this survey.

(b) What is the size of the sample?

6. (a) Calculate the response rate in this survey.

(b) Explain why this survey was subject to nonresponse bias.

7. (a) Can you explain the big difference in the data from interviewer to interviewer?

(b) One of the interviewers conducted the interviews at a street corner downtown. Which interviewer? Explain.

(c) Do you think the survey was subject to selection bias? Explain.

(d) Was the sampling method used in this survey the same as quota sampling? Explain.

8. (a) Do you think this was a good survey? If you were a consultant to the Cleansburg Planning Department, could you suggest some improvements? Be specific.

Exercises 9 through 12 refer to the following survey. The dean of students at Tasmania State University wants to determine the percent of undergraduates living at home during the current semester. There are 15,000 undergraduates at TSU, so it is decided that the cost of checking with each and every one would be prohibitive. The following method is proposed to choose a representative sample of undergraduates to interview: Start with the registrar's alphabetical listing containing the names of all undergraduates. Pick randomly a number between 1 and 100 and count that far down the list, taking the name and every 100th name after it. (For example, if the random number chosen is 73, then pick the 73rd, 173rd, 273rd, etc., names on the list.) Assume the survey has a response rate of 0.95.

9. (a) Describe the population for this survey.

(b) Give the exact value of N.

10. (a) Find the size n of the sample.

(b) Find the sampling rate.

11. (a) Was this survey subject to selection bias? Explain.

(b) Explain why the method used for choosing the sample is not simple random sampling.

12. Do you think the results of this survey will be reliable? Explain.

Exercises 13 through 16 refer to the following hypothetical study. The manufacturer of a new vitamin (vitamin X) decides to sponsor a study to determine its effectiveness in curing the common cold. Five hundred college students in the San Diego area who are suffering from colds are paid to participate as subjects in this study. They are all given two tablets of vitamin X a day. Based on information provided by the subjects themselves, 457 out of the 500 subjects are cured of their colds within 3 days. The average number of days a cold lasts is 4.87 days. As a result of this study, the manufacturer launches an advertising campaign claiming that "vitamin X is more than 90% effective in curing the common cold."

13. **(a)** Describe as specifically as you can the population for this study.

(b) How was the sample selected?

(c) What was the size n of the sample?

(d) Was this health study a controlled experiment?

14. **(a)** Do you think the placebo effect could have played a role in this study?

(b) List three possible causes other than the effectiveness of vitamin X itself that could have confounded the results of this study.

15. List four different problems with this study that indicate poor design.

16. Make some suggestions for improving the study.

Exercises 17 through 20 refer to the following. A study by a team of Harvard University scientists [Science News, 138, no. 20 (November 17, 1990), 308] found that regular doses of beta carotene (a nutrient common in carrots, papayas, and apricots) may help prevent the buildup of plaque-produced arteriosclerosis (clogging of the arteries), which is the primary cause of heart attacks. The subjects in the study were 333 volunteer male doctors, all of whom had shown some early signs of coronary artery disease. The subjects were randomly divided into two groups. One group was given a 50-milligram beta carotene pill every other day for six years, and the other group was given a similar-looking placebo pill. The study found that the men taking the beta carotene pills suffered 50% fewer heart attacks and strokes than the men taking the placebo pills.

17. Describe as specifically as you can the population for this study.

18. **(a)** Describe the sample.

(b) What was the size n of the sample?

(c) Was the sample chosen by random sampling? Explain.

19. **(a)** Explain why this study can be described as a controlled placebo experiment.

(b) Describe the treatment group in this study.

(c) Explain why this study can be described as a randomized controlled experiment.

20. **(a)** Mention two possible confounding variables in this study.

(b) Carefully state what a legitimate conclusion from this study might be.

Exercises 21 through 24 refer to the following hypothetical situation. A college professor has a theory that a dose of about 10 milligrams of caffeine a day can actually improve students' performance in their college courses. To test his theory, he chooses the 13 students in his Psychology 101 class that got an "F" in the first midterm and asks them to come to his office three times a week for "individual tutoring." When the students come to his office, he engages them in friendly conversation, while at the same time pouring them several cups of strong coffee. After a month of doing this, he observes that of the 13 students, 8 show significant improvement in their second midterm scores; 3 show some improvement, and 2 show no improvement at all. Based on this, he concludes that is theory about caffeine is correct.

21. Which of the following terms best describes the professor's study: (i) randomized controlled experiment, (ii) double-blind experiment, (iii) controlled placebo experiment, or (iv) clinical study? Explain your choice and why you ruled out the other choices.

22. **(a)** Describe the population and the sample of this study.

(b) What was the value of n?

(c) Which of the following percentages best describes the sampling rate for this study: (i) 10%, (ii) 1%, (iii) 0.1%, (iv) 0.01%, or (v) less than 0.01%? Explain.

23. List at least three possible causes other than caffeine that could have confounded the results of this study.

24. Make some suggestions to the poor professor as to how he might improve the study.

Jogging

25. **Informal surveys.** In everyday life, we are constantly involved in activities that can be described as *informal surveys,* often without even realizing it. Here are some examples:

(i) Al gets up in the morning and wants to know what kind of day it is going to be, so he peeks out the window. He doesn't see any dark clouds, so he figures it's not going to rain.

(ii) Betty takes a sip from a cup of coffee and burns her lips. She concludes the coffee is too hot and decides to add a tad of cold water to it.

(iii) Carla goes to the doctor to have a checkup. The nurse draws 5 ml of blood from Carla's right arm and sends it to the lab. The lab report comes out negative for all diseases tested.

For each of the above examples:

(a) Describe the population.

(b) Discuss whether the sample is random or not.

(c) Discuss the validity of the conclusions drawn. (There is no right or wrong answer to this question, but you should be able to make a reasonable case for your position.)

26. Read the examples of informal surveys given in Exercise 25. Give three more examples of your own. Make them as different as possible from the ones given in Exercise 25. (Changing coffee to tea or soup in (ii) is not acceptable.)

27. **Leading-question bias.** In many surveys, the way the questions in the survey are phrased can itself be a source of bias. When a question is worded in such a way as to predispose the respondent to provide a particular response, the results of the survey are tainted by a special type of bias called leading-question bias. The following is an extreme hypothetical situation intended to drive the point home.

The American Self-Righteous Institute is a conservative think tank. In an effort to find out how the American taxpayer feels about a tax increase, the institute conducts a "scientific" poll. The main question in the poll is phrased as follows:

Are you in favor of paying higher taxes to bail the federal government out of its disastrous economic policies and its mismanagement of the federal budget? Yes _____. No _____.

Ninety-five percent of the respondents answered no. The results of the survey are announced by the sponsors with the statement:

Public opinion polls show that 95% of American taxpayers oppose a tax increase.

(a) Explain why the results of this survey might be invalid.

(b) Rephrase the question in a neutral way. Pay particular attention to "highly charged" words.

(c) Make up your own (more subtle) example of leading-question bias. Analyze the critical words that are the cause of bias.

28. Consider the following hypothetical survey designed to find out what percentage of people cheat on their income taxes: Fifteen hundred taxpayers are randomly selected from the Internal Revenue Service (IRS) rolls. These individuals are then interviewed in person by representatives of the IRS and read the following statement:

This survey is for information purposes only. Your answer will be held in strict confidence. Have you ever cheated on your income taxes? Yes _____. No _____.

Twelve percent of the respondents answered yes.

(a) Explain why the above figure might be unreliable.

(b) Can you think of ways in which a survey of this type might be designed so that more reliable information could be obtained? In particular, discuss who should be sponsoring the survey and how the interviews should be carried out.

29. **Listing bias.** Today, most consumer marketing surveys are conducted by telephone. In selecting a sample of households that are representative of all the households in a given geographical area the two basic techniques used are (i) randomly selecting telephone numbers to call from the local telephone directory or directories, and (ii) using a computer to randomly generate 7-digit numbers to try that are compatible with the local phone numbers.

(a) Briefly discuss the advantages and disadvantages of each technique. In your opinion, which of the two will produce the more reliable data? Explain.

(b) Suppose that you are trying to market burglar alarms in New York City. Which of the two techniques for selecting the sample would you use? Explain your reasons.

30. The following two surveys were conducted in January 1991 in order to assess how the American public viewed media coverage of the Persian Gulf war.

Survey 1 was an Area Code 900 telephone poll survey conducted by "ABC News." Viewers were asked to call a certain 900 number if they felt the media was doing a good job of covering the war, and a different 900 number if they felt the media was not doing a good job in covering the war. Each call cost 50 cents. Of the 60,000 respondents, 83% felt the media was not doing a good job.

Survey 2 was a telephone poll of 1500 randomly selected households across the United States conducted by the *Times-Mirror* survey organization. In this poll 80% of the respondents indicated that they approved of the press coverage of the war.

(a) Briefly discuss survey 1, indicating any possible types of bias.

(b) Briefly discuss survey 2, indicating any possible types of bias.

(c) Can you explain the discrepancy between the results of the two surveys?

(d) In your opinion, which of the two surveys gives the more reliable data?

31. (a) You have a small pond stocked with fish and want to estimate how many fish there are in the pond. Let's suppose that you capture $n_1 = 500$ fish, tag them, and throw them back in the pond. After a couple of days you go back to the pond and capture $n_2 = 120$ fish, of which $k = 30$ are tagged. Give an estimate of the N-value of the fish population in the pond.

(b) The following real example is based on data given in D. G. Chapman and A. M. Johnson, "Estimation of Fur Seal Pup Populations by Randomized Sampling," *Transactions of the American Fisheries Society,* 97 (July 1968), 264–270. To estimate the population in a rookery, 4965 fur seal pups were captured and tagged in early August. In late August, 900 fur seal pups were captured. Of these, 218 had been tagged. Based on these figures, estimate the population of fur seal pups in the rookery to the nearest hundred.

32. (a) For the capture-recapture method to give a reasonable estimate of N, list all the assumptions you think should be made about the two samples.

(b) Give reasons why in many situations the assumptions in (a) may not hold true.

33. (Open-ended question) Consider the following hypothetical situation. A potentially effective new drug for treating AIDS patients must be tested by means of a clinical study. Based on experiments conducted with laboratory animals, the drug appears to be extremely effective in treating the more serious effects of AIDS, but it also appears to have caused many side effects, including serious kidney disorders in about 20% of the laboratory animals tested.

(a) Discuss the ethical and moral issues you think should be considered in designing a clinical study to test this drug.

(b) Taking into account the issues discussed in (a), describe how you would design a clinical study for this new drug? (In particular, how would you choose the participants in the study, the treatment and the control groups, etc.?)

References and Further Readings

1. Francis, Thomas, Jr., et al., "An Evaluation of the 1954 Poliomyelitis Vaccine Trials—Summary Report, *American Journal of Public Health,* 45 (1955), 1–63.
2. Freedman, D., R. Pisani, R. Purves, and A. Adhikari, *Statistics,* 2d ed. New York: W. W. Norton, Inc., 1991, chaps. 19 and 20.
3. Gallup, George, *The Sophisticated Poll Watchers Guide.* Princeton, NJ: Princeton Public Opinion Press, 1972.
4. Glieck, James, "The Census: Why We Can't Count," *New York Times Magazine* (July 15, 1990), 22–26, 54.
5. Hansen, Morris, and Barbara Bailar, "How to Count Better: Using Statistics to Improve the Census," in *Statistics: A Guide to the Unknown,* 3d ed., ed. Judith M. Tanur et al. Belmont, CA: Wadsworth, Inc., 1989, 208–217.
6. Meier, Paul, "The Biggest Public Health Experiment Ever: The 1954 Field Trial of the Salk Poliomyelitis Vaccine," in *Statistics: A Guide to the Unknown,* 3d ed., ed. Judith M. Tanur et al. Belmont, CA: Wadsworth, Inc., 1989, 3–14.
7. Mosteller, F., et al., *The Pre-election Polls of 1948.* New York: Social Science Research Council, 1949.
8. Paul, John, *A History of Poliomyelitis.* New Haven, CT: Yale University Press, 1971.
9. Scheaffer, R. L., W. Mendenhall, and L. Ott, *Elementary Survey Sampling.* Boston: PWS-Kent, 1990.
10. Utts, Jessica M., *Seeing Through Statistics.* Belmont, CA: Wadsworth, Inc., 1996.
11. Warwick, D. P., and C. A. Lininger, *The Sample Survey: Theory and Practice.* New York: McGraw-Hill Book Co., 1975.
12. Yates, Frank, *Sampling Methods for Censuses and Surveys.* New York: Macmillan Publishing Co., Inc., 1981.

It is a proof of high culture to say the greatest matters in the simplest way.

Ralph Waldo Emerson

Descriptive Statistics

Graphing and Summarizing Data

CHAPTER 14

Data[1] are the building blocks in the language of statistics. The primary purpose of collecting data is to give meaning to a statistical story, to uncover some new fact about our world, and—last but certainly not least—to make a point, no matter how outlandish. But how is this done?

Imagine asking a mother the ages of her children. The answer in most cases is simple, a short list such as 15, 12, and 8. Imagine now asking for the ages of all people in the United States.[2] An analogous list would be incomprehensible—a huge babble of numbers. There comes a point when a list of numbers is too long for the human mind to digest or comprehend, and that point comes surprisingly early (at five or at the most six numbers, psychologist believe). What do we do when we have too much data?

One important purpose of statistics is to describe large amounts of data in a way that is understandable and, if need be, convincing—as Emerson put it, "to say the greatest matters in the simplest way." This area, called **descriptive statistics**, is the subject of this chapter.

---

[1] In current usage the word *data* can represent both the singular and plural forms of the word.

[2] Why? you might ask. For insurance companies, the U.S. Government, McDonald's, Nike, and hundreds of other corporations, this information is an essential part of doing business.

There are two strategies for describing data: One is to present the data in the form of pictures or graphs; the other is to use numerical summaries that serve as "snapshots" of the data. Sometimes we even combine the two strategies, using pictures and numerical summaries together.

Graphical Descriptions of Data

Statisticians call individual pieces of data **data values** or **data points** and a list of data values a **data set**. For a large data set, an effective way to describe it is by means of a "picture," usually called a graph. There are many different ways of doing this, the best choice often depending on the nature of the data we are dealing with. In this section we will discuss several ways to describe data sets graphically, and the situations in which each is appropriate.

We begin with a visit to a hypothetical statistics class the day after an exam, a routine scene on any college campus.

Example 1. The Stat 101 Midterm Scores

The day after the Stat 101 midterm exam, Professor Blackbeard posts the scores (Table 14-1).

Table 14-1 Stat 101 Midterm Exam Scores (25 Points Possible); $N = 75$.

ID	Score	ID	Score	ID	Score	ID	Score	ID	Score
1257	12	2651	10	4355	8	6336	11	8007	13
1297	16	2658	11	4396	7	6510	13	8041	9
1348	11	2794	9	4445	11	6622	11	8129	11
1379	24	2795	13	4787	11	6754	8	8366	13
1450	9	2833	10	4855	14	6798	9	8493	8
1506	10	2905	10	4944	6	6873	9	8522	8
1731	14	3269	13	5298	11	6931	12	8664	10
1753	8	3284	15	5434	13	7041	13	8767	7
1818	12	3310	11	5604	10	7196	13	9128	10
2030	12	3596	9	5644	9	7292	12	9380	9
2058	11	3906	14	5689	11	7362	10	9424	10
2462	10	4042	10	5736	10	7503	10	9541	8
2489	11	4124	12	5852	9	7616	14	9928	15
2542	10	4204	12	5877	9	7629	14	9953	11
2619	1	4224	10	5906	12	7961	12	9973	10

Each student is identified by a student ID number (rights of privacy prohibit the use of names), and the *data points* are the midterm scores (whole numbers between 0 and 25) shown to the right of each ID number. The data set for this exam consists of the 75 scores. (We will use this data set several times in the chapter. For ease of reference we will call it the *Stat 101 data set.*) ■

Like students everywhere, Professor Blackbeard's students have two questions foremost on their minds regarding the exam: (1) How did I do? and (2) How did the class as a whole do? The answer to question 1 can be found directly in

Table 14-1, but the answer to question 2 requires a little extra effort. How can all the information given by Table 14-1 be packaged into a single intelligible whole? Let us count the ways.

Bar Graphs and Variations Thereof

Our first approach is to put the scores into a **frequency table**, as shown in Table 14-2.

Table 14-2 Frequency Table for the Stat 101 Data Set

Exam score	1	6	7	8	9	10	11	12	13	14	15	16	24
Frequency	1	1	2	6	10	16	13	9	8	5	2	1	1

The number below each score represents the **frequency** of that score—that is, the number of students getting that score. In this example there are 16 students with a score of 10, 2 students with a score of 15, and so on. Note that when no students get a particular score (i.e., the frequency is 0), we can omit the score from the table.

While Table 14-2 is a considerable improvement over Table 14-1, we can do even better. Figure 14-1 shows the same information in a much more visual way called a **bar graph**, with the possible test scores listed in increasing order on a horizontal axis and the frequencies of each test score displayed by the *height* of the column above that test score. Notice that in the bar graph even the missing test scores show up—there simply is no column above these scores.

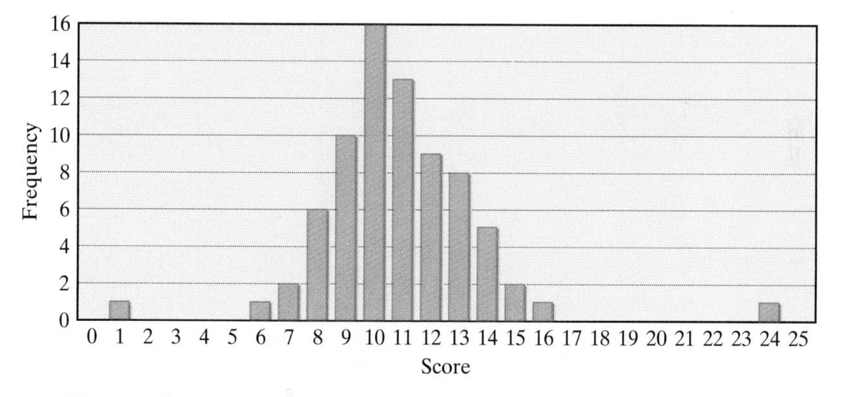

FIGURE 14-1 Bar graph for the Stat 101 data set.

Bar graphs are easy to read and they are a nice way to present an overall "picture" of the data. With a bar graph, for example, it is easy to detect **outliers**. (An outlier is a data value that stands out from the crowd, that is to say, a value that is noticeably larger or smaller than the rest of the data.) In Fig. 14-1 there are two outliers, one being the abnormally low score of 1, the other the abnormally high score of 24.

When the values of the frequencies are large numbers, it is customary to describe the bar graph in term of **relative frequencies**—that is, the frequencies expressed as percentages of the total population. Figure 14-2 shows a *relative-frequency bar graph* for the Stat 101 data set. Note that we indicated on the graph that we are dealing with percentages rather than total counts, and that the size of the data set is $N = 75$. [Letting the viewer know the size of the data set in a relative-frequency bar graph is important, because it allows anyone who wishes to do so to compute the actual frequencies (actual frequency = percentage ×

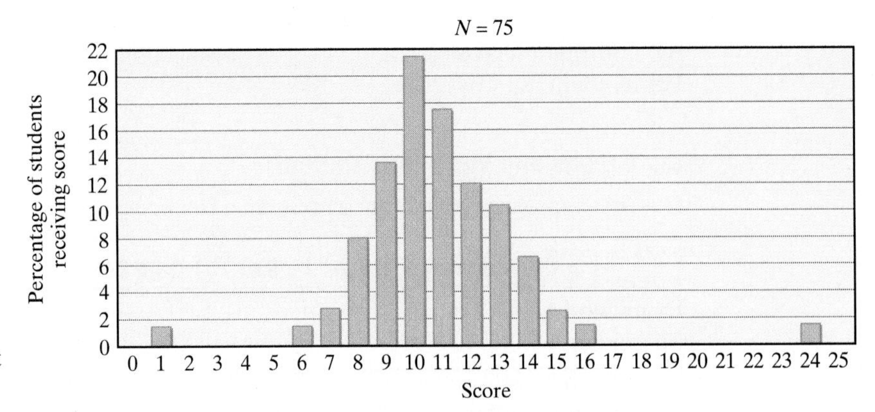

FIGURE 14-2 Relative frequency bar graph for the Stat 101 data set.

N/100).] The change from actual frequencies to percentages does not change the shape of the graph—it is basically a change of scale.

While the term *bar graph* is most commonly used for graphs like the ones in Figs. 14-1 and 14-2, devices other than "bars" can be used to add a little extra flair or to subtly influence the content of the information given by the raw data. Professor Blackbeard, for example, could have chosen to display the midterm data using a graph like the one shown in Fig. 14-3, which conveys all the information given by the more staid version (Fig. 14-1) and at the same time sends a subtle individual message to each student.

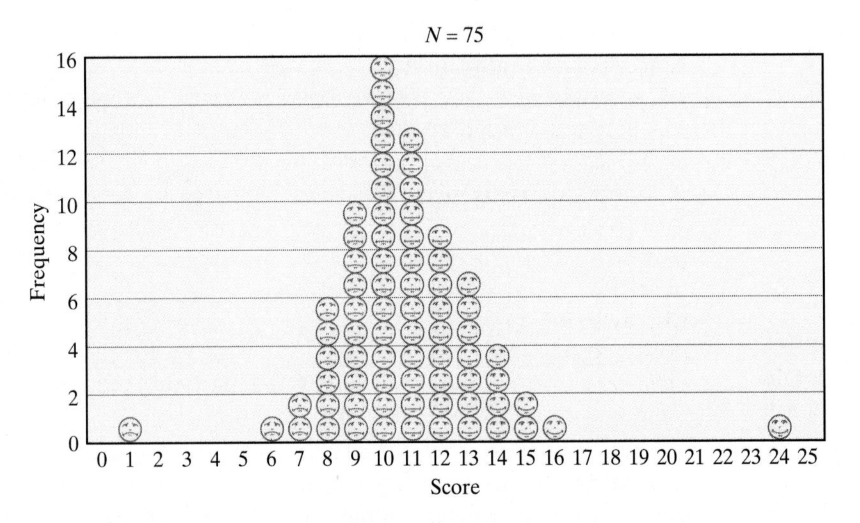

FIGURE 14-3 Frequency chart for the Stat 101 data set.

The general point here is that a bar graph is often used not only to inform but also to impress and persuade, and in such cases a clever design for the frequency columns can be more effective than just a bar. Graphs such as those in Figs. 14-3, 14-4, and 14-5, which are just fancy bar graphs using gadgets other than bars to show the frequencies, are commonly referred to as **pictograms**.

Example 2.

Figure 14-4 is a pictogram showing the growth in yearly sales of the XYZ Corporation over the period from 1991 to 1996. It looks very impressive, but the picture is actually quite misleading. Figure 14-5 shows a pictogram for exactly the same data, a much more accurate and sobering picture of how well XYZ corporation had been doing.

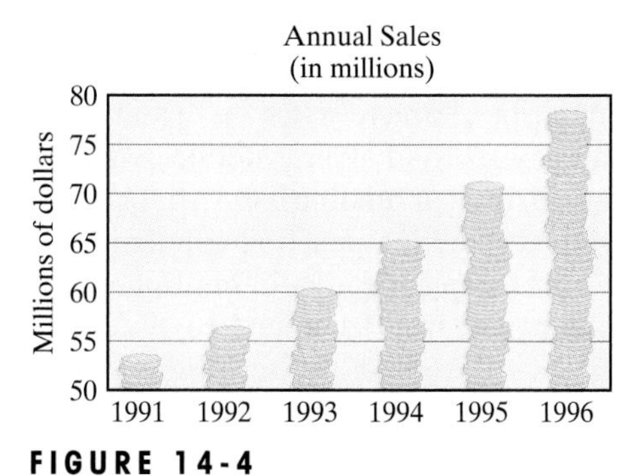

FIGURE 14-4
XYZ Corp. annual sales.

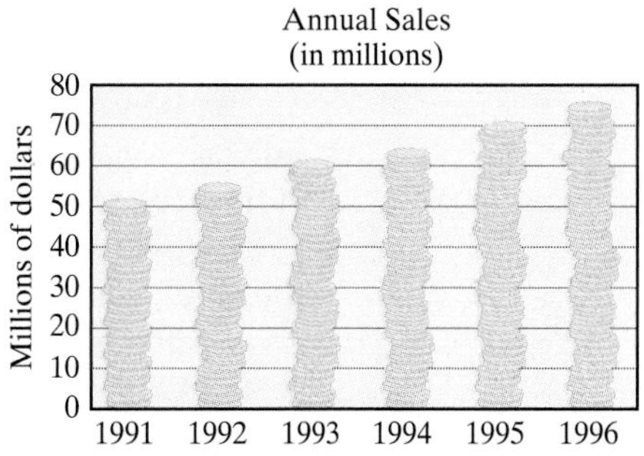

FIGURE 14-5
XYZ Corp. annual sales.

The difference between the two pictograms can be attributed to a couple of standard tricks in the creative-chart-making business: (1) stretching the scale of the vertical axis, and (2) "cheating" on the choice of starting value on the vertical axis. As an educated consumer, you should always be on the lookout for these tricks. In graphical descriptions of data, only a fine line separates objectivity from propaganda. ■

Variables: Quantitative and Qualitative; Continuous and Discrete

Before we continue with our discussion of graphs, we need to briefly discuss the concept of a **variable**. In statistical usage, a variable is any characteristic that varies with the members of a population. The students in Professor Blackbeard's Stat 101 course (the population) do not all perform equally on the exam. Thus, the *test score* is a variable, which in this particular example is a whole number between 0 and 25. In some instances, such as when the instructor gives partial credit or when there is subjective grading, a test score may take on a fractional value, such as 18.5 or even 18.25. Even in these cases, however, the possible increments for the values of the variable are given by some minimum amount: a quarter-point, a half-point, whatever. In contrast to this situation, consider a different variable, the *length of time* it takes a student to complete the exam. In this case the variable can take on values that differ by arbitrarily small increments—a second, a tenth of a second, a hundredth of a second, and so on.

When a variable represents a measurable quantity, it is called a **numerical** (or **quantitative**) variable. When the difference between the values of a numerical variable can be arbitrarily small, we call the variable **continuous**; when possible values of the numerical variable change by minimum increments, the variable is called **discrete**. Examples of discrete variables are IQ, pulse, shoe size, family size, number of automobiles owned, and points scored in a basketball game. Examples of continuous variables are height, weight, foot size (as opposed to shoe size), and the time it takes to run a mile.

Sometimes in the real world the distinction between continuous and discrete variables is blurred. Height, weight, and age are all continuous variables in theory, but in practice they are frequently rounded off to the nearest inch, ounce, and year (or month in the case of babies), respectively, at which point they become discrete variables. On the other hand, money, which is in theory a discrete variable (the difference between two values cannot be less than a penny), is almost always thought of as continuous, because in most real-life situations a penny can be thought of as an infinitesimally small amount of money.

Variables can also describe characteristics that cannot be measured numerically: nationality, sex, hair color, brand of automobile owned, and so on. Variables of this type are called **categorical** (or **qualitative**) variables.

In some ways, categorical variables must be treated differently from numerical variables: They cannot, for example, be added, multiplied, or averaged. In other ways, categorical variables can be treated much like discrete numerical variables, particularly when it comes to graphical descriptions such as bar graphs and pictograms.

Example 3. Enrollments (by School) at Tasmania State University

Table 14-3 shows undergraduate enrollments in each of the 5 schools at Tasmania State University. A sixth category ("Other") includes undeclared students, interdisciplinary majors, and so on.

Table 14-3 Undergraduate Enrollments at TSU

School	Enrollment
Agriculture	2400
Business	1250
Education	2840
Humanities	3350
Science	4870
Other	290

The bar graph in Fig. 14-6 is very similar to the one in Fig. 14-1, except that here the variable being described is categorical, the categories being the 5 schools plus the catch-all category "Other." Figure 14-7 shows the same data using relative frequencies.

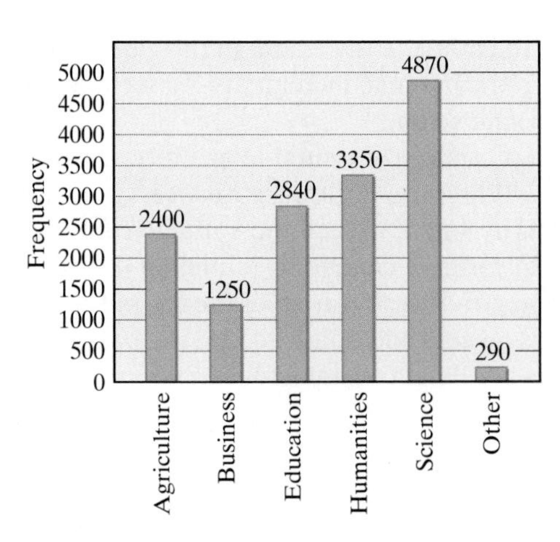

FIGURE 14-6
Undergraduate enrollments at TSU (by school). Bar graph.

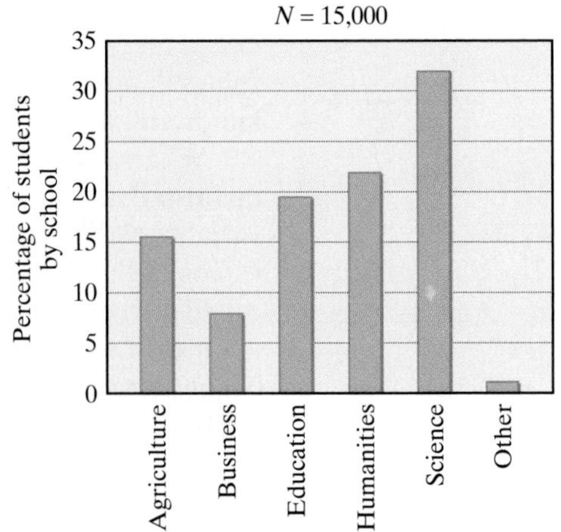

FIGURE 14-7
Undergraduate enrollments at TSU (by school). Relative frequency chart. ■

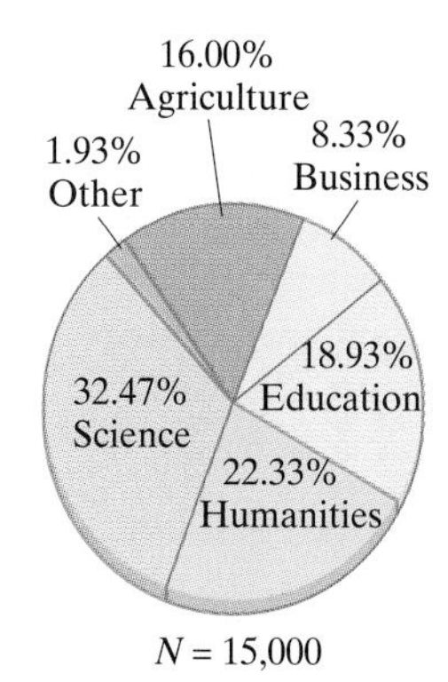

FIGURE 14-8 Undergraduate enrollments at TSU (by school). Pie chart.

When the number of categories is small, as it is in Example 3, another commonly used way to describe relative frequencies of a population by categories is the **pie chart**. The "pie" represents the entire population (100%), and the "slices" represent the categories or classes, with the size (area) of each slice being proportional to the relative frequency of the corresponding category. Some relative frequencies, such a 50% and 25%, are very easy to describe; but how do we accurately draw the slice corresponding to a more complicated frequency—say 32.47%? Here a little high school geometry comes in handy; Since 100% equals 360°, 1% corresponds to an angle of 360°/100 = 3.6°. It follows that a frequency such as 32.47% is given by $32.47 \times 3.6° = 117°$ (rounded to the nearest degree, which is generally good enough for most practical purposes). Figure 14-8 shows the school-enrollment data in Example 3 described by a pie chart.

Bar graphs and pie charts are an excellent way to graphically display categorical data, but, as always, we should be wary of jumping to hasty conclusions based on what we see on a graph. Our next example illustrates this point.

Example 4. Who's Watching the Boob Tube Tonight?

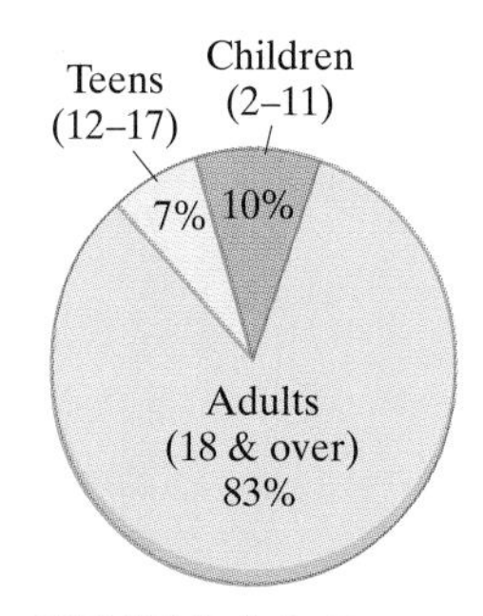

FIGURE 14-9 Audience composition for prime-time TV viewship by age group. (Source: Nielsen Media Research.)

According to Nielsen Media Research data, the percentages of TV audience watching TV during prime time (8 P.M. to 11 P.M.), broken up by age group, is: adults (18 years and over), 83%; teenagers (12–17 years), 7%; children (2–11 years), 10%.[3] The pie chart in Fig. 14-9 shows this breakdown of audience composition by age group.

When looking at this pie chart, one is tempted to conclude that, at least during prime time, children and teenagers do not watch much TV. Could all the reports we read about how much TV young people watch be wrong?

The problem with this pie chart is that, while accurate, it is also very misleading: Children (2–11 years) make up only 15% of the population at large, teens (12–17) make up only 8%, and adults make up the rest. Given that there are more than 5 times as many adults as there are children, is it any wonder that there are more prime-time TV-viewing adults than there are prime-time TV-viewing children? Likewise, in absolute terms there are more TV-viewing children than teenagers, but that is only because there are almost twice as many children as there are teenagers. In relative terms, a higher percentage of teenagers (taken out of the total teenage population) watch prime-time TV than do children. (This is not all that surprising, given that most children's bedtime is around 8 P.M.). ■

The moral of this example is that using absolute percentages, as we did in Fig. 14-9, can be quite misleading. When comparing characteristics of a population that is broken up into categories, it is essential to take into account the relative sizes of the various categories.

Class Intervals

While the distinction between qualitative and quantitative data is important in many aspects of statistics, when it comes to deciding how best to display graphi-

[3] These figures are rough approximations based on information taken from the *World Almanac* and averaged over several years. The exact figures vary from year to year.

cally the frequencies of a population, a critical issue is the number of categories into which the data can fall. When the number of categories is too big (say, for example, in the dozens) a bar graph or pictogram can become muddled and ineffective. This generally is not a problem with qualitative data, but often is with quantitative data: Both continuous and discrete variables can take on infinitely many values, and even when they don't, the number of values can be too large for any reasonable graph.

Example 5. SAT Scores

Suppose that, as part of a special research project, we want to look at the cumulative SAT test scores for the population of students discussed in Example 1 (those in Professor Blackbeard's Stat 101 course). Just as in Example 1, our data represents a discrete quantitative variable (in this case, cumulative SAT scores). While in theory the situation is no different from that in Example 1, in practice, because of the extremely large number of possible SAT scores (they are given in 10-point increments and range between 400 and 1600), we must deal with such data differently. The standard way to display bar graphs in this situation is to break up the range of scores into **class intervals**. The decision as to how the class intervals are defined and how many there should be is a matter of personal choice. (As a general rule of thumb, the number of class intervals should be somewhere between 5 and 20.) In this example, a sensible thing to do might be to break up the SAT scores into 12 class intervals. In this case our bar graph would look something like Fig. 14-10.

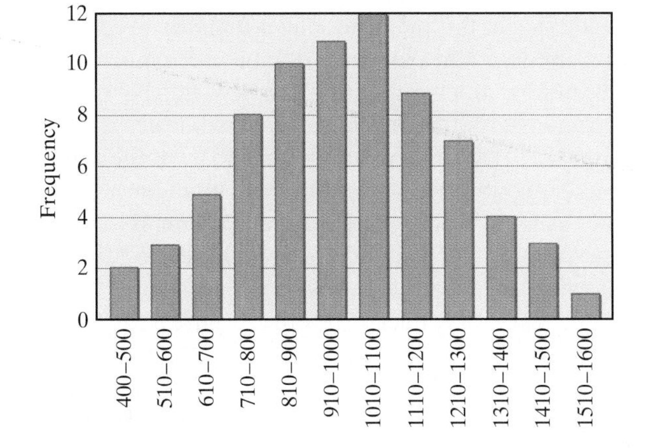

FIGURE 14-10 Cumulative SAT scores for the students in Prof. Blackbeard's class.

■

Note that in Example 5 we made it a point to create class intervals of the same size,[4] and this should be done as much as possible. Sometimes, however, it might make more sense to define class intervals of different lengths, as illustrated by our next example.

Example 6. Midterm Grades

Imagine now that Professor Blackbeard wants to convert the test scores in the Stat 101 data set into letter grades. In our terminology, this means converting a numerical variable (test score) into a categorical one (letter grade) by defining class inter-

[4] A tiny exception was made for the class interval 400–500, which has one more possible test score than the others.

vals associated with each grade category (A, B, C, D, and F). In this case there is a good reason not to use class intervals of equal length. Following his own mysterious way of doing things, Professor Blackbeard defines the class intervals for this particular exam according to the breakdown shown in Table 14-4.

Table 14-4

Class interval	Grade
18–25	A
14–17	B
11–13	C
9–10	D
0–8	F

If we combine the Stat 101 exam scores in Table 14-2 with the class intervals for grades as defined in Table 14-4, we get a new frequency table (Table 14-5) and a corresponding bar graph for the grade distribution in the exam (Fig. 14-11).

Table 14-5

Grade	Frequency	Percentage
F	10	13.33%
D	26	34.67%
C	30	40%
B	8	10.67%
A	1	1.33%

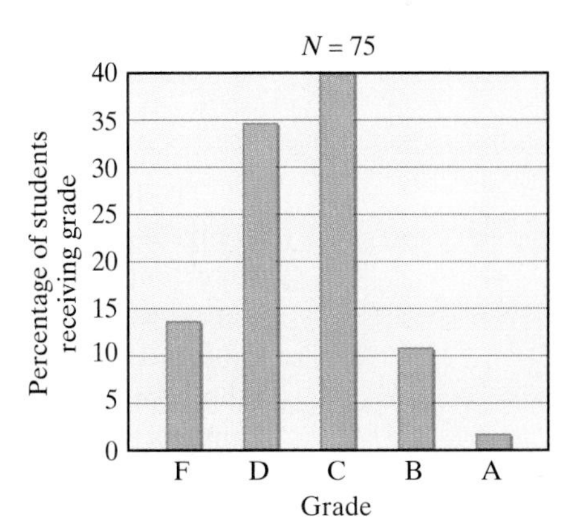

FIGURE 14-11
Grade distribution for the Stat 101 test based on Tables 14-4 and 14-5. ■

Histograms

When a numerical variable is continuous, then its possible values can vary by infinitesimally small increments. As a consequence, there are no gaps between the class intervals, and our old way of doing things (using columns or stacks separated by gaps) will no longer work. In this case we use a variation of a bar graph called a **histogram**. We illustrate the concept of a histogram in the next example.

Example 7. Starting Salaries of TSU Graduates

Suppose we want to use a graph to display the distribution of starting salaries for last year's graduating class at Tasmania State University.

The starting salaries of the $N = 3258$ graduates ranged from a low of $20,350 to a high of $54,800. Based on this range and the amount of detail we want to show, we must decide on the length of the class intervals. A reasonable choice would be to use class intervals defined in increments of $5000. Table 14-6 is a frequency table for the data based on these class intervals. We chose a starting value of $20,000 for convenience. (The third column in the table shows the data as a percentage of the population.)

Table 14-6 Starting Salaries of First-Year TSU Graduates

Salary	Number of students	Percentage
20,000–25,000	228	7%
25,000$^+$–30,000	456	14%
30,000$^+$–35,000	1043	32%
35,000$^+$–40,000	912	28%
40,000$^+$–45,000	391	12%
45,000$^+$–50,000	163	5%
50,000$^+$–55,000	65	2%
Total	3258	100%

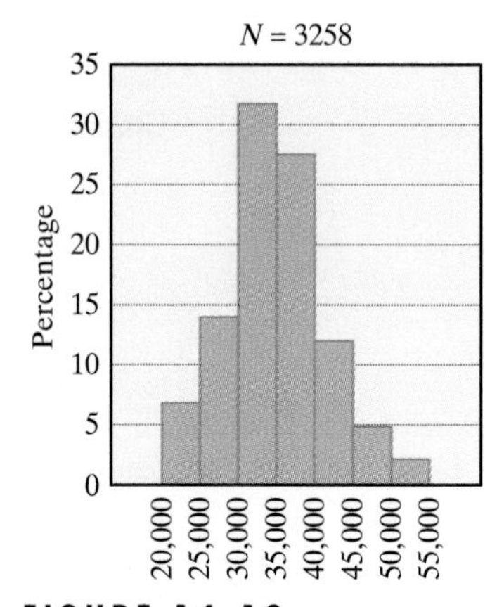

FIGURE 14-12
Histogram for starting salaries for first-year graduates of TSU. (class intervals of $5000.)

The histogram showing the relative frequency of each class interval is shown in Fig. 14-12. As we can see, a histogram is very similar to a bar graph. Several important distinctions must be made, however. To begin with, because a histogram is used with continuous variables, there can be no gaps between the class intervals, and it follows therefore that the columns of a histogram must touch each other. Among other things, this forces us to make an arbitrary decision as to what happens to a value that falls exactly on the boundary between two class intervals. Should it always belong to the class interval to the left or to the one to the right? This is called the *endpoint convention.* The superscript "plus" marks in Table 14-6 indicate how we chose to deal with the endpoint convention in Fig. 14-12: A starting salary of exactly $30,000, for example, should be assigned to the second rather than to the third class interval. ■

As with regular bar graphs, in creating histograms we should try, as much as possible, to define class intervals of equal length. When the class intervals are of unequal length, the rules for creating a histogram are considerably more complicated, since it is no longer appropriate to use the heights of the columns to indicate the frequencies of the class intervals. We will not discuss the details of this situation here but refer the interested reader to Exercises 43 and 44 at the end of the chapter.

Numerical Summaries of Data

As we have seen, pictures can be an excellent tool for summarizing large data sets. Unfortunately, circumstances do not always lend themselves equally well to the use of pictures, and bar graphs and pie charts cannot be readily used in everyday conversation. A different and very important approach is to use a few well-chosen numbers as a "summary" of the entire data set.

In this section we will discuss several of the most commonly used *numerical summaries* of a data set. First we will draw an important distinction. Numerical summaries of data fall into two categories—numbers that tell us something about where the values of the data fall, and numbers that tell us something about how spread out the values of the data are. The former are called **measures of location**, the latter **measures of spread**. The most important measures of location are the *average* (or *mean*), the *median,* and the *quartiles.* The most important measures of spread are the *range,* the *interquartile range,* and the *standard deviation.* We will discuss each of these in order.

The Average

The best known of all numerical summaries of data is the *average,* sometimes also called the *mean.* (As much as possible, we will stick to the good down-to-earth word "average.") The **average** of a set of N numbers is obtained by adding the numbers and dividing by N. When the set of numbers is small, one can often calculate the average in one's head; for larger data sets, pencil and paper or a calculator can be helpful. In either case the idea is very straightforward.

Example 8. Average Points per Game

In 10 playoff games a basketball player scores 8, 5, 11, 7, 15, 0, 7, 4, 11, and 14 points, respectively. The total is 82 points in 10 games. The average is 8.2 points per game. Note that it is actually impossible for a player to score 8.2 points. As it often happens, the average taken by itself can be an impossible data value. ■

Example 9. The Average Test Score in the Stat 101 Test

Table 14-7 is the same as Table 14-2, shown again for the reader's convenience. We want to calculate the average test score.

Table 14-7 Frequency Table for the Stat 101 Data Set

Exam score	1	6	7	8	9	10	11	12	13	14	15	16	24
Frequency	1	1	2	6	10	16	13	9	8	5	2	1	1

The 75 data values can be totaled by taking each score, multiplying it by its corresponding frequency, and adding. In this case we get

$$\text{Total} = (1 \times 1) + (6 \times 1) + (7 \times 2) + (8 \times 6) + (9 \times 10) + (10 \times 16) + (11 \times 13) + (12 \times 9) + (13 \times 8) + (14 \times 5) + (15 \times 2) + (16 \times 1) + (24 \times 1) = 814$$

The average score on the midterm exam (rounded off to two decimal places) is

$$814 \div 75 \approx 10.85 \text{ points.}$$

Intuitively, we think of this average as representing a "typical" student's score. If all test scores had been about the same, then, given the same total, each score would have been "about" 10.85 points. ■

Table 14-8 shows a generic frequency table. To find the average of the data we do the following:

Table 14-8

Data value	Frequency
s_1	f_1
s_2	f_2
...	...
s_k	f_k

- **Step 1.** Calculate the total of the data:

  $\text{total} = (s_1 \times f_1) + (s_2 \times f_2) + \cdots + (s_k \times f_k).$

- **Step 2.** Calculate N:

  $N = f_1 + f_2 + \cdots + f_k.$

- **Step 3.** Calculate the average:

  $\text{average} = \text{total} \div N.$

Sometimes averages can be quite deceiving, as illustrated in our next example.

Example 10. Starting Salaries of Philosophy Majors

The average annual starting salary for the 75 philosophy majors who recently graduated from Tasmania State University is \$56,400. This is an impressive figure, but before we all rush out to change majors, consider the fact that one of these graduates is professional basketball star "Hoops" Tallman, whose starting salary is a whopping \$2.5 million a year.

If we were to disregard this one outlier, the average annual starting salary for the remaining 74 philosophy majors could be computed as follows:

$$75 \times \$56{,}400 = \$4{,}230{,}000 \quad \text{(the total of all 75 salaries combined)}$$

$$\$4{,}230{,}000 - \$2{,}500{,}000 = \$1{,}730{,}000 \quad \text{(the total of all salaries excluding "Hoops" Tallman's)}$$

$$\frac{\$1{,}730{,}000}{74} = \mathbf{\$23{,}378} \quad \text{(the average of the other 74 salaries)}$$

■

Example 10 underscores the point that even a single outlier can have a big impact on the average. We must always be alert to the possibility that an average may have been distorted by one or more outliers. On the other hand, if we know that the data set does not have outliers, we can rely on the average as a useful numerical summary.

So far, all our examples have involved data values that are positive, but negative data values are also possible, and when both negative and positive data values are averaged, the results can be a little misleading.

Example 11.

The monthly savings (monthly income minus monthly spending) of a college student over a one-year period is shown in Table 14-9. A negative amount indicates that, rather than saving money, the student spent more that month than his monthly income.

Table 14-9

Month	Savings (in $)	Comment
Jan.	−732	(Christmas bills)
Feb.	−158	
Mar.	−71	
Apr.	−238	
May	1839	($2000 lottery winnings)
Jun.	−103	
Jul.	−148	
Aug.	−162	
Sep.	−85	
Oct.	−147	
Nov.	−183	
Dec.	500	(Christmas present from mom)

The average monthly savings of this student over the year is

$$\frac{-732 - 158 - 71 - 238 + 1839 - 103 - 148 - 162 - 85 - 147 - 183 + 500}{12}$$
$$= \$26,$$

which is an accurate but deceptive figure. The true picture is that of a student living beyond his means and bailed out by a lucky lottery ticket.

The Median

The median is another important, commonly used numerical summary of a set of data. The **median** is a number that separates the data set into two equal halves: half of the numbers are *smaller than or equal to* the median (this is called the **lower half** of the data set) and half of the numbers are *bigger than or equal to* the median (the **upper half** of the data set).

To find the median of a set of numbers *we must first sort the numbers by size*—that is to say, we must rewrite the numbers in increasing order from left to right (or right to left—it makes no difference). The median is the number in the "middle" of the sorted list. Where the "middle" is depends on whether the size of the data set is odd or even.

Example 12.

The data set below shows the yards gained by a high school running back over a nine-game football season:

$$48, -12, 31, 85, 16, -5, 42, 61, 39.$$

To find the median we first sort the numbers. The *sorted data set* is

−12, −5, 16, 31, 39, 42, 48, 61, 85.

Of these nine, the "middle" number is the fifth number counting from left to right (or right to left), so the median number of yards gained by the player over the nine games is 39. The two halves of the data set are −12, −5, 16, 31, 39 (the *lower* half) and 39, 42, 48, 61, 85 (the *upper* half). Notice that in this example the median (39) belongs to both halves. ■

Example 13.

The annual profits (and losses) of the XYZ Corporation over an 8-year period are given (in millions of dollars) in the following data set:

2.2, −1.1, −2.7, 4.4, 6.2, −2.4, 3.8, 1.6.

The sorted data set is

−2.7, −2.4, −1.1, 1.6, 2.2, 3.8, 4.4, 6.2.

With 8 numbers, none of them can be designated as the "middle" number. The closest to the middle would be the fourth and fifth numbers counting from left to right. Since we don't want to choose between them, we split the difference and declare that the median in this case is the number halfway between the fourth number (1.6) and the fifth number (2.2), namely 1.9.

Notice that here, again, the median (even though it is not part of the data set) splits the data set into two equal halves: the lower half −2.7, −2.4, −1.1, 1.6 and the upper half 2.2, 3.8, 4.4, 6.2. ■

Examples 12 and 13 illustrate the two possible scenarios for calculating the median of a set of numbers, which can be generalized as follows:

Finding the Median of N Numbers

1. Sort the data set.
2. (a) When N is odd, the *median* is the number in position $(N + 1)/2$ (counting from left to right) in the sorted data set.

   (b) When N is even, the *median* is the number halfway between the numbers in position $N/2$ and $(N/2) + 1$ (counting from left to right) in the sorted data set.

Example 14. The Median Test Score for the Stat 101 Test

We will now find the median score for the Stat 101 data set given in Table 14-10.

Having the frequency table available eliminates the need for sorting the scores—the frequency table has in fact done this for us. The total number of scores is $N = 75$, which means that the median can be found in the 38th position, counting

Table 14-10 Frequency Table for the Stat 101 Data Set

Exam score	1	6	7	8	9	10	11	12	13	14	15	16	24
Frequency	1	1	2	6	10	16	13	9	8	5	2	1	1

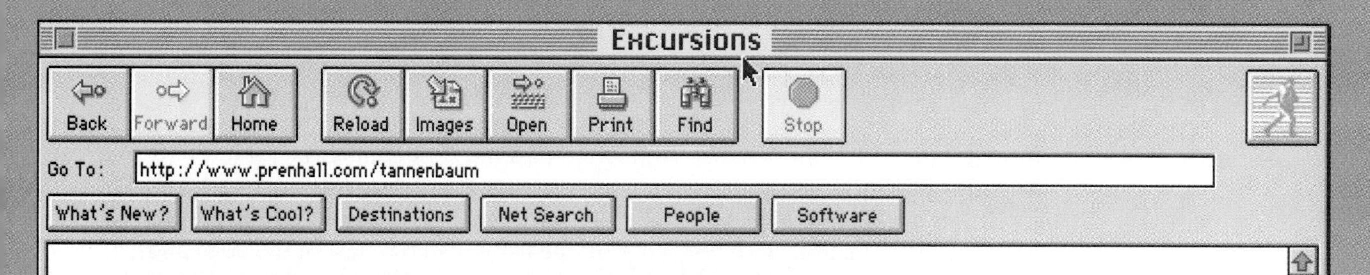

The Big One

Disasters occur every day, but the worst are those that strike unexpectedly. In various parts of the world, an earthquake of magnitude 7.0 or greater on the Richter scale can be disastrous. Studying past data and visually representing them in terms of illustrations, charts, and graphs can help us in finding patterns and predicting such occurrences.

Information on earthquakes can be found through the United States Geological Survey's National Earthquake Information Center located in Golden, Colorado. Go to: http://www.prenhall.com/tannenbaum.

Questions:

1. Study the graphic on the page. What information is given in the graphic? How can you use this in predicting occurrences of earthquakes?
2. Is it possible to present the information given in the graphic in a more numerical way? If not, what additional information would you need in order to do this?
3. Next, select the link for *Number of Earthquakes per Year, Magnitude 7 or Greater, 1990–1995.* Construct a box plot for the frequency of the number of earthquakes per year.
4. What information does this provide? How can you use this in predicting the number of earthquakes per year?
5. Is there some other way to graphically present the data to show if there is a trend in the number of earthquakes? If so, make an appropriate chart or graph using the data.
6. Suppose you are interested only in California? What would you require of the given data to be meaningful?
7. Explore NEIC's Website and find any additional information that will be helpful in predicting earthquakes in California. Construct appropriate charts and graphs to illustrate your predictions.

from left to right in the frequency table. To find the 38th number in Table 14-10, we tally frequencies as we move from left to right: $1 + 1 = 2$; $1 + 1 + 2 = 4$; $1 + 1 + 2 + 6 = 10$; $1 + 1 + 2 + 6 + 10 = 20$; $1 + 1 + 2 + 6 + 10 + 16 = 36$. At this point we know that the 36th test score on the list is a 10 (the last of the 10's) and the next 13 scores are all 11's. We can conclude that the 38th test score (which is the median test score) is 11. ■

Finding a median is not nearly as complicated as it seems, and with a little practice the reader will find it quite easy. Surprisingly, with large data sets the most-time consuming part is sorting the numbers. Once this is done, finding the median is just a matter of knowing where to look for it.

A fairly common mistake is to confuse the median and the mean. The two words are quite similar, and they both define related concepts. This is one more good reason why (as much as possible) the term "average" should be used instead of "mean." Even those who can keep the two concepts straight often assume, mistakenly, that the median and the average must be close in value. While this is indeed the case in many types of real-life data, it is not true in general. Take, for example, the numbers 1, 1, 1, and 97. The median of these numbers is 1, while the average is 25, a much large number. On the other hand, if we take the numbers 1, 1, 100, 101, and 102, then the median (100) is much larger than the average (61). We can see that it is a mistake to assume, as many people do, that the median and the average are "about the same."

The Quartiles

Sometimes it is useful to know how the data set splits up into quarters (not just halves). The **quartiles** are the numbers that tell us this. There are three quartiles: the *first quartile* (Q_1), the second quartile (Q_2), and the *third quartile* (Q_3), but only the first and third quartiles are new concepts—the second quartile is our old friend the *median.* The **first quartile** is a number such that *one-quarter of the numbers in the data set are smaller than or equal to it and three-quarters of the numbers in the data set are bigger than or equal to it.* The **third quartile** is the mirror twin of the first quartile: *three-quarters of the numbers are smaller than or equal to it and one-quarter of the numbers are bigger than or equal to it.*

Now that we know how to find medians of data sets, finding the quartiles is quite easy: *the first quartile can be found by finding the median of the lower half of the data set;* likewise, *the third quartile is the median of the upper half of the data set.*

Finding the Quartiles of a Data Set

- Sort the data set.
- Find the median M.
- Find the *lower* and *upper* halves of the data set.
- Find the median of the lower half. This is the first quartile Q_1.
- Find the median of the upper half. This is the third quartile Q_3.

Example 15.

Consider the sorted data set

$$-12, -5, 16, 31, 39, 42, 48, 61, 85.$$

The lower half is $-12, -5, 16, 31, 39$ (see Example 12). Now think of these 5 numbers as a new data set. The median of these five numbers is the third number (from left to right), namely 16. This makes 16 the first quartile of the original data set ($Q_1 = 16$).

The upper half of the data set is 39, 42, 48, 61, 85, and the median of these five numbers is 48. It follows that the third quartile of the original data set is 48 ($Q_3 = 48$). ■

Example 16.

Consider the sorted data set

$$-2.7, -2.4, -1.1, 1.6, 2.2, 3.8, 4.4, 6.2. \text{ (see Example 13).}$$

Here the lower half of the data set is $-2.7, -2.4, -1.1, 1.6$. The median of these 4 numbers (halfway between -2.4 and -1.1) is -1.75. Likewise, the upper half of the data set is 2.2, 3.8, 4.4, 6.2, and the median of these 4 numbers is 4.1. Thus, $Q_1 = -1.75$ and $Q_3 = 4.1$. ■

The Five-Number Summary

A good summary for a large data set can be provided by giving the lowest value of the data (**Min**), the first quartile (Q_1), the median (M), the third quartile (Q_3), and the largest value of the data (**Max**). These five numbers constitute the **five-number summary** of the data set.

Example 17.

Let's find the five-number summary of the Stat 101 data set given once again in Table 14-11.

Table 14-11 Frequency Table for the Stat 101 Data Set

Exam score	1	6	7	8	9	10	11	12	13	14	15	16	24
Frequency	1	1	2	6	10	16	13	9	8	5	2	1	1

We already found the median score ($M = 11$) in Example 14, and a quick look at the table tells us that Min = 1 and Max = 24. All we have left to do is to find the first and third quartiles. Since $N = 75$ (odd), we know that the first half of the data consists of the first 38 numbers. Since 38 is even, the median of the first 38 numbers is halfway between the 19th and 20th scores. To find the 19th score we go to the frequency table and start counting frequencies beginning on the left. We leave it to the reader to verify that the 19th and 20th scores are both 9. It follows that $Q_1 = 9$.

The upper half of the test scores consists of the 38th through the 75th scores, and the third quartile is the median of this set. The easiest way to locate it is to count back *from right to left* until we find the 19th and 20th numbers from the end. The median is halfway between these two numbers. Since the 19th and 20th scores counting right to left in the frequency table are both 12, it follows that $Q_3 = 12$.

The five-number summary for the Stat 101 data set is given by

$$\text{Min} = 1, \quad Q_1 = 9, \quad M = 11, \quad Q_3 = 12, \quad \text{Max} = 24.$$

Note that without Q_1 and Q_3 we would have a very distorted picture of the Stat 101 data set, since both Min = 1 and Max = 24 are outliers. The scores were not evenly spread out in the range between 1 and 24; in fact just the opposite was true. With the quartiles we can get a much better idea of what happened: The middle half of the Stat 101 test sores were bunched up in a very narrow range (between 9 and 12 points); only one-fourth of the test scores were 9 or less and only one-fourth of the test scores fell between 12 and 25. Let's give Prof. Blackbeard an "F" in test writing! ■

Box Plots

A *box plot* is a picture of the 5-number summary of a data set. The **box plot** consists of a rectangular box that goes from the first quartile Q_1 to the third quartile Q_3. A vertical line crosses the box, indicating the position of the median M. On both sides of the box are "whiskers" extending to the smallest value Min and largest value Max of the data.

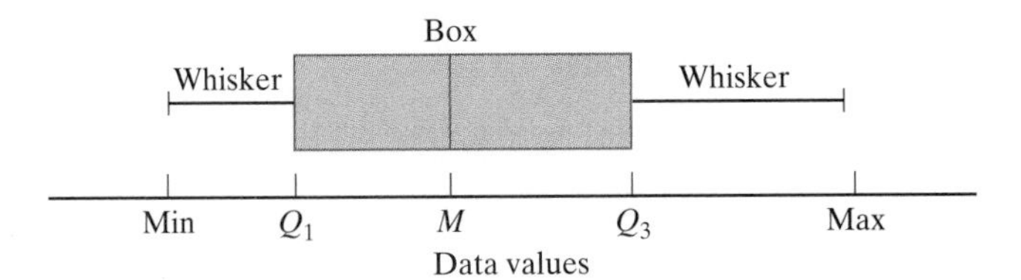

FIGURE 14-13
Box plot.

Figure 14-13 shows a generic box plot for a data set. Figure 14-14(a) shows a box plot for the Stat 101 data set. The long whiskers in this box plot are largely due to the outliers 1 and 24. Figure 14-14(b) shows a variation of the same box plot but with the two outliers separated from the rest of the data (they are shown by the 2 crosses). This last box plot is a much more accurate picture of the data set.

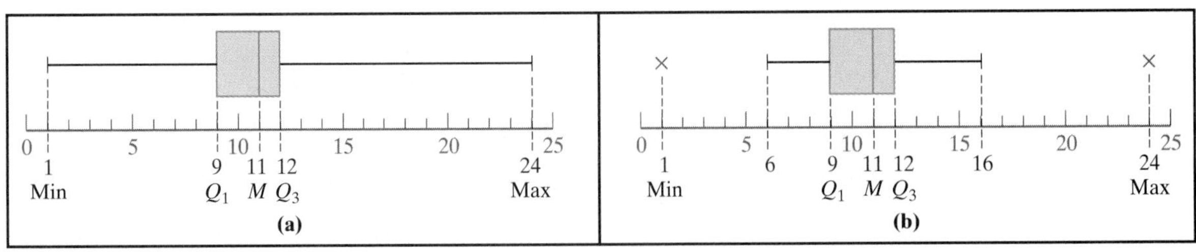

FIGURE 14-14
(a) Box plot for the Stat 101 data set. (b) Same box plot with the outliers separated from the rest of the data.

Box plots are particularly useful when comparing similar data for two or more populations. This is illustrated in the next example.

Example 18.

Figure 14-15 shows box plots for the starting salaries of two different populations: first-year agriculture and engineering graduates of Tasmania State University. Superimposing the two box plots on the same scale allows us to make some useful comparisons. It is clear, for instance, that engineering graduates are doing better overall than agriculture graduates, even though at the very top levels agriculture graduates are better paid. Another interesting point is that the median salary of agriculture graduates is less than the first quartile of the salaries of engineering graduates. The very short whisker on the left side of the agriculture box plot tells us that the bottom 25% of agriculture salaries are concentrated in a very narrow salary range. We can also see that agriculture salaries are much more spread out than engineering salaries, even though most of the spread occurs at the higher end of the salary scale.

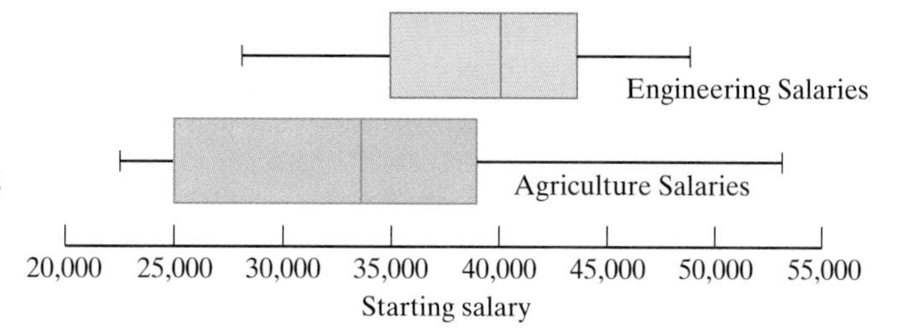

FIGURE 14-15 Comparison of starting salaries of first-year graduates in agriculture and engineering.

We can see that the old chestnut "a picture is worth a thousand words" applies well to statistics. There is a lot being told even in a simple picture like the one in Fig. 14-15—if we know how to read it. ■

Measures of Spread

An important aspect of summarizing numerical data is to give an idea of how *spread out* the data values are. Consider, for example, the following two data sets:

set 1 = {45, 46, 47, 48, 49, 51, 52, 53, 54, 55},

set 2 = {1, 11, 21, 31, 41, 59, 69, 79, 89, 99}.

We leave it to the reader to verify that for both data sets the average is 50 and the median is 50. If we used only the average (or the median) to summarize these sets, there would appear to be no significant difference between them, which is clearly not the case. It is obvious to the naked eye that the two data sets differ in their "spread": the numbers in set 2 are much more spread out than those in set 1.

There are several different ways to describe the spread of a data set; in this section we will describe the three most commonly used ones. An obvious approach is to take the difference between the highest and lowest values of the data (Max − Min). This difference is called the **range**. For set 1 the range is 55 − 45 = 10, and for set 2 the range is 99 − 1 = 98.

As a measure of spread, the range is useful only if there are no outliers, since outliers can significantly affect the range. For example, for the Stat 101 data set the range of the exam scores is 24 − 1 = 23 points, but without the outliers the range would be 16 − 6 = 10 (see Fig. 14-14).

To eliminate the possible distortion caused by outliers, a common practice is to use the **interquartile range** (IQR). The interquartile range is the difference between the third quartile and the first quartile (IQR $= Q_3 - Q_1$), and it tells us how spread out the middle 50% of the data values are. For many types of real-world data the interquartile range is a useful measure of spread. When the five-number summary is used, both the range and the interquartile range come, essentially, free in the bargain.

Example 19.

A total of 500 students take an entrance exam. Rather than give all 500 test scores, we summarize the information by means of the five-number summary, which is Min = 34, $Q_1 = 53$, $M = 71$, $Q_3 = 82$, Max = 96. For this exam the range is Max − Min = 62, and the interquartile range is IQR $= Q_3 - Q_1 = 29$. In addition, the five-number summary tells us the partial story of what happened: the lowest 125 scores (the bottom fourth) fell between 34 and 53; the next 125 scores fell between 53 and 71; the next 125 scores after that fell between 71 and 82; and the top 125 scores fell between 82 and 96. ■

The Standard Deviation

The most important and most commonly used measure of spread for a data set is the *standard deviation.* The key concept for understanding the standard deviation is the concept of *deviation from the mean.* The idea is to measure spread by looking at how far each data point is from a fixed reference point. If we pick a good reference point, the "distances" between it and each data point could give a good description of the spread of the data. The reference point we will use is the mean (average) of the data set. Imagine that we plant a flag there, and we measure how far each data point is from the flag by taking the difference (*data value − mean value*). These numbers are called the **deviations from the mean.**

Example 20. Deviations from the Mean in the Stat 101 Test

Once again we return to the Stat 101 data set. We calculated (way back) in Example 9 the average (mean) test score, which came out to be 10.85. For each

Table 14-12 Stat 101 Data: Deviations from the Mean

Test score (x)	Deviation from the mean ($x - 10.85$)	Frequency
1	−9.85	1
6	−4.85	1
7	−3.85	2
8	−2.85	6
9	−1.85	10
10	−0.85	16
11	0.15	13
12	1.15	9
13	2.15	8
14	3.15	5
15	4.15	2
16	5.15	1
24	13.15	1

possible test score, we can now calculate "how far" that score is from the average score of 10.85 (middle column of Table 14-12). ■

The deviations from the mean are themselves a data set, which we would like to summarize. One way would be to average them, but if we do that, the negative deviations and positive deviations will always cancel each other out, so that we end up with an average of 0 (Exercises 30, 31, and 48). This, of course, makes the average useless in this case. The cancellation of positive and negative deviations can be avoided by squaring each of the deviations. The squared deviations are never negative, and if we average them out we get an important measure of spread called the *variance.*[5] If we take the square root of the variance, we get the **standard deviation.**[6] The process is complicated, but not necessarily difficult, if we take it one step at a time.

Example 21.

Let's find the standard deviation of the 10 numbers 45, 46, 47, 48, 49, 51, 52, 53, 54, 55. The first step is to find the mean (average) of the data set, which we will call A. Here $A = 50$. The second step is to calculate the *deviations from the mean.* They are $-5, -4, -3, -2, -1, 1, 2, 3, 4, 5$. The third step is to square each of the preceding deviations. This gives the set of *squared deviations*: 25, 16, 9, 4, 1, 1, 4, 9, 16, 25. Next, we average these numbers. This gives the *variance*: 11. Finally, we take the square root of the variance to get the standard deviation: $\sqrt{11} \approx 3.317$. ■

Finding the Standard Deviation of a Data Set

- **Step 1.** Find the *average* (mean) of the data set. Call it A.
- **Step 2.** For each number x in the data set, find $x - A$, the *deviation from the mean.*
- **Step 3.** Square each of the deviations found in step 2. These are the *squared deviations.*
- **Step 4.** Find the average of the squared deviations. This number is called the *variance.*
- **Step 5.** Take the square root of the variance. This is the *standard deviation.*

---

[5] In many statistics books and statistical computer programs, the variance is defined by dividing the squared deviations by $N - 1$ (instead of by N, as one would in an ordinary average). There are reasons why this definition is appropriate in some circumstances, but a full explanation would take us beyond the purpose and scope of this chapter. In any case, except for small values of N the difference between the two definitions tends to be very small.

[6] Taking the square root of the variance makes the standard deviation have the same units as the original data. Thus, if the data represents dollars, then the standard deviation will also be given in dollars.

Standard deviations of large data sets are not fun to calculate by hand, and it is rarely done that way. Standard procedure for calculating standard deviations is to use a computer or a good scientific or business calculator, which often are preprogrammed to do all the steps automatically. Be that as it may, it is still important to know what the steps are in calculating a standard deviation, even when the actual grunt work is farmed out to a machine.

As a measure of spread, the standard deviation is particularly useful when analyzing real-life data. We will come to appreciate its importance in this context in Chapter 16.

Conclusion

Whether we like to or not, as we navigate through life in the information age, we are awash in a sea of data. Today, data is the common currency of scientific, social, and economic discourse. Powerful satellites constantly scan our planet, collecting prodigious amounts of weather, geological, and geographical data. Government agencies, such as the Census Bureau and the Bureau of Labor Statistics, collect millions of numbers a year about our living, working, spending, and dying habits. Even in our less serious pursuits, such as sports, we are flooded with data, not all of it great.

Faced with the common problem of data overload, statisticians and scientists of all kinds have devised many ingenious ways to organize, display, and summarize large amounts of data. In this chapter we discussed some of the basic concepts in this area of statistics.

Graphical summaries of data can be produced by bar graphs, pictograms, pie charts, histograms, and so on. (There are many types of graphical descriptions that we did not discuss in the chapter.) Which kind of graph is the most appropriate for which situation depends on many factors, and creating good "pictures" of a data set is as much an art as a science.

Numerical summaries of data, when properly used, help us understand the overall pattern of a data set without getting bogged down in the details. They fall into two categories: (1) *measures of location,* such as the *average,* the *median,* and the *quartiles,* and (2) *measures of spread,* such as the *range,* the *interquartile range,* and the *standard deviation.* Sometimes we even combine numerical summaries and graphical displays, as in the case of the *box plot.* We touched upon all of these in this chapter, but the subject is a big one, and by necessity we only scratched the surface.

In this day and age we are all consumers of data, and at one time or another we are likely to also be providers of data. As we enter the twenty-first century, understanding the basics of how data is organized and summarized has become an essential requirement for "efficient citizenship." The concepts that you learned in this chapter are to statistical literacy what a basic vocabulary is to ordinary language.

Key Concepts

average (mean)
bar graph
box plot
categorical (qualitative) variable
category (class)
class interval
continuous (variable)
data set
data values (data points)
deviations from the mean
discrete (variable)
five-number summary
frequency table
histogram
interquartile range
lower half
measures of location
measures of spread

median	**quartiles**
numerical (quantitative) variable	**range**
outlier	**standard deviation**
pictogram	**upper half**
pie chart	

Exercises

Walking

Exercises 1 through 4 refer to the scores in a Chem 103 final exam which consisted of 10 questions worth 10 points each. The scores on the exam are given in Table 14-13.

Table 14-13 Chem 103 Final Exam Scores

Student ID	Score	Student ID	Score	Student ID	Score	Student ID	Score
1362	50	2877	80	4315	70	6921	50
1486	70	2964	60	4719	70	8317	70
1721	80	3217	70	4951	60	8854	100
1932	60	3588	80	5321	60	8964	80
2489	70	3780	80	2872	100	9158	60
2766	10	3921	60	6433	50	9347	60

1. (a) Make a frequency table for the Chem 103 final exam scores.

(b) Make a bar graph showing the actual frequencies of the scores on the exam.

(c) Make a bar graph showing the relative frequencies of the scores on the exam (i.e., the percentage of students receiving each score).

2. For the Chem 103 final exam scores, find:

(a) The range.

(b) The median.

(c) The first and third quartiles.

(d) The interquartile range.

3. For the Chem 103 final exam scores:

(a) Find the five-number summary.

(b) Draw a box plot.

4. For the Chem 103 final exam scores, find:

(a) The average.

(b) The standard deviation, including the units.

Exercises 5 through 8 refer to a midterm exam in History 3B. The percentage scores on the exam are given in Table 14-14.

Table 14-14 History 3B Midterm Exam Scores

Student ID	Score	Student ID	Score	Student ID	Score	Student ID	Score	Student ID	Score
1075	74%	1998	75%	3491	57%	4713	83%	6234	77%
1367	83%	2103	59%	3711	70%	4822	55%	6573	55%
1587	70%	2169	92%	3827	52%	5102	78%	7109	51%
1877	55%	2381	56%	4355	74%	5381	13%	7986	70%
1946	76%	2741	50%	4531	77%	5717	74%	8436	57%

5. For the data set in Table 14-14 find:

(a) The average.

(b) The median.

(c) The first and third quartiles.

(d) The interquartile range.

6. For the data set in Table 14-14:

(a) Find the five-number summary.

(b) Draw a box plot.

7. For the data set in Table 14-14, find:

(a) The variance.

(b) The standard deviation, including the units.

8. Suppose the class intervals for letter grades on the History 3B midterm exam are:

90%–100%: A

80%–89%: B

70%–79%: C

60%–69%: D

Below 60%: F

(a) Make a frequency table for the letter grades on the exam.

(b) Draw a pie chart for the percentage of students receiving each letter grade.

9. The percentage of the U.S. population enrolled in HMOs for the years 1989 to 1995 is given in the following table (*Source: The World Almanac and Book of Facts 1997,* p. 973).

Year	Percent in HMOs
1989	13.0
1990	13.4
1991	13.6
1992	14.3
1993	15.1
1994	16.1
1995	17.7

Using the ideas of Example 2, make two different-looking pictograms showing the growth in the percentage of the population enrolled in HMOs from 1989 to 1995. In the first pictogram you are trying to convince your audience that HMOs are growing very fast. The second pictogram should give a more accurate picture.

10. The percentage sales of recorded music on compact disc from 1991 to 1995 is given in the following table (*Source: The World Almanac and Book of Facts 1997,* p. 291).

Year	Percent CDs
1991	38.9
1992	46.5
1993	51.2
1994	58.4
1995	65.0

Using the ideas of Example 2, make two different-looking pictograms showing the growth in the percentage of recorded music sold on compact discs from 1991 to 1995. In the first pictogram you are trying to convince your audience that CDS are taking over the market. The second pictogram should give a more accurate picture.

Exercises 11 and 12 refer to Table 14-15, which gives the distance from home to school (measured to the closest half-mile) for each kindergarten student at Cleansburg Elementary School.

Table 14-15 Distance from Home to School for Cleansburg Elementary School Kindergarten Students

Student ID	Distance to school (miles)	Student ID	Distance to school (miles)	Student ID	Distance to school (miles)	Student ID	Distance to school (miles)
1362	1.5	2877	1.0	4355	1.0	6573	0.5
1486	2.0	2964	0.5	4454	1.5	8436	3.0
1587	1.0	3491	0.0	4561	1.5	8592	0.0
1877	0.0	3588	0.5	5482	2.5	8854	0.0
1932	1.5	3711	1.5	5533	1.0	8964	2.0
1946	0.0	3780	2.0	5717	8.5		
2103	2.5	3921	5.0	6307	1.5		

11. **(a)** Make a frequency table for the data set in Table 14-15.

(b) Draw a bar graph showing the relative frequencies for the data set in Table 14-15.

12. Suppose that class intervals for the distances from home to school for the kindergartners at Cleansburg Elementary School are defined by:

Very close: Less than 1 mile.

Close: 1 mile up to and including 1.5 miles.

Nearby: 2 miles up to and including 2.5 miles.

Not too far: 3 miles up to and including 4.5 miles.

Far: 5 miles or more.

(a) Make a frequency table for the class intervals.

(b) Draw a pie chart for the percentage of students in each category.

13. The bar graph below shows the frequencies of various scores received by students in Math A on a 10-point pop quiz.

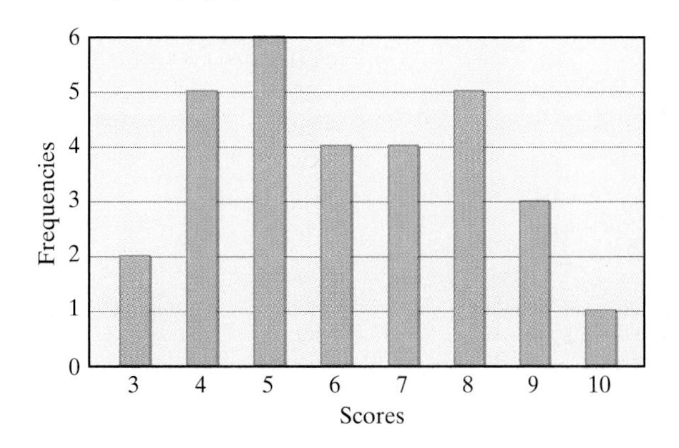

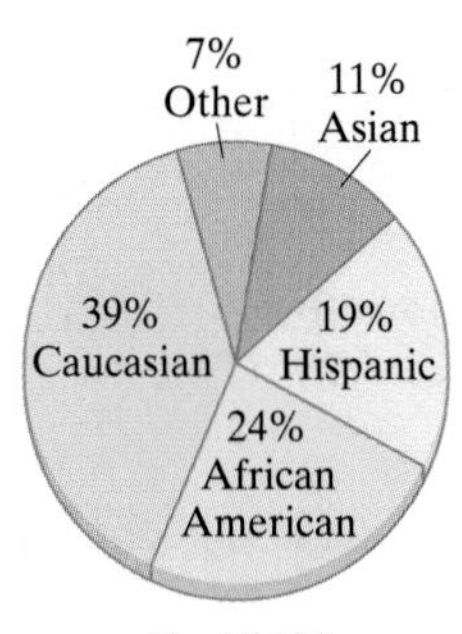

$N = 15{,}000$

(a) How many students took the quiz?

(b) Make a frequency table for the scores on the pop quiz.

(c) What is the average score?

(d) What is the median score?

Exercises 14 and 15 refer to the pie chart in the margin. The pie chart shows the percentage of the undergraduate student body at Tasmania State University for each ethnic group.

14. **(a)** Give a frequency table showing the actual frequencies for each category.

(b) Draw the bar graph corresponding to the frequency table in (a).

15. Calculate the size of the angle (to the nearest degree) for each of the slices shown in the pie chart.

16. A pie chart is made up of 4 slices representing 4 different categories in a population (A, B, C, and D).

Slice A has an angle of 52°.
Slice B has an angle of 108°.
Slice C has an angle of 125°.
Slice D has an angle of 75°.

Find the percentage of the population in each category.

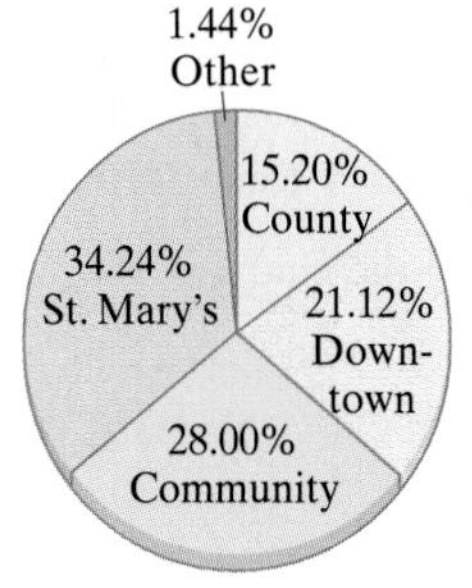

$N = 625$

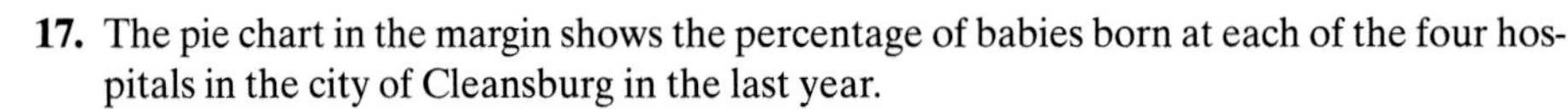

17. The pie chart in the margin shows the percentage of babies born at each of the four hospitals in the city of Cleansburg in the last year.

(a) How many babies were born at Downtown Hospital?

(b) How many babies were born outside one of the four hospitals (at home, on the way to the hospital, and so on)?

(c) Draw a bar graph showing the frequency for each category.

18. Calculate the size of the angle (in degrees) for each of the slices shown in the pie chart used in Exercise 17.

Exercises 19 through 22 refer to the data in Table 14-16, which shows the weights (in ounces) of the 625 babies born in the city of Cleansburg in the last year.

Table 14-16

Weight (in ounces)		
More than	**Less than or equal to**	**Frequencies**
48	60	15
60	72	24
72	84	41
84	96	67
96	108	119
108	120	184
120	132	142
132	144	26
144	156	5
156	168	2

19. **(a)** Give the length of each class interval (in ounces).

(b) Suppose a baby weighs exactly 5 pounds 4 ounces. What class interval does she belong to? Describe the endpoint convention.

20. Write a new table for these data values using class intervals of length equal to 24 ounces.

21. Draw the histogram corresponding to these data values using the class intervals as shown in the original table.

22. Draw the histogram corresponding to the same data when class intervals of 24 ounces are used.

23. Find the average and the standard deviation for each of the following three sets of numbers: $\{5, 5, 5, 5\}$, $\{0, 5, 5, 10\}$, $\{-5, 0, 0, 25\}$. Note the similarities and differences in your answers, and explain.

24. Find the average and the standard deviation for each of the three data sets: $\{10, 10, 10, 10\}$, $\{1, 6, 13, 20\}$, $\{1, 1, 18, 20\}$. Note the similarities and differences in your answers, and explain.

25. For the data set $\{0, 1, 2, 3, 4, 5, 6, 7, 8, 9\}$ find:

(a) The average.

(b) The median.

(c) The first quartile.

(d) The third quartile.

(e) The interquartile range.

(f) The standard deviation.

26. For the data set $\{1, 2, 3, 4, 5, 6, 7, 8, 9, 10\}$ find:

(a) The average.

(b) The median.

(c) The first quartile.

(d) The third quartile.

(e) The interquartile range.

(f) The standard deviation.

27. For the data set $\{1, 2, 3, \ldots, 99, 100\}$ find:

(a) The average.

(b) The median.

28. For the data set $\{1, 2, 3, \ldots, 99, 100\}$ find:

(a) The first quartile.

(b) The third quartile.

(c) The interquartile range.

29. **(a)** Find the five-number summary, the average, and the standard deviation for the data given in the following frequency table.

Value	9	10	11	12	13	14	15	16	17	18	19	20	21	22	23	24	25
Frequency	3	5	7	4	12	10	13	11	15	13	11	4	3	0	0	0	1

(b) Draw a box plot for this data.

30. For the data set $\{25, 13, 18, 37, 11, 16\}$ find:

(a) The average (mean).

(b) The deviations from the mean.

(c) The average of the deviations from the mean.

31. For the data set $\{2.2, -1.1, -2.7, 4.4, 6.2, -2.4, 3.8, 1.6\}$ (Example 13) find:

(a) The average (mean).

(b) The deviations from the mean.

(c) The average of the deviations from the mean.

Exercises 32 and 33 refer to the following figure describing the starting salaries for Tasmania State University first-year graduates in agriculture and engineering. (These are the two box plots discussed in Example 18.)

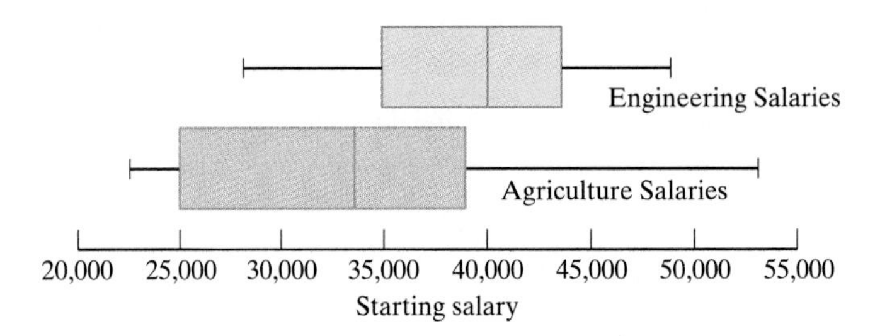

32. (a) Approximately how much is the median salary for agriculture majors?

(b) Approximately how much is the median salary for engineering majors?

(c) Explain how we can tell that the median salary for engineering majors is more than the third quartile of the salaries for agriculture majors.

33. (a) If the number of engineering majors was 612, how many of them had a starting salary of $35,000 or more?

(b) If the number of agriculture majors was 960, approximately how many of them made less than $25,000?

34. Mike's average on the first five exams in Econ 1A is 88. What must he earn on the next exam in order to raise his overall average to 90?

35. Sarah's overall average in Physics 101 was 93%. Her average was based on four exams each worth 100 points and a final exam worth 200 points. What is the lowest possible score she could have made on the first exam?

36. Find the standard deviation of the Stat 101 test scores (see Example 20).

Jogging

37. Josh and Ramon each have an 80% average on the five exams given in Psychology 4. Ramon, however, did better than Josh on all of the exams except one. Give an example that illustrates this situation.

38. Kelly and Karen each have an average of 75 on the six exams given in Botany 1. Kelly's scores have a small standard deviation and Karen's scores have a large standard deviation. Give an example that illustrates this situation.

39. (a) Give an example of 10 numbers with an average less than the median.

(b) Give an example of 10 numbers with a median less than the average.

(c) Give an example of 10 numbers with an average less than the first quartile.

(d) Give an example of 10 numbers with an average more than the third quartile.

40. Suppose that the average of 10 numbers is 7.5 and that the smallest of them is Min = 3.

(a) What is the smallest possible value of Max?

(b) What is the largest possible value of Max?

41. What happens to the five-number summary of the Stat 101 data set (Example 17) if:

(a) Two points are added to each score?

(b) Ten percent is added to each score?

42. A data set is called **constant** if every value in the data set is the same. A constant data set can be described by $\{a, a, a, \ldots, a\}$.

(a) Show that the standard deviation of a constant data set is 0.

(b) Show that if the standard deviation of a data set is 0, it must be a constant data set.

Exercises 43 and 44 refer to histograms with unequal class intervals. When drawing such histograms, the columns must be drawn so that the frequencies or percentages are proportional to the area of the column. The figure illustrates what we mean. If the column over class interval 1 represents 10% of the population, then the column over class interval 2, also representing 10% of the population, must be one third as high, because the class interval is three times as large.

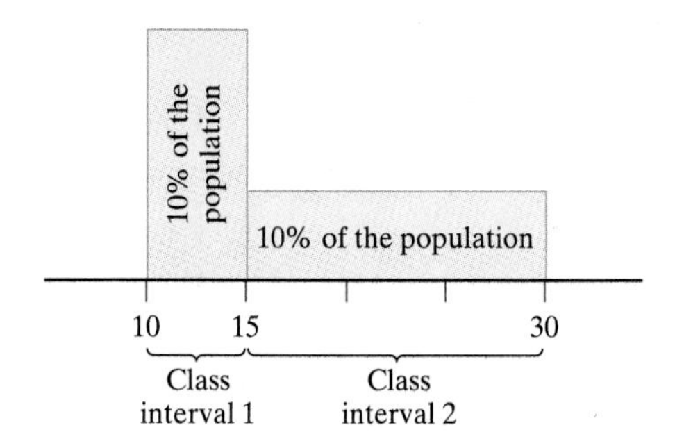

43. If the height of the column over the class interval 20–30 is 1 unit and the column represents 25% of the population, then:

(a) How high should the column over the interval 30–35 be if 50% of the population falls into this class interval?

(b) How high should the column over the interval 35–45 be if 10% of the population falls into this class interval?

(c) How high should the column over the interval 45–60 be if 15% of the population falls into this class interval?

44. Two hundred senior citizens are tested for fitness and rated on their times on a 1-mile walk. These ratings and associated frequencies are given in the following table.

Time	Rating	Frequency
6^+–10 minutes	Fast	10
10^+–16 minutes	Fit	90
16^+–24 minutes	Average	80
24^+–40 minutes	Slow	20

Draw a histogram for these data based on the categories given by the ratings in the table.

45. News media have accused Tasmania State University of discriminating against women in the admission policies in its schools of architecture and engineering. The *Tasmania*

Gazette states, "68% of all male applicants to the schools of architecture or engineering are admitted while only 51% of the female applicants to these same schools are admitted." The actual data is given in the following table.

	School of Architecture		School of Engineering	
	Applied	**Admitted**	**Applied**	**Admitted**
Male	200	20	1000	800
Female	500	100	400	360

(a) What percent of the male applicants to the School of Architecture were admitted? What percent of the female applicants to this same school were admitted?

(b) What percent of the male applicants to the School of Engineering were admitted? What percent of the female applicants to this same school were admitted?

(c) How did the *Tasmania Gazette* come up with its figures?

(d) Explain how it is possible for (a), (b), and the *Tasmania Gazette* statement all to be true.

46. Given that the numbers $x_1, x_2, x_3, \ldots, x_N$ have average a and median M, explain why the numbers $x_1 + c, x_2 + c, x_3 + c, \ldots, x_N + c$ have:

(a) Average $A + c$.

(b) Median $M + c$.

47. Show that the standard deviation of any set of numbers is always less than or equal to the range of the set of numbers.

Running

48. (a) Calculate the average of the deviations from the mean for the Stat 101 data set (see Example 20).

(b) Show that if $\{x_1, x_2, x_3, \ldots, x_N\}$ is a data set with average A, then the average of $x_1 - A, x_2 - A, x_3 - A, \ldots, x_N - A$ is 0.

49. (a) Show that if $\{x_1, x_2, x_3, \ldots, x_N\}$ is a data set with average A and standard deviation s, then $s\sqrt{N} \geq |x_i - A|$ for every data value x_i.

(b) Use (a) to show that

$A - s\sqrt{N} \leq x_i \leq A + s\sqrt{N}$

for every data value x_i.

50. (a) Find two numbers whose average is A and standard deviation is s.

(b) Find three equally spaced numbers whose average is A and standard deviation is s.

(c) Generalize the preceding by finding N equally spaced numbers whose average is A and standard deviation is s. (*Hint:* Consider N even and N odd separately.)

51. Show that the median and the average of the numbers $1, 2, 3, \ldots, N$ are always the same.

52. Suppose that the average of the numbers $x_1, x_2, x_3, \ldots, x_N$ is A and that the variance of these same numbers is V. Suppose also that the average of the numbers $x_1^2, x_2^2, x_3^2, \ldots, x_N^2$ is B. Show that $V = B - A^2$. (In other words, for any data set, if we take the average of the squared data values and subtract the square of the average of the data values, we get the variance.)

53. Given that the numbers $x_1, x_2, x_3, \ldots, x_N$ have standard deviation s, explain why the numbers $x_1 + c, x_2 + c, x_3 + c, \ldots, x_N + c$ also have standard deviation s.

54. Given that the numbers $x_1, x_2, x_3, \ldots, x_N$ have standard deviation *s*, explain why the numbers $ax_1, ax_2, ax_3, \ldots, ax_N$ (where *a* is a positive number) have standard deviation *as*.

55. Using the formula $1^2 + 2^2 + 3^2 + \ldots + N^2 = N(N + 1)(2N + 1)/6$,

(a) Find the standard deviation of the data set $\{1, 2, 3, \ldots, 98, 99\}$. (*Hint:* Use Exercise 52.)

(b) Find the standard deviation of the data set $\{1, 2, 3, \ldots, N\}$.

56. (a) Find the average and standard deviation of the data set $\{315, 316, \ldots, 412, 413\}$. (*Hint:* Use Exercises 46 and 53.)

(b) Find the average and standard deviation of the data set $\{k + 1, k + 2, \ldots, k + N\}$.

57. (Open-ended question) The following table gives the number of violent crimes committed in the United States and the population of the United States for the years 1985 to 1995 (*Source: The World Almanac and Book of Facts 1997,* p. 958).

Year	Population	Violent Crimes
1985	238,740,000	1,328,800
1986	241,077,000	1,489,170
1987	243,400,000	1,484,000
1988	245,807,000	1,566,220
1989	248,239,000	1,646,040
1990	248,709,873	1,820,130
1991	252,177,000	1,911,770
1992	255,082,000	1,932,270
1993	257,908,000	1,926,020
1994	260,341,000	1,857,670
1995	262,755,000	1,798,790

Using the knowledge you have acquired in this chapter, summarize, display graphically, and discuss this data.

References and Further Readings

1. Cleveland, W. S., *The Elements of Graphing Data,* rev. ed. New York: Van Nostrand Reinhold Co., 1994.
2. Freedman, D., R. Pisani, R. Purves, and A. Adhikari, *Statistics,* 2d ed. New York: W. W. Norton, Inc., 1991, chaps. 3 and 4.
3. Mosteller, F., W. Kruskal, et al., *Statistics by Example: Exploring Data.* Reading, MA: Addison-Wesley Publishing Co., Inc., 1973.
4. Sincich, Terry, *Statistics by Example,* 5th ed. New York: Macmillan Publishing Co., 1993, chaps. 2 and 3.
5. Tanner, Martin, *Investigations for a Course in Statistics.* New York: Macmillan Publishing Co., Inc., 1990.
6. Tufte, Ed, *Envisioning Information.* Cheshire, CT: Graphics Press, 1990.
7. Tufte, Ed, *The Visual Display of Quantitative Information.* Cheshire, CT: Graphics Press, 1983.
8. Utts, Jessica, *Seeing Through Statistics.* Belmont, CA: Wadsworth Publishing Co., 1996.
9. Wainer, H., "How to Display Data Badly," *The American Statistician,* 38 (1984), 137–147.
10. Wildbur, Peter, *Information Graphics.* New York: Van Nostrand Reinhold Co., 1986.

Ninety percent chance of rain tomorrow.
(Weather report.)
The probability of rolling a 7 or 11 with a pair of dice is 1/6.
(Gambler's manual.)
The odds of dying in an automobile crash are 1 in 140.
(National Highway Safety Board.)

Chances, Probability, and Odds

Measuring Uncertainty

CHAPTER 15

"Probability," "chance," "odds"—these words are as much a part of our everyday vocabulary as "mother," "baseball," and "apple pie." While we all use these words in everyday conversation and probably (there it is again!) have a rough idea of what each of them means, giving a precise definition of *probability* (or *chance,* or *odds*—they are just different versions of the same concept) is surprisingly difficult.

In this chapter we will learn how to interpret and calculate probabilities, chances, and odds using mathematical methods. This will be our very brief introduction to the mathematical theory of probability—a relatively young branch of mathematics which has become of fundamental importance to many aspects of modern life. Anywhere there is uncertainty—and that is practically everywhere—probability theory plays a role: insurance, health, biology, the economy, sports.

When we talk about the *probability* of something happening, we always include a number—for example, "The probability of rolling a 7 or an 11 with a pair of dice is 2/9." Likewise, when we talk about the *chances* of something happening, we include a number given in percentage form. Thus, we could say that "The chances of rolling a 7 or an 11 with a pair of dice are 22.22%."[1] When we talk about *odds,* we usually give two numbers—for example, "The odds of the Chicago Bulls winning the NBA championship are 3 to 2"—but we are still describing an equivalent idea. Later in this chapter we will explain how to convert odds to probabilities and vice versa.

Our discussion in this chapter is broken up into two parts. In the first part we lay down the basic concepts needed for a meaningful discussion of probability; in the second part we define and calculate probabilities using a mathematical approach.

Random Experiments and Sample Spaces

In broad terms, probability is the *quantification of uncertainty.* To understand what that means, we may start by formalizing the notion of uncertainty.

We will use the term **random experiment** to describe an activity or process *whose outcome cannot be predicted ahead of time.* Typical examples of random experiments are tossing a coin, rolling a pair of dice, shooting a free throw, drawing a number out of a hat, having a baby (in the sense of "Will it be a boy or a girl?"), and predicting the outcome of a basketball game. As these examples show, random experiments do not require elaborate setups or fancy equipment.

Associated with every random experiment is the *set* of all its possible outcomes, called the **sample space** of the experiment. For the sake of simplicity we will concentrate on experiments for which there is only a finite set of outcomes, although experiments with infinitely many outcomes are both possible and important.

We illustrate the importance of the sample space by means of several examples. Since the sample space of any experiment is a set of outcomes, we will use set notation to describe it. We will consistently use the letter S to denote a sample space, and N to denote its size (that is, the number of outcomes in S).

Example 1. A coin toss

Our random experiment is to *toss a quarter.* The sample space can be described by $S = \{H, T\}$ (where H stands for heads, and T for tails). Here $N = 2$. ■

A couple of comments about coins are in order here. First, the fact that the coin in Example 1 was a quarter is essentially irrelevant. Practically all coins have an obvious "head" side (and thus a "tail" side), and even when they don't—as in a "buffalo" nickel—we can agree ahead of time which side is which. Second, we all know (and if we don't, we should) that there are fake coins out there on which both sides are "heads." Tossing such a coin does not fit our definition of a random experiment, so from now on we will assume that all coins used in our experiments have two different sides, which we will call H and T.

---

[1] It is customary to express probabilities in decimals (or fractions) and chances in percentages, and we will follow that custom in this chapter.

Example 2. A double coin toss

(a) Suppose we *toss a coin twice.* The sample space now is $S = \{HH, HT, TH, TT\}$, where HT means the first toss came up H and the second toss came up T, which is a different outcome from TH (first toss T, second toss H). As the reader can see, we are being very meticulous about the details. For this sample space, $N = 4$.

(b) Suppose now we *toss two coins—say a nickel and a quarter—at the same time.* This random experiment appears to be different from the one in (a), but the sample space is the same: $S = \{HH, HT, TH, TT\}$. Here we must agree what the order of the symbols is (for example, the first symbol describes the quarter and the second the nickel). ■

Example 3. Free throws

Suppose *a basketball player shoots a pair of free throws.* This is a random experiment with sample space $S = \{ss, sf, fs, ff\}$, where s means success and f means failure. Here again $N = 4$. ■

Notice the similarities between Examples 2 and 3. In fact, if we were to identify H with success and T with failure (an arbitrary decision), the sample spaces would be exactly the same. Examples 2 and 3 illustrate the fact that very different random experiments (tossing a pair of coins, shooting a pair of free throws) can turn out to have essentially the same sample space (the symbols may be different, but the idea is the same).

We will now discuss a few examples of random experiments involving dice. A die[2] is a cube, usually made of plastic, whose six faces are marked with dots (from 1 to 6) called "pips." Random experiments using dice have a long-standing tradition in our culture, and are a part of both gambling and recreational games (Monopoly, Yahtzee, etc.).

Example 4. Rolling a die

Suppose we *roll a single die.* The sample space for this experiment is S = {⚀, ⚁, ⚂, ⚃, ⚄, ⚅}. Here $N = 6$. ■

Example 5. Rolling a pair of dice: Part I

Suppose we *roll a pair of dice.* The sample space now is a little bigger ($N = 36$):

Notice that, as we did with the coins, we are treating the dice as *distinguishable* objects (as if one were white and the other red), so that ⚂⚄ and ⚄⚂ are considered different outcomes. ■

[2] Singular, die; plural, dice.

In many games (such as Monopoly, craps, etc.), we roll a pair of dice and what matters is the total rather than the actual numbers rolled. In this case the outcomes that really matter are the possible sums (2 through 12), so our interpretation of the sample space can change accordingly.

Example 6. Rolling a pair of dice: Part II

Suppose we *roll a pair of dice and add the points on the two dice.* Now the sample spaces is $S = \{2, 3, 4, 5, 6, 7, 8, 9, 10, 11, 12\}$ and $N = 11$. ■

There is no contradiction between Examples 5 and 6. A random experiment is a process and, as such, it can consist of more than one step. The random experiments described in Examples 5 and 6 are indeed different.

Example 7. Ranking the top 3 candidates in an election

Five candidates (*A, B, C, D,* and *E*) are running in an election. The top 3 candidates are chosen President, Vice-President, and Secretary, in that order. The election can be considered a random experiment with sample space $S = \{ABC, ACB, BAC, BCA, CAB, CBA, ABD, ADB, \ldots\}$ (where the outcome ABC signifies that candidate A is elected President, B is elected Vice-President, and C is elected Treasurer). The "..." at the end of the sample space is another way of saying "and so on—hopefully you got the picture." ■

What happened in Example 7 is commonplace. Once we realize that the sample space S is big, we decide against writing each and every outcome down. The critical thing will be to find the actual size N of the sample space without having to list each individual outcome, and this can be done by using a few basic rules of counting. We will learn how to do this next.

Counting: The Multiplication Rule

Example 8. Triple coin toss

Suppose we *toss a coin three times.* Here the sample space is $S = \{HHH, HHT, HTH, HTT, THH, THT, TTH, TTT\}$, with $N = 8$. ■

Example 8 sets the stage for the next example:

Example 9. Multiple coin toss

Suppose we *toss a coin 8 times.* This sample space is too big to fully write down. Nonetheless, we can find its size in a relatively painless way.

The first thing we should ask is: What does a random outcome look like? Taking our cue from Example 8, we can say that a random outcome can be described by a string of 8 consecutive letters, where the letters can be either H's or T's. For example, the string $THHTHTHH$ represents a *single* outcome in our sample space—the one in which the first toss came up T, the second toss came up H, the third toss came up H, and so on. To count *all* the outcomes, we will argue as follows: (1) the number of possibilities for the first letter is 2 (H or T); (2) the number of possibilities for the second letter is also 2, . . . ; (8) the number of possibilities for the last letter is 2. The total number of outcomes is given by *multiplying* all of these numbers.

Total number of outcomes: $N = 2 \times 2 \times 2 \times 2 \times 2 \times 2 \times 2 \times 2 = 256$. ■

The basic rule we used in Example 9 is called (for obvious reasons) the **multiplication rule**. Informally stated, the multiplication rule says that *when something takes place in several stages, to find the total number of ways it can occur we multiply the number of ways each individual stage can occur.*

The easiest way to understand the multiplication rule is through examples.

Example 10. Buying ice cream

Imagine you want to buy a *single* scoop of ice cream. There are two types of cones available (sugar and regular) and three flavors to choose from (vanilla, chocolate chip, and strawberry). Figure 15-1 shows all the possible combinations.

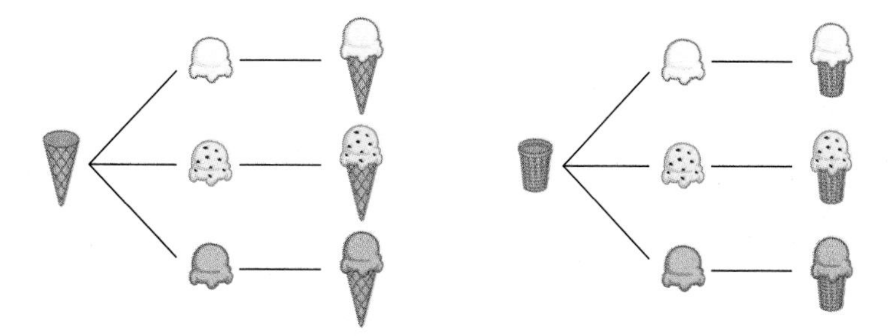

FIGURE 15-1
2 cones and 3 flavors make $2 \times 3 = 6$ combinations.

■

The multiplication rule is undoubtedly the most important tool used in solving *counting problems* (that is, problems that ask in how many different ways can one thing or another happen). These are exactly the kinds of questions one needs to answer to carry out the basic probability calculations that we will want to do later in this chapter. Before we get to that, we will take a brief detour to explore a few of the subtleties of "counting." This is an important and rich subject, full of interesting twists and turns, but our detour necessarily will be brief.

Our detour starts with a straightforward application of the multiplication rule. As we move on, the level of sophistication will gradually increase, with each successive example showing some variation of the original theme.

Example 11. The making of a wardrobe: Part I

Dolores is a young saleswoman planning her next business trip. She is thinking about packing 3 different pairs of shoes, 4 skirts, 6 blouses, and 2 jackets. If all the items are color coordinated, how many different *outfits* will she be able to make out of these items?

To answer this question we must first define what we mean by an outfit. Let's assume that an "outfit" consists of a pair of shoes, a skirt, a blouse, and a jacket. Here we can use the multiplication rule directly: The total number of possible outfits Dolores can make is $3 \times 4 \times 6 \times 2 = 144$. (Color coordination obviously pays: Dolores can be on the road for over 4 months and never have to wear the same outfit twice!) ■

The next example is a more subtle variation of Example 11.

Example 12. The making of a wardrobe: Part II

Once again, Dolores is packing for a business trip. This time she packs 3 pairs of shoes, 4 skirts, 3 pairs of slacks, 6 blouses, 3 turtlenecks, and 2 jackets. As before, we can assume that she coordinates the colors so that everything goes with everything else. This time we will define an "outfit" as consisting of a pair of shoes, a choice of "lower wear" (either a skirt *or* a pair of slacks), a choice of "upper wear" (it could be a blouse *or* a turtleneck *or both*), and finally, she may or may not choose to wear a jacket (it's early spring and the weather is unpredictable).

Once again, we want to count how many different such outfits are possible. Our strategy will be to think of an "outfit" as being put together in stages and to draw a box for each of the stages. We then separately count the number of choices at each stage and enter that number in the corresponding box. (Some of these calculations can themselves be mini-counting problems.) The last step is to multiply the numbers in each box. The details are illustrated in Fig. 15-2. The final count for the number of different outfits is $N = 3 \times 7 \times 27 \times 3 = 1701$.

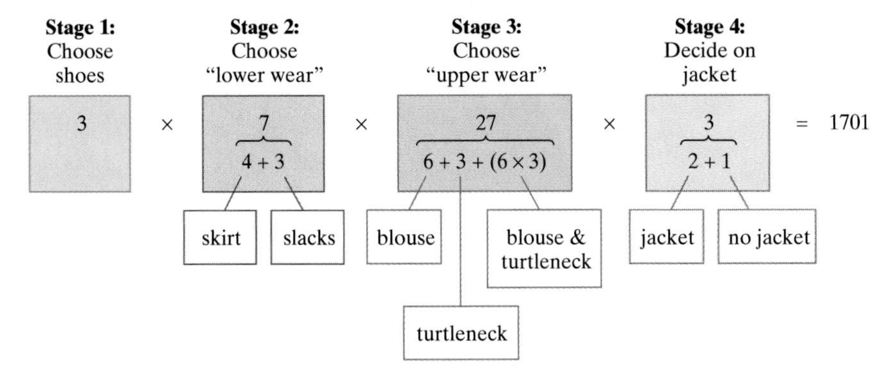

FIGURE 15-2 Counting all possible "outfits" using a box model.

■

The method of drawing boxes representing the successive stages in a process, and putting the number of choices for each stage inside the box is a convenient device that often helps clarify one's thinking. Silly as it may seem, we strongly recommend it. For ease of reference, we will call it the *box model for counting.*

Example 13. Ranking the top 3 candidates in a 5-person election

We are back to the question raised in Example 7: Five candidates are running in an election, with the top 3 getting elected (in order) President, Vice-President, and Secretary. We want to know how big the sample space is. Using a box model, this becomes a reasonably easy counting problem, as illustrated in Fig. 15-3.

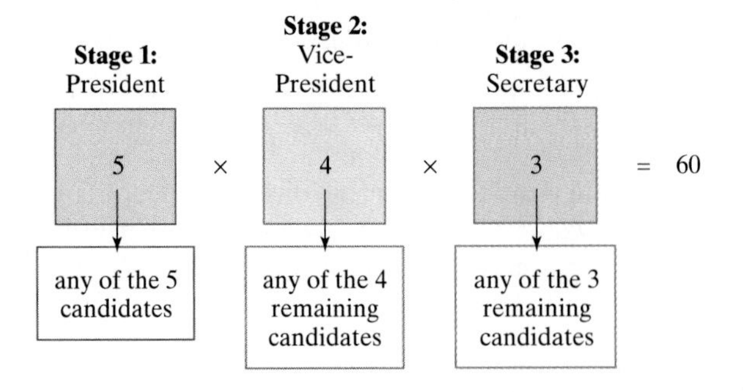

FIGURE 15-3 Ranking 3 out of 5 candidates using a box model.

■

Permutations and Combinations

In many counting problems the multiplication rule and the box model are by themselves not enough, and we need to add some new tools to our toolbox. Take, for example, the question of ordering ice cream at Baskin-Robbins.

Example 14. "True-doubles" at Baskin-Robbins

Baskin-Robbins offers 31 different flavors of ice cream. A "true-double" is the name some kids use for two scoops of ice cream of two *different* flavors. How many different "true-doubles" are possible?

It would appear at first glance that this is a simple variation of Example 13, and that the total number of possible "true-doubles" is $31 \times 30 = 930$, as shown in Fig. 15-4. But if we give the matter a little careful thought, we will realize that we have double counted. Double counting "true-doubles"? Why? Most people would agree that the order in which the scoops of ice cream are put in a bowl or a cone is irrelevant, and that picking strawberry first and chocolate second is no different from picking chocolate first and strawberry second. But in a box model, *there is always a definite order to things,* and strawberry-chocolate is counted separately from chocolate-strawberry, so our count of 930 is wrong. Fortunately, we also know exactly how and why the count of 930 is wrong—it is double what is should be! Thus, dividing 930 by 2 gives us the correct count: a total of 465 "true-doubles" are possible at Baskin-Robbins.

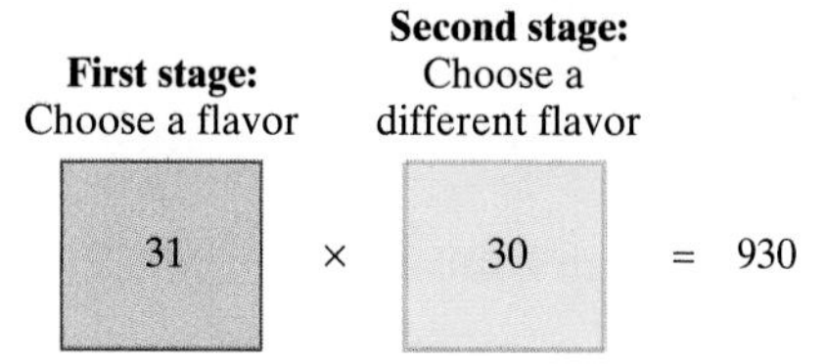

FIGURE 15-4 "True-doubles" at Baskin-Robbins are only half of this count.

■

Example 14 is an important one. It warns us that we have to be careful about how we use the multiplication rule and box models, especially in problems where changing the order in which we choose the parts does not change the whole.

Example 15. "True-triples" at Baskin-Robbins

Let's carry the ideas of Example 14 one step further. Let's say that a "true-triple" consists of 3 scoops of ice cream, each of the 3 scoops being a different flavor. How many different "true-triples" can be ordered at Baskin-Robbins?

Starting with a box model, we have 31 choices for the "first" flavor, 30 choices for the "second" flavor, and 29 choices for the "third" flavor, for a grand total of $31 \times 30 \times 29 = 26{,}970$ combinations. But once again, this is not the correct answer—in fact, the correct answer is the above number divided by 6 (Fig 15-5), giving a total count of 4495 "true-triples."

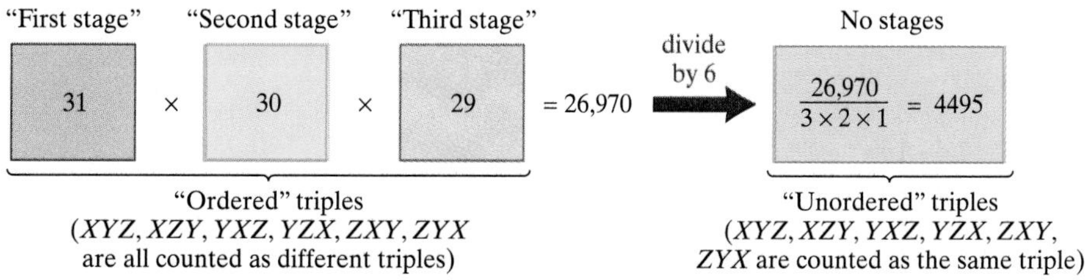

"First stage" "Second stage" "Third stage"

$$31 \times 30 \times 29 = 26{,}970 \xrightarrow{\text{divide by 6}} \frac{26{,}970}{3 \times 2 \times 1} = 4495$$

No stages

"Ordered" triples ($XYZ, XZY, YXZ, YZX, ZXY, ZYX$ are all counted as different triples)

"Unordered" triples ($XYZ, XZY, YXZ, YZX, ZXY, ZYX$ are counted as the same triple)

FIGURE 15-5 Counting "true-triples" at Baskin-Robbins.

The key question is: Why did we divide by 6? For any combination of three flavors (call them X, Y, and Z), there are 6 orders in which these flavors can be listed (XYZ, XZY, YXZ, YZX, ZXY, and ZYX). We can call these "ordered" triples. Each one of these is counted as a different triple under the multiplication rule, but when we consider them as "unordered" triples, they are all the same. *Changing from "ordered" triples to "unordered" triples was accomplished through the division by 6.* ■

It is helpful to think of the final answer to Example 15 (4495) in terms of its pedigree: $4495 = (31 \times 30 \times 29)/(3 \times 2 \times 1)$. The numerator $(31 \times 30 \times 29)$ comes from counting "ordered" triples using the multiplication rule; the denominator $(3 \times 2 \times 1)$ comes from the fact that there are that many ways to shuffle around 3 things (in this case the 3 flavors in a triple). The denominator $3 \times 2 \times 1$ is already familiar to us—it is the *factorial* of 3. The factorial of a positive integer N is denoted by $N!$ and is the number $N \times (N-1) \times (N-2) \times \cdots \times 2 \times 1$. It represents, among other things, the number of ways in which N objects can be ordered or, if you will, shuffled. We discussed the factorial in Chapters 2 and 6, so we won't dwell on it here.

Our next example will deal with poker, a popular card game and the source of many interesting counting questions. Poker is played with a standard deck of 52 cards (the cards are divided into 4 *suits* and there are 13 *values* in each suit). Many variations of poker are played (5-card, 7-card, draw poker, stud poker, etc.). We will dispense with most of the details; what will be most important to us is the distinction between *down* cards (which only the player who gets them sees) and *up* cards (which are dealt face up and everyone gets to see).

Example 16. Five-card poker hands

(a) Let's start by counting the number of possible 5-card *stud poker hands*. In stud poker, the first card is dealt down, and the remaining 4 cards are dealt up, one at a time. In between successive cards there is a round of betting. In this situation the order in which the cards are dealt is important. If you get "good cards" as your second and third cards, then the other players know you have a good hand and are not likely to stick around. If your best card is the first card, which is a down card, that is much better—your opponents won't know that you have a strong hand. In any case, since the order of the 5 cards matters, we can make direct use of the multiplication rule, as shown in Fig. 15-6(a). The total number of 5-card stud poker hands is an enormous number: 311,875,200.

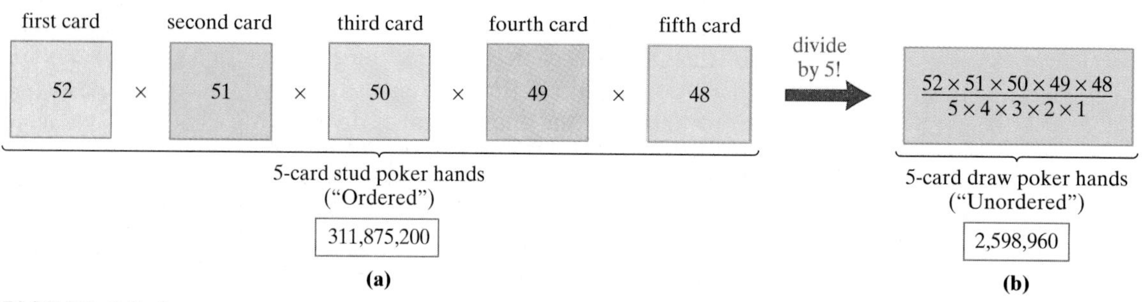

FIGURE 15-6
(a) Number of 5-card stud poker hands, (b) number of 5-card draw poker hands.

(b) Now let's consider the number of possible 5-card hands in draw poker. In draw poker, all cards are down cards. This means that the order in which the cards are dealt is irrelevant—the only person seeing the cards is the player getting them. Once the cards are in her hand, she can shuffle them around any way she sees fit. In fact, we can be more specific: 5 cards can be shuffled in 120 $(5! = 5 \times 4 \times 3 \times 2 \times 1)$

ways. Using the idea of Example 14, if we divide the total number of "ordered" hands by 5!, we get the total number of "unordered" (draw poker) hands: 311,875,200/120 = 2,598,960 [Fig. 15-6(b)]. ■

We should now be able to generalize the ideas we learned in Examples 14, 15, and especially 16. Imagine that we have a set of n distinct objects and we want to select r different objects from this set. The number of ways that this can be done depends on whether the selections are "ordered" or "unordered." Ordered selections are the generalization of stud poker hands—change the order of selection of the same objects and you get something different; unordered selections are the generalization of draw poker hands—change the order of selection of the same objects and you are just "spinning your wheels," you get nothing new. This distinction is of fundamental importance in counting, and mathematicians have a name for each scenario: "ordered" selections are called **permutations**, "unordered" selections are called **combinations**. For a given number of objects n, and for a given selection size r, we can talk about the "number permutations of n objects taken r at a time" and the "number of combinations of n objects taken r at a time," and these two extremely important families of numbers are denoted $\boldsymbol{{}_nP_r}$ and $\boldsymbol{{}_nC_r}$, respectively. Essentially, the numbers ${}_nP_r$ can be computed directly using the multiplication rule (see Example 13); the numbers ${}_nC_r$ can be computed by dividing the corresponding ${}_nP_r$ by $r!$ (see Examples 14, 15, and 16).[3] A summary of the essential facts about these numbers is given in Table 15-1.

Table 15-1 Selecting r different objects out of n objects ($r \le n$)

	"Ordered" selections	**"Unordered selections"**
Name	Permutations	Combinations
Symbol	${}_nP_r$	${}_nC_r$ $\left[\text{also } \binom{n}{r}\right]^4$
Formula (with factorials)	$\dfrac{n!}{(n-r)!}$	$\dfrac{n!}{(n-r)!r!}$
Formula (without factorials)	$n \times (n-1) \times \cdots \times (n-r+1)$	$\dfrac{n \times (n-1) \times \cdots \times (n-r+1)}{r \times (r-1) \times \cdots \times 1}$
Applications	Rankings; Stud poker hands; Committees (when each member has a different job).	Subsets; Draw poker hands; Lottery tickets; coalitions.

[3] Most business and scientific calculators have built-in ${}_nP_r$ and ${}_nC_r$ keys. The standard sequence of keystrokes is to first enter the value of n, then push the ${}_nP_r$ (or ${}_nC_r$) key, then enter the value of r and finally press the equals key.

[4] This number is also known as a *binomial coefficient*.

Example 17. The California Lottery

To play the California Lottery, a person has to pick 6 out of 51 numbers (after paying $1 for the privilege). If the pigeon, er, person picked exactly the same 6 numbers as the ones drawn by the lottery, he or she can win mountains of money (usually a few million but it can be as much as 40 or 50 million). How many different lottery tickets are possible?

The key question is: Are lottery tickets ordered or unordered selections of 6 out of 51 numbers? In a lottery the order in which the numbers come up is irrelevant, so lottery draws are unordered. The rest is easy: The number of California Lottery ticket combinations is ${}_{51}C_6 = 18{,}009{,}460$. ■

We now leave the wonderful world of counting problems and return to the main theme of the chapter—calculating probabilities mathematically.

What Is a Probability?

Suppose that we toss a coin in the air. What is the probability that it will land heads? That is not a deep mathematical question, and almost everybody agrees on the answer, although not necessarily for the same reason. The standard answer given is 1/2 or (or 50%, or 1 out of 2). But why is the answer 1/2, and what does such an answer mean?

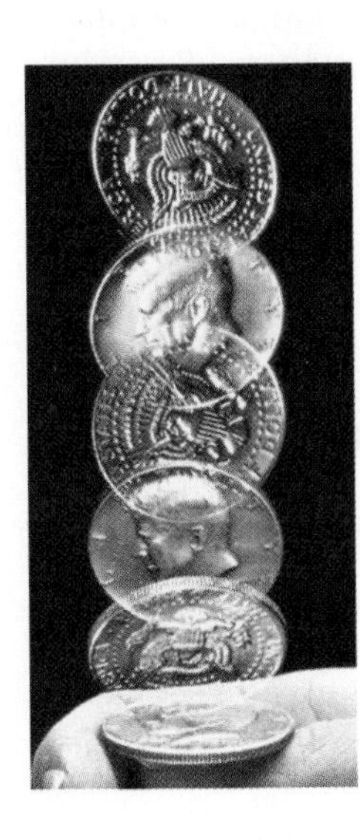

One common argument given for the probability of 1/2 is that when we toss a coin there are two possible outcomes (H and T), and since H represents one of the two possibilities, the probability of an H outcome must be 1 out of 2 or 1/2. This logic, while correct in the case of an honest coin, has a lot of holes in it.

Consider how the same argument would sound in a different scenario: We are in the last few seconds of an important basketball game and Michael Jordan is at the free-throw line, shooting a free throw that can decide the outcome of the game. Once again, there are two possible outcomes to the free throw (success or failure), but it would be absurd to conclude in this case that the probability of success is 1/2. The two outcomes, while both possible, are not both equally likely, and thus their probabilities should be different. What then is the probability that Michael makes the winning free throw? This is a much harder question to answer, and many people might argue that in fact it has no real answer. But, where there is a will there is a way. So we checked in the *Sports Almanac* and found out that over his professional career, Michael Jordan has been at the free-throw line more than 4000 times and has made 84% of the free throws he attempted. A reasonable answer to this question may be, then, that the probability that Michael Jordan makes the free throw is *approximately* .84.

This last argument leads us to a different way to think about probabilities and to make a different argument about the coin toss: The probability of an H when tossing an honest coin is 1/2 because, if we toss the coin over and over again, hundreds, possibly thousands of times, in the long run about half of the tosses will turn out to be heads and about half will turn out to be tails. We don't actually have to do it—we instinctively believe that this is true, and that's why we give the answer of 1/2.

The argument as to exactly how to interpret the statement "the probability of X is such and such" goes back to the late 1600s, and it wasn't until the 1930s that a formal mathematical theory of probabilities was developed by the Russian math-

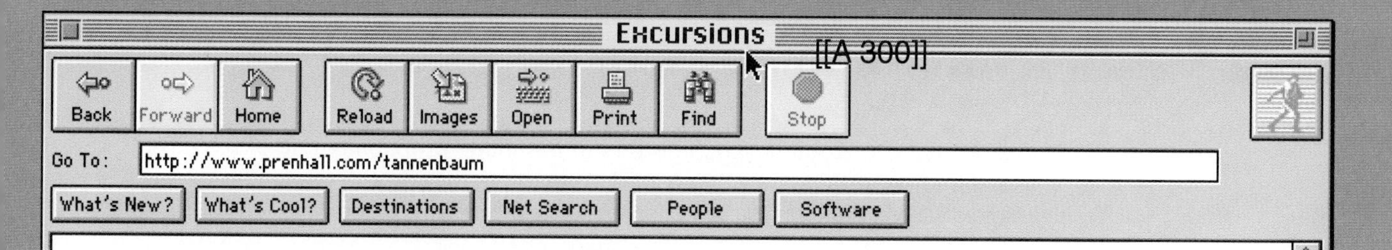

It's Your Birthday

Suppose there are 35 people in the room. What do you think the probability is that at least two people share the same birthday? The answer to such a question is not necessarily obvious. It would seem that the probability would be fairly small. One way to get an idea is to take a survey of groups of thirty-five. Another way would be to run a simulation. (We will assume an equal likelihood of each birthday and not include February 29.)

Questions:

1. Run the simulation ten times for 35 people and tabulate the results. What is the experimental probability based on the simulation?
2. Repeat the simulation for groups of 20, 25, 40, 45, and 50 people. Does the probability appear to increase rapidly or slowly?
3. Compute the theoretical probability directly using a graphing calculator or computer. (*Hint*: Compute the probability that no one has the same birthday.)
4. The simulation randomly generates one of 365 days for each person. How might you take into account a person born on February 29?
5. How would you use the simulation to estimate the probability that no one has the same birthday as you?

ematician A. N. Kolmogorov. This theory has made probability one of the most useful and important concepts in modern mathematics. In the remainder of this chapter we will develop the key concepts of Kolmogorov's theory.

Probability Spaces

Let's return to free-throw shooting—it is a useful metaphor for many probability questions.

Example 18. The Probability of Making a Free Throw

A person shoots a free throw. We know nothing about his or her abilities (for all we know, the person could be Michael Jordan or it could be Joe Schmoe). What is the probability that he or she will make the free throw?

It seems that there is no way to answer this question, since we know nothing about the shooter. One could argue that the probability could be just about any number, as long as it is not a negative number and it is not bigger than 1 (a perfect 100% free-throw shooter). We learned in high school algebra how to handle situations like this: we just make our unknown probability a variable—say p.

What can we say about the probability that our shooter misses the free throw? A lot. Since there are only two possible outcomes, their probabilities are related by the fact that they must add up to 1 (the free throw is either successful or unsuccessful—there are no other alternatives!). This means that the probability of missing the free throw must be $1 - p$.

Table 15-2 is a summary of the line on a generic free-throw shooter.

Table 15-2 A generic free-throw shooter

Outcome	Probability
Success (s)	p
Failure (f)	$1 - p$

Table 15-2, humble as it may seem, gives a complete model of free-throw shooting: It works when the free-throw shooter is Michael Jordan (make $p = .84$) or Joe Schmoe (make $p = .25$) or any other Tom, Dick, or Shaq in between. Each one of the choices results in a different assignment of numbers to the outcomes in the sample space. Such an assignment is called a *probability assignment.* The combination of the full description of the same space and the assignment of specific probabilities to the outcomes is called a **probability space**. ■

Example 19.

Five players, Boris, Steffi, Andre, Gabriela, and Monica enter a tennis tournament. We are interested in who is going to win the tournament. The sample space is S = {Boris, Steffi, Andre, Gabriela, Monica}. According to one expert, the probabilities of victory assigned to each of the players [denoted by Pr(*name*)] are Pr(Boris) = .25, Pr(Steffi) = .22, Pr(Andre) = .14, Pr(Gabriela) = .18. The value of Pr(Monica) is not given, but we can determine that Pr(Monica) = .21, because the total sum of the probabilities must be 1. ■

Examples 18 and 19 illustrate the concept of a probability assignment: A **probability assignment** for a sample space S is a set of numbers that is assigned to the outcomes in the sample space and that satisfies the following two conditions:

1. Each number in the set is between 0 and 1 (inclusive).
2. The numbers add up to 1.

Any set of numbers that satisfies conditions 1 and 2 is a legal probability assignment.

Events

So far we have talked about the probabilities of the individual outcomes in a sample space, specifically described by a probability assignment. We will also want to talk about the probabilities of events. An **event** is any subset of the sample space, in other words, a set of individual outcomes.

Example 20.

In Example 8 we considered the random experiment of tossing a coin 3 times and saw that the sample space was $S = \{HHH, HHT, HTH, HTT, THH, THT, TTH, TTT\}$. There are many possible events for this sample space. Table 15-3 shows just a few of them.

There are many ways of combining outcomes in a sample space to make an event, and the same event can be described (in English) in more than one way (e.g., events 2 and 4 in Table 15-3). The actual number of individual outcomes in an event can be as low as 0 and as high as N (the size of the sample space). In the case in which the number of outcomes is 0 (as in event 8 in Table 15-3), the event is called the **impossible event**; in the case in which the event is the whole sample space S (as in event 9 in Table 15-3), it is called the **certain event**.

Table 15-3 Some of the Many Possible Events in a Sample Space

Event	Set of Outcomes	Size of Event
1. Toss 2 or more heads	$\{HHT, HTH, THH, HHH\}$	4
2. Toss more than 2 heads	$\{HHH\}$	1
3. Toss 2 heads or less	$\{TTT, TTH, THT, HTT, THH, HTH, HHT\}$	7
4. Toss no tails	$\{HHH\}$	1
5. Toss exactly 1 tail	$\{HHT, HTH, THH\}$	3
6. Toss exactly 1 head	$\{HTT, THT, TTH\}$	3
7. First toss is heads	$\{HHH, HHT, HTH, HTT\}$	4
8. Toss same number of heads as tails	$\{\ \}$	0
9. Toss at most 3 heads	S	8
10. First toss is heads and at least 2 tails are tossed	$\{HTT\}$	1

■

Once a probability assignment is made on the sample space, *we can find the probability of any event by simply adding the probabilities of the individual outcomes that make up that event.* In addition, there are two special rules: the probability of the *impossible event* is always 0 [$\Pr(\{\ \}) = 0$] and the probability of the *certain event* is always 1 [$\Pr(S) = 1$].

At this point, it might be a good idea to summarize the different elements that make up a probability space.

The Elements of a Probability Space

1. A finite *sample space* $S = \{o_1, o_2, \ldots, o_N\}$ (the o's are the individual outcomes).
2. A *probability assignment* for S. To each individual outcome o_i we assign a number $\Pr(o_i)$. The two rules for a probability assignment are: $0 \leq \Pr(o_i) \leq 1$ and $\Pr(o_1) + \Pr(o_2) + \cdots + \Pr(o_N) = 1$.
3. *Events.* Any subset of S is an event. Two special events are $\{\ \}$ (called the *impossible* event) and S itself (called the *certain* event).
4. *Probabilities of events.* The probability of an event is obtained by adding the probabilities of the individual outcomes that make up the event. In particular, $\Pr(\{\ \}) = 0$ and $\Pr(S) = 1$.

Probability Spaces with Equally Likely Outcomes

An important special case of a probability space is the one in which every individual outcome has an equal probability assigned to it. This is the case when we toss an honest coin, or roll an honest die, or draw a card from a well-shuffled deck of cards.

When the probability of each individual outcome in the sample space is the same, then calculating probabilities becomes simply a matter of counting. For a sample space of size N, the probability of each individual outcome must be $1/N$ (remember that these probabilities must add up to 1), and the probability of an event is the number of outcomes in the event divided by N.

Computing Probabilities When All Outcomes Are Equally Likely

- Size of sample space = N.
- Pr(*individual outcome*) = $1/N$.
- If E is an event, $\Pr(E) = \dfrac{\text{number of outcomes in } E}{N}$.

Example 21. The Probability of Drawing an Ace

(a) The top card is drawn from a well-shuffled deck of 52 cards. What is the probability of drawing an ace?

Here $N = 52$, and the event

ace = {A♥, A♣, A♦, A♠}

is made up of 4 outcomes, each of which has probability 1/52. It follows that

$$\Pr(\text{ace}) = \frac{4}{52} = \frac{1}{13} \approx .077.$$

(b) Suppose now that we want to know the probability that the tenth card in the deck is an ace. Is it different than for the top card? When the deck is well shuffled, the aces can be anywhere in the deck—there is nothing about the top position that makes it special or different from the tenth position or for that matter any of the other positions in the deck. The probability of the tenth card in the deck being an ace is still 1/13. ■

Example 22.

Suppose that we roll a pair of honest dice. **(a)** What is the probability of rolling a total of 11? **(b)** What is the probability of rolling a total of 7? **(c)** What is the probability of rolling a total of 7 or 11?

Here $N = 36$ (see Example 5), and since the dice are honest, each of the 36 outcomes has probability 1/36.

(a) There are 2 ways of rolling a total of 11 ("roll 11" = {⚄⚅, ⚅⚄}). Thus,

$$\Pr(\text{"roll 11"}) = \frac{2}{36} = \frac{1}{18} \approx .056.$$

(b) There are 6 ways of rolling a total of 7:

"roll 7" = {⚀⚅, ⚁⚄, ⚂⚃, ⚃⚂, ⚄⚁, ⚅⚀}

$$\Pr(\text{"roll 7"}) = \frac{6}{36} = \frac{1}{6} \approx .167.$$

(c) The event "roll 7 or 11" has 8 possible outcomes (the 6 in "roll 7" and the 2 in "roll 11"), so

$$\Pr(\text{"roll 7 or 11"}) = \frac{8}{36} = \frac{2}{9} \approx .222.$$ ■

Example 23.

If we roll a pair of honest dice, what is the probability that at least one of them is a ⚀?

We know that each individual outcome in the sample space has probability of 1/36. We will show three different ways to solve this problem.

Solution 1 (the brute-force approach). If we just write down the event E, which is "we will roll at least one ⚀," we have

E = {⚀⚀, ⚀⚁, ⚀⚂, ⚀⚃, ⚀⚄, ⚀⚅, ⚁⚀, ⚂⚀, ⚃⚀, ⚄⚀, ⚅⚀}.

It follows that $\Pr(E) = 11/36$.

Solution 2 (the roundabout approach). Let's say for the sake of argument that we will win if at least one of the two dice comes up a ⚀ and we will lose otherwise. This means that we will lose if both dice come up with a number other than ⚀. Let's calculate first the probability that we will lose (this is called the roundabout way of doing things). Using the multiplication principle, we can calculate the number of individual outcomes in the event "we lose":

- Number of ways first die can come up (not a ⚀) = 5.
- Number of ways the second die can come up (not a ⚀) = 5.
- Total number of ways both dice can come up (neither a ⚀) = 5× 5 = 25.

$$\text{Probability that we will lose: } \Pr(\text{lose}) = \frac{25}{36}.$$

$$\text{Probability that we will win: } \Pr(\text{win}) = 1 - \frac{25}{36} = \frac{11}{36}.$$

Solution 3 (independent events). In this solution we consider each die separately. In fact, we will find it slightly more convenient to think of rolling a single honest die twice (mathematically it is exactly the same thing as rolling a pair of honest dice once).

Let's start with the first roll. The probability that we won't roll a ⚀ is 5/6 (there are 6 possible outcomes, 5 of which are not a ⚀). For the same reason, the probability that the second roll will not be a ⚀ is also 5/6.

Now comes a critical observation: The probability that neither of the first two rolls will be a ⚀ is $5/6 \times 5/6 = 25/36$. The reason we can multiply the probabilities of the two events ("first roll is not a ⚀" and "second roll is not a ⚀") is that these two events are **independent**: The outcome of the first roll does not in any way affect the outcome of the second roll.

We finish the problem exactly as in solution 2:

$$\Pr(\text{lose}) = \frac{25}{36} \quad \text{and therefore} \quad \Pr(\text{win}) = \frac{25}{36} = \frac{11}{36}.$$ ■

Of the three solutions to Example 23, solution 3 appears to be the most complicated, but in fact it shows us the most useful approach. It is based on what we will call the **multiplication principle for independent events**.

Independent events. Two events are said to be independent if the outcome of one event does not affect the outcome of the other.
The multiplication principle for independent events. When a complex event E can be broken down into a combination of two simpler events that are *independent* (call them F and G), then we can calculate the probability of E by multiplying the probabilities of F and G.

The multiplication principle for independent events in an important and useful rule, but it works only when the parts are independent. The next two examples illustrate the usefulness of this principle.

Example 24.

If we roll an honest die 4 times, what is the probability that at least once we will roll a ⚀?

Let's try the same approach we used in Example 23, solution 3. We will win if we roll a ⚀ at least once, and we will lose if none of the four rolls comes up ⚀. We know that

Pr(roll 1 not a ⚀) = 5/6.

Pr(roll 2 not a ⚀) = 5/6.

Pr(roll 3 not a ⚀) = 5/6.

Pr(roll 4 not a ⚀) = 5/6.

Because each roll is independent of the preceding ones, we can use the multiplication principle for independent events.

Pr(lose) = Pr(not rolling any ⚀'s in four rolls) = $(5/6)^4 \approx .482$.

It follows that:

Pr(win) = Pr(rolling at least one ⚀ in four rolls) $\approx .518$. ■

Example 25. The Probability of Four Aces

What is the probability of getting four aces in a 5-card draw poker hand? We computed the size of this sample space in Example 16 ($N = 2{,}598{,}960$). The event E = "draw four aces" has 48 different outcomes (four of the cards are aces, the fifth card can be any one of the 48 other cards). Thus,

$$\Pr(E) = \frac{48}{2{,}598{,}960} = \frac{1}{54{,}145}.$$

The probability of drawing 4 aces in a 5-card draw pokerhand is 1 in 54,145. ■

Example 26.

If we toss an honest coin 10 times, what is the probability of getting 5 H's and 5 T's? (This is an important question, and you might find the answer surprising—before you read on you are encouraged to make a rough guess.)

The size of the sample space when tossing 10 coins is $N = 2^{10} = 1024$. How many of the 1024 possible strings of 10 H's and T's have exactly 5 H's and 5 T's? To count these, we count the possible ways in which we can choose the 5 "slots" for the H's. These are unordered selections, and thus the answer is ${}_{10}C_5 = 252$. Thus, the possibility of tossing 5 H's and 5 T's is $252/1024 \approx .246$. ■

Odds

Dealing with probabilities as numbers that are always between 0 and 1 is the mathematician's way of having a consistent terminology. To the everyday user consistency is not that much of a concern, and we know that people talk about *chances* (probabilities expressed as percentages) and *odds,* which are most frequently used to describe probabilities associated with gambling situations. In this section we will briefly discuss how to interpret and calculate odds. To simplify our discussion we will consider only the situation in which all outcomes are equally likely.

Odds in favor of an event. The odds in favor of event E are given by the ratio of the number of ways event E can occur to the number of ways in which event E cannot occur.

Example 27.

If we roll a pair of honest dice, what are the odds in favor of rolling a total of 7?

We saw in Example 22(b) that of the 36 different outcomes that are possible when rolling a pair of dice, 6 are favorable (result in a total of 7) and the other 30 are unfavorable (result in a total that is not 7). It follows that the odds in favor of rolling a 7 are 6 to 30 or, equivalently, 1 to 5. ■

Example 28.

If we roll a pair of honest dice, what are the odds *against* rolling a total of 7?

This question is essentially the opposite of the one asked in Example 27. Of the 36 possible outcomes, 30 are favorable (result in a total that is not 7) and 6 are unfavorable (result in a total of 7). Thus, the odds against rolling a 7 are 30 to 6 or 5 to 1, the same numbers as in Example 27 but reversed. ■

Odds against an event. If the odds in favor of event E are m to n, then the odds against event E are n to m.

Sometimes we want to calculate the odds in favor of an event, but all we have to go on is the probability of that event.

Example 29.

When Michael Jordan shoots a free throw, the probability that he will make it is .84. In other words, on the average, out of every 100 free throws he attempts, he will make 84 and miss 16. It follows that the odds in favor of his making a free throw are 84 to 16 or, reduced to simplest form, 21 to 4. ■

The general rule for converting probabilities into odds is

If $\Pr(E) = a/b$, the odds in favor of E are a to $b - a$ and the *odds against* E are $b - a$ to a.

The general rule for converting odds to probabilities is

If the *odds* in *favor* of an event E are m to n, then $\Pr(E) = \dfrac{m}{m+n}$.

A word of caution: There is a difference between odds as discussed in this section, and the *payoff odds* posted by casinos or bookmakers in sports gambling situations. Suppose we read in the newspaper, for example, that the Las Vegas sports books have established that "the odds that the Chicago Bulls will win the NBA championship are 3 to 2." What this means is that if you want to bet in favor of the Bulls, for every \$2 that you bet, you can win \$3 if the Bulls win. This ratio may be taken as some indication of the actual odds in favor of the Bulls winning, but several other factors affect payoff odds, and the connection between payoff odds and actual odds is tenuous at best.

Conclusion

While the average citizen thinks of probabilities, chances, and odds as vague, informal concepts that are useful primarily when discussing the weather or playing the lottery, scientists and mathematicians think of probability as a formal framework within which the laws that govern chance events can be understood. The basic elements of this framework are a *sample space* (which represents a precise mathematical description of all the possible outcomes of a *random experiment*) and a *probability assignment* (which associates a numerical value to each of these outcomes).

Of the many ways in which probabilities can be assigned to outcomes, a particularly important case is the one in which all outcomes have the same probability. When this happens, the critical steps in calculating probabilities revolve around two basic (but not necessarily easy) questions: (1) given a sample space, what is its size? and (2) given an event, what is its size? To answer these kinds of questions, knowing how to "count" large sets is critical.

When one stops to think how much of life is ruled by fate and chance, the importance of probability theory in almost every area of modern society is hardly surprising. In this chapter we only scratched the surface of a deep and important mathematical theory.

Key Concepts

certain events
event
impossible event
independent events
multiplication rule
odds
probability assignment
probability space
multiplication principle for independent events
random experiment
sample space

Exercises

Walking

1. Consider the random experiment of tossing a coin 4 times.
   (a) Write out the sample space for this random experiment.
   (b) What is the size of the sample space?
   (c) Write out the event E: "exactly 2 of the coin tosses come out heads" as a set.
   (d) Assuming that the coin is honest, what is the probability of the event E in (c)?

2. Consider the random experiment of drawing 1 card out of an ordinary deck of 52 cards.
   (a) What is the size of the sample space?
   (b) Describe the event E: "the card drawn is a spade."
   (c) What is the probability of the event E described in (b)? (Assume the deck is well shuffled.)
   (d) Describe the event F: "the card drawn is a face card" (jack, queen, or king).
   (e) What is the probability of the event F described in (d)? (Assume the deck is well shuffled.)

3. There are 7 players (call them $P_1, P_2, \ldots, P_7$) entered in a tennis tournament. According to one expert, P_1 is twice as likely to win as any of the other players. and $P_2, P_3, \ldots, P_7$ all have an equal chance of winning.
   (a) Write down the sample space and find the probability assignment for the sample space based on this expert's opinion.
   (b) What are the odds that P_1 will win the tournament? How about P_2?

4. There are 8 players (call them $P_1, P_2, \ldots, P_8$) entered in a chess tournament. According to an expert P_1 has a 25% chance of winning the tournament, P_2 a 15% chance, P_3 a 5% chance of winning, and all the other players an equal chance.
   (a) Write down the sample space and find the probability assignment for the sample space based on this expert's opinion.
   (b) What are the odds that P_1 will win the tournament? How about P_2?

5. **(a)** How many license plates can be made using 3 letters followed by 3 digits (0 through 9)?
   (b) How many license plates can be made using 3 letters followed by 3 digits, if the first digit cannot be a zero?
   (c) How many license plates can be made using 3 letters followed by 3 digits, if no license plate can have a repeated letter or digit?

6. **(a)** How many 4-letter code words are there? (A code word is any string of letters—it doesn't have to mean anything.)
   (b) How many 4-letter code words are there that start with the letter A?
   (c) How many 4-letter code words are there that have no repeated letters?

7. A set of reference books consists of 8 volumes numbered 1 to 8.
   (a) In how many ways can the 8 books be arranged on a shelf?
   (b) In how many ways can the 8 books be arranged on a shelf so that at least 1 book is out of order?

8. Four men and 4 women line up at a checkout stand in a grocery store.
   (a) In how many ways can they line up?
   (b) In how many ways can they line up if the first person in line must be a woman?
   (c) In how many ways can they line up if they must alternate woman, man, woman, man, and so on. (A woman is first in line.)

9. A child's spinner has 5 outcomes; red, blue, yellow, purple, and orange. Experience shows that each primary color (red, blue, or yellow) comes up about 100 times in every 1000 spins and that each nonprimary color has an equal probability of occurring.

(a) What is the sample space for the random experiment consisting of a single spin?

(b) Write down the event that the outcome is a primary color.

(c) Write down the event that the outcome is not a primary color.

(d) Give a reasonable probability assignment for the sample space given in (a).

10. A child's spinner has 5 outcomes: 1, 2, 3, 4, and 5.

(a) What is the sample space for the random experiment consisting of a single spin?

(b) Write down the event that the outcome of a spin is an odd number.

(c) Write down the event that the outcome of a spin is an even number.

(d) If the probability of the event described in (b) is .3, find the probability of the event described in (c).

(e) If the probability of the event described in (b) is .3 and $\Pr(1) = \Pr(3) = \Pr(5)$ and $\Pr(2) = \Pr(4)$, find the probability assignment for the sample space.

11. Consider the sample space $S = \{A, B, C\}$. Make a list of all the possible events for this sample space. (Remember that an event is any subset of S including $\{\ \}$ and S itself.)

12. Consider the sample space $S = \{A, B, C, D\}$. Make a list of all the possible events for this sample space. (Remember than an event is any subset of S including $\{\ \}$ and S itself.)

In Exercises 13 through 18 a pair of honest dice are rolled and the points are added.

13. (a) What is the probability that the sum is 10?

(b) What is the probability of *not* rolling a 10?

(c) What are the odds in favor of rolling a 10?

(d) What are the odds against rolling a 10?

14. (a) What is the probability of rolling snake eyes (i.e., both dice coming up 1)?

(b) What are the odds in favor of rolling snake eyes?

(c) What is the probability of not rolling snake eyes?

(d) What are the odds against rolling snake eyes?

15. (a) Describe the event E: "the sum is less than 6."

(b) What is the probability of the event E in (a)?

(c) What are the odds in favor of the event in E in (a)?

16. (a) What is the probability of the event E: "at least one of the two dice comes up 1?"

(b) What are the odds in favor of the event E in (a)?

17. Suppose that if we roll a 7 we win and for any other total we lose. If we roll the dice twice:

(a) Find the probability that we win in both rolls.

(b) Find the probability that we lose in both rolls.

(c) Find the probability that we win once and lose once.

18. Suppose that if we roll any number less than 7 we win and for any other total we lose. If we roll the dice twice:

(a) Find the probability that we win in both rolls.

(b) Find the probability that we lose in both rolls.

(c) Find the probability that we win once and lose once.

19. **(a)** An honest coin is tossed three times. What is the probability of getting at least one H?

(b) An honest coin is tossed ten times. What is the probability of getting at least one H?

20. Assuming that boys and girls have an equal chance of being born, what is the probability that in a family of three children:

(a) All three will be girls?

(b) There will be 2 girls and 1 boy?

21. The Greens and the Browns each have 2 children. The Greens have at least 1 boy and the Browns' oldest child is a boy.

(a) What is the probability that the Greens have 2 boys? (*Hint*: Write out the sample space for the Greens' children.)

(b) What is the probability that the Browns have 2 boys? (*Hint*: Write out the sample space for the Browns' children.)

22. A box contains 25 red marbles and 50 white marbles.

(a) What is the probability that if 2 marbles are drawn at random from the box, both are red?

(b) What is the probability that if 2 marbles are drawn at random from the box, both are white?

(c) What is the probability that if 2 marbles are drawn at random from the box, they are of different colors?

23. A committee of 3 people (chair, vice-chair, and treasurer) must be chosen from a board consisting of 12 people. In how many ways can this be done?

Exercises 24 through 28 refer to a club that has 15 members. A delegation of 4 members (all of equal standing) must be chosen to represent the club at a convention.

24. How many different 4-person delegations are possible?

25. How many different 4-person delegations are there that include Mary?

26. How many different 4-person delegations are there that do not include Mary?

27. Anna and Bob are a pair, and each has refused to be in the delegation unless the other one is. How many different 4-person delegations are possible if Anna and Bob refuse to be separated?

28. Carol and Luis are not speaking to each other, and each refuses to be in the delegation if the other one is. How many 4-person delegations are possible if Carol and Luis will not serve on the same delegation?

Jogging

29. If we toss an honest coin 12 times, what is the probability of

(a) Getting 5 H's and 7 T's?

(b) Getting 6 H's and 6 T's?

(c) Getting 7 H's and 5 T's?

Exercises 30 through 34 refer to 5-card draw poker hands [see Example 16(b)].

30. What is the probability of getting "4 of a kind" (4 cards of the same value)?

31. What is the probability of getting all 5 cards of the same color?

32. What is the probability of getting a "flush" (all 5 cards of the same suit)?

33. What is the probability of getting an "ace-high straight" (10, J, Q, K, A of any suit but not all of the same suit)?

34. What is the probability of getting a "full house" (3 cards of equal value and 2 other cards of equal value)?

35. A study group of 10 students is to be split into 2 groups of 5 students each. In how many ways can this be done?

36. Eight points are taken on a circle.

(a) How many chords can be drawn by joining all possible pairs of the points?

(b) How many triangles can be made using these points as vertices?

37. Dolores wants to walk from point A to point B (a total of 6 blocks) as shown on the street map. Assuming that she always walks toward B, how many different ways can she take this walk?

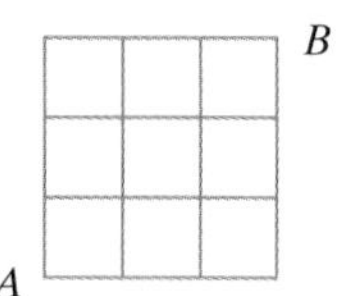

38. Consider the following game. We roll a pair of honest dice 25 times. If we roll "boxcars" (i.e., ⚅⚅) at least once, we will win; otherwise we will lose. What is the probability that we will win?

39. Consider the following game: We roll a pair of honest dice 5 times. If we roll a total of 7 at least once, we will win; otherwise we will lose. What is the probability that we will win?

40. A factory assembles car stereos. From random testing at the factory it is known that on the average 1 out of every 50 car stereos will be defective (which means that the probability that a car stereo randomly chosen from the assembly line will be defective is .02.) After manufacture, car stereos are packaged in boxes of 12 for delivery to the stores.

(a) What is the probability that in a box of 12 there are no defective car stereos? What assumptions are you making?

(b) What is the probability that in a box of 12 there is at most 1 defective car stereo?

41. In the game of craps, a pair of dice is rolled. A person betting "on the field" is betting that on the next roll the total rolled will be a 2, 3, 4, 9, 10, 11, or 12. What is the probability of winning a field bet?

42. A pizza parlor offers 6 toppings—pepperoni, Canadian bacon, sausage, mushroom, anchovies, and olives—that can be put on their basic cheese pizza. How many different pizzas can be made? (A pizza can have anywhere from no toppings to all 6 toppings.)

43. **(a)** In how many different ways can 10 people form a line?

(b) In how many different ways can 10 people hold hands and form a circle? [*Hint*: The answer to (b) is much smaller than the answer to (a). There are many different ways in which the same circle of 10 people can be broken up to form a line. How many?]

44. Two teams (call them X and Y) play against each other in the World Series. The World Series is a best-of-7 series. This means that the first team to win 4 games wins the series and the series is over. (Games cannot end in a tie.) We can describe an outcome for the World Series by writing a string of letters that indicate (in order) the winner of each game. For example, the string $XYXXYX$ represents the outcome: X wins game 1, Y wins game 2, X wins game 3, and so on.

(a) Using the notation described above, write the sample space S for the World Series.

(b) Describe the event "X wins in 5 games."

(c) Describe the event "the series lasts 7 games."

Running

45. If an honest coin is tossed N times, what is the probability of getting the same number of H's as T's? (*Hint*: Consider two cases—N even and N odd.)

46. How many different "words" (they don't have to mean anything) can be formed using all the letters in

(a) The word PARSLEY. (*Note*: This one is easy!)

(b) The word PEPPER. (*Note*: This one is much harder!) [*Hint*: Think about the difference between (a) and (b).]

47. In the game of craps, the player's first roll of a pair of dice is very important. If the first roll is 7 or 11, the player wins. If the first roll is 2, 3, or 12, the player loses. If the first roll is any other number (4, 5, 6, 8, 9, 10), this number is called the player's "point." The player then continues to roll until either the point reappears, in which case the player wins, or a 7 shows up before the point, in which case the player loses. What is the probability that the player will win? (Assume the dice are honest.)

48. The birthday problem. There are 30 people in a room. What is the probability that at least 2 of these people have the same birthday—that is, have their birthdays on the same day and month?

49. (Open-ended question) You have just been chosen to be on a game show in which there are 3 doors, behind 1 of which is a new sports car. There is nothing behind the other 2 doors. You know in advance that the game show host will ask you to pick 1 of the doors, but before the door you picked is opened, the host will open 1 of the other 2 doors with nothing behind it. You will then be given the option of keeping the door you picked or switching to the other closed door. Discuss why the best strategy is to always switch your choice to the other closed door.

50. (Open-ended question) A large box contains 100 tickets. Each ticket has a different number written on it. Other than that the numbers are different, we know nothing about them. The numbers can be positive or negative, integers or decimals, large or small—anything goes. Of all the tickets in the box there is, of course, one that bears the biggest of all the numbers. That's the winning ticket. If we turn that ticket in, we will win a $1000 prize. If we turn any other ticket in, we will get nothing. The ground rules are that we can draw a ticket out of the box, look at it, and, if we think it's the winning ticket, turn it in. If we don't, we get to draw again, but first we must tear the other ticket up—once we pass on a ticket, we can't use it again! We can continue drawing tickets this way until we find one we like or run out of tickets.

(a) What is the probability that the first ticket we draw will be the winning ticket?

(b) What is the probability that after we have drawn 50 tickets, the winning ticket will still be in the box?

(c) Describe a strategy for playing this game that will give a better than 25% chance of winning.

References and Further Readings

1. Bernstein, Peter, *Against the Odds: The Remarkable Story of Risk.* New York: John Wiley & Sons, 1996.
2. di Finetti, B., *Theory of Probability.* New York: John Wiley & Sons, Inc., 1970.
3. Epstein, Richard A., *The Theory of Gambling and Statistical Logic.* San Diego, CA: Academic Press, 1995
4. Gnedenko, B. V., and A. Y. Khinchin, *An Elementary Introduction to the Theory of Probability.* New York: Dover Publications, Inc., 1962.
5. Keynes, John M., *A Treatise of Probability.* New York: Harper and Row, 1962.
6. Krantz, Les, *What the Odds Are.* New York: HarperCollins Publishers, Inc., 1992.
7. Levinson, Horace, *Chance, Luck and Statistics: The Science of Chance.* New York: Dover Publications, Inc., 1963.
8. McGervey, John D., *Probabilities in Everyday Life.* New York: Ivy Books, 1986.
9. Mosteller, Frederick, *Fifty Challenging Problems in Probability.* New York: Dover Publications, Inc., 1965.
10. Mosteller, F., R. Rourke, and G. Thomas, *Probability and Statistics.* Reading, MA: Addison-Wesley Publishing, 1961.
11. Packel, Edward, *The Mathematics of Games and Gambling.* Washington DC: Mathematical Association of America, 1981.
12. Weaver, Warren, *Lady Luck: The Theory of Probability.* New York: Dover Publications, Inc., 1963.

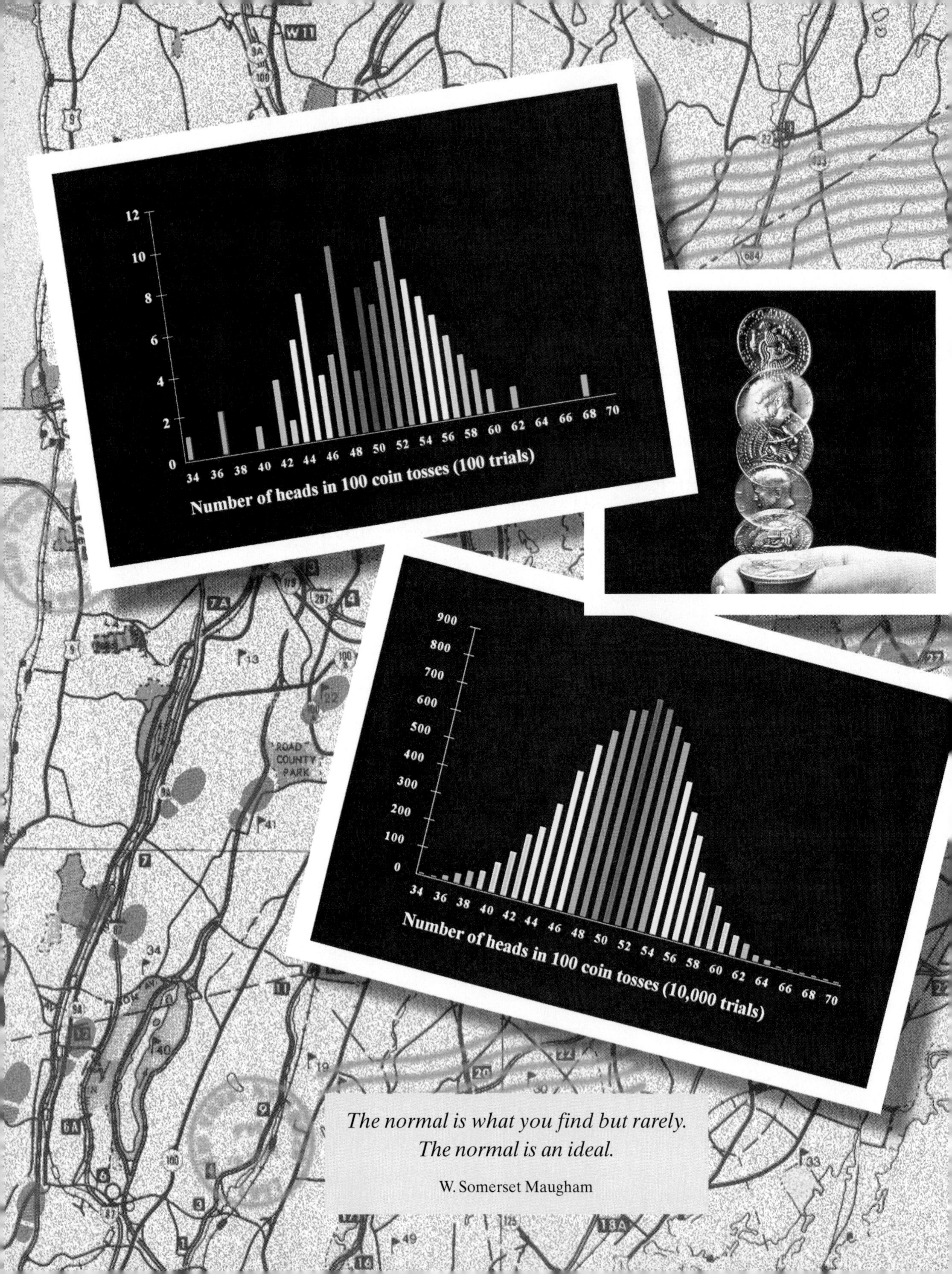

The normal is what you find but rarely.
The normal is an ideal.

W. Somerset Maugham

Normal Distributions

Everything Is Back to Normal (Almost)

CHAPTER 16

What does a scientist do when he has nothing but time on his hands? Some 60 years ago, the South African mathematician John Kerrich spent five years as a German prisoner of war. To pass the time, Kerrich decided to try a coin-tossing experiment. He tossed a coin 100 times and recorded the number of heads. He got 44 heads. He decided to do it again. The second time he got 54 heads. Undaunted, he repeated his coin-tossing experiment (tossing 100 times and recording the number of heads) again, and again. By the time he was done, he had tossed the coin 10,000 times and had meticulous records of the number of heads in every 100 tosses. The bar graph at the top left of the opposite page shows the result of Kerrich's experiment, displaying the head counts for 100 sets of 100 tosses.[1] The picture does not show anything to get particularly excited about—amazingly, Kerrich had quit his coin tossing experiment too soon! Had Kerrich continued tossing his coin just a little longer he would have come across a truly remarkable picture. With the aid of a computer we reproduced and extended Kerrich's experiment, but this time "we" repeated the 100 "coin tosses" *ten thousand times.* (It is easy to have a computer toss a pretend coin, and the results are just as valid as those we would get tossing a real coin. It is also a lot easier on the thumb!) What we got is the bar graph at the bottom of the opposite page—a smooth, consistent, beautiful bell-shaped graph!

---

[1] Source: John Kerrich, *An Experimental Introduction to the Theory of Probability,* 1964.

Bell-shaped patterns of data show up in more than just coin-tossing experiments—they are pervasive throughout the natural world. For a homogeneous population, a surprising number of measurements—heights, weights, IQ's, test scores—consistently fit a bell-shaped pattern (when the population is large enough). In addition, many other random phenomena besides coin tossing follow mathematical laws that guarantee that after enough repetitions the patterns that will emerge are bell shaped.

More than any other type of regular pattern, bell-shaped patterns rule the statistical world. The purpose of this chapter is to gain an understanding of these patterns and how they can be used to draw inferences about the way things are, and the way things ought to be.

Approximately Normal Distributions of Data

We start with a pair of examples.

Example 1. Heights of NBA Players

Table 16-1 shows a frequency table for the heights of National Basketball Association players listed on team rosters at the start of the 1996–1997 season.

Table 16-1 Heights of NBA Players 1996–1997
Source: *National Basketball Association (http://www.nba.com)*

Height	Frequency	Height	Frequency	Height	Frequency	Height	Frequency
5-3	1	6-3	22	6-8	44	7-2	5
5-10	2	6-4	24	6-9	45	7-3	2
5-11	6	6-5	25	6-10	44	7-4	2
6-0	9	6-6	29	6-11	30	7-6	1
6-1	13	6-7	39	7-0	21	7-7	1
6-2	15			7-1	6		

A corresponding bar graph for the data is shown in Fig. 16-1. One can see a distinctive bell-shaped pattern to the bar graph, with a mathematical idealization of what a perfect bell distribution would be like (the red curve) superimposed on the bar graph. We can see that the fit isn't perfect, with a few too many players in the 6 ft 8 in. to 7 ft range,[2] but it is still a reasonably bell-shaped pattern. The bar graph also shows an obvious outlier: Tyrone "Muggsy" Bogues, who, at 5 ft 3 in. is not only the shortest player in the NBA by a wide margin, but also one of the great success stories of our time.

Example 2. 1996 SAT Scores (Verbal)

Table 16-2 is a relative-frequency table for scores on the 1996 SAT examination (verbal). The scores range from 200 to 800 and are grouped in class intervals of 50 points. The population for this data are 1996 college-bound seniors, and the size of the data

[2] The excessive number of players in the 6 ft 8 in. to 7 ft range is a recent phenomenon—a consequence of modern NBA playing styles and not a quirk of nature.

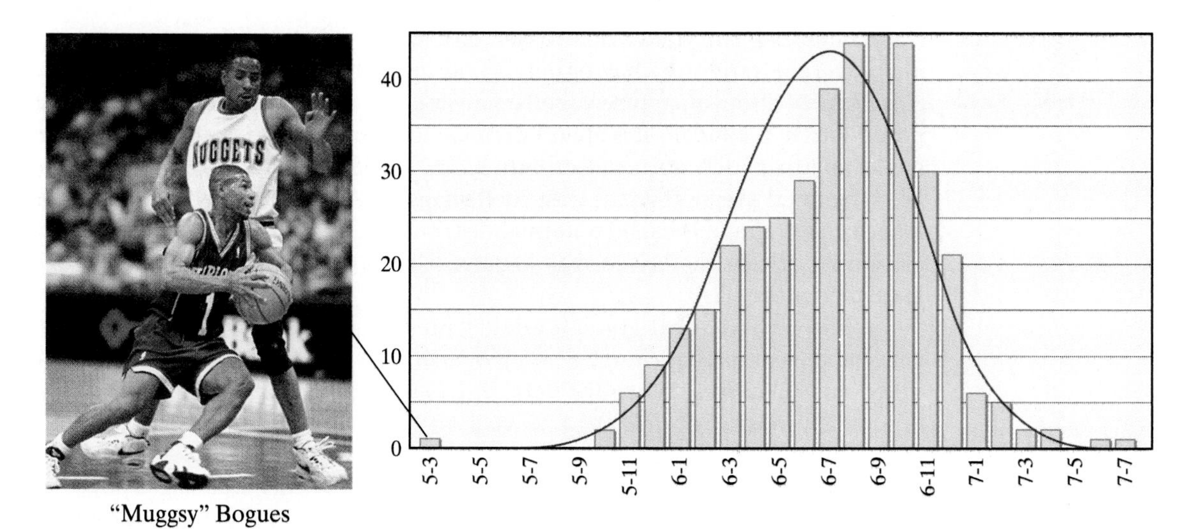

"Muggsy" Bogues

FIGURE 16-1
Heights of NBA players (1996–1997 season). [Source: National Basketball Association (http://www.nba.com)]

Table 16-2 1996 SAT Scores (Verbal). $N = 1{,}084{,}725$. (Source: The College Board National Report, 1996)

Score Ranges	Frequency	Percentage
750–800	16,857	1.6%
700–740	30,503	2.8%
650–690	66,066	6.1%
600–640	115,401	10.6%
550–590	158,138	14.6%
500–540	187,773	17.3%
450–490	193,470	17.8%
400–440	143,630	13.2%
350–390	94,037	8.7%
300–340	46,583	4.3%
250–290	20,874	1.9%
200–240	11,393	1.1%

set is $N = 1{,}084{,}725$. A bar graph for the data is shown in Fig. 16-2. Once again, a smooth bell-shaped curve is superimposed on the bar graph, showing a mathematical idealization of the data. In this example the data fit the curve quite well.

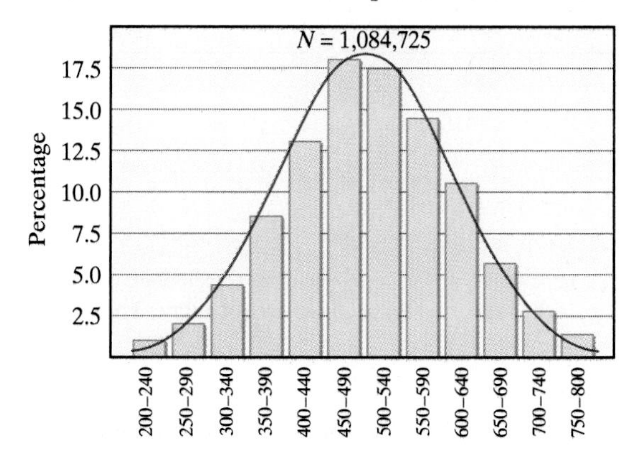

FIGURE 16-2
Test scores SAT-Verbal (1996)

■

Examples 1 and 2 illustrate two very different situations (different populations and different variables) having one thing in common: Both data sets can be described as fitting an approximately bell-shaped pattern. In Example 1 the fit is borderline; in Example 2 it is almost perfect. In either case, we say that the data set has an **approximately normal distribution**. The world "normal" in this context is to be interpreted as synonymous with "bell-shaped." A distribution of data that has a perfect bell shape is called a **normal distribution**. Real-world bell-shaped data are always approximately normal—sometimes, as we have seen, more approximate than others.

When we have a bar graph for data with a normal distribution, we can connect the tops of the bars into a smooth bell-shaped curve. Perfect bell curves are called **normal curves**. When the data set has an approximately normal distribution, an appropriate normal curve (such as the ones shown in red in Figs. 16-1 and 16-2) represents an idealization of the data (what things would look like in a perfect world). This is not wishful thinking—it is mathematical modeling, a powerful tool for understanding and describing the data. Thus, to fully understand real-world data sets with an approximately normal distribution, we first need to learn some of the mathematical properties of normal curves.[3]

Normal Curves and Their Properties

In this section we will briefly discuss some of the important properties of normal curves and then use these properties to analyze and describe data sets with an approximately normal distribution.

Normal curves (Fig. 16-3) can come in many different "looks," even as they all have the same shape: Some are tall and skinny, others are short and squat, others fall somewhere in between. Mathematically speaking, however, they are essentially all the same. In fact, whether a normal curve is skinny and tall or short and squat or somewhere in between depends on the way we scale the units on the axes. With the proper choice of scale on the axes, any two normal curves can be made to look the same.

What follows is a summary of some of the essential facts about normal curves. These facts are going to help us greatly later on in the chapter.

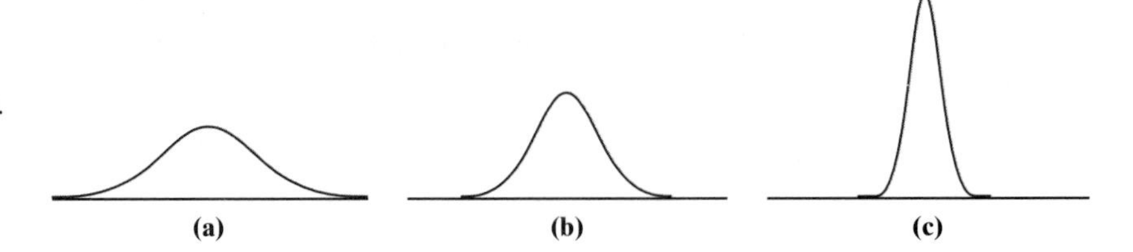

FIGURE 16-3
Three normal curves.
(a) Short and squat.
(b) In between.
(c) Tall and skinny.

[3] The study of normal curves can be traced back to work of the great German mathematician Karl Friedrich Gauss, and these curves are sometimes known as *Gaussian curves.*

1. Symmetry. Every normal curve is symmetric about a vertical axis (Fig. 16-4). The axis of symmetry splits the bell-shaped region outlined by the curve into two identical halves.

2. Mean = Median = Center. The point where the vertical axis of reflection symmetry cuts the horizontal (data) axis is called the **center** of the normal distribu-

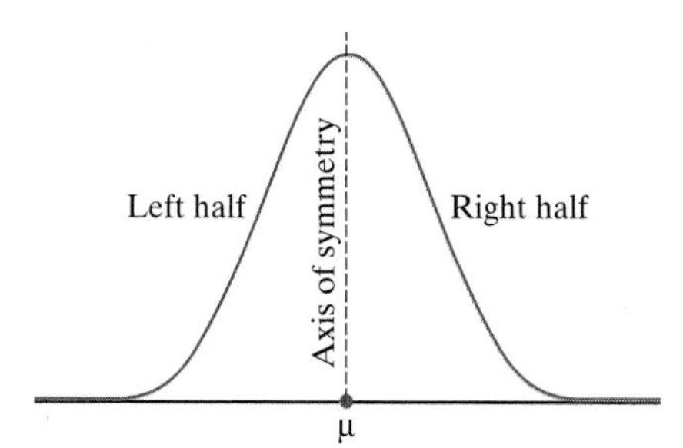

FIGURE 16-4 A normal curve has one axis of reflection symmetry. μ is the center (median and mean of the data.

tion. The data value corresponding to this point is both the *median* and the *mean* (*average*) of the distribution. Thus, for a normal distribution, the mean always equals the median, and both are denoted by the same symbol—the Greek letter μ (mu).

One consequence is that for a population that follows a normal distribution, half of the population measurements are smaller than or equal to the mean and half are bigger than or equal to the mean. For a population with an approximately normal distribution, all we can predict is that the mean and the median are close.

3. Standard Deviation. We discussed the standard deviation—traditionally denoted by the Greek letter σ (sigma)—of a data set in Chapter 14. The standard deviation is an important measure of spread in general, but it is particularly important when dealing with normal (or approximately normal) distributions. In these cases, we use the standard deviation as a yardstick with which to measure "distances" between data values, or the position of a data value relative to the center (mean, median) of the data.

For example, suppose that a normal distribution has a standard deviation of 10 ($\sigma = 10$). Then we could describe the data values 32 and 42 as being "1 standard deviation apart" and the data values 20 and 50 as being "3 standard deviations apart." We could even describe the data values 53 and 67 along the same lines—they would be "1.4 standard deviations apart" (Fig. 16-5).

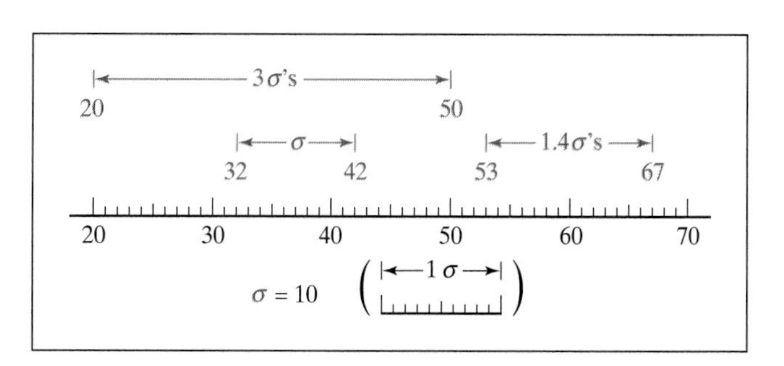

FIGURE 16-5 Measuring distances using the standard deviation σ as a yardstick.

Along similar lines, it is a common practice to **standardize** the values of the data by measuring how many standard deviations a data value is from the mean. For example, suppose the mean of the data set is 45, with the standard deviation still 10. Then, the data value 55 is said to be "one standard deviation *above* the mean" or, equivalently, to have a "standardized value of 1." For a data value like 35, which is "one standard deviation *below* the mean," we say that it has a "standardized value of -1." Likewise, the data value 50 has a standardized value of 0.5 and the data value 20 has a standardized value of -2.5 (Fig. 16-6).

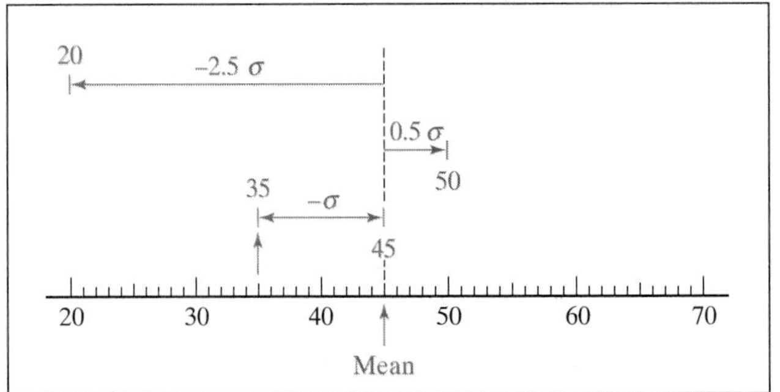

FIGURE 16-6 Standardized values are given by number of standard deviations above (positive) or below (negative) the mean.

How does one find the standard deviation of a data set with a normal distribution? The easiest way is to describe the standard deviation of the corresponding normal curve in geometric terms. Pretend you want to bend a piece of wire into the shape of a normal curve. At the very top you must bend the wire downward [see Fig. 16-7(a)], and at the bottom you must bend the wire upward [Fig. 16-7(b)]. As we move our hands shaping the wire, the curvature gradually changes, and there is one point on each side of the curve where the transition from being bent downward to being bent upward takes place. Such a point [P in Fig. 16-7(c)] is called an **inflection point** of the curve. Every normal curve has two inflection points (P and P' as shown in Fig. 16-8), and *the horizontal distance between the axis of the curve and either of these points is the standard deviation.*

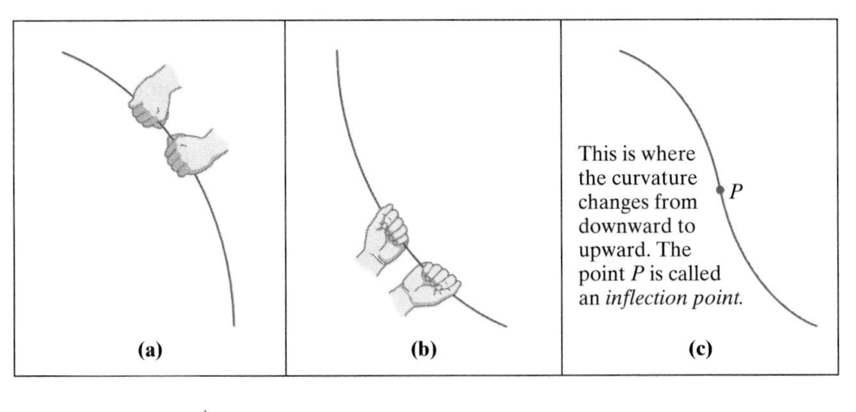

FIGURE 16-7 (a) At the top, the wire has "downward" curvature, (b) at the bottom, the wire has "upward" curvature, (c) at P the transition takes place.

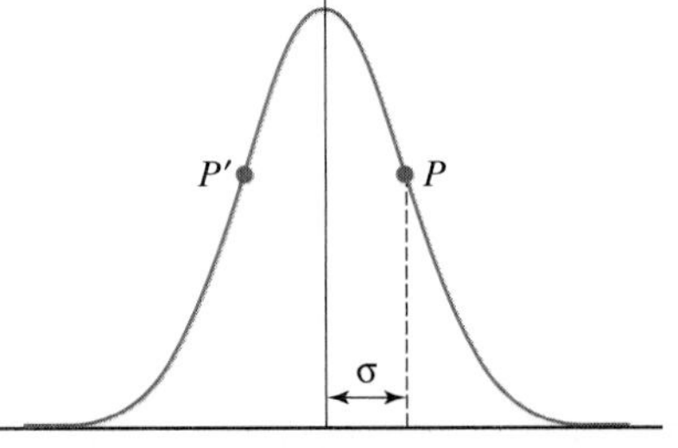

FIGURE 16-8 The horizontal distance between an inflection point and the axis of symmetry equals the standard deviation (σ).

4. The 68-95-99.7 Rule. For any data set with a normal distribution the following are standardized facts:

- **Standardized values between −1 and 1.** Data whose standardized values are between −1 and 1 represent 68% of all the data. (In other words, 68% of all the data falls within one standard deviation below and above the mean.) The remaining 32% of the data have a standardized value bigger than 1 or smaller than −1. By symmetry, there is an equal amount of each [Fig. 16-9(a)].

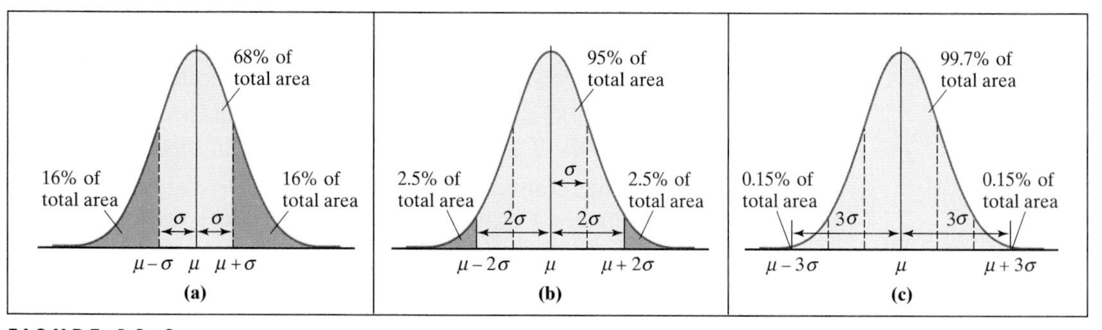

FIGURE 16-9
The 68-95-99.7 Rule.

- **Standardized values between −2 and 2.** Data whose standardized values are between −2 and 2 represent 95% of all the data [Fig. 16-9(b)]. The remaining 5% of the data is divided equally between data with standardized values bigger than 2 and smaller than −2.
- **Standardized values between −3 and 3.** Data whose standardized values are between −3 and 3 represent 99.7% (which is practically 100%) of all the data [Fig 16-9(c)]. The practical consequence is that essentially all the data can be considered to fall within 3 standard deviations above and below the mean, and thus, *one can estimate the range of a data set with an approximately normal distribution to be 6 standard deviations.*

As a practical fact, the 68-95-99.7 rule is one of the most useful properties of normal distributions. Figure 16-9 illustrates its details, and the reader is advised to study it carefully.

5. Quartiles. For every normal distribution, the third quartile has a standardized value of 0.675,[4] and by symmetry it follows that the first quartile has a standardized value of −0.675. These two important facts can be restated by the two formulas given below:

$$Q_3 \approx \mu + (0.675)\sigma,$$

$$Q_1 \approx \mu - (0.675)\sigma.$$

Example 3.

Let's suppose that we have to analyze a data set having a normal distribution with mean $\mu = 505$ and standard deviation $\sigma = 110$. Here are all the things we can now say about the data:

---

[4] For the purposes of memorization, one might think of 0.675 as being approximately 2/3 = 0.666 . . . and estimate the first and third quartiles as having standardized values of −2/3 and 2/3, respectively.

- The median is 505. Thus, we know that half of the data are smaller than or equal to 505 and half are bigger than or equal to 505.
- The first quartile is $Q_1 \approx 505 - 0.675 \times 110 = 430.75$. This means that one-fourth of the data are smaller than or equal to 430.75 and another one-fourth fall between 430.75 and 505.
- The third quartile is $Q_3 \approx 505 + 0.675 \times 110 = 579.25$. This means that one-fourth of the data fall between 505 and 579.25, and one-fourth of the data are bigger than or equal to 579.25.
- Data values between 395 and 615 correspond to standardized values between -1 and 1. By the 68-95-99.7 rule, 68% of the data fall in this range.
- Data values between 285 and 725 correspond to standardized values between -2 and 2. By the 68-95-99.7 rule, 95% of the data fall in this range.
- 99.7% of the data has standardized values between -3 and 3, which correspond to values between 175 and 835. For most practical purposes, this can be considered all of the data. ■

Normal Curves As Models of Real-Life Data Sets

The reason we like to idealize a real-life, approximately normal data set by means of a normal distribution is that we can use many of the properties we just learned about normal curves to draw useful conclusions about our data. For example, the 68-95-99.7 rule for normal curves can be reinterpreted in the context of an approximately normal data set as follows:

1. About 68% of the data values fall within (plus or minus) 1 standard deviation from the mean.
2. About 95% of the data values fall within (plus or minus) 2 standard deviations from the mean.
3. About 99.7% of the data values fall within (plus or minus) 3 standard deviations from the mean.

Example 4. Analyzing SAT Scores

In 1996, a total of 1,084,725 college-bound high school seniors took the SAT. (Let's call it 1 million to keep it simple.) The scores in the verbal part fit an approximately normal distribution with mean of 505 and standard deviation of 110 points.[5]

Without even looking at the data (which is given in Table 16-2, Example 2), we can estimate the median and the quartiles.

- Since the median should be about the same as the mean, we can estimate that the median score was 505. The actual median score was 510. Thus, about half of the million or so students taking the SAT in 1996 scored 510 points or less on the verbal part.
- The first quartile should be approximately $0.675 \times 110 = 74.25$ points below the mean of 505. Rounding 74.25 to 75 gives us an estimate of 430 points for the first quartile. This turns out to be right on the money—the actual value

[5] *Source:* The College Board, National Report, 1996.

of the first quartile was 430 points. This tells us that about 250,000 students taking the SAT in 1996 scored 430 points or less on the verbal part.

- The third quartile should be approximately $505 + 75 = 580$ points ($0.675 \times 110 \approx 75$). Once again, this estimate is exactly right—the third quartile was indeed 580 points. Thus, about 750,000 students scored 580 points or less.

Additional information about the distribution of scores can be obtained using the 68-95-99.7 rule. To wit:

- The percentage of students scoring within 1 standard deviation of the mean should be about 68%. In this case that means scores between $505 - 110 = 395$ and $505 + 110 = 615$ points. Since SAT scores can only come in multiples of 10, this really means scores between 400 and 610.
- The percentage of students scoring within 2 standard deviations of the mean should be about 95%. In this case that means scores between $505 - 220 = 285$ and $505 + 220 = 725$ points. Since SAT scores can come only in multiples of 10, this really means scores between 290 and 720.
- The 99.7 part of the 68-95-99.7 rule is not much help in this example. Essentially, it would tell us that approximately 99.7% of the students had test scores between 175 and 835, which does not tell us anything useful, since everybody's score has to fall between 200 and 800. ■

Normal Distributions of Random Events

We are now ready to take up another important aspect of normal curves—their connection with random events and, through that, their critical role in margins of error of public opinion polls. Our starting point is the following important example.

Example 5. A coin-tossing experiment

In opening this chapter we discussed the coin-tossing experiments performed by John Kerrich while he was a prisoner of war during World War II. Kerrich tossed a coin 10,000 times and kept records of the number of heads in groups of 100 tosses.

With modern technology, one can repeat Kerrich's experiment and take it much further. Practically any computer can imitate the tossing of a coin by means of a random-number generator. If we use this technique, it isn't hard to "toss" the coin many (hundreds, thousands, millions) of times.

We will start modestly. We will toss our make-believe coin 100 times and count the number of heads, which we will denote by X. Before we do that, let's say a few words about X. Since we cannot predict ahead of time its exact value—we are tempted to think that it should be 50, but in principle it could be anything from 0 to 100—we call X a **random variable**. The possible values of the random variable X are governed by the laws of probability: some values of X are extremely unlikely ($X = 0$, $X = 100$), others are much more likely ($X = 50$), although the likelihood of $X = 50$ is not as great as one would think. It also seems reasonable that (assuming the coin is fair and heads and tails are equally likely) the likelihood of $X = 49$ should be the same as the likelihood of $X = 51$; the likelihood of $X = 48$ should be the same as the likelihood of $X = 52$, and so on.

While all of the above statements are true, we still don't have a clue as to what is going to happen when we toss the coin 100 times. One way to get a sense of the probabilities of the different values of X is to repeat the experiment a few times and check the frequencies of the various outcomes. Finally, we are ready to do some experimenting!

Our first trial results in 46 heads out of 100 tosses ($X = 46$). The first 10 trials give in order $X = 46, 49, 51, 53, 49, 52, 47, 46, 53, 49$. Figure 16-10(a) shows a bar graph for this data.

Continuing this way, we collect data for the values of X in 100, 500, 1000, 5000, and 10,000 trials. The bar graphs are shown in Figs. 16-10(b) through (f), respectively.

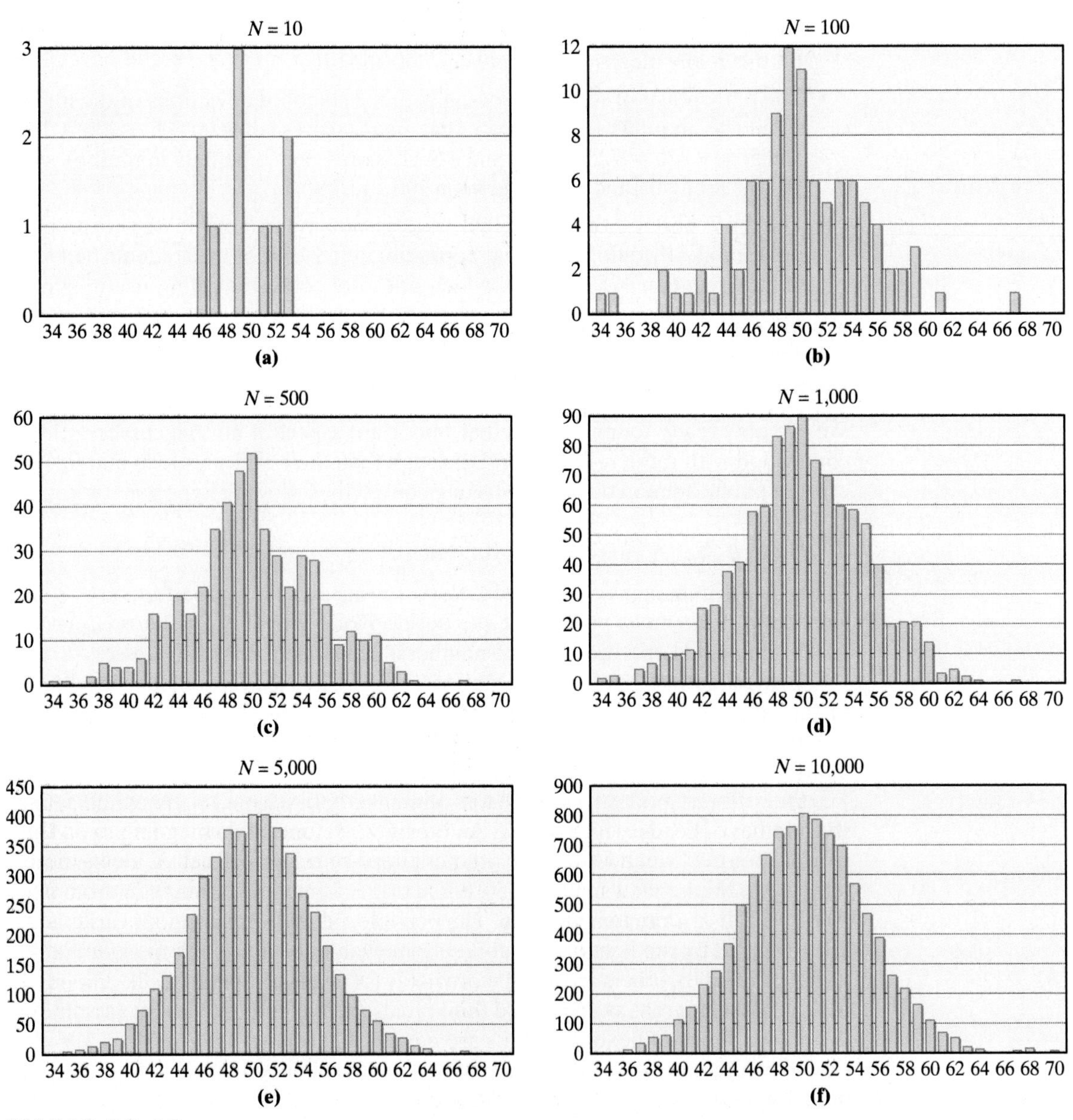

FIGURE 16-10
Distribution of random variable X (number of heads in 100 coin tosses). (a) 10 times; (b) 100 times; (c) 500 times; (d) 1000 times; (e) 5000 times; (f) 10,000 times.

Figure 16-10 paints a pretty clear picture of what happens: As the number of trials increases, the distribution of the data becomes more and more bell shaped. At the end we have data from 10,000 trials, and the bar graph gives an almost perfect normal distribution!

What would happen if someone else decided to repeat what we did—toss an honest coin (be it by hand or by computer) 100 times, count the number of heads, and repeat this experiment a few times? The first 10 trials are likely to produce results very different from ours, but as the number of trials increases, the results will begin to look more and more alike. After 10,000 trials, the corresponding bar graph will be almost identical to the bar graph in Fig 16-10(f). In a sense, this says that doing the experiments a second time is a total waste of time—in fact, it was even a waste the first time! *Everything that happened at the end could have been predicted without ever tossing a coin!*

Knowing that the random variable X has an approximately normal distribution is, as we have seen, quite useful. The clincher would be to find out the values of the mean μ and the standard deviation σ of this distribution. Looking at Fig. 16-10(f), we can pretty much see where the mean is—right at 50. This is not surprising, since the axis of symmetry of the distribution has to pass through 50 as a simple consequence of the fact that the coin is an honest coin. The value of the standard deviation is less obvious: For now, let's accept the fact that it is $\sigma = 5$. We will explain how we got this value shortly.

Let's summarize what we now know: An honest coin is tossed 100 times. The number of heads in the 100 tosses is a random variable, which we call X. If we repeat this experiment a large number of times (call it N), the random variable X will have an approximately normal distribution with mean $\mu = 50$ and standard deviation $\sigma = 5$, and the larger the value of N, the better this approximation will be.

The real significance of these facts is that they are true not because we took the trouble to toss a coin a million times. Even if we did not toss a coin at all, all of the above would still be true: *For a sufficiently large number of repetitions of the experiment of tossing an honest coin 100 times, the number of heads X is a random variable that has an approximately normal distribution with center $\mu = 50$ heads and standard deviation $\sigma = 5$ heads.* This is a mathematical, rather than an experimental fact. ■

Statistical Inference

Next we are going to take our first tentative leap into statistical inference. Suppose that we have an honest coin and plan to toss it 100 times. We are going to do this just once—period, and the resulting number of heads we will call X. Been-there, done-that! What's new now is that we a have a solid understanding of the statistical behavior of this random variable—it has an approximately normal distribution with mean $\mu = 50$ and standard deviation $\sigma = 5$—and this allows us to make some very accurate predictions about what is to happen.

For starters, we can predict the chances that the number of heads will fall somewhere between 45 and 55 (1 standard deviation below and above the mean)—they are 68%. Likewise, we know that the chances that the number of heads will fall somewhere between 40 and 60 are 95%, and between 35 and 65 are a whopping 99.7%.

What if, instead of tossing the coin 100 times, we were to toss it 500 times? Or 1000 times? Or n times? Not surprisingly, the bell-shaped distribution we saw in

Example 5 would still be there—only the values of μ and σ would change. Specifically, the number of heads X would be a random variable with an approximately normal distribution with mean $n/2$ heads and standard deviation $\sigma = \sqrt{n}/2$ heads. This is an important fact for which we have coined the name the *honest-coin principle:*

The Honest-Coin Principle

Suppose an honest coin is tossed n times, and X denotes the number of heads that come up. The random variable X has an approximately normal distribution with mean $\mu = n/2$ and standard deviation $\sigma = \sqrt{n}/2$.

When we apply the above principle with $n = 100$ we get the mean $\mu = 100/2 = 50$ heads and the standard deviation $\sigma = \sqrt{100}/2 = 10/2 = 5$ heads.)

Example 6. Betting on the outcome of 256 coin tosses

An honest coin is going to be tossed 256 times. Before this is done, we have the opportunity to make some bets. Let's say that we can make a bet (with even odds) that if the number of heads falls somewhere between 120 and 136, we will win, otherwise we will lose. Should we make such a bet?

By the honest-coin principle we know that the number of heads in 256 tosses of an honest coin is a random variable having a distribution that is approximately normal with mean $\mu = 128$ heads and standard deviation $\sigma = \sqrt{256}/2 = 8$ heads. The values 120 to 136 are exactly 1 standard deviation below and above the mean of 128, which means that there is a 68% chance that the number of heads will fall somewhere between 120 and 136. We should indeed make this bet! A similar calculation tells us that there is a 95% chance that the number of heads will fall somewhere between 112 and 144, and the chance that the number of heads will fall somewhere between 104 and 152 is 99.7%. ■

What happens when the coin being tossed is not a fair coin? Surprisingly, the distribution of the number of heads X in n tosses of such a coin is still approximately normal, as long as the number n is not too small.[6] All we need now is a **dishonest-coin principle** to tell us how to find the mean and the standard deviation:

The Dishonest-Coin Principle

Suppose an arbitrary coin is tossed n times ($n \geq 30$), and X denotes the number of heads that come up. Suppose also that p is the probability of the coin landing heads, and $1 - p$ is the probability of the coin landing tails. Then, the random variable X has an approximately normal distribution with mean $\mu = n \cdot p$ and standard deviation $\sigma = \sqrt{n \cdot p \cdot (1 - p)}$.

[6] An accepted rule of thumb in statistics is that n should be at least 30.

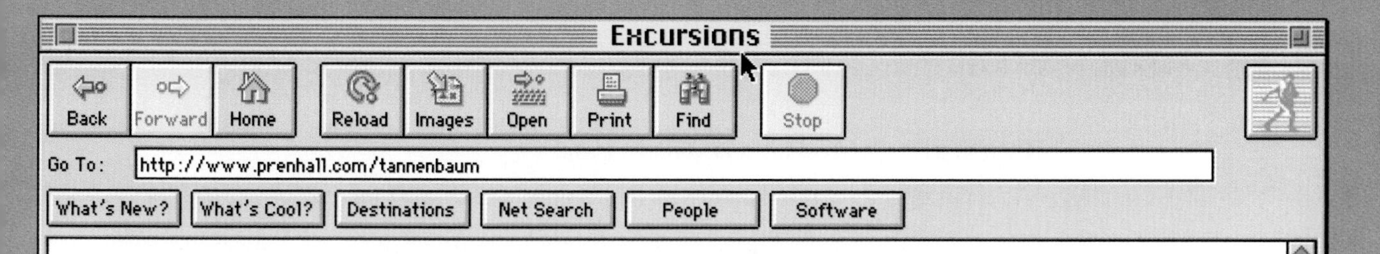

How Much Under The Normal

The *standard normal* curve has a mean of 0 and a standard of deviation of 1. Suppose you wish to find the percentage of the values between -2.5 and 1. How would you do this? This happens to be the area under the normal curve. Statisticians can use tables, computational software, or advanced mathematics to compute this.

To compute the area under a standard normal curve go to Chapter 16 in the Tannenbaum website, and use the applet on the page.

We can also use the standard normal curve to compute the estimate the percentage between two values of some normal curve with a given mean and standard deviation. For example, suppose we wish to estimate the percentage of 1996 seniors in Example 4 that scored between 550 and 650. First, we must "standardize" (find the number of standard deviations) each score is from the mean. The expression $\frac{x - \sigma}{\mu}$ gives the number of standard deviations x is away from the mean μ. Thus, 550 and 650 are standardized to $\frac{550 - 505}{110} \approx 0.4091$ and $\frac{650 - 505}{110} \approx 1.3182$.

Questions:

1. Intuitively, how does the computation of the area under the curve relate to using a histogram to find the number of data between two values?
2. Find the percentage of the values between -2.5 and 1.
3. Estimate the percentage of seniors that scored between 550 and 650.
4. Imagine you are a student in the class of 1996 who scored 650 on the SAT-Verbal. Estimate the percentile of the population this score puts you in.

Example 7. Tossing a Dishonest Coin

A coin is rigged so that it comes up heads only 20% of the time (i.e., $p = .20$). The coin is tossed 100 times ($n = 100$), and X is the number of heads in the 100 tosses. What can we say about X?

According to the dishonest-coin principle, the distribution of the X is approximately normal with center $\mu = 100 \times .20 = 20$ heads and standard deviation $\sigma = \sqrt{100 \times .20 \times .80} = 4$ heads

(Note that in this case heads and tails are no longer symmetric, but the dishonest-coin principle will work just as well for tails as it does for heads. The distribution for the number of tails is approximately normal with center $\mu = 100 \times .80 = 80$ and standard deviation $\sigma = \sqrt{100 \times .80 \times .20} = 4$. Note that σ is still the same.)

Based on these facts we can now make the following assertions:

- There is a 68% chance that the number of heads will fall somewhere between 16 and 24, which represents one standard deviation below and above the mean.
- There is a 95% chance that the number of heads will fall somewhere between 12 and 28, which represents standardized values between -2 and 2.
- The number of heads is almost guaranteed (a 99.7% chance) to fall somewhere between 8 and 32. ■

The dishonest-coin principle can be applied to any coin, even one that is fair ($p = 1/2$). In the case $p = 1/2$ the honest- and dishonest-coin principles say the same thing (Exercise 30).

The dishonest-coin principle is a down-to-earth version of one of the most important facts in statistics, known by the somewhat intimidating name of the *central limit theorem.* We will now briefly illustrate why the importance of the dishonest-coin principle goes beyond the tossing of coins.

Example 8. Sampling for Defective Light Bulbs

An assembly line produces 100,000 light bulbs a day, 20% of which generally turn out to be defective. Suppose we draw a random sample of $n = 100$ light bulbs. Let X represent the *number of defective light bulbs* in the sample. What can we say about X?

A moment's reflection will show that statistically this example is almost identical to Example 7—each light bulb chosen has an approximate probability of .20 of being defective (like the coin did of coming up heads in Example 7). We can use the dishonest-coin principle to infer that the number of defective light bulbs in the sample is a random variable having an approximately normal distribution with a mean of 20 light bulbs and standard deviation of 4. Thus,

- There is a 68% chance that the number of defective light bulbs in the sample will fall somewhere between 16 and 24.
- There is a 95% chance that the number of defective light bulbs in the sample will fall somewhere between 12 and 28.
- The number of defective light bulbs in the sample is practically guaranteed (a 99.7% chance) to fall somewhere between 8 and 32.

Probably the most important point here is that each of the above facts can be rephrased in terms of sampling errors, a concept we first discussed in Chapter 13.

For example, say we had 24 defective light bulbs in the sample, in other words, 24% of the sample (24 out of 100) is made of defective light bulbs. If we use this statistic to estimate the percent of defective light bulbs overall, then the sampling error would be 4% (the estimate is 24% and the value of the parameter is 20%). By the same token, if we had 16 defective light bulbs in the sample, the sampling error would be −4%. Coincidentally, the standard deviation is $\sigma = 4$ light bulbs, or 4% of the sample (we computed it in Example 7). Thus, we can rephrase our previous assertions about sampling errors as follows:

- When estimating the proportion of defective light bulbs coming out of the assembly line by using a sample of 100 light bulbs, there is a 68% chance that the sampling error will fall somewhere between −4% and 4%.
- When estimating the proportion of defective light bulbs coming out of the assembly line by using a sample of 100 light bulbs, there is a 95% chance that the sampling error will fall somewhere between −8% and 8%.
- When estimating the proportion of defective light bulbs coming out of the assembly line by using a sample of 100 light bulbs, there is a 99.7% chance that the sampling error will fall somewhere between −12% and 12%. ■

Example 9. Sampling With Larger Samples

Suppose we have the same assembly line as in Example 8, but this time we are going to take a really big sample of $n = 1600$ light bulbs. Before we even count the number of defective light bulbs in the sample, let's see how much mileage we can get out of the dishonest-coin principle. The standard deviation for the distribution of defective light bulbs in the sample is $\sqrt{1600 \times .2 \times .8} = 16$, which just happens to be exactly 1% of the sample (16/1600 = 1%). This means that when we estimate the proportion of defective light bulbs coming out of the assembly line using this sample, we can have some sort of a handle on the sampling error.

- We can say with some confidence (68%) that the sampling error will fall somewhere between −1% and 1%.
- We can say with a lot of confidence (95%) that the sampling error will fall somewhere between −2% and 2%.
- We can say with tremendous confidence (99.7%) that the sampling error will fall somewhere between −3% and 3%. ■

Our last example shows how a variation of the dishonest coin principle can be used to give reasonable estimates on the margin of error in a public opinion poll, an issue of considerable importance in modern statistics.

Example 10. Measuring the Margin of Error of a Poll

In California, school bond measures require a 66.67% vote for approval. Suppose that an important school bond measure is on the ballot in the upcoming election. In the most recent poll of 1200 randomly chosen voters, 744 of the 1200 voters sampled, or 62%, indicated that they would vote for the school bond measure. Let's assume that the poll was properly conducted, and that the 1200 voters sampled represent an unbiased sample of the entire population. What are the chances that the 62% statistic is the result of sampling variability and that the actual vote for the bond measure will be 66.67% or more?

Here we will use a variation of the dishonest-coin principle, with each voter being likened to a coin toss. Voting for the bond measure is like the coin coming up heads; against is tails. The probability (p) of "heads" for this "coin" will turn out to be the proportion of voters in the population that support the bond measure: If p turns out to be .667 or more the bond measure will pass. Our problem is that we don't know p, so How can we use the dishonest-coin principle to estimate the mean and standard deviation of the sampling distribution?

The idea here is to use the 62% (.62) statistic from the sample as an estimate for the actual value of p in the formula for the standard deviation given by the dishonest-coin principle. (Even though we know that this is only a rough estimate for p, this generally turns out to give us a good estimate for the standard deviation.) In our example, the approximate standard deviation for the number of "heads" in the sample turns out to be $\sqrt{1200 \times .62 \times .38} = 16.8$ voters. When we convert this number to percentages we get a standard deviation that is approximately 1.4% of the sample ($16.8/1200 = .014$).

The standard deviation for the sampling distribution expressed as a percentage of the entire sample is called the **standard error**. (For our example, we have found that the standard error is approximately 1.4%). In sampling and public opinion polls, it is customary to express the information about the population in terms of **confidence intervals**, which are themselves based on standard errors: a 95% confidence interval is given by 2 standard errors below and above the statistic obtained from the sample; a 99.7% confidence interval is given by going 3 standard errors below and above the sample statistic.

In our example, we have a 95% confidence interval of 62% plus or minus 2.8%, which means that we can say with 95% confidence (we would be right 95 out of 100 times) that the actual vote for the bond measure will fall somewhere between 59.2% ($62 - 2.8$) and 64.8% ($62 + 2.8$), and thus, that the bond measure will lose. Want even more certainty? Take a 99.7% confidence interval of 62% plus or minus 3.6%—it is almost certain that the actual vote will turn out somewhere in that range. Even in the most optimistic scenario, the vote will not reach the 66.7% needed to pass the bond measure. ■

Conclusion

Bell-shaped curves are everywhere around us. In this chapter we studied bell-shaped curves, some of their mathematical properties, and how these properties can be translated into analyzing real-life data that often approximate a bell-shaped pattern. It was a brief introduction to what is undoubtedly one of the most widely used and sophisticated tools of modern mathematical statistics.

In this chapter we also got a brief glimpse at the concept of statistical inference. The process of drawing conclusions and inferences based on limited data is an essential part of statistics. It gives us a way not only to analyze what has already taken place but to make reasonably accurate large scale predictions of what will happen in certain random situations. Casinos know, without any shadow of a doubt, that in the long run, they will make a profit—it is a mathematical law! A similar law gives us the confidence to trust the results of surveys and public opinion polls (up to a point!), the quality of the products we buy, and even the statistical data our government uses to make many of its decisions. In all of these cases, bell-shaped distributions of data and the laws of probability come together to give us insight into what was, is, and most likely will be.

Key Concepts

68-95-99.7 rule
approximately normal distribution
center (mean, median)
confidence interval
dishonest-coin principle
honest-coin principle
inflection point
normal curve
normal distribution
standard deviation
standard error
standardized value

Exercises

Walking

1. Suppose that a normal distribution has a standard deviation of 15. How far apart (measured in standard deviations) are the following pairs of numbers?
   (a) 45 and 75.
   (b) 30 and 51.
   (c) 50 and 89.
   (d) 20 and 32.

2. Suppose that a normal distribution has a standard deviation of 4.8. How far apart (measured in standard deviations) are the following pairs of numbers?
   (a) 15.3 and 20.1.
   (b) 24.7 and 27.1.
   (c) 12.4 and 13.6.
   (d) 30.2 and 43.4.

3. Suppose that a normal distribution has a mean of 30 and a standard deviation of 15. Find the standardized value of each of the following numbers.
   (a) 45.
   (b) 54.
   (c) 0.
   (d) 3.

4. Suppose that a normal distribution has a mean of 47.3 and a standard deviation of 4.8. Find the standardized value of each of the following numbers.
   (a) 56.9.
   (b) 38.9.
   (c) 58.1.
   (d) 31.7.

Exercises 5 through 8 refer to the following: 250 students take a college entrance exam. The scores on the exam have an approximately normal distribution with mean $\mu = 52$ points and the standard deviation $\sigma = 11$ points.

5. **(a)** Estimate the average score on the exam.
   (b) Estimate what percent of the students scored 52 points or more.
   (c) Estimate what percent of the students scored between 41 and 63 points.
   (d) Estimate what percent of the students scored 63 points or more.

6. **(a)** Estimate how many students scored between 30 and 74 points.
   (b) Estimate how many students scored 74 points or more.
   (c) Estimate how many students scored 85 points or more.

7. **(a)** Estimate the first-quartile score for this exam.

(b) Estimate the third-quartile score for this exam.

(c) Estimate the interquartile range for this exam

8. For each of the following scores, estimate in what percentile of the students taking the exam the score would place you.

(a) 51.

(b) 64.

(c) 60.

(d) 85.

Exercises 9 through 12 refer to the following: As part of a research project, the blood pressures of 2000 patients in a hospital are recorded. The systolic blood pressures (given in millimeters) have an approximately normal distribution with mean $\mu = 125$ and $\sigma = 13$.

9. **(a)** Estimate the number of patients whose blood pressure was between 99 and 151 millimeters.

(b) Estimate the number of patients whose blood pressure was 99 millimeters or less.

10. **(a)** Estimate the third quartile (Q_3) for the distribution of blood pressures.

(b) Estimate the interquartile range for the distribution of blood pressures.

11. For each of the following blood pressures, estimate the percentile of the patient population to which they correspond.

(a) 100 millimeters.

(b) 112 millimeters.

(c) 115 millimeters.

(d) 138 millimeters.

(e) 164 millimeters.

12. **(a)** Estimate the value of the lowest (Min) and the highest (Max) blood pressures. (Assume there were no outliers and use the 68-95-99.7 rule.)

(b) Assuming there were no outliers, given an estimate of the five-number summary (Min, Q_1, μ, Q_3, Max) for the distribution of blood pressures.

Exercises 13 through 16 refer to the following: Packaged foods sold at supermarkets are not always the weight indicated on the package. Variability always crops up in the manufacturing and packaging process. Suppose that the exact weight of a "12-ounce" bag of potato chips is a random variable that has an approximately normal distribution with mean $\mu = 12$ ounces and standard deviation $\sigma = 0.5$ ounce.

13. If a "12-ounce" bag of potato chips is chosen at random, what are the chances that:

(a) It weighs somewhere between 11 and 13 ounces?

(b) It weighs somewhere between 12 and 13 ounces?

(c) It weighs more than 11 ounces?

14. If a "12-ounce" bag of potato chips is chosen at random, what are the chances that:

(a) It weighs somewhere between 11.5 and 12.5 ounces?

(b) It weighs somewhere between 12 and 12.5 ounces?

(c) It weighs more than 12.5 ounces?

15. Suppose that 500 bags of potato chips are chosen at random. Estimate the number of bags with weight:

(a) 11 ounces or less.
(b) 11.5 ounces or less.
(c) 12 ounces or less.
(d) 12.5 ounces or less.
(e) 13 ounces or less.
(f) 13.5 ounces or less.

16. Suppose that 1500 bags of potato chips are chosen at random. Estimate the number of bags of potato chips with weight:
(a) Between 11 and 11.5 ounces.
(b) Between 11.5 and 12 ounces.
(c) Between 12 and 12.5 ounces.
(d) Between 12.5 and 13 ounces.
(e) Between 13 and 13.5 ounces.

17. Find μ and σ for the normal curve shown in the figure. P and P' are the inflection points of the curve—(*Hint*: see Fig. 16-8.)

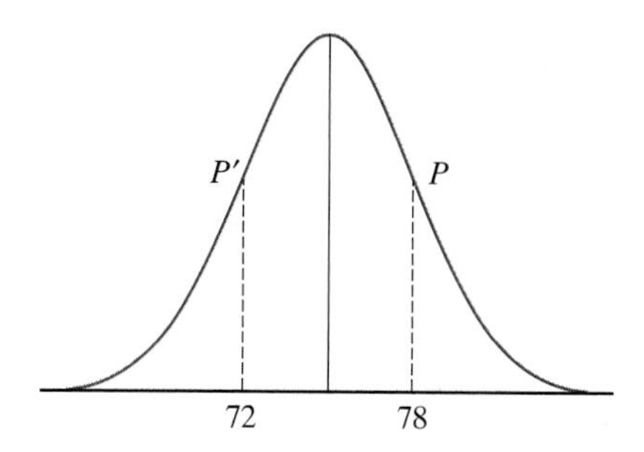

18. Find μ and σ for the normal curve shown in the figure.

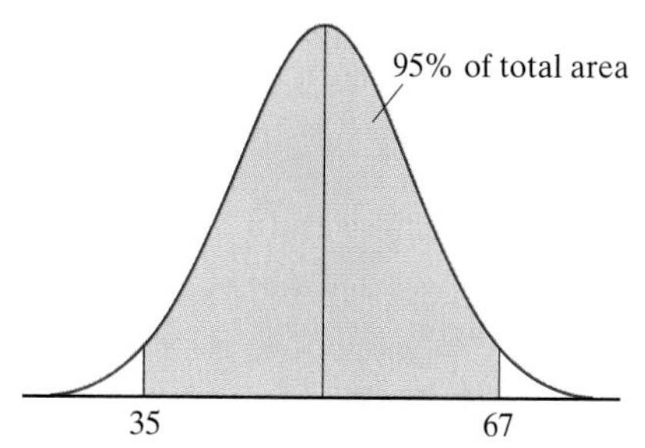

19. Find μ and σ for the normal curve shown in the figure.

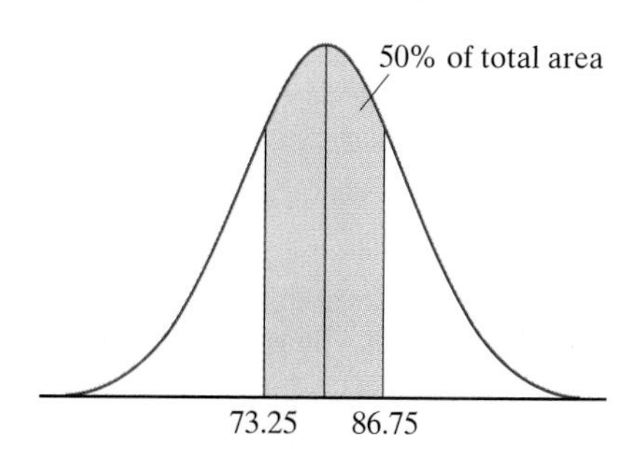

20. Find μ and σ for the normal curve shown in the figure.

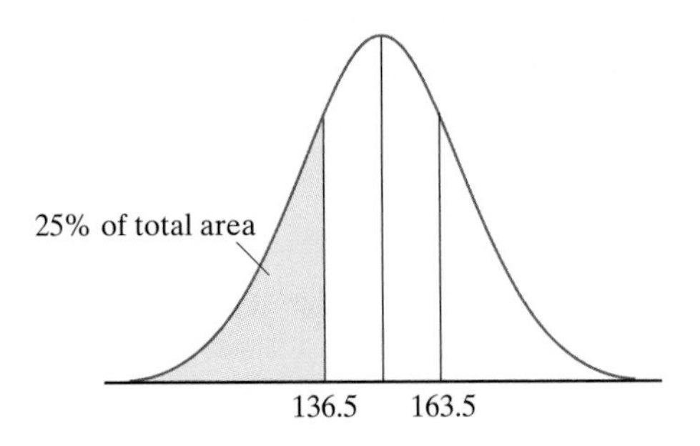

Exercises 21 through 24 refer to the following: The distribution of weights for children of a given age and sex is approximately normal. This fact allows a doctor or nurse to find from a child's weight the weight percentile of the population (all children of the same age and sex) to which the child belongs. Typically, this is done using special charts provided to the doctor or nurse, but these percentiles can also be computed using facts about approximately normal distributions such as the ones we learned in this chapter. (The figures in these examples are approximate values taken from tables produced by the National Center for Health Statistics, U.S. Department of Health and Human Services.)

21. The distribution of weights for 6-month-old baby boys is approximately normal with mean $\mu = 17.25$ pounds and standard deviation $\sigma = 2$ pounds.

(a) Suppose that a 6-month-old boy weighs 15.25 pounds. Approximately what weight percentile is he in?

(b) Suppose that a 6-month-old boy weighs 21.25 pounds. Approximately what weight percentile is he in?

(c) Suppose that a 6-month-old boy is in the 75th percentile in weight. Estimate his weight.

22. The distribution of weights for 12-month-old baby girls is approximately normal with mean $\mu = 21$ pounds and standard deviation $\sigma = 2.2$ pounds.

(a) Suppose that a 12-month-old girl weighs 16.6 pounds. Approximately what weight percentile is she in?

(b) Suppose that a 12-month-old girl weighs 18.8 pounds. Approximately what weight percentile is she in?

(c) Suppose that a 12-month-old girl is in the 75th percentile in weight. Estimate her weight.

23. The distribution of weights for 1-month-old baby girls is approximately normal with mean $\mu = 8.75$ pounds and standard deviation $\sigma = 1.1$ pounds.

(a) Suppose that a 1-month-old girl weighs 11 pounds. Approximately what weight percentile is she in?

(b) Suppose that a 1-month-old girl weighs 12 pounds. Approximately what weight percentile is she in?

(c) Suppose that a 1-month-old girl is in the 25th percentile in weight. Estimate her weight.

24. The distribution of weights for 12-month-old baby boys is approximately normal with mean $\mu = 22.5$ pounds and standard deviation $\sigma = 2.2$ pounds.

(a) Suppose that a 12-month-old boy weighs 24 pounds. Approximately what weight percentile is he in?

(b) Suppose that a 12-month-old boy weighs 21 pounds. Approximately what weight percentile is he in?

(c) Suppose that a 12-month-old boy is in the 84th percentile in weight. Estimate his weight.

25. This exercise is based on Example 1 (heights of NBA players 1996–97). The frequency table without the outlier is reproduced below.

Height	Frequency	Height	Frequency	Height	Frequency	Height	Frequency
5-10	2	6-3	22	6-8	44	7-1	6
5-11	6	6-4	24	6-9	45	7-2	5
6-0	9	6-5	25	6-10	44	7-3	2
6-1	13	6-6	29	6-11	30	7-4	2
6-2	15	6-7	39	7-0	21	7-6	1
						7-7	1

(a) Estimate the value of the standard deviation.

(b) Estimate the value of the mean.

(c) Using your estimates in parts (a) and (b), estimate the number of players whose heights fall between 2 standard deviations below and above the mean.

(d) How does your answer in (c) compare with the actual data?

Jogging

26. Over the years, the eggs produced by the Fibonacci Egg Company have been known to have weights with an approximately normal distribution whose mean is $\mu = 1.5$ ounces and standard deviation is $\sigma = 0.35$ ounce. Eggs are classified by weight as follows:

Extra large: more than 2.2 ounces.

Large: between 1.5 and 2.2 ounces.

Discount (sold to wholesalers): less than 1.5 ounces.

Out of 5000 eggs, approximately how many would you expect would fall in each of the three categories?

27. An honest coin is tossed $n = 3600$ times. Let the random variable Y denote the number of tails tossed.

(a) Find the center μ and the standard deviation σ for the distribution of the random variable Y.

(b) What are the chances that the value of Y will fall somewhere between 1770 and 1830?

(c) What are the chances that the value of Y will fall somewhere between 1800 and 1830?

(d) What are the chances that the value of Y will fall somewhere between 1830 and 1860?

28. An honest die is rolled. If the roll comes out even (2, 4, or 6), you will win \$1; if the roll comes out odd (1, 3, or 5), you will lose \$1. Suppose that in one evening you play this game $n = 2500$ times in a row.

(a) What is the probability that by the end of the evening you will not have lost any money?

(b) What is the probability that the number of even rolls will fall between 1250 and 1300?

(c) What is the probability that you will win \$100 or more?

(d) What is the probability that you will win exactly \$101?

29. A dishonest coin with probability of heads $p = .4$ is tossed $n = 600$ times. Let the random variable X represent the number of times the coin comes up heads.

(a) Find the center and standard deviation for the distribution of X.

(b) Find the first and third quartiles for the distribution of X.

(c) Suppose that you could choose between one of the following bets (\$1 to win \$1):

(1) The number of heads will fall somewhere between 230 and 250.

(2) The number of heads will be less than or equal to 230 or more than or equal to 250.

Which of the two bets would you choose? Explain your answer.

30. Explain why when the dishonest-coin principle is applied with an honest coin, we get the honest-coin principle.

Exercises 31 and 32 refer to the following table. It is a simplified version of a more elaborate statistical table that gives, for every value of x, the approximate location of the xth *percentile for a normal distribution with mean* μ *and standard deviation* σ.

Percentile	Approximate location	Percentile	Approximate location
99th	$\mu + 2.33\sigma$	1st	$\mu - 2.33\sigma$
95th	$\mu + 1.65\sigma$	5th	$\mu - 1.65\sigma$
90th	$\mu + 1.28\sigma$	10th	$\mu - 1.28\sigma$
80th	$\mu + 0.84\sigma$	20th	$\mu - 0.84\sigma$
75th	$\mu + 0.675\sigma$	25th	$\mu - 0.675\sigma$
70th	$\mu + 0.52\sigma$	30th	$\mu - 0.52\sigma$
60th	$\mu + 0.25\sigma$	40th	$\mu - 0.25\sigma$
50th	μ		

31. The distribution of weights for 6-month-old baby boys is approximately normal with mean $\mu = 17.25$ pounds and standard deviation $\sigma = 2$ pounds.

(a) Suppose that a 6-month-old baby boy weighs in the 95th percentile of his age group. Find his weight in pounds approximated to 2 decimal places.

(b) Suppose that a 6-month-old boy weighs 17.75 pounds. Determine the percentile of the weight distribution for his age group this child is in.

(c) Suppose that a 6-month-old baby boy weighs 15.2 pounds. Estimate the percentile of the weight distribution for his age group this child is in.

32. Five thousand students took a college entrance exam. The scores on the exam have an approximately normal distribution with mean $\mu = 55$ points and standard deviation $\sigma = 12$ points.

(a) Suppose a student's score places her in the 60th percentile. Find her score on the exam.

(b) Suppose that a student scored 45 points on the exam. Estimate the pecentile in which this score places him.

(c) Suppose that a student scored 83 points on the exam. Estimate the percentile in which this score places her.

(d) Approximately how many students scored 83 points or more?

(e) Approximately how many students scored 52 points or less?

Running

33. On an American roulette wheel, there are 18 red numbers and 18 black numbers, plus 2 green numbers (0 and 00). Thus, the probability of a red number coming up on a spin of the wheel is $p = 18/38 \approx .47$. Suppose that we go on a binge and we bet \$1 on red 10,000 times in a row. (A \$1 bet wins \$1 if red comes up; otherwise we lose the \$1.)

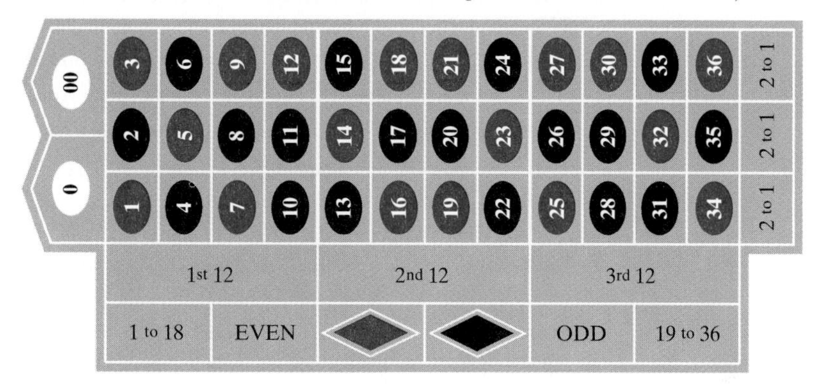

(a) Let Y represent the number of times we lose (i.e., red does not come up). Use the dishonest-coin principle to describe the distribution of the random variable Y.

(b) Approximately what are the chances that we will lose 5300 times or more?

(c) Approximately what are the chances that we will lose somewhere between 5150 and 5450 times?

(d) Explain why the chances that we will break even or win in this situation are essentially zero.

34. An urn contains 10,000 beads, of which 20% are red and the rest white. Suppose that we draw a sample of size $n = 400$ (drawing a bead and each time replacing it in the urn before drawing again).

(a) If Y is the number of white beads in the sample, find the center and standard deviation for the distribution of Y.

(b) What are the chances that the number of white beads in the sample will fall somewhere between 304 and 332?

(c) When estimating the proportion of white beads in the urn using a sample of 400 beads, what are the chances that the sampling error will fall somewhere between -4% and 4%?

35. An urn contains a large number of beads which are either red or white. Suppose that we draw a random sample of 1200 beads, of which 300 (25%) are red. We want to use the 25% statistic to estimate the percentage of red beads in the urn.

(a) Use the dishonest coin principle with $p = .25$ to estimate the standard error.

(b) Compute a 95% confidence interval for the percentage of red beads in the urn.

(c) Compute a 99.7% confidence interval for the percentage of red beads in the urn.

References and Further Readings

1. Converse, P. E., and M. W. Traugott, "Assessing the Accuracy of Polls and Surveys," *Science,* 234 (1986), 1094–1098.
2. Frankel, Max, "Margins of Error," *New York Times Magazine,* December 15, 1996, 34.
3. Fraser, Steven, ed., *The Bell Curve Wars.* New York: Basic Books, 1995.
4. Freedman, D., R. Pisani, R. Purves, and A. Adhikari, *Statistics,* 2d ed. New York: W. W. Norton, Inc., 1991, chaps. 16 and 18.
5. Herrnstein, R. J., and C. Murray, *The Bell Curve: Intelligence and Class Structure in American Life.* New York: Free Press, 1994.
6. Kerrich, John, *An Experimental Introduction to the Theory of Probability.* Witwatersrand, South Africa: University of Witwatersrand Press, 1964.
7. Larsen, J., and D. F. Stroup, *Statistics in the Real World.* New York: Macmillan Publishing Co., Inc., 1976.
8. Mosteller, F., W. Kruskal, et al., *Statistics by Example: Detecting Patterns.* Reading, MA: Addison-Wesley Publishing Co., Inc., 1973.
9. Mosteller, F., R. Rourke, and G. Thomas, *Probability and Statistics.* Reading, MA.: Addison-Wesley Publishing Co., Inc., 1961.
10. Tanner, Martin, *Investigations for a Course in Statistics.* New York: Macmillan Publishing Co., Inc., 1990.

Answers to Selected Problems

Chapter 1

Walking

1. **(a)**

Number of voters	5	3	1	3
1st choice	*A*	*C*	*B*	*C*
2nd choice	*B*	*B*	*D*	*B*
3rd choice	*C*	*A*	*C*	*D*
4th choice	*D*	*D*	*A*	*A*

(b) No **(c)** The Country Cookery (*C*) **(d)** Blair's Kitchen (*B*)

3. **(a)** 55 **(b)** *B*

5. *D*

7. *C*

9. Winner: *D*. Second place: *A*. Third place: *C*. Fourth place: *B*. Last place: *E*.

11. Winner: *D*. Second place: *B* and *C* (tied). Last place: *A* and *E* (tied).

13. **(a)**

Number of voters	5	3	5	5	3
1st choice	*A*	*A*	*C*	*D*	*B*
2nd choice	*B*	*D*	*D*	*C*	*D*
3rd choice	*C*	*B*	*A*	*B*	*C*
4th choice	*D*	*C*	*B*	*A*	*A*

(b) *D*

15. *R*

17. Winner: *H*. Second place: *R*. Third place: *C*. Fourth place: *O*, *S* (tied).

19. Winner: *O*. Second place: *C*. Third place: *H*. Fourth place: *R*. Last place: *S*.

21. **(a)** B

(b) Yes. D has a majority (13) of the first-place votes but does not win the election.

23. Winner: D. Second place: A. Third place: C. Fourth place: E. Last place: B.

25. Winner: A. Second place: B and C (tied). Last place: D and E (tied).

27. **(a)** Winner: E. Second place: C. Third place : A. Fourth place: D. Last place: B.

(b) Yes. A is a Condorcet candidate but does not win the election.

29. **(a)** A

(b) No. No candidate has a majority of the first-place votes.

31. Winner: C. Second place: D. Third place: A. Last place: B.

33. Winner: C. Second place: B. Third place: D. Last place: A.

35. Winner: E. Second place: A. Third place: B. Fourth place: D. Last place: C.

37. Winner: A. Second place: B. Third place: D. Fourth place: E. Last place: C.

39. Winner: E. Second place: A. Third place: C. Fourth place: B. Last place: D.

Jogging

41. **(a)** 1225 **(b)** 2450

43. **(a)** Since there are only *two* columns in the preference schedule, one of the columns must represent the votes of 11 or more voters and so the first choice in that column is the first choice of more than half of the voters.

(b) The argument given in **(a)** can be repeated provided there are an odd number of voters since then one of the *two* columns must represent the votes of more than half of the voters (there cannot be a tie).

45. Suppose that the results of the election under the Borda count method (first place: 5 points, second place: 4 points, third place: 3 points, fourth place: 2 points, and fifth place: 1 point) were

Candidate	A	B	C	D	E
Points	a	b	c	d	e

Under the revised scheme (first place: 4 points, second place: 3 points, third place: 2 points, fourth place: 1 point, and fifth place: 0 points), A loses a point for every voter, so does B, etc. It follows that the result of the election under the revised schedule must be

Candidate	A	B	C	D	E
Points	$a - 21$	$b - 21$	$c - 21$	$d - 21$	$e - 21$

Since these numbers have the same values relative to each other as the original numbers, the outcome of the election is still the same.

47. If there is a candidate that is the first choice of a majority of the voters, then using the plurality-with-elimination method, that candidate will be declared the winner of the election in the first round.

49. If there is a candidate that is the first choice of a majority of the voters, then that candidate will win every head-to-head comparison with any other candidate and so will be a Condorcet candidate. If a voting method violates the majority criterion, then there is an election for which a candidate is the first choice of a majority of the voters (and hence a Condorcet candidate), and yet is not the winner of the election. Consequently, the voting method also violates the Condorcet criterion.

51. **(a)** No. With 27 voters, 14 votes are necessary for a majority.

(b) Yes. (C is a Condorcet candidate.) **(c)** B **(d)** C

(e) The Condorcet criterion. (C is a Condorcet candidate but B wins the election under the plurality-with-elimination method.) The independence of irrelevant alternatives criterion. (B wins the election under the plurality-with-elimination method, but when D drops out, C wins the election.)

53. **(a)** $36 + 25 + 1 = 62$, so all of the first-place votes are accounted for.

(b)

	2nd-place votes	**3rd-place votes**
Florida State	25	1
Notre Dame	18	19
Nebraska	19	42

CHAPTER 2

Walking

1. **(a)** 4 **(b)** 10 **(c)** 10
(d) $\{P_1, P_2\}, \{P_1, P_3\}, \{P_1, P_2, P_3\}, \{P_1, P_2, P_4\}, \{P_1, P_3, P_4\}, \{P_2, P_3, P_4\}, \{P_1, P_2, P_3, P_4\}$
(e) P_1 only **(f)** $P_1: \frac{5}{12}; P_2: \frac{1}{4}; P_3: \frac{1}{4}; P_4: \frac{1}{4}$

3. **(a)** $P_1: \frac{3}{5}; P_2: \frac{1}{5}; P_3: \frac{1}{5}$ **(b)** $P_1: \frac{3}{5}; P_2: \frac{1}{5}; P_3: \frac{1}{5}$; The systems are equivalent.

5. **(a)** $P_1: \frac{1}{3}; P_2: \frac{1}{4}; P_3: \frac{1}{6}; P_4: \frac{1}{6}; P_5: \frac{1}{12}$ **(b)** $P_1: \frac{7}{19}; P_2: \frac{5}{19}; P_3: \frac{3}{19}; P_4: \frac{3}{19}; P_5: \frac{1}{19}$

7. **(a)** $\langle P_1, P_2, P_3\rangle, \langle P_1, P_3, P_2\rangle, \langle P_2, P_1, P_3\rangle, \langle P_2, P_3, P_1\rangle, \langle P_3, P_1, P_2\rangle, \langle P_3, P_2, P_1\rangle$
(b) $\langle P_1, \underline{P_2}, P_3\rangle, \langle P_1, \underline{P_3}, P_2\rangle, \langle P_2, \underline{P_1}, P_3\rangle, \langle P_2, P_3, \underline{P_1}\rangle, \langle P_3, \underline{P_1}, P_2\rangle, \langle P_3, P_2, \underline{P_1}\rangle$
(c) $P_1: \frac{2}{3}; P_2: \frac{1}{6}; P_3: \frac{1}{6}$

9. **(a)** $P_1: 1; P_2: 0; P_3: 0$ **(b)** $P_1: \frac{2}{3}; P_2: \frac{1}{6}; P_3: \frac{1}{6}$
(c) $P_1: \frac{2}{3}, P_2: \frac{1}{6}, P_3: \frac{1}{6}$ **(d)** $P_1: \frac{1}{2}, P_2: \frac{1}{2}; P_3: 0$ **(e)** $P_1: \frac{1}{3}; P_2: \frac{1}{3}; P_3: \frac{1}{3}$

11. **(a)** $P_1: \frac{1}{2}; P_2: \frac{3}{10}; P_3: \frac{1}{10}; P_4: \frac{1}{10}$ **(b)** $P_1: \frac{7}{12}; P_2: \frac{1}{12}; P_4: \frac{1}{12}$

13. **(a)** P_1 and P_2 have veto power; P_3 is a dummy.
(b) P_1 is a dictator; P_2 and P_3 are dummies.
(c) There is no dictator, no one has veto power, and no one is a dummy.

15. **(a)** P_1 and P_2 have veto power; P_5 is a dummy.
(b) P_1 is a dictator; P_2, P_3, P_4 are dummies.
(c) P_1 and P_2 have veto power; P_3 and P_4 are dummies.
(d) All 4 players have veto power.

17. **(a)** 14 **(b)** 27 **(c)** 31 **(d)** 120

19. **(a)** $P_1: 1; P_2: 0; P_3: 0$
(b) $P_1: \frac{2}{3}; P_2: \frac{1}{6}; P_3: \frac{1}{6}$
(c) $P_1: \frac{1}{2}; P_2: \frac{1}{2}; P_3: 0$
(d) $P_1: \frac{1}{2}; P_2: \frac{1}{2}; P_3: 0$
(e) $P_1: \frac{1}{3}, P_2: \frac{1}{3}; P_3: \frac{1}{3}$

21. **(a)** 720 **(b)** 3,628,800
(c) $11! = 10! \times 11 = 3{,}628{,}800 \times 11 = 39{,}916{,}800$
(d) $9! = \frac{10!}{10} = \frac{3{,}628{,}800}{10} = 362{,}880$
(e) $x = \frac{12!}{12} = 11! = 39{,}916{,}800$

23. $A: \frac{1}{3}; B: \frac{1}{3}; C: \frac{1}{3}; D: 0$

25. $A: \frac{7}{17}; B: \frac{7}{17}; C: \frac{1}{17}; D: \frac{1}{17}; E: \frac{1}{17}$

Jogging

27. $P_1: \frac{15}{52}; P_2: \frac{1}{4}; P_3: \frac{11}{52}; P_4: \frac{9}{52}; P_5: \frac{3}{52}; P_6: \frac{1}{52}$

29. **(a)** 720

(b) The player must be the last (sixth) player in the sequential coalition.

(c) 120 **(d)** $\frac{120}{720} = \frac{1}{6}$

(e) $\frac{1}{6}$ (Each player is the last player in 120 of the 720 sequential coalitions.)

(f) If the quota equals the sum of all the weights $(q = w_1 + w_2 + \cdots + w_N)$ then the only way a player can be pivotal is for the player to be the last player in the sequential coalition. Since every player will be the last player in the same number of sequential coalitions, all players must have the same Shapley-Shubik power index. It follows that each of the N players has Shapley-Shubik power index of $\frac{1}{N}$.

31. (a) [7: 6,3,2,1,1]. Shapley-Shubik power distribution: P_1: $\frac{3}{5}$; P_2: $\frac{1}{10}$; P_3: $\frac{1}{10}$; P_4: $\frac{1}{10}$; P_5: $\frac{1}{10}$

(b) [9: 6,3,1,1,1]. Shapley-Shubik power distribution: P_1: $\frac{11}{20}$; P_2: $\frac{6}{20}$; P_3: $\frac{1}{20}$; P_4: $\frac{1}{20}$; P_5: $\frac{1}{20}$

(c) [10: 6,3,2,1,1]. Shapley-Shubik power distribution: P_1: $\frac{1}{2}$; P_2: $\frac{1}{4}$; P_3: $\frac{1}{12}$; P_4: $\frac{1}{12}$; P_5: $\frac{1}{12}$

(d) [13: 6,3,2,1,1]. Shapley-Shubik power distribution: P_1: $\frac{1}{5}$; P_2: $\frac{1}{5}$; P_3: $\frac{1}{5}$; P_4: $\frac{1}{5}$; P_5: $\frac{1}{5}$

33. (a) $7 \le q \le 13$

(b) For $q = 7$ or $q = 8$, P_1 is a dictator since $w_1 = 8 \ge q$.

(c) For $q = 9$, only P_1 has veto power since P_1 is the only player that can single-handedly prevent the rest of the players from passing a motion.

(d) For $10 \le q \le 12$, both P_1 and P_2 have veto power since no motion can pass without both of their votes. For $q = 13$, all three players have veto power.

(e) For $q = 7$ or $q = 8$, both P_2 and P_3 are dummies. For $10 \le q \le 12$, P_3 is a dummy since all winning coalitions contain $\{P_1, P_2\}$ which is itself a winning coalition.

35. (a) Both have Banzhaf power distribution: P_1: $\frac{2}{5}$; P_2: $\frac{1}{5}$; P_3: $\frac{1}{5}$; P_4: $\frac{1}{5}$.

(b) In the weighted voting system $[q: w_1, w_2, \ldots, w_N]$, P_k is critical in a coalition means that the sum of the weights of all the players in that coalition (including P_k) is at least q but the sum of the weights of all the players in the coalition except P_k is less than q. Consequently, if the weights of all the players are multiplied by $c > 0$ ($c \le 0$ would make no sense), then the sum of the weights of all the players in the coalition (including P_k) is at least cq but the sum of the weights of all the players in the coalition except P_k is less than cq. Therefore P_k is critical in the same coalition in the weighted voting system $[cq: cw_1, cw_2, \ldots, cw_N]$. Since the critical players are the same in both weighted voting systems, the Banzhaf power distributions will be the same.

37. (a) If a player X has Banzhaf power index 0 then X is not critical in any coalition and so the addition or deletion of X to or from any coalition will never change the coalition from losing to winning or winning to losing. It follows that X can never be pivotal in any sequential coalition and so X must have Shapley-Shubik power index 0.

(b) If a player X has Shapley-Shubik power index 0 then X is not pivotal in any sequential coalition and so X can never be added to a losing coalition and turn it into a winning coalition. It follows that X can never be critical in any coalition so X has Banzhaf power index 0.

39. You should buy your vote from P_1. The following table explains why.

Buying a vote from	Resulting weighted voting system	Resulting Banzhaf power distribution	Your power
P_1	[6: 3, 2, 2, 2, 2]	$P_1: \frac{1}{5}; P_2: \frac{1}{5}; P_3: \frac{1}{5}; P_4: \frac{1}{5}; P_5: \frac{1}{5}$	$\frac{1}{5}$
P_2	[6: 4, 1, 2, 2, 2]	$P_1: \frac{1}{2}; P_2: 0; P_3: \frac{1}{6}; P_4: \frac{1}{6}; P_5: \frac{1}{6}$	$\frac{1}{6}$
P_3	[6: 4, 2, 1, 2, 2]	$P_1: \frac{1}{2}; P_2: \frac{1}{6}; P_3: 0; P_4: \frac{1}{6}; P_5: \frac{1}{6}$	$\frac{1}{6}$
P_4	[6: 4, 2, 2, 1, 2]	$P_1: \frac{1}{2}; P_2: \frac{1}{6}; P_3: \frac{1}{6}; P_4: 0; P_5: \frac{1}{6}$	$\frac{1}{6}$

41. (a) You should buy your vote from P_2. The following table explains why.

Buying a vote from	Resulting weighted voting system	Resulting Banzhaf power distribution	Your power
P_1	[18: 9, 8, 6, 4, 3]	$P_1: \frac{4}{13}; P_2: \frac{3}{13}; P_3: \frac{3}{13}; P_4: \frac{2}{13}; P_5: \frac{1}{13}$	$\frac{1}{13}$
P_2	[18: 10, 7, 6, 4, 3]	$P_1: \frac{9}{25}; P_2: \frac{1}{5}; P_3: \frac{1}{5}; P_4: \frac{3}{25}; P_5: \frac{3}{25}$	$\frac{3}{25}$
P_3	[18: 10, 8, 5, 4, 3]	$P_1: \frac{5}{12}; P_2: \frac{1}{4}; P_3: \frac{1}{6}; P_4: \frac{1}{12}; P_5: \frac{1}{12}$	$\frac{1}{12}$
P_4	[18: 10, 8, 6, 3, 3]	$P_1: \frac{5}{12}; P_2: \frac{1}{4}; P_3: \frac{1}{6}; P_4: \frac{1}{12}; P_5: \frac{1}{12}$	$\frac{1}{12}$

(b) You should buy 2 votes from P_2. The following table explains why.

Buying a vote from	Resulting weighted voting system	Resulting Banzhaf power distribution	Your power
P_1	[18: 8, 8, 6, 4, 4]	$P_1: \frac{7}{27}; P_2: \frac{7}{27}; P_3: \frac{7}{27}; P_4: \frac{1}{9}; P_5: \frac{1}{9}$	$\frac{1}{9}$
P_2	[18: 10, 6, 6, 4, 4]	$P_1: \frac{5}{13}; P_2: \frac{2}{13}; P_3: \frac{2}{13}; P_4: \frac{2}{13}; P_5: \frac{2}{13}$	$\frac{2}{13}$
P_3	[18: 10, 8, 4, 4, 4]	$P_1: \frac{11}{25}; P_2: \frac{1}{5}; P_3: \frac{3}{25}; P_4: \frac{3}{25}; P_5: \frac{3}{25}$	$\frac{3}{25}$
P_4	[18: 10, 8, 6, 2, 4]	$P_1: \frac{9}{25}; P_2: \frac{7}{25}; P_3: \frac{1}{5}; P_4: \frac{1}{25}; P_5: \frac{3}{25}$	$\frac{3}{25}$

(c) Buying a single vote from P_2 raises your power from $\frac{1}{25} = 4\%$ to $\frac{3}{25} = 12\%$. Buying a second vote from P_2 raises your power to $\frac{2}{13} \approx 15.4\%$. The increase in power is less with the second vote, but if you value power over money, it might still be worth it to you to buy that second vote.

CHAPTER 3

Walking

1. (a) \$9.00 **(b)** \$3.00 **(c)** \$1.00 **(d)** \$2.00 **(e)** \$3.00

3. (a) \$6.00 **(b)** \$4.00 **(c)** \$2.00

(d) piece 1: \$1.00; piece 2: \$1.50; piece 3: \$2.00; piece 4: \$2.50; piece 5: \$3.00; piece 6: \$2.00

5. Ana: s_2, s_3; Ben: s_3; Cara: s_1, s_2, s_3

7. Abe: s_1, s_4; Betty: s_1, s_2; Cory: s_1, s_2; Dana: s_4

9. (a) only (iii) **(b)** either

11. (a) Three answers are possible as shown in the following table:

Chooser 1	Chooser 2	Divider
s_2	s_1	s_3
s_2	s_3	s_1
s_3	s_1	s_2

(b) Two answers are possible as shown in the following table:

Chooser 1	Chooser 2	Divider
s_3	s_1	s_2
s_2	s_1	s_3

(c) The only possible fair division is

Chooser 1	Chooser 2	Divider
s_1	s_2	s_3

(d) The divider can pick between s_2 and s_3—let's say the divider picks s_2. Then s_1 and s_3 can be combined again into a cake that may then be divided between Chooser 1 and Chooser 2 using the divider-chooser method.

13. (a) One possible fair division of the cake is

Chooser 1	Chooser 2	Chooser 3	Divider
s_2	s_3	s_1	s_4

(b) Another possible fair division of the cake is

Chooser 1	Chooser 2	Chooser 3	Divider
s_3	s_1	s_2	s_4

(c) Since none of the choosers have chosen s_4, s_4 can only be given to the divider.

15. (a) A fair division of the cake is

Chooser 1	Chooser 2	Chooser 3	Chooser 4	Divider
s_2	s_4	s_3	s_5	s_1

(b) Another fair division of the cake is

Chooser 1	Chooser 2	Chooser 3	Chooser 4	Divider
s_4	s_2	s_3	s_5	s_1

(c) Since none of the choosers chose s_1, s_1 can only be given to the divider.

17. (a) A fair division of the cake is

Chooser 1	Chooser 2	Chooser 3	Chooser 4	Chooser 5	Divider
s_5	s_1	s_6	s_2	s_3	s_4

(b) Chooser 5 must get s_3 which forces chooser 4 to get s_2. This leaves only s_5 for chooser 1 which in turn leaves only s_6 for chooser 3. Consequently only s_1 is left for chooser 2 and the divider gets the piece left over—s_4.

19. (a) Chooser 1: $\{s_3, s_4\}$; Chooser 2: $\{s_1, s_3, s_4\}$; Chooser 3: $\{s_3\}$

(b) A fair division of the land is

Chooser 1	Chooser 2	Chooser 3	Divider
s_4	s_1	s_3	s_2

21. The value of the parts of the cake (as a percentage of the total) as seen by each person is shown in the figure.

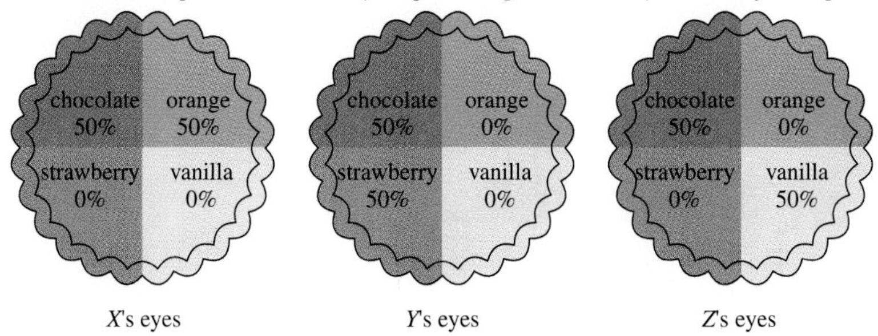

(a) One possible fair division is

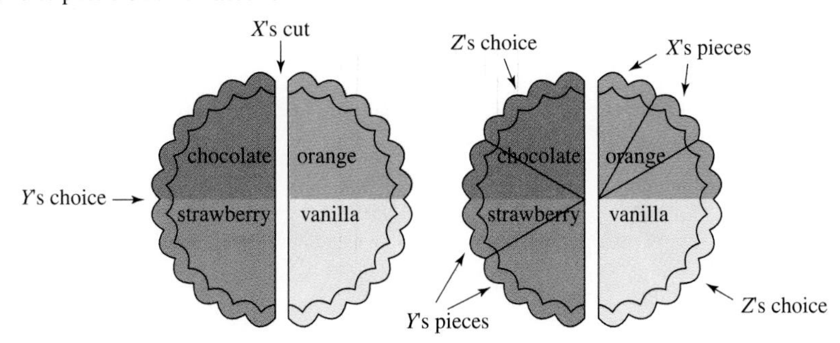

In this division the value of *X*'s final share in *X*'s own eyes is $\frac{2}{3} \cdot 50\% + 50\% = 33\frac{1}{3}\%$, the value of *Y*'s final share in *Y*'s own eyes is $\frac{2}{3} \cdot 100\% = 66\frac{2}{3}\%$, and the value of *Z*'s final share in *Z*'s own eyes is $\frac{2}{3} \cdot 50\% + 50\% = 83\frac{1}{3}\%$.

(b) One possible fair division is

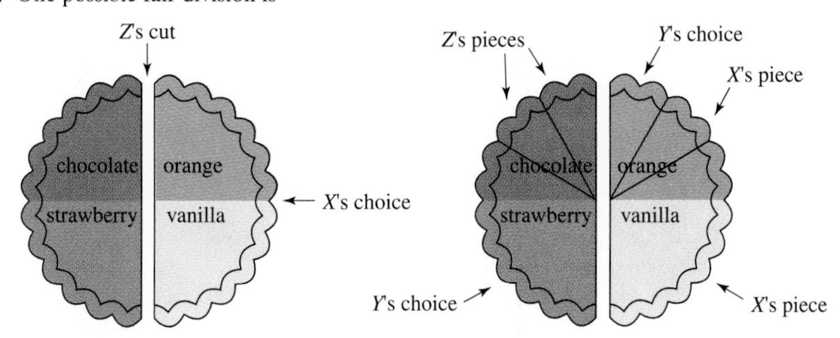

In this division the value of *X*'s final share in *X*'s own eyes is $\frac{2}{3} \cdot 50\% = 33\frac{1}{3}\%$, the value of *Y*'s final share in *Y*'s own eyes is $\frac{1}{3} \cdot 50\% + 50\% + 0\% = 66\frac{2}{3}\%$, and the value of *Z*'s final share in *Z*'s own eyes is $\frac{2}{3} \cdot 50\% = 33\frac{1}{3}\%$.

(c) One possible fair division is

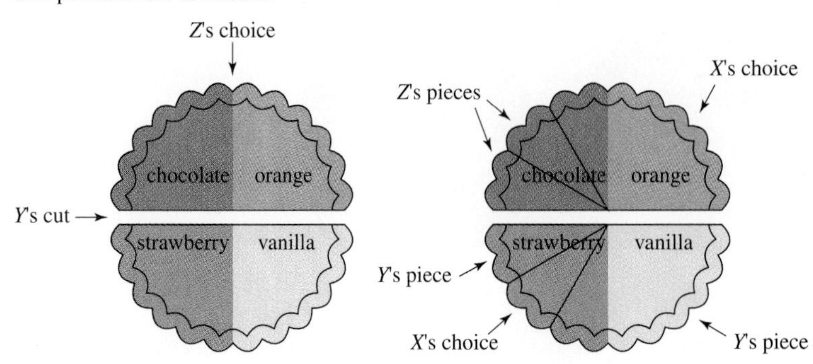

In this division the value of X's final share in X's own eyes is $\frac{1}{3} \cdot 50\% + 50\% + 0\% = 66\frac{2}{3}\%$, the value of Y's final share in Y's own eyes is $\frac{2}{3} \cdot 50\% = 33\frac{1}{3}\%$, and the final value of Z's final share in Z's own eyes is $\frac{2}{3} \cdot 50\% = 33\frac{1}{3}\%$.

23. **(a)** P_9 **(b)** P_1 **(c)** P_{12} **(d)** P_5 **(e)** P_1 **(f)** P_2 **(g)** P_{12} **(h)** 11

25. A gets the desk and receives \$200 in cash; B gets the dresser and receives \$80; C gets the vanity and the tapestry and pays \$280.

27. **(a)** Bob gets the partnership and pays \$155,000.
(b) Jane gets \$80,000 and Ann gets \$75,000.

29. A ends up with items 1, 2, and 4 and must pay \$170,666.66; B ends up with \$90,333.33; C ends up with item 3 and \$80,333.33.

31. A ends up with items 4 and 5 and pays \$739; B ends up with \$608; C ends up with items 1 and 3 and pays \$261; D ends up with \$632; E ends up with items 2 and 6 and pays \$240.

33. **(a)** A gets items 10, 11, 12, 13; B gets items 1, 2, 3; C gets items 5, 6, 7.
(b) Items 4, 8, and 9 are left over.

35. **(a)** A gets items 1, 2; B gets items 10, 11, 12; C gets items 4, 5, 6, 7.
(b) Items 3, 8, and 9 are left over.

37. **(a)** A gets items 19, 20; B gets items 15, 16, 17; C gets items 1, 2, 3; D gets items 11, 12, 13; E gets items 5, 6, 7, 8.
(b) Items 4, 9, 10, 14, and 18 are left over.

39. **(a)** A gets items 4, 5; B gets item 10; C gets item 15; D gets items 1, 2.
(b) Items 3, 6, 7, 8, 9, 11, 12, 13, and 14 are left over.

Jogging

41. Paul would choose the larger portion, worth \$2.70 to him.

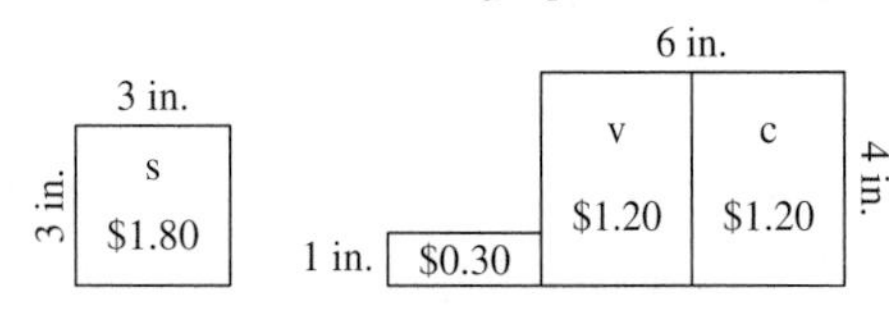

Since Paul likes all flavors equally, he would divide his piece into 3 equal (in volume) pieces, each worth \$0.90 to him. Peter would also divide his pieces into 3 equal (in volume) pieces.

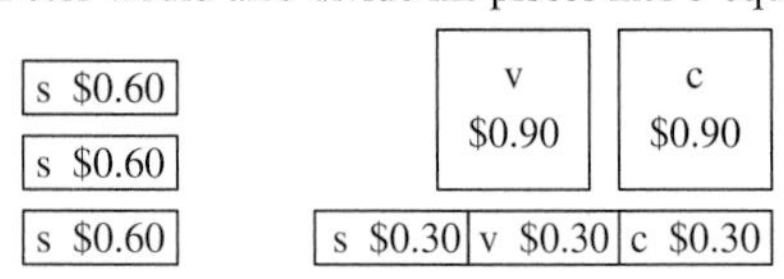

Mary would choose any one of Peter's pieces (each worth \$0.10 to her) and the vanilla piece from Paul (worth \$1.35 to her).

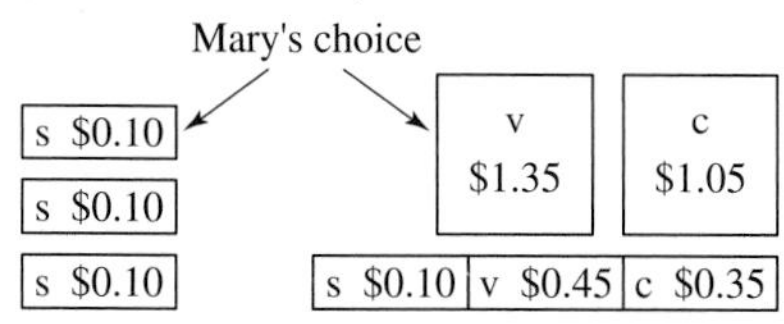

Mary's choice

After all is said and done, Peter values his portion at \$1.20, Paul values his portion at \$1.80, and Mary values her portion at \$1.45.

43. **(a)** The total area is 30,000 m^2 and the area of C is only 8000 m^2. Since P_2 and P_3 value the land uniformly, each thinks that a fair share must have an area of at least 10,000 m^2.

(b) Since there are 22,000 m^2 left, any cut that divides the remaining property in parts of 11,000 m^2 will work. For example

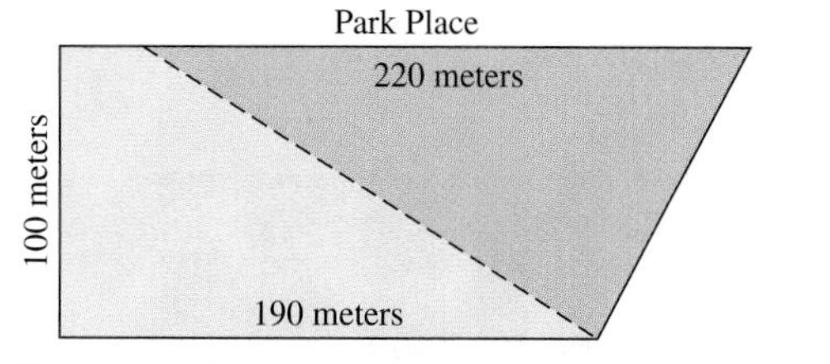

(c) The cut parallel to Baltic Avenue which divides the remaining parcel in half is

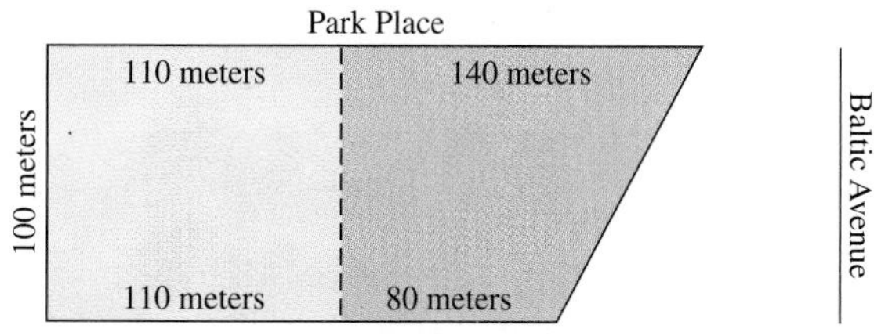

45. **(a)** H **(b)** G **(c)** F **(d)** F **(e)** F **(f)** G

47. **Step 1 (The Bids).** Each player makes a sealed bid giving his or her honest assessment of the dollar value of each of the items in the estate.

Step 2 (The Allocation). Each item goes to the highest bidder for that item. (In case of ties, a predetermined tie-breaking procedure should be invoked.)

Step 3 (The Payments). Each player's fair share is calculated by multiplying the total of that player's bids by the percentage that player is entitled to. $\left(\text{Multiply the total of } P_1\text{'s bids by } \frac{r_1}{100}, \text{ etc.}\right)$ Each player puts in or takes out from a common pot (*the estate*) the difference between his or her fair share and the total value of the items allocated to that player.

Step 4 (Dividing the surplus). After the original allocations are completed there may be a surplus of cash in the estate. This surplus is divided among the players according to the percentage each is entitled to. (P_1 gets $\frac{r_1}{100}$, of the surplus, etc.)

49. Ruth cleans bathrooms and pays \$11.67 per month; Sarah cooks and pays \$11.67 per month; Tamara washes dishes, vacuums, mows the lawn and receives \$23.34 per month.

51. **(a)**

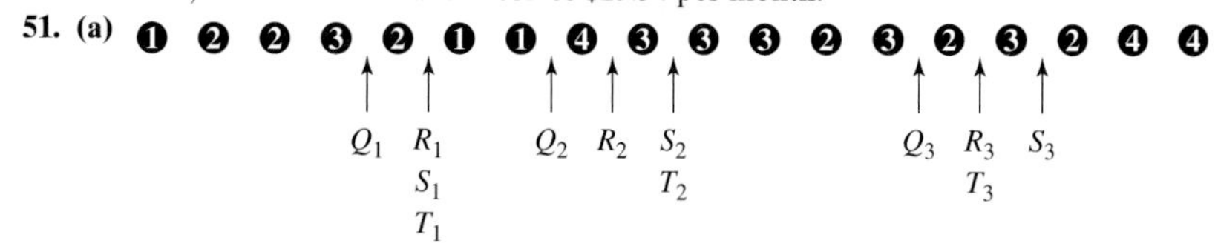

(b)

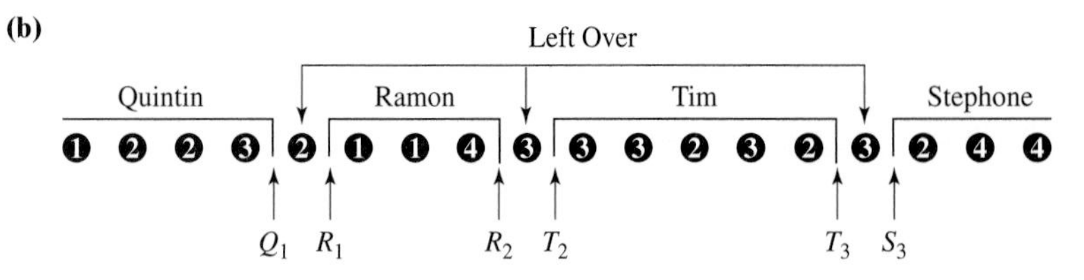

(c) These are discrete, indivisible items, so the only available methods are the method of markers and the sealed bids method. With only a few items to share, the sealed bid method is appropriate.

53. (a) One possible answer is given below.

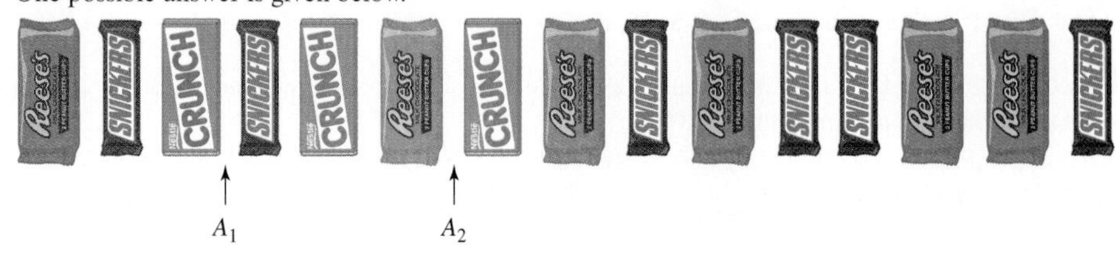

(b) One possible answer is given below.

(c) One possible answer is given below.

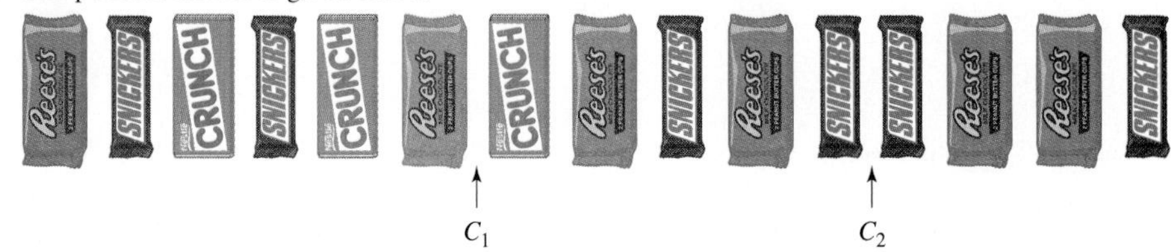

(d) *A* gets 1 Nestle Crunch Bar, 1 Reese's Peanut Butter Cups, and 1 Snickers Bar; *B* gets 1 Snickers Bar, 2 Nestle Crunch Bars, and 2 Reese's Peanut Butter Cups; *C* gets 2 Snickers Bars and 2 Reese's Peanut Butter Cups.

(e) 2 Snickers Bars and 1 Reese's Peanut Butter Cups.

55. This problem can be thought of as a fair division problem with 20 players in which 5 of the players are clones of *A*, 7 of the players are clones of *B*, and 8 of the players are clones of *C*.

CHAPTER 4

Walking

1. (a) 50,000 **(b)** *A*: 66.2; *B*: 53.4; *C*: 26.6; *D*: 13.8 **(c)** *A*: 66; *B*: 53; *C*: 27; *D*: 14

3. (a) *A*: 65.545; *B*: 52.871; *C*: 26.337; *D*: 13.663 **(b)** *A*: 66; *B*: 53; *C*: 27; *D*: 14

5. (a) The standard divisor (50,000) works. **(b)** *A*: 66; *B*: 53; *C*: 27; *D*: 14

7. *A*: 46; *B*: 31; *C*: 21; *D*: 14; *E*: 10; *F*: 8

9. *A*: 45; *B*: 31; *C*: 21; *D*: 14; *E*: 10; *F*: 9

11. *A*: 72; *B*: 86; *C*: 51; *D*: 16

13. *A*: 72; *B*: 86; *C*: 51; *D*: 16

15. (a) 119 **(b)** 200,000
(c) *A*: 81,000,000; *B*: 5,940,000; *C*: 4,730,000; *D*: 2,920,000; *E*: 2,110,000

17. *A*: 41; *B*: 30; *C*: 24; *D*: 14; *E*: 10

19. *A*: 40; *B*: 30; *C*: 24; *D*: 15; *E*: 10

21. Agriculture: 33; Business: 15; Education: 42; Humanities: 21; Science: 139.

23. Agriculture: 33; Business: 15; Education: 42; Humanities: 21; Science: 139.

25. **(a)** Bob: 0; Peter: 3; Ron: 8
(b) Bob: 1; Peter: 2; Ron: 8
(c) Yes. For studying an extra 2 minutes (an increase of 3.70%), Bob gets a piece of candy while Peter, who studies an extra 12 minutes (an increase of 4.94%), has to give up a piece. This is an example of the population paradox.

Jogging

27. Answers will vary. One such example is: Apportion 10 seats among the four states A, B, C, and D with populations given in the following table. Hamilton's method and Adams' method result in the same apportionment: A: 2; B: 3; C: 2; and D: 3.

State	*A*	*B*	*C*	*D*
Population (in millions)	2.24	2.71	2.13	2.92

29. **(a)** The standard quotas add up to 100. Rounding these standard quotas in the usual way gives A: 11; B: 24; C: 8; D: 36; and E: 20. These integers add up to 99. Consequently, we must choose a divisor that is *smaller* than the standard divisor so that we can obtain modified quotas that are slightly larger.
(b) The standard quotas add up to 100. Rounding these standard quotas in the usual way gives A: 12; B: 25; C: 8; D: 36; and E: 20. These integers add up to 101. Consequently, we must choose a divisor that is *bigger* than the standard divisor so that we can obtain modified quotas that are slightly smaller.
(c) If the standard quotas rounded in the conventional way (to the nearest integer) add up to M, then the standard divisor works as an appropriate divisor for Webster's method.

31. Answers will vary. One such example is: Apportion 90 seats among the four states A, B, C, and D with populations given in the following table. Webster's method results in the apportionment A: 72; B: 7; C: 6; D: 5. Adams' method results in the apportionment
A: 69; B: 8; C: 7; D: 6.

State	*A*	*B*	*C*	*D*
Population	70,800	7400	6400	5400

33. Answers will vary. One such example is: Apportion 90 seats among the four states A, B, C, and D with populations given in the following table. Webster's method apportions 72 seats to state A although state A has a standard quota of 70.8.

State	*A*	*B*	*C*	*D*
Population	70,800	7400	6400	5400

35. **(a)** A: 5, B: 10; C: 15; D: 21
(b) For $D = 100$ the modified quotas are A: 5, B:10, C: 15, D: 20. For $D < 100$, each of the modified quotas above will increase, so rounding upward will give at least A: 6, B; 11; C: 16, D: 21 for a total of at least 54. For $D > 100$, each of the modified quotas above will decrease, so rounding upward will give at most A: 5, B: 10, C: 15, D: 20 for a total of at most 50.
(c) From part (b) we see that there is no divisor such that after rounding the modified quotas upward, the total is 51.

37. **(a)**

CT	DE	GA	KY	MD	MA	NH	NJ	NY	NC	PA	RI	SC	VT	VA
7	2	2	2	8	14	4	5	10	10	13	2	6	2	18

(b)

CT	DE	GA	KY	MD	MA	NH	NJ	NY	NC	PA	RI	SC	VT	VA
7	1	2	2	8	14	4	5	10	10	13	2	6	2	19

(c) Virginia; Delaware

39. **(a)** In Jefferson's method the modified quotas are larger than the standard quotas and so rounding downward will give each state at least the integer part of the standard quota for that state.
(b) In Adams' method the modified quotas are smaller than the standard quota and so rounding upward will give each state at most one more than the integer part of the standard quota for that state.
(c) If there are only two states, an upper quota violation for one state results in a lower quota violation for the other state (and vice versa). Since neither Jefferson's nor Adams' method can have both upper and lower violations of the quota rule, neither can violate the quota rule when there are only two states.

41. **(a)** Take for example $q_1 = 3.9$ and $q_2 = 10.1$ ($M = 14$). Under both Hamilton's method and Lowndes' method, A gets 4 seats and B gets 10 seats.
(b) Take for example $q_1 = 3.4$ and $q_2 = 10.6$ ($M = 14$). Under Hamilton's method, A gets 3 seats and B gets 11 seats. Under Lowndes' method, A gets 4 seats and B gets 10 seats.

(c) If $f_1 > f_2$, then under Hamilton's method the surplus seat goes to A. Under Lowndes' method, the surplus seat would go to B if $\dfrac{f_2}{q_2 - f_2} > \dfrac{f_1}{q_1 - f_1}$.

CHAPTER 5

Walking

1. (a) Vertices: A, B, C, D; Edges: AB, AC, AD, BD; deg$(A) = 3$, deg$(B) = 2$, deg$(C) = 1$, deg$(D) = 2$
(b) Vertices: A, B, C; Edges: none; deg$(A) = 0$, deg$(B) = 0$, deg$(C) = 0$
(c) Vertices: V, W, X, Y, Z; Edges: $XX, XY, XZ, XV, XW, WY, YZ$; deg$(V) = 1$, deg$(W) = 2$, deg$(X) = 6$, deg$(Y) = 3$, deg$(Z) = 2$

3. (a)

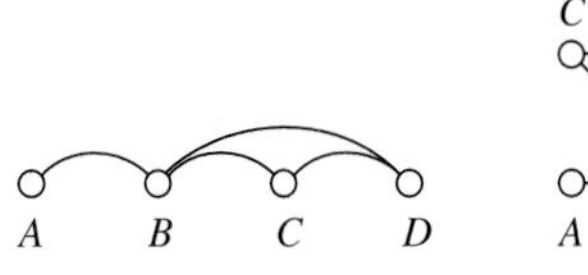

C D

A B

(b)

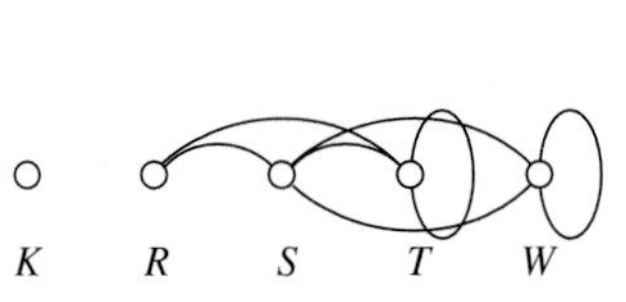

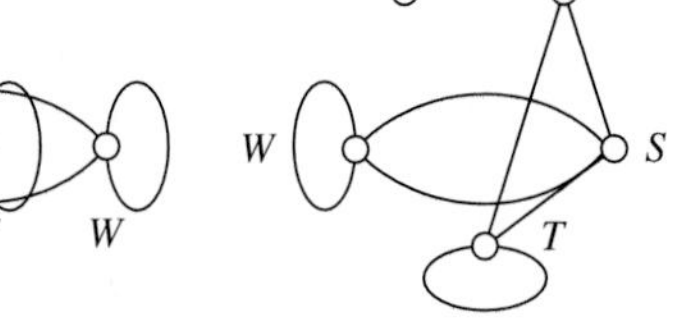

5. (a) Both graphs have four vertices A, B, C, and D and (the same) edges AB, AC, AD, BD.
(b)

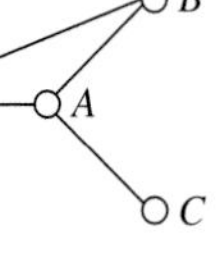

7. (a) **(b)**

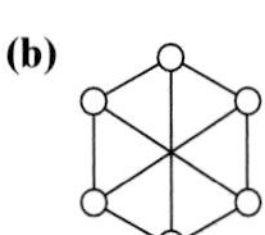

9. (a)

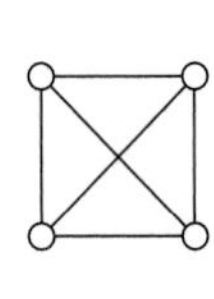

(b) **(c)**

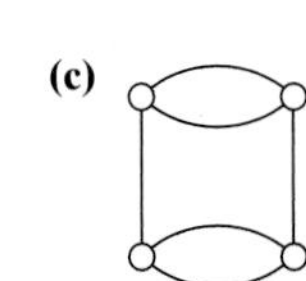

(d)

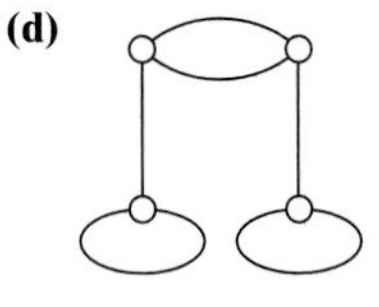

11. (a) C, B, A, H, F **(b)** C, B, D, A, H, F **(c)** 4 (C, B, A; C, D, A; C, B, D, A; C, D, B, A)
(d) 3 (H, F; H, G, F; H, G, G, F)
(e) 12 [Any one of the paths in (c) followed by AH, followed by any one of the paths in (d).]

13. (a) D, C, B, A, D
(b) 6 (D, C, B, D; D, B, C, D; D, A, B, D; D, B, A, D; D, C, B, A, D; D, A, B, C, D)
Note: Is the "same" circuit read backward a "different" circuit or not? This is always a matter of some controversy but for several reasons it is somewhat more convenient to start with the presumption that they are indeed different. This is consistent with the definition on page 157.
(c) HA and FE

15. (a) None of them. (b) o———o———o———o

17. (a) Has an Euler circuit since all vertices have even degree.
(b) Has no Euler circuit, but has an Euler path since there are exactly two vertices of odd degree.
(c) Has neither an Euler circuit nor an Euler path since there are four vertices of odd degree.

19. (a) Has an Euler circuit since all vertices have even degree.
(b) Has no Euler circuit, but has an Euler path since there are exactly two vertices of odd degree.
(c) Has no Euler circuit, but has an Euler path since there are exactly two vertices of odd degree.

21. (a) Neither since there are more than two vertices of odd degree.

(b) Open unicursal tracing

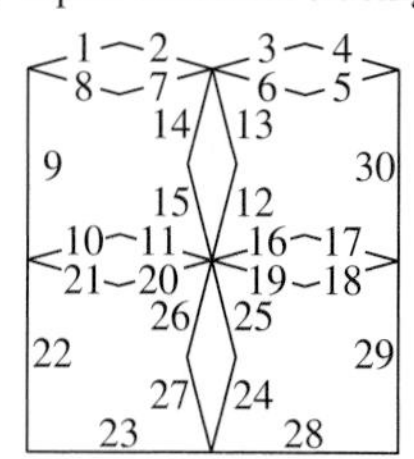

(c) Open unicursal tracing

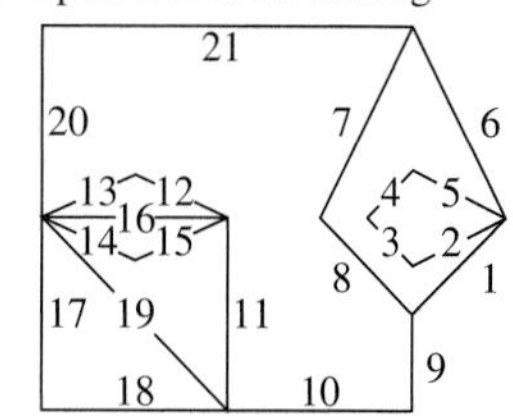

23. (a)

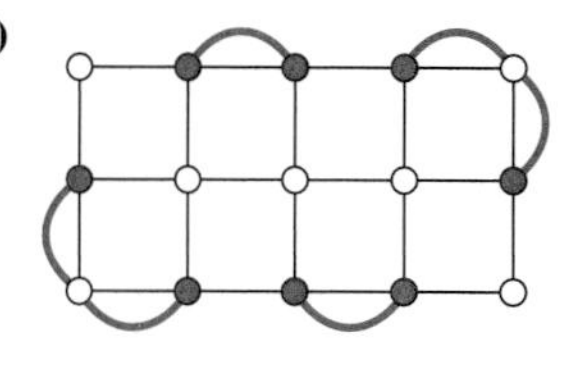

(b)

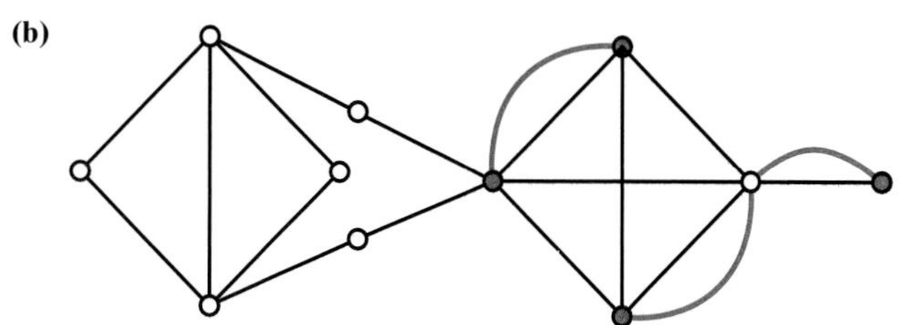

25. (a)

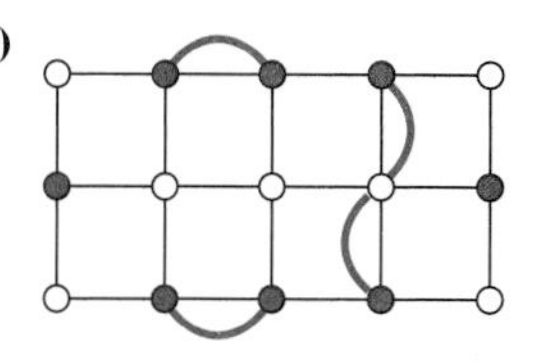

(b)

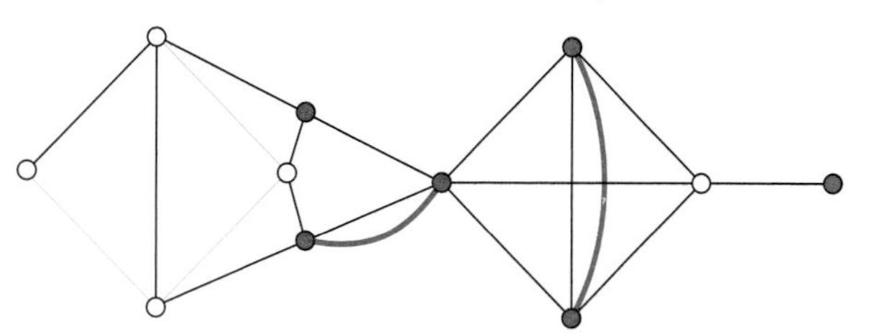

27.

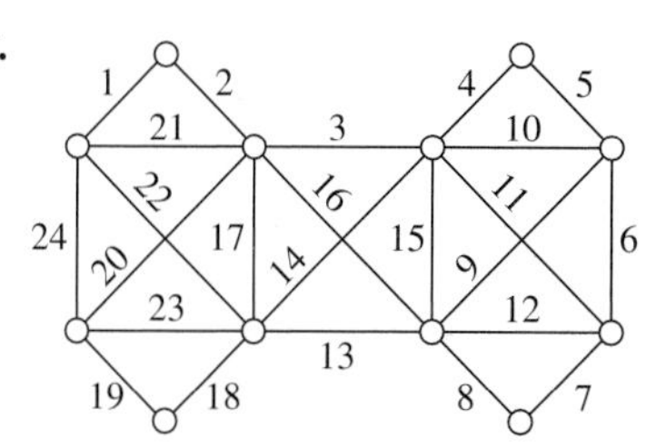

29. $A, B, C, D, E, F, G, A, C, E, G, B, D, F, A, D, G, C, F, B, E, A$

31.

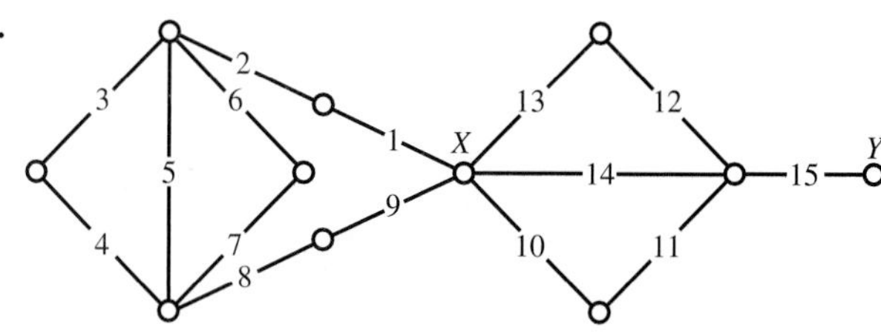

33. (a)

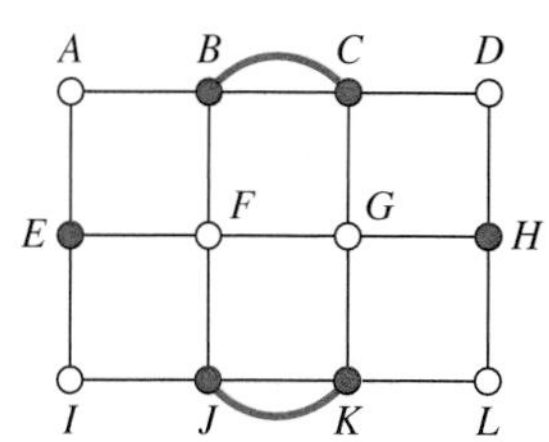

(b) BC and JK

35. 3 times

37.

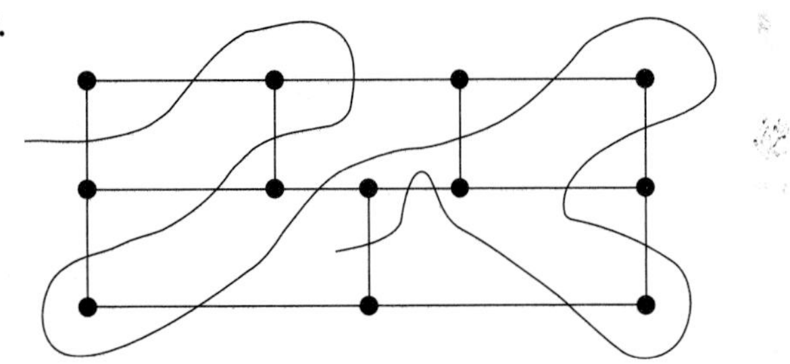

Jogging

39. (a)

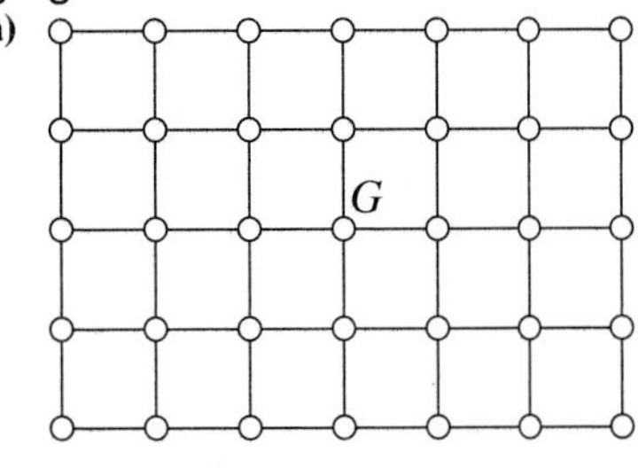

(b)

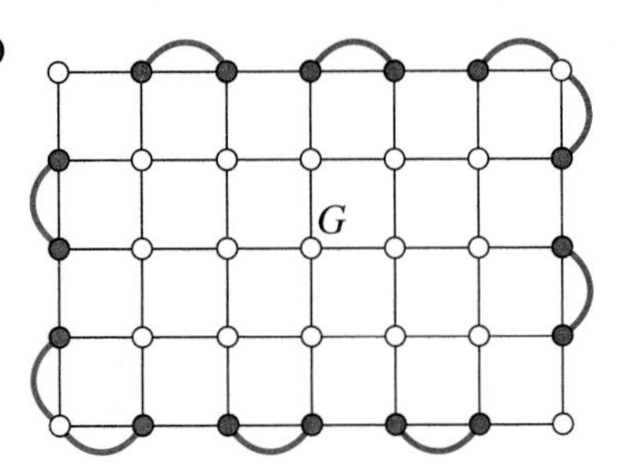

(c)

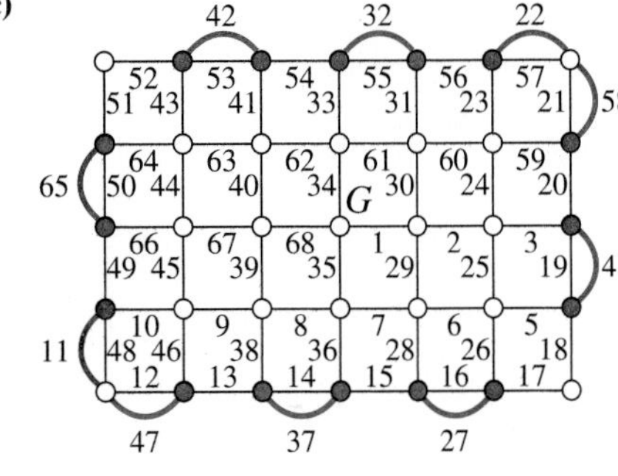

41. (a)

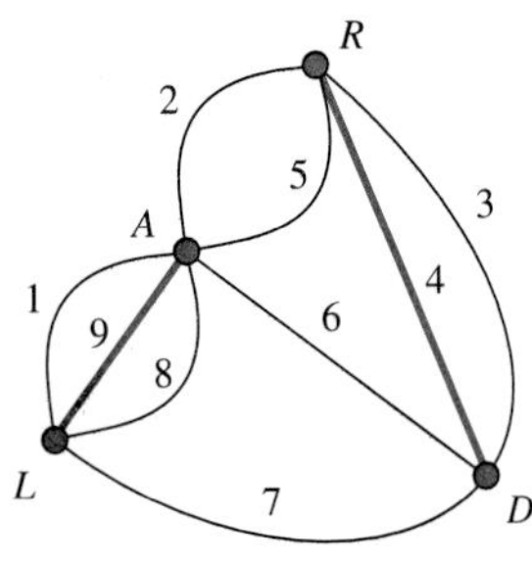

Eulerizing the graph shown in Fig. 5–18(b) requires the addition of two edges so the cheapest walk will cost $9.00. One possible such walk is shown in the figure.

(b)

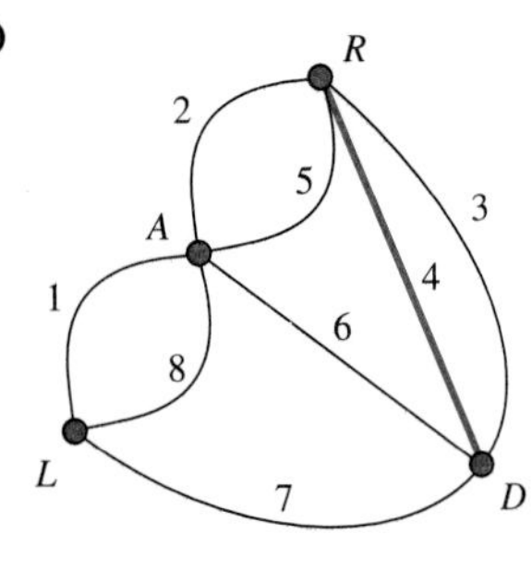

Semi-eulerizing the graph shown in Fig. 5–18(b) requires the addition of one edge so the cheapest walk will cost $8.00. One possible such walk (starting at L and ending at A) is shown in the figure.

43. **(a)**

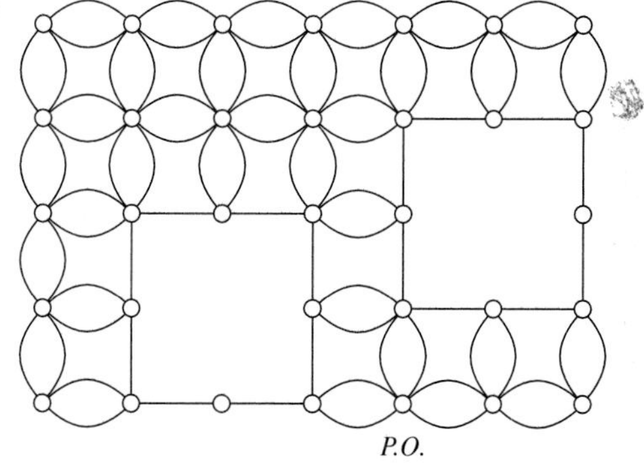

(b) The graph is already eulerized.

(c)

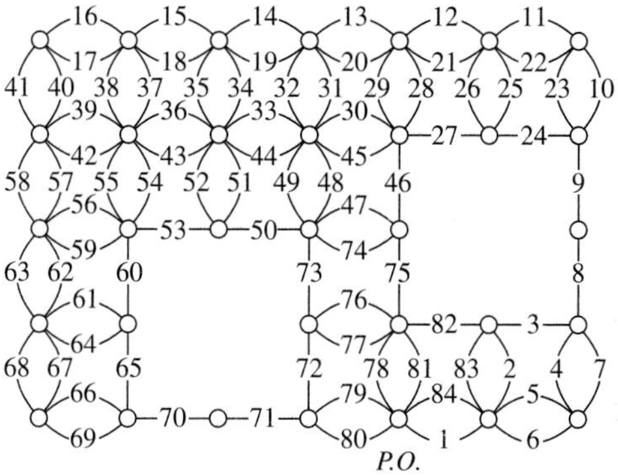

45.

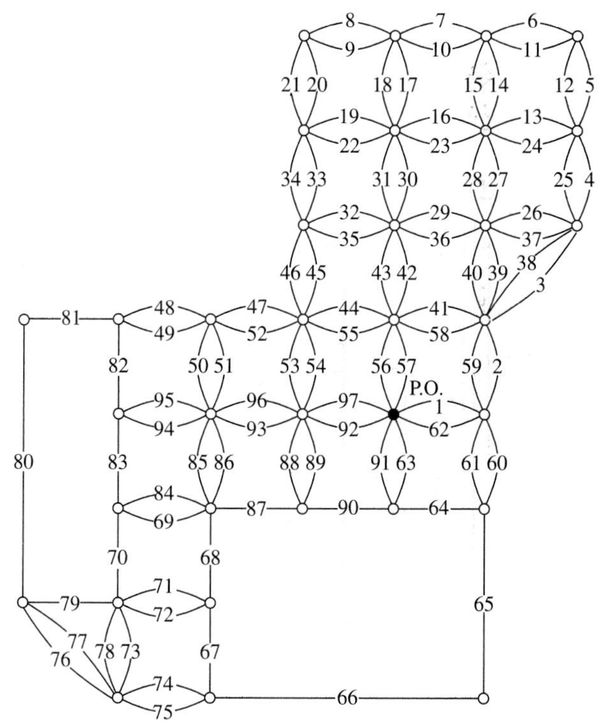

47. **(a)** 12

(b)

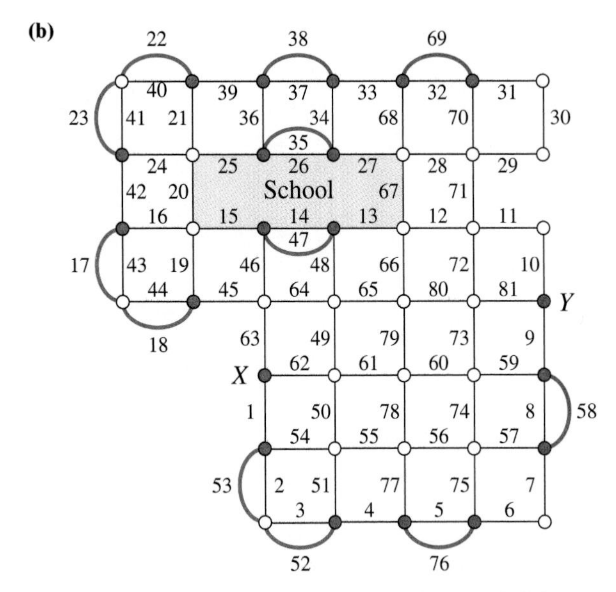

49. (a) The office complex can be represented by a graph (where each vertex represents a location and each edge a door).

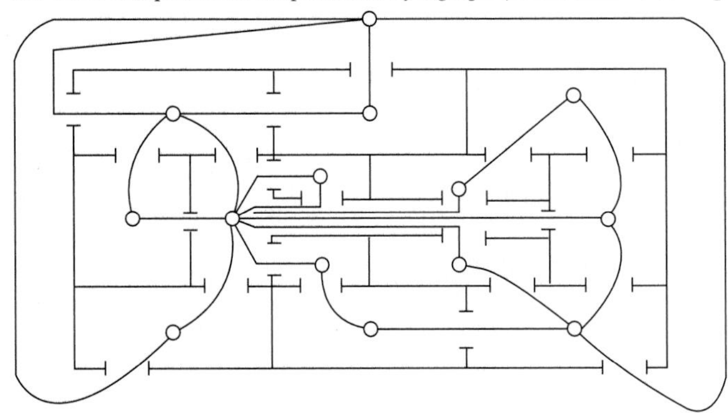

Since there are vertices of odd degree, (for example the secretary's office has degree 3), there is no Euler circuit.

(b) Since there are exactly 2 vertices of odd degree (the secretary's office has degree 3 and the hall has degree 9), there is an Euler path starting at either the secretary's office or the hall and ending at the other.

(c) If the door from the secretary's office to the hall is removed (i.e., the edge between the secretary's office and the hall is removed), then every vertex will have even degree and so there will be an Euler circuit. Consequently, it would be possible to start at any location, walk through every door exactly once and end up at the starting location.

51. (a)

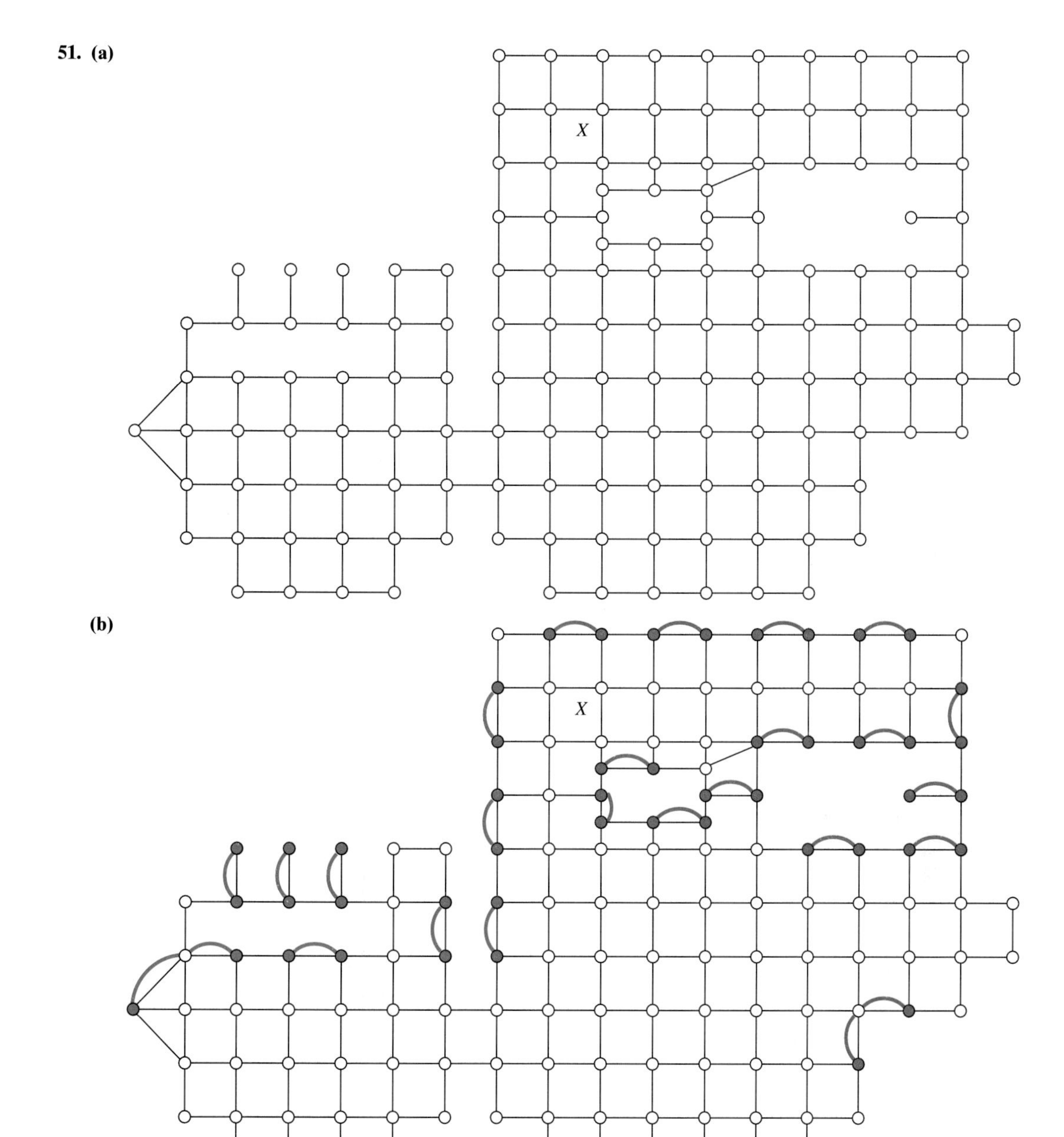

(c)

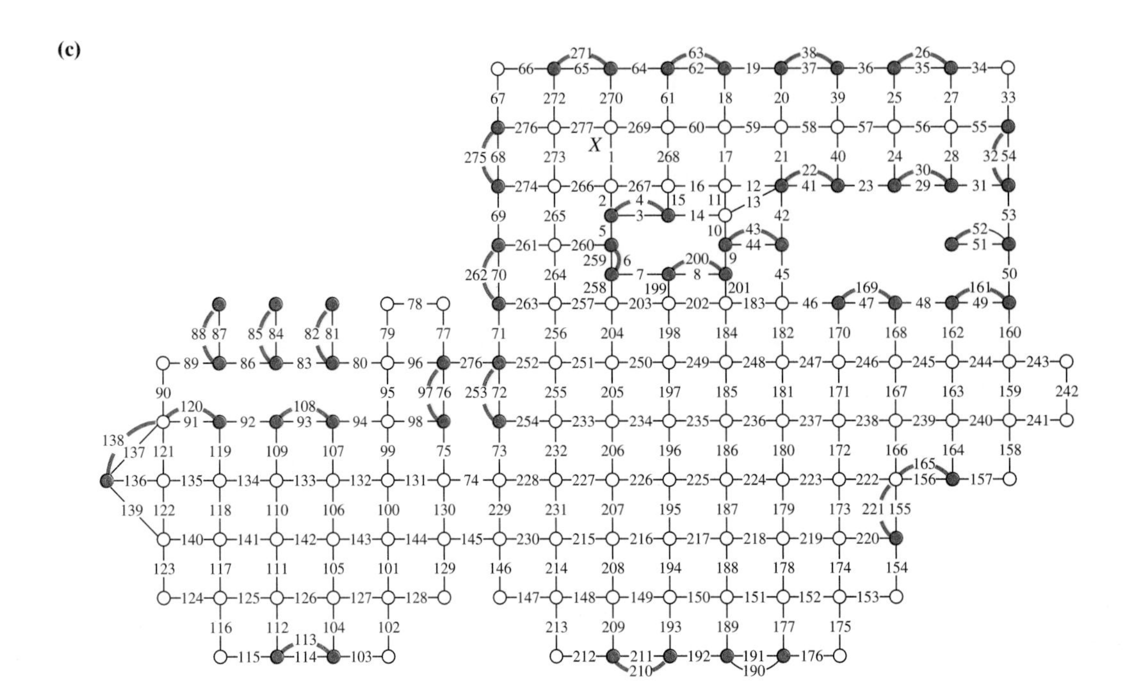

Chapter 6

Walking

1. (a) A, D, B, E, C, F, G, A **(b)** A, G, F, E, C, D, B **(c)** D, A, G, B, C, E, F

3.
A, B, C, D, E, F, G, A | A, G, F, E, D, C, B, A
A, B, E, D, C, F, G, A | A, G, F, C, D, E, B, A
A, F, C, D, E, B, G, A | A, G, B, E, D, C, F, A
A, F, E, D, C, B, G, A | A, G, B, C, D, E, F, A

5. (a) A, F, B, C, G, D, E, A and A, F, B, E, C, G, D, A **(b)** A, F, B, C, G, D, E **(c)** A, F, B, E, D, G, C
(d) F, A, B, E, D, C, G

7. (a)
A, B, C, D, E, F, A | A, F, E, D, C, B, A
A, B, E, D, C, F, A | A, F, C, D, E, B, A

(b)
D, E, F, A, B, C, D | D, C, B, A, F, E, D
D, C, F, A, B, E, D | D, E, B, A, F, C, D

9. The degree of every vertex in a graph with a Hamilton circuit must be at least 2 since the circuit must "pass through" every vertex. A graph with Hamilton path can have at most 2 vertices (the starting and ending vertices of the path) of degree 1 since the path must "pass through" the remaining vertices. This graph has 4 vertices of degree 1.

11. (a) 6 **(b)** 4 **(c)** A, B, C, D, E, A (weight 32) **(d)** A, D, B, C, E, A (weight 27)

13. (a) 11 **(b)** A, B, C, F, E, D, A (weight 37) **(c)** A, D, F, E, B, C, A (weight 41)

15. (a) $6! = 5! \times 6 = 120 \times 6 = 720$
(b) $9! = \dfrac{10!}{10} = 362{,}880$
(c) $9! = 362{,}880$

17. (a) A, C, B, D, A (weight 62)
(b) A, D, C, B, A (weight 80)
(c) A, B, D, C, A (weight 74)

(d) The solution obtained in part (b) is 29.0% more expensive than the optimal solution found in part (a)—relative error = 0.290. The solution obtained in part (c) is 19.4% more expensive than the optimal solution—relative error = 0.194.

19. **(a)** B, C, A, E, D, B; $121 + 119 + 133 + 199 + 150 = 722$
(b) C, A, E, D, B, C; $119 + 133 + 199 + 150 + 121 = 722$
(c) D, B, C, A, E, D; $150 + 121 + 119 + 133 + 199 = 722$
(d) E, C, A, D, B, E; $120 + 119 + 152 + 150 + 200 = 741$

21. A, D, B, E, C, A; $1250

23. A, D, B, C, E, A; $1220

25. E, P, C, H, T, G, E; 12.5 years

27. **(a)** A, E, B, C, D, A
(b) No. The 5 edges used in the Hamilton circuit are the 5 cheapest edges in the graph.

29. A, E, C, D, B, A (weight 91)

31. A, B, E, D, C, A (weight 9.9)

33. C, D, E, A, B, C (weight 9.8)

35. A, B, F, C, D, E, A (weight (203)

37. **(a)**

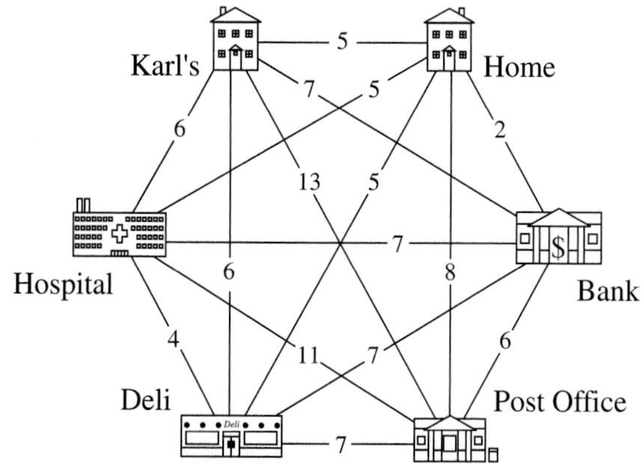

(b) Home, Bank, Post Office, Deli, Hospital, Karl's, Home. The total length of the trip is 30 miles.

Jogging

39. Each vertex is adjacent to each of the other vertices, so each vertex has degree $N - 1$. Since there are N vertices, the sum of the degrees of all the vertices is $N(N - 1)$. But the sum of the degrees of all the vertices in a graph is always equal to twice the number of edges. Therefore, the number of edges in a complete graph with N vertices is $\frac{N(N-1)}{2}$.

41.

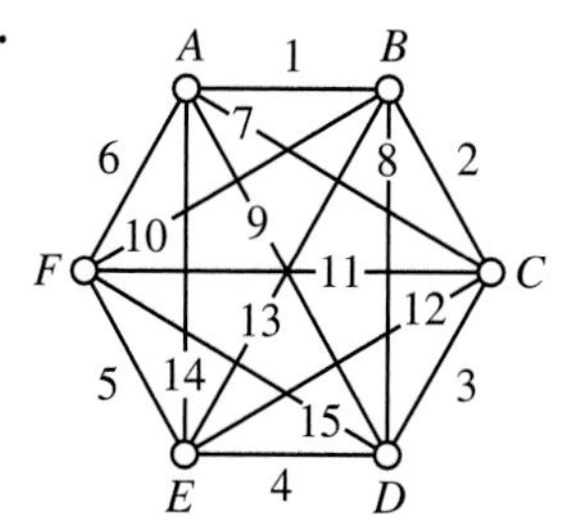

43. $A, J, D, H, B, F, K, G, C, I, E, A$

45. $A, B, C, D, J, I, F, G, E, H$

47. **(a)** $2^3 = 8 > 6 = 3!$ **(b)** $2^4 = 16 < 24 = 4!$

(c) $N!$ is bigger.

$2^5 = 2 \cdot 2^4 < 2 \cdot 4! < 5 \cdot 4! = 5!$
$2^6 = 2 \cdot 2^5 < 2 \cdot 5! < 6 \cdot 5! = 6!$
$2^7 = 2 \cdot 2^6 < 2 \cdot 6! < 7 \cdot 6! = 7!$
$\vdots$
$2^{k+1} = 2 \cdot 2^k < 2 \cdot k! < (k+1) \cdot k = (k+1)!$
$\vdots$

In other words, as k increases by 1, 2^k increases by a factor of 2, but $k!$ increases by a factor of $(k+1)$.

49. Dallas, Houston, Memphis, Louisville, Columbus, Chicago, Kansas City, Denver, Atlanta, Buffalo, Boston, Dallas.

CHAPTER 7

Walking

1. (a) tree **(b)** not a tree (has a circuit, is not connected) **(c)** not a tree (has a circuit) **(d)** tree

3.

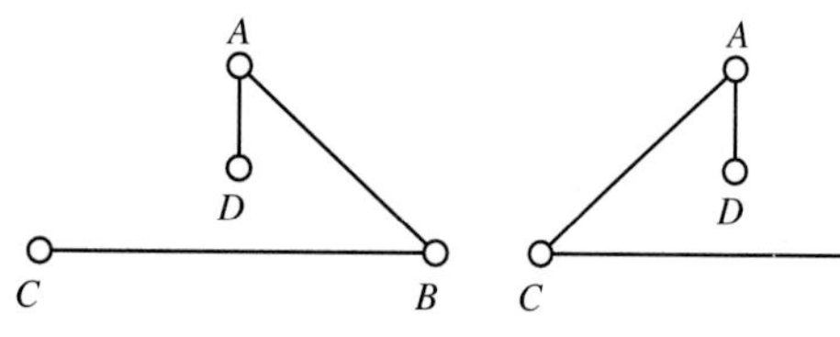

5. (a) **(b)** **(c)** **(d)**

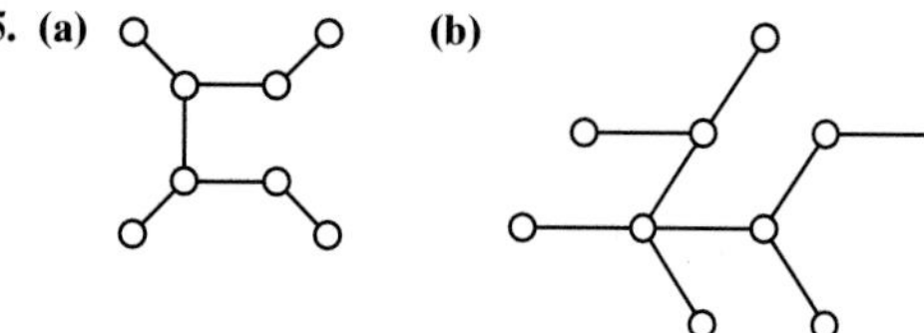

7.

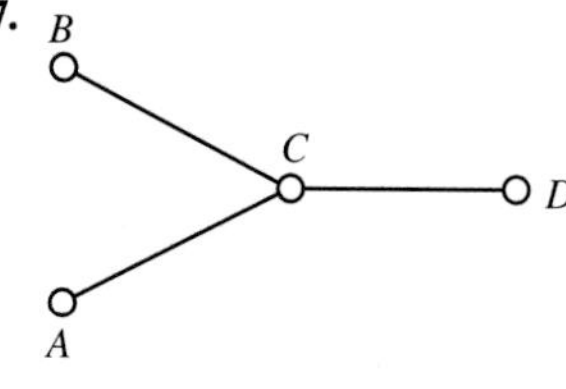

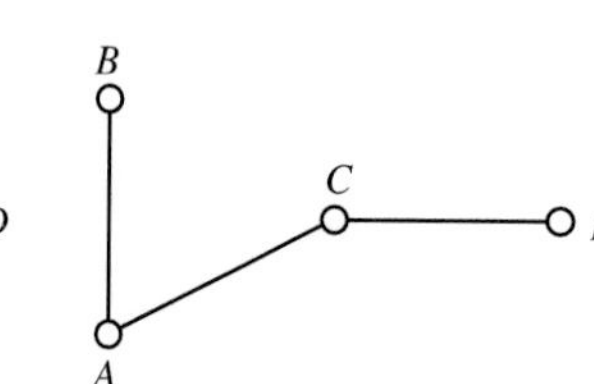

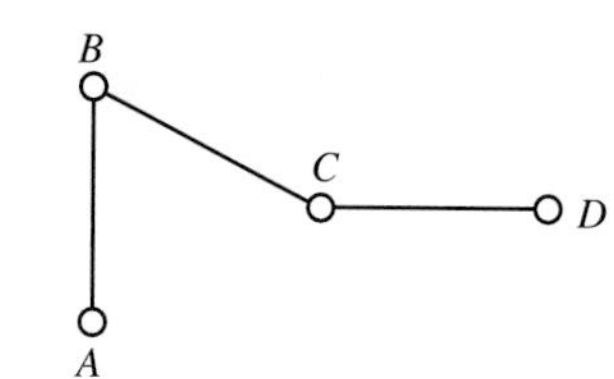

9. (a) 3 **(b)** 1

11.

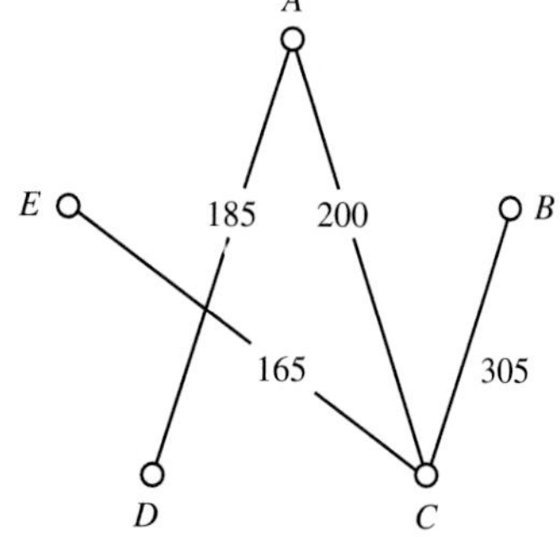

13.

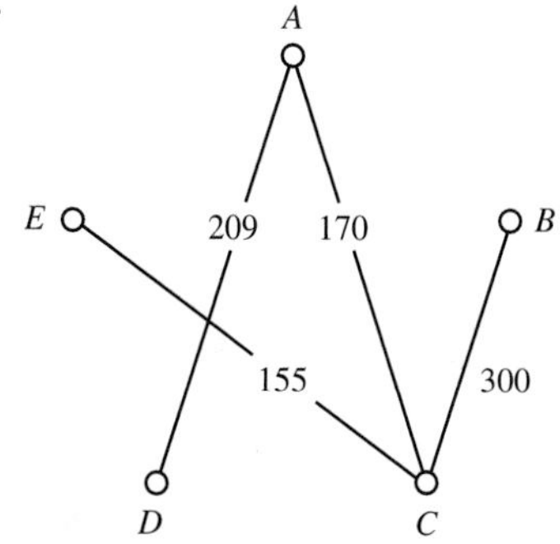

15.

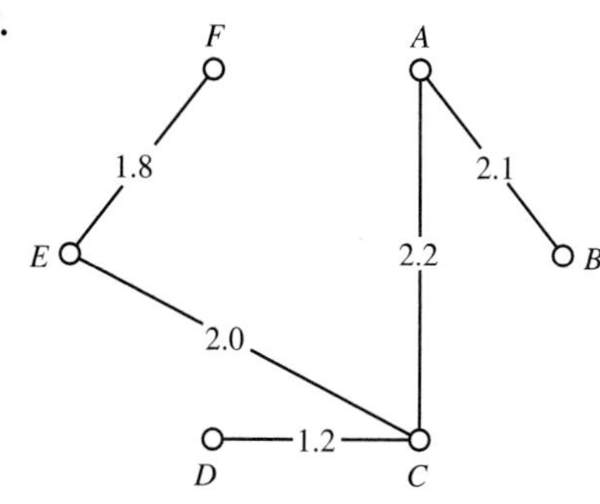

17. **(a)** $CE + ED + EB$ is larger since $CD + DB$ is the shortest network connecting the cities C, D, and B.
 (b) $CD + BD$ is the shortest network connecting the cities C, D, and B since angle CDB is 120° and so the shortest network is the same as the minimum spanning tree
 (c) $CE + EB$ is the shortest network connecting the cities C, E, and B since angle CEB is more than 120° and so the shortest network is the same as the minimum spanning tree.

19. 88.2

21. 50.6

23. **(a)** BC **(b)** $AB + AC$
 (c) No. Since all the angles of the triangle are less than 120°, the shortest network is a Steiner tree of length less than that of the minimum spanning tree.

25. $\sqrt{3}AB$

Jogging

27. The minimum cost network connecting the 4 cities has a 3-way junction point at A and has a total cost of 205 million dollars.

29. In a tree there is one and only one path joining any two vertices. Consequently, the only path joining two adjacent vertices is the edge connecting them and so if that edge is removed, the graph will become disconnected.

31. **(a)** 27 **(b)** 18 **(c)** 42

33. (a) According to Cayley's theorem, there are $3^{3-2} = 3$ spanning trees in a complete graph with 3 vertices and $4^{4-2} = 16$ spanning trees in a complete graph with 4 vertices.

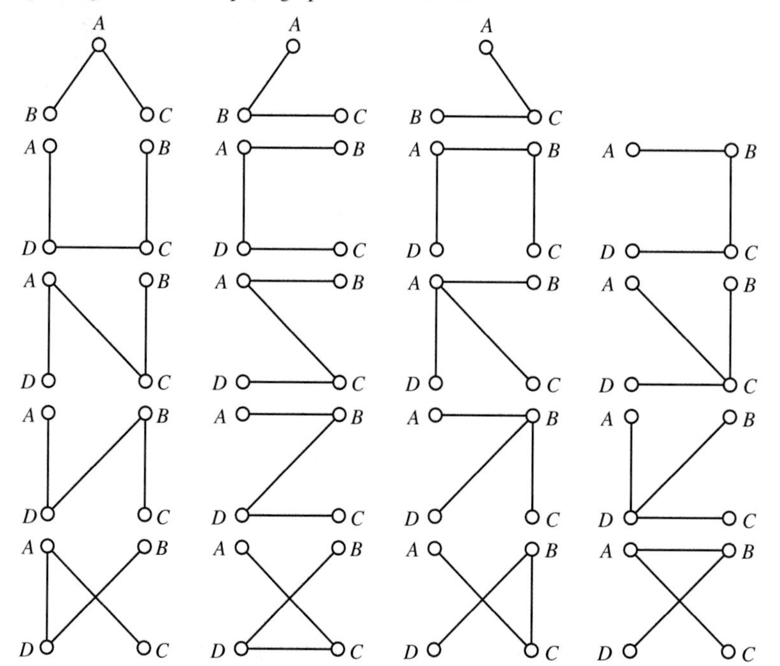

(b) There are $(N-1)!$ Hamilton circuits and N^{N-2} spanning trees in a complete graph with N vertices. Since

$$\frac{(N-1)!}{N^{N-2}} = \frac{2}{N} \times \frac{3}{N} \times \frac{4}{N} \times \cdots \times \frac{N-1}{N} < \frac{2}{N} < 1 \text{ for } N \geq 3,$$

there are more spanning tress than there are Hamilton circuits.

35. (a)

(b)

(c)

37. (a) 2. The solution of Exercise 35(c) shows an example of a graph with N vertices with only 2 vertices of degree 1. To show that there cannot be any fewer vertices of degree 1, let v be the number of vertices in the graph, e the number of edges, and k the number of vertices of degree 1. Recall that in a tree $v = e + 1$ and in any graph the sum of the degrees of all the vertices is $2e$. Now, since we are assuming there are exactly k vertices of degree 1, the remaining $v - k$ vertices must have degree at least 2. Therefore the sum of the degrees of all the vertices must be at least $k + 2(v - k)$. Putting all this together we have

$$2e \geq k + 2(v-k) = k + 2(e + 1 - k),$$
$$2e \geq k + 2e + 2 - 2k,$$
$$k \geq 2.$$

(b) $N - 1$. The solution of Exercise 36(c) shows an example of a graph with N vertices with $N - 1$ vertices of degree 1. It is clear that there cannot be any more vertices of degree 1, since the graph must be connected and so there must be at least one vertex of degree more than 1 if there are more than 2 vertices.

39. If some edge had weight more than weight of e, deleting that edge would result in a spanning tree with total weight less than that of the minimum spanning tree.

41. (a) If J is a Steiner point then $\angle BJC = 120°$.
But since $\angle BJC > \angle BAC = 130°$, this is impossible.
[See part (b) for a proof.]

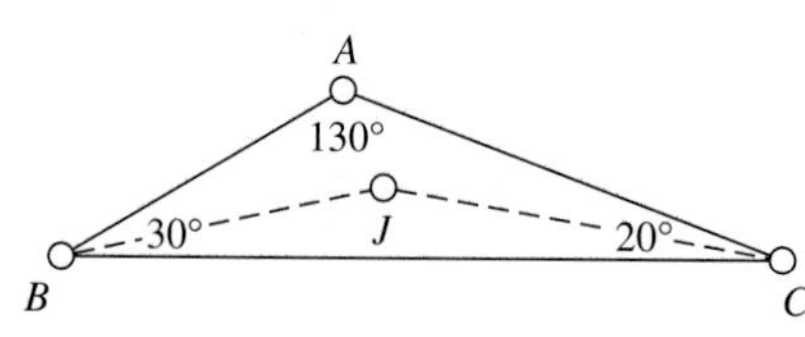

(b) In $\triangle ABC$, $\angle BAC + \angle ABC + \angle BCA = \angle BAC + u + v + r + s = 180°$ (see figure). In $\triangle BJC$, $\angle BJC + v + s = 180°$.
Therefore, $\angle BAC + u + v + r + s = \angle BJC + v + s$ and so $\angle BAC + u + r = \angle BJC$. Consequently, $\angle BAC < \angle BJC$ and so if $\angle BAC > 120°$ then $\angle BJC > 120°$.
Thus, J cannot be a Steiner point since $\angle BJC \neq 120°$.

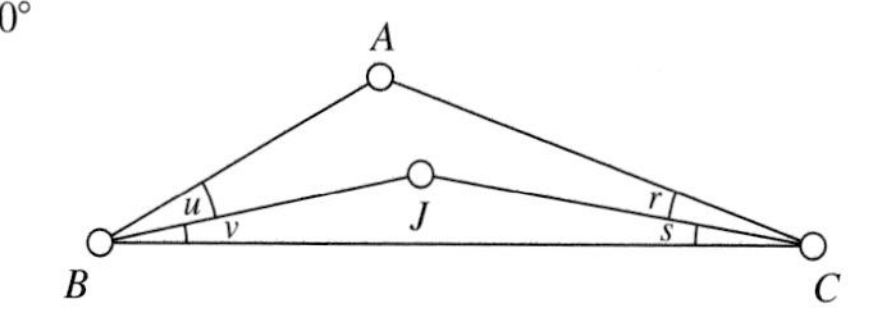

43. The length of the network is $4x$ (see figure). Since the diagonals of a square are perpendicular, by the Pythagorean theorem we have $x^2 + x^2 = 500^2$ which gives $2x^2 = 500^2$ or $x = \frac{500}{\sqrt{2}} = \frac{500\sqrt{2}}{2} = 250\sqrt{2}$.

Thus $4x = 1000\sqrt{2} \approx 1414$.

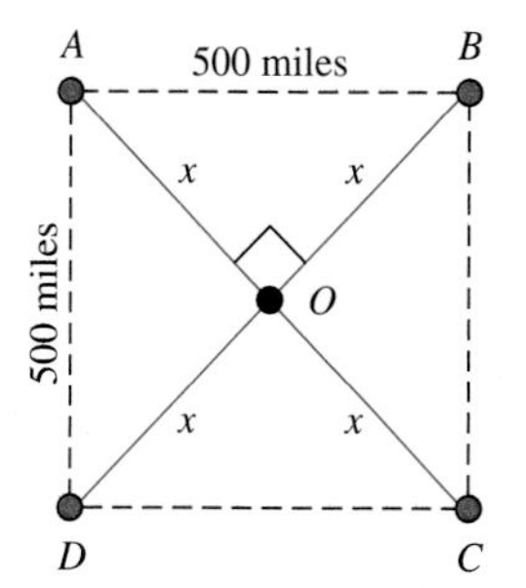

45. (a) The length of the network is $4x + (300 - x) = 3x + 300$, where $200^2 + \left(\frac{x}{2}\right)^2 = x^2$. (See figure.)

Solving the equation gives $x = \frac{400\sqrt{3}}{3}$, and so

the length of the network is $400\sqrt{3} + 300 \approx 993$.

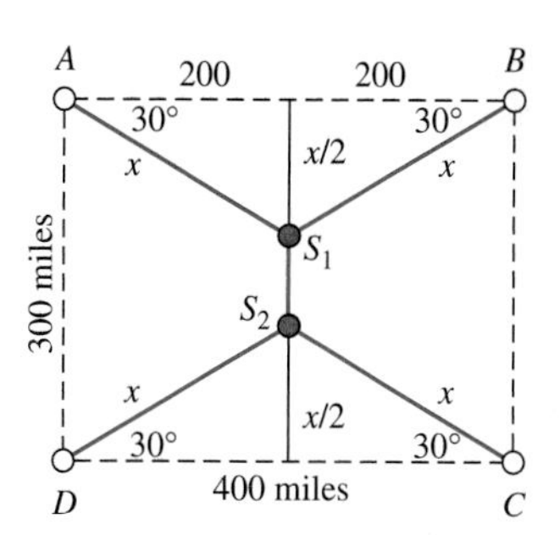

(b) The length of the network is $4x + (400 - x) = 3x + 400$, where $150^2 + \left(\frac{x}{2}\right)^2 = x^2$. (See figure.)
Solving the equation gives $x = 100\sqrt{3}$, and so

the length of the network is $300\sqrt{3} + 400 \approx 919.6$.

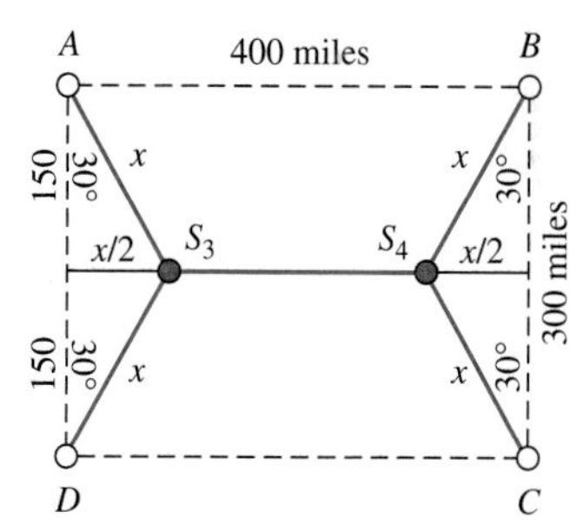

47. Because of Exercise 37(a), if all the vertices had the same degree, they would all have to have degree 1 and the sum of the degrees of all the vertices would be N. This is impossible when $N \geq 3$ since in a tree with N vertices the sum of the degrees of all the vertices is $2N - 2$, and for $N \geq 3$, $2N - 2 > N$.

CHAPTER 8

Walking

1. (a)

Vertex	Degree	Indegree	Outdegree	Vertex is incident to	Vertex is incident from
A	3	2	1	C	B, D
B	2	0	2	A, D	–
C	1	1	0	–	A
D	2	1	1	A	B

(b)

Vertex	Degree	Indegree	Outdegree	Vertex is incident to	Vertex is incident from
A	4	2	2	B, C	C, E
B	2	1	1	D	A
C	4	1	3	A, D, E	A
D	3	3	0	–	B, C, E
E	3	1	2	A, D	C

(c)

Vertex	Degree	Indegree	Outdegree	Vertex is incident to	Vertex is incident from
A	1	1	0	–	B
B	3	2	1	A	E
C	2	1	1	F	E
D	1	1	0	–	E
E	5	0	5	B, C, D, F	–
F	2	2	0	–	C, E

3. (a) **(b)**

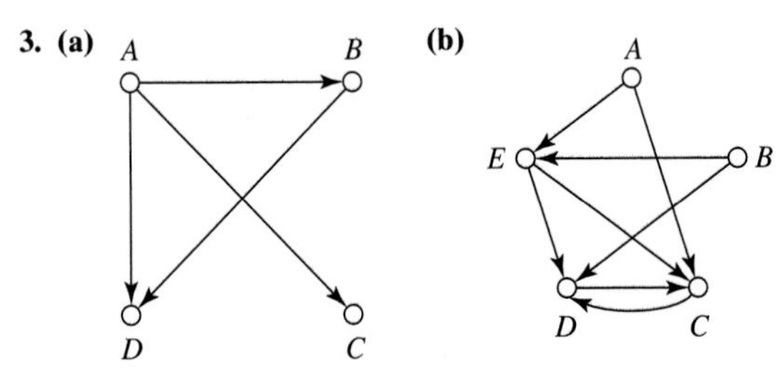

5.

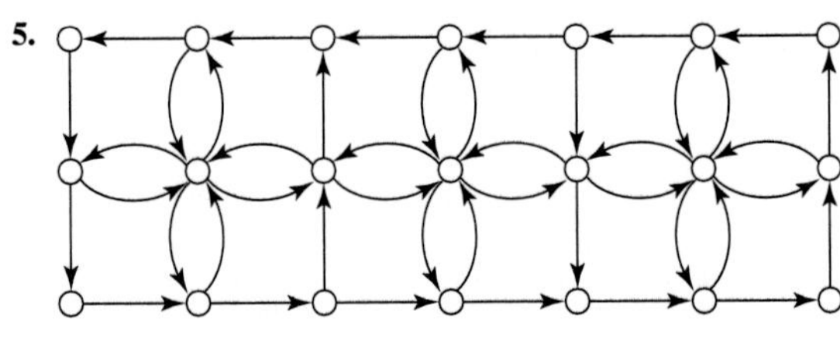

7. (a) **(b)**

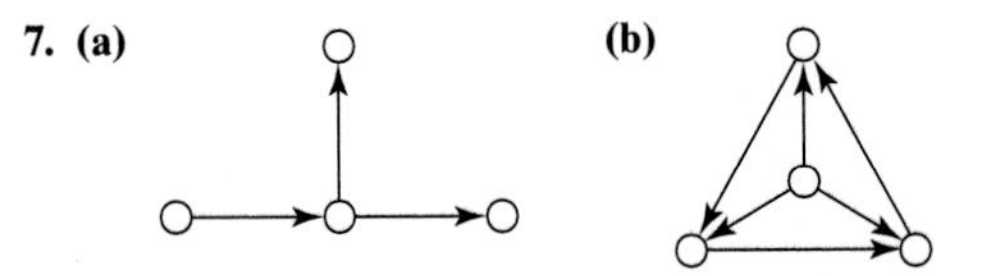

9. (a)

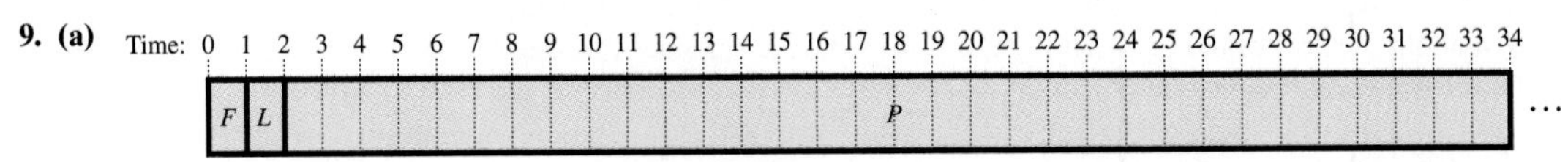

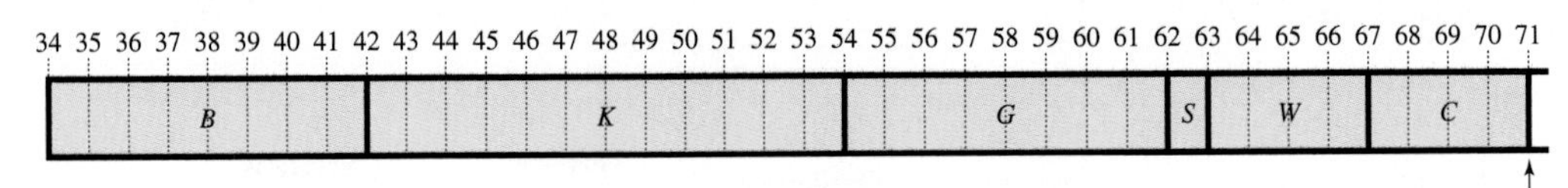

(b)

Time: 0 1 2 3 4 5 6 7 8 9 10 11 12 13 14 15 16 17 18 19 20 21 22 23 24 25 26 27 28 29 30 31 32 33
W_1
W_2
F
P
L
Idle
...
33 34 35 36 37 38 39 40 41 42 43 44 45 46 47 48 49 50 51 52 53 54 55 56 57 58 59 60 61
B
Idle
G
S
C
K
Idle
W
Idle
Finishing time = 61

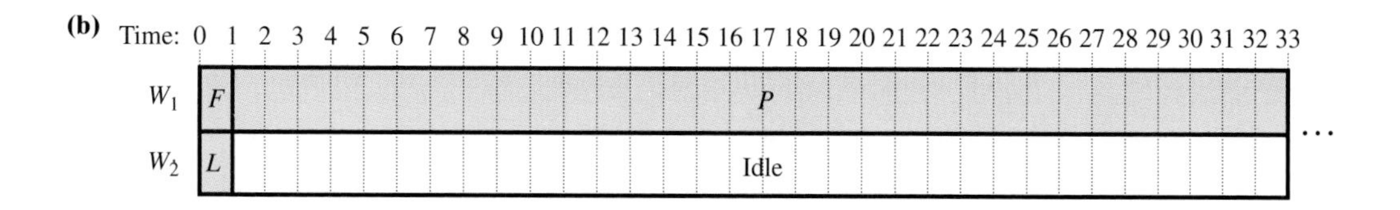

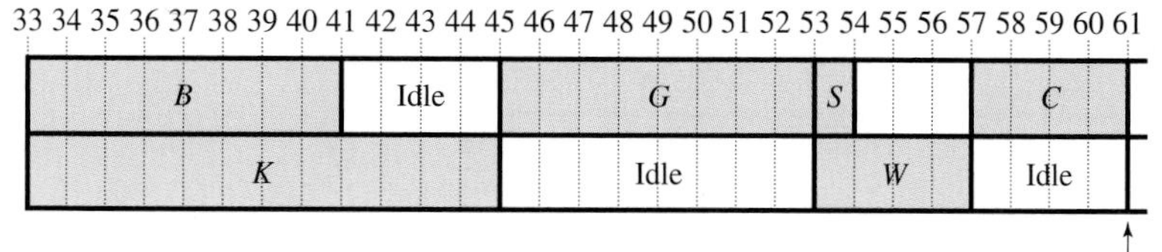

11. According to the precedence relations, G cannot be started until K is completed.

13. According to the precedence relations, G cannot be started until both K and B are completed.

15.

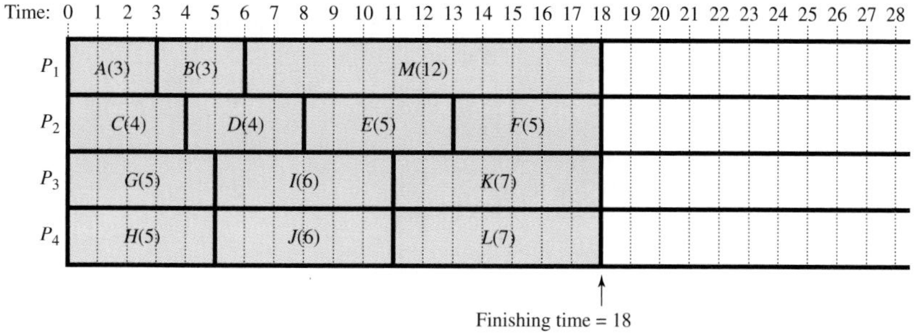

Time: 0 1 2 3 4 5 6 7 8 9 10 11 12 13 14 15 16 17 18 19 20 21 22 23 24 25 26 27 28
P_1
A(3)
B(3)
M(12)
P_2
C(4)
D(4)
E(5)
F(5)
P_3
G(5)
I(6)
K(7)
P_4
H(5)
J(6)
L(7)
Finishing time = 18

17.

Time: 0 1 2 3 4 5 6 7 8 9 10 11 12 13 14 15 16 17 18 19 20 21 22 23 24 25 26 27 28
P_1
M(12)
F(5)
Idle
P_2
L(7)
H(5)
E(5)
Idle
P_3
K(7)
G(5)
D(4)
B(3)
P_4
J(6)
I(6)
C(4)
A(3)
Finishing time = 19

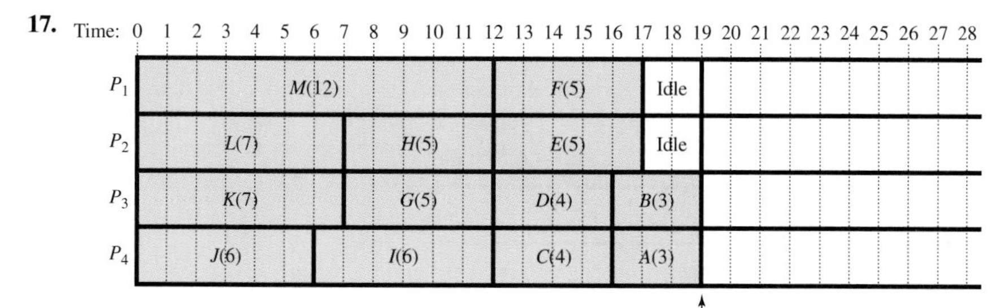

19.

Time: 0 1 2 3 4 5 6 7 8 9 10 11 12 13 14 15 16 17 18 19 20 21 22 23 24
P_1
A(3)
L(7)
M(12)
P_2
B(3)
K(7)
Idle
P_3
C(4)
J(6)
Idle
P_4
D(4)
I(6)
Idle
P_5
E(5)
H(5)
Idle
P_6
F(5)
G(5)
Idle
Finishing time = 22

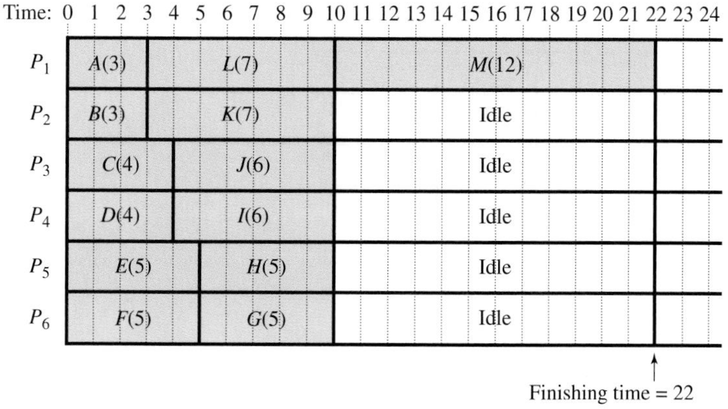

21. No. The total length of the 13 copying jobs is 72 minutes. If 5 processors are going to do 72 minutes of work with no idle time, each processor must do $\frac{72}{5} = 14.4$ minutes of work. Since all of the jobs require a whole number of minutes to complete, a completion time of 14.4 minutes for a processor is impossible.

23.

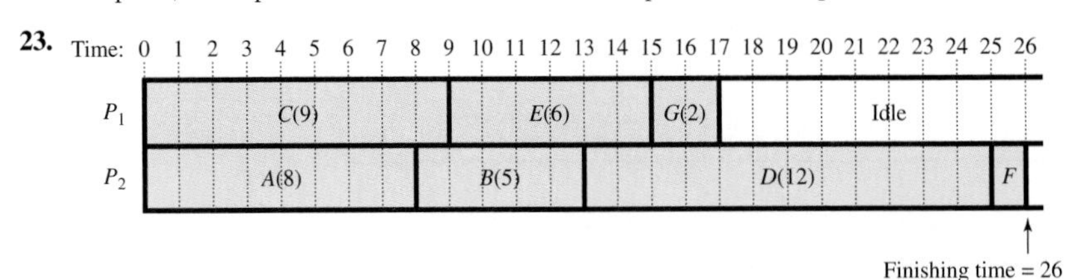

25.

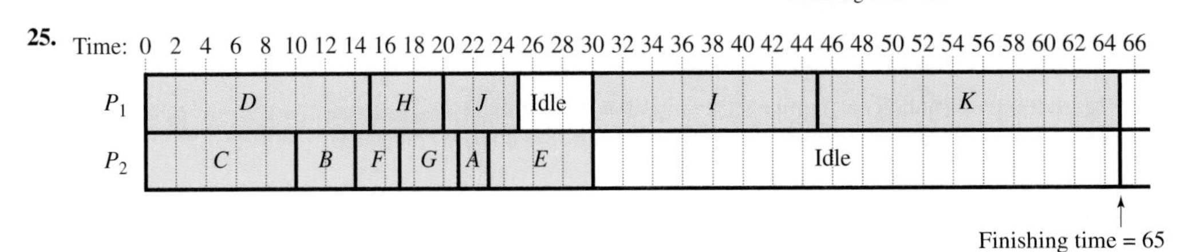

27.

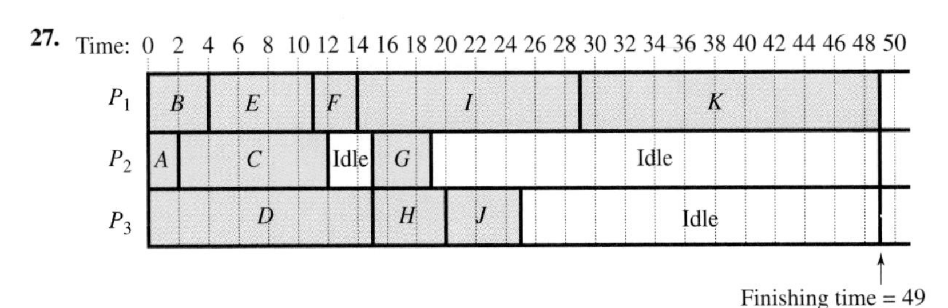

29. Since there is a total of 25 hours of work to be done by 2 processors, the work cannot be completed in less than 12.5 hours. But the times for all the jobs are whole numbers, so the work cannot be completed in less than 13 hours.

31. Time: 0 2 4 6 8 10 12 14 16 18 20 22 24 26 28 30 32 34 36 38 40 42 44 46 48 50

P_1	AP	IF	IW	IP	HU	IL	FW	EU	
P_2	AF	AW	AD	PL	ID	PU	PD	Idle	

Finishing time = 36

33. (a)

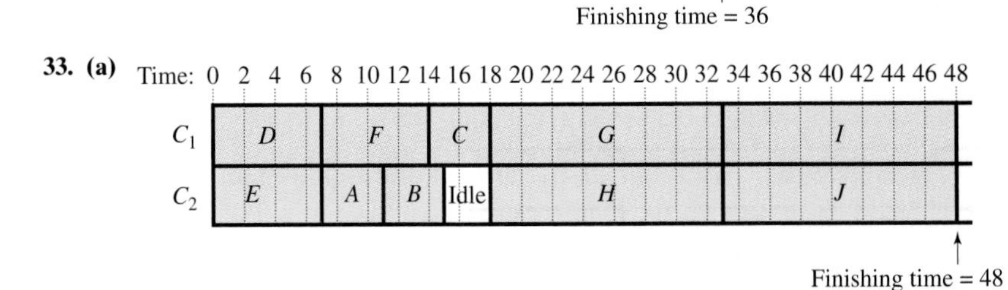

(b)

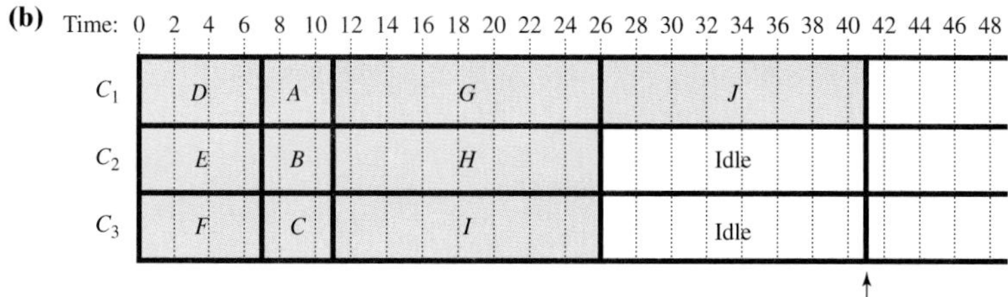
Time: 0 2 4 6 8 10 12 14 16 18 20 22 24 26 28 30 32 34 36 38 40 42 44 46 48
C_1
D
A
G
J
C_2
E
B
H
Idle
C_3
F
C
I
Idle
Finishing time = 41

(c)

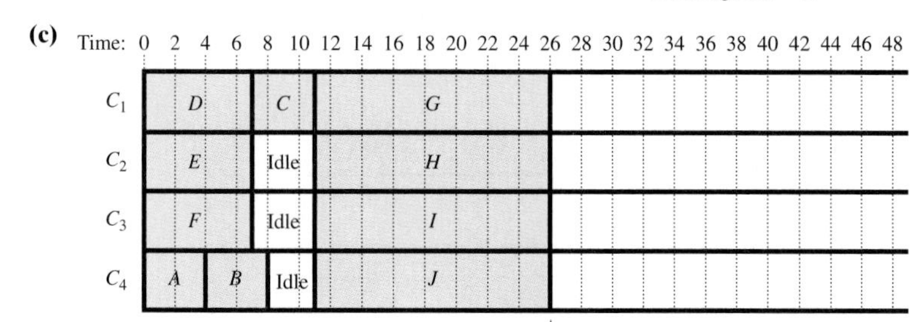
Time: 0 2 4 6 8 10 12 14 16 18 20 22 24 26 28 30 32 34 36 38 40 42 44 46 48
C_1
D
C
G
C_2
E
Idle
H
C_3
F
Idle
I
C_4
A
B
Idle
J
Finishing time = 26

35.

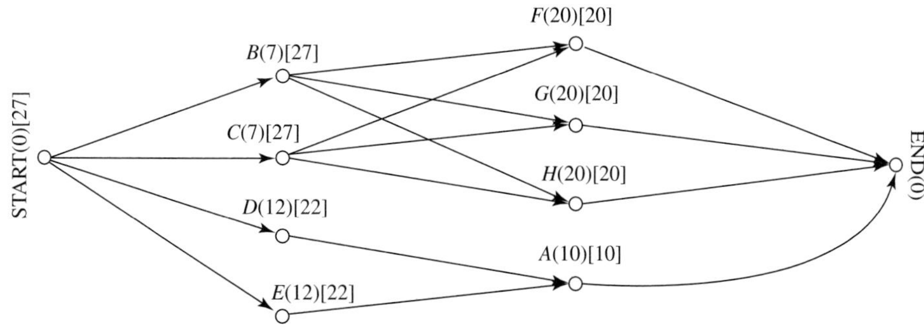
START(0)[27]
B(7)[27]
C(7)[27]
D(12)[22]
E(12)[22]
F(20)[20]
G(20)[20]
H(20)[20]
A(10)[10]
END(0)

37. (a)

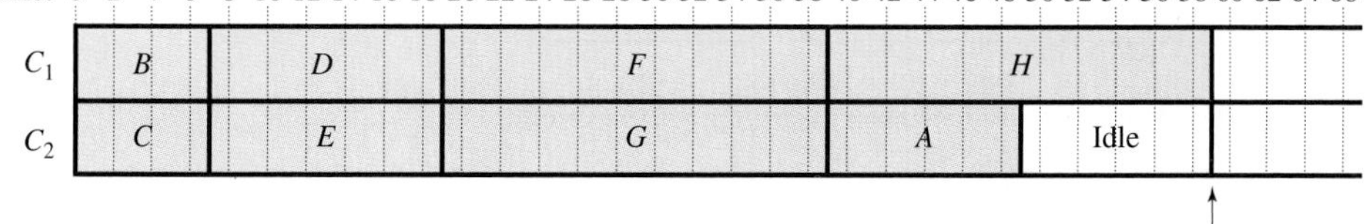
Time: 0 2 4 6 8 10 12 14 16 18 20 22 24 26 28 30 32 34 36 38 40 42 44 46 48 50 52 54 56 58 60 62 64 66
C_1
B
D
F
H
C_2
C
E
G
A
Idle
Finishing time = 59

(b)

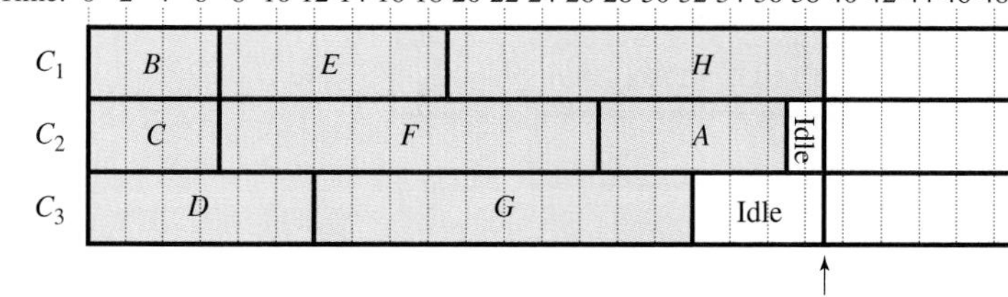
Time: 0 2 4 6 8 10 12 14 16 18 20 22 24 26 28 30 32 34 36 38 40 42 44 46 48
C_1
B
E
H
C_2
C
F
A
Idle
C_3
D
G
Idle
Finishing time = 39

39. (a)

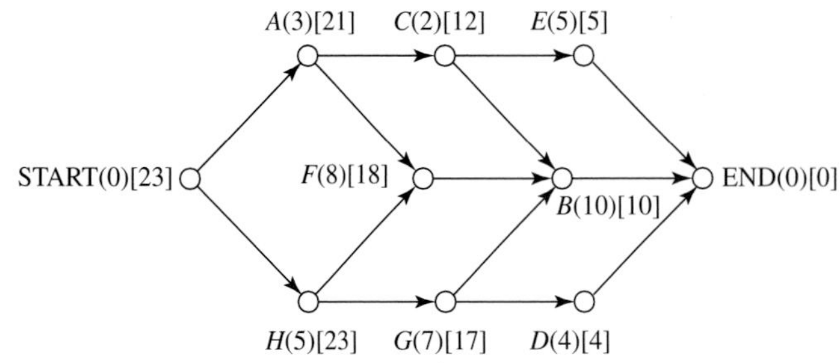

Note: The critical path times shown inside the square brackets are not required for part (a), but are included for part (b).

(b)

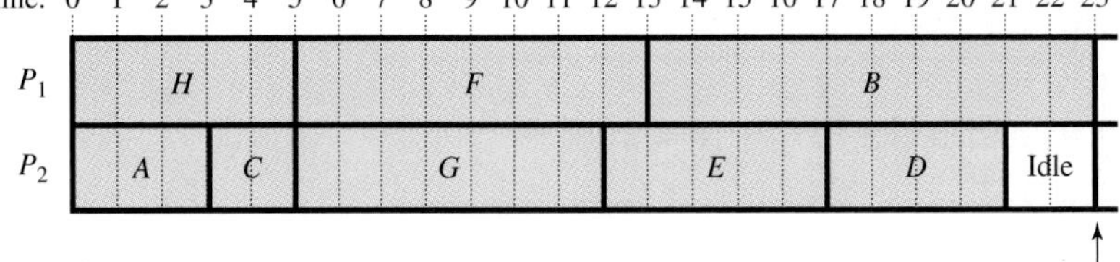

Jogging

41. Each arc of the graph contributes 1 to the sum of the indegrees and 1 to the sum of the outdegrees.

43. (a) True. If no processor is idle, then the total of the processing times of all the tasks is the same as the complete time of the schedule and so there can be no shorter schedule.

(b) False. Precedence relations may force idle time for one or more processors.

45. (a) $\frac{4}{15}, \frac{5}{18}, \frac{6}{21}, \frac{7}{24}, \frac{8}{27}, \frac{9}{30}$

(b) Since $M - 1 \leq M$, we have $\frac{M-1}{3M} \leq \frac{M}{3M} = \frac{1}{3}$.

47. (a) Time: 0 1 2 3 4 5 6 7 8 9 10 11 12 13 14 15 16 17 18 19 20

	0–7	7–11	11–15
P_1	E	A	F
P_2	I	B	Idle
P_3	D (0–6)	C (6–11)	Idle
P_4	H (0–6)	G (6–11)	Idle

Finishing time = 15

(b)

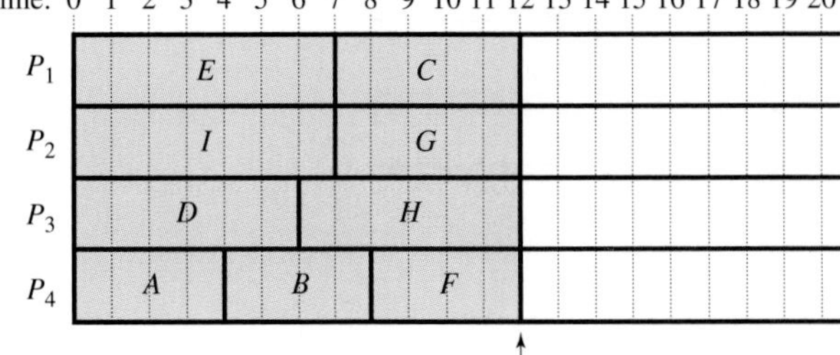

49.

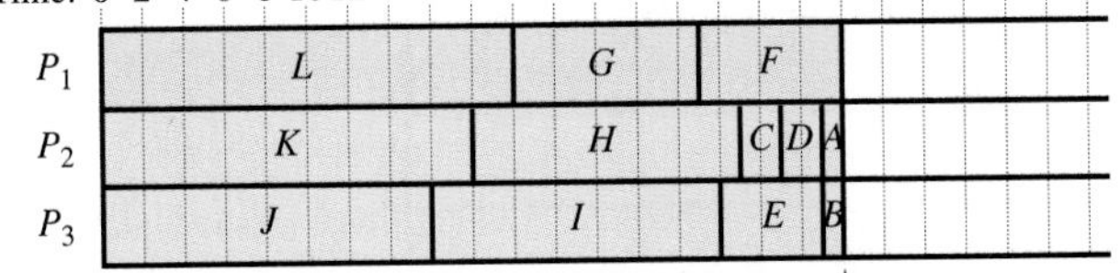

Finishing time = 36

This schedule is obviously optimal, since the processors are always busy.

51. (a)

Time: 0 1 2 3 4 5 6 7 8 9 10 11 12 13 14 15 16 17 18 19 20 21 22 23 24

P_1: A(3) C(4) E(5)

P_2: B(3) D(4) F(5)

P_3: G(5) K(7)

P_4: H(5) L(7)

P_5: I(6) J(6)

P_6: M(12)

Finishing time = 12

(b) See the answer to Exercise 19.

53. (a)

Time: 0 5 10 15 20 25 30 35 40 45 50 55 60 65

P_1: A F G H I

P_2: D B Idle

P_3: E C Idle

Finishing time = 65

(b)

Time: 0 5 10 15 20 25 30 35 40 45 50 55 60 65

P_1: A F I

P_2: B G D

P_3: C H E

Finishing time = 45

55. (a)

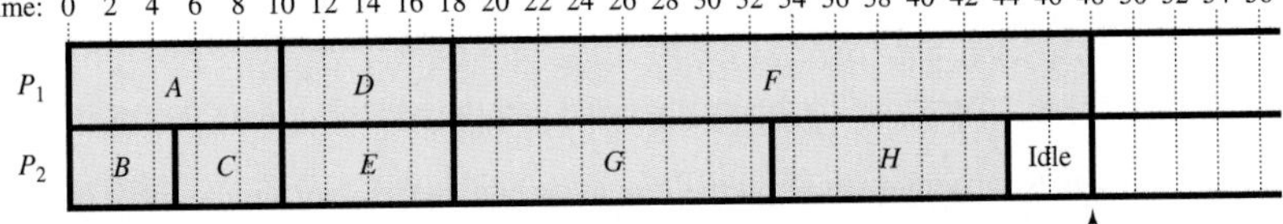

Finishing time = 48

(b)

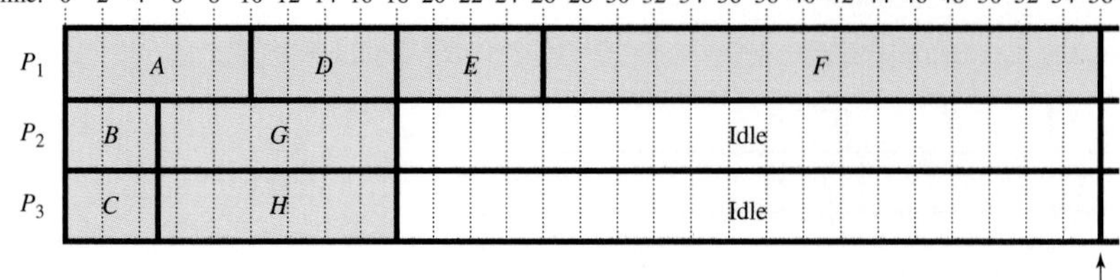

(c) The critical path for this project has length 48 and so the job cannot be completed any sooner. Consequently, the critical path algorithm with 2 processors produced an optimal schedule. With 3 processors the critical path algorithm produced a longer schedule. This paradoxical situation is a consequence of the fact that in the list processing model, a processor cannot choose to be idle if there is a ready task to be executed. In this example tasks D and E need to be completed early so that the long task F can be started, but processor 2 and processor 3 were forced to start tasks G and H and so were not available to start D or E when they were available. We can see from the two schedules that it is the very addition of processor 3 that actually ended up messing up the timing of things.

CHAPTER 9

Walking

1. $F_{15} = 610, F_{16} = 987, F_{17} = 1597, F_{18} = 2584$

3. **(a)** $F_{10} = 55$ **(b)** $F_{10} + 2 = 55 + 2 = 57$ **(c)** $F_{10+2} = F_{12} = 144$ **(d)** $F_{10-8} = 55 - 8 = 47$ **(e)** $F_{10-8} = F_2 = 1$
(f) $3F_4 = 3 \times 3 = 9$ **(g)** $F_{3\times 4} = F_2 = 1$

5. **(a)** $F_{38} = F_{37} + F_{36} = 24{,}157{,}817 + 14{,}930{,}352 = 39{,}088{,}169$
(b) $F_{35} = F_{37} - F_{36} = 24{,}157{,}817 - 14{,}930{,}352 = 9{,}227{,}465$

7. **(a)** Both equations express the fact that each term of the Fibonacci sequence is the sum of the two preceding terms. However, in the first equation ($F_N = F_{N-1} + F_{N-2}$) we must have $N \geq 3$, whereas in the second equation ($F_{N+2} = F_{N+1} + F_N$) we must have $N \geq 1$.
(b) $F_1 = 1, F_2 = 1$, and $F_N = F_{N-1} + F_{N-2}$ for $N \geq 3$; $F_1 = 1, F_2 = 1$ and $F_{N+2} = F_{N+1} + F_N$ for $N \geq 1$.

9. $N = 12$ and $M = 12$. ($F_{12} = 12^2$)

11. **(a)** $47 = 13 + 34$ **(b)** $48 = 1 + 13 + 34$ **(c)** $207 = 8 + 55 + 144$ **(d)** $210 = 3 + 8 + 55 + 144$

13. **(a)** $(F_1 + F_2 + F_3 + F_4) + 1 = (1 + 1 + 2 + 3) + 1 = 8 = F_6 = F_{4+2}$
(b) $(F_1 + F_2 + F_3 + F_4 + F_5) + 1 = (1 + 1 + 2 + 3 + 5) + 1 = 13 = F_7 = F_{5+2}$
(c) $(F_1 + F_2 + F_3 + \cdots + F_{10}) + 1 = (1 + 1 + 2 + 3 + 5 + 8 + 13 + 21 + 34 + 55) + 1 = 144 = F_{12} = F_{10+2}$
(d) $(F_1 + F_2 + F_3 + \cdots + F_{11}) + 1 = (1 + 1 + 2 + 3 + 5 + 8 + 13 + 21 + 34 + 55 + 89) + 1 = 233 = F_{13} = F_{11+2}$

15. **(a)** $2F_{N+2} - F_{N+3} = F_N$
(b) $2F_{1+2} - F_{1+3} = 2F_3 - F_4 = 2 \cdot 2 - 3 = 1 = F_1$
(c) $2F_{4+2} - F_{4+3} = 2F_6 - F_7 = 2 \cdot 8 - 13 = 3 = F_4$
(d) $2F_{8+2} - F_{8+3} = 2F_{10} - F_{11} = 2 \cdot 55 - 89 = 21 = F_8$

17. $x = 1 + \sqrt{2} \approx 2.414, x = 1 - \sqrt{2} \approx -0.414$

19. $x = -1, x = \frac{8}{3} \approx 2.66667$

21. $\frac{F_{14}}{F_{13}} = \frac{377}{233} \approx 1.6180258, \frac{F_{16}}{F_{15}} = \frac{987}{610} \approx 1.6180328,$
$\frac{F_{18}}{F_{17}} = \frac{2584}{1597} \approx 1.6180338$. These numbers are increasing and getting closer to the golden ratio Φ.

23. **(a)** True. Two polygons are similar if their corresponding angles are equal and the lengths of their corresponding sides are proportional.
(b) False. A 10 by 20 rectangle and a 10 by 10 square have all their angles equal, but are not similar since their sides are not proportional.

25. 20 by 30

27. $c = 24$

29. $x = 4$

31. $x = 12, y = 10$

33. 10 by approximately 6.18

Jogging

35. Using the hint, the sum of the two solutions to the equation $x^2 - x - 1 = 0$ must be 1. But we know one of the solutions is $\Phi \approx 1.618$ and so the other solution must be $1 - \Phi$ which is just the negative of the decimal part of Φ.

37. $x = 6, y = 12, z = 10$

39. $x = 3, y = 5$

41. If $\Phi^N = a\Phi + b$ then $\Phi^{N+1} = (a\Phi + b)\Phi$
$= a\Phi^2 + b\Phi = a(\Phi + 1) + b\Phi = (a + b)\Phi + a$. (Remember $\Phi^2 = \Phi + 1$.)

43. We must have $\dfrac{b+y}{b} = \dfrac{h+x}{h}$ or, equivalently, $1 + \dfrac{y}{b} = 1 + \dfrac{x}{h}$. This gives $\dfrac{y}{b} = \dfrac{x}{h}$ or, equivalently, $\dfrac{y}{x} = \dfrac{b}{h}$.

45. (a) Since we are given that $AB = BC = 1$, we know that $\angle BAC = 72°$ and so $\angle BAD = 180° - 72° = 108°$.

This makes $\angle ABD = 180° - 108° - 36° = 36°$ and so $\triangle ABD$ is isosceles with $AD = AB = 1$. Therefore,

$AC = x - 1$. Using these facts and the similarity of $\triangle ACB$ and $\triangle BCD$ we have $\dfrac{x}{1} = \dfrac{1}{x-1}$ or $x^2 = x + 1$ for which we

know the solution is $x = \Phi$.

(b) 36°-36°-108°

(c) $\dfrac{\text{Longer side}}{\text{Shorter side}} = \dfrac{x}{1} = x = \Phi$

47. $A_N = 5F_N$

49. (a) $T_1 = aF_2 + bF_1 = a + b$

(b) $T_2 = aF_3 + bF_2 = 2a + b$

(c) $T_N = aF_{N+1} + bF_N$
$= a(F_N + F_{N-1}) + b(F_{N-1} + F_{N-2})$
$= (aF_N + bF_{N-1}) + (aF_{N-1} + bF_{N-2})$
$= T_{N-1} + T_{N-2}$

51. Follows from Exercise 45 and the following figure.

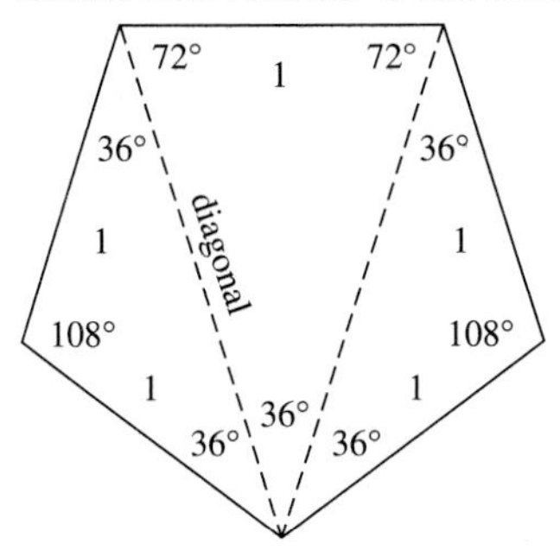

53. From the Pythagorean theorem, $CA^2 = r^2 + (2r)^2 = 5r^2$ and so $CA = r\sqrt{5}$. Therefore,

$AD = CA - r = r\sqrt{5} - r = r(\sqrt{5} - 1)$ and $AP = r\left(\dfrac{\sqrt{5}-1}{2}\right) = r\left(\dfrac{\sqrt{5}+1}{2} - 1\right) = r(\Phi - 1) = \dfrac{r}{\Phi}$ which is the length of

the side of the decagon. This length of the side of the decagon follows from Exercise 45 and the following figure.

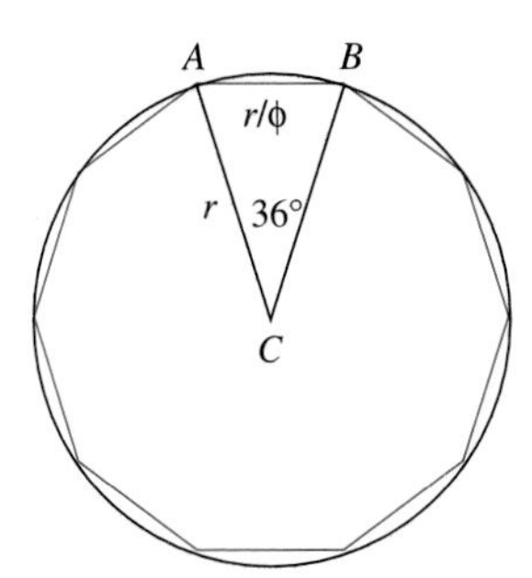

CHAPTER 10

Walking

1. **(a)** $P_4 = 26$ **(b)** $P_6 = 68$ **(c)** No. The sum of two even numbers is even.
3. **(a)** $P_N = P_{N-1} + 5; P_1 = 3$ **(b)** $P_N = 3 + 5(N-1) = 5N - 2$ **(c)** $P_{300} = 3 + 5 \times 299 = 1498$
5. **(a)** $P_2 = 205, P_3 = 330, P_4 = 455$ **(b)** $P_{100} = 12{,}455$ **(c)** $P_N = 80 + 125(N-1) = 125N - 45$
7. **(a)** $d = 3$ **(b)** $P_{51} = 158$ **(c)** $P_N = 8 + 3(N-1) = 3N + 5$
9. 24,950
11. 16,050
13. **(a)** 3,519,500 **(b)** 3,482,550
15. **(a)** 213 **(b)** $137 + 2N$ **(c)** \$7124 **(d)** \$2652
17. **(a)** \$40.50 **(b)** 40.5% **(c)** 40.5%
19. 39.15%
21. \$4587.64
23. **(a)** \$9083.48 **(b)** 12.6825%
25. The Great Bulldog Bank: 6%; The First Northern Bank: ≈ 5.9%; The Bank of Wonderland: ≈ 5.65%
27. ≈ \$1133.56
29. **(a)** \$6209.21 **(b)** \$6102.71 **(c)** \$6077.89
31. **(a)** $P_2 = 11 \times 1.25 = 13.75$ **(b)** $P_{10} = 11 \times 1.25^9 \approx 81.956$ **(c)** $P_N = 11 \times 1.25^{N-1}$
33. **(a)** $P_{100} = 3 \times 2^{99}$ **(b)** $P_N = 3 \times 2^{N-1}$ **(c)** $3 \times (2^{100} - 1)$ **(d)** $3 \times 2^{49} \times (2^{51} - 1)$
35. **(a)** $p_2 = 0.357$ **(b)** $p_3 \approx 0.64274$ **(c)** $p_5 \approx 0.64278$
37. $p_2 = 0.3825, p_3 = 0.70858125, p_4 \approx 0.619481586, p_5 \approx 0.707172452, p_6 \approx 0.621238726, p_7 \approx 0.705903514, p_8 \approx 0.622811229, p_9 \approx 0.704752206, p_{10} \approx 0.624229602$ (Answers rounded to 9 decimal places after each transition.)

Jogging

39. 100%
41. \$10,737,418.23
43. No. This would require $p_1 = p_2 = 0.8(1 - p_1)p_1$ and so $1 = 0.8(1 - p_1)$ or $p_1 = -\frac{1}{4}$.
45. ≈ 14,619 snails
47. ≈ \$105,006
49. 6425

CHAPTER 11

Walking

1.

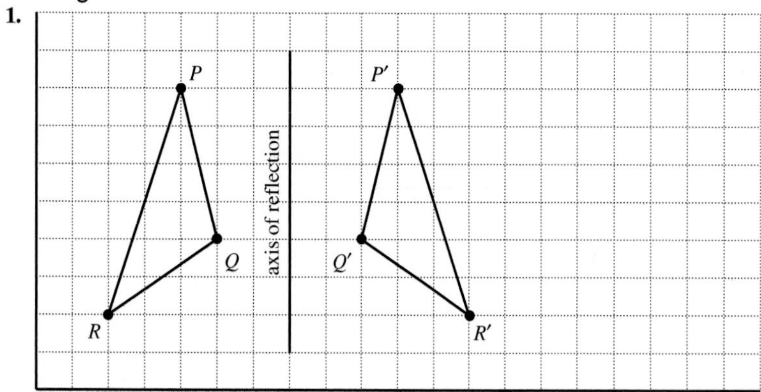

3. **(a)** 140° **(b)** 279°

5.

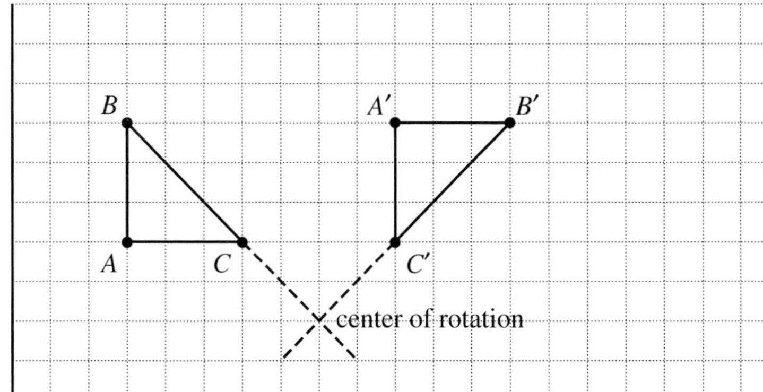

7.

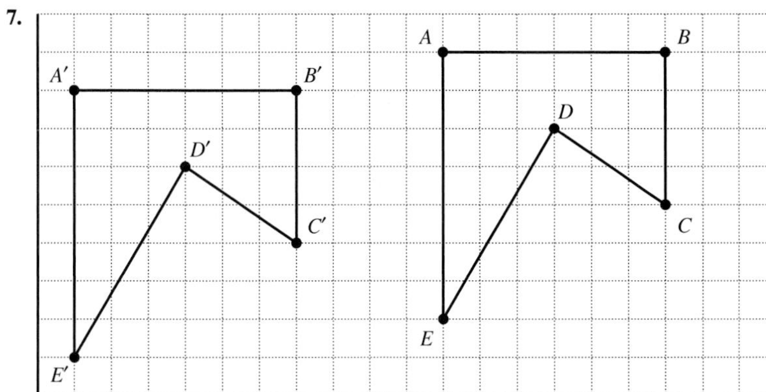

9.

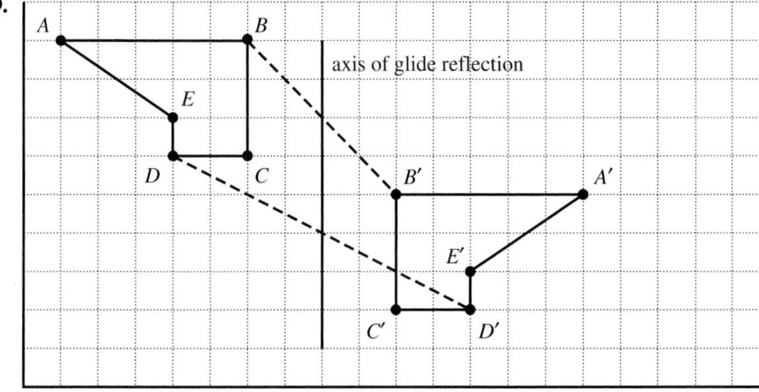

11. **(a)** b A T **(b)** M I T **(c)** b A d **(d)** b o Y

13. **(a)** D_1 **(b)** D_1 **(c)** D_4 **(d)** Z_2 **(e)** Z_1

15. **(a)** J **(b)** T **(c)** S **(d)** I **(e)** O

17. **(a)** **(b)**

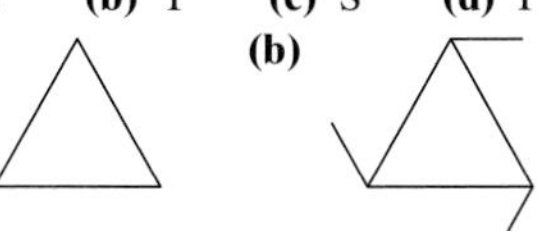

19. **(a)** D_2 **(b)** D_4 **(c)** D_3 **(d)** D_8

21. **(a)** Translation, vertical reflection, identity
(b) Translation, horizontal reflection, identity
(c) Translation, 180° rotation, identity
(d) Translation, identity

23. (a) Translation, identity
(b) Translation, vertical reflection, identity
(c) Translation, horizontal reflection, identity
(d) Translation, 180° rotation, identity

25. Since every rigid motion is equivalent to either a reflection, rotation, translation, or glide reflection, and a rotation has only one fixed point while translations and glide reflections have no fixed points, the specified rigid motion must be equivalent to a reflection.

Jogging

27. A reflection is an improper rigid motion and hence reverses the left-right orientation. The propeller blades have a protrusion on the right (as you face the blade) and so after a reflection the protrusion will be on the left.

29. (a)

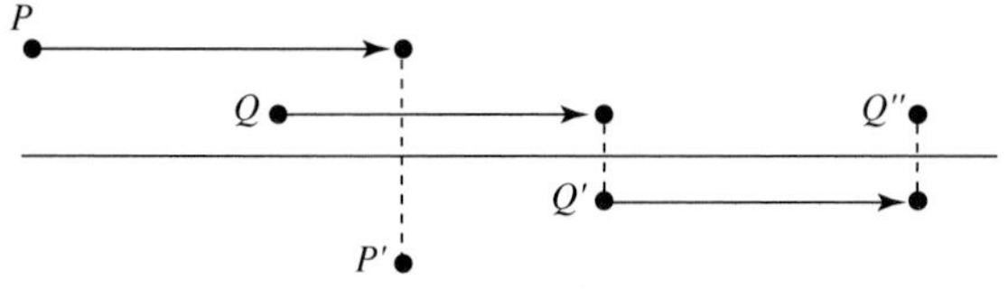

(b) The result of applying the same glide reflection twice is equivalent to a translation in the direction of the glide of twice the amount of the original glide. [See figure in part (a).]

31. (a) P′, Q′, and R′ are the result of applying reflection 1; P″, Q″, and R″ are the result of applying reflection 2 to P′, Q′, and R′, respectively.

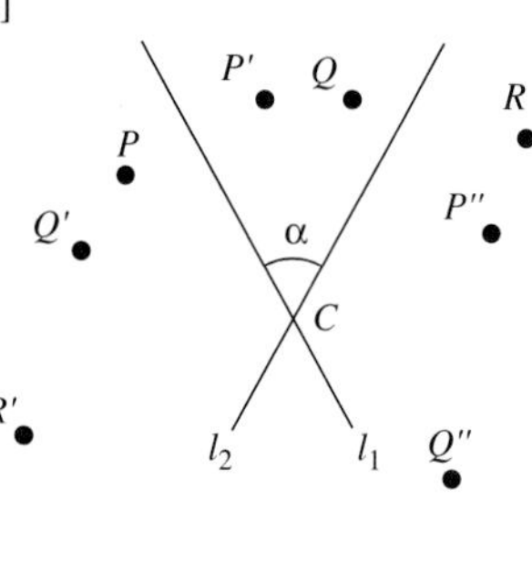

(b) The result of applying reflection 1 followed by reflection 2 is a clockwise rotation with center *C* and angle of rotation $\gamma + \gamma + \beta + \beta = 2(\gamma + \beta) = 2\alpha$. (See figure.)

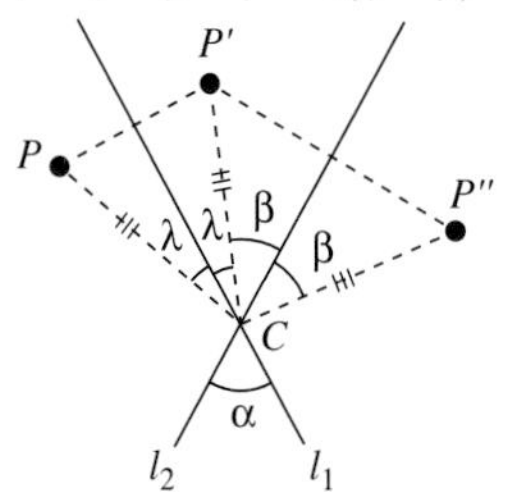

(c) The result of applying reflection 2 followed by reflection 1 is a counter-clockwise rotation with center *C* and angle 2α.

33. (a) By definition, a border pattern has translation symmetries in exactly one direction (let's assume the horizontal direction). If the pattern had a reflection symmetry along an axis forming a 45° with the horizontal direction, there would have to be a second direction of translation symmetry (vertical).

(b) If a border pattern had a reflection symmetry along an axis forming an angle of $\alpha°$ with the horizontal direction, it would have to have translation symmetry in a direction that forms an angle of $2\alpha°$ with the horizontal. This could only happen for $\alpha = 90°$ or $\alpha = 180°$ (since the only allowable direction for translation symmetries is horizontal).

35. (a) Translation, 180° rotation, identity
(b) Translation, 180° rotation, identity
(c) Translation, vertical reflection, 180° rotation, glide reflection, identity

37. (a) Rotations and translations are proper rigid motions, and hence preserve clockwise-counterclockwise orientations. The given motion is an improper rigid motion (it reverses the clockwise-counterclockwise orientation).

(b) If the rigid motion was a reflection, then PP', RR', and QQ' would all be perpendicular to the axis of reflection and hence would all be parallel.

(c) It must be a glide reflection (the only rigid motion left).

39. A rotation (See Exercise 31.)

41. A glide reflection

43. Answers will vary.

CHAPTER 12

Walking

1. (a), (b), (c) The line segment joining the midpoints of two sides of a triangle is parallel to the third side and has length one-half the third side. This implies that all the triangles are congruent and therefore the area of each one is $\frac{1}{4}X$.

(d) $\frac{3}{4}X$. [See (a), (b), (c), above.]

3. (a) $\frac{9}{16}X$ **(b)** $\frac{27}{64}X$ **(c)** $\left(\frac{3}{4}\right)^N X$

(d) $\left(\frac{3}{4}\right)^N$ gets closer and closer to 0 as N gets bigger and bigger.

5. $36 = 6^2, 6^9, 6^{N-1}$

7. (a) $\frac{1}{3}P$ **(b)** $2P$ **(c)** $4P$

9.

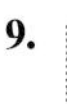

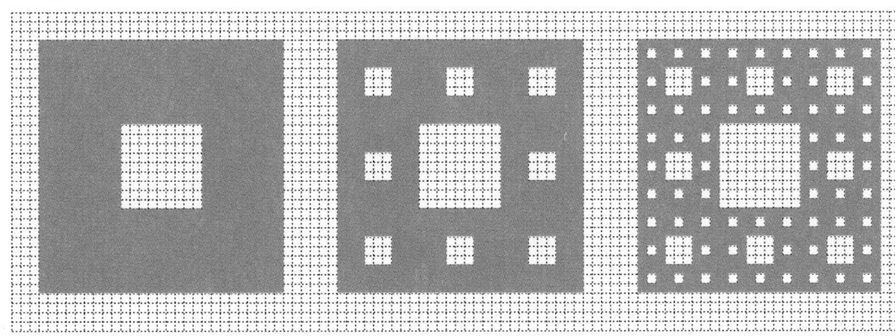

Step 1 Step 2 Step 3

11. (a) 512 **(b)** $\left(\frac{8}{9}\right)^3 X = \frac{512}{729}X$ **(c)** 8^N **(d)** $\left(\frac{8}{9}\right)^N X$

(e) $\left(\frac{8}{9}\right)^N$ gets closer and closer to 0 as N gets bigger and bigger.

13. Start with a ■. Wherever you see a ______, replace it with a _■_.

15. (a) $\frac{20}{3}, \frac{100}{9}$ **(b)** $4\left(\frac{5}{3}\right)^N$

17.

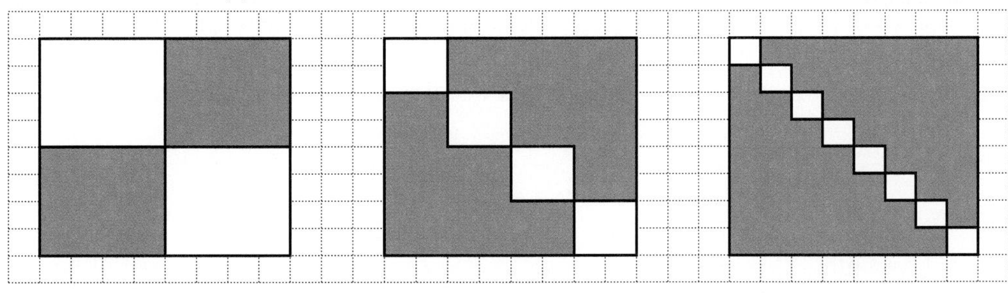

Step 1 Step 2 Step 3

19. $\left[1 - \left(\frac{1}{2}\right)^N\right]X$

21.

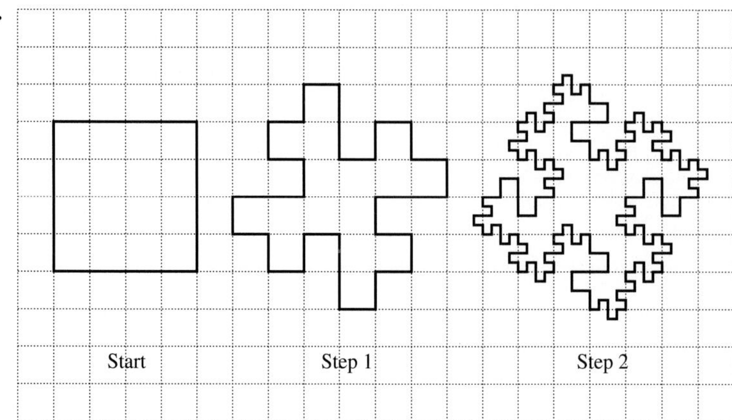

23. **(a)** $2P, 4P$ **(b)** $2^N P$

25.

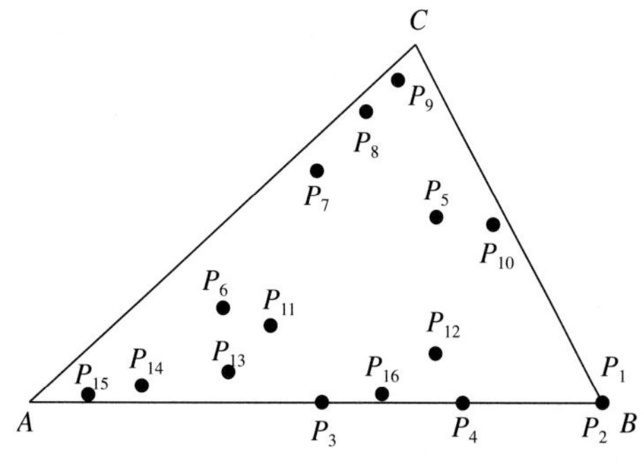

27.

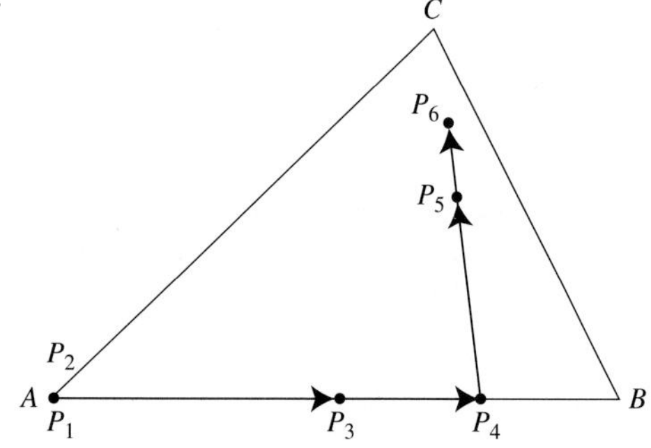

29. **(a)** 2, 2, 2, 2, 2 **(b)** attracted

Jogging

31. At each step, one new white triangle is introduced for each black triangle and the number of black triangles is tripled.

	White △'s	**Black △'s**
Step 1	1	3
Step 2	$1 + 3$	3^2
Step 3	$1 + 3 + 3^2$	3^3
Step 4	$1 + 3 + 3^2 + 3^3$	3^4

We see that at the Nth step there will be

$$1 + 3 + 3^2 + 3^3 + \cdots + 3^{N-1} = \frac{1 - 3^N}{1 - 3} = \frac{3^N - 1}{2}$$

white triangles.

33. There will be infinitely many points left. For example, the 3 vertices of the original triangle will be left as well as the vertices of every black triangle that occurs at each step of the construction.

35. $-0.1875, -0.71484375, -0.238998413, -0.692879759, -0.26991764, -0.677144468, -0.29147537, -0.665042109, -0.307718994, -0.655309021, -0.320570087, -0.647234819, -0.331087089, -0.64038134, -0.33991174, -0.634460009, -0.347460497, -0.629271203, -0.354017753, -0.624671431$
Attracted to -0.5 (although it takes many more terms to see this). This is one of the solutions of the equation $x^2 - 0.75 = x$.

37. This sequence is attracted to 0.5, a solution of the equation $x^2 + 0.25 = x$.

39. This sequence is escaping.

CHAPTER 13

Walking

1. **(a)** Answer 1: All married people. Answer 2: All married people who read Dear Abby's column. Note: While both answers are acceptable, it is clear that Dear Abby was trying to draw conclusions about married people at large so answer 1 is better. Also note that from the wording of her conclusion ("... there are far more faithfully wed couples than I had surmised.") one can assume that she was primarily interested in currently married people as opposed to divorcees, widows and widowers.
(b) 210,336 **(c)** self-selection **(d)** 85% is a statistic, since it is based on data taken from a sample.

3. **(a)** 74.0% **(b)** 81.8%
(c) Not very accurate. The sample was far from being representative of the entire population.

5. **(a)** The citizens of Cleansburg (b) 475

7. **(a)** The choice of street corner could make a great deal of difference in the responses collected.
(b) *D*. (We are making the assumption that people who live or work downtown are much more likely to answer yes than people in other parts of town.)
(c) Yes, for two main reasons: (i) people out on the street between 4 P.M. and 6 P.M. are not representative of the population at large. For example, office and white collar workers are much more likely to be in the sample than homemakers and school teachers. (ii) The five street corners were chosen by the interviewers and the passersby are unlikely to represent a cross section of the city.
(d) No. No attempt was made to use quotas to get a representative cross section of the population.

9. **(a)** All undergraduates at Tasmania State University
(b) $N = 15{,}000$

11. **(a)** No. The sample was chosen by a random method.
(b) In simple random sampling, any two members of the population have as much chance of both being in the sample as any other two. But in this sample, two people with the same last name—say Len Euler and Linda Euler—have no chance of both being in the sample. (By the way, the sampling method described in this exercise is frequently used and goes by the technical name of **systematic sampling**.)

13. **(a)** Anyone who could have a cold and would consider buying vitamin X (i.e., pretty much all adults).
(b) Presumably they volunteered. (We could infer this from the fact that they are being paid.)
(c) $n = 500$ **(d)** No. There was no control group.

15. (i) Using college students. (College students are not a representative cross section of the population in terms of age and therefore in terms of how they would respond to the treatment.)
(ii) Using subjects only from the San Diego area.
(iii) Offering money as an incentive to participate.
(iv) Allowing self-reporting (the subjects themselves determine when their colds are over) is a very unreliable way to collect data and is especially bad when the subjects are paid volunteers.

17. Anyone who could potentially suffer from arterioscelerosis (clogging of the arteries)

19. **(a)** There was a treatment group (the ones getting the beta-carotene pill) and there was a control group. The control group received a placebo pill. These two elements make it a controlled placebo experiment.
(b) The group that received the beta-carotene pills.
(c) Both the treatment and control groups were chosen by random selection.

21. The professor was conducting a clinical study because he was, after all, trying to establish the connection between a cause (10 milligrams of caffeine a day) and an effect (improved performance in college courses). Other than that, the experiment had little going for it: it was not controlled (no control group); not randomized (the subjects were chosen because of their poor grades); no placebo was used and consequently the study was not double-blind.

23. (i) A regular visit to the professor's office could in itself be a boost to a student's self-confidence and help improve his or her grades.

(ii) The "individualized tutoring" that took place during the office meetings could also be the reason for improved performance.

(iii) The students selected for the study all got F's on their first midterm making them likely candidates to show some improvement.

Jogging

25. (a) (i) the entire sky; (ii) all the coffee in the cup; (iii) all the blood in Carla's body

(b) The sample is not random in any of the three examples.

(c) (i) In some situations one can have a good idea as to whether it will rain or not by seeing only a small section of the sky, but in many other situations rain clouds can be patchy and one might draw the wrong conclusions by just peeking out the window. (ii) If the coffee is burning hot on top, it is likely to be pretty hot throughout, so Betty's conclusion is likely to be valid. (iii) Because of the constant circulation of blood in our system, the 5 ml of blood drawn out of Carla's right arm is representative of all the blood in her body, so the lab's results are very likely to be valid.

27. (a) The question was worded in a way that made it almost impossible to answer yes.

(b) "Will you support some form of tax increase if it can be proven to you that such a tax increase is justified?" is better, but still not neutral. "Do you support or oppose some form of tax increase?" is bland but probably as neutral as one can get.

29. (a) Under method 1, people whose phone numbers are unlisted are automatically ruled out from the sample. At the same time, method 1 is cheaper and easier to implement than method 2.

(b) For this particular situation, method 2 is likely to produce much more reliable data than method 1. The two main reasons are: (i) People with unlisted phone numbers are very likely to be the same kind of people that would seriously consider buying a burglar alarm, and (ii) the listing bias is more likely to be significant in a place like New York City. (People with unlisted phone numbers make up a much higher percentage of the population in a large city such as New York than in a small town or rural area. Interestingly enough, the largest percentage of unlisted phone numbers for any American city is in Las Vegas, Nevada.)

31. (a) 2000

$$N = \frac{n_2}{k} \cdot n_1 = \frac{120}{30} \times 500 = 2000$$

(b) 20,5000

$$N = \frac{n_2}{k} \cdot n_1 = \frac{900}{218} \times 4965 \approx 20{,}497.7 \approx 20{,}500$$

33. (a) Issue 1. Who should be getting the treatment? (Individuals who are HIV positive but otherwise appear healthy? Individuals who are already at a very advanced stage of AIDS? Those in between? Some of each group?) How about money? (Should anyone who can afford the treatment be allowed to get it?)

Issue 2. Is the remedy worse than the disease? How serious are the possible side effects in humans? (The drug has been tested only on laboratory animals.) Who should be making the decision as to whether the risks justify the benefits? (Patient? Doctor? Insurance Company?)

Issue 3. Should there be a placebo group? (In matters of life and death is it fair to tell a patient that he/she may be getting a "fake" treatment?)

(b) This is an open ended question and appropriate for class discussion and/or a report. At present, there are no well established protocols for dealing with some of these issues.

CHAPTER 14

Walking

1. (a)

Score	10	50	60	70	80	100
Frequency	1	3	7	6	5	2

(b)

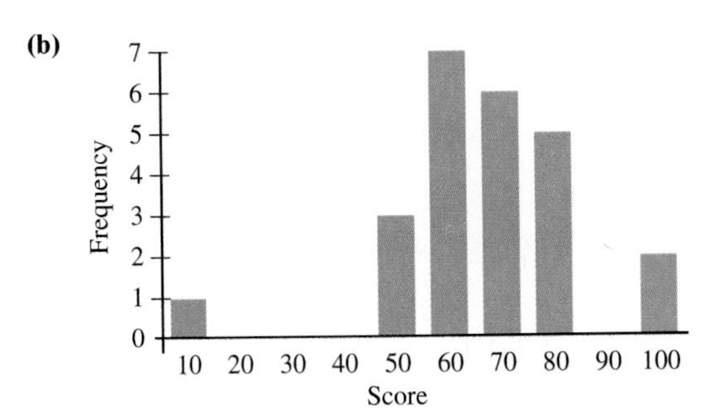

(c)

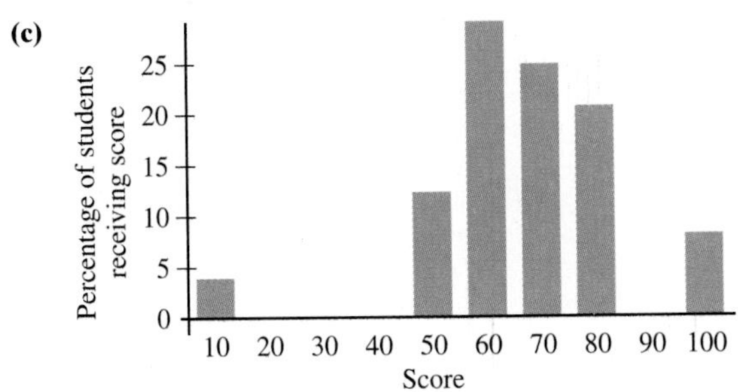

3. (a) Min = 10, Q_1 = 60, M = 70, Q_3 = 80, Max = 100

(b)

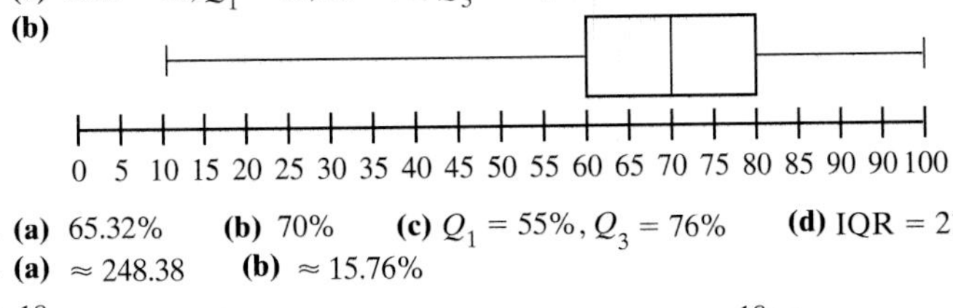

5. (a) 65.32% **(b)** 70% **(c)** $Q_1 = 55\%, Q_3 = 76\%$ **(d)** IQR = 21%

7. (a) ≈ 248.38 **(b)** $\approx 15.76\%$

9.

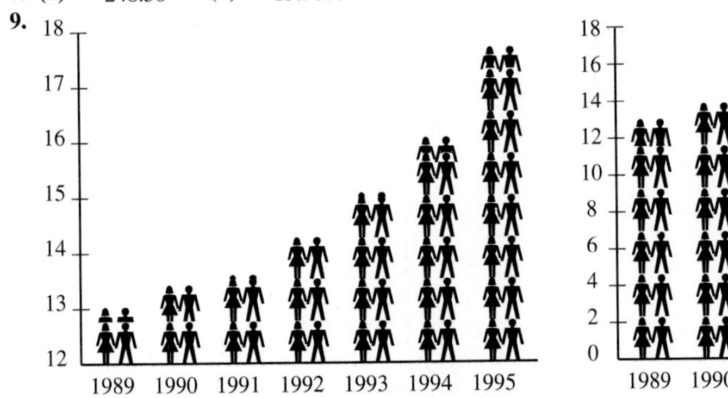

11. (a)

Distance to School (miles)	0.0	0.5	1.0	1.5	2.0	2.5	3.0	5.0	8.5
Frequency	5	3	4	6	3	2	1	1	1

(b)

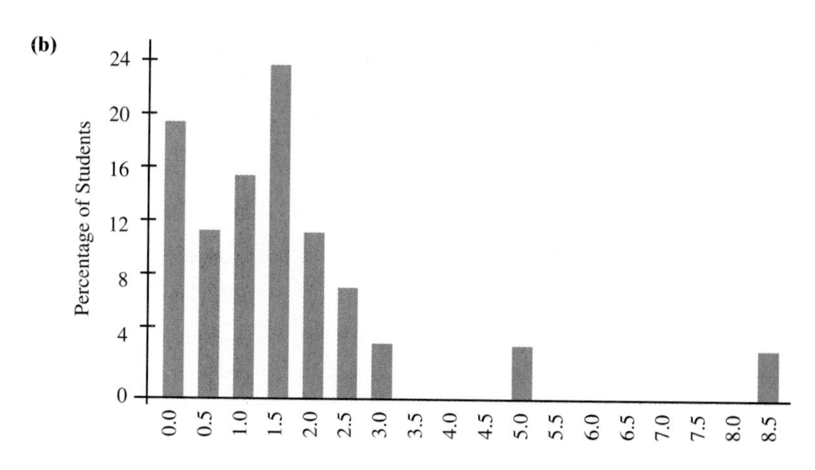

13. **(a)** 30

(b)

Score	3	4	5	6	7	8	9	10
Frequency	2	5	6	4	4	5	3	1

(c) 6.17 **(d)** 6

15. Asian: 40°; Hispanic: 68°; African American: 86°; Caucasian: 140°; Other: 25°

17. **(a)** 132 **(b)** 9

(c)

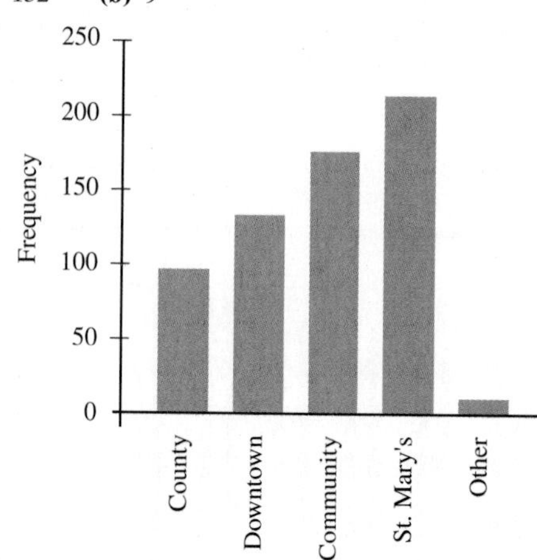

19. **(a)** 12 ounces

(b) The third class interval: "more than 72 ounces and less than or equal to 84 ounces." Values that fall exactly on the boundary between two class intervals belong to the class interval to the left.

21.

Percentage

35 30 25 20 15 10 5 0

48 60 72 84 96 108 120 132 144 156 168

Ounces

23. Average = 5, SD = 0; Average = 5, SD ≈ 3.54; Average = 5, SD ≈ 11.73.
The averages are all the same but the standard deviations get larger as the numbers are more spread out.

25. **(a)** 4.5 **(b)** 4.5 **(c)** 2 **(d)** 7 **(e)** 5 **(f)** ≈ 2.87

27. **(a)** 50.5 **(b)** 50.5

29. **(a)** Min = 9, Q_1 = 13, M = 16, Q_3 = 18, Max = 25. Average ≈ 15.49; standard deviation ≈ 3.11

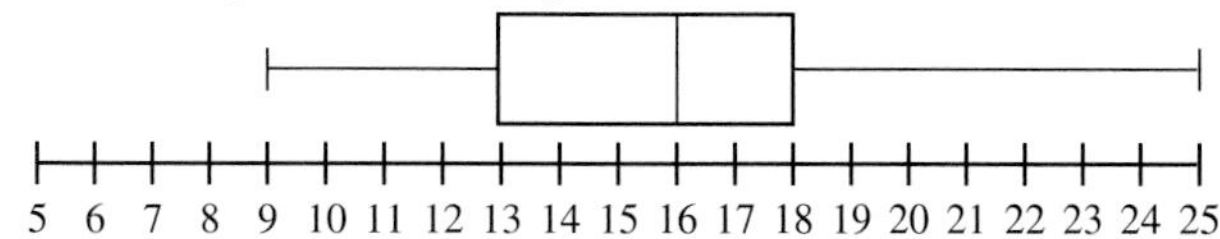

31. **(a)** 1.5 **(b)** {0.7 −2.6, −4.2, 2.9, 4.7, −3.9, 2.3, 0.1} **(c)** 0

33. **(a)** 459 **(b)** 240

35. 58 points

Jogging

37. An example is Ramon gets 85 out of 100 on each of the first four exams and 60 out of 100 on the fifth exam while Josh gets 80 out of 100 on all 5 of the exams.

39. **(a)** {1, 1, 1, 1, 6, 6, 6, 6, 6, 6} Average = 4; Median = 6
(b) {1, 1, 1, 1, 1, 1, 6, 6, 6, 6} Average = 3; Median = 1
(c) {1, 1, 6, 6, 6, 6, 6, 6, 6, 6} Average = 5; Q_1 = 6
(d) {1, 1, 1, 1, 1, 1, 1, 1, 6, 6} Average = 2; Q_3 = 1

41. **(a)** The five-number summary for the original scores was Min = 1, Q_1 = 9, M = 11, Q_3 = 12, and Max = 24. When 2 points are added to each test score, the five-number summary will also have 2 points added to each of the its numbers (i.e., Min = 3, Q_1 = 11, M = 13, Q_3 = 14, and Max = 26).
(b) When 10% is added to each score (i.e., each score is multiplied by 1.1), then each number in the five-number summary will also be multiplied by 1.1 (i.e., Min = 1.1, Q_1 = 9.9, M = 12.1, Q_3 = 13.2, and Max = 26.4).

43. **(a)** 4 **(b)** 0.4 **(c)** 0.4

45. **(a)** 10%, 20% **(b)** 80%, 90%
(c) The figures for both schools were combined. A total of 820 males were admitted out of a total of 1200 that applied—giving approximately 68.3% of the total number of males that applied were admitted. Similarly, a total of 460 females were admitted out of a total of 900 that applied—giving approximately 51.5% of the total number of females that applied were admitted.
(d) In this example, females have a higher percentage $\left(\frac{100}{500} = 20\%\right)$ than males $\left(\frac{20}{200} = 10\%\right)$ for admissions to the School of Architecture and also a higher percentage $\left(\frac{360}{400} = 90\%\right)$ than males $\left(\frac{800}{1000} = 80\%\right)$ for admissions to the School of Engineering. When the numbers are combined, however, females have a lower percentage $\left(\frac{100 + 360}{500 + 400} \approx 51.1\%\right)$ than males $\left(\frac{20 + 800}{200 + 1000} \approx 68.3\%\right)$ in total admissions. The reason that this apparent paradox can occur is purely a matter of arithmetic: Just because $\frac{a_1}{a_2} > \frac{b_1}{b_2}$ and $\frac{c_1}{c_2} > \frac{d_1}{d_2}$, it does not necessarily follow that $\frac{a_1 + c_1}{a_2 + c_2} > \frac{b_1 + d_1}{b_2 + d_2}$.

47. Consider a data set $\{x_1, x_2, x_3, \ldots, x_N\}$ and assume Min $= x_1 \le x_2 \le x_3 \le \ldots \le x_N =$ Max. Then since the average A satisfies the inequality Min $\le A \le$ Max, we see that $(x_k - A)^2 \le (\text{Max} - \text{Min})^2 = \text{Range}^2$ for each $k = 1, 2, 3, \ldots, N$. Therefore,

$$SD = \sqrt{\frac{(x_1 - A)^2 + (x_2 - A)^2 + \ldots + (x_N - A)^2}{N}} \le \sqrt{\frac{N \times \text{Range}^2}{N}} = \text{Range}.$$

CHAPTER 15

Walking

1. **(a)** {*HHHH*, *HHHT*, *HHTH*, *HHTT*, *HTHH*, *HTHT*, *HTTH*, *HTTT*, *THHH*, *THHT*, *THTH*, *THTT*, *TTHH*, *TTHT*, *TTTH*, *TTTT*}
(b) 16 **(c)** {*HHTT*, *HTHT*, *HTTH*, *THHT*, *THTH*, *TTHH*} **(d)** $\frac{6}{16} = \frac{3}{8}$

3. (a) $\{P_1, P_2, P_3, P_4, P_5, P_6, P_7\}$; $\Pr(P_1) = \frac{1}{4}$; $\Pr(P_2) = \Pr(P_3) = \Pr(P_4) = \Pr(P_5) = \Pr(P_6) = \Pr(P_7) = \frac{1}{8}$

(b) The odds for P_1 winning the tournament are 1 to 3. The odds for P_2 winning the tournament are 1 to 7.

5. (a) 17,576,000 **(b)** 15,818,400 **(c)** 11,232,000

7. (a) 40,320 **(b)** 40,319

9. (a) {red, blue, yellow, purple, orange} **(b)** {red, blue, yellow} **(c)** {purple, orange}
(d) Pr(red) = Pr(blue) = Pr(yellow) = 0.1; Pr(purple) = Pr(orange) = 0.35

11. { }, $\{A\}$, $\{B\}$, $\{C\}$, $\{A, B\}$, $\{A, C\}$, $\{B, C\}$, $\{A, B, C\}$

13. (a) $\frac{1}{12}$ **(b)** $\frac{11}{12}$ **(c)** 1 to 11 **(d)** 11 to 1

15. (a) E = { ⚀⚀, ⚀⚁, ⚀⚂, ⚀⚃, ⚁⚀, ⚁⚁, ⚁⚂, ⚂⚀, ⚂⚁, ⚃⚀ }

(b) $\frac{10}{36} = \frac{5}{18}$ **(c)** 5 to 13

17. (a) $\frac{1}{36}$ **(b)** $\frac{25}{36}$ **(c)** $\frac{5}{18}$

19. (a) $\frac{7}{8}$ **(b)** $\frac{1023}{1024}$

21. (a) $\frac{1}{3}$ **(b)** $\frac{1}{2}$

23. 1320

25. 364

27. 793

Jogging

29. (a) $\frac{99}{512} \approx 0.19$ **(b)** $\frac{231}{1024} \approx 0.226$ **(c)** $\frac{99}{512} \approx 0.19$

31. $\frac{253}{4998} \approx 0.05$

33. $\frac{1}{2548} \approx 0.00039$

35. 252

37. 20

39. $1 - \left(\frac{5}{6}\right)^5 \approx 0.6$

41. $\frac{4}{9}$

43. (a) 3,628,800 **(b)** 362,880

CHAPTER 16

Walking

1. (a) 2 **(b)** 1.4 **(c)** 2.6 **(d)** 0.8

3. (a) 1 **(b)** 1.6 **(c)** −2.0 **(d)** −1.8

5. (a) 52 **(b)** 50% **(c)** 68% **(d)** 16%

7. (a) 44.6 **(b)** 59.4 **(c)** 14.8

9. (a) 1900 **(b)** 50

11. (a) approximately the 3rd percentile **(b)** the 16th percentile **(c)** around the 22nd or 23rd percentile
(d) the 84th percentile **(e)** the 99.85th percentile

13. (a) 95% **(b)** 47.5% **(c)** 97.5%

15. (a) 13 **(b)** 80 **(c)** 250 **(d)** 420 **(e)** 488 **(f)** 499

17. $\mu = 75, \sigma = 3$

19. $\mu = 80, \sigma = 10$

21. (a) 16th percentile **(b)** 97.5th percentile **(c)** 18.6 lbs

23. **(a)** 97.5th percentile **(b)** 99.85th percentile **(c)** 8 lbs

25. **(a)** $\sigma \approx 3.5$ inches **(b)** 6-8 **(c)** 366

(d) The actual number of players that fall with ± 7 inches (2 estimated standard deviations) of the estimated mean (6-8) is $13 + 15 + 22 + 24 + 25 + 29 + 39 + 44 + 45 + 44 + 30 + 21 + 6 + 5 + 2 = 364$.

Jogging

27. **(a)** $\mu = 1800, \sigma = 30$ **(b)** 68% **(c)** 34% **(d)** 13.5%

29. **(a)** $\mu = 240, \sigma = 12$ **(b)** $Q_1 \approx 232, Q_3 \approx 248$

(c) From (b) we know that 50% of the time the number of heads will fall between 232 and 248 so bet (1) is the better bet.

31. **(a)** 20.55 pounds **(b)** 60th percentile **(c)** Slightly below the 16th percentile

Index

Photo Credits

CHAPTER 1 **CO (clockwise, from top right)** Mark Lennihan/AP/Wide World Photos; Barr/Gamma-Liaison, Inc.; Reprinted with permission of *Automobile Magazine.* Copyright 1997 by K-III Magazines Corporation; Courtesy Peter Tannenbaum **p. 36** Phillip Hayson/Photo Researchers, Inc. **p.38** Barr/Gamma-Liaison, Inc.

CHAPTER 2 **CO** Maria Melin/ABC, Inc. **p. 52** Emile Wamsteker/AP/Wide World Photos

CHAPTER 3 **CO** All photos courtesy Peter Tannenbaum **p. 74** Poussin, Nicolas. "The Judgment of Solomon." Louvre, Paris, France. Giraudon/Art Resource, NY. **p. 87** Margo Taussig Pinkerton/Gamma-Liaison, Inc.

CHAPTER 4 **CO** Corbis-Bettmann **p. 117** The Granger Collection **p. 119** The New York Historical Society **p. 125** PH College Archives **p. 126** The White House Photo Office **p. 129** Corbis-Bettmann **p. 132** Corbis-Bettmann **p. 145 (left)** The Ohio Historical Society **(right)** The Granger Collection

CHAPTER 5 **CO** Map by Martin Zeiller (1652)

CHAPTER 6 **CO** Michael Carroll **p. 186 (top)** Reprinted by permission of *Newsweek.* All rights reserved. **(bottom)** Chris Butler/Science Photo Library/Photo Researchers, Inc.

CHAPTER 7 **p. 244** Courtesy of Peter Tannenbaum **p. 258** All photos courtesy of Peter Tannenbaum.

CHAPTER 8 **CO** All photos courtesy of Peter Tannenbaum **p. 262 (left)** Lawrence Migdale/Photo Researchers, Inc. **(right)** Will & Deni McIntyre/Photo Researchers, Inc. **p. 263 (left)** Jonathan Pite/Gamma-Liaison, Inc. **(right)** Ed Lallo/Gamma-Liaison, Inc. **p. 264 (top left, right)** Etienne de Malglaive/Gamma-Liaison, Inc. **(bottom left)** William R. Sallaz/Gamma-Liaison, Inc. **(bottom right)** Master Sgt. Bill Thompson/U.S. Department of Defense/Corbis **p. 266** Christian Grzimek/OKAPIA/Photo Researchers, Inc. **p. 267** Courtesy of NASA **p. 285** Tom Campbell/Gamma-Liaison, Inc.

CHAPTER 9 **CO (top)** Courtesy of Peter Tannenbaum **(bottom)** Martine Franck/Magnum Photos, Inc. **p. 302** Courtesy of Peter Tannenbaum **p. 305 (top, from left to right)** Richard Parker/Photo Researchers, Inc.; Courtesy of Peter Tannenbaum; Courtesy of Peter Tannenbaum; John Kaprielian/Photo Researchers, Inc.; Bonnie Sue/Photo Researchers, Inc. **p. 305 (center)** "Fibonacci Numbers in Nature," Trudi Garland and Edith Allgood. Copyright 1988 by Dale Seymour Publications. Reprinted by permission. **p. 307** John G. Ross/Photo Researchers, Inc. **p. 308** Erich Lessing/Art Resource, NY; **p. 316** Courtesy of Peter Tannenbaum **p. 317 (left to right)** Art Resource, NY; G. Buttner/Naturbild/OKAPIA/Photo Researchers, Inc.; Photo by Gherandi-Fiorelli. Photo supplied by The Cement and Concrete Association, a division of British Cement Association, Berkshire England

CHAPTER 10 **CO** Frans Lanting/Minden Pictures **p. 330 (left to right)** Frans Lanting/Minden Pictures; Stephen Ferry/Gamma-Liaison, Inc.; Courtesy of Peter Tannenbaum **p. 333** Stephen Ferry/Gamma-Liaison, Inc.

CHAPTER 11 **CO (top left)** Corbis-Bettmann **(top right)** Science Source/Photo Researchers, Inc. **(bottom)** M. C. Escher, "Day and Night." Copyright 1938 M. C. Escher/Cordon Art B. V., Baarn, Holland **p. 363** Robert Winslow/Animals Animals/Earth Scenes **p. 366** Nigel J. Dennis/Photo Researchers, Inc. **p. 370** Photo Researchers, Inc. **p. 372 (left to right)** Science Source/Photo Researchers, Inc.; Courtesy of Peter Tannenbaum; Courtesy of Peter Tannenbaum **p. 374 (left margin)** Top and bottom photos courtesy of Peter Tannenbaum **(center of page, from left to right)** Courtesy of Peter Tannenbaum; Giraudon/Art Resource, NY; Courtesy of Peter Tannenbaum

CHAPTER 12 **CO (top left)** Carr Clifton/Minden Pictures **(top right)** John Buitenkant/Photo Researchers, Inc. **(bottom)** Courtesy Pete Watterberg and Gary Mastin **p. 398 (left)** Christopher Burke/Quesada/Burke Photography **(right)** Science Photo Library/Photo Researchers, Inc. **p. 405** Photos courtesy of Peter Tannenbaum **p. 406 (left to right)** Robert Finken/Photo Researchers; Inc.; Robert A. Isaacs/Photo Researchers, Inc.; Jeffrey M. Hamilton/Gamma-Liaison, Inc. **p. 408** All computer images by Rollo Silver **p. 409** All computer images by Rollo Silver **p. 414 (top row, left to right)** Cindy Charles/Gamma-Liaison, Inc.; Sylvain Grandadam/Photo Researchers, Inc.; Cindy Lewis/Cindy Lewis Photography **(bottom row, left to right)** John Buitenkant/Photo Researchers, Inc.; Robert A. Isaacs/Photo Researchers, Inc.; Robert Finken/Photo Researchers, Inc **p. 415 (left)** NASA **(right)** Copyright F. Kenton Musgrave

CHAPTER 13 **CO (top left)** AP/Wide World Photos **(top right)** Catherine Ursillo/Photo Researchers, Inc.; **(bottom)** Bill Brown/Gamma-Liaison, Inc. **p. 433** UPI/Corbis-Bettmann **p. 436** Courtesy of Peter Tannenbaum

CHAPTER 14 **CO** Courtesy of Peter Tannenbaum

CHAPTER 15 **CO (top left)** Gene Peach/Gamma-Liaison, Inc. **(top right)** Ken Whitmore/Tony Stone Images **(bottom)** Andrea Krause/Photo Researchers, Inc. **p. 492** Photos courtesy of Peter Tannenbaum **p. 494 (top margin)** Courtesy of Peter Tannenbaum **(bottom margin)** Al Francekevich/The Stock Market **p. 496 (left)** William R. Sallaz/Duomo Photography **(right)** Courtesy of Peter Tannenbaum

CHAPTER 16 **CO** Graphs courtesy of Peter Tannenbaum; Photo by Al Francekevich/The Stock Market **p. 513** William R. Sallaz/Duomo Photography